AF800314

Fungal Endophytes Volume II

Kamel A. Abd-Elsalam • Amr H. Hashem
Editors

Fungal Endophytes Volume II

Applications in Agroecosystems and Plant Protection

Springer

ARC

Editors
Kamel A. Abd-Elsalam
Plant Pathology Research Institute
Agricultural Research Center (ARC)
Giza, Egypt

Amr H. Hashem
Botany and Microbiology Department
Faculty of Science
Al-Azhar University
Cairo, Egypt

ISBN 978-981-97-8803-3 ISBN 978-981-97-8804-0 (eBook)
https://doi.org/10.1007/978-981-97-8804-0

© The Editor(s) (if applicable) and The Author(s), under exclusive license to Springer Nature Singapore Pte Ltd. 2025

This work is subject to copyright. All rights are solely and exclusively licensed by the Publisher, whether the whole or part of the material is concerned, specifically the rights of translation, reprinting, reuse of illustrations, recitation, broadcasting, reproduction on microfilms or in any other physical way, and transmission or information storage and retrieval, electronic adaptation, computer software, or by similar or dissimilar methodology now known or hereafter developed.
The use of general descriptive names, registered names, trademarks, service marks, etc. in this publication does not imply, even in the absence of a specific statement, that such names are exempt from the relevant protective laws and regulations and therefore free for general use.
The publisher, the authors and the editors are safe to assume that the advice and information in this book are believed to be true and accurate at the date of publication. Neither the publisher nor the authors or the editors give a warranty, expressed or implied, with respect to the material contained herein or for any errors or omissions that may have been made. The publisher remains neutral with regard to jurisdictional claims in published maps and institutional affiliations.

This Springer imprint is published by the registered company Springer Nature Singapore Pte Ltd.
The registered company address is: 152 Beach Road, #21-01/04 Gateway East, Singapore 189721, Singapore

If disposing of this product, please recycle the paper.

Preface

In recent years, the exploration and utilization of fungal endophytes have received much attention in the field of agriculture and plant protection. These fascinating microorganisms have emerged as promising allies in plant growth promotion, enhancing nutrient uptake, improving stress tolerance, and offering sustainable alternatives to conventional agrochemicals. The diverse and intricate interactions between fungal endophytes and their host plants have opened up new avenues for innovative and eco-friendly approaches in agroecosystems.

This book, *Fungal Endophytes: Applications in Agroecosystems and Plant Protection*, delves into the multifaceted world of fungal endophytes and their potential applications in sustainable agriculture. It brings together a collection of insightful chapters contributed by experts and researchers from various disciplines, shedding light on the latest advancements and discoveries in this exciting field.

The chapters in this book cover a wide range of topics, providing a comprehensive overview of the diverse roles and applications of fungal endophytes. From their isolation and identification to their mechanisms of action and their effects on plant growth and protection, each chapter unravels a specific aspect of these intriguing symbiotic relationships. Also, fungal endophytes have various applications in various fields, including plant growth promotion, sustainable agriculture, biosynthesis of metal nanoparticles, bioremediation, pesticides, and antifungal and antiviral compounds. They have been found to be multifunctional, eco-friendly, and have potential in wheat, blight diseases, and postharvest diseases. They can also be used as biocontrol agents for pathogenic fungi, soil-borne fungi, and vascular wilt diseases. Trichoderma species can be used as fungal inoculants to treat agro-industrial crop diseases and promote plant growth. Mycorrhizal fungi also play a role in plant health. Secondary metabolites of endophytic fungi have antimicrobial properties, making them a valuable source of biological control mechanisms. The book also delves into the exploration of novel bioactive compounds produced by fungal endophytes, their potential for biocontrol of plant pathogens, and their contribution to the overall sustainability and resilience of agroecosystems. These examples highlight the promising potential of fungal endophytes in contributing to sustainable agriculture practices, but further studies are needed to fully explore and harness their benefits in real-world agricultural systems.

While the primary focus of this book is on fungal endophytes and their applications in agroecosystems, it also emphasizes the importance of interdisciplinary collaborations and knowledge exchange. By bringing together experts from the fields of microbiology, plant pathology, agronomy, ecology, and biotechnology, this book seeks to foster a holistic understanding of fungal endophytes and their potential impact on sustainable agriculture.

It is our sincere hope that this book serves as a valuable resource for researchers, academicians, and professionals seeking to explore the untapped potential of fungal endophytes in agroecosystems and plant protection. By disseminating the latest research findings, sharing practical insights, and stimulating further investigations, we aspire to contribute to the advancement of knowledge and the development of innovative strategies for sustainable agriculture.

I would like to express my deepest gratitude to all the authors who have contributed their expertise and valuable insights to this compilation. Their dedication and commitment have made this book possible. I would also like to extend my appreciation to the editorial team, reviewers, and everyone involved in the publication process for their invaluable contributions.

We sincerely thank the Springer staff, especially Dr. Naren Aggarwal Editorial Director—Books, Springer, Asia, Associate Editor Sanchi Bhimrajka, PhD., and Production Editor Mrs. Nandhini Viswanathan, for their kind assistance and hard work in making this volume possible.

Finally, I extend my heartfelt thanks to the readers for their interest in this book. It is my sincere belief that the contents presented within these pages will inspire further exploration and spark new ideas that lead to the development of sustainable and environmentally friendly solutions for agroecosystems and plant protection.

Giza, Egypt — Kamel A. Abd-Elsalam
Cairo, Egypt — Amr H. Hashem

Contents

Contributors

Amer M. Abdelaziz Botany and Microbiology Department, Faculty of Science, Al-Azhar University, Cairo, Egypt

Salah M. Abdel-Momen Plant Pathology Research Institute, Agricultural Research Center (ARC), Giza, Egypt

Kamel A. Abd-Elsalam Plant Pathology Research Institute, Agricultural Research Center (ARC), Giza, Egypt

Faijuddin Ahammad Department of Plant Pathology, Sam Higgingbottom University of Agriculture, Technology and Science, Prayagraj, Uttar Pradesh, India

Hira Akhtar Institute of Plant Protection, MNS-University of Agriculture Multan, Multan, Pakistan

Sk Hasibul Alam Department of Plant Pathology, Bidhan Chandra Krishi Viswavidyalaya (BCKV), Mohanpur, West Bengal, India

Muhammad Ashfaq Plant Pathology, Institute of Plant Protection, Muhammad Nawaz Shareef University of Agriculture, Multan, Punjab, Pakistan

Mohamed S. Attia Botany and Microbiology Department, Faculty of Science, Al-Azhar University, Cairo, Egypt

Julekha Bagum Centre for Life Sciences, Vidyasagar University, Midnapore, India

Arghya Banerjee Department of Plant Pathology, Bidhan Chandra Krishi Viswavidyalaya (BCKV), Mohanpur, West Bengal, India

Department of Plant Pathology, School of Agriculture and Allied Science, The Neotia University, Jhinga, West Bengal, India

Debdulal Banerjee Centre for Life Sciences, Vidyasagar University, Midnapore, West Bengal, India

Microbiology and Microbial Biotechnology Laboratory, Department of Botany and Forestry, Vidyasagar University, Midnapore, West Bengal, India

Debora Luiza Costa Barreto Departamento de Microbiologia, Instituto de Ciências Biológicas, Universidade Federal de Minas Gerais, Belo Horizonte, Minas Gerais, Brazil

Henrique Oliveira Chuman Borges Departamento de Microbiologia, Instituto de Ciências Biológicas, Universidade Federal de Minas Gerais, Belo Horizonte, Minas Gerais, Brazil

Charles Lowell Cantrell Natural Products Utilization Research Unit, Agricultural Research Service, United States Department of Agriculture, Oxford, MS, USA

Jaime Alioscha Cuervo-Parra Escuela Superior de Apan, Universidad Autónoma del Estado de Hidalgo, Apan, Hidalgo, Mexico

Camila Rodrigues de Carvalho René Rachou Institute, FIOCRUZ-MG, Belo Horizonte, Minas Gerais, Brazil

Patrick Brice Defo Deeh Department of Bio-Health Convergence, Kangwon National University, Chuncheon, Republic of Korea

Láuren Machado Drumond de Souza Departamento de Microbiologia, Instituto de Ciências Biológicas, Universidade Federal de Minas Gerais, Belo Horizonte, Minas Gerais, Brazil

Sonia Claudia do Nascimento de Queiroz Laboratório de Resíduos e Contaminantes, Embrapa Meio Ambiente, Jaguariúna, São Paulo, Brazil

Stephen Oscar Duke National Center for Natural Products Research, School of Pharmacy, Oxford, MS, USA

Riya Dutta Centre for Life Sciences, Vidyasagar University, Midnapore, West Bengal, India

Dounia Elfadil Biology and Chemistry Department, Faculty of Sciences and Techniques, Hassan II University of Casablanca, Mohammedia, Morocco

Gharieb S. El-Sayyad Medical Laboratory Technology Department, Faculty of Applied Health Sciences Technology, Badr University in Cairo (BUC), Cairo, Egypt

Department of Microbiology and Immunology, Faculty of Pharmacy, Galala University, New Galala City, Suez, Egypt

Drug Microbiology Lab., Drug Radiation Research Department, National Center for Radiation Research and Technology (NCRRT), Egyptian Atomic Energy Authority (EAEA), Cairo, Egypt

Paula Cecilia Guadarrama-Mendoza Universidad Tecnológica de la Mixteca, Huajuapan de León, Oaxaca, Mexico

Akhtar Hameed Institute of Plant Protection, MNS-University of Agriculture Multan, Multan, Pakistan

Kiseok Han Department of Bio-Health Convergence, Kangwon National University, Chuncheon, Republic of Korea

Amr H. Hashem Botany and Microbiology Department, Faculty of Science, Al-Azhar University, Cairo, Egypt

Saidul Islam Nadia Krishi Vigyan Kendra, Bidhan Chandra Krishi Viswavidyalaya (BCKV), Gayeshpur, Nadia, West Bengal, India

Sania Javeed Plant Pathology, Institute of Plant Protection, Muhammad Nawaz Shareef University of Agriculture, Multan, Punjab, Pakistan

Yachana Jha N.V. Patel College of Pure & Applied Sciences, CVM University, V.V. Nagar, Anand, (Gujarat), India

Samah Bashir Kayani Govt. Graduate College (W) Mumtazabad, Multan, Punjab, Pakistan

Feba Sara Koshy Department of Biotechnology, Karpagam Academy of Higher Education, Coimbatore, Tamil Nadu, India

Simon Law CSIRO Agriculture and Food, Black Mountain Science and Innovation Park, Acton, ACT, Australia

Zhang Leena Department of Bio-Health Convergence, Kangwon National University, Chuncheon, Republic of Korea

Kumudini Mangalika Meepagala The National Center for Agricultural Utilization Research, Crop Bioprotection Research Unit, Peoria, IL, USA

Mirza Abid Mehmood Plant Pathology, Institute of Plant Protection, Muhammad Nawaz Shareef University of Agriculture, Multan, Punjab, Pakistan

Bholanath Mondal Department of Plant Pathology, Palli Siksha Bhavana, Visva Bharati University, Sriniketan, Bolpur, West Bengal, India

Sahar Murmu Department of Plant Pathology, Bidhan Chandra Krishi Viswavidyalaya (BCKV), Mohanpur, Nadia, West Bengal, India

Iqra Mushtaq Plant Pathology, Institute of Plant Protection, Muhammad Nawaz Shareef University of Agriculture, Multan, Punjab, Pakistan

Kumar Vishven Naveen Department of Bio-Health Convergence, Kangwon National University, Chuncheon, Republic of Korea

M. D. Oak Genetics and Plant breeding group, Agharkar Research Institute, Pune, India

Prajakta Y. Pachorkar P.G. Department of Microbiology, K. R. T. Arts, B. H. Commerce and A. M. Science College, Nashik, Maharashtra, India

Vaishnavi R. Pawar P.G. Department of Microbiology, K. R. T. Arts, B. H. Commerce and A. M. Science College, Nashik, Maharashtra, India

Tikka Dewage Chamarika Priyadarshani Department of Plant Sciences, Faculty of Agriculture, Rajarata University of Sri Lanka, Puliyankulama, Anuradhapura, Sri Lanka

Institute of Plant Protection, MNS-University of Agriculture Multan, Multan, Pakistan

Muzzamil Qazi Institute of Plant Protection, MNS- University of Agriculture, Multan, Multan, Pakistan

Muhammad Umair Rafiq Institute of Plant Protection, MNS-University of Agriculture Multan, Multan, Pakistan

J. M. Rajwade Nanobioscience group, Agharkar Research Institute, Pune, India

Heriberto Fortino Ramírez-Cariño Universidad de la Sierra Juárez, Ixtlán de Juárez, Oaxaca, Mexico

Areeba Rauf Plant Pathology, Institute of Plant Protection, Muhammad Nawaz Shareef University of Agriculture, Multan, Punjab, Pakistan

Victor Pena Ribeiro Natural Products Utilization Research Unit, Agricultural Research Service, United States Department of Agriculture, Oxford, MS, USA

Teresa Romero-Cortes Escuela Superior de Apan, Universidad Autónoma del Estado de Hidalgo, Apan, Hidalgo, Mexico

Luiz Henrique Rosa Departamento de Microbiologia, Instituto de Ciências Biológicas, Universidade Federal de Minas Gerais, Belo Horizonte, Minas Gerais, Brazil

Salem S. Salem Botany and Microbiology Department, Faculty of Science, Al-Azhar University, Cairo, Egypt

Hiran Kanti Santra Department of Botany and Forestry, Vidyasagar University, Midnapore, West Bengal, India

Anbazhagan Sathiyaseelan Department of Bio-Health Convergence, Kangwon National University, Chuncheon, Republic of Korea

Mohamed T. Selim Botany and Microbiology Department, Faculty of Science, Al-Azhar University, Cairo, Egypt

Maksud Hasan Shah Department of Agronomy, Bidhan Chandra Krishi Viswavidyalaya (BCKV), Mohanpur, Nadia, West Bengal, India

Mohamed K. Y. Soliman Botany and Microbiology Department, Faculty of Science, Al-Azhar University, Cairo, Egypt

Elisa Amorim Amâncio Teixeira Departamento de Microbiologia, Instituto de Ciências Biológicas, Universidade Federal de Minas Gerais, Belo Horizonte, Minas Gerais, Brazil

Rogelio Valadez-Blanco Universidad Tecnológica de la Mixteca, Huajuapan de León, Oaxaca, Mexico

Myeong-Hyeon Wang Department of Bio-Health Convergence, Kangwon National University, Chuncheon, Republic of Korea

Prathiba Aruni Weerasinghe Department of Plant Sciences, Faculty of Agriculture, Rajarata University of Sri Lanka, Puliyankulama, Anuradhapura, Sri Lanka

Zhang Xin Department of Bio-Health Convergence, Kangwon National University, Chuncheon, Republic of Korea

Sangilimuthu Alagar Yadav Department of Biotechnology, Karpagam Academy of Higher Education, Coimbatore, Tamil Nadu, India

Heba Yousef Plant Pathology Research Institute, Agricultural Research Center (ARC), Giza, Egypt

Lu Yuting Department of Bio-Health Convergence, Kangwon National University, Chuncheon, Republic of Korea

Mohamed Cassim Mohamed Zakeel CSIRO Agriculture and Food, Black Mountain Science and Innovation Park, Acton, ACT, Australia

About the Editors

Kamel A. Abd-Elsalam, Ph.D., is currently a Research Professor and Deputy Director for Research at the Plant Pathology Research Institute, Agricultural Research Center, Giza, Egypt. Dr. Kamel's research interests include developing and improving plant biosecurity diagnostic tools, understanding fungal pathogen genomes, and creating eco-friendly hybrid nanomaterials for controlling toxigenic fungi and plant diseases as well as applications in agroecosystems. Kamel earned his Ph.D. in Molecular Plant Pathology from Christian Albrechts University of Kiel (Germany) and Suez Canal University (Egypt). In 2008, he was awarded a postdoctoral fellowship from the same institution. Dr. Kamel served as a Visiting Associate Professor at Mae Fah Luang University in Thailand, the Institute of Microbiology at TUM in Germany, the Laboratory of Phytopathology at Wageningen University in the Netherlands, and the Plant Protection Department at Sassari University in Italy.

He has published 35 books related to nanobiotechnology applications in agriculture and plant protection with major publishing houses such as Springer, Taylor & Francis, and Elsevier. Since 2019, he has served as the Editor-in-Chief of the Elsevier book series Nanobiotechnology for Plant Protection and as the Series Editor for Applications of Genome Modified Plants and Microbes in Food and Agriculture. Additionally, he has been an active member of the Elsevier Advisory Panel since 2020, providing feedback and suggestions to improve Elsevier's products and services. He has published more than 302 scientific research articles in international and regional specialized journals with high impact factors, and has an H-index of 49, an i-10 index of 155, and over 9292 citations. He served as a Guest Editor for the *Journal of Fungi, Plants, and Microorganisms*, and as a Review Editor for *Frontiers in Genomic Assay Technology*, *Agrochemicals*, and *Plant Nano Biology*. He was ranked in the top 2% of the most influential scientists in the world in nanobiotechnology for 2021, 2022, 2023 and 2024 by Elsevier and Stanford University. In 2014, he was awarded the Federation of Arab Scientific Research Councils Prize for excellent scientific research in biotechnology (fungal genomics), ranking first. Additionally, he received the Egyptian State Award for Agricultural Scientific Excellence from ASRT, Egypt, in 2022 with a silver medal.

Amr H. Hashem, Ph.D., is currently an Associate Professor at the Botany and Microbiology Department, Faculty of Science, Al-Azhar University, Cairo, Egypt. Hashem granted his Ph.D. in Applied Mycology from Al-Azhar University in 2018. Hashem's research interests include fungal endophytes and their biomedical and agricultural applications, control of fungal plant pathogens using safe biofungicides, biosynthesis of metal nanoparticles and their different biological applications, and lipid production from oleaginous fungi.

He has published more than 100 articles in high-impact international peer-reviewed journals as well as reviewed more than 100 articles in impacted international journals. He has an H-index of 47, an i10 index 90, and over 5164 citations. He was awarded the "State Encouragement Award" in Basic Biological Sciences for the year 2021. Also, he received Abdel Aal Mubasher prize for excellent Mycologist (2022), granted from Adel Aal Mubasher Center, Assiut University, Egypt. His name is included in the list of the 2% of the most distinguished and influential scientists in the world according to the classification of the international publishing house Elsevier and Scopas with Stanford University of America in 2022. He is an Associate Editor of *Frontiers in Microbiology and BMC Microbiology Journals*. Also, he is a review editor of both *Frontiers in Nanotechnology* and *Frontiers in Bioengineering and Biotechnology*.

Part I

Agricultural Applications

1 Enhancing Agroecosystem Productivity with Endophytic Fungi: Unveiling Their Role

Kamel A. Abd-Elsalam and Amr H. Hashem

Abstract

Endophytic fungi are organisms that live in plant tissues and contribute to improved resistance to diseases, tolerance to environmental stresses, and accelerated plant growth. The molecular intermediates they produce also inhibit the emergence of shoots on the stem itself, so the use of natural biopesticides can replace synthetic chemical insecticide and herbicide compounds from an environmentally friendly point of view, but is not cost-effective. Therefore, higher productivity is required. They can also improve plant nutrient uptake, plant growth, and agricultural yields. Based on these general data, applications are being developed for biological control of diseases caused by Fusarium weevils in sugarcane fields. Endophytic fungi lead to a more stratified energy intake in these improved ecosystems and support a richer population of herbivores. Although difficult to quantify, biodiversity has a very real impact on plant health and yield. Another approach to plant protection using biotechnology is using endophytes. This chapter provides an view of the understanding work in recent years on the use of endophytic fungi for biological control of agricultural pests.

Keywords

Endophytic fungi · Agroecosystems · Biocontrol · Plant growth promotion

K. A. Abd-Elsalam (✉)
Plant Pathology Research Institute, Agricultural Research Center (ARC), Giza, Egypt

A. H. Hashem
Botany and Microbiology Department, Faculty of Science, Al-Azhar University, Cairo, Egypt

© The Author(s), under exclusive license to Springer Nature Singapore Pte Ltd. 2025
K. A. Abd-Elsalam, A. H. Hashem (eds.), *Fungal Endophytes Volume II*,
https://doi.org/10.1007/978-981-97-8804-0_1

1.1 Introduction

A cluster of organisms known as fungal endophytes resides within plant tissues without harming their host. Researchers have invested effort in studying these beneficial relationships due to the various advantages they offer plants, such as improved growth, heightened resilience to environmental pressures and bolstered immunity, against diseases. Recently, the exploration of fungal endophytes has broadened to explore their applications in safeguarding plants and enhancing environments (Dutta et al. 2014; Latz et al. 2018).

Many bioactive substances produced by endophytes help plants protect themselves from pests and diseases offering an alternative to chemical pesticides. Studies have shown that these compounds not only improve absorption and support plant growth but also boost overall crop yield (Zhao et al. 2010). Fungal endophytes play a role in agriculture by promoting biodiversity reducing reliance on artificial inputs and increasing ecosystem resilience. By harnessing the benefits of these microorganisms farmers can enhance crop productivity maintain soil health and minimize impacts (Khan 2020; Verma et al. 2022; Abo Nouh 2019). Such fungal endophytes could further lead to innovative technologies geared at protecting plants. Manipulation in such microbes can easily take place to enhance their beneficial traits, and they would be able to colonize a wide variety of plant species. With the advancement in biotechnology and genomics research, mechanisms related to endophytes' plant interaction are now understood better, and practical methods have been formulated using these microorganisms in plant protection (Baron and Rigobelo 2022; Collinge et al. 2022; Sharma et al. 2023).

This chapter aims to present an overview of the very latest research into fungal endophytes, their applications in agroecosystems, and plant protection. We consider mechanisms by which endophytes promote plant health and defense, potential benefits of utilizing endophytes within agriculture, and challenges with research opportunities for further fundamental and applied investigations. We, therefore hope that by putting forward the potential of fungal endophytes as environmentally benign and sustainable tools in enhancing crop productivity and safeguarding crops against pests and diseases, more research and development relating to these critical agricultural microorganisms will be prompted.

1.2 Role of Fungal Endophytes in Agroecosystems

Fungal endophytes are one of the most diversified groups of fungi that reside within plant tissues without inducing disease symptoms. Symbiotic organisms have been described to exert various positive effects on plant growth, development, and health. In this paper, we will demonstrate how fungal endophytes function and act to sustain agroecosystems in many ways (Khan 2020).

1.2.1 Enhanced Nutrient Uptake and Efficiency

Fungal endophytes have an essential ecological role in agroecosystems, interfering with the nutrient acquisition and use efficiency of host plants. It has been shown that endophytic colonization can enhance the availability and uptake of certain essential nutrients required for plant growth through the enhancement of their absorption from the soil. They do this by employing a variety of mechanisms, which extend from producing enzymes that degrade organic matter and make extracted nutrients available to forming associations with the roots of plants such as mycorrhizae (Alves et al. 2021). The mycorrhizal association of fungal endophytes with plant roots expands the effective surface area for acquiring nutrients; they develop tools in their nutrient uptake that would certainly otherwise go unavailable. Improved nutrient uptake will thus not only increase plant growth and yield, but it also benefits the environment economically by limiting the use of synthetic fertilizers (García-Latorre et al. 2021).

1.2.2 Improved Abiotic Stress Tolerance

Agroecosystems commonly suffer from abiotic stresses such as drought, salinity, extreme temperatures, and others that negatively affect plants' growth and productivity. Fungal endophytes have been documented to be critical players in promoting plant resilience to various abiotic stresses (Khan et al. 2015). Fungal endophytes can address abiotic stress through the following mechanisms: production of scavenging chemicals against reactive oxygen species, maintenance of cell membranes' integrity, and regulation of plant hormone concentrations. In this way, plants stay physiologically functioning and can grow in unfavorable conditions. Similarly, fungal endophytes increase water-use efficiency through enhanced water intake efficacy and reduced water loss due to transpiration. Especially in water-scarce environments, this is highly advantageously designed for water conservation to be productive for sustainable agriculture (Singh et al. 2011).

1.2.3 Biocontrol of Plant Pathogens

Pathogenic microorganisms significantly threaten crop output and food safety and cause plant diseases. Fungal endophytes have been considered as potential biocontrol agents against various plant diseases caused by bacteria, fungi, and nematodes (De Silva et al. 2019). These endophytes can adopt various ways of defending plants from disease-causing agents through synthesizing antifungal chemicals, competing for resources and space, and triggering plant defensive responses. Colonization of plant tissues with fungal endophytes creates a physical barrier to the invasion of pathogens and can further suppress the growth and development of the pathogen within the plant. Fungal endophytes have several advantages as biocontrol agents over conventionally used chemical pesticides (Berg and Hallmann 2006). Another good thing about them is that they do not leave any dangerous residues on the environment and have a diminished ability to develop resistance to target diseases. Therefore, tapping into the potential of fungal endophytes as biocontrol agents is consistent with sustainable agriculture principles, seeking to work according to natural and ecological plant protection methods (Berg and Hallmann 2006).

1.2.4 Induction of Plant Systemic Resistance

The ability of fungal endophytes to induce systemic resistance in plants is just incredible. Systemic resistance is a mechanism of defense that enhances the possibility of surviving attacks from various diseases in plants. This systemic resistance does not remain specific to the pathogen the endophyte protects from but can also give broad-spectrum resistance to other infections. Fungal endophytes, by inducing resistance, turn on protection routes in plants that produce defense-related chemicals like phytoalexins and pathogenesis-related proteins. It means they offer a better defense with the aid of such chemicals against several diseases (Fontana et al. 2021). Systemic resistance induced by fungal endophytes relies on activating plant defense pathways, which in turn leads to the synthesis of defense-related chemicals, including phytoalexins and pathogenesis-related proteins. These chemicals serve as a sort of chemical defense weaponry system, providing better plant protection against any potential diseases. Production of the plant-induced resistance through fungal endophytes offers long-term protection for plants; simultaneously, it reduces the need for chemical pesticides, making this approach environmentally sustainable for plant protection.

Critically, fungal endophytes enhance nutrient uptake and use efficiency, improve abiotic stress tolerance, biocontrol plant diseases, and induce systemic resistance in plants within agroecosystems. Such beneficial interactions between the fungal endophytes and plants will promote the sustainability and productivity of agricultural systems with reduced reliance on synthetic inputs. Information on fungal endophyte roles in agroecosystems is thus instrumental for their full exploitation and integration into sustainable agriculture practices (Akram et al. 2023).

1.3 Importance of Fungal Endophytes in Plant Protection

Fungal endophytes have received a lot of attention in recent years because of their ability to protect plants. These beneficial bacteria have been shown to significantly improve plant defense mechanisms against a variety of diseases, pests, and environmental challenges. Because fungal endophytes can increase plant health and reduce reliance on chemical pesticides, they are a viable and environmentally friendly crop protection alternative. This makes them vital for plant protection (Sodhi and Saxena 2023).

1.3.1 Induction of Systemic Resistance

Fungal endophytes help plants protect themselves by building systemic resistance. When these endophytes enter plants, they can cause a variety of biochemical and physiological changes, resulting in increased resistance to infections (Waqar et al. 2023). This systemic resistance can be achieved by activating a number of defense pathways, including as antimicrobial chemical production, cell wall reinforcement, and defense-related gene expression.

1.3.2 Biocontrol of Plant Pathogens

It has been discovered that fungal endophytes have biocontrol properties against a variety of plant diseases. These endophytes can create secondary metabolites with antimicrobial characteristics, such as antifungal chemicals, antibiotics, and enzymes that break down pathogens' cell walls (Malarvizhi et al. 2023). Colonizing plant tissues allows them to directly compete with diseases for space and resources, reducing their establishment and spread. Furthermore, some fungal endophytes can cause systemic resistance in plants, making them more resistant to pathogen attacks.

1.3.3 Suppression of Plant Diseases

Fungal endophytes have been shown to reduce a variety of plant illnesses caused by fungal, bacterial, and viral pathogens. They can prevent disease growth and progression by creating antifungal chemicals, enzymes that degrade pathogen cell walls, and activating plant defensive responses (Latz et al. 2018). In some conditions, fungal endophytes can create volatile organic molecules with antibacterial characteristics, which aids in disease prevention. These endophytes are an important component of integrated pest control systems because of their capacity to reduce plant diseases.

1.3.4 Enhancement of Plant Growth and Development

In addition to plant protection, fungal endophytes have been demonstrated to promote plant growth and development. These endophytes can improve nutrient uptake and digestion, increase water efficiency, and boost plant tolerance to abiotic stressors like drought, salt, and heavy metal toxicity. Fungal endophytes indirectly contribute to plant protection by improving plant health and vigor, making plants more resistant to pathogens and environmental challenges (Fadiji and Babalola 2020).

1.3.5 Reduction of Chemical Inputs

The use of chemical pesticides in agriculture has sparked worries about their possible harm to human health and the environment. Fungal endophytes provide a more sustainable alternative to chemical inputs in crop cultivation. Farmers can reduce their usage of synthetic pesticides and fertilizers by leveraging the biocontrol and plant growth-promoting characteristics of these endophytes, resulting in lower chemical residues in food crops and a healthier agroecosystem (Lugtenberg et al. 2016).

1.3.6 Preservation of Biodiversity

Fungal endophytes help in the preservation of biodiversity in agroecosystems by diversifying the microbial populations that coexist with plants. These endophytes can increase the diversity and number of beneficial bacteria in the rhizosphere and phyllosphere, resulting in cascading impacts on ecosystem health. Fungal endophytes contribute to a healthy and robust agroecosystem that is less vulnerable to pest and disease outbreaks (Rana et al. 2019).

1.3.7 Sustainable and Eco-friendly Agriculture

The role of fungal endophytes in plant protection goes beyond their direct impact on crop health. These endophytes promote sustainable and environmentally friendly farming practices, ensuring the long-term viability of agroecosystems (Abo Nouh 2019). The use of fungal endophytes can assist minimize agriculture's environmental effect by lowering chemical inputs, reducing soil erosion, increasing soil fertility, and conserving water. Furthermore, introducing these beneficial microbes can make farming systems more economically sustainable by lowering production costs and increasing crop yields (Khan 2020). Lastly, by promoting plant development and growth, biocontrolling plant pathogens, building systemic resistance, and averting disease, fungal endophytes greatly enhance plant protection (Table 1.1). Their application will have a significant positive impact on crop production in the future since it can lessen the need for chemical inputs, protect biodiversity, and advance sustainable agriculture. It is anticipated that additional research in this field will broaden the range of possible uses for fungal endophytes in plant protection, creating new opportunities for environmentally conscious and sustainable agriculture.

Table 1.1 The role of fungal endophytes in agroecosystems

Role	Description	Benefits
Plant Growth Promotion	Endophytes can solubilize nutrients and produce plant hormones such as gibberellins and auxins, promoting root formation, plant growth, and nutrient uptake.	Crop yield rose, and plant health improved.
Stress Tolerance	Endophytes can help plants cope with abiotic stresses such drought, salinity, and heavy metals by producing appropriate solutes, boosting nutrient absorption, and increasing antioxidant activity.	Crops are more resilient to environmental conditions.
Disease Resistance	Endophytes can compete with and reduce harmful fungi and bacteria through a range of tactics, including antibiosis, activation of plant defense mechanisms, and niche occupancy.	Reduced reliance on chemical pesticides and improved agricultural disease control.
Biocontrol of Insect Pests	Some endophytes produce toxic secondary metabolites or insect repellents that aid in natural pest control in agroecosystems.	Sustainable and environmentally friendly insect pest management.
Improved Nutrient Cycling	Endophytes can help with nutrient cycling by decomposing organic detritus, which makes nutrients more available to plants.	Improved soil fertility and ecological function.

1.4 Fungal Endophytes and Sustainable Agriculture

1.4.1 Promotion of Plant Health and Productivity

Fungal endophytes have emerged as powerful tools for enhancing plant health and productivity in agroecosystems. These beneficial bacteria can create symbiotic relationships with plants, providing a variety of advantages that boost the host plant's overall health. This section will discuss how fungal endophytes increase plant health and productivity, as well as their potential applications in sustainable agriculture (Abo Nouh 2019).

1.4.1.1 Enhanced Nutrient Uptake and Efficiency

The capacity of fungal endophytes to improve plant nutrient uptake and efficiency is one of its main advantages. Le Cocq et al. (2017) report that these endophytes have the ability to create mutualistic connections with plant roots, facilitating the absorption of vital minerals like nitrogen, phosphate, and potassium. Fungal endophytes have the ability to explore a larger area of soil and gather nutrients that the plant would not otherwise have due to the size of their hyphal networks. Better plant development, more biomass production, and ultimately larger crop yields can all be attained through improved nutrient uptake (Lugtenberg et al. 2016). Furthermore, fungal endophytes can help plants absorb nutrients. They can improve nutrient uptake and utilization by changing the root architecture, boosting nutrient absorption capacity, and promoting nutrient translocation within the plant. This greater efficiency not only reduces the need for synthetic fertilizers, but it also reduces nutrient losses, hence reducing the environmental impact of fertilizer use.

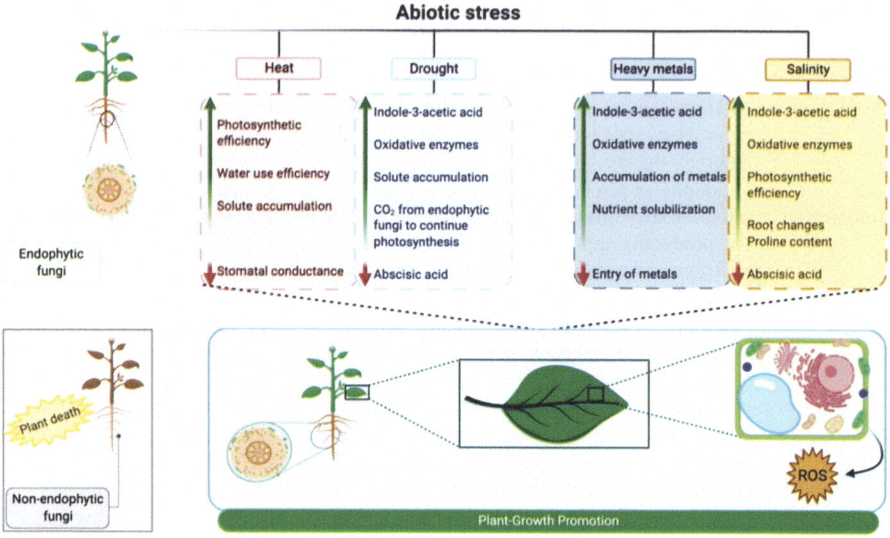

Fig. 1.1 Potential reactions to abiotic stresses evidenced by plants when interacting with endophytic fungi. The green arrow represents the increase, and the red arrow represents the reduction of the listed characteristics. (Reprinted from Fontana et al. 2021)

1.4.1.2 Alleviation of Abiotic Stress

The development and yield of plants are significantly affected by abiotic stressors like drought, salt, and extreme temperatures. Nevertheless, fungal endophytes have shown great potential in reducing the adverse effects of these stresses on plants. Through symbiotic interactions, fungal endophytes can enhance plant tolerance to abiotic stress in various ways. For instance, they produce osmolytes and other stress-related chemicals that help plants maintain cellular osmotic balance, protect cellular structures, and eliminate reactive oxygen species, thereby alleviating the impact of abiotic stress. Additionally, fungal endophytes can boost the production of stress-responsive proteins and enzymes in plants, enabling them to better cope with harsh environmental conditions (Vardharajula et al. 2017; Verma et al. 2022).

Furthermore, fungal endophytes can improve plant water relations by increasing water intake and lowering transpiration (Mosaddeghi et al. 2021). They can also regulate stomatal opening and closure to maximize water efficiency. These adaptations help plants survive and produce more when they are under water stress. Endophytic fungi can enter plant tissues and act as control agents, such as biological agents or elicitors, in the process of induced resistance and abiotic stressor attenuation. Figure 1.1 displays the many processes influenced by abiotic stress and contact with endophytic fungi.

1.4.1.3 Biocontrol of Plant Pathogens

Plant diseases caused by pathogenic microorganisms are a threat to agriculture output worldwide. Despite all this, fungal endophytes have a very bright future as biocontrol agents of plant diseases. The endophytes can synthesize a variety of bioactive compounds like antibiotics, antifungal metabolites, and enzymes responsible for the inhibition of growth as well as sporulation by pathogens (Shahrtash and Brown 2023). Among these, the fungal endophytes potentially are a source of microbial inoculants that colonize plant tissues and represent an initial-level resistance against invading diseases. Competition for space and nutrients, anti-microbial compounds that block pathogen growth (antibiotics) or even induce systemic resistance in plants. Moreover, the presence of fungal endophytes can enhance immunity in plants by upregulating disease resistance-related genes and signaling pathways paving way for strengthened mechanisms against pathogen infestations (Collinge et al. 2022). The biocontrol role of fungal endophytes goes far beyond the direct antagonism against pathogens. They may also prime plants to exhibit systemic resistance in order for the hosts to mount a more robust defense response against pathogen attacks. Here are several methods by which plant resistance conferred via Art Of Nature can lead to long-term protection, reducing the use of chemicals and promoting sustainable disease management.

1.4.1.4 Induction of Plant Growth-Promoting Traits

In addition to their direct effects on nutrient intake, stress tolerance, and disease management, fungal endophytes can promote a variety of plant development features. These include the production of phytohormones such as auxins, cytokinins, and gibberellins, which regulate plant growth and development. They can also promote root development, leading to better nutrient absorption and plant anchoring in the soil (Harman and Uphoff 2019). Furthermore, fungal endophytes can stimulate the production of secondary metabolites in plants, such as phenolics and flavonoids, which have antioxidant and allelopathic properties. These metabolites can increase plant defense systems, protect against oxidative stress, and inhibit competitive plant development, so improving the host plant's general health and productivity (Angeeva et al. 2020).

1.4.1.5 Contribution to Soil Health and Microbial Diversity

Soil health and microbial diversity are maintained by fungal endophytes. Creating symbiotic partnerships with plants makes the soil microbiome more diverse and robust. This microbial community, in turn, enhances nutrient transport, enhances soil structure, and deters the emergence of soil-borne diseases (Devi et al. 2020). The presence of fungal endophytes in the rhizosphere can encourage the growth of beneficial soil microbes like mycorrhizal fungi and nitrogen-fixing bacteria. These interactions improve nutrient availability, improve soil fertility, and reduce the need for further inputs. Fungal endophytes can help break down organic materials in the soil, which promotes nutrient release and cycling. Furthermore, the presence of fungal endophytes in plants could result in the production of rhizosphere fluid, which attracts beneficial flora and fauna. The resulting dynamic and interconnected network of plant–microbe interactions promotes a healthy and productive agroecosystem (Nasslahsen et al. 2022).

1.4.1.6 Integration of Fungal Endophytes in Agroecosystems

The successful integration of fungal endophytes into agricultural systems necessitates careful consideration of several criteria. These include selecting suitable endophyte–host plant pairings, optimizing inoculation processes, and regulating environmental variables. Furthermore, it is critical to adopt sustainable and cost-effective production procedures for fungal endophytes in large-scale agricultural operations (Toju et al. 2018). The application of fungal endophytes as biofertilizers, biocontrol agents, and plant growth boosters has important implications for sustainable agriculture. Endophytes are a feasible and environmentally beneficial agricultural production alternative because they reduce the need for chemicals, improve plant health and productivity, and promote soil health and microbial variety (Paveda et al. 2021).

1.5 Contribution to Soil Health and Microbial Diversity

Fungal endophytes play an important function in improving soil health and increasing microbial diversity in agroecosystems. These helpful fungi form symbiotic connections with plants, which improves soil health and productivity. In this part, we will study how fungal endophytes assist to soil health and microbial diversity (Zhou et al. 2017).

1.5.1 Nutrient Cycling and Soil Fertility

Fungal endophytes play an important role in soil health by cycling nutrients and increasing soil fertility. These fungi have the potential to solubilize minerals including phosphate and micronutrients, increasing their availability to plants. Fungal endophytes can gain access to nutrients that plants would otherwise be unable to obtain, hence improving nutrient absorption and use. This not only enhances the plant's nutrient status but also enriches the soil, resulting in higher soil fertility (Chhabra and Dowling 2017). Moreover, fungal endophytes can fix atmospheric nitrogen and transform it to a form that plants can use. This nitrogen fixation capability decreases dependency on synthetic nitrogen fertilizers, which can harm the environment. Fungal endophytes promote natural nitrogen fixation, contributing to sustainable agriculture methods and reducing the environmental impact of traditional farming (Lacava et al. 2022).

1.5.2 Soil Structure and Aggregation

Fungal endophytes also play a significant role in improving soil structure and aggregation. The hyphal networks formed by these fungi create a complex web-like structure that binds soil particles together, resulting in the formation of stable aggregates. These aggregates improve soil porosity, water infiltration, and water-holding capacity, thereby enhancing soil structure and reducing the risk of soil erosion (Hartmann and Six 2023). Additionally, the presence of fungal endophytes in the soil stimulates the growth of beneficial soil microbes including bacteria and fungus. These microbes aid in the breakdown of organic materials, releasing vital nutrients and organic compounds into the soil. The enhanced microbial activity promoted by fungal endophytes improves soil structure and nutrient availability, creating an ideal environment for plant growth.

1.5.3 Disease Suppression and Biocontrol

It has been demonstrated that fungal endophytes exhibit biocontrol characteristics against a variety of plant diseases, which helps to reduce disease in agroecosystems. These endophytes have the ability to create secondary metabolites such as antibiotics and antifungal chemicals, which restrict the growth and activity of pathogens. Fungal endophytes colonies plant tissues and provide a protective barrier against diseases, inhibiting their entry and establishment (Latz et al. 2018). The presence of fungal endophytes in soil can cause systemic resistance in plants, making them more disease-resistant. This systemic tolerance response is triggered by a contact between endophytes and the plant's immune system, which activates defense mechanisms. The generated resistance not only protects the host plant but also has a good impact on the surrounding soil microbial population, which promotes overall disease suppression in the agroecosystem (Fontana et al. 2021). In a variety of ways, fungal endophytes interact with fungal plant diseases. They produce antimicrobial compounds that inhibit pathogen growth and development, compete for resources, promote systemic resistance, mycoparasitism, and indirect competition by altering the plant's physiology and metabolism (Fig. 1.2). These interactions lead to increased plant health, growth, and disease resistance. Fungal endophytes can also fight illnesses directly, parasitizing them to get nutrients and eventually destroying them (Akram et al. 2023). Fungal endophytes have a significant impact on plant health and resistance.

1.5.4 Enhancing Microbial Diversity

They interact with other soil microorganisms to increase overall microbial diversity in agroecosystems. The development of symbiotic interactions between fungal endophytes and plants results in a dynamic microbial population in the rhizosphere and surrounding soil. This diversified microbial community is essential for nutrient cycling, disease suppression, and general ecosystem health. The presence of fungal endophytes in the soil also stimulates the growth and activity of beneficial soil bacteria. These bacteria can form mutualistic relationships with the endophytes, further enhancing nutrient cycling and disease suppression. Additionally, the interactions between fungal endophytes and other soil microorganisms can lead to the

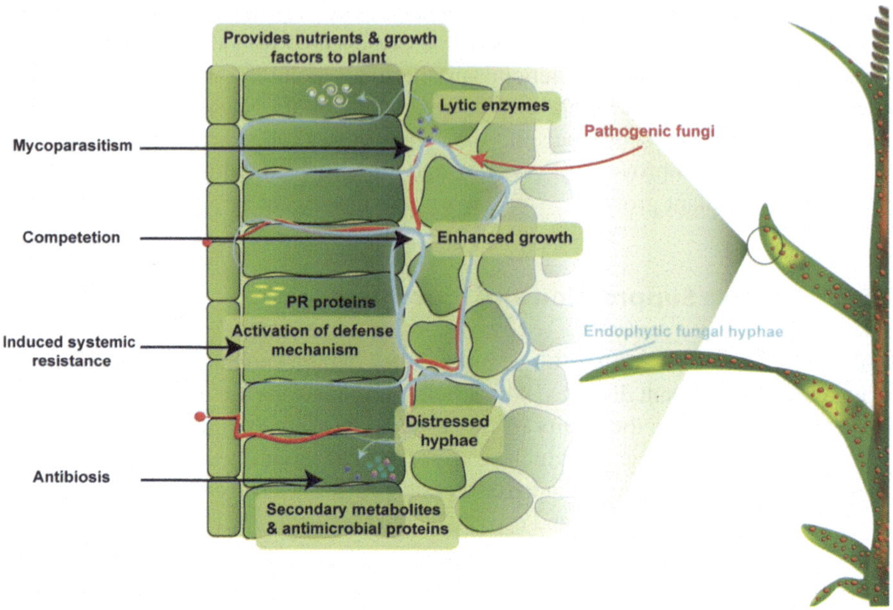

Fig. 1.2 Mycoparasitism, competition, and antibiosis are a few of the defense mechanisms that endophytic fungi use to both directly and indirectly reduce infections and induce resistance in plant hosts (Akram et al. 2023)

production of novel compounds and metabolites, contributing to the overall chemical diversity of the soil ecosystem (Gakuubi et al. 2021).

1.5.5 Conservation of Soil Biodiversity

Soil biodiversity conservation is critical for the long-term viability of agricultural systems. Fungal endophytes help to preserve soil biodiversity by fostering the growth and survival of beneficial soil microbes. Symbiotic connections between fungal endophytes and plants promote the creation of different microbial communities, ensuring the soil ecosystem's stability and resilience. It can also help plants withstand environmental challenges like drought and salinity. This improved resilience helps to the overall stability of the agroecosystem and decreases the need for chemical inputs, supporting more sustainable agriculture practices (Devi et al. 2020).

In summary, fungal endophytes contribute significantly to soil health and microbial diversity in agroecosystems. Their potential to raise nutrient cycling, soil structure, disease suppression, and microbial interactions makes them useful partners in sustainable agriculture. By incorporating fungal endophytes into agroecosystems, we can improve soil fertility, preserve soil biodiversity, and improve the general health and productivity of agricultural systems.

1.6 Applications of Fungal Endophytes in Crop Production

1.6.1 Enhancement of Nutrient Uptake and Efficiency

Fungus endophytes have been found to be helpful allies for improving nitrogen uptake and efficiency in plants. These helpful bacteria aid in the intake and application of essential nutrients by forming symbiotic interactions with their host plants. Fungal endophytes enhance plant development, growth, and production through many methods (Baron and Rigobelo 2022).

1.6.1.1 Nutrient Acquisition by Fungal Endophytes

They play a crucial role in helping plants absorb nutrients. Nutrients in the soil may be mobilized and solubilized by these bacteria, increasing their availability to their host plants. Enzymes like phosphatases and proteases, which decompose complex organic molecules into simpler forms that plants may readily absorb, are produced by fungal endophytes. In addition to releasing nutrients from organic materials, this enzyme activity also makes essential elements like potassium, phosphate, and nitrogen more readily available (Verma et al. 2021). Furthermore, they have the ability to form symbiotic partnerships with mycorrhizal fungi, which are widely recognized for their role in the uptake of nutrients. Because of these connections, plants can obtain nutrients from a wider range of sources, including minerals, organic matter, and even nitrogen from the atmosphere. By extending the reach of plant roots, the mycorrhizal network created by fungal endophytes increases the surface area that is available for nutrient absorption. According to Lanfranco et al. (2016), this enhanced nutrient uptake supports the growth and development of plants by improving their overall nutritional status.

1.6.1.2 Nutrient Mobilization and Translocation

The transportation and mobilization of nutrients inside plants is greatly aided by fungi endophytes. These microbes transfer nutrients from one plant organ to another, enabling effective nutrient utilization. For example, fungus endophytes can transfer nutrients from older to younger leaves, where they are required for growth and development. According to Haro and Benito (2019), "nutrient remobilization" aids plants in allocating nutrients as effectively as possible by focusing resources where they are most required. They can facilitate the transfer of nutrients from a plant's roots to its surface. By creating a network of hyphae that penetrates the soil, they expand the amount of surface area that can be used to absorb nutrients. Plants can absorb nutrients from a larger area of soil thanks to the network's role as a conduit for nutrient transfer. According to Chhabra and Dowling (2017), fungal endophytes increase the effectiveness of nutrient transfer, improving the general nutritional condition of plants as well as their vigor and productivity.

1.6.1.3 Nutrient Use Efficiency

Enhancing plant nutrient consumption efficiency is one of the main benefits of fungal endophytes in agroecosystems. The ability of plants to efficiently use the nutrients that are available while reducing losses and raising output is referred to as nutrient utilization efficiency. Fungal endophytes increase the efficiency of nutrient use in a number of ways (Chitnis et al. 2020). First off, micronutrients and other nutrients that plants would not otherwise be able to take are made possible by fungal endophytes. The aforementioned bacteria produce chelating compounds and organic acids that have the ability to solubilize and mobilize micronutrients, hence improving plant absorption. Fungal endophytes ensure that plants have an adequate supply of micronutrients, which are necessary in small quantities but play crucial roles in many physiological processes. They do this by improving the uptake of micronutrients (Sarkar and others 2021). In agroecosystems, they can improve nutrient cycling. These microorganisms aid in the breakdown of organic materials, liberating nutrients that had been previously trapped in complex compounds. Organic components are broken down by fungus endophytes, which releases nutrients back into the soil for plant uptake. By ensuring the continuous availability of essential components, this nutrient cycle mechanism reduces reliance on outside inputs and promotes sustainable agricultural practices (Zhou et al. 2017).

1.6.2 Improvement of Abiotic Stress Tolerance

Global agricultural productivity is significantly hampered by abiotic factors such as salt, drought, extreme heat, and heavy metal toxicity. Lower yields, reduced plant development, and even crop collapse might result from these stressors. It is essential for food security and sustainable agriculture to develop methods to increase plants' resistance to abiotic stresses. As a potential method of improving plant resistance to abiotic stressors, fungal endophytes have recently come to light (Fontana et al. 2021).

1.6.2.1 Drought Stress

Abiotic stressors that significantly affect agricultural productivity include drought. It depletes plants of water, which hinders photosynthesis, impedes nutrient intake, and damages cells. Plant tolerance to drought stress has been reported to be increased by fungus endophytes in a number of ways. First of all, fungi endophytes have the ability to improve plant water uptake. They achieve this by raising soil water input, decreasing transpiration, and regulating stomatal closure. Additionally, these endophytes can promote root formation and expansion, enabling plants to reach deeper soil layers with higher water availability (Decunta et al. 2021). Plants utilize them to make soluble carbohydrates and proline, two osmoprotectants. Plant cells are kept from drying out by these osmoprotectants, which also maintain the water equilibrium within cells. In addition, fungal endophytes have the ability to enhance the production of stress-related proteins and antioxidants, both of which are crucial in mitigating the deleterious effects of drought stress (Dastogeer et al. 2020).

1.6.2.2 Salinity Stress

Agricultural development is severely hampered in many locations by salinity stress, which is caused by high salt levels in the soil. It causes water deprivation and ion toxicity in plants by upsetting their osmotic balance. Significant promise for improving plant resilience to salt stress has been shown by fungus endophytes. As stated by Gupta et al., by a number of techniques, fungal endophytes can increase plants' resistance to salt. By reducing the uptake and accumulation of detrimental ions like sodium (Na+) and chloride (Cl–), they can first control ion homeostasis in plant tissues. Potassium (K+) and calcium (Ca2+), two essential ions needed for the best possible growth and development of plants, are balanced out in part via this regulation. They can promote the synthesis of proline, glycine, betaine, and other plant osmolytes. By controlling cellular water balance and guarding against ion toxicity, these osmolytes serve as osmoprotectants. Moreover, Cosoveanu et al. (2021) found that fungal endophytes can increase the activity of antioxidant enzymes like catalase and superoxide dismutase, which scavenge reactive oxygen species (ROS) created under salinity stress.

1.6.2.3 Temperature Stress

The growth and development of plants can be harmed by extreme temperatures, both hot and cold. Stress brought on by heat produces denaturation of proteins, damage to membranes, and oxidative stress; stress brought on by cold causes dehydration and interferes with cellular function. Several strategies are employed by fungal endophytes to increase plant resistance to temperature stress (Verma et al. 2022). By modifying antioxidant defense mechanisms and heat shock proteins (HSPs), fungal endophytes can enhance plant thermotolerance in heat-stressed environments. Protein folding and the prevention of protein aggregation under heat stress are crucial roles played by HSPs. Plant resistance to high temperatures can be enhanced by fungal endophytes by enhancing the expression of HSP genes. Additionally, plant cells are shielded from oxidative stress by antioxidant enzymes like glutathione reductase and peroxidase, which are activated by fungal endophytes. These enzymes scavenge reactive oxygen species (ROS). By modifying cold-responsive gene expression and increasing the synthesis of antifreeze proteins, fungal endophytes can enhance plant cold tolerance in conditions of cold stress. These proteins shield plant cells from the damaging effects of freezing by preventing the development of ice crystals. According to Adhikari et al. (2022) fungi endophytes can also boost the synthesis of sugars and polyols, two compounds known as cryoprotectants that aid in preserving cellular water balance and preventing cellular dehydration during cold stress.

1.6.2.4 Heavy Metal Stress

A serious environmental issue that endangers human health and negatively impacts plant growth is heavy metal pollution in soil. In a variety of ways, fungi endophytes have shown promise in enhancing plant resistance to heavy metal stress (Domka and associates 2019). Because fungi endophytes may store and immobilize heavy metals in their tissues, they can help plants endure exposure to them. They have the capacity to retain heavy metals in their cells, preventing them from migrating to plant parts above ground. Heavy metals' detrimental impacts on plant growth and development are lessened by this sequestration. It can also aid in the production of molecules like metallothioneins and phytochelatins that assist plants chelate metals. These substances lessen the toxicity of heavy metals by binding to them to form stable complexes. Antioxidant enzyme activity can also be increased by fungi endophytes, helping to lessen the oxidative stress brought on by heavy metal toxicity (Priyadarshini and associates 2021).

1.6.3 Biocontrol of Plant Pathogens

Because of their ability to act as biocontrol agents against plant diseases, fungi endophytes have drawn a lot of attention recently. Fungal infections can cause plant diseases that can seriously reduce yields and cause financial harm to the agricultural industry. Chemical pesticides, which can be hazardous to both humans and the environment, are widely used in traditional disease management techniques. On the other hand, a sustainable and environmentally safe substitute for treating plant diseases is the use of fungal endophytes as biocontrol agents (De Silva et al. 2019; Collinge et al. 2022).

1.6.3.1 Mechanisms of Biocontrol

Numerous mechanisms can be employed by fungal endophytes to demonstrate biocontrol effect. The synthesis of antifungal compounds is one of the most well-known processes. By impeding the establishment and development of plant pathogens, these substances can reduce the spread of illness. Antibiotics with broad-spectrum activity against a range of plant diseases, such as streptomycin and penicillin, are produced by certain fungal endophytes. VOCs, which are produced by other endophytes, have the ability to impede pathogen growth by interfering with their biological processes (Adeleke et al. 2022). Numerous mechanisms can be employed by fungal endophytes to demonstrate biocontrol effect. The synthesis of antifungal compounds is one of the most well-known processes. By impeding the establishment and development of plant pathogens, these substances can reduce the spread of illness. Antibiotics with broad-spectrum activity against a range of plant diseases, such as streptomycin and penicillin, are produced by certain fungal endophytes. VOCs, which are produced by other endophytes, have the ability to impede pathogen growth by interfering with their biological processes (Adeleke et al. 2022).

1.6.3.2 Selection and Application of Biocontrol Agents

It is necessary to carefully select and evaluate potential candidates in order to use fungal endophytes as biocontrol agents. It is necessary to take into consideration a number of factors, such as the endophyte's capacity to colonize the target plant, its antagonistic activity against the particular disease, and its compatibility with other beneficial bacteria. Promising endophytes with biocontrol potential can be identified using screening techniques like as greenhouse research and in vitro testing (Raymaekers and associates 2020). Various methods can be employed to apply suitable biocontrol agents to plants once they have been identified. One common technique is seed treatment, which involves treating the seeds with endophytes prior to sowing. Endophytes are able to colonize immature seedlings as a result, shielding them from early disease attacks. Foliar application is an additional method that entails misting endophytes into the leaves of established plants. This can protect against foliar infections and help slow the spread of illness (Yan and others 2019).

1.6.3.3 Benefits and Limitations of Biocontrol

Fungal endophytes offer several benefits over conventional chemical pesticides when used as biocontrol agents. They are not harmful to non-target species, like pollinators and beneficial insects, and they are environmentally friendly. Fungal endophytes provide no harmful residues and are biodegradable, in contrast to chemical pesticides that can linger in the environment and build up in food chains (Segaran and Sathiavelu 2019). Utilizing biocontrol agents can aid in reducing the rate at which infection resistance spreads. Chemical pesticides can be rendered ineffective over time by fungal infections that develop resistance to them. Conversely, it is challenging for illnesses to develop resistance due to the intricate relationships that exist between plants and fungal endophytes. According to Köhl et al. (2019), endophytes have a complex biocontrol activity that includes both the production of several antifungal compounds and the induction of plant defense mechanisms. Fungal endophytes can be used as biocontrol agents, however there are some restrictions. One problem is that the environment affects how effective they are. The effectiveness of biocontrol agents can be affected by soil conditions, temperature, and humidity. The specific circumstances of the target agroecosystem must therefore be taken into consideration when choosing and applying biocontrol chemicals. The dearth of approved methods and commercial products utilizing fungal endophytes is another disadvantage. The potential of endophytes for biocontrol is being studied more and more; however, future application method development and enhancement are still needed. To ensure the safe and responsible use of biocontrol agents in agriculture, regulatory concerns and safety assessments also need to be addressed (Dos Reis et al. 2022).

1.6.3.4 Future Perspectives

For plant protection and sustainable agriculture, the application of fungal endophytes as biocontrol agents holds great promise. Some issues need further investigation as this field of inquiry develops. One area of study is the potential benefits of mixing several endophyte strains or using them in combination with other biocontrol agents. This has the potential to enhance the overall effectiveness of biocontrol and offer a more comprehensive approach to managing diseases. Understanding the molecular mechanisms governing plant–endophyte–pathogen interactions is another area of attention for future research. Gaining knowledge of these pathways at the molecular and genetic levels may help design treatments tailored to individual diseases. Furthermore, the diversity and functional potential of fungal endophytes in agroecosystems can be better understood by the application of cutting-edge technologies like transcriptomics and metagenomics. And lastly, using fungal endophytes for biocontrol of plant diseases offers an environmentally sound and long-lasting substitute for managing disease in agroecosystems. Effective allies in the battle against plant diseases, endophytes can produce compounds that are antifungal, develop systemic resistance, and enhance plant health. To guarantee their safe and efficient application in agriculture, better understanding of their interactions with plants and pathogens, and improved application techniques, more study is necessary.

1.7 Induction of Plant Systemic Resistance

Plant systemic resistance is an important defense mechanism that enables plants to protect themselves from a variety of diseases. It entails the activation of a variety of physiological and biochemical mechanisms that improve the plant's ability to resist and overcome pathogens. Fungal endophytes have been revealed to play a substantial role in developing plant systemic resistance, making them useful allies in agroecosystems for plant protection (Hossain et al. 2017; Fontanta et al. 2021).

1.7.1 Mechanisms of Induced Systemic Resistance

Plant systemic resistance is induced by fungal endophytes through complex interactions with the plant. The plant is able to withstand potential pathogen attacks because of these interactions, which trigger a number of defense mechanisms. The mechanisms by which fungal endophytes induce systemic resistance in plants have been elucidated.

1.7.1.1 Priming of Defense Responses

Plant defense mechanisms may be heightened by fungus endophytes, which can improve the plant's resistance to disease and increase its responsiveness. Defensive substances including phytohormones and secondary metabolites are produced as a result of this priming effect, which also activates several genes linked to defense. Under pathogen attack, the primed condition enables the plant to mount a more robust and rapid defense response (Mengistu 2020).

1.7.1.2 Activation of Systemic Acquired Resistance (SAR)

In addition, fungal endophytes have the ability to boost a plant's systemic acquired resistance (SAR), which is a long-term defense mechanism that guards against a variety of illnesses. Cell wall reinforcement, antibacterial compounds, and pathogenesis-related (PR) proteins are produced during SAR. By triggering the signaling pathways that result in the activation of SAR, fungi endophytes can increase a plant's resistance to disease (Poveda et al. 2020).

1.7.1.3 Genes Associated with Systemic Resistance Induction

Plants can produce genes for systemic resistance in response to fungal endophytes. These genes are involved in defense mechanisms that include defense signaling pathway activation, cell wall reinforcement, and the manufacture of antimicrobial chemicals. These genes are expressed more frequently by fungi endophytes, which strengthens the plant's defenses against infections (Yu et al. 2022).

1.7.2 Enhanced Disease Resistance

Fungal endophytes induce plant systemic resistance, which enhances disease resistance in a variety of crops. Numerous studies have shown that fungal endophytes reduce the prevalence and severity of plant illnesses caused by bacterial, fungal, and viral pathogens. Fungal endophytes give better disease resistance, which is a sustainable and environmentally benign way to plant protection.

1.7.2.1 Suppression of Pathogen Growth and Development

Fungal endophytes have the power to directly control disease growth and development in plants. They generate antimicrobial compounds that are fungicidal, bactericidal, and virucidal, effectively preventing pathogen development. Furthermore, fungal endophytes can compete with pathogens for nutrients and space, limiting their ability to reproduce and cause illness (Latz et al. 2018).

1.7.2.2 Activation of Defense Signaling Pathways

Plants' defensive signaling pathways can be triggered by fungus endophytes, which causes the synthesis of defense chemicals and the bolstering of physical barriers. The plant can detect and react to pathogen assaults more effectively as a result of this activation. Fungal endophytes enhance the plant's defensive mechanisms, making it possible for the plant to fight off infections more quickly and effectively, which lessens the severity of sickness (Mattoo and Nonzom 2021)

1.7.2.3 Induction of Systemic Resistance in Non-Infected Tissues

Plant systemic resistance's ability to shield both diseased and healthy tissues is among its most astounding features. Systemic resistance to potential pathogen assaults can be induced in non-infected tissues by fungal endophytes. This provides the plant with an early line of defense. Reactive tactics like chemical pesticides are less necessary because to this systemic protection, which provides a proactive approach to disease prevention (Shankar 2019) (Table 1.2).

1.7.3 Field Applications of Fungal Endophytes for Inducing Plant Systemic Resistance

The capacity of fungal endophytes to develop systemic resistance in plants has garnered a great deal of attention in recent years. To take use of fungal endophytes' advantages for plant protection in agricultural settings, scientists and farmers are investigating a variety of strategies (Sudheep et al. 2017; Murphy et al. 2018).

Some of the field applications of fungal endophytes for inducing plant systemic resistance include:

Table 1.2 Mechanisms of induced systemic resistance using endophytic fungi

Mechanism	Description	Examples
Signaling pathway activation	Endophytes initiate defensive signaling pathways in plants, such as the ethylene (ET) and jasmonic acid (JA) pathways, which in turn cause the synthesis of defense chemicals and the induction of immunological responses.	*JA/ET pathway*: Piriformosporaindica, Serendipitaindica
Production of elicitors	Plant defense responses are stimulated by endophytes through the production of chemicals known as elicitors. Different kinds of molecules, including enzymes, secondary metabolites, and pieces of cell walls, can function as these elicitors.	Chitinases, glucanases, small peptides
Priming of plant defense	When faced with a pathogen, endophytes have the ability to "prime" the plant's defensive mechanisms, increasing their responsiveness and speed of activation. Long-term protection can be obtained from this priming effect, which has a lengthy half-life.	Unspecified fungal elicitors
Phytoalexin production	Phytoalexins are antimicrobial molecules that plants make in response to stress or illness, and they can be induced by endophytes.	Several endophytic fungi from various plant species
Antimicrobial metabolite production	Certain endophytes manufacture their own antimicrobial substances that either directly inhibit or reduce the development of infections. This may strengthen the plant's entire defensive mechanism.	Lolitrem B (antifungal) by Epichloetyphina
Competition for nutrients and space	Pathogenic microorganisms may face competition from endophytes for nutrients and available space in plant tissue, which might hinder the pathogen's ability to establish and spread.	Various endophytic fungi

1.7.3.1 Seed Treatment

Fungal endophyte seed treatment has had positive results in enhancing plant systemic resistance. It is ensured that endophytes take root in the plant from the beginning by covering seeds with endophytic fungus before to sowing (Rétif et al. 2023). The plant's defenses are strengthened by this early colonization, which offers long-term disease protection.

1.7.3.2 Foliar Sprays

To create systemic resistance, plant aerial sections can be sprayed with foliar solutions containing fungal endophytes. When endophytes penetrate the leaves, the plant as a whole is protected by defense mechanisms activated. When seed treatment is not an option or additional protection is required throughout the growth season, foliar sprays are particularly helpful.

1.7.3.3 Crop Rotation and Intercropping

Agroecosystems can benefit from the planned use of crop rotation and intercropping to increase the abundance of beneficial fungal endophytes. Farmers may help these helpful bacteria develop and remain in the soil by cultivating crops that are compatible with particular endophytes. Long-term pathogen protection is offered by this strategy, which also lessens the need for chemical treatments.

1.7.4 Challenges and Future Directions

Even though fungal endophytes have a promising potential to cause plant systemic resistance, a number of issues must be resolved before they can be successfully used in agroecosystems. Among the principal difficulties are:

1.7.4.1 Identification and Selection of Effective Endophytes

It is a challenging task to assess effective fungal endophytes for the development of plant systemic resistance. The host plant, target pathogen, and environmental factors all affect endophyte efficiency. To find and characterize endophytes that consistently and reliably protect plants, more research is required.

1.7.4.2 Integration with Other Plant Protection Strategies

For optimal results, it is recommended to combine fungal endophytes with other plant protection strategies. In order to give a thorough approach to disease control, endophytes can be employed in conjunction with cultural techniques such as crop rotation, intercropping, and the adoption of resistant cultivars. Moreover, the total effectiveness of plant protection techniques can be increased by combining endophytes with biopesticides and biological control agents.

1.7.4.3 Regulatory Considerations and Safety Assessment

It is crucial to consider the safety and regulatory implications before using fungal endophytes in commercial agriculture. It is important to carefully assess the risks associated with using endophytes that have unknown ecological repercussions or releasing genetically modified endophytes into the environment. It is necessary to create regulatory frameworks to guarantee the appropriate and safe application of fungal endophytes in agroecosystems.

1.8 Fungal Endophytes in Horticulture

The science and practice of cultivating plants for their aesthetic value as well as to yield fruits, vegetables, and other plant-based goods is known as horticulture. By promoting plant development, enhancing crop quality, and providing defense against a variety of biotic and abiotic challenges, fungal endophytes have shown to have significant promise for advancing horticultural practices (Hanaka et al. 2021).

1.8.1 Promotion of Plant Growth and Development

It has been demonstrated that fungi endophytes play a significant role in promoting plant growth and development in horticultural crops. These endophytes can enhance the absorption and use of nutrients, which will raise the vigor and production of the plants. They may solubilize micronutrients and minerals like phosphate, making them more accessible to plants. Additionally, phytohormones and enzymes produced by fungi endophytes can stimulate the development of roots and shoots, improve nutrient absorption, and boost overall plant performance. Fungal endophytes can also aid horticultural crops in resisting a range of abiotic stresses, such as salt, drought, and severe temperatures. By producing stress-related proteins and enzymes, regulating osmotic balance, and improving water and nutrient absorption efficiency, they can help plants cope with these difficulties (Abo Nouh et al. 2021; Gupta et al. 2021). Fungal endophytes can enhance the resilience and sustainability of horticultural systems by enhancing their ability to withstand stress, particularly in areas where environmental problems are common.

1.8.2 Future Prospects and Areas for Further Investigation

As more is known about fungal endophytes, so is their potential use for stabilizing agro-ecosystems and plant protection. The investigations done pro tempore have provided valuable insights into the multifaceted roles that fungal endophytes may play in enhancing plant health and productivity. Future studies need to be aimed in order to exploit fully the potential of these exciting microbes. In this section, we will take a closer look at some of the prospects and topics of the research in endophytic fungi.

1.8.2.1 Unexplored Diversity of Fungal Endophytes

While much has been done towards an understanding of endophytic fungal diversity, in truth, much about them remains underexploited. Most studies in this regard have been done on only those culturable endophytes, hence representing a very minute proportion of total fungal diversity. Future research using culture-independent investigations and the ongoing revolution in genomics at the community level will further untangle the hidden complexity of fungal endophyte diversity, yielding insights into distribution and abundance; the ecological functions of these endophytes play in a wide range of agroecosystems (Savi et al. 2019).

1.8.2.2 Functional Characterization of Fungal Endophytes

While a few aspects of their various functional potentials have been detailed, much still needs to be understood on how these fungal endophytes confer benefits or tolerate stress. Research on bioactive chemicals beyond the general characterization needs to be done to discover precise bioactive chemicals produced and their input potentials in agriculture. The fungal endophytes produce many chemicals likely to inhibit the growth of other microorganisms like bacteria, fungi, and viruses. Fungal

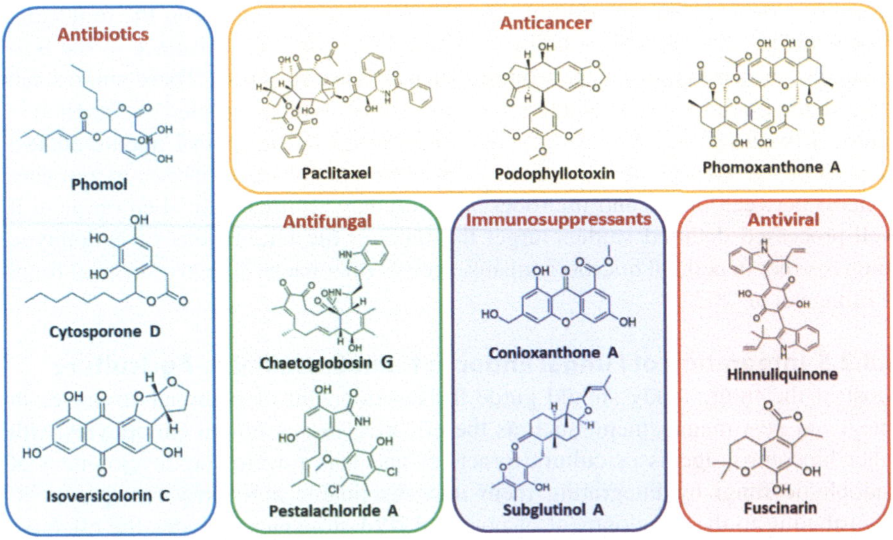

Fig. 1.3 Graphic representations of some bioactive substances derived from fungus endophytes that are categorized according to their bioactivities. It gives a summary of the wide variety of compounds that make up this group and emphasizes some of their possible uses in various industries (Gakuubi et al. 2021)

endophytes contain several chemicals that can hinder the growth of microorganisms like bacteria, fungus, and viruses (Fig. 1.3). Since these chemicals exhibit anticancer activities, scavenge damaging free radicals and reduce inflammation, modulate the immune system, prevent injury to brain cells, and suppress virus replication. Anti-parasitic chemicals would be limiting the growth or activity of bacteria and viruses. Generally, the fungal endophytes find immense health significance, and they can be applied in treating the available diseases.

1.8.2.3 Optimization of Fungal Endophyte Inoculation Strategies

Inoculation procedures need to be developed so that the full potentials of fungal endophytes in agriculture can be realized. Research needs to concentrate on developing efficient and cost-effective mass production and delivery techniques to plants. This may involve investigations on new formulations of the inocula that enhance survival and establishment during the colonization of fungi in the plant host, such as encapsulation and seed coating. Furthermore, research on the time and frequency of inoculation is justified to show the best conditions in which maximal plant growth promotion and disease suppression could occur (Sharma et al. 2023).

1.8.2.4 Understanding the Role of Fungal Endophytes in Plant Defense

While fungal endophytes have been shown to influence plant growth and development positively, their value as plant defense mechanisms against disease has yet to be entirely determined. In addition, future studies need to focus on the molecular events mediated by fungal endophytes that induce systemic resistance in the host plant and, in general, that condition the plant defense response. These studies can confirm the likelihood of valuable information that might be targeted and applied in future activities concerned with fungal endophytes as biocontrol agents against plant diseases. These dense, complex webs of interactions that happen in agroecosystems between plants and microbes will only have further light shed on them if well-processed detailed studies target the links in the interactions of endophytic fungi to other beneficial microorganisms, such as rhizobacteria or mycorrhizal fungi (Collinge et al. 2022).

1.8.2.5 Integration of Fungal Endophytes in Sustainable Agriculture

Most of the future study should guide the development of potential strategies in integrated pest management, such as the effective use of fungal endophytes with other biocontrol agents or cultural practices and lean toward the development of endophytic fungi by integrating them into sustainable agroecosystem practices. Contributing to the development of potential IPM strategies includes the effective use of bio-agents in conjunction with other bio-control agents and with acknowledged cultural practices in this field. Such a strategy will reduce the depreciation on inputs made on the tank of crop protection chemicals and have more environmentally friendly ways of plant protection. Therefore, long-term experiments are needed not only to test them on their effect on soil health and determine whatever sustainable benefits are accruing in the process but also to have subsequent implications on microbial diversity so that these mycorrhizal and endophyte inoculants, amongst others, based on this concept are used sustainably in agro-ecosystems (Baron and Rigobelo 2022).

1.8.2.6 Commercialization and Large-scale Application of Fungal Endophytes

The real challenge for fungal endophytes lies in their commercialization and realizing their potential for application in agricultural land. Further research must include methods of low-cost production for fungal endophytes and the development of quality control measures to assure their efficacy. Furthermore, regulatory considerations and pre- and post-registration safety studies can also help in registering and licensing products based on fungal endophytes. Collaboration among researchers, industries, and the regulatory bodies of this product will be a critical ingredient for addressing the challenges associated with commercialization and large-scale deployment of fungal endophytes (Ganeshan et al. 2021). Many more research efforts should be directed toward commercializing and applying them in agriculture on a much larger scale for the potential of endophytes to be realized. The most crucial thing in low-cost production is quality.

There should be inbuilt measures of controlling the quality of these fungal endophytes by upholding high standard measures and an upsurge of the endophyte. More so, regulatory consideration and safety studies must be conducted to make product registration and licensing easier about fungal endophytes. It is only a collaboration between the researchers, industries, and regulatory authorities that can provide a formidable solution to overcoming the fact that there are sundry challenges involved in the commercialization and large-scale deployment of fungal endophytes. Future activities in fungal endophyte research look promising. It is expected that the unknown diversity and functional characterization of inoculum protocol optimization in further experimental depth will bring us closer to understanding possible applications of fungal entomopathogens in the context of agroecosystems and plant protection. More enlightenment on molecular mechanisms of plant–fungus interaction unraveled how fungal endophytes could be hosted into sustainable agricultural practices for the development of environmentally friendly crop production that is also economically viable. More and improved data on the FEs and combined efforts in advanced research collaboration promised to bring about the changes needed to renovate the face of agriculture into a replicable food system. The past few decades have greatly emphasized and witnessed the influential role of fungal endophytes as potential biocontrol agents against many infections, which post an eco-friendly solution against chemical pesticides. Fungal endophytes reduce the development and activity of plant pathogens by nutrient and space competition, antibiotic chemical synthesis with fungicidal activity, and other modes, including induction of responses for plant defense.

Studies establish the effectiveness of fungal endophytes out of many diseases of horticultural crops, which are probably involved in the defense mechanisms of horticultural crops: A host of different studies establish successful protection against a wide ex vivo pathogen-induced spectrum of diseases found on several fruits and vegetables using different endophytic fungi, namely Botrytis c. cinerea, Colletotrichum spp., etc. Interestingly enough, these endophytes live within the plant tissues and produce antifungal compounds to limit the growth and development of pathogens; therefore, they reduce the disease incidence on the plants and protect horticultural produce through shelf life.

1.8.2.7 Enhancement of Crop Quality and Nutritional Value

Moreover, the fungal endophytes may influence the synthesis of and metabolism of plants in secondary metabolites by producing phytochemicals of medicinal and industrial values. There are several reports on the stimulation of some fungal endophytes participating in the synthesis of essential oils in aromatic plants, which use these fairly apply in fragrance and cosmetic industries. In such a context, it is promising that the potential of fungal endophytes can enhance resistance in horticultural crops against the ever-increasing demand for high-quality, nutritious, and enhanced-value horticultural crops. An AM fungus (Funneliformis mosseae) and an endophytic fungus (Piriformospora indica) were inoculated either singly or in combination into Polygonum cuspidatum in a time-course experiment designed to explore effective inoculation procedures for enhancing disease-suppressive activity and biomass production in Boerugie medic: growth and development, physiological activities—chlorophyll contents, soluble proteins, gas exchange, and their contents of medicinal ingredients—coupled with changes in the expression of six genes related to the resveratrol synthesis pathway (Sun et al. 2022) (Fig. 1.4).

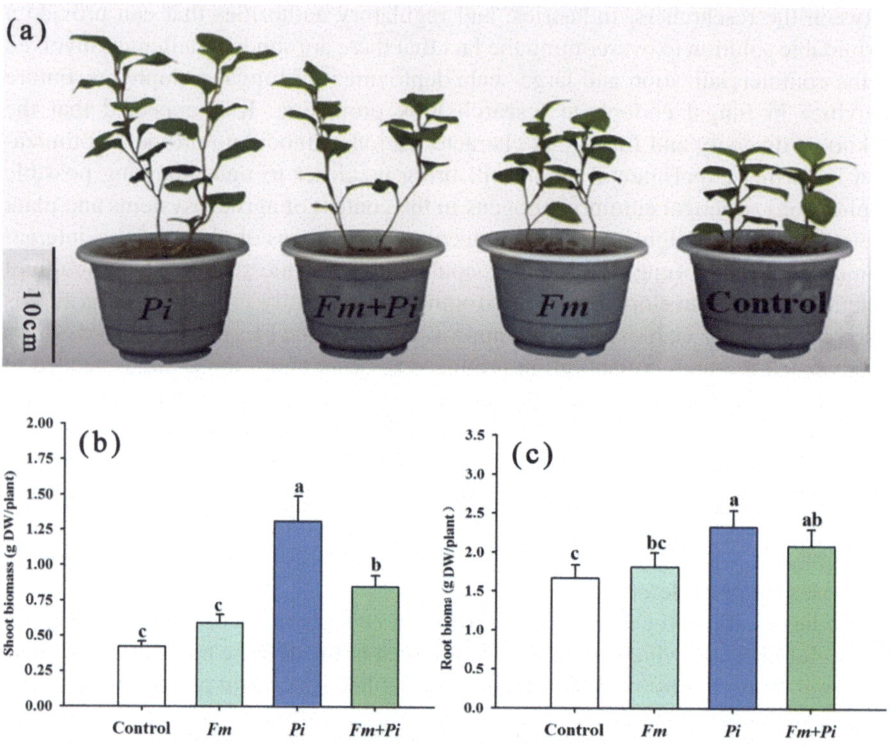

Fig. 1.4 Effects of single and co-inoculation with *Funneliformis mosseae* (*Fm*) and *Piriformospora indica* (*Pi*) on plant growth (**a**), shoot biomass (**b**), and root biomass (**c**) in *Polygonum cuspidatum*. Data (means ± SD, $n = 5$) followed by different letters above the bars indicate significant ($p < 0.05$) differences among treatments. (Reprinted from Sun et al. 2022)

1.8.2.8 Integration of Fungal Endophytes in Horticultural Systems

The integration of such fungal endophytes into the horticultural system must take into account specific inputs in aspects such as host specificity, application techniques, and environmental conditions. In this respect, the critical stage is selection: the endophytes must be capable of establishing symbiotic linkages with the selected target crops without the ensuing detrimental consequences. Two following leading stages involve isolation and identification of appropriate endophytes from healthy plants or natural habitats (Samal et al. 2023). Once the right endophytes have been identified, interest application methods can be implemented in horticultural crops. Endophytes can be introduced through seed treatment, foliar spraying, root drenching, and soil applications. The treatment time and frequency should be modulated with the correct timing for maximum colonization and efficacy of the endophytes. In addition, the ambient conditions of temperature, humidity, and nutrient availability should be conducive.

1.9 Conclusion

This chapter discusses the major role of endophytes to increase the agroecosystem productivity. They help enhance plant growth and nutrient uptake while also improving stress tolerance; meanwhile, they are indeed capable of safeguarding these plants from pathogens and pests. These fungi ease nutrient uptake through augmented roots and release growth-promoting compounds. Endophytic fungi confer abiotic stress tolerance. Moreover, they contribute to controlling certain plant pathogens by producing secondary antimicrobial metabolites. The key interactions, primary metabolites involved, and factors affecting the efficiency of the mentioned interactions under different agricultural systems remain to be further elaborated on in future work. Therefore, endophytic fungi present one of the most important potentials for promoting sustainable agriculture, minimizing the synthetic inputs usage, and plant health enhancement and agricultural systems productivity.

References

Abo Nouh FA (2019) Endophytic fungi for sustainable agriculture. Microb Biosyst 4(1):31–44

Abo Nouh F, Abu-Elsaoud A, Abdel-Azeem A (2021) The role of endophytic fungi in combating abiotic stress on tomato. Microb Biosyst 6(1):35–48

Adeleke BS, Ayilara MS, Akinola SA, Babalola OO (2022) Biocontrol mechanisms of endophytic fungi. Egypt J Biol Pest Control 32(1):1–17

Adhikari L, Baral R, Paudel D, Min D, Makaju SO, Poudel HP et al (2022) Cold stress in plants: strategies to improve cold tolerance in forage species. Plant Stress 4:100081

Akram S, Ahmed A, He P, He P, Liu Y, Wu Y et al (2023) Uniting the role of endophytic fungi against plant pathogens and their interaction. J Fungi 9(1):72

Alves GS, Bertini SCB, Barbosa BB, Pimentel JP, Junior VAR, de Oliveira Mendes G, Azevedo LCB (2021) Fungal endophytes inoculation improves soil nutrient availability, arbuscular mycorrhizal colonization and common bean growth. Rhizosphere 18:100330

Ancheeva E, Daletos G, Proksch P (2020) Bioactive secondary metabolites from endophytic fungi. Curr Med Chem 27(11):1836–1854

Baron NC, Rigobelo EC (2022) Endophytic fungi: a tool for plant growth promotion and sustainable agriculture. Mycology 13(1):39–55

Berg G, Hallmann J (2006) Control of plant pathogenic fungi with bacterial endophytes. In: Microbial root endophytes. Springer Berlin Heidelberg, Berlin, Heidelberg, pp 53–69

Chhabra S, Dowling DN (2017) Endophyte-promoted nutrient acquisition: phosphorus and iron. In: Functional importance of the plant microbiome: implications for agriculture, forestry and bioenergy. Springer, Cham, pp 21–42

Chitnis VR, Suryanarayanan TS, Nataraja KN, Prasad SR, Oelmüller R, Shaanker RU (2020) Fungal endophyte-mediated crop improvement: the way ahead. Front Plant Sci 11:561007

Collinge DB, Jensen B, Jørgensen HJ (2022) Fungal endophytes in plants and their relationship to plant disease. Curr Opin Microbiol 69:102177

Cosoveanu A, Chowdhary K, Cabrera R, Sharma S (2021) Role of phytohormones-producing fungal endophytes in plant–microbial interactions under stress. In: Endophytes: potential source of compounds of commercial and therapeutic applications. Springer, pp 195–223

Dastogeer KMG, Chakraborty A, Sarker MSA, Akter MA (2020) Roles of fungal endophytes and viruses in mediating drought stress tolerance in plants. Int J Agric Biol 24(6):1497–1512

De Silva NI, Brooks S, Lumyong S, Hyde KD (2019) Use of endophytes as biocontrol agents. Fungal Biol Rev 33(2):133–148

Decunta FA, Pérez LI, Malinowski DP, Molina-Montenegro MA, Gundel PE (2021) A systematic review on the effects of Epichloë fungal endophytes on drought tolerance in cool-season grasses. Front Plant Sci 12:644731

Devi R, Kaur T, Kour D, Rana KL, Yadav A, Yadav AN (2020) Beneficial fungal communities from different habitats and their roles in plant growth promotion and soil health. Microb Biosyst 5(1):21–47

Domka AM, Rozpaądek P, Turnau K (2019) Are fungal endophytes merely mycorrhizal copycats? The role of fungal endophytes in the adaptation of plants to metal toxicity. Front Microbiol 10:371

Dos Reis JBA, Lorenzi AS, do Vale HMM (2022) Methods used for the study of endophytic fungi: a review on methodologies and challenges, and associated tips. Arch Microbiol 204(11):675

Dutta D, Puzari KC, Gogoi R, Dutta P (2014) Endophytes: exploitation as a tool in plant protection. Braz Arch Biol Technol 57:621–629

Fadiji AE, Babalola OO (2020) Exploring the potentialities of beneficial endophytes for improved plant growth. Saudi J Biol Sci 27(12):3622–3633

Fontana DC, de Paula S, Torres AG, de Souza VHM, Pascholati SF, Schmidt D, Dourado Neto D (2021) Endophytic fungi: biological control and induced resistance to phytopathogens and abiotic stresses. Pathogens 10(5):570

Gakuubi MM, Munusamy M, Liang ZX, Ng SB (2021) Fungal endophytes: a promising frontier for discovery of novel bioactive compounds. J Fungi 7(10):786

Ganeshan S, Kim SH, Vujanovic V (2021) Scaling-up production of plant endophytes in bioreactors: concepts, challenges and perspectives. Bioresour Bioprocess 8:1–16

García-Latorre C, Rodrigo S, Santamaría O (2021) Endophytes as plant nutrient uptake-promoter in plants. In: Endophytes: mineral nutrient management, Volume 3. Springer, pp 247–265

Gupta S, Schillaci M, Walker R, Smith PM, Watt M, Roessner U (2021) Alleviation of salinity stress in plants by endophytic plant-fungal symbiosis: current knowledge, perspectives and future directions. Plant Soil 461:219–244

Hanaka A, Ozimek E, Reszczyńska E, Jaroszuk-Ściseł J, Stolarz M (2021) Plant tolerance to drought stress in the presence of supporting bacteria and fungi: an efficient strategy in horticulture. Horticulturae 7(10):390

Harman GE, Uphoff N (2019) Symbiotic root-endophytic soil microbes improve crop productivity and provide environmental benefits. Scientifica 2019:9106395

Haro R, Benito B (2019) The role of soil fungi in K+ plant nutrition. Int J Mol Sci 20(13):3169

Hartmann M, Six J (2023) Soil structure and microbiome functions in agroecosystems. Nat Rev Earth Environ 4(1):4–18

Hossain MM, Sultana F, Islam S (2017) Plant growth-promoting fungi (PGPF): phytostimulation and induced systemic resistance. In: Plant-microbe interactions in agro-ecological perspectives: Volume 2: Microbial interactions and agro-ecological impacts. Springer, Singapore, pp 135–191

Khan AAH (2020) Endophytic fungi and their impact on agroecosystems. In: Medicinal plants: biodiversity, sustainable utilization and conservation. Springer, Singapore, pp 443–499

Khan AL, Hussain J, Al-Harrasi A, Al-Rawahi A, Lee IJ (2015) Endophytic fungi: resource for gibberellins and crop abiotic stress resistance. Crit Rev Biotechnol 35(1):62–74

Köhl J, Kolnaar R, Ravensberg WJ (2019) Mode of action of microbial biological control agents against plant diseases: relevance beyond efficacy. Front Plant Sci 845

Lacava PT, Bogas AC, Cruz FDPN (2022) Plant growth promotion and biocontrol by endophytic and rhizospheric microorganisms from the tropics: a review and perspectives. Front Sustain Food Syst 6:796113

Lanfranco L, Bonfante P, Genre A (2016) The mutualistic interaction between plants and arbuscular mycorrhizal fungi. Microbiol Spectr 4(6):4–6

Latz MA, Jensen B, Collinge DB, Jørgensen HJ (2018) Endophytic fungi as biocontrol agents: elucidating mechanisms in disease suppression. Plant Ecol Diversity 11(5–6):555–567

Le Cocq K, Gurr SJ, Hirsch PR, Mauchline TH (2017) Exploitation of endophytes for sustainable agricultural intensification. Mol Plant Pathol 18(3):469–473

Lugtenberg BJ, Caradus JR, Johnson LJ (2016) Fungal endophytes for sustainable crop production. FEMS Microbiol Ecol 92(12):fiw194

Malarvizhi K, Murali TS, Kumaresan V (2023) Fungal endophytes of crop plants: diversity, stress tolerance and biocontrol potential. Egypt J Biol Pest Control 33(1):1–7

Mattoo AJ, Nonzom S (2021) Endophytic fungi: understanding complex cross-talks. Symbiosis 83(3):237–264

Mengistu AA (2020) Endophytes: colonization, behaviour, and their role in defense mechanism. Int J Microbiol 2020:6927219

Mosaddeghi MR, Hosseini F, Hajabbasi MA, Sabzalian MR, Sepehri M (2021) *Epichloë* spp. and *Serendipita indica* endophytic fungi: functions in plant-soil relations. Adv Agron 165:59–113

Murphy BR, Doohan FM, Hodkinson TR (2018) From concept to commerce: developing a successful fungal endophyte inoculant for agricultural crops. J Fungi 4(1):24

Nasslahsen B, Prin Y, Ferhout H, Smouni A, Duponnois R (2022) Management of plant beneficial fungal endophytes to improve the performance of agroecological practices. J Fungi 8(10):1087

Poveda J, Abril-Urias P, Escobar C (2020) Biological control of plant-parasitic nematodes by filamentous fungi inducers of resistance: trichoderma, mycorrhizal and endophytic fungi. Front Microbiol 11:992

Poveda J, Eugui D, Abril-Urías P, Velasco P (2021) Endophytic fungi as direct plant growth promoters for sustainable agricultural production. Symbiosis 85(1):1–19

Priyadarshini E, Priyadarshini SS, Cousins BG, Pradhan N (2021) Metal-fungus interaction: review on cellular processes underlying heavy metal detoxification and synthesis of metal nanoparticles. Chemosphere 274:129976

Rana KL, Kour D, Sheikh I, Yadav N, Yadav AN, Kumar V et al (2019) Biodiversity of endophytic fungi from diverse niches and their biotechnological applications. In: Advances in endophytic fungal research: present status and future challenges. Springer, pp 105–144

Raymaekers K, Ponet L, Holtappels D, Berckmans B, Cammue BP (2020) Screening for novel biocontrol agents applicable in plant disease management–a review. Biol Control 144:104240

Rétif F, Kunz C, Calabro K, Duval C, Prado S, Bailly C, Baudouin E (2023) Seed fungal endophytes as biostimulants and biocontrol agents to improve seed performance. Front Plant Sci 14:1260292

Samal I, Bhoi TK, Majhi PK, Murmu S, Pradhan AK, Kumar D et al (2023) Combatting insects mediated biotic stress through plant associated endophytic entomopathogenic fungi in horticultural crops. Front Plant Sci 13:1098673

Sarkar S, Dey A, Kumar V, Batiha GES, El-Esawi MA, Tomczyk M, Ray P (2021) Fungal endophyte: an interactive endosymbiont with the capability of modulating host physiology in myriad ways. Front Plant Sci 12:701800

Savi DC, Aluizio R, Glienke C (2019) Brazilian plants: an unexplored source of endophytes as producers of active metabolites. Planta Med 85(08):619–636

Segaran G, Sathiavelu M (2019) Fungal endophytes: a potent biocontrol agent and a bioactive metabolites reservoir. Biocatal Agric Biotechnol 21:101284

Shahrtash M, Brown SP (2023) Interactions between fungal endophytes and pathogens isolated from the invasive plant kudzu (Pueraria montana var. lobata). Arch Phytopathol Plant Protect 56(2):87–107

Shankar Naik B (2019) Functional roles of fungal endophytes in host fitness during stress conditions. Symbiosis 79(2):99–115

Sharma I, Raina A, Choudhary M, Kaul S, Dhar MK (2023) Fungal endophyte bioinoculants as a green alternative towards sustainable agriculture. Heliyon 9:e19487

Singh LP, Gill SS, Tuteja N (2011) Unraveling the role of fungal symbionts in plant abiotic stress tolerance. Plant Signal Behav 6(2):175–191

Sodhi GK, Saxena S (2023) Role of endophytic fungi in promoting plant health. In: Fungal resources for sustainable economy: current status and future perspectives. Singapore, Springer Nature Singapore, pp 319–345

Sudheep NÁ, Marwal A, Lakra N, Anwar K, Mahmood S (2017) Fascinating fungal endophytes role and possible beneficial applications: an overview. In: Plant-microbe interactions in agro-ecological perspectives: Volume 1: Fundamental mechanisms, methods and functions. Springer, Singapore, pp 255–273

Sun RT, Zhang ZZ, Feng XC, Zhou N, Feng HD, Liu YM et al (2022) Endophytic fungi accelerate leaf physiological activity and resveratrol accumulation in Polygonum cuspidatum by up-regulating expression of associated genes. Agronomy 12(5):1220

Toju H, Peay KG, Yamamichi M, Narisawa K, Hiruma K, Naito K et al (2018) Core microbiomes for sustainable agroecosystems. Nature Plants 4(5):247–257

Vardharajula S, SkZ A, Shiva Krishna Prasad Vurukonda S, Shrivastava M (2017) Plant growth promoting endophytes and their interaction with plants to alleviate abiotic stress. Curr Biotechnol 6(3):252–263

Verma SK, Sahu PK, Kumar K, Pal G, Gond SK, Kharwar RN, White JF (2021) Endophyte roles in nutrient acquisition, root system architecture development and oxidative stress tolerance. J Appl Microbiol 131(5):2161–2177

Verma A, Shameem N, Jatav HS, Sathyanarayana E, Parray JA, Poczai P, Sayyed RZ (2022) Fungal endophytes to combat biotic and abiotic stresses for climate-smart and sustainable agriculture. Front Plant Sci 13:953836

Waqar S, Bhat AA, Khan AA (2023) Endophytic fungi: unravelling plant-endophyte interaction and the multifaceted role of fungal endophytes in stress amelioration. Plant Physiol Biochem 206:108174

Yan L, Zhu J, Zhao X, Shi J, Jiang C, Shao D (2019) Beneficial effects of endophytic fungi colonization on plants. Appl Microbiol Biotechnol 103:3327–3340

Yu Y, Gui Y, Li Z, Jiang C, Guo J, Niu D (2022) Induced systemic resistance for improving plant immunity by beneficial microbes. Plan Theory 11(3):386

Zhao J, Zhou L, Wang J, Shan T, Zhong L, Liu X, Gao X (2010) Endophytic fungi for producing bioactive compounds originally from their host plants. Curr Res Tech Educ Trop Appl Microbiol Microbial Biotechnol 1:567–576

Zhou Y, Zhu H, Yao Q (2017) Improving soil fertility and soil functioning in cover cropped agroecosystems with symbiotic microbes. In: Agro-environmental sustainability: Volume 1: Managing crop health. Springer, pp 149–171

Multifunctional Efficacy and Eco-friendly Applications of Fungal Endophytes

2

Riya Dutta, Hiran Kanti Santra, and Debdulal Banerjee

Abstract

Recent advancements in agricultural practices lead to rapid production of agricultural commodities but with some serious long-term irrepressible consequences for natural resources. The problems of pathogenic fungal, bacterial, and nematode attacks and the issue of increased production of crops could also be managed by sustainable techniques other than using hazardous chemicals with detrimental side effects. Out of many possibilities, endophytic microorganisms are also one of the most valued solutions in this respect. Applying the plant's integral microbial entities would be a possible way to regain rhizospheric fertility under sustainable agricultural practices to restore the healthy upbringing of Mother Nature. In this aspect, fungal endophytes represent a potential group of genera that can modulate local and systemic disease resistance as an integral part of their host plant. The beneficial effects can be obtained by modifying the biochemical and genetic expression of fungal inhabitants. Many pioneering scientific studies have pointed out the positive kinship of endophytic fungal interactions such as the acquisition of soil-bound minerals, production of bioactive secondary metabolites, phytohormones, siderophores, osmolytes, and induction of acquired and systemic resistance against plant pathogens. They are highly assorted in

R. Dutta
Centre for Life Sciences, Vidyasagar University, Midnapore, West Bengal, India

H. K. Santra
Department of Botany and Forestry, Vidyasagar University, Midnapore, West Bengal, India

D. Banerjee (✉)
Centre for Life Sciences, Vidyasagar University, Midnapore, West Bengal, India

Microbiology and Microbial Biotechnology Laboratory, Department of Botany and Forestry, Vidyasagar University, Midnapore, West Bengal, India
e-mail: db@mail.vidyasagar.ac.in

© The Author(s), under exclusive license to Springer Nature Singapore Pte Ltd. 2025

K. A. Abd-Elsalam, A. H. Hashem (eds.), *Fungal Endophytes Volume II*,
https://doi.org/10.1007/978-981-97-8804-0_2

terms of tissue-specific occurrence and ecological diversity with respect to host plants. In recent years, proteomics and meta-genomics approaches have given an extra highlight on the motion to understand the nature of complexity and compatibility of plant-microbe symbiosis. Endophytic fungi also significantly influence the activation of stress-mitigating genes to safeguard the plant cell against biotic and abiotic environmental stimuli. Findings reveal the beneficial application of endophytic fungal genera, including Piriformospora, Colletotrichum, Trichoderma, Epicoccum, Paecilomyces, Aureobasidium, and so on, displaying intimate linkage with their host plant to certify the ultimate supremacy of existence under stress condition. There is an urgent need to manifest the hidden interaction of plants and fungal flora that can be used as a supplement for establishing green technology. This chapter summarizes the current scenario on the diversity of fungal endophytes and deciphers their diverse roles to maintain agricultural sustainability despite scientific barriers.

Keywords

Fungal endophytes · Bioactive secondary metabolites · Green technology · Sustainable agriculture

2.1 Introduction

Endophytes define a very promising and complex system of relationships with a diverse group of phytotaxa worldwide. They are naturally occurring microflora of plant integral tissues that manifest their activity through co-evolution with their host defense mechanisms. Pioneering research on plant–microbe interactions was most widely focused on the rhizospheric microbial community, with special emphasis on nodule formation in legumes and mycorrhizal associations. Endophytes modified their micro-environmental conditions in equilibrium with their host's internal defense mechanisms. They act as a mutualistic protectant against environmental stresses by reducing biotic and abiotic pressure. From the first decade, fungal species were considered before bacterial species as a part of the endophyte community. However, omics approaches open up new dimensions for exploring bioactive metabolites of endophytic origin. Endophytes produced a group of bioactive compounds to manage the biometric as well as disease resistance capabilities of agricultural commodities more than 100 years ago (Kumar et al. 2020). They have been well characterized and provide significant benefits for maintaining plant growth and development under climatic challenges. Research studies have always focused on plant–microbe interaction to better understand the complexities of the environment. This dynamic interaction involves a complex exchange of signals, nutrients, and support, contributing to the overall well-being of the plant. The most frequent candidate in the endophytic community is fungi. A large number of fungal endophytes have also been reported from plants growing in extremely cold or hot regions, and their enormous diversity may play a significant role in the host's life cycle (Kapoor et al. 2022). The range of fungal endophytes differs within different plant tissues according to time and function.

The growing population density has caused environmental concerns, leading to excessive agricultural land agrochemical inputs. Whereas, huge accessibility and a lack of knowledge about commercially available inorganic products cause hazardous consequences in terms of environmental as well as human health. As biofertilizers, endophytes contribute to sustainable agriculture by reducing the reliance on synthetic fertilizers and promoting a healthier plant ecosystem. Reports about the integrity of fungal endophytic species have been manifesting their more intense impact than rhizospheric microorganisms. Numerous studies have been conducted to explore the beneficial effect of endophytes as biological control agents in a bioformulating way. For example, various fungal endophytes, including *Trichoderma* sp., *Phoma* sp., *Piriformospora* sp., *Colletotrichum* sp., entomopathogenic fungi, etc., can successfully colonize different dicot and monocot plant species and possess the essential function to accelerate the growth and yield of agriculture crops under stress conditions (Santra and Banerjee 2022a, b). Recognizing the need to counter the adverse effects of commercial fertilizers, researchers have acknowledged the importance of exploring crop-specific bioinoculants. The gradual investigation has opened new dimensions to overcome environmental barriers by applying modern molecular techniques.

2.2 Fungal Endophytes and Their Multifunctional Efficacy

Endophytes are gradually gaining interest among the new era's microbiological researchers due to their different habitat occurrences tied to multifaceted activities. It represents a dynamic field of research with implications for agriculture, ecology, and biotechnology. Modern molecular techniques have revolutionized the study of fungal endophytes, allowing for a deeper understanding of their genetics, functional attributes, and ecological roles. Though root colonization by rhizospheric microbes immensely helps the host plant survive environmental stresses, rather than the intrinsic participants of plant tissues, they possess more significant interaction. Fungal endophytic species occupy a novel group that differs in space, time, and function within a diverse range of host plants. They can spread from one generation to another by spreading their mycelium within meristematic tissues along the longitudinal axis and through seeds (vertical transmission) or by transmitting plant decay materials from nature to a closely related species (horizontal transmission). In the balanced antagonistic interaction of plant–microbe equilibrium, fungal endophytes play a recognized role. Host defensive systems activate numerous mechanisms to tolerate biotic and abiotic stimuli, endophytes ultimately contributing to the eco-friendly establishment of sustainable agriculture (Fig. 2.1). Fungal endophytes have been extensively studied for their potential as biocontrol agents against plant pathogens. Akram et al. (2023) highlight the role of these endophytes in enhancing plant growth, inducing systemic resistance, and producing antifungal secondary metabolites. Breeyen et al. (2022) introduce the concept of mycoparasitism, where

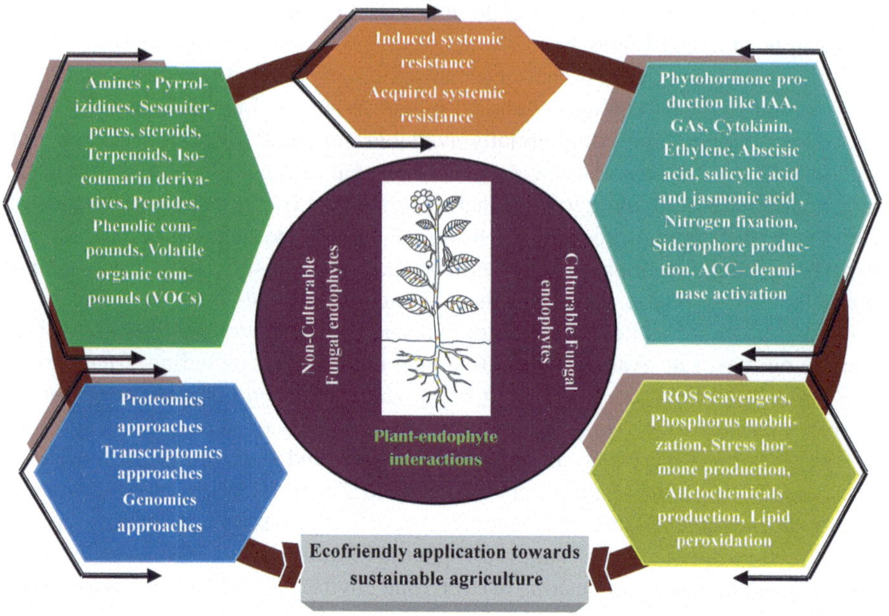

Fig. 2.1 Multifunctional interactions of plant–fungal endophytes to achieve crop sustainability

endophytes parasitize other fungi, potentially affecting the success of classical biological control agents. Their multifunctional efficacy also provides a suitable atmosphere to cope with environmental challenges caused by the interactions of xenobiotic factors. Fungal endophytes consecutively evolve as per the requirement of evolutional changes day by day. These studies collectively underscore the diverse mechanisms through which fungal endophytes can serve as effective biocontrol agents in agriculture.

There are several bases, such as the mode of transmission, ecological habitat, host range and diversity, colonization, and mode of reproduction, to classify fungal endophytes into different classes. According to the classification of Rodrigeuz et al. (2009), fungal endophytes are divided into two major groups of classes clavicipitaceous or C-endophytes (especially reside within grasses of tropical and temperate regions) and non-clavicipitaceous or NC-endophytes (especially residing within vascular and non-vascular plants species) (Table 2.1). However, non-clavicipitaceous endophytic fungi are further categorized into 3 different classes according to their diverse host range, species diversity, and colonization rate. Class 2 belongs to the genera of Ascomycota and Basidiomycota, and is generally associated with green tissues of algae, bryophytes, pteridophytes, conifers, and angiosperms via vertical and horizontal transmission. Class 3 is also rich in diversity of fungal genera occupying leaves of trees in tropical regions. Endophytes of Class 4 generally consisted of dark septate fungal species transmitted through the horizontal mode of transmission and colonizing the root tissues of the host plant, basically of cold geographic

Table 2.1 Characteristics of fungal endophytes according to the classification of Rodriguez et al. (2009)

Class of endophytes		Specificity	Mode of transmission	Biodiversity
Clavicipitaceous endophytes		Shoot and root tissues of grasses	Both Vertical and horizontal	Low
Non-Clavicipitaceous endophytes	Class 2	Root, stem, and Leaves of green plants	Both Vertical and horizontal	Low
	Class 3	Leaves of tropical trees	Only horizontal	High
	Class 4	Mainly roots of cold region's plants	Only horizontal	Unknown

regions. Recently, Lugtenberg et al. (2016) proposed including an additional class of endophytic entomopathogens due to their ability to grow as asymptomatic endophytes in multiple plant species and their unique ability to infect and colonize insects. About their host specificity and wide range of biodiversity, many aspects remain unknown to scientists. Endophytic fungi have already been discovered in many ecological habitats, including arctic environments, deserts, mangroves, and tropical and temperate forests (Baron and Rigobelo 2022). Considering the universal availability of fungal endophytes with multipotent beneficial actions are gaining importances among plant scientists in the field of biotechnology all over the world.

2.2.1 Fungal Metabolites as Bio-control Agents

Fungi are prolific producers of secondary metabolites with their diverse chemical compositions and biological activities. They have emerged as promising candidates for bio-control agents in sustainable agricultural and pest management. Endophytes remain a mine of bioactive compounds, including antiviral, antibacterial, antifungal, antioxidant, anticancer, anti-insect, and immunosuppressant, with their vast application area in different global industries. The compounds are produced either for signaling in the defense system or as a process of integral interaction with their host tissues, recognized as secondary metabolites. It has been summarized that approximately four thousand types of secondary metabolites have been listed as having a fungal origin. Most secondary metabolites with wider applicability in therapeutic applications were identified in endophytic fungi. They may interfere with the physiological processes of pests, disrupt their reproductive cycles, or act as toxins affecting the target organism selectively. The majority of the bioactive metabolites derived from endophytes are found to be affected by pathogenic attacks. Fungal metabolites directly affect pests and play a vital role in plant defense mechanisms. Antimicrobial resistance (AMR) and multidrug resistance (MDR) are continuously chasing the existence of the modern world's civilization due to the rapid modification of the structural complexities of microbes. A range of antibacterial compounds have been identified from endophytic fungi. To defeat microbial infections, antibiotics of endophytic origin also became a trustworthy arena of interest in the medical field. Widjajanti et al. (2022) isolated ten endophytic fungi from *Syzygium zeylanicum* and found that *Lasiodiplodia pseudotheobromae* showed strong antibacterial activity. Similarly, Haung et al. (2021) identified three new hydroxylphenylacetic acid derivatives and a new alkaloid from *Mortierella* sp., with compounds 7 and 10 exhibiting selective antibacterial activity. Liu et al. (2022) isolated endophytic *Aspergillus polyporicola*, producing two nucleoside derivatives, exhibiting inhibitory activity against various bacteria.

Several studies revealed the origin of endophytic fungal metabolites, which are derivatives of alkaloids, peptides, steroids, phenols, flavonoids, terpenoids, and quinines. Echinocandin A is a well-known antifungal peptide produced by the endophytic fungus *Cryptosporiopsis* sp., isolated from the plant *Pinus sylvestris* (Nisa and Kamili 2019). Zang et al. (2022) isolated two novel heterodimeric tetrahydroxanthones from the mangrove endophytic fungus *Aspergillus flavus* QQYZ, which exhibited broad-spectrum antifungal and moderate antibacterial activities. Similarly, Song et al. (2023) identified seven new compounds from the endophytic fungus *Fusarium decemcellulare* F25, some showing inhibitory activities against *Colletotrichum musae*. The endophytic fungi from *Artemisia argyi*, namely *Diaporthe* sp. AC1 was also discovered by Gu et al. (2022) to have considerable antifungal activity against *F. graminearum*, *F. moniliforme*, and *Botrytis cinerea*.

Human diseases like heart disease, carcinoma, and age-related diseases are caused by the imbalance between reactive oxygen species and cellular changes regarding oxidative stress. Antioxidants of microbial origin are considered novel candidates to nullify the bad effects of ROS. Not only fruits and vegetables but endophytic fungi are also recognized as a great source of potent antioxidant compounds. Endophytic fungi from various medicinal plants, including *Polygonatum sibiricum*, *Peronema canescens* (Sungkai), and *Syzygium zeylanicum*, have been investigated for their antioxidant compounds. In the study of Zhao et al. (2023), the endophytic fungus *Fusarium oxysporum* HJ.15 from *Peronema sibiricum* showed significant antioxidant activity, with the presence of various compounds responsible for this property. Additionally, endophyte from Sungkai leaves was found to produce a compound with strong antioxidant activity, identified as 4-hydroxybenzoic acid (Oktiansyah et al. 2023). These findings suggest that endophytic fungi from various medicinal plants have the potential to produce antioxidant compounds, which can be further explored as therapeutic agents. The experimental work of Billar de Almeida et al. (2020) on the biocontrol activity of endophytic isolates *Epicocum nigrum* against the causal agent of grapevine trunk diseases proved their antagonistic activity against mycelia growth. Bertini et al. (2022) studied the biodiversity and bio-prospecting activity of the fungal endophytic strains *Tremetes* sp., *Lenzites* sp., *Sistotrema* sp., and *Peniophora* sp. from *Colobanthus quitensis*, which showed anti-tumor activity with secondary root development in *Arabidopsis*. This study emphasizes the biotechnological perspective of the Antarctic ecosystem. Recent research on the biocontrol potentiality of ethyl acetate crude extract from endophytic *Pestalotiopsis neglecta* revealed the existence of various bioactive chemicals such as tolycaine, eugenol, 1H-pyrazole, and bis(2-Ethylhexyl) phthalate, all of which have important biological applications (Almustafa and Yehia 2023).

Recent studies have highlighted the potential of epigenetic manipulation in activating silent biosynthetic gene clusters in endophytic fungi, producing novel and diverse secondary metabolites (Verma et al. 2023; Pillay et al. 2022). This strategy effectively enhances the production of specialized metabolites and inducing the synthesis of previously undescribed compounds (Toghueo et al. 2020). These findings enlighten the potential of epigenetic manipulation as a promising approach for discovering new drug candidates from endophytic fungi. The potential fungal endophytes have been listed in Table 2.2.

2.2.2 Fungal Metabolites as a Plant Growth Promoter

Fungal metabolites encompass a rich reservoir of bioactive compounds that influence various aspects of plant physiology. This is achieved through mechanisms such as nutrient solubilization, which leads to improved nutrient and water absorption. It also plays a crucial role in alleviating the impact of abiotic stresses such as drought, salinity, and temperature extremes. The activation and inactivation of stress factor-regulating genes eventually mediate growth-regulating factors. As a result of co-evolution, the fungal endophytic community offers a multiphase barrier system among the many mechanisms that facilitate mutual interaction and yield promotion. Despite their poorly understood genetic combination, they can diversify within a wide number of plant species and promote the balanced activation of plant defense genes. The crosstalk between secondary metabolites and plant hormones influences hormonal balance and signaling pathways in stress responses. The benefits of endophytic contact are classified as either direct or indirect. The immediate benefits include the acquisition of insoluble nutrients, the generation of phytohormones, the action of ACC deaminase, and nitrogen fixation. The indirect advantages are the production of chelating agents like siderophores and the evaluation of HCN with extracellular enzyme synthesis to improve systemic resistance to diseases and pests.

Endophytic fungi have been identified as key players in sustainable agriculture, offering a range of benefits including improved nutrient uptake, stress tolerance, and disease resistance in plants. These microorganisms can also reduce the need for synthetic fertilizers and pesticides, minimizing the associated risks (Watts et al. 2023). However, the loss of endophyte diversity in domesticated plants and the reduced effectiveness of some endophyte strains in conferring plant benefits are challenges that must be addressed.

Plant growth-promoting endophyte fungi have been studied in various crops, including halophytic rice, tomatoes, and sugarcane. In the case of halophytic rice, three different fungal endophytes were isolated and characterized, showing various plant growth-promoting abilities (Airin et al. 2023). Similarly, in tomatoes, five different entamopathogenic fungi were evaluated for their effects on tomato growth promotion and pest protection, with all five isolates successfully colonizing the plants and promoting growth (Zhang et al. 2022). Many research studies have

Table 2.2 Bio-control activities of endophytic fungal genera

Source	Endophytic fungi	Activity	Reference
Camellia oleifera	*Pestalotiopsis* sp.	Antifungal activity against *C. oleifera* anthracnose pathogen	Yu et al. (2018)
Cornus florida	*Nigrospora sphaerica*	Effectively reduced the growth of *Phytopthora capsici* root rot	Mmbaga et al. (2018)
Litsea cubeba	*Chaetomium globossum*	Suppress the growth of *Ceratosphaeria phyllostachydis*	Wu et al. (2019)
Theobroma cocao	*Colletotrichum gloeosporioides*	Antagonist activity against *Phytopthora palmivora*	Chhipa and Desmukh (2019)
Cucurbita pepo	*Dichotomopilus* sp. and *Schizothecium* sp.	Antagonistic potential against *Verticillium dahilae*, a significant cucurbit pathogen	Tymon et al. (2020)
Quercus suber	*Simplicillium aogashimaense*	Suppressing the growth of cork oak causing pathogens *Biscogniauxia mediterranea* and *Diplodia corticola*	Costa et al. (2020)
Cucurbita maxima	*Chaetomium*, *Colletotrichum* and *Acrocalymma*	Effectively inhibiting the growth of soil borne fungal pathogens related to cucumbers	Haung et al. (2020)
Cunninghamia lanceolata	*Epicoccum dendrbii*	Significantly suppressing the mycelia growth of anthracnose causing phytopathogen *Colletotrichum gloeosporoides*	Bian et al. (2021)
Olea europa	*Epiccocum nigrum*	Control the phytopathogenic fungi *Colletotrichum acutatum*	Poveda and Baptista (2021)
Barleria prionitis	*Daldinia eschscholtzii*	Biocontrol activity agnist *Colletotrichum acutatum* and its post-harvest infections	Khruengsai et al. (2021)
Buchanania axillaris	*Diaporthe caatingaensis*	Producing camptothecin like anticancer compounds	Dhashinamoorthy et al. (2021)
Triticum aestivum	*Penicillium olsonii*	Effective against *Fusarium graminearum*	Rojas et al. (2022)

(continued)

Table 2.2 (continued)

Source	Endophytic fungi	Activity	Reference
Solanum lycopersicum	*Sarocladium terricola*	Potential biocontrol agent against *Meloidogyne incognita*	Tian et al. (2022)
Zingiber officinale	*Aspergillus terreus*	Effective against *Colletotrichum gloeosporioides*	Gupta et al. (2022)
Aloe dhufarensis Lavranos	*Sarocladium liliense* and *Penicillium oxalicum*	Inhibitory activity against several phytopathogens like *Fusarium* sp. and *Cladosporium* sp.	Al-Rashdi et al. (2022)
Triticum aestivum	*Prietia aryabhattai*	Multi-stress reducer and growth promoter in agriculturally important crops including cereals	Shahid et al. (2022)
Solanum lycopersicum	*Acremonium sclerotigenum*	In vivo and in vitro biocontrol potency against root knot nematodes *Meloidogyne* spp.	Yao et al. (2023)
Garcinia atroviridis	*Nigrospora sphaerica*	Effective against anthracnose disease causing pathogen *Colletotrichum gloeosporoides*	Zanudin and Mansor (2023)
Solanum lycopersicum	*Paecilomyces formosus*	Growth inhabiting activity against *Rhizoctonia solani*, *Fusarium oxysporium*, *Rhizopus oryzae*, *R. stolonifer*, *Fusarium verticillioides*, and *Botrytis cinerea*	Arasu and Al-Dhabi (2023)
Dioscorea bulbifera	*Clonostachys pseudochroleucha*	Antagonistic against *Parathyridaria percutanea* and *Curvularia lunata*	Sharma et al. (2023)
Anthemis altissima, *Martricaria parthenium* and *Achillea millefolium*	*Prussia africana*, *Alternaria embellisia*, *Septoria malagutii*, *Fusarium avencearum* and *Tranetes versicolor*	Plant growth promoting ability with IAA producing activity and significantly improved the seed germination of maize hybrid cultivar	Hatamzadeh et al. (2023)

discovered evidence of fungal endophytes as biologically active growth inducers in a well-known category of crops. The invasion of endophytes mediated the uptake of both micro- and macro-nutrients such as nitrogen, phosphorus, potassium, magnesium, zinc, iron, and copper. Many studies have highlighted the potential of nutrient-solubilizing endophytic fungi in promoting plant growth and improving soil fertility. Alves et al. (2021) found that fungal endophytes, including *Beauveria bassiana* and *Trichoderma asperellum*, can enhance root colonization by arbuscular mycorrhizal fungi and increase the concentration of certain nutrients in the rhizospheric soil. Similarly, Howard et al. (2022) highlighted the potential role of mucoromycotina "fine root endophytes" in plant nitrogen nutrition, suggesting their importance in nutrient uptake and transfer. These studies collectively expose the potential of nutrient-solubilizing endophytic fungi in sustainable agricultural practices.

Endophytic fungi can also upregulate auxin-sensitive genes in response to environmental stressors. IAA is the primary phytohormone that has attracted the greatest interest to date, along with the synthesis of gibberellin by fungal endophytes. The endophytic fungus *Rhizophagus irregularis* has been found to produce phytohormones, including cytokinin, auxins, and gibberellins, which can influence plant growth and development (Pons et al. 2020). The community structure of fungal endophytes in tomato roots is influenced by host genotype and phytohormones, with changes in phytohormone balance affecting species richness (Manzotti et al. 2020). Recent research on the fungal consortium ability of *Purpureocilium lilacium*, *P. lavenduluma*, and *M. marquandii* demonstrated the yield-enhancing characteristic tested in soybean, bean, and maize plants with high auxin production and phosphate solubilization capability (Baron et al. 2020). Numerous studies are being conducted to determine the potentiality of various fungal endophytic species to comprehend the true mechanism of plant growth stimulation. To demonstrate this point, Al-Harthi et al. (2023) attempted to extract, purify, and describe plant growth-promoting fungal species such as *Trichoderma*, *Fusarium*, *Colletotrichum*, *Alternaria*, *Phoma*, and others from the three aromatic plants *Jasminum sambac*, *Camellia sinensis*, and *Ocimum basilicum,* respectively. Fungal endophytes can boost plant growth through multi-domain methods (Table 2.3).

2.2.3 Fungal Metabolites in Stress Management

As one of the primary invaders populating the earth's surface, green plants have faced major ecological tragedies since their origin, creating survival instincts with the assistance of their internal endophytic colonizers. They aid in biotic and abiotic stress adaptation by creating more soluble carbohydrates, amino acids, alkaloids, and osmotic solutes. Fungal metabolites maintain cellular osmotic balance, prevent oxidative damage, and enhance plants' ability to withstand adverse environmental stress. Endophytes can exchange their genetic makeup with the host genomic architecture to overcome environmental hurdles in adverse situations. The fungal endophytic community is classified based on its habitat and the types of interactions it has with host plants. In recent decades, some microbial populations have dominated the rhizosphere and within the plant tissues to protect against various biological processes driven by abiotic factors. In addition, parasitism or competition may arise to evade disease and herbivory. Fungal endophytes activate induced systemic resistance (ISR) and acquired systemic resistance (ASR) to protect plants against environmental stresses.

Table 2.3 Plant growth-promoting activities of endophytic fungal genera

Source	Endophytic fungi	Crops	Activity	Reference
Cenchrus americans	*Trichoderma asperellum* and *Acremonium* sp.	*Cenchrus americans*	Plant growth-promoting activity with downy mildew disease suppression	Nandhini et al. (2018)
Dendrobium moniliforme	*Colletotrichum alatae*	Protocorms of *Rhynchostylis retusa*	Production of Indole acetic acid with a higher root/shoot ratio	Shah et al. (2018)
Withenia somnifera	*Aspergillus awamori*	*Zea mays*	Produced indole-3-acetic acid and significantly enhanced the growth of maize	Mehmood et al. (2019)
Salicornia europea	*Arthrinium gamsii* and *Stereum gausapatum*	*Lolium perenne*	Promote the growth of ryegrass with IAA, siderophore-producing activity, phosphate solubilizing activity, and extracellular enzyme production	Furtado et al. (2019)
Sophora flavescens	*Alternaria* sp., *Didymella* sp., *Fusarium* sp. and *Xylogone* sp.	*Arabidopsis thaliana*	Increasing different photosynthetic pigments of the plant with high IAA	Turbat et al. (2020)
Sorghum bicolor	*Trichoderma asperellum*, *Epicoccum nigrum* and *Alternaria longipes*	*Sorghum bicolor*	Significant increase in seed vigor index	Rajini et al. (2020)
Vanda cristata	*Fusarium tricinctum*, *Alternaria tenuissima* andF. *oxysporum*	*Cymbidium aloifolium* protocorms	Increased the root and shoot length	Chand et al. (2020)
Ephedra pachyclada	*Penicillium commune*, *P. crustosum*, *P. caseifulvum* and *Alternaria tenuissima*	*Zea mays*	Increased root lengths with other plant growth-promoting activity	Khalil et al. (2021)

(continued)

Table 2.3 (continued)

Source	Endophytic fungi	Crops	Activity	Reference
Phaseolus vulgaris	*Alternaria sorghii* and *Penicillium commune*	*Phaseolus vulgaris*	Improved plant biomass, photosynthetic pigments, and endogenous hormone content	Ismail et al. (2021)
Oryza coarctata	*Talaromyces adpressus* and *T. argentinensis*	*O. coarctata*	Plant growth-promoting abilities in vitro, such as IAA, ACC-deaminase and siderophore production, phosphate and zinc solubilization	Khan et al. (2021)
Justicia brandegeana	*Colletotrichum siamense* and *Diaporthe masirevicii*	*Lycopersicon esculentum*	Production of IAA and phosphate solubilization	Silva et al. (2022)
Dactylis glomerata	*Trichoderma harzianum*	*D. glomerata*	Nitrogen-fixing capacity with the production of IAA and nutrient solubilization	Zhang et al. (2022)
Cheilanthes vellea	*Penicillium chrysogenum and Aspergillus fumigatiaffinis*	*Triticum aestivum*	Solubilize phosphorus in alkaline medium with IAA and siderophore production	Mohamed et al. (2022)
Pistacia vera	*Clonostachys rosea*	*Piper nigrum*, *Solanum lycopersicum*, and S. *melongena*	Showing higher germination rate with vigorous seedling growth	Türkölmez et al. (2023)
Oryza sativa	*Aspergillus welitschiae*	*Oryza sativa*	Enhanced metabolite and photosynthetic pigment content at the seedling stage	Airin et al. (2023)
Cremastra appendiculata	*Colletotrichum gloeosporoides*	*Glycine max*	Increased in soybean radicle growth and plant biomass with IAA and protease-producing activity	Wang et al. (2023)
Triticum aestivum	*Penicillium citrinum*	*Triticum aestivum*	Exhibiting growth-promoting activities like IAA and GA production, siderophore production with extracellular enzymatic activities	Kaur et al. (2023)

Endophytic fungi are key to plant oxidative stress management, particularly in abiotic stresses such as salinity and drought. These fungi interact with their host plants to enhance magnification, increment fitness, and reinforce tolerance to stress while promoting the accumulation of secondary metabolites that provide immunity (Chandran et al. 2020). The potential of endophytic fungi in sustainable agriculture is paramount, with their application as biofertilizers, biocontrol agents, and stress mitigators becoming increasingly apparent. Oxidative damage in plant tissues is becoming a significant problem regarding stress-relieving strategies. Fungal metabolites stimulate the activation of antioxidant defense mechanisms within plants. This includes the activation of ROS-suppressing enzyme genes such as catalase (CAT), superoxide dismutase (SOD), peroxidase (POD), and ascorbate peroxidase (APX). Published results in this domain of effective endophyte-mediated tolerance to ROS in plants have raised the curiosity of researchers looking for novel bioactive modulators. This interaction plays a crucial role in orchestrating physiological adjustments that contribute to plant stress tolerance. In this regard, endophytic strains *Alternaria* and *Trichoderma harzarium* inoculation have been demonstrated to be effective reducers of ROS accumulation within tomato seedlings (Azad and Kaminsky 2016). In the work of Ahmad et al. (2020), endophytic colonization of *M. robertsii* in maize plants was able to boost plant growth, increase plant biomass, regulate the expression of defense genes, and decrease the development of insect larvae supplied to plants that received fungal inoculation. Significant changes in the expression of genes related to the jasmonic acid (JA) and salicylic acid (SA) biosynthesis pathways show the "priming effect" of the defense system, which ensures a more effective response in future exposures to potential pathogens and stress situations. Gul et al. (2019) has described a favorable endophytic interaction of *Yarrowia lipolytica* that reduced the impact of salinity on maize plants. Chlorophyll content, electrolyte leakage, leaf-relative water, and levels of oxidative enzymes and phytohormones were all enhanced in inoculated plants, suggesting that these fungi might be employed as biofertilizers in a salty environment. Ismail et al. (2020) demonstrated the activity of the endophytic strain *Aspergillus niger* in increasing plant height, biomass, and pigment concentration, along with lipid peroxidation capacity, against high-temperature stress.

Secondary metabolites are created as defense mechanisms to establish their connection with the host plant. Genetic and metabolic engineering strategies are also being explored to enhance the production of stress-alleviating compounds for agricultural applications. Nodulisporic acid, derived from Nodulisporium sp., is one of the secondary metabolites of endophytic origin that activates glutamate in the chloride channel to limit insect herbivory by causing flaccid paralysis (Demain 2000). The feasible effect of cucumber plants on the endophytic fungus *Paecilomyces formosus* (=*P. maximus*) strain LHL10 has been discovered. Compared to non-inoculated plants, inoculated plants adapted to salt conditions demonstrated increased vegetative development (Khan et al. 2012). Studies on the production of an antifungal resistance gene cluster in the endophytic fungal strain *Burkholderia vietnamiensis* have shown the strain's ability to generate auxin and siderophores and to be used as a biocontrol agent against plant infections (Doty et al. 2022). The symbiotic relationship improved nutrient absorption, particularly potassium, calcium, and magnesium, and reduced the consequences of sodium toxicity during salinity stress. Moreover, it was shown that the fungus reprogrammed plant growth in response to stress by regulating the production of salicylic, jasmonic, and abscisic acids.

Recent studies by Farhat et al. (2023) demonstrated the biocontrol potential of the endophytic strain *Talaromyces assiutensis* against root rot pathogens in sunflowers. Pot tests revealed that the biomass increased, and biochemical indicators like polyphenolic content and salicylic acid increased. Many additional recent studies can be used as examples of the benefits outlined above, demonstrating that research has been undertaken to investigate more than one conceivable benefit that might be supplied by fungal endophytic colonization. *Phoma glomerata* (LWL2) and *Penicillium* sp. (LWL3) have been characterized as having the ability to form endophytic contacts with cucumber plants, resulting in significantly increased biomass and improved growth under water and salt stress conditions (Waqes et al. 2012).

Plants are exceptional in managing internal and external difficulties with the support of beneficial microbes such as bacteria, actinomycetes, and fungi, which provide essential nutrients while also strengthening defenses against biotic and abiotic stresses (Table 2.3). Among the many stressors, salinity and drought circumstances have a significant impact on various croplands as well as lands adjacent to the coast, resulting in stunted growth of crops with a broad range of pathogenic interactions causing loss of key grains worldwide. Colonization of endophytic fungi on several crops has been shown to induce systemic resistance to pathogens, mitigate stress by increasing levels of protective metabolites and osmoprotectants, and activate antioxidant systems to prevent ROS damage, decrease salt-induced root respiration, and modulate the phytohormone profile to minimize salt effects on plant growth. The fungal isolate *Trichoderma longibrachiatum* T6 effectively promoted the growth of wheat seedlings to NaCl stress through ACC-deaminase and auxin-producing activity by minimizing intracellular ionic homeostasis within plant cells (Zhang et al. 2019). An effective isolate *Bipolaris* sp. CSL-1 was discovered as a potent growth promoter that significantly decreased endogenous ABA concentration with respect to Salicylic acid level increased to combat the effect of high NaCl concentration within *Glycine max*. The non-inoculated plants showed a higher level of antioxidant enzymes which minimize the harmful consequences of oxidative damage (Khan et al. 2021). Yuan et al. (2023) extensively studied the osmotic stress tolerance ability of the fungal endophytic strain *Chaetomimum globossum* D5, increasing the nutrients uptake and accumulating carbon and photosynthetic pigment to mitigate salt causing damage to the *Paeonia lactiflora* plant.

Root exudates influence the successful interaction of plant's internal microorganisms which aids in overall growth by promoting healthy root, shoot, and leaf development under dry conditions. In this context, it is hypothesized that ABA serves as a signal that prepares the plant to withstand a water deficit, mostly by modulating stomata closure and water loss. The current study revealed that the co-inoculation of endophytic strain *Penicillium funiculosum* LHL6 and *Paecilomyces formosus* LHL10 promoted plant growth attributes by enhancing micronutrient uptake under heavy metal and drought stress conditions (Bilal et al. 2020). Aside from using a single endophytic inoculum as a vector to mitigate stress consequences, fungal consortiums, or the combined activity of endophytes with specific bioactive metabolites, are emerging as another potential option for modern science in establishing the concept of sustainable agriculture. A recent study demonstrated the efficiency of the endophytic isolate *Trichoderma virens* SB10 with glycine betaine as a redox homeostasis regulator against heat and salinity stress (Bilal et al. 2023) (Table 2.4).

2.2.4 Volatile Organic Compounds of Endophytic Fungal Origin Serve a Variety of Functions

The term "volatile organic compounds" refers to organic, low-molecular-weight carbon-containing small compounds (up to C20), which are typically released in small or significant amounts from fungal endophytes and may have a wide range of specific bioactivities, including antimicrobial, antioxidant, anti-proliferative, anti-larval, immune-enhancer, anticancer, and antifungal. Microbial volatiles are now widely used as a source of bio-control agents' myco-diesel and as a bio-toilet component. Production of mVOCs is greatly influenced by cultural circumstances, including the medium's pH and composition, ideal temperature and aeration for fungal growth, and the source of the isolated fungus (Kanchiswamy et al. 2015). Endophytes have recently been shown to be powerful makers of bioactive chemicals, with the majority of these molecules coming from sources that are known to be volatile (Fig. 2.2). Because endophytes are hidden microbial floras of green plants that have existed since the beginning of land plants on Earth, there is a complex relationship between the host and endophyte. It is well known that volatiles function as allelochemicals, clearing agricultural fields of unwanted plants and crops. The word "allelochemical" refers to substances that interact with a plant's many physiological processes and/or fully impair its normal biochemical processes, which can impede, delay, or halt a plant's growth. Controlling weeds and eliminating collateral and alternative hosts of disease-causing pathogens are serious issues.

Therefore, to control an organism's growth or eradicate its population from a particular ground or agricultural field, agriculturalists and plant scientists are working to develop a novel system in which one biological organism or its secondary metabolites will interact with another organism. In an effort to maintain sustainable growth while simultaneously controlling weeds and unwanted plants, researchers are now looking for allelochemicals from microbial sources rather than only plant sources, with a particular emphasis on microbial volatiles. *Muscodor* species are among the most diverse genera in the endophytic community since they have been isolated from various hosts around the globe. Every *Muscodor* strain produces unique combinations of volatiles that set them apart from other *Muscodor* isolates worldwide.

The number of volatile compounds previously reported to be widespread was only about five. The main volatile mixtures are effective against pathogenic fungi such as *Fusarium oxysporum* and *Xylaria* sp. as allelochemical. However, they do

Table 2.4 Environmental stress-preventing activities of endophytic fungal genera

Source	Endophytic fungi	Biotic/Abiotic factor	Activity	Reference
Zingiber officinale	*Rhizopycnis vagum*	Soft rot pathogen *Phythium myriotylum*	Antibiosis	Anisha et al. (2018)
Triticum aestivum	*Piriformospora indica*	Drought	Activation of stress responsive hormonal genes	Zhang et al. (2018)
Calotropis gigantea	*Alternaria destruens*	Pest *Spodoptera litura*	Insecticidal with α-glycosidase inhibitors	Kaur et al. (2019)
Euphorbia indica	*Aspergillus flavus*	Temperature	Induced IAA and Salicylic acid production	Hamayun et al. (2019)
Glycyrrhiza uralensis	*Acrocalymma vagum, Fusarium acuminatum*	Drought	Antioxidative enzyme activation with phosphate solubilization	He et al. (2019)
Pennisetum glaucum	*Aspergillus terreus*	Salinity	Increased IAA content	Khushdil et al. (2019)
Lolium perene	*Epichloe festucae* var. *lolii*	Pasture pest *Wiseana cervinata*	Anti-insect activity by Epoxyjanthitrem	Finch et al. (2020)
Vitis vinifera	*Trichoderma* sp. strainT154	Pathogen *Phaeoacremonium minimum*	Antibiosis	Carro-Huerga et al. (2020)
Solanum lycopersicum	*Penicillium* sp.	Drought	Phytohormone production	Morsy et al. (2020)
Coffea Arabica	*Indurantia coffeana*	*Colletotrichum lindemuthianum*, *Sclerotinia aclerotiorum* and *Pseudocercospora griseola*	The higher percentage of growth inhibition	Mota et al. (2021)
Pinus tobulaeformis	*Phoma* species	Drought	Anti-oxidative enzyme increased	Zhou et al. (2018)
Oryza sativa	*Phomopsis liqudambaris*	Salinity	Decreasing ACC deaminase activity	Siddikee et al. (2022)

(continued)

Table 2.4 (continued)

Source	Endophytic fungi	Biotic/Abiotic factor	Activity	Reference
Dysoxylum gotadhora	*Aspergillus* sp. *and Porostereum* sp.	Soil-borne phytopathogen *Verticilium dahilae* and *Fusarium oxysporum*	Antagonist against the pathogen's growth	Kapoor et al. (2022)
Solanum lycopersicum	*Serenpidita indica*	Drought	Anti-oxidant enzymes increased with low H_2O_2 and MDA content	Ghabooli and Kaboosi (2022)
Dactylis glomerata	*Trichoderma harzianum, Aspergillus fumigatus*	Pathogen *Phoma exima, Fusarium tricinctum, Colletotrichum cliviicola*	Antibiosis	Zhang et al. (2022)
Glycine max	*Bipolaris* sp.	Salinity	Decreased endogenous ABA content with an increase in IAA and GAs content	Khan et al. (2022)
Vitis vinifera	*Piriformospora indica*	Temperature	Abscisic acid and proline content increased	Karimi et al. (2022)
Glycine max	*Aspergillus welwitschiae*	Heavy metal	Increased Ascorbic acid oxidase and peroxidase activity	Husna et al. (2021)
Mangifera indica	*Alternaria* sp., *Daldinia* sp., *Rosalinia* sp.	*Botrytic cinerea* and *Penicillium digitatum*	Antagonist against phytopathogens	Yang et al. (2023)
Zea mays	*Serendipita indica, Rhizophagus intraradices* and *Azotobactor chroococcum*	Drought	Induced the activity of Peroxidase, Catalase, Polyphenol oxidase and Superoxide dismutase	Tyagi et al. (2023)
Ecballium elaterium	*Fusarium clavum*	Salinity	Reduced the level of H_2O_2, produced hydrolytic enzyme and IAA	Meshram et al. (2023)
Solanum lycopersicum	*Curvularia lunata*	Heavy metal	Elevated the level of Jasmonic acid, ABA, and ethylene	Asaf et al. (2023)

not show any effectiveness against *Colletotrichum* sp., *Phomopsis* sp., *Alternaria solani*, *Rhizoctonia* sp., *Phytophthora capsici*, and *P. parasitica*. The main substances that prevent the elongation of the roots of the plants include octane, 2-methyl butyl acetate, 2-pentyl furan, caryophyllene, aromadendrene, and several naphthalene derivatives. The culture medium extract and mycelial extract share 12 allelochemicals, including benzene derivatives, phenolic compounds, cyclopentadienes,

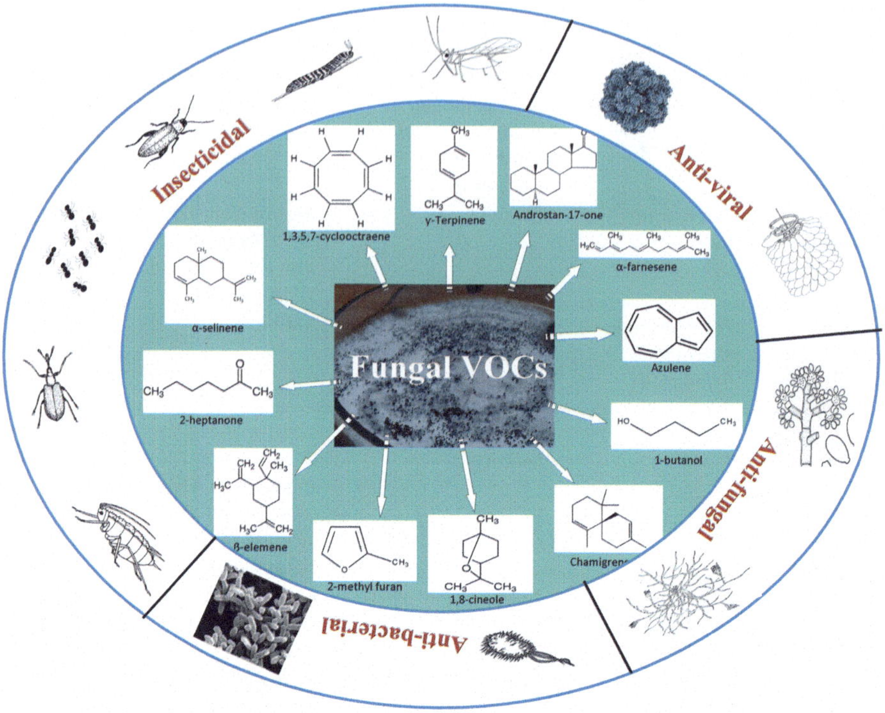

Fig. 2.2 Fungal VOCs and their efficacy to control biotic factors

esters, lactones, alkanes, aldehydes, and carboxylic acids. These VOCs are crucial because they alter how the host defends it from certain rivals. This work thus supports the idea that endophytic fungi that produce VOCs can be used as effective biocontrol agents, green herbicides, bio-herbicides, and weed controllers. Volatiles are also effective for controlling infections with harmful nematodes and moths. PTM (Potato Tuber Moth) and *Pthorimaea poerculella* are controlled by *Muscodor* strains' biofumigation. According to reports, PTM primarily affects tropical and subtropical nations and is a devastating pest of potato plants and tubers. *Bacillus thuringiensis*, granulovirus, and chemical fumigants like methyl bromide are some of the traditional methods for controlling this issue, but all of these have some safety concerns for humans and the environment. In that context, *M. albus* volatiles would be a possible method for controlling pest infection. There was suppression in the growth of PTM neonate larvae and pupal stages when exposed to 15–30 grams of prepared *M. albus* mycelia. It has been demonstrated that potato tubers treated with *M. albus* fungal volatiles are less colonized by moth larvae and are not a good environment for their survival. In addition to being powerful antimicrobials, substances like 1-butanol, 3-methyl acetate, and isobutyric acid also cause PTM mortality and development retardation. Generally speaking, *Muscodor albus* is a physiologically promising creature with a wide range of uses. A powerful bio-fumigant agent against the codling moth (*Cydia pomonella* L. (Lepidoptera: Totricidae), *Muscodor albus, is* a member of the Xylariaceae family. At various phases of its life cycle, *Cydia pomonella*, a dangerous pest of pome fruit plants, was totally and partially inhibited when treated with VOC mixes of *M. albus*, according to Lacey and Neven (2006), Lacey et al. (2009) (Larval and adult). In that context, *M. albus* volatiles would be a possible method for controlling pest infection. There was suppression in the growth of PTM neonate larvae. When exposed to *M. albus* VOC for a random three-day period and subsequently incubated at room temperature (25 °C) for a whole day (24 hours), adult moths exhibit a modest mortality rate of 81%, but longer exposure times result in higher mortality rates ranging from 84% to 100%. The newborn Lepidoptera larvae used by the researchers to infect the apple species *Malus* sp. were then continuously exposed to VOCs for 36 hours, incubated for a lengthy period of 7 days, and ultimately died 86% of them. In contrast to the control or untreated one (85%), it progressed to the second and third larval instars progressively; even the larvae with volatile treatment retained the first instars. This issue could be resolved by employing VOC combinations of *M. albus* as an alternate source of prohibited chemicals. This could also help to reduce the systemic loss of *Malus* fruits during exportation and possibly even enhance storage conditions. VOCs have, therefore, been shown to be effective. A polyphagous plant parasitic nematode like *Meloidogyne javanica*, an obligatory parasite, preys on various hosts (root-knot nematodes). This results in a significant loss of profitable agricultural plants like tomatoes. The second-stage juveniles' survivability is reduced by up to 99% when volatiles such as 3-methyl-1-butanol, 2-methyl-1-butanol, 4-heptanone, and isoamyl acetate are applied. The volatiles can even reduce egg hatching by as much as 87%. Liarzi et al. (2016) infected the nematode after treating the roots with synthetic volatile mixtures, which significantly decreased the number of eggs produced and the galling index while astonishingly leaving the root weight unaffected. As a result, endophytic *Daldinia* cf. *concentrica* provides an alternative source of strong chemicals capable of suppressing the parasite known as root-knot.

Other than these, endophyte-originated volatiles are mainly reported for their ability to control post-harvest phytopathogens in cash crops. In this case, the *Muscodor* species, a fungus that lives as an endophyte on plants in tropical and temperate rainforests, releases various volatile compounds with antibacterial properties. The genus *Muscodor* is widely dispersed in many host species and has a wide range of bioactivity (anti-bacterial, anti-fungal, biocontrol agent, postharvest control agent). The growing medium significantly regulates the bioactivity of this genus. Research has indicated that using sucrose as a carbon source in the growth medium is the only way to limit the emission of volatile organic compounds (VOCs) such as acetic acid, butyl ester, and methyl isobutyl ketone worldwide (Ezra and Strobel 2003). The bioactivity is increased directly by how nutrient-rich the growth medium compositions are. Up to this point, reports of *Muscodor* species from countries around the world, including Australia, Central and South America, and Central and Southeast Asia, total about twelve (Suwannarach et al. 2010; Zhang et al. 2010) and the several plant species of these tropical and subtropical rain forests have been studied for *Muscodor* species isolation. *Muscodor albus* is referred to as an endophytic fungus that produces antibiotics because it creates a vast number of distinctive volatile combinations with antibacterial and antifungal properties. Butanoic acid, 2-methyl-butanoic acid, 3-methyl-2-butanol, and 3-methyl butyl esters are the main volatiles of *Muscodor albus* that were identified from *Guazuma ulmifolia,* a member of the Sterculiaceae family growing in a dry tropical forest in south-west Ecuador. The pathogenic fungus *Pythium ultimum*, *Phytophthora cinnamomi*, and Gram-positive and Gram-negative bacteria are partially or completely inhibited by the volatiles. All *Muscodor* isolates produce volatiles; however, each one's organic component composition is different. The first *Muscodor* species to be documented to eradicate Gram-positive bacteria is this strain of *M. albus* (E-6). The discovery that *M. albus* VOCs cause bacterial cell death by destroying their DNA results from research into the likely mechanism of action of *M. albus* volatiles for suppressing the growth of dangerous fungi. As a result, mycofumigation with volatiles causes problems with DNA repair, DNA metabolism, and increased membrane fluidity. Finally, the alkylation of the DNA bases renders the organism incapable of surviving (Alpha et al. 2015). *Muscodor* species effectively control and inhibit pathogenic fungi associated with seeds like *Aspergillus*, *Penicillium*, *Fusarium*, and *Leptosphaeria.* They can stop the development of disease in foods that have been preserved, such as fruits, vegetables, and other food products, in addition to producing mVOCs in vitro. According to a study done by Corcuff et al. (2011), the M. *albus* CZ-620 strain was found to produce a variety of volatile chemicals, including alcohols (2-, 3-, and ethanol is the most prevalent), aldehydes, organic acids (isobutyric acid), terpenoids (sesquiterpene), and esters. These bioactive volatiles can completely eradicate silver scurf and considerably limit the development of dry and soft rot in potato tubers. This establishes *M. albus* (strain CZ-620) as a highly effective postharvest control agent. From *Cinnamomum bejolghota*, Suwannarach and his colleagues (2013) and (2015) have isolated new species of *Muscodor*. The antifungal activity of the *M. cinnamoni* isolate against the disease-causing pathogen *Rhizoctonia solani* AG-2 was then investigated. Rye grains infected with *Muscodor cinnamonum* are useful myco-fumigating agents; the most successful inoculum included 30 g of *M. cinnamonum*.

The endophytic fungal flora of *Cinnamomum camphora*, *Cinnamomum zeylanicum*, and *Aegle marmelos* was investigated by Meshram et al. (2015) and Meshram (2017) to isolate *Muscodor* species that will produce volatile antibiotics. Finally, five species of *Muscodor* were isolated and tested for volatile emissions and antifungal activity, including *M. ghoomensis*, *M. indica*, *M. camphora*, and *M. khashaynum*. Tetraoxapropellan, terpinolene, octadec-9-enoic acid, aspidofractinine-3-methanol, 3-cyclohexen-1-ol, 1-(1,5-dimethyl-4-hexenyl)-4-methyl, 1,6-dioxacyclododecane-7,12-dione, 1,4-dimethyl-7-prop-1-en-2-yl-2. Numerous plant species, human pathogenic fungi, yeast species, and Gram-positive and Gram-negative bacterial species all experience growth inhibition from these volatiles, and some even experience mortality. The antifungal, antibacterial, and volatile emissions of the isolates *M. musae* (derived from *Musaacuminata*), *M. equiseti* (derived from *Equisetum debile* and *Hevea brasiliensis*), and *M. suthepensis* (derived from *C. bejolghata*) are also tested. The study reveals that a variety of volatile organic compounds are reported, including 2-methylpropanoic acid, 3,7-dimethyl-1,6. These species can therefore manage postharvest disease just as *M. suthepensis* can manage *Penicillium digitatum*, making them powerful biofumigants that cause tangerine fruit rot. The fungi *Cercospora beticola*, *Penicillium marneffei*, *Rhizoctonia solani*, *Candida albicans*, and *C. vishwanathi*, as well as the bacterial pathogen *Staphylococcus aureus*, can all grow more slowly when these two new species of *Muscodor*, *M. darjeelingensis,* and *M. tigerii*, are present. Saxena et al. (2015) isolated these two Musc 2, 6-Bis (1, 1-dimethylethyl)-4-(1-oxopropyl) phenol, 1, 6-Dioxacyclododecane-7, 12-dione, 4-octadecylmorpholine, 1-Tetradecanamine, N, N-dimethyl, and 1,2-Benzenedicarboxylic acid, mono(2-ethylhexyl) ester, which are stated to be the main volatile organic compounds.

To isolate a fungus that produces volatile antibiotics, *Prestonia trifidi* and *Ananas ananassoides* (a wild variety of pineapples) were chosen. The result is *M. sutura* and *M. crispans*, two species that both produce significant amounts of VOCs and can kill a variety of volatile compounds. Phuzopsene, chamigrene, isocaryophyllene, butanoic acid, and 2-methyl are all compounds produced by *M. sutura* that are effective against bacteria but not against fungi. *M. crispans* is a potent endophytic isolate that can stop the development of various bacterial and fungal species. Apart from *Muscodor* species endophytic isolates *Curvularia eragrostidis*, *Daldinia* cf *concentrica*, *Oxyporus latemarginatus*, *Hypoxylon anthochroum*, *Gliocladium* sp., and *Nodulisporium* sp. (Strain GS4d2II1a), *Myrothecium inundatum* produce various mixtures of volatiles with multidimensional bioactivities and can be useful a biocontrol agent, especially in postharvest conditions of fruit and vegetables (Cherry, tomato fruits, apple decay, *Rhizoctonia* root rot of moth orchid) (Santra and Banerjee 2020, 2022a, b). Numerous pathogenic fungi and bacteria are either entirely or partially inhibited by the antimicrobial volatile emissions of the endophytic isolates, according to in vitro and in vivo research.

2.3 Conclusion

As a member of the most populous country with vast biodiversity, it has become extremely important to develop new generational techniques or employ modern scientific approaches to sustain the expansion of green. In this regard, plants and their associated microbiota never cease to amaze us with their multifaceted activities. They are the epitome of an active candidate for environmental sustainability. Recently, agricultural food resources have been severely depleted due to excessive use of chemical fertilizers, uncontrolled infection of phytopathogens, and insect pests. Fungal endophytes are one of the candidates that contribute to plant growth and yield development by producing bioactive metabolites that either directly or indirectly limit the pathogen's growth by activating stress-related genes. Omics approaches will be used as a new route technology for better understanding the complexity of host–endophyte genome interactions. They can be studied for multifunctional applications such as biocontrol, bioremediation, and plant growth-stimulating activity. Furthermore, using microbial consortiums with beneficial outcomes will increase the possibility of dynamic interactions between the host and environmental microbiome. This might be used to develop crop-specific management as a valuable tool for future generations of farmers.

References

Ahmad I, María del Mar JG, Dawn SL, Samina NS, Mary EB (2020) Endophytic *Metarhizium robertsii* promotes maize growth, suppresses insect growth, and alters plant defense gene expression. Biol Control 144:104167

Airin AA, Arafat MI, Begum RA, Islam MR, Seraj ZI (2023) Plant growth-promoting endophytic fungi of the wild halophytic rice *Oryza coarctata*. Ann Microbiol 73(1):36

Akram S, Ahmed A, He P, He P, Liu Y, Wu Y, He Y (2023) Uniting the role of endophytic fungi against plant pathogens and their interactions. J Fungi 9(1):72

Al-Harthi HF, Elgorgan AM, Ahmed B, Bahkali AH, ElSheshtawi M, Shaik JP et al (2023) Identification, molecular characterization, and plant growth promoting activities of endophytic fungi of *Jasminum sambac*, *Camellia sinensis*, and *Ocimum basilicum*. J King Saud Univ-Sci 35(3):102558

Almustafa HI, Yehia RS (2023) Antioxidant, cytotoxic, and DNA damage protection activities of endophytic fungus Pestalotiopsis neglecta isolated from *Ziziphus spina-christi* medicinal plant. Microorganisms 11(1):117

Alpha CJ, Campos M, Jacobs-Wagner C, Strobel SA (2015) Mycofumigation by the volatile organic compound-producing fungus *Muscodor albus* induces bacterial cell death through DNA damage. Appl Environ Microbiol 81(3):1147–1156

Al-Rashdi FKH, Al-Sadi AM, Al-Riyamy BZ, Maharachchikumbura SS, Al-Sabahi JN, Velazhahan R (2022) Endophytic fungi from the medicinal plant *Aloe dhufarensis* Lavranos exhibit antagonistic potential against phytopathogenic fungi. S Afr J Bot 147:1078–1085

Alves GS, Bertini SCB, Barbosa BB, Pimentel JP, Junior VAR, de Oliveira Mendes G, Azevedo LCB (2021) Fungal endophytes inoculation improves soil nutrient availability, arbuscular mycorrhizal colonization and common bean growth. Rhizosphere 18:100330

Anisha C, Jishma P, Bilzamol VS, Radhakrishnan EK (2018) Effect of ginger endophyte *Rhizopycnis vagum* on rhizome bud formation and protection from phytopathogens. Biocatal Agric Biotechnol 14:116–119

Arasu MV, Al-Dhabi NA (2023) Biological control of root rot disease-causing *Rhizoctonia solani* in tomato plant by an endophytic fungus and analysis of growth promoting activities in greenhouse and field. Physiol Mol Plant Pathol 127:102080

Asaf S, Jan R, Khan MA, Khan AL, Asif S, Bilal S et al (2023) Unraveling the mutualistic interaction between endophytic *Curvularia lunata* CSL1 and tomato to mitigate cadmium (Cd) toxicity via transcriptomic insights. Sci Total Environ 861:160542

Azad K, Kaminskyj S (2016) A fungal endophyte strategy for mitigating the effect of salt and drought stress on plant growth. Symbiosis 68:73–78

Baron NC, Rigobelo EC (2022) Endophytic fungi: a tool for plant growth promotion and sustainable agriculture. Mycology 13(1):39–55

Baron NC, de Souza Pollo A, Rigobelo EC (2020) *Purpureocillium lilacinum* and *Metarhizium marquandii* as plant growth-promoting fungi. PeerJ 8:e9005

Bertini L, Perazzolli M, Proietti S, Capaldi G, Savatin DV, Bigini V et al (2022) Biodiversity and bioprospecting of fungal endophytes from the antarctic plant *Colobanthus quitensis*. J Fungi 8(9):979

Bian JY, Fang YL, Song Q, Sun ML, Yang JY, Ju YW et al (2021) The fungal endophyte *Epicoccum dendrobii* as a potential biocontrol agent against *Colletotrichum gloeosporioides*. Phytopathology 111(2):293–303

Bilal S, Shahzad R, Imran M, Jan R, Kim KM, Lee IJ (2020) Synergistic association of endophytic fungi enhances *Glycine max* L. resilience to combined abiotic stresses: heavy metals, high temperature and drought stress. Ind Crop Prod 143:111931

Bilal S, Shahzad R, Asaf S, Imran M, Al-Harrasi A, Lee IJ (2023) Efficacy of endophytic SB10 and glycine betaine duo in alleviating phytotoxic impact of combined heat and salinity in *Glycine max* L. via regulation of redox homeostasis and physiological and molecular responses. Environ Pollut 316:120658

Billar de Almeida A, Concas J, Campos MD, Materatski P, Varanda C, Patanita M et al (2020) Endophytic fungi as potential biological control agents against grapevine trunk diseases in Alentejo region. Biology 9(12):420

Carro-Huerga G, Compant S, Gorfer M, Cardoza RE, Schmoll M, Gutiérrez S, Casquero PA (2020) Colonization of *Vitis vinifera* L. by the endophyte *Trichoderma* sp. strain T154: biocontrol activity against Phaeoacremonium minimum. Front Plant Sci 11:1170

Chand K, Shah S, Sharma J, Paudel MR, Pant B (2020) Isolation, characterization, and plant growth-promoting activities of endophytic fungi from a wild orchid *Vanda cristata*. Plant Signal Behav 15(5):1744294

Chandran V, Shaji H, Mathew L (2020) Endophytic microbial influence on plant stress responses. In: Microbial endophytes. Woodhead Publishing, pp 161–193

Chhipa H, Deshmukh SK (2019) Diversity of endophytic fungi and their role in artificial agarwood production in aquilaria tree. In: Advancing frontiers in mycology and mycotechnology: basic and applied aspects of fungi. Springer, pp 479–494

Corcuff R, Mercier J, Tweddell R, Arul J (2011) Effect of water activity on the production of volatile organic compounds by *Muscodor albus* and their effect on three pathogens in stored potato. Fungal Biol 115(3):220–227

Costa D, Tavares RM, Baptista P, Lino-Neto T (2020) Cork oak endophytic fungi as potential biocontrol agents against *Biscogniauxia mediterranea* and *Diplodia corticola*. J Fungi 6(4):287

Demain AL, (2000) Microbial natural products: a past with a future. Cambridge: The Royal Society of Chemistry. 3–16

Den Breeyen A, Lange C, Fowler SV (2022) Plant pathogens as introduced weed biological control agents: could antagonistic fungi be important factors determining agent success or failure? Front Fungal Biol 3:959753

Dhakshinamoorthy M, Packiam KK, Kumar PS, Saravanakumar T (2021) Endophytic fungus *Diaporthe caatingaensis* MT192326 from *Buchanania axillaris*: an indicator to produce biocontrol agents in plant protection. Environ Res 197:111147

Doty SL, Joubert PM, Firrincieli A, Sher AW, Tournay R, Kill C et al (2022) Potential biocontrol activities of Populus endophytes against several plant pathogens using different inhibitory mechanisms. Pathogens 12(1):13

Ezra D, Strobel GA (2003) Effect of substrate on the bioactivity of volatile antimicrobials produced by *Muscodor albus*. Plant Sci 165(6):1229–1238

Farhat H, Urooj F, Irfan M, Sohail N, Majeed S, Ullah S, Shafique HA (2023) Biological control potential of endophytic fungi with amelioration of systemic resistance in sunflower and GC–MS metabolic profiling of *Talaromyces assiutensis*. Curr Microbiol 80(2):61

Finch SC, Prinsep MR, Popay AJ, Wilkins AL, Webb NG, Bhattarai S et al (2020) Identification and structure elucidation of epoxyjanthitrems from *Lolium perenne* infected with the endophytic fungus *Epichloë festucae* var. lolii and determination of the tremorgenic and anti-insect activity of epoxyjanthitrem I. Toxins 12(8):526

Furtado BU, Szymańska S, Hrynkiewicz K (2019) A window into fungal endophytism in *Salicornia europaea*: deciphering fungal characteristics as plant growth promoting agents. Plant Soil 445:577–594

Gu H, Zhang S, Liu L, Yang Z, Zhao F, Tian Y. (2022) Antimicrobial potential of endophytic fungi from *Artemisia argyi* and bioactive metabolites from *Diaporthe* sp. AC1. Frontiers in Microbiol. 23;13:908836

Ghabooli M, Kaboosi E (2022) Alleviation of the adverse effects of drought stress using a desert adapted endophytic fungus and glucose in tomato. Rhizosphere 21:100481

Gul Jan F, Hamayun M, Hussain A, Jan G, Iqbal A, Khan A, Lee IJ (2019) An endophytic isolate of the fungus *Yarrowia lipolytica* produces metabolites that ameliorate the negative impact of salt stress on the physiology of maize. BMC Microbiol 19(1):1–10

Gupta S, Choudhary M, Singh B, Singh R, Dhar MK, Kaul S (2022) Diversity and biological activity of fungal endophytes of *Zingiber officinale* Rosc. with emphasis on *Aspergillus terreus* as a biocontrol agent of its leaf spot. Biocatal Agric Biotechnol 39:102234

Hamayun M, Hussain A, Afzal Khan S, Iqbal A, Lee IJ (2019) *Aspergillus flavus* promoted the growth of soybean and sunflower seedlings at elevated temperature. BioMed Res Int 2019:1295457

Hatamzadeh S, Rahnama K, White JF, Oghaz NA, Nasrollahnejad S, Hemmati K (2023) Investigation of some endophytic fungi from five medicinal plants with growth promoting ability on maize (*Zea mays* L.). J Appl Microbiol 134(1):lxac015

He C, Wang W, Hou J. (2019) Plant growth and soil microbial impacts of enhancing licorice with inoculating dark septate endophytes under drought stress. Front Microbiol 10:2277

Howard N, Pressel S, Kaye RS, Daniell TJ, Field KJ (2022) The potential role of Mucoromycotina 'fine root endophytes' in plant nitrogen nutrition. Physiol Plant 174(3):e13715

Huang LQ, Niu YC, Su L, Deng H, Lyu H (2020) The potential of endophytic fungi isolated from cucurbit plants for biocontrol of soilborne fungal diseases of cucumber. Microbiol Res 231:126369

Huang ZD, Wang WJ, Han XL, Yang XL (2021) Three new hydroxyphenylacetic acid derivatives and a new alkaloid from endophytic fungus *Mortierella* sp. in *Epimedium acuminatum* Franch. and their antibacterial activity. Chem Biodivers 18(12):e2100741

Husna, Hussain A, Shah M, Hamayun M, Qadir M, Iqbal A (2021) Heavy metal tolerant endophytic fungi *Aspergillus welwitschiae* improves growth, ceasing metal uptake and strengthening antioxidant system in *Glycine max* L. Environ Sci Pollut Res 29:1–15

Ismail HM, Hussain A, Iqbal A, Khan SA, Lee IJ (2020) *Aspergillus Niger* boosted heat stress tolerance in sunflower and soybean via regulating their metabolic and antioxidant system. J Plant Interact 15(1):223–232

Ismail MA, Amin MA, Eid AM, Hassan SED, Mahgoub HA, Lashin I et al (2021) Comparative Study between exogenously applied plant growth hormones versus metabolites of microbial endophytes as plant growth-promoting for *Phaseolus vulgaris* L. Cells 10(5):1059

Kanchiswamy CN, Malnoy M, Maffei ME (2015) Chemical diversity of microbial volatiles and their potential for plant growth and productivity. Front Plant Sci 6:151

Kapoor N, Ntemafack A, Chouhan R, Gandhi SG (2022) Anti-phytopathogenic and plant growth promoting potential of endophytic fungi isolated from *Dysoxylum gotadhora*. Arch Phytopathol Plant Protect 55(4):454–473

Karimi R, Amini H, Ghabooli M (2022) Root endophytic fungus *Piriformospora indica* and zinc attenuate cold stress in grapevine by influencing leaf phytochemicals and minerals content. Sci Hortic 293:110665

Kaur R, Saxena S (2023) Penicillium citrinum, a drought-tolerant endophytic fungus isolated from wheat (*Triticum aestivum* L.) leaves with plant growth-promoting abilities. Curr Microbiol 80(5):184

Kaur J, Sharma A, Sharma M, Kumari Manhas R, Kaur S, Kaur A (2019) Effect of α-glycosidase inhibitors from endophytic fungus *Alternaria destruens* on survival and development of insect pest Spodoptera litura Fab. and fungal phytopathogens. Sci Rep 9(1):11400

Khalil AMA, Hassan SED, Alsharif SM, Eid AM, Ewais EED, Azab E et al (2021) Isolation and characterization of fungal endophytes isolated from medicinal plant *Ephedra pachyclada* as plant growth-promoting. Biomol Ther 11(2):140

Khan AL, Hamayun M, Kang SM, Kim YH, Jung HY, Lee JH, Lee IJ. (2012) Endophytic fungal association via gibberellins and indole acetic acid can improve plant growth under abiotic stress: an example of*Paecilomyces formosus* LHL10. BMC Microbiol. 12:1–4

Khan MS, Gao J, Munir I, Zhang M, Liu Y, Xue J, Zhang X (2021) Characterization of endophytic fungi, *Acremonium* sp., from *Lilium davidii* and analysis of its antifungal and plant growth-promoting effects. BioMed Res Int 2021:9930210

Khruengsai S, Pripdeevech P, Tanapichatsakul C, Srisuwannapa C, D'Souza PE, Panuwet P (2021) Antifungal properties of volatile organic compounds produced by *Daldinia eschscholtzii* MFLUCC 19-0493 isolated from *Barleria prionitis* leaves against *Colletotrichum acutatum* and its post-harvest infections on strawberry fruits. PeerJ 9:e11242
Khushdil F, Jan FG, Jan G, Hamayun M, Iqbal A, Hussain A, Bibi N (2019) Salt stress alleviation in *Pennisetum glaucum* through secondary metabolites modulation by *Aspergillus terreus*. Plant Physiol Biochem 144:127–134
Kumar A, Droby S, Singh VK, Singh SK, White JF (2020) Entry, colonization, and distribution of endophytic microorganisms in plants. In: Microbial endophytes. Woodhead Publishing, pp 1–33
Lacey LA, Neven LG (2006) The potential of the fungus, *Muscodor albus*, as a microbial control agent of potato tuber moth (Lepidoptera: Gelechiidae) in stored potatoes. J Invertebr Pathol 91(3):195–198
Lacey LA, Horton DR, Jones DC, Headrick HL, Neven LG (2009) Efficacy of the biofumigant fungus *Muscodor albus* (Ascomycota: Xylariales) for control of codling moth (Lepidoptera: Tortricidae) in simulated storage conditions. J Econ Entomol 102(1):43–49
Liarzi O, Bucki P, Braun Miyara S, Ezra D (2016) Bioactive volatiles from an endophytic *Daldinia* cf. *concentrica* isolate affect the viability of the plant parasitic nematode *Meloidogyne javanica*. PLoS One 11(12):e0168437
Liu SS, Huang R, Zhang SP, Xu TC, Hu K, Wu SH (2022) Antimicrobial secondary metabolites from an endophytic fungus *Aspergillus polyporicola*. Fitoterapia 162:105297
Lugtenberg BJ, Caradus JR, Johnson LJ (2016) Fungal endophytes for sustainable crop production. FEMS Microbiol Ecol 92(12):fiw194
Manzotti A, Bergna A, Burow M, Jørgensen HJ, Cernava T, Berg G et al (2020) Insights into the community structure and lifestyle of the fungal root endophytes of tomato by combining amplicon sequencing and isolation approaches with phytohormone profiling. FEMS Microbiol Ecol 96(5):fiaa052
Mehmood A, Hussain A, Irshad M, Hamayun M, Iqbal A, Khan N (2019) In vitro production of IAA by endophytic fungus *Aspergillus awamori* and its growth promoting activities in *Zea mays*. Symbiosis 77:225–235
Meshram V (2017) *Muscodor camphora*, a new endophytic species from *Cinnamomum camphora*. Mycosphere 8(4):568–582
Meshram V, Mahiti G, Sanjai S (2015) *Muscodor ghoomensis* and *Muscodor indica*: new endophytic species based on morphological features and molecular and volatile organic analysis from Northeast India. Sydowia 67:133–146
Meshram V, Elazar M, Maymon M, Sharma G, Shawahna R, Belausov E et al (2023) Endophytic *Fusarium clavum* confers growth and salt tolerance in *Cucumis melo*. Environ Exp Bot 206:105153
Mmbaga MT, Gurung S, Maheshwari A (2018) Screening of plant endophytes as biological control agents against root rot pathogens of pepper (*Capsicum annum* L.). J Plant Pathol Microbiol 9(3):1000435
Mohamed AH, Abd El-Megeed FH, Hassanein NM, Youseif SH, Farag PF, Saleh SA, Abdel-Wahab BA, Alsuhaibani AM, Helmy YA, Abdel-Azeem AM. (2022) Native rhizospheric and endophytic fungi as sustainablesources of plant growth promoting traits to improve wheat growth under low nitrogen input. Journal of Fungi. 19;8(2):94
Morsy M, Cleckler B, Armuelles-Millican H (2020) Fungal endophytes promote tomato growth and enhance drought and salt tolerance. Plan Theory 9(7):877
Mota SF, Pádua PF, Ferreira AN, Gomes LDBW, Dias MA, Souza EA et al (2021) Biological control of common bean diseases using endophytic Induratia spp. Biol Control 159:104629
Nandhini M, Rajini SB, Udayashankar AC, Niranjana SR, Lund OS, Shetty HS, Prakash HS (2018) Diversity, plant growth promoting and downy mildew disease suppression potential of cultivable endophytic fungal communities associated with pearl millet. Biol Control 127:127–138
Nisa H, Kamili AN (2019) Fungal endophytes from medicinal plants as a potential source of bioactive secondary metabolites and volatile organic compounds: an overview. In: Jha S

(ed) Endophytes and secondary metabolites, Reference series in phytochemistry. Springer, pp 527–537

Oktiansyah R, Elfita E, Widjajanti H, Setiawan A, Mardiyanto M, Nasution SS (2023) Antioxidant and antibacterial activity of endophytic fungi isolated from the leaves of Sungkai (*Peronema canescens*). Trop J Nat Prod Res 7(3)

Pillay LC, Nekati L, Makhwitine PJ, Ndlovu SI (2022) Epigenetic activation of silent biosynthetic gene clusters in endophytic fungi using small molecular modifiers. Front Microbiol 13:815008

Pons S, Fournier S, Chervin C, Bécard G, Rochange S, Frei Dit Frey N, Puech Pagès V (2020) Phytohormone production by the arbuscular mycorrhizal fungus *Rhizophagus irregularis*. PLoS One 15(10):e0240886

Poveda J, Baptista P (2021) Filamentous fungi as biocontrol agents in olive (*Olea europaea* L.) diseases: mycorrhizal and endophytic fungi. Crop Prot 146:105672

Rajini SB, Nandhini M, Udayashankar AC, Niranjana SR, Lund OS, Prakash HS (2020) Diversity, plant growth-promoting traits, and biocontrol potential of fungal endophytes of *Sorghum bicolor*. Plant Pathol 69(4):642–654

Rodriguez RJ, White JF Jr, Arnold AE, Redman ARA (2009) Fungal endophytes: diversity and functional roles. New Phytol 182(2):314–330

Rojas EC, Jensen B, Jørgensen HJ, Latz MA, Esteban P, Collinge DB (2022) The fungal endophyte *Penicillium olsonii* ML37 reduces *Fusarium* head blight by local induced resistance in wheat spikes. J Fungi 8(4):345

Santra HK, Banerjee D (2022a) Broad-spectrum antimicrobial action of cell-free culture extracts and volatile organic compounds produced by endophytic fungi *Curvularia eragrostidis*. Front Microbiol 13:920561

Santra HK, Banerjee D (2022b) Production, optimization, characterization and drought stress resistance by β-Glucan-Rich heteropolysaccharide from an endophytic fungi *Colletotrichum alatae* LCS1 isolated from clubmoss (*Lycopodium clavatum*). Front Fungal Biol 2:796010

Santra HK, Banerjee D (2020) Natural products as fungicide and their role in crop protection. *Natural Bioactive Products in Sustainable Agriculture*. 131–219

Saxena S, Meshram V, Kapoor N (2015) *Muscodor tigerii* sp. nov.-Volatile antibiotic producing endophytic fungus from the Northeastern Himalayas. Ann Microbiol 65(1):47–57

Shah S, Shrestha R, Maharjan S, Selosse MA, Pant B (2018) Isolation and characterization of plant growth-promoting endophytic fungi from the roots of *Dendrobium moniliforme*. Plan Theory 8(1):5

Shahid M, Zeyad MT, Syed A, Singh UB, Mohamed A, Bahkali AH et al (2022) Stress-tolerant endophytic isolate *Priestia aryabhattai* BPR-9 modulates physio-biochemical mechanisms in wheat (*Triticum aestivum* L.) for enhanced salt tolerance. Int J Environ Res Public Health 19(17):10883

Sharma S, Dhar MK, Kaul S (2023) Antagonistic, plant growth promoting and extracellular hydrolytic enzyme activity of fungal endophytes of *Dioscorea bulbifera* L. Biocatal Agric Biotechnol 50:102694

Siddikee MA, Zereen MI, Wu M, Zhang W, Chuan-Chao D (2022) Correction to: *Phomopsis liquidambaris* reduces ethylene biosynthesis in rice under salt stress via increasing the activity of 1-aminocyclopropane-1-carboxylate deaminase. Arch Microbiol 204(4)

Silva Santos SD, Silva AAD, Polonio JC, Polli AD, Orlandelli RC, Oliveira JADS et al (2022) Influence of plant growth-promoting endophytes *Colletotrichum siamense* and *Diaporthe masirevici* on tomato plants (*Lycopersicon esculentum* Mill.). Mycology 13(4):257–270

Song Z, Sun YJ, Xu S, Li G, Yuan C, Zhou K. (2023) Secondary metabolites from the Endophytic fungi *Fusarium decemcellulare* F25 and their antifungal activities. Frontiers in Microbiol. 1;14:1127971

Suwannarach N, Bussaban B, Hyde KD, Lumyong S (2010) *Muscodor cinnamomi*, a new endophytic species from *Cinnamomumbejolghota*. Mycotaxon 114(1):15–23

Suwannarach N, Kumla J, Bussaban B, Hyde KD, Matsui K, Lumyong S (2013) Molecular and morphological evidence support four new species in the genus *Muscodor* from northern Thailand. Ann Microbiol 63:1341–1351

Suwannarach N, Kumla J, Matsui K, Lumyong S (2015) Characterization and efficacy of *Muscodor cinnamomi* in promoting plant growth and controlling Rhizoctonia root rot intomatoes. Biol Control. 1;90:25–33

Tian X, Sun T, Lin R, Liu R, Yang Y, Xie B, Mao Z (2022) Genome sequence resource of *Sarocladium terricola* TR, an endophytic fungus as a potential biocontrol agent against *Meloidogyne incognita*. Mol Plant-Microbe Interact 35(6):505–508

Toghueo RMK, Sahal D, Boyom FF (2020) Recent advances in inducing endophytic fungal specialized metabolites using small molecule elicitors including epigenetic modifiers. Phytochemistry 174:112338

Turbat A, Rakk D, Vigneshwari A, Kocsubé S, Thu H, Szepesi Á et al (2020) Characterization of the plant growth-promoting activities of endophytic fungi isolated from *Sophora flavescens*. Microorganisms 8(5):683

Türkölmez Ş, Özer G, Derviş S (2023) *Clonostachys rosea* Strain ST1140: an endophytic plant-growth-promoting fungus, and its potential use in seedbeds with wheat-grain substrate. Curr Microbiol 80(1):36

Tyagi J, Mishra A, Kumari S, Singh S, Agarwal H, Pudake RN et al (2023) Deploying a microbial consortium of *Serendipita indica*, *Rhizophagus intraradices*, and *Azotobacter chroococcum* to boost drought tolerance in maize. Environ Exp Bot 206:105142

Tymon LS, Morgan P, Gundersen B, Inglis DA (2020) Potential of endophytic fungi collected from *Cucurbita pepo* roots grown under three different agricultural mulches as antagonistic endophytes to *Verticillium dahliae* in western Washington. Microbiol Res 240:126535

Verma A, Tiwari H, Singh S, Gupta P, Rai N, Kumar Singh S et al (2023) Epigenetic manipulation for secondary metabolite activation in endophytic fungi: current progress and future directions. Mycology:1–17

Wang S, Chen S, Wang B, Li Q, Zu J, Yu J et al (2023) Screening of endophytic fungi from *Cremastra appendiculata* and their potential for plant growth promotion and biological control. Folia Microbiol 68(1):121–133

Waqas M, Khan AL, Kamran M, Hamayun M, Kang SM, Kim YH, Lee IJ (2012) Endophytic fungi produce gibberellins and indoleacetic acid and promotes host-plant growth during stress. Molecules 17(9):10754–10773

Watts D, Palombo EA, Jaimes Castillo A, Zaferanloo B (2023) Endophytes in agriculture: potential to improve yields and tolerances of agricultural crops. Microorganisms 11(5):1276

Widjajanti H, Setiawan A, Kurniawati AR (2022) Antioxidant and antibacterial activities of endophytic fungi extracts of *Syzygium zeylanicum*. Sci Technol Indones 7(3):303–312

Wu F, Yang D, Zhang L, Chen Y, Hu X, Li L, Liang J (2019) Diversity estimation and antimicrobial activity of culturable endophytic fungi from *Litsea cubeba* (Lour.) Pers. in China. Forests 10(1):33

Yang D, Shi H, Zhang K, Liu X, Ma L. (2023) The antifungal potential of the chelating agent EDTA against postharvest plant pathogen *Botrytis cinerea*. Int J Food Microbiol 388:110089

Yao Y, Huo J, Ben H, Gao W, Hao Y, Wang W, Xu J (2023) Biocontrol efficacy of endophytic fungus, *Acremonium sclerotigenum*, against *Meloidogyne incognita* under in vitro and in vivo conditions. Biologia 78(11):3305–3313

Yu J, Wu Y, He Z, Li M, Zhu K, Gao B (2018) Diversity and antifungal activity of endophytic fungi associated with *Camellia oleifera*. Mycobiology 46(2):85–91

Yuan Y, Zu M, Zuo J, Li R, Yang L, Tao J (2023) *Chaetomium globosum* D5 confers salinity tolerance on Paeonia lactiflora Pall. J Plant Physiol 280:153878

Zanudin NAM, Mansor P (2023) Identification and characterization of endophytic fungi from *Garcinia atroviridis* for potential antagonistic against phytopathogenic, Colletotrichum gloeosporioides. Malays Appl Biol 52(4):97–106

Zhang CL, Wang GP, Mao LJ, Komon-Zelazowska M, Yuan ZL, Lin FC et al (2010) *Muscodor fengyangensis* sp. nov. from southeast China: morphology, physiology and production of volatile compounds. Fungal Biol 114(10):797–808

Zhang W, Wang J, Xu L, Wang A, Huang L, Du H et al (2018) Drought stress responses in maize are diminished by *Piriformospora indica*. Plant Signal Behav 13(1):e1414121

Zhang S, Gan Y, Xu B (2019) Mechanisms of the IAA and ACC-deaminase producing strain of *Trichoderma longibrachiatum* T6 in enhancing wheat seedling tolerance to NaCl stress. BMC Plant Biol 19(1):1–18

Zhang C, Cai K, Li M, Zheng J, Han Y (2022) Plant-growth-promoting potential of PGPE isolated from *Dactylis glomerata* L. Microorganisms 10(4):731

Zhao L, Liu Y, Sun H, Li Y, Pachaiyappan SK, Fang X et al (2023) Metabolomics-guided identification of compounds with antibacterial and antioxidant activities from *Polygonatum sibiricum*-derived endophytic fungi. Beni-Suef Univ J Basic Appl Sci 12(1):1–10

Zhou W, Wheeler TA, Starr JL, Valencia CU, Sword GA (2018) A fungal endophyte defensive symbiosis affects plant-nematode interactions in cotton. Plant Soil 422:251–266

Role of Endophytic Fungi in Induced Resistance to Abiotic Stresses

3

Yachana Jha

Abstract

Crop productivity is constantly challenged by a changing environment. To feed the growing global population, there is a requirement to increase crop productivity. Initially, this was achieved through uncontrolled application of chemicals in the fields, but over time this has become ineffective and is responsible for pollution and loss of biodiversity. To meet the needs of the growing population, exponential urbanization and industrialization are occurring, which use our farming lands and are responsible for various types of abiotic stress on crop production as well as limiting available croplands. To enhance crop productivity without harming the ecosystem, utilizing plant-specific and environment-specific symbiotic endophytic fungi is one of the best tools. These fungi possess a variety of bioactive metabolites that not only help plants develop tolerance against abiotic and biotic stress but also increase plant biomass and productivity in an eco-friendly manner.

Keywords

Endophytic fungi · Abiotic stress · Nutrient · Drought · Salinity heat

3.1 Introduction

The group of microbes which resides intercellularly in the healthy plant cell without having any adverse effect in it is called endophytes. Endophytes colonized in the living plant cell and are ubiquitous. Endophytes get enter in the plant mainly through

Y. Jha (✉)
N.V. Patel College of Pure & Applied Sciences, CVM University, V.V. Nagar, Anand (Gujarat), India

© The Author(s), under exclusive license to Springer Nature Singapore Pte Ltd. 2025

K. A. Abd-Elsalam, A. H. Hashem (eds.), *Fungal Endophytes Volume II*,
https://doi.org/10.1007/978-981-97-8804-0_3

the roots, as plants roots are directly in contact with soil (Mushtaq et al. 2023). However, aerial parts of the plants can be also used for entry and it can either spread throughout the plant or remains localized at the point of entry. Endophytes can reside within healthy cells or within the vascular tissues or in the intercellular spaces and establish different types of relationship with plants and other microorganisms, which including mutualistic or supportive. The types of endophytes associates with the plants show great variation and depend on plant root exudates, plant age, plant type, and surrounding environment (Ogbe et al. 2020). So, endophytes are asymptomatic microorganisms residing inside plants, can be fungi or bacteria. Among the two, the fungi are one of the important components of soil ecosystem, efficiently associated with many processes in the soil, such as the decomposition of dead living organism, recycling of organic and inorganic component, and conversion of complex compound to simple one to be used as nutrients (Corbu et al. 2023). More than millions of different fungal species are expected on the Earth, but only few have been identified till date. The association of bacteria with the plant are well studied, but very little has been explored about the fungi. But recently it has been studied that endophytic bacteria coexist with endophytic fungi and can regulate the distribution of minerals, ecology of residing plant cell, biochemistry of associated plant cell, and physiology of the host plants (Jha et al. 2023). Endophytic fungi have a rich source of bioactive compounds such as hydrolytic enzymes, antimicrobial agents, phytohormones required by the endophytic fungi for its establishment in the associated plants as well as for the growth and development of host plant. The association of endophytic fungi with the host plant involves a complex interaction and plant stringently limits the number of endophytic fungi and associated endophytic fungi apply several different mechanisms to progressively adapt in its living environments. To maintain stable association with its host plant, endophytic fungi produce versatile compound to sustain their association; this is responsible for enhanced growth and better adaptability of host plant under adverse environmental state (Fig. 3.1) (Wen et al. 2022).

So, recognition of endophytic fungal resources with effective, specific, and efficient bioactive compounds according to different environmental condition is highly helpful in developing effective microbial consortium to sustain plant growth and development in different environmental condition. As such environment and plant-specific bioactive compounds cannot be available by chemical synthesis and precise dose also not be predicted properly. Thus, understanding mechanism of plants and associated fungal ecology in different stressful condition is important for sustainable yield of crop to its full genetic potential (Verma et al. 2022). The aim of the chapter is to review the effect and mechanism of endophytic fungi in mitigating abiotic stress in the crop plant to sustain crop productivity under stress.

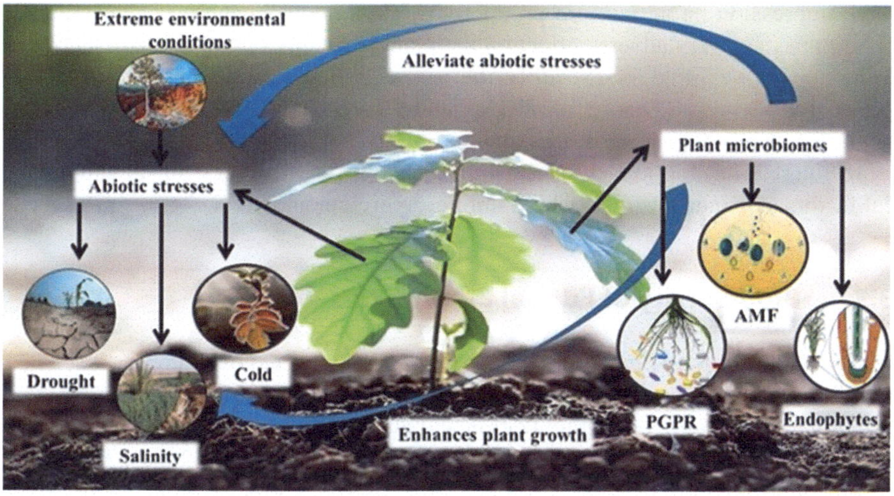

Fig. 3.1 Function of endophytes for plant under stress (Munir et al. 2022)

3.2 Need of Focus on Endophytic Microbes

World population is continuously increasing and corresponding increase in food production is demand of the hour, to feed the growing population. The hasty growth in population is one of the main reasons for the degradation of environment globally, which badly affects the natural resources and creating natural imbalance. The rising population and the environmental degradation is reason behind the challenge for sustainable development. To feed the growing population there is huge pressure on crop land, to increase crop yield uncontrolled use of chemicals- are cause for land degradation, to increase infrastructure for the growing population- are cause for deforestation, habitat destruction of many other living organism and loss of biodiversity (Adla et al. 2022). To fulfil the energy demand of growing population, there is increased pressure on natural resources, responsible for water scarcity, pollution, climate change, and the global warming. The uncontrolled urbanization, industrialization, and the destruction of natural habitats is main outcome of population, which remarkably altered land-use practices. The increasing food demand and reducing land resources put immense pressure on land extensification. The horizontal addition of land has limited or no scopes, so, it totally relies on vertical enhancement, which includes technical improvement in the agriculture field, like development of hybrid/improved seeds with less requirements of chemical fertilizers/pesticides or use of plant-associated endophytic microbes like fungi with novel bioactive compound for sustainable plant growth. Such endophytic fungi potentially increased crop production, serve as an eco-friendly approach for the developing sustainable farming practice to increase crop yield under different environmental conditions as well as stress. Water availability, soil fertility, and variety of seed have large impression on agriculture productivity. Continuous changing environmental conditions have very adverse effect on the plant productivity and to challenge such limitations; for assured food security, there is need for development of new climate-smart crop varieties (Zhao et al. 2023). Under such changing climatic conditions, plants frequently face several stresses like salinity, cold, drought, waterlogging, and temperature. The abiotic factors like light intensities, UV rays, chemical and physical factors also induce stress, which limit crop growth and yield.

3.3 Biodiversity of Fungal Endophytes

The wide biodiversity of endophytic fungi has been reported by many plant science researchers from almost all crops in all climatic zones worldwide. On the basis of site of the endophytic fungi colonization and mode of transmittance, the endophytic fungi has been categorized into four groups by Rodriguez et al. (2009). The endophytic fungi inhabits in grasses are classified as Class I—Clavicipitaceous fungi with vertical transmittance; the endophytic fungi resides in rhizomes, roots, and shoots are classified as Class II—Non-Clavicipitaceous fungi with both vertical and horizontally transmittance through seeds or rhizomes: the endophytic fungi resides in leaves/shoots are classified as Class III—Non-Clavicipitaceous fungi with horizontal transmittance only and the endophytic fungi resides in roots forming conidia are classified as Class IV—Non-Clavicipitaceous typical ascomycetes fungi with horizontally transmittance. Class I endophytic fungi is free-living symbiotic fungi include *Neotyphodium tembladerae, Epichloe festucae*, etc., help associated plant to develop resistance against pathogen by producing anti-pathogenic compound as well as provide tolerance against abiotic stress. Class II endophytic fungi include *Arthrobotrys* species, *Mycophycia ascophylli*, etc., produce phytohormone and enhance plant biomass and also confer tolerance against both biotic and abiotic stress. Class III endophytic fungi include *Colletotrichum species, Ustilago maydis*, etc., are not habitat-specific and reside in above-ground tissues, which initiate plant microbes' interaction. Class IV ascomycetous endophytic fungi are root-associated *Phialocephala dimorphosphora, Chloridium paucisporum*, etc., produce melanized tissues under high-stress environments (Table 3.1). The endophytic fungi predominantly produce several secondary metabolites likes, terpenes, phenolic acids, quinines, alkaloids, flavonoids, chinones, steroids, and phytohormones in host plants in response to environmental stress, to confer resistance in host plants (Jha 2020).

3.4 Fungi in Agriculture Production

In the last century, agriculture production has been enormously increased by the use of synthetic fertilizers and pesticides, but it comes with many drawbacks for the agriculture farms, biodiversity, and ecosystem. Uncontrolled use of chemical inputs in the crop field is responsible for pollution, global warming, and resistance in pathogen towards such chemical inputs.

Table 3.1 Endophytic fungal diversity in different abiotic stress climatic zone to develop tolerance in plant against abiotic stress

S. No	Stress	Endophytic fungi	Reference
1	Temperature	*Thermomyces* sp. *Aspergillus niger. Rhizopus oryzae*	(Ali et al. 2019)
2	Cold	*Paecilomyces formosus* LHL10 *Penicillium janthinellum* LK5	(Baazeem et al. 2021)
3	Drought	*Trichoderma atroviride, Arbuscular mycorrhizal fungi* *Penicillium citrinum, Aureobasidium pullulans* *Dothideomycetes* sp. *Piriformospora indica, Nectria haematococca*	(Al-Arjani et al. 2020)
4	Salinity	*Yarrowia lipolytica, Streptomyces sp, Trichoderma longibrachiatum, Fusarium oxysporum* sensu lato *Arbuscular mycorrhizal fungi, Aspergillus flavus* *Piriformospora indica, Epichloë bromicola Phoma glomerata, Penicillium* sp.	(Siddiqui et al. 2022)
5	Heavy Metal	*Penicillium janthinellum, Trametes hirsute, Chaetomium cupreum, Phialophora mustea, Penicillium roqueforti* Curvularia geniculate, *Paecilomyces formosus* and*Penicillium funiculosum,Sordariomycetes* sp.	(Pietro-Souza et al. 2020)
6.	Nutrient	*Mucor hiemalis, Fusarium equiseti, Byssochlamys spectabilis, Curvularia geniculate, Sordariomycetes* sp	(García-Latorre et al. 2021)

Fungi and bacteria are the groups of microorganism in the soil have inherent ability to enhance the soil nutrient content by solubilization/decomposition of organic matter and also use extra nutrient for its own growth and development (Wang et al. 2022). These microbes are generally remain either associated with plant roots or remain in its rhizosphere. These groups of microorganism can be used to substitute the uncontrolled input of chemicals in the agriculture field, to maintain or enhance the soil fertility as well as to overcome the disadvantage related with the chemical inputs. The augmented advantage of fungi is the production of unique secondary metabolites like alkaloids, phenolics, terpines, steroids, etc., in the host plant cell which is bioactivity effective against several invertebrates/vertebrates, phytopathogens, and can also provide tolerance against various abiotic stresses (Jha 2022).

The association of endophytic fungi is not an abrupt phenomenon, but it is predetermined chemotactic response, which will decide the types of endophytic fungi associated with which climatic zone and with which plants. The association of endophytic fungal modes and strategies are considerably different and endophytic fungi colonized intra and inter-cellularly with the complete root system (Alam et al. 2021). The endophytic fungi follow different mechanisms for entry and colonization with host plants like host cell surface receptor recognition and presence of specific plant protein, which facilitates efficient interaction. Efficient colonization of endophytic fungi depends on multiple factors like climatic condition, plant genotype, plant age, type of plant tissue, plant growth condition, and types of abiotic/biotic stress. The presence of endophytic fungi has already been reported in a wide range of habitats, starting from artic environments to hot deserts, and from mangrove, temperate to tropical forests. The endophytic fungi can serve as microbial factories for the production of novel bioactive compounds with a wide degree of structural variation and biological activity diversity (Hashem et al. 2023). Synthesis of such plant-derived secondary novel bioactive compounds by the endophytic fungi is due to establishment of unique mechanisms during the co-evolution of endophytic fungi with plants host, for the development of symbiotic relationships. The endophytic fungi able to synthesize the novel bioactive compounds by highly regulated mechanism depends on the developmental requirement of fungus and environmental changes, as change in growth parameter causes corresponding change in metabolic profile of endophytic fungi. The endophytic fungi have been acting as the source of the chemical diversity due to activation of the expression of silent gene clusters in competitive environment (Tiwari and Bae 2022). The population of endophytic fungi is highly variable in plant roots and depends on numerous factors as plant genotype and environmental conditions. The endophytic fungi are highly host-specific, which requires an adjacent adaptation between the host plant and its fungal partner. The interaction specificity between plants and endophytic fungi results in production of varieties of unique metabolites, which will take place by different methods: the endophytic fungi-induced production of unique metabolites of the host, the host plant also induce specific metabolic pathway in the endophytic fungi. The host plant and the endophytic fungi participate in parts of specific metabolic pathway for the production of novel bioactive compound (Fig. 3.2).

The host plant and the endophytic fungi metabolize products of each other and the endophytic fungi metabolize secondary compounds of host plants. The endophytic fungi not only initiate production of bioactive metabolites, but also stimulate the biochemical pathways of host plant for the production/accumulation of unique secondary metabolites (Toppo et al. 2023). These unique secondary metabolites utilized by the endophytic fungi for the production of novel compound, which enhance the resistance potential of the plant to ensure smooth adaption in its habitat. The interactions between plant and endophytic fungi are complex, involving multi-species communications and vary from host to host and from endophyte to endophyte. When endophytic fungi physically encounter a host plant, fungi have to cross several chemical and physical barriers to establish as efficient association. Different types of secondary metabolites produced by the plant to protect itself from the harmful pathogen interaction, which is harmful to endophytic fungi and act as chemical barrier for its entry. The endophytic fungi have to produce enzymes to hydrolyze these secondary metabolites/chemical barrier for its entry in the host plants (Chaudhary et al. 2022). The hydrolyzing enzyme produced by the endophytic fungi at the entry site, open the way for other microorganism also. The colonization of endophytic fungi can occur inter or intra-cellularly and involves several stages, including recognition of the host, spore germination, penetration of the epidermis, and tissue multiplication (Dutta et al. 2014). The transmission of endophytic fungi may be horizontal (when takes place by airborne spores) or vertical (when takes place by seeds from one generation of plants to other). The vertically transmitted endophytic fungi has systemic interaction, growing inside the seeds are

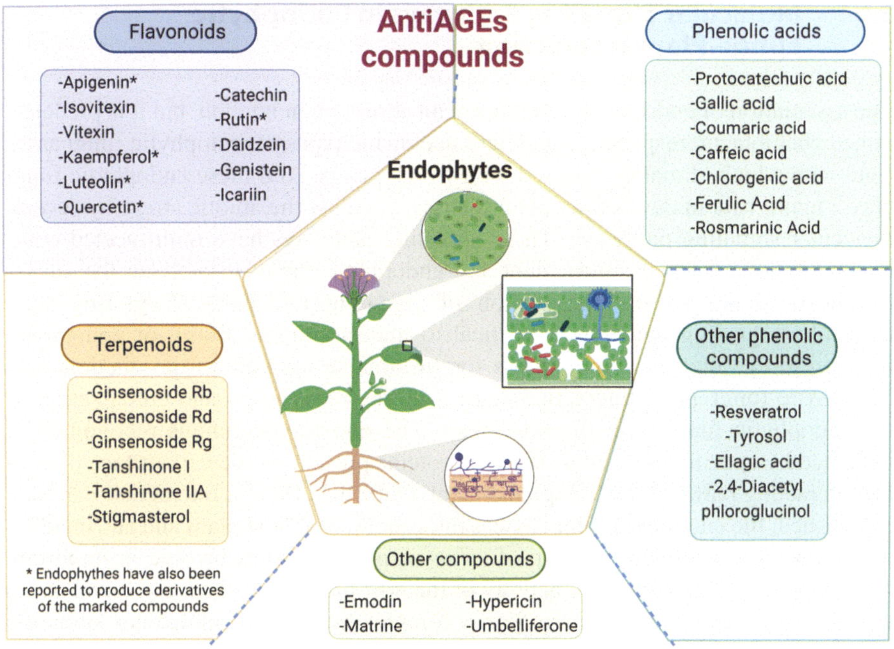

Fig. 3.2 Different novel bioactive compound produced by endophytic fungi (Rochín-Hernández et al. 2022)

more likely to be mutualistic, while horizontally transmitted endophytic fungi via spores has non-systemic interaction, to be more antagonistic to the host. So, the interaction of endophytic fungi and plant are flexible. The pathogenesis and symbiosis has been directly determined by the fungal gene expression and response to fungus for the host plant and recognition of host cell receptor.

The response of interaction is based on the genotypes of both interacting partners and can be determined as neutral, positive, or negative. However, symbiotic endophytic fungi maintain a sensible antipathy without the progression of disease in the host plant. Instead, they induce the production of defense metabolites potentially effective against a wide range of phytopathogens. Thus, a perfect balance must be established by the endophytic fungi in the host plant: they block the activation of the host's defense mechanism while protecting themselves from the host's toxic metabolites by disabling them. If plant defense systems counteract endophytic fungal virulence factors, it can result in the death of the endophytic fungi (Pasrija et al. 2022).

3.5 Molecular Signaling Pathways in Endophytic Fungi–Plant Interaction

The association of endophytic fungi is not an abrupt phenomenon, but it is predetermined chemotactic response, which will decide the types of endophytic fungi associate with which climatic zone and with which plants. And these endophytic fungi play a major role in developing plant tolerance against the abiotic stress by several molecular signaling pathways. These signaling pathways have multifaceted communications between the host plant and endophytic fungi, to activate the stress-responsive genes for the production of signaling molecules. Exploring such molecular signaling pathways is critical to understand the ability of endophytic fungi in mitigating the abiotic stress for sustainable agricultural production. The endophytic fungi–plant interaction starts with the recognition and colonization of the endophytic fungi with the host plant. The endophytic fungus recognition is mediated by specific molecular patterns proteins present on the cell surface of both the interacting partners, i.e., fungi and plant. Once the efficient interaction has been established, the endophytic fungi assess the condition of host plant and surrounding environment and activate a group of stress-responsive genes involve in important biochemical and physiological activity of the host plant for its better survival under stress. These gene activation regulate a complex network of molecular signaling pathways like the abscisic acid signaling pathway to develop tolerance against abiotic stress in the host plant. The abscisic acid is a phytohormone, having remarkable role in modulating the plant responses to abiotic stress. Moreover, endophytic fungi also induce the expression of genes involve in osmotic regulation, ion homeostasis as well as antioxidant defense in host plants. Not only that the molecular signaling pathways of endophytic fungi–plant interaction also cross-talk with other signaling pathways involved in important physiological pathway for plant growth and development. The endophytic fungi itself induce production of as indole-3-acetic acid like plant growth-promoting hormones as well as modulate the genes expression involved in mineral nutrient solubilization, nitrogen fixation for improved nutrient status of the host plant. Sometime there will be exchange of signaling molecules taking place between the plant and the fungi, which may act as communication signals for the establishment of efficient symbiotic relationship. The strigolactones and flavonoids like signaling molecules produced by the plant are utilized by the endophytic fungi for the activation of fungal genes to overcome the abiotic stress. So, understanding of such complex and specific interactions is critical for exploring the ability of the of endophytic fungi for its use in specific climatic zone to mitigate adverse effect of stress for the development of sustainable agricultural practices.

3.6 Role of Endophytic Fungi in the Mitigation of Abiotic Stress

The continuous changes in the surrounding environmental condition are frequently faced by the plant, but being sessile in nature plant do not able to escape as animals. Therefore, plants develop several internal mechanisms to overcome the adverse effect of changing surrounding environmental condition. These variations result in stress in plants and are major limiting factor for reduced crop productivity (Fig. 3.3).

Plant tries to counteract the changing surrounding environmental condition by corresponding modification in its internal environment to maintain its growth and development (Seleiman et al. 2021). All adverse impact of inert factors on living organism is designated as abiotic stress. The human activity is responsible for incessant climate change and environmental deterioration, which is major threat to food security for growing population. Cold stress, salt stress, nutrient imbalances, drought, extremes of temperature, flooding and many more in the list directly affect the growth and yield of crops. These stresses affect at all growth stages of the plant starting from germination to flowering and alter plant biochemical pathways, plant physiology as well as plants molecular events. Impact of these stresses can be visible at all stages of plant growth (Table 3.2).

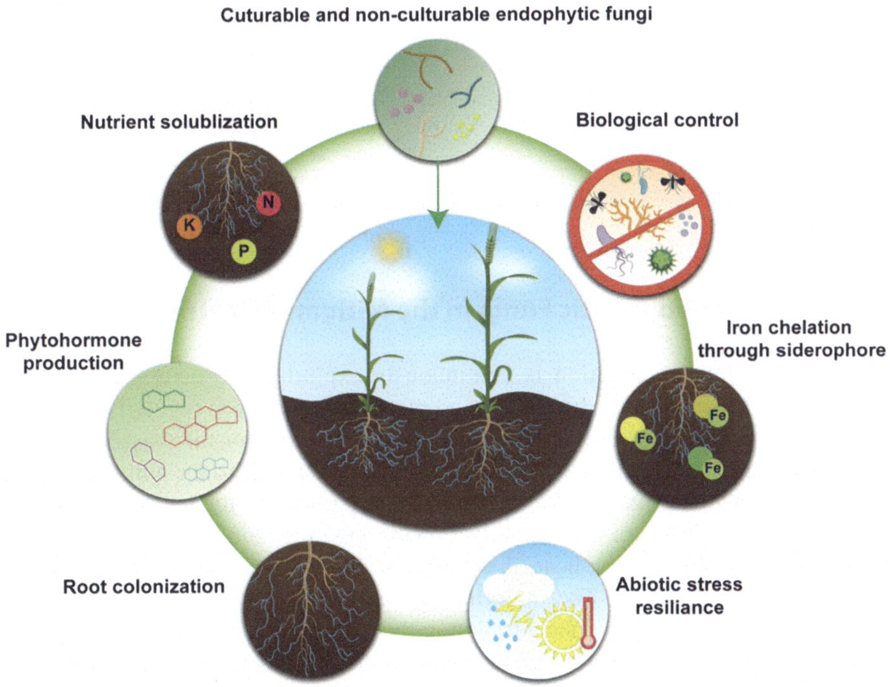

Fig 3.3 Different abiotic stresses and their effects on plants (Kumari et al. 2022)

Table 3.2 Mechanism of endophytic fungi to develop tolerance in plant against abiotic stress

S. No	Stress	Plants	Endophytic fungi	Mechanism	Reference
1	Nutrient	Wheat Rice	Curvularia geniculata *Sordariomycetes* sp.	IAA production;phosphate solubilization;siderophore production; Maintain photosystem II function	(Baron and Rigobelo 2021)
2	Cold	Soybean Tomato	*Paecilomyces formosus* LHL10 *Penicillium janthinellum* LK5	Enhance chlorophyll content;Reduce lipid peroxidation;Antioxidant production(LNA, GSH, PPO, CAT, SOD)Enhance the translocation of Ni from the root to the shoot	(Li et al. 2021)
3	Drought	Wheat Soybean	*Arbuscular mycorrhizal fungi* *Piriformosporaindica*	Upregulate antioxidant defense system;synthesis of osmolytes;maintain phytohormone levels;	(Al-Arjani, et al. 2020)
4	Salinity	*Arabidopis thaliana* Wheat Rice	*Streptomyces sp* *Arbuscular mycorrhizal fungi* *Piriformosporaindica*	Increase expression of the ion channels;increase plant biomass, lateral roots density, and chlorophyll content Increase osmoprotectant content	(Gupta et al. 2021)
5	Heavy Metal	Soybean Tomato	*Penicillium janthinellum* Curvularia geniculata *Sordariomycetes* sp.	decrease heavy metal translocation factor; downregulation of heavy metal transporter genes;activate signaling network of stress-responsive hormones and antioxidant systems	(Pietro-Souza et al. 2020)

3.6.1 Role of Endophytic Fungi in the Mitigation Nutrient Stress

The growing global population is continuously exploring the natural resources, which cause climate change drastically and plant being sessile is regularly exposed to biotic and abiotic stresses. All these stress is responsible for the massive threats to crop productivity. Most common abiotic stress corresponds to the entire environmental factor, which is responsible for reduced crop productivity and is great threat to global food security. In a constantly varying environment, all the unfavorable factors like drought, salinity, heat, flood, and cold are common factors for abiotic stress (Zhang et al. 2022). The majority of abiotic stress results in water stress, which results in higher ions concentration in the surrounding soil and water, which limits the uptake of both water and nutrients by the plants. Lower nutrient uptake reduced the plant growth and productivity, ultimately resulting in lower crop yield. The most common soil-related abiotic stress is lower mineral nutrient (phosphorus, potassium, and zinc) availability. To overcome the lower availability of inorganic nutrient and to sustain crop productivity, farmers applying uncontrolled amount of chemical fertilizers in the crop field to sustain crop yield, to feed the growing population globally. Water deficiency causes nutrient imbalance or deficiency for the crop due to Na+ and Cl– ion, will compete with nutrient ions Ca++, NO3– and K+, and also causes altered distribution of mineral nutrient as well as altered its root absorption (Long et al. 2023). The suboptimal availability of necessary nutrients or the excessive amounts of essential/non-essential macronutrients is designated as nutrient stress.

The endophytic fungi has the enzymes for the transformation and solubilization of mineral macronutrients like potassium, phosphorous, zinc, etc., and various other micronutrients, thus making such insoluble mineral nutrients easily available to associated host plants. The endophytic fungi like *Piriformospora indica, chroococcum,* and *Azotobacter are reported to increase* uptake of mineral nutrient and improve nutrient acquisition ability of the host plant (Singh et al. 2022). This suggests that characterization of endophytic fungi from different climatic zone may help plant scientist to develop novel endophytic fungi species consortium, which will help crop plant for improved production under various abiotic stress.

3.6.2 Role of Endophytic Fungi in the Mitigation Heat Stress

The growing global population continuously requires enhanced energy sources to fulfill their energy requirement and consume natural resources in uncontrolled way, which is responsible for global warming. The global warming is responsible for environmental high temperature due to different gaseous emissions due to human-related activities, which considerably add greenhouse gasses like methane, chlorofluorocarbons, CO_2, nitrous oxide and isresponsible for elevated environmental temperature. For all living organism, cellular activity is maintenance specifically at optimum temperature and elevated environmental temperature apparent as heat stress by the plants is responsible for irreparable damage in plant activity (El-Remaly 2023). The effect of elevated environmental temperature on the plant is a complex process accounting for many factors including temperature intensity, temperature duration, as well as the rate of increase in temperature above optimum levels. The elevated environmental temperature altered the cellular homeostasis, eventually the foremost cause that retarded plant growth and finally plant death, which ultimately results in reduced crop production worldwide. The elevated environmental temperatures shorten the life cycle of most crop plants by reducing the duration of the vegetative stage inducing senescence. Most of the stable crops are temperature sensitive and are able to tolerate very narrow range of temperature only. The elevated environmental temperatures range if exceed during floral transition is responsible for improper seed setting and production (Hampton et al. 2016).

The plant metabolic activity and physiology have been badly altered by elevated environmental temperature, but the plant-associated endophytic fungi help the host plant to mitigate the effect of elevated environmental temperature by altering,

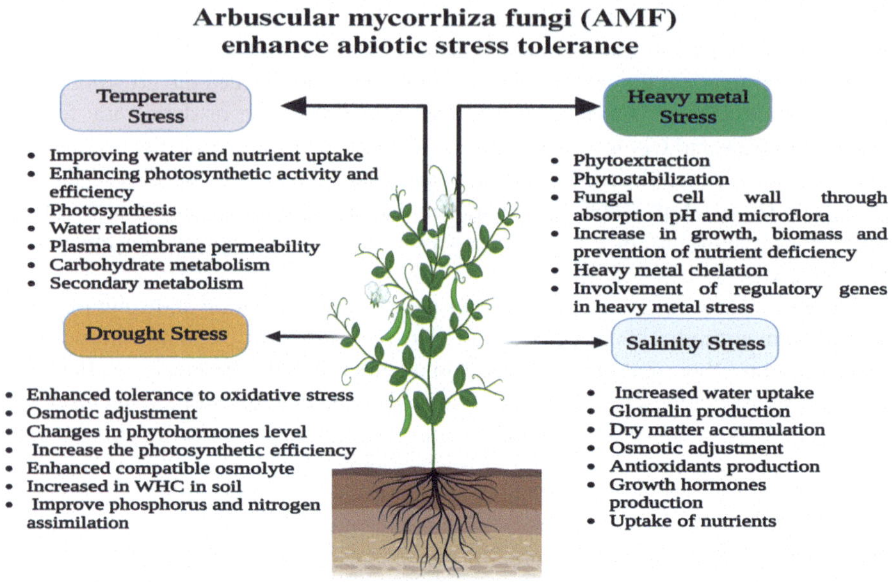

Fig. 3.4 Different mechanisms imparting abiotic stress tolerance in plants by endophytic fungi (Wahab et al. 2023)

regulating, or adjusting the metabolic activity, metabolites, and ultimately plant physiological activity according to its own biology as well as the plant developmental stage and plant genotype. According to the environmental condition, the corresponding alteration will take place in the endophytic fungi community, which enhanced the production of a variety of condition-specific metabolites for environmental acclimatization, to develop tolerance in host plant. Morsy (2020) reported production of heat shock protein 90 inhibitor in tomato by an endophytic fungi *Curvularia protubrate* to confer heat tolerance in tomato (Fig. 3.4).

3.6.3 Role of Endophytic Fungi in the Mitigation Drought Stress

Drought is most common abiotic stress, which remarkably reduced agricultural productivity and drastically challenging food security to feed the growing population globally. The continuous changing environmental conditions due to the growing global population, which continuously exploit excessively the natural resources, cause climate change drastically (Wiedmann et al. 2020). The climate change due to the growing global population requires growing water demand for their household activity as well as for crop productivity. Excessive human interference cause variably impacting rainfall pattern, elevated environmental temperature, which collectively cause water-deficit condition in the crop field. The water-deficit condition affects all living organisms, but predominantly affect plants, due to its inability to move away from stress condition. Plant develops several mechanisms to migrate the effect of water deficit, thereby impacting crop production. The water deficit is driven by many factors like the water holding capacity of the soil, amount of water losses by evapo-transpiration, and altered rainfall pattern, which cause drought stress (Seleiman et al. 2021). The adverse effect of drought include retarded seedling growth, low rate of seed germination, poor availability of soil nutrients, altered photosynthesis due to reduced leaves size and number, altered stomatal conduction, and poor gaseous exchange. Plants' response to drought depends on many factors like growth stage of the plant and genotype of plant, which stimulate altered gene expression followed by corresponding change in biochemical pathway, metabolites, and finally physiological changes as closure of stomata, reduction in rate of transpiration, reduced rate of photosynthesis. Plants also induce production of phytohormone like abscisic acid under drought stress and enhanced concentration of stress phytohormone like abscisic acid stimulates a cascade reaction for the modified physiological and transcriptional changes, which induce reduced water loss via reduced transpiration by stomata, reduced cell size, altered membrane integrity, enhanced leaf senescence, and induce ROS production, responsible for reduced agricultural productivity (EL Sabagh et al. 2022).

There are several endophytic fungi like *Trichoderma* sp., *Piriformospora indica, Aspergillus* sp., *Alternaria* sp. that are predominantly present in desert land are reported for their positive impact on the plant to develop tolerance against the drought stress by improving the water holding and absorbing ability of associated root system; it also produces and induces production of phytohormones and responsible for higher chlorophyll content in host plant under drought stress (Jha and Yadav 2021). There are several endophytic fungi which get associated with diverse group of crop plants like wheat, rice, corn, and tomato under drought condition and help the host plant utilize less water for its survival as well as increase biomass. The drought tolerance features induced by such endophytic fungi include reduced stomatal conductance and lower rate of transpiration by the associated host plant by formation of thicker cuticle. Ripa et al. (2019) reported that *Trichoderma* sp. induce drought tolerance ability of host plant by altering gene expression patterns of drought stress-inducible genes, which results in production of tolerance target proteins and other associated regulatory proteins for stimulation of specific signal transduction pathway in host plant to overcome the adverse effect of drought stress.

3.6.4 Role of Endophytic Fungi in the Mitigation of Salinity Stress

Salinity is one of the main limiting factors for crop productivity and due to our farming method and global warming this problem is increasing constantly worldwide. The agricultural land located in arid and semi-arid regions are also affected by salinity. Salinity stress badly affects plant growth and productivity due to reduced uptake of water, nutrients, and production of ROS. This ROS react with important biomolecules like DNA, lipid, proteins and hamper important molecular and biochemical pathways, which ultimately resulted in altered plant physiology like

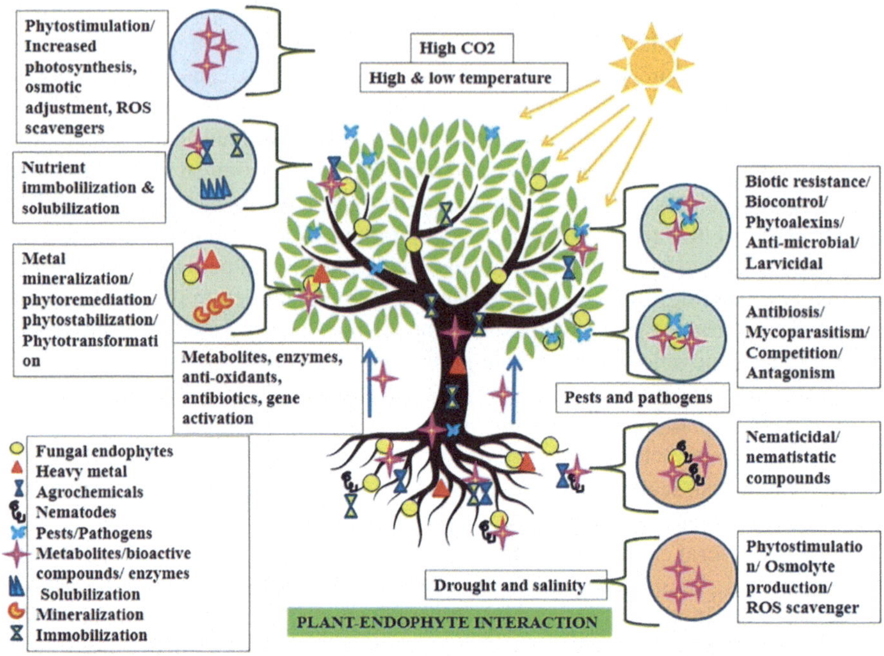

Fig. 3.5 Fungal endophyte–plant interaction expediting phytostimulation conferring stress mitigation (Verma et al. 2022)

reduced plant growth, low biomass, lower rate of photosynthesis, and altered redox homeostasis (Hasanuzzaman et al. 2021). At the same time, plants also use multilevel mechanisms to overcome the adverse effect of salinity to sustain their survival. Plants develop various mechanisms to sense such alteration in their surrounding and accordingly respond and try to adapt to changing environment by changing their morpho-physiological feature, by regulating the ionic concentration, biochemical and molecular mechanisms. Under salinity stress, plants feel osmotic stress and exhibited a wide range of responses at molecular, cellular, tissue, and whole-plant level. Plants induce enlargement of root growth, reduced shoot growth, modulate ions uptake, extrusion and compartmentalization, modification in ion transport systems, regulation of metabolites, altered carbon metabolism, and the production and accumulation of compatible solutes. Plants also divert primary metabolites for osmotic regulation and induce secondary stress signals, for activation of plant wide response against stress (Jha and Mohamed 2022). Such secondary signals are production of stress phytohormones like ABA and ethylene, induction of ROS generation, and intracellular second messengers like phospholipids (Singhal et al. 2021).

The endophytic fungi develop many strategies in host plant to overcome salinity stress, which is similar to drought tolerance mechanism. The endophytic fungi induce the production of antioxidant enzymes to encounter different reactive radicals to preserve the regular function of the cell under salinity. Endophytic fungi apply multiple mechanisms to develop tolerance against the salinity by producing enhanced quantity of organic osmolytes like proline, glycine, betaine, and trehalose as an adaptive response against salinity (Jha 2018). Some endophytic fungi improve salinity tolerance ability of the host plant not only by reprogramming the antioxidant enzyme profile, but also modulate hormonal profile of the host plant (Fig. 3.5).

Hosseyni et al. (2022) have reported association of *Curvularia* sp. induces production of antioxidant enzymes, like superoxide dismutase and ascorbate peroxidase in the associated plant. The endophytic fungi induce enhanced production and accumulation of osmolytes and contribute to the plant cell matrix potential to maintain water potential, necessary for retaining water in associated plant under salinity. The osmotic potential is determined mainly by matrix potential and solute potential. The root endophytic fungi *Piriformospora indica* are reported to enhance the plant biomass by stimulating the production of antioxidant enzymes, catalase, glutathione reductase, dehydroascorbate reductase, lipid peroxidase, alteration of composition of fatty acid of plasma membrane as well as antioxidative capacity in the associated host plant. Endophytic fungi *Exopiala* sp., *Fusarium culmorum,* and *S. vermifera* are reported to increase plant biomass in associated plant under saline stress by higher osmolyte accumulation. However, there is a great need to unveil how fungal endophytes stimulate plant defense response against abiotic stress (Ameen et al. 2024).

3.7 Molecular Interactions Between Endophytic Fungi and Plants

Plants being sessile regularly interact with various changing environment conditions as abiotic stress including extreme temperatures, salinity, drought, flood, heavy metal toxicity, etc., which have deleterious effect on plant growth and productivity. At the same time, plants also develop multilayered defense mechanism to overcome all such deleterious effects to ensure its survival. One such eco-friendly mechanism is development of symbiotic association among endophytic fungi and plant. The endophytic fungi are group of microbes which resides intercellularly in the healthy plant cell without having any adverse effect. Endophytic fungi colonized in the living plant cell and play a major role in developing plant tolerance against the abiotic stress by improving water uptake efficiency, enhancing nutrient uptake ability, regulating plant defense responses to promote plant growth response.

3.7.1 Induced Systemic Tolerance

Plants response to abiotic stress depends on many factors like type of abiotic effector, genotype of plant, and growth stage of the plant, which stimulate altered gene expression followed by corresponding change in biochemical pathway, metabolites, and finally physiological changes. The entire response of microbes-mediated tolerance ability in the plants can protect plant against abiotic stress in an eco-friendly manner (Jha, 2024). In this one of the main component is use of endophytic fungi with plants to cope with abiotic stress is by the induction of systemic tolerance. The endophytic fungi induced change in biochemical pathway, metabolites, and finally physiological changes in the host plant to develop stress tolerance by activating various defense pathways and the production of stress-related compounds (Chaubey et al. 2023). Endophytic fungi also induce production of stress-related hormones, phytohormone like jasmonic acid, abscisic acid under abiotic stress to modulate plant response. These stress-related hormones, phytohormone, can induce stress-responsive signaling pathways as well as genes for the synthesis of stress-related metabolites and tolerance proteins. Furthermore, endophytic fungi also stimulate the antioxidant defense system of plants, to scavenge reactive oxygen species to protect important plant biomolecules under stress conditions.

3.7.2 Gene Expression in Endophytic Fungi–Plant Interaction

The endophytic fungi–plant interaction starts with the recognition and colonization of the endophytic fungi with the host plant, involve the suppression/activation of specific genes in the plant/fungi. The suppression/activation of specific genes is a complex process that involves different molecular mechanisms. The regulation of gene expression in endophytic fungi is by transcription factors, which bind to specific gene for its controlled expression. In response to abiotic stress, certain transcription factors in endophytic fungi are activated, leading to the upregulation of stress-responsive genes. Moreover, some epigenetic modifications like histone acetylation, DNA methylation, also meditated by endophytic fungi for the repression or activation of precise genes to mitigate stress response in plant, which takes place due to altered expression of genes at the transcriptional level (Jha and Subramanian 2015). The regulation of gene expression in endophytic fungi/by the endophytic fungi in the host plant also takes place at post-transcriptional and post-translational level, such as modifications of effector protein, phosphorylation of target protein, acetylation of regulatory protein alternative splicing, and RNA processing. The endophytic fungi are responsible for the fine-tuning of the gene in the host plant to optimize the adverse effect of abiotic stress.

3.8 Conclusion

In the current scenario, to fulfill the food demand for the growing population, increase in the crop productivity is need of hour, without affecting the health of soil and environment for our future generations. But crop productivity is badly affected by adverse environmental conditions or different types of abiotic stress. To overcome all these limitations, it requires a non-toxic, cost-effective, easily applicable and eco-friendly approach for sustainable development in crop field to increase the crop yield. One of the best such tool is use of endophytic fungi as a substitute to fertilizers or pesticide for sustainable agriculture. Endophytic fungi and their metabolites need to be explored rigorously up to the multiomics level to design proper endophytic fungal combination as per crop and environment.

References

Adla K, Dejan K, Neira D, Dragana Š (2022) Chapter 9 - Degradation of ecosystems and loss of ecosystem services. In: Prata JC, Ribeiro AI, Rocha-Santos T (eds) One Health. Academic Press, pp 281–327

Alam B, Lǐ J, Gě Q, Khan MA, Gōng J, Mehmood S, Yuán Y, Gǒng W (2021) Endophytic fungi: from symbiosis to secondary metabolite communications or vice versa? Front Plant Sci 12:791033

Al-Arjani AF, Hashem A, Abd Allah EF (2020) Arbuscular mycorrhizal fungi modulate dynamics tolerance expression to mitigate drought stress in Ephedra foliata Boiss. Saudi J Biol Sci 27:380–394

Ali AH, Radwan U, El-Zayat S (2019) The role of the endophytic fungus, *Thermomyces lanuginosus*, on mitigation of heat stress to its host desert plant *Cullen plicata*. Biol Futur 70:1–7

Ameen M, Mahmood A, Sahkoor A, Zia MA, Ullah MS (2024) The role of endophytes to combat abiotic stress in plants. *Plant Stress,12*:100435.

Baron NC, Rigobelo EC (2021) Endophytic fungi: a tool for plant growth promotion and sustainable agriculture. Mycology 13(1):39–55

Baazeem A, Alorabi M, Manikandan P, Alotaibi SS, Almanea A, Abdel-Hadi A, Vijayaraghavan P, Raj SRF, Kim YO, Kim HJ (2021) *Paecilomyces formosus* MD12, a Biocontrol Agent to Treat *Meloidogyne incognita* on Brinjal in Green House. J Fungi (Basel). 7(8):632.

Chaubey AK, Sharma V, Prajapati PK, Mishra S, Panday R, Dwivedi SV, Shing A, Soni R (2023) Induction of stress tolerance in plants by metabolic secretions of endophytes for sustainable development. In: Soni R, Suyal DC, Morales-Oyervides L (eds) Microbial bioactive compounds. Springer, Cham

Chaudhary P, Agri U, Chaudhary A, Kumar A, Kumar G (2022) Endophytes and their potential in biotic stress management and crop production. Front Microbiol 13:933017

Corbu VM, Gheorghe-Barbu I, Dumbravă AŞ, Vrâncianu CO, Şesan TE (2023) Current insights in fungal importance—a comprehensive review. Microorganisms 11(6):1384

Dutta D, Puzari KC, Gogoi R, Dutta P (2014) Endophytes: exploitation as a tool in plant protection. Braz Arch Biol Technol 57(5):621–629

El-Remaly E (2023) Morphological, physio-biochemical, and molecular indications of heat stress tolerance in cucumber. Sci Rep 13:18729

García-Latorre C, Rodrigo S, Santamaria O (2021) Effect of fungal endophytes on plant growth and nutrient uptake in *Trifolium subterraneum* and *Poa pratensis* as affected by plant host specificity. Mycol Prog 20:1217–1231

Gupta S, Schillaci M, Walker R (2021) Alleviation of salinity stress in plants by endophytic plant-fungal symbiosis: current knowledge, perspectives and future directions. Plant Soil 461:219–244

Hampton JG, Conner AJ, Boelt B, Chastain TG, Rolston P (2016) Climate change: seed production and options for adaptation. Agriculture 6(3):33

Hasanuzzaman M, Raihan MRH, Masud AAC, Rahman K, Nowroz F, Rahman M, Nahar K, Fujita M (2021) Regulation of reactive oxygen species and antioxidant defense in plants under salinity. Int J Mol Sci 22(17):9326

Hashem AH, Attia MS, Kandil EK (2023) Bioactive compounds and biomedical applications of endophytic fungi: a recent review. Microb Cell Factories 22:107

Hosseyni Moghaddam MS, Safaie N, Rahimlou S, Hagh-Doust N (2022) Inducing tolerance to abiotic stress in *Hordeum vulgare* L. by halotolerant endophytic fungi associated with salt lake plants. Front Microbiol 13:906365

Jha Y (2018) Effects of salinity on growth physiology, accumulation of osmo-protectant and autophagy-dependent cell death of two maize variety. Russ Agric Sci 44:124–130

Jha Y (2020) Plant microbiomes with phytohormones: attribute for plant growth and adaptation under the stress conditions. In: Yadav A, Rastegari A, Yadav N, Kour D (eds) Advances in plant microbiome and sustainable agriculture. Microorganisms for sustainability, vol 20. Springer, Singapore

Jha Y (2022) Enhanced cell Viability with Induction of Pathogenesis Related Proteins Against Aspergillus niger in Maize by Endo-Rhizospheric Bacteria. Jordan J Biol Sci 15(1):139–147.

Jha Y (2024) Chapter 23 - Differential fungal metabolite accumulation in response to abiotic and biotic stresses. In: Abd-Elsalam KA, Mohamed HI (eds) Nanobiotechnology for plant protection, fungal secondary metabolites. Elsevier, pp 457–467

Jha Y, Mohamed HI (2022) Plant secondary metabolites as a tool to investigate biotic stress tolerance in plants: a review. Gesunde Pflanzen 74:771–790

Jha Y, Subramanian RB (2015) Reduced cell death and improved cell membrane integrity in rice under salinity by root associated bacteria. Theor Exp Plant Physiol 27:227–235

Jha Y, Yadav AN (2021) *Piriformospora indica*: biodiversity, ecological significances, and biotechnological applications for agriculture and allied sectors. In: Abdel-Azeem AM, Yadav AN, Yadav N, Usmani Z (eds) Industrially important fungi for sustainable development. Fungal biology. Springer, Cham

Jha P, Kaur T, Chhabra I, Panja A, Paul S, Kumar V, Malik T (2023) Endophytic fungi: hidden treasure chest of antimicrobial metabolites interrelationship of endophytes and metabolites. Front Microbiol 14:1227830

Kumari VV, Banerjee P, Verma VC, Sukumaran S, Chandran MAS, Gopinath KA, Venkatesh G, Yadav SK, Singh VK, Awasthi NK (2022) Plant nutrition: an effective way to alleviate abiotic stress in agricultural crops. Int J Mol Sci 23:8519

Li D, Bodjrenou DM, Zhang S, Wang B, Pan H, Yeh KW, Lai Z, Cheng C (2021) The endophytic fungus *Piriformospora indica* reprograms banana to cold resistance. Int J Mol Sci 22(9):4973

Long F, Hu M-F, Chen S, Bao G-S, Dan H, Chen S-H (2023) Endophytic fungi regulate *HbNHX1* expression and ion balance in *Hordeum bogdanii* under alkaline stress. J Fungi 9(3):331

Morsy M, Cleckler B, Armuelles-Millican H (2020) Fungal endophytes promote tomato growth and enhance drought and salt tolerance. Plants (Basel) 9(7):877

Munir N, Hanif M, Abideen Z, Sohail M, El-Keblawy A, Radicetti E, Mancinelli R, Haider G (2022) Mechanisms and strategies of plant microbiome interactions to mitigate abiotic stresses. Agronomy 12(9):2069

Mushtaq S, Shafiq M, Tariq MR, Sami A, Nawaz-ul-Rehman MS, Bhatti MHT, Haider MS, Sadiq S, Abbas MT, Hussain M, Shahid MA (2023) Interaction between bacterial endophytes and host plants. Front Plant Sci 13:1092105

Ogbe AA, Finnie JF, Van Staden J (2020) The role of endophytes in secondary metabolites accumulation in medicinal plants under abiotic stress. S Afr J Bot 134:126–134

Pasrija P, Girdhar M, Kumar M, Arora S, Katya A (2022) Endophytes: an unexplored treasure to combat multidrug resistance. Phytomed Plus 2(2):100249

Pietro-Souza W, de Campos PF, Mello IS, Stachack FFF, Terezo AJ, Cunha CND, White JF, Li H, Soares MA (2020) Mercury resistance and bioremediation mediated by endophytic fungi. Chemosphere 240:124874

Ripa FA, Cao WD, Tong S, Sun JG (2019) Assessment of plant growth promoting and abiotic stress tolerance properties of wheat endophytic fungi. Biomed Res Int 2019:6105865

Rochín-Hernández LS, Rochín-Hernández LJ, Flores-Cotera LB (2022) Endophytes, a potential source of bioactive compounds to curtail the formation–accumulation of advanced glycation end products: a review. Molecules 27(14):4469

Rodriguez RJ, White JF Jr, Arnold AE, Redman RS (2009) Fungal endophytes: diversity and functional roles. New Phytol 182:314–330

Sabagh ELA, Islam MS, Hossain A, Iqbal MA, Mubeen M, Waleed M, Reginato M, Battaglia M, Ahmed S, Rehman A, Arif M, Athar H-U-R, Ratnasekera D, Danish S, Raza MA, Rajendran K, Mushtaq M, Skalicky M, Brestic M, Soufan W, Fahad S, Pandey S, Kamran M, Datta R, Abdelhamid MT (2022) Phytohormones as growth regulators during abiotic stress tolerance in plants. Front Agron 4:765068

Seleiman MF, Al-Suhaibani N, Ali N, Akmal M, Alotaibi M, Refay Y, Dindaroglu T, Abdul-Wajid HH, Battaglia ML (2021) Drought stress impacts on plants and different approaches to alleviate its adverse effects. Plants (Basel) 10(2):259

Siddiqui ZS, Wei X, Umar M, Abideen Z, Zulfiqar F, Chen J, Hanif A, Dawar S, Dias DA, Yasmeen R (2022) Scrutinizing the application of saline endophyte to enhance salt tolerance in rice and maize plants. Front Plant Sci 12:770084

Singh S, Annapurna SK, Shrivastava N, Varma A (2022) Symbiotic interplay of *Piriformospora indica* and *Azotobacter chroococcum* augments crop productivity and biofortification of Zinc and Iron. Microbiol Res 262:127075

Singhal RK, Saha D, Skalicky M, Mishra UN, Chauhan J, Behera LP, Lenka D, Chand S, Kumar V, Dey P, Indu PS, Vachova P, Gupta A, Brestic M, El Sabagh A (2021) Crucial cell signaling compounds crosstalk and integrative multi-omics techniques for salinity stress tolerance in plants. Front Plant Sci 12:670369

Tiwari P, Bae H (2022) Endophytic fungi: key insights, emerging prospects, and challenges in natural product drug discovery. Microorganisms 10(2):360

Toppo P, Kagatay LL, Gurung A (2023) Endophytic fungi mediates production of bioactive secondary metabolites via modulation of genes involved in key metabolic pathways and their contribution in different biotechnological sector. 3 Biotech 13(6):191

Verma A, Shameem N, Jatav HS, Sathyanarayana E, Parray JA, Poczai P, Sayyed RZ (2022) Fungal endophytes to combat biotic and abiotic stresses for climate-smart and sustainable agriculture. Front Plant Sci 13:953836

Wahab A, Muhammad M, Munir A, Abdi G, Zaman W, Ayaz A, Khizar C, Reddy SPP (2023) Role of arbuscular mycorrhizal fungi in regulating growth, enhancing productivity, and potentially influencing ecosystems under abiotic and biotic stresses. Plan Theory 12(17):3102

Wang G, Ren Y, Bai X, Su Y, Han J (2022) Contributions of beneficial microorganisms in soil remediation and quality improvement of medicinal plants. Plants (Basel) 11(23):3200

Wen J, Okyere SK, Wang S, Wang J, Xie L, Ran Y, Hu Y (2022) Endophytic fungi: an effective alternative source of plant-derived bioactive compounds for pharmacological studies. J Fungi (Basel) 8(2):205

Wiedmann T, Lenzen M, Keyßer LT (2020) Scientists' warning on affluence. Nat Commun 11:3107

Zhao J, Liu D, Huang R (2023) A review of climate-smart agriculture: recent advancements, challenges, and future directions. Sustain For 15(4):3404

Zhang H, Zhu J, Gong Z, Zhu JK (2022) Abiotic stress responses in plants. Nat Rev Genet. 23(2):104–119.

Endophytic Fungi: A Natural Approach to Plant Growth Promotion and Sustainable Agriculture

4

Muhammad Ashfaq, Iqra Mushtaq, Kamel A. Abd-Elsalam, Mirza Abid Mehmood, and Samah Bashir Kayani

Abstract

Endophytic fungi play a crucial role in plant habitat adaptation, enhancing productivity and protecting against biotic and abiotic stresses. The benefits of endophytic colonization arise from both direct effects, such as phytohormone synthesis, and indirect effects, including the production of secondary metabolites and antibiotics, induced resistance, and protection from various stressors. In addition to protecting plants from pests and diseases, certain fungal endophytes have been utilized to reduce animal toxicity associated with these fungi in temperate grasses, enhance the cultivation of maize and rice plants tolerant to a range of biotic and abiotic stresses, and improve post-harvest control management. The loss of fungal endophyte diversity in domesticated plants can result from various practices, such as plant breeding and seed treatments, often due to a lack of awareness of the importance of fungal endophytes. This may explain why some endophyte strains are less effective at conferring benefits to plants. We offer recommendations for mitigating these negative effects in modern agriculture.

Keywords

Biotic stress · Fungal endophyte · Growth promotion · Modern agriculture

M. Ashfaq (✉) · I. Mushtaq · M. A. Mehmood
Plant Pathology, Institute of Plant Protection, Muhammad Nawaz Shareef University of Agriculture, Multan, Punjab, Pakistan
e-mail: mashfaq@mnsuam.edu.pk

K. A. Abd-Elsalam
Plant Pathology Research Institute, Agricultural Research Center (ARC), Giza, Egypt

S. B. Kayani
Govt. Graduate College (W) Mumtazabad, Multan, Punjab, Pakistan

© The Author(s), under exclusive license to Springer Nature Singapore Pte Ltd. 2025

K. A. Abd-Elsalam, A. H. Hashem (eds.), *Fungal Endophytes Volume II*,
https://doi.org/10.1007/978-981-97-8804-0_4

4.1 Introduction

The relationship between fungi and plants, specifically the association between mycorrhizas and endophytes, is prevalent and has been documented in fossil records for over 400 million years (Krings et al. 2007; Chadha et al. 2015; Card et al. 2016). This association initiated when plants began to colonize the soil which highlights its importance in this process (Rai et al. 2014; Anjum et al. 2019). The advantages of this distinctive relationship have been recognized and discussed. However, comprehensive investigations about the actual advantages provided by these fungi have only been conducted recently (Vega et al. 2009; Card et al. 2016; Quesada Moraga 2020). To fully understand fungal endophytic interaction, further research is required in numerous areas. Fortunately, this field of study is progressing swiftly (Chadha et al. 2015; Khan et al. 2015; Collinge et al. 2022). The main driving force behind current scientific research is the heightened awareness of the need for more sustainable agriculture. Developing sustainable agricultural practices should help to mitigate and reduce harmful environmental impacts in the future (Baron et al. 2019). The unregulated and inadequate use of fertilizers and pesticides adversely affects the health of humans and the environment. Food contains residues from various processes, including grains, cereals, fruits, and vegetables, as well as byproducts such as wines and juices. These residues depend on the manufacturing practices used in this process (Zikankuba et al. 2019; Maji et al. 2020; Briffa et al. 2020). Sabarwal et al. (2018) have described several studies that link pesticide exposure to an increased risk of numerous health disorders in both children and adults. These diseases include Hodgkin's disease, endocrine disorders, reproductive and respiratory problems, Parkinson's disease, cancer, and lymphoma (Sabarwal et al. 2018). Furthermore, non-target organisms in both aquatic and terrestrial ecosystems are continuously affected. This includes fish, crustaceans, zooplankton in the aquatic environment, and birds, livestock, natural pollinators, and beneficial microscopic organisms in the soil (Chagnon et al. 2015; Sánchez-Bayo 2021). Excessive use of fertilizers also results in heavy metal accumulation in soil, soil acidification, contamination of water reservoirs and aquifers, pollution of lakes, and production of greenhouse gases contributing to environmental pollution (Khan et al. 2018; Rashmi et al. 2020). Understanding the co-occurrence patterns that represent positive interactions between soil microbiota and plant is important for exploring different alternative pathways. These alternatives have the potential to reduce or replace the use of fertilizers and pesticides, thereby promoting the sustainable agriculture (Baron Rigobelo 2022; Ortiz and Sansinenea 2022). In this area further research has been conducted, by considering the possible application of classical anti-bacterial agents (BCAs) through large-scale release of pathogens in crops. *Beauveria bassiana* and *Metarhizium anisopliae* (entomopathogenic fungi) are frequently used as biological control agents to control arthropods that are vectors or pests (Garido-Jurado et al. 2020; Quesada Moraga 2020; Miranda-Fuentes et al. 2020). Fungi can act as antagonists against plant pathogens through a variety of mechanisms, including the synthesis of various metabolites such as enzymes, allergens, pesticides, bacteria, nitrogen, carbon sources, minerals and compete for space, they can also kill parasites, promote plant growth, and reduce disease activity by strengthening plant immunity (Heimpel and Mills 2017; Quesada Moraga 2020; Bahram and Netherway 2022). The study of endophytes sheds light on the ecology and various interactions between bacteria and plants, as well as contributing to the understanding of the benefits derived from these interventions. It draws attention to factors that should be carefully considered in order to create sustainable human practices, particularly in agriculture. The preparation of endophytic fungi as biofertilizer, biocontrol agent, and biostimulant is interesting due to their properties (Busby et al. 2016). In this chapter, we discuss different aspects of endophyte life as well as the progress in scientific research demonstrating the use of endophytes in the production of crops and agricultural bio products.

4.2 Definition, Classification, Biodiversity, and Distribution of Fungal Endophytes

The term "endophytic" has been extensively discussed and defined in various ways. Initially proposed by De Bary in 1886, it was used to describe any organism that lived inside (éndon) a plant (phyton). Endophytic microorganisms encompass viruses, fungi, bacteria, protozoa, and even microalgae. These organisms exist within the different tissues of host plant without inducing disease symptoms (Baron and Rigobelo 2022; Elnahal et al. 2022). However, the usage of the term "endophyte" is currently under debate among scientists. Some argue that it should refer solely to habitat rather than function (Hardoim et al. 2015; Hyde and Soytong 2008). Others suggest the term "mutualistic endophytes" should be used to describe endophytic microbes that are responsible for benefit to their host. This term specifically excludes latent or inactive pathogens as well as saprophytes (Card et al. 2016). According to Le Cocq et al. (2017), endophytes are microorganisms that spend at least some of their life cycle in interior plant tissue and do not harm the host plant under any circumstances. For colonization, host plant's function requires initiation of virulence mechanisms and activation of host defense. In interchange of benefits between host plant and endophytic fungi, included resistance to biotic and abiotic stresses in host plant as well as fungi takes nutrients from the host plant (Bamisile et al. 2018). Endophytic fungi are categorized as either clavicipitaceous (which is found in grasses) or nonclavicipitaceous (which is not found in grasses) (Hyde and Soytong 2008). In 2009, both types of fungi were classified into four classes as shown in Table 4.1. Class 1 includes all clavicipitaceous fungi that specify in grass colonizing and can be found in the aerial parts or roots of their host plants. These fungi are transmitted both vertically and horizontally (Rodriguez et al. 2009). On the other hand, Classes 2, 3, and 4 consist of non-clavicipitaceous fungi (Rodriguez et al. 2009). Class 2 endophytes can colonize both the aerial section and the roots, with vertical and horizontal transmission capabilities as shown in Table 4.1. Class 3 is a diverse group (transmitted horizontally) which is often linked with the leaves of tropical trees. Class 4 consists of fungi having melanin in their septa, often known as "dark septate endophytes." These fungi live inside the host roots for their whole

Table 4.1 Four fungal endophytic classes

Endophyte class		Host range	Host	Localized tissue	Transmission type	References
Clavicipitaceous (which is found in grasses)	Class 1	Narrow	Cool season grasses	Leaf sheath, culms, shoots and rhizome	Horizontal and vertical	Rodriguez et al. (2009), Tadych et al. (2014)
Nonclavicipitaceous (which is found in grasses)	Class 2	Broad	Great host range	Stem, leaves, roots and low abundance in rhizosphere	Horizontal and vertical	Rodriguez et al. (2009), Abo Nouh (2019), Sharma et al. (2018)
	Class 3	Broad	Many tropical trees	Leaves, shoots	Horizontal	Abo Nouh (2019), Rodriguez et al. (2009)
	Class 4	Broad	Abundant host range	Roots	Horizontal	Rodriguez et al. (2009)

life and transmit horizontally. According to Lugtenberg et al. (2016), the addition of a new class occurs as endophytic entomopathogenic fungi. These fungi have the unusual capacity to colonize and infect various insects in addition to being able to survive as symptomless endophytes in a range of plant types. Bamisile et al. (2018) organized data from numerous studies, suggesting that endophytic fungi can be categorized based on function, ecology, and diversity. Based on their method of reproduction, spores are categorized as either sexual or asexual (Bamisile et al. 2018). Additionally, their transmission can be either vertical or horizontal. Horizontal transmission involves the distribution of spores or vegetative propagules to the plant population by a vector or air. On the other hand, vertical transmission occurs when fungi are transferred to plant offspring through seeds (Lugtenberg et al. 2016; Franco et al. 2021; Baron and Rigobelo 2022).

Depending on the plant they infect, endophytic fungi are categorized as either asymptomatic or symptomatic, leaf endophytes or root endophytes. Depending on their nutrition, they are also classified as biotrophic or necrotrophic. Biotrophic endophytic fungi take nutrients from living tissues of host plant but the necrotrophic fungi induce necrosis to get nutrients from dead tissues (Kemen and Jones 2012). Studies on fungal endophyte was comprehensive, focusing on their relationships with grasses, particularly the genus *Epichlöe* (Lugtenberg et al. 2016; Card et al. 2016). The best thing about fungi for plants is that they produce secondary metabolites, mostly alkaloids. These compounds accumulate in tissues and exhibit biological activity against a variety of vertebrates, invertebrates, fungi, and bacteria. They also increase the plant's resistance against abiotic diseases (Nisa et al. 2015; Card et al. 2016). *Epichlo* species are now used commercially worldwide to cultivate (Johnson et al. 2019). Among non-clavical fungi, representatives of group 2 fungi are common in research. This study focuses on their relationship with plants and their benefits to agriculture (Rodriguez et al. 2009). However, much is still unknown about their distribution and biodiversity. Endophytic fungi have been found from many places, including hot deserts, cold climates, mangroves, tropical and subtropical forests (Abo Nouh 2019; El Mansy et al. 2020). Chadha et al. (2015) and Lugtenberg et al. (2016) showed that the distribution and diversity of endophytic fungi over large areas is still in its infancy. However, some general changes are also recognized such as there are many species of tropical endophytic fungi compared to those at higher latitudes. Additionally, small populations attempt to offset the diversity of endobiotic organisms found in tropical environments.

Endophytic fungal diversity is plant-associated and can vary between different environments (Wani et al. 2015; Chen et al. 2020) even the abundance of endophytic fungi in plant tissues can vary within a plant species. Additionally, plant age can influence the diversity of fungal endophytic groups (Sieber 2007). Some endophytic fungi can colonize a wide variety of plants, but others are more specific and are found only in certain plants. On the basis of plant colonized area, plant specificity has been observed (Bamisile et al. 2018; Hartmann et al. 2009; Aly et al. 2011). As compared to horizontally transmitted diseases, vertically transmitted diseases appear to have better relationships between plants which are considered antagonistic (Aly et al. 2011). Genetic communication among different endophytes and host plants is important for the evolution of the relationship but is a complex process that is still poorly understood. It involves the selection of genes which are responsible for the synthesis of secondary chemicals and enzymes that facilitate colonization (Yan et al. 2019). The attachment of endophytic germination tubes with roots causes the root to lose its apical dominance and stimulates the development of hyphal penetrating cells (hyphal spores). It initiates the colonization process by penetrating the hyphae in the root cortex (Khan et al. 2015), after it hyphae penetrates in the cell wall of plant and continue to colonize the surrounding tissues (such as pericycle, endoderm, phloem, and xylem) roots and soil (Yan et al. 2019; Brundrett and Tedersoo 2020). Some endophytes have the ability to produce the specific type of host metabolites and also have the unique characteristics that the interaction occur between the fungal endophyte and host plant without any defense response against them indicate that this interaction is beneficial for both of them. The capability of certain endophytes to produce host metabolites and the unique characteristics of plants linked to fungal endophytes, such as the absence of defense responses against them, indicate that the interaction is beneficial (Wani et al. 2015). This relationship may even extend to simultaneous reproductive processes in both the host plant and fungi during vertical transmission via seeds (Aly et al. 2011). By analyzing the plant fungal endophytic community presence, a significant challenge occurs for scientists. It is difficult to detect the presence of microorganism in tissue due to abnormal hyphae and lack of differentiation (Rashmi et al. 2019). Traditionally, the endophytic population is evaluated by isolating it from surface-sterilized plant tissues, aiming to recover fungal strains exclusively found inside the plant. To avoid osmotic barriers and facilitate exploratory hyphae passage, traditional culture media are typically modified as necessary, often by adding a higher proportion of water, usually double, to the medium. Additionally, plants can also be used in culture media (Murphy et al. 2018). Many biotic and abiotic factors can influence the properties of endophytic fungi using cultivation (Chen et al. 2020; Rashmi et al. 2019). Therefore, scientific research has greatly improved with the introduction of genetic methods to identify endophytic populations in various plant species. Microbiome studies can reveal a wide range of endobiotic pathogens using DNA sequencing of plant material (Deyett and Rolshausen 2020; Brader et al. 2017).

4.3 Transmission of Endophytic Fungi

Although they can reproduce vertically or horizontally, the mode of reproduction or transmission of many endophytic organisms remains unknown. *Epichlo* species is an example of an endophyte that can reproduce in two ways (sexual and asexual) as long as the host retains the ability to produce ascospore as shown in Fig. 4.1. For asexual *Epichlo* species, vertical transmission via seed is a good example of study. The endophytic fungi can be established by growing in the embryos of developing seeds and colonizing small seedlings after germination (Card et al. 2016). In the lab conditions vertically transmitted *Epichlo* endophytes can be transferred from one plant to another plant (Latch and Christensen 1985), allowing the cultivation of different types of grasses to create grass–endophyte combinations for future work (Johnson et al. 2013). In category 2, another type of vertical transmission is found, in which the endophyte spreads throughout the seed coat.

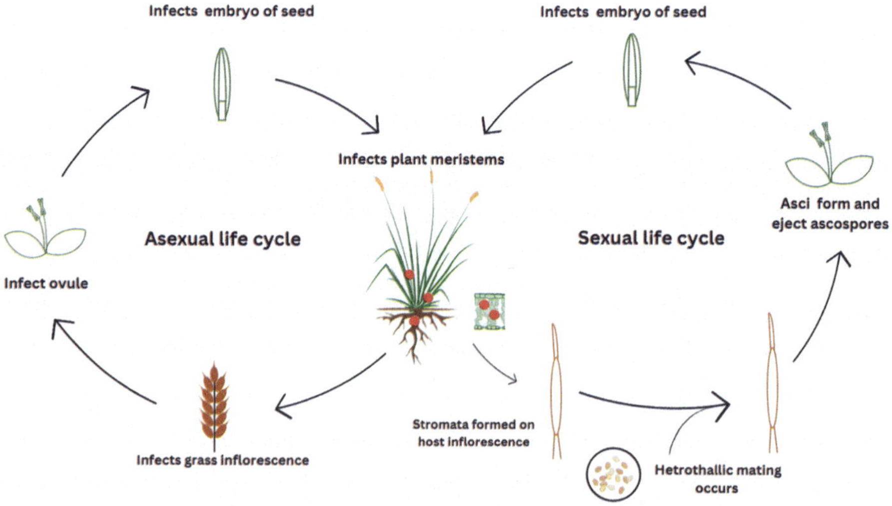

Fig. 4.1 Epichloae fungal endophytes sexual and asexual lifecycle

Rodriguez et al. (2009) revealed *Curvularia protuberate* characteristics which provides high temperature tolerance to tropical plants in underground habitats (Singh et al. 2021). In contrast, horizontal transmission occurs mainly in the distribution of reproductive endophytic structures (spores). These are transferred from one plant to another by vectors such as wind, rain, air, soil, and movement of any person or animal. Colonization of plants by type 2 and 3 endophytes is usually achieved through fungal epiphytes infection or direct hyphal penetration into tissues (Schardl et al. 2008). For the sexual type of *Epichlo* (category 1), ascospores are generally considered a horizontal transmission method (Schardl et al. 2004). Their presence leads to the formation of external structures and florets, thus causing inhibition of seed production and flower growth (strangulation disease). This type of mating is heterothallic and competition is necessary for reproduction. Genus *Phytophthora* spores (mitotic spores or conidia) carried by a vector fly from one mating species to other. Ascospores are produced and released into the air if mating partners are compatible. They are then spread by wind to spread infectious diseases (Shardl et al. 2008). The life cycle of endophytes directly affects the reproduction of the host. Importantly, in sexual *Epichlo* species, the choice of transmission method can influence whether symbiotic effects are beneficial or hostile to the host (Schardl 1996).

4.4 Role in the Ecosystem

In an ecosystem, endophytic fungus plays an important role by protecting plants from various stresses (biotic and abiotic), boosting their resilience, and supporting them in adapting to new habitats (Lu et al. 2021; Verma et al. 2022; Madouh and Quoreshi 2023; George et al. 2024). Endophytes can provide protection against biological pathogens such as insects, nematodes, and plant pathogens as shown in Fig. 4.2. Abiotic stresses include temperature changes, drought, salinization, nutrient limitation, and pH extremes shown in Fig. 4.2. Through vertical transport, plants become responsible for spatial structure as well as protection from drought and nutrients. In addition, this process facilitates transmission to the next generation. The relationship between *Ustilago zeae* and the maize endophyte *Fusarium verticillioides* demonstrates the ability of the endophyte to reduce fungal growth. This can be achieved by secretion of plant-damaging metabolites that limit corn starch (Lugtenberg et al. 2016). Additionally, some endophytic fungi initiate recycling processes through biodegradation of dead plants (Aly et al. 2011; Nair and Padmavathy 2014). Vazquez de Aldana et al. (2013) revealed that the predominant endophyte species in grasses are also airborne pathogens, which supports the previous results (Vazquez de Aldana et al. 2013). For example, in some species, the transition from endophytes to saprophytes requires sporulation of the old host. During the growing season, endophytes acquire nutrients from plant tissues (Rodriguez et al. 2008; Aly et al. 2011).

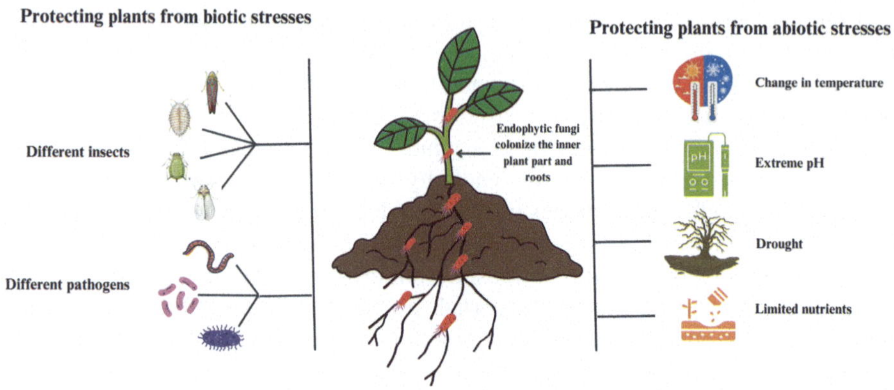

Fig. 4.2 Endophytic fungi protecting plants from biotic and abiotic stresses

4.4.1 Improving Plant Performance with Fungus Endophytes

Endophytic fungi recommended for promoting the plant adaptation against abiotic and biotic stresses (Rodriguez et al. 2008; Aly et al. 2011; Johnson et al. 2013; Chadha et al. 2015). These fungi produce a number of secondary metabolites, including bioactive molecules that work to protect plants from pathogens and pests. They also produce other metabolites that promote association with the host (Hussain et al. 2019; Keswani et al. 2019). Rodríguez et al. (2008) introduced the term "habitat-adapted symbiosis" to describe the specific response of fungal endophytes (category 2) against stress in one habitat in a different environment. However, the biochemical and molecular mechanisms of how habitat variation symbiosis leads to plant tolerance under severe stress conditions are still unknown. This will be achieved through altering and accelerating nutrient supply and regulating phytohormones involved in plant growth and development (Lugtenberg et al. 2016).

4.4.2 Fungal Endophytes for Resistance to Diseases and Pests

Type 1, a common feature of many grass species is the presence of *Epichlo* endophytes, which causes plants to produce bioactive compounds that inhibit herbivores and pests. A toxic endophyte is found in perennial grasses known as *Epichlo festucae* var. *lolii*. This pathogen enters from Europe to New Zealand and produces three major metabolites. Fifteen years ago, a special *Epichlo* strain (AR1) was introduced. These fungi do not produce ergovaline or lolitrem B, but they produce peramine, which provides resistance in plants against the Argentine rootworm (Johnson et al. 2013). The subsequent discovery of other *Epichlo* endophyte species led to the identification of another endophyte alkaloid, epoxy–janthitrems (Philippe 2016). These indolediterpenes are found only in annual ryegrass and cause disease in the AR37 strain (Ludlow et al. 2019). AR37-grass virus has broad-spectrum insecticide activity against 5–6 major pests in New Zealand. On the other hand, the effect of epoxy–janthitrem compounds in resistance of insect remains unknown (Johnson et al. 2013). Farmers are rapidly adopting these new endophytes due to their benefits, such as increasing the persistence of ryegrass without negative consequences for animals (Johnson and Caradus 2019). Johnson et al. (2013) provided a more detailed analysis of the importance and use of secondary metabolites found in *Epichlo* species. *Epichlo* endophytes are recognized worldwide for their importance in agriculture, providing economic and agricultural solutions. They also serve as good models for studying many aspects of endophyte biology, including the molecular basis of symbiosis and its ecological implications. Especially, *Epicchlo festucae* culture in sterile medium is suitable for molecular studies. A comprehensive study on *Epicchlo festucae* mutants shows that regulation of reactive oxygen species and iron homeostasis are important for a healthy relationship with grasses (Takemoto et al. 2011). Facultative endophytes such as *Pyroplasma indica* associate with agricultural plants by colonizing roots. This can be used in agriculture, horticulture, and floriculture (Ansari et al. 2013; Unnikumar et al. 2013; Johnson et al. 2013; Das et al. 2021; Kumar et al. 2021). *P. indica* was first discovered in the Indian Thar desert's rhizosphere of various xerophytic plants in 1998 (Verma et al. 1998). This filamentous fungus, a member of the Basidiomycota, promotes plant growth, especially when nutrients are scarce, and enhances tolerance to a variety of biotic stresses such as foliar pathogens and root diseases, as well as abiotic stresses including salinity, high temperature, drought and heavy metals (Weiss et al. 2004). *P. indica* colonized plant roots which demonstrates host resistance to a range of root and foliar diseases and serve as a biocontrol agent for treating various root diseases in numerous crops, such as tomato (Fakhro et al. 2010), wheat (Rabiey et al. 2015), maize (Kumar et al. 2009), and barley (Waller et al. 2005). *P. indica* interaction has also exhibited systemic resistance to numerous foliar diseases (Roylawar et al. 2021). Fakhro et al. (2010) discovered that *P. indica*, when colonizing tomato roots, reduces the severity of *Verticillium* wilt disease by more than 30% in soil conditions. It also enhances leaf biomass by up to 20% and increases the tomatoes fresh fruit biomass up to 100% in hydroponics. This results in a greater quantity of fruits and an increase of up to 20% in dry matter content. Moreover, it suppresses the *Pepino Mosaic Virus* in high light densities. This virus has been extensively identified in tomato greenhouses across various European countries, China, North and South America, and Morocco (Fakhro et al. 2010). Numerous fungal endophytes have been found to protect host plants against harmful fungi (Badal and Brown 2024). They are especially well known for their capacity to control diseases including *Epicoccum nigrum* in sugarcane, *Pythium* in cotton, *Monilinia* spp. in peaches, *Sclerotinia sclerotiorum* in sunflowers, and phytoplasma in apples (Singh et al. 2021). Murphy et al. (2015) revealed that fungal endophytes which were isolated from wild barley plants, efficiently inhibited seed-borne fungal infections. Several dangerous barley diseases, including *Cochiolobus*, *Pyrenophora*, *Rhynchosporium*, and *Fusarium*, were found in the tested seeds. The soilborne pathogen *Gaeumannomyces graminis* var. *tritici* was inhibited by the *Penicillium brevicompactum* strain, which completely reduced these infections (Murphy et al. 2015). According to Meja et al. (2008), key cacao pathogens such as *Phytophthora palmivora*, which causes black pod rot, *Moniliophthora perniciosa*, which causes witches broom, and *Moniliophthora roreri*, which causes frosty pod rot, were studied. In vitro, from healthy *Theobroma cacao* tissues, these pathogens were examined for their ability to inhibit fungal endophytes isolated. *Colletotrichum gloeosporioides* treatment significantly reduced pod loss induced by black pod rot (Meja et al. 2008). Finally, entomopathogenic fungi, a type of endophyte, deserve consideration due to their capacity to manage insects through biocontrol (Quesada Moraga 2020). These fungi are remarkable in their capacity to colonize and infect insects, potentially affecting insect survival and reproduction (Vidal and Jaber 2015). Notably, numerous commercially available entomopathogenic fungi, including *Isaria*, *Metarrhizium*, *Lecanicillium*, and *Beauveria*, can exhibit endophytic characteristics for a portion of their life cycle (Vidal and Jaber 2015). Wagner and Lewis (2000) were the first to thoroughly describe the entomopathogenic fungus *Beauveria bassiana's* penetration and colonization of plant tissues. Vidal and Jaber (2015) present a more comprehensive evaluation of plant colonization by entomopathogenic fungi, including the efficacy and range of colonization rates seen with diverse strain/cultivar combinations. Entomopathogenic fungi are efficient against numerous insect orders that feed on roots. These include the white grubs *Melolontha melolontha*, *Tetanops myopaeformis*, *diabroticine* rootworms, sugar beet root maggot, and *Cataenococcus ensete*. The most detrimental issue to coffee is thought to be the coffee berry borer. Vega et al. (2010) collected many fungal endophytes from coffee trees in Puerto Rico, Mexico, Colombia, and Hawaii. Numerous fungal entomopathogens genera have been discovered, including Clonostachys, Beauveria, Paecilomyces, Cladosporium, and Acremonium (Vega et al. 2010). According to Vega et al. (2010), *Clonostachys rosea* and *B. bassiana* are pathogenic to the coffee berry borer. These fungal endophytes that are entomopathogenic also contribute to the nitrogen cycle in the soil. Behie et al. (2012) investigated *M. robertsii's* capacity to transfer nitrogen obtained from insects to plants in further detail. The scientists introduced nitrogen labeled with 15N into wax moth larvae, then placed the labeled insects on haricot beans and switch grass. The incorporation of 15N into plant amino acids was found; however, it was only evident when *M. robertsii* was present (Behie et al. 2015).

4.4.3 Fungal Endophytes Causes Abiotic Stress, with the Effects of Climate Change

Mutualistic fungal endophytes contribute to or are responsible for plant adaptation for both biotic and abiotic stress (Bharadwaj 2024; Srivastava et al. 2024). They adapt to water scarcity, drought, salt water, and heat (Aly et al. 2011; Tiwari et al. 2020). It is thought that the domain-shifting stress resistance of habitat-specific symbionts promotes plant output under high stress circumstances (Rodriguez et al. 2008). Many type 2 endophytes have habitat adaptations that allow them to tolerate certain environment selection, such as soil pH, salinity, and temperature (Rodriguez et al. 2009). Fungal endophytes (symbiotic) that enhance temperature tolerance are present in grass species that are well suited for tropical and coastal environments, respectively (Rodriguez et al. 2008). When introduced into salt-sensitive plants, endophytes can give salt tolerance (Rodriguez et al. 2008). Furthermore, the tropical panicle plant *Dichanthelium lanuginosum*, found in Yellowstone National Park's geothermal soils, is closely related to the type 2 endophyte *Curvularia protuberata*. Symbiotic plants with endophytes may endure heat when the root zone is heated to 65 °C. When the soil temperature rises above 38 °C, no microbes or plants can grow on their own (Márquez et al. 2007). Symbionts native to these areas are exceedingly diverse and have been found to provide distinctive habitats for many plant species. In coastal plants, endophytes shows the unique characteristics as they can invade monocot rice plants and induce salt tolerance. In other ecosystems (such as geothermal, coastal, and agricultural) endophytes can only sustain dicotyledonous tomato specimens and cannot withstand salt or heat (Govindasamy et al. 2018; Rodriguez et al. 2008). Interestingly, regardless of the plant's initial habitat, all of these endophytes are drought tolerant (Rodriguez et al. 2008). Crop production is becoming increasingly problematic in many parts of the world due to increase in soil salinization process and decrease in water availability due to many reasons (Egamberdieva and Lugtenberg 2014). Climate change is predicted to make this issue worse. According to Redman et al. (2011), the ability of endophytic bacteria to stress plants may provide solutions to reduce the impact of climate change on plant communities. Some endophytes, particularly those belonging to class 2, improved the tolerance of two types of commercial rice to drought and salt, which they could not tolerate previously (Rodriguez et al. 2008). In green houses, by using these endophytes in grown plants also reduce the water need of plants (20–30%), while increasing their productivity, growth, and biomass. They also be responsible for cold tolerance to plants grown in greenhouses and grow rooms. These findings suggest that in scenario of climate change, incorporation of fungal symbionts may be a suitable strategy to reduce its impact on major crops and increase agricultural productivity (Redman et al. 2011). *P. Indica* is known for its ability to stimulate plant growth, stimulate plant growth, and promote early growth, fruiting, and seed germination. Many phytohormones and their signaling networks appear to regulate plant growth-promoting activities, helping to improve early root growth stimulation and ultimately increase biomass. The growth rate generally hovers around 50% but may vary due to differences between plants and experiments (Hossain et al. 2017). Many studies have been conducted on the salt stress tolerance performance in tobacco, rice, wheat, and barley, which plays a role in stimulating a high antioxidant environment to increase the photosynthetic efficiency and detoxify reactive oxygen species (Johnson et al. 2014). To enhance the cultivation and yield of rice or any other crop in Bangladesh's highly salinized coastal regions as shown in Fig. 4.3, halotolerant Plant Growth-Promoting Endophytic Fungi (PGPEF), like *Aspergillus welwitschiae* Ocstreb1 strain (AwOcstreb1), isolated from *Oryza coarctata*, are a promising option for an environmentally friendly biofertilizer formula (Airin et al. 2023).

P. indica combined with its suitability for basic biological research makes it an attractive research topic. The ability to get a good product is very exciting. The interaction between *P. indica* and barley resulted in increased yield (Gill et al. 2016).

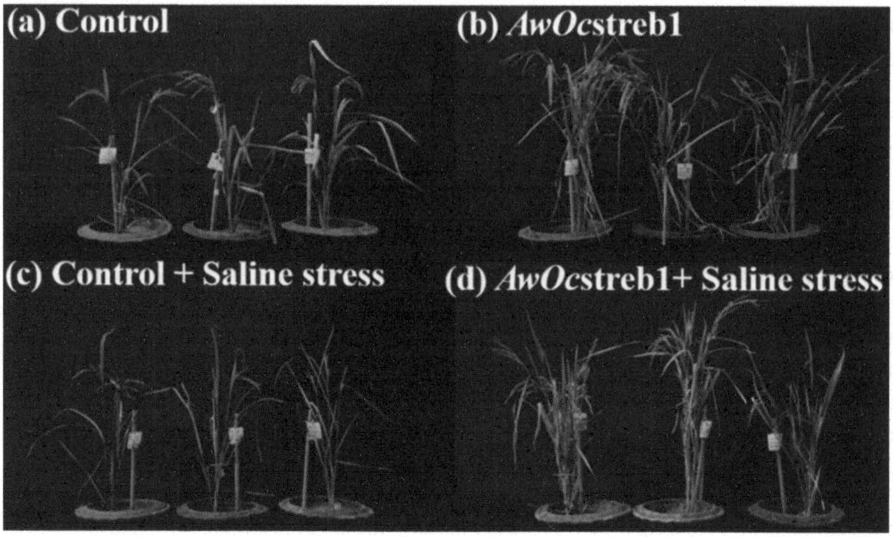

Fig. 4.3 Impact of AwOcstreb1, a halotolerant endophytic fungal strain, on rice plant reproduction in both normal and salinized stress environments (**a**) Control (**b**) *AwOc*streb1 (**c**) Control + Saline stress (**d**) *AwOc*streb1+ Saline stress (Cited from Airin et al. 2023)

This increase was observed at higher and lower nutrient levels. Rice can also flower earlier in the growing season when colonized by *P. indica* (Gill et al. 2016). These findings suggest that *P. indica* has the potential to be used as an effective treatment agent for barley crops grown at low temperatures. Crops grown in cold climates may show best yields if nutrients are available (Murphy et al. 2015). Franken (2012), P. Indicators can be difficult for two main reasons. Therefore, business potential depends on the patent owner's willingness to manufacture or license production. Secondly, sometimes he does not know what is wrong with growing plants. The indica trade name "ROOTONIC" has been developed and is currently being tested in India (Franken 2012; Shrivastava and Varma 2014). Other endophytes in barley may be useful in agriculture. Murphy et al. (2015) isolated endophytic fungi from wild barley and found that, in addition to their health benefits, one or more of these isolates has the potential to boost the yield (up to 29%) in barley. One of these endophytic fungi has the greatest impact on crop yield and shoot dry weight and has the lowest nutritional content. As a result, this species is capable of producing enough food with less chemical fertilizer input. Continuing the topic of improving plant performance, dark compartment endophytes (DSEs) are characterized by dark, melanized compartments and represent a group of endophytes. They occur in many terrestrial environments, especially polar and montane environments (Rodriguez et al. 2009; Murphy et al. 2015). Arbuscular mycorrhizal fungi, often associated with low-elevation and low-latitude grass roots, generally cannot survive in cold and water-poor environments (Newsham 2011). In certain areas, it has been proposed that DSE species might act similarly to other mycorrhizae (Bledsoe et al. 1990). Plants do not appear to benefit from DSE species when inorganic nitrogen is accessible through their roots (Newsham 2011; Kumar et al. 2021).

4.5 Direct Benefits of Interaction with Endophytic Fungi

Increased nutrient and phytohormone uptake in the plant is one of the direct advantages of interacting with endophytic fungus as shown in Fig. 4.4. This is directly linked to higher biomass production, expanded weight, plant height, development of root system, reproduction, and yield. These advantages allow them to be stated to as "biofertilizers" (Bamisile et al. 2018).

4.5.1 Acquisition of Nutrients

Endophytic fungus may improve the plant host's intake of macronutrients like nitrogen, phosphorus, magnesium, and potassium or micronutrients like copper, iron, and zinc by improving their uptake from the organic matter and soil (Rana et al. 2020; Bhusal 2024). The entomopathogenic fungus *Metarhizium robertsii* was first reported to have colonized switchgrass (*Panicum virgatum*) and bean plants (*Phaseolus vulgaris*) endophytically. Behie et al. (2012) also reported this fungal invasion's mechanism for transferring nitrogen (N) by which the fungus infects and kills the *Galleria mellonella* larvae. Since then, more research has been done. For example, Behie and Bidochka (2014) assessed that seven endophytic fungal species transfer the nitrogen from insects, which are *B. bassiana*, *M. acridum*, *M. flavoviridae*, *M. brunneum*,

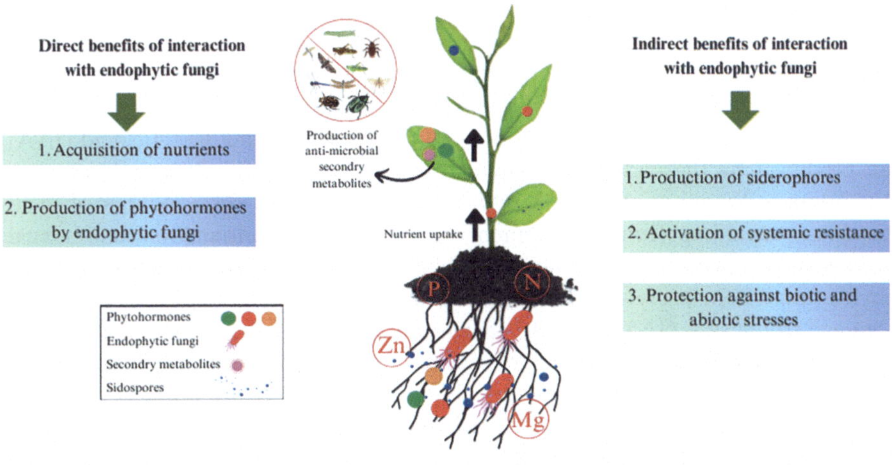

Fig. 4.4 Direct and indirect benefits of host plant having interaction with endophytic fungi

M. guizhouense, *Akanthomyces* (=Lecanicillium) *lecanii*, *M. robertsii*. These fungal species were grown in four cultures: two dicots, soybean and common bean and two monocots, switchgrass, and wheat.

Consequently, the scientists discovered that five species of *B. bassiana* and *Metarhizium* are capable of colonizing plants endophytically, transferring nitrogen from insects to these plants and killing insect larvae. A well-characterized endophyte, *Serendipita* (=Piriformospora) *indica*, has been shown in the literature to be capable of transporting nutrients, including phosphates to host plants (Card et al. 2016). Although by endophytic fungi phosphorus (P) transference has not been the subject of many investigations, a number of reports have shown that fungal inoculation improves P acquisition, and these findings assume that endophytic fungi play a role in this process. According to Ortega-Garcia et al. (2015), phosphorus fertilization of onions (*Allium cepa*) was substantially reduced by inoculating the plant with *Trichoderma asperellum*. Similarly, Baron et al. (2019) conducted a field investigation and introduced *Aspergillus sydowii* into maize (*Zea mays*). The plants that have the interaction with the fungus gained considerably larger amounts of P in their tissues, even at lower fertilization dosages. The precise molecules and metabolic pathways, which involved in many activities, have not yet been thoroughly characterized as how they relate to the mechanisms of nutrient transport.

4.5.2 Production of Phytohormones by Endophytic Fungi

Endophytic fungi are able to generate gibberellins, cytokinins, and auxins as shown in Fig. 4.4. These fungi have the unique characteristics to produce the phytohormones, particularly gibberellins, is still little known, and despite their critical function in chemical signaling for plant growth under a variety of environmental situations (Khan et al. 2015) .The primary auxin produced by this fungus is Indole-3-acetic acid (IAA). The main plant growth regulators, auxins, have a variety of advantageous impacts on the development of shoots and roots. According to Jaroszuk-Ściseł et al. (2014), these impacts include reactions to tropism, vascular tissue differentiation, cell division, elongation, and the start of root development. According to Waqas et al. (2012) Gibberellin (GA) and Indole-3-acetic acid (IAA) are produced by the endophytic fungus *Penicillium* ssp. and *Phoma glomerata*. According to Waqas et al. (2012), gibberellins are essential for a number of plant responses, including fruit development, senescence, stem elongation, and sexual expression. By using the mevalonic acid (MVA), gibberellins are produced by endophytic fungi from acetyl-CoA. The two main end products, GA1 and GA3, are obtained from GA4, GA5, and GA7 (Khan et al. 2015). Afandhi et al. (2019) investigated whether *Beauveria bassiana*, an entomopathogenic fungus, might act as an endophyte and promote the development of *Gossypium*. In this work, *B. bassiana* was isolated, identified, and its potential role as a helpful endophyte in fostering common bean development was examined. Two different techniques to inoculate

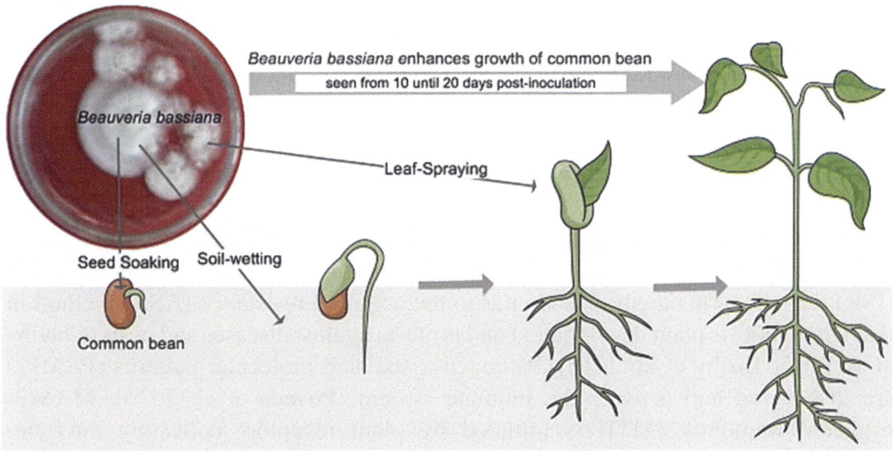

Fig. 4.5 The method of soil-wetting to introduce *B. bassiana* conidia into *Phaseolus vulgaris* L. exhibited the most potential for boosting plant development at 10 days post-inoculation. In this experiment, *B. bassiana* was inoculated by seed soaking, leaf spraying, and soil wetting, demonstrating the fungus's capacity to act as an endophytic agent that promotes growth (Afandhi et al. 2019)

B. bassiana: soil-wetting and leaf-spraying were evaluated. Day 10 post-inoculation observation of common bean development demonstrated the efficacy of both approaches. These results show that the technique of *B. bassiana* inoculation and the encouragement of common bean development are positively correlated as shown in Fig. 4.5.

Compared to extracts of the recognized GA producer from the pathogen *Gibberella fujikuroi* which causes bakanae disease in rice, fungal extracts exhibited greater concentrations of the bioactive gibberellins GA1, GA3, GA4, and GA7. In study, this fungus used as a positive control although it did not boost growth as much as *P. citrinum* IR-3-3. *Cladosporium sphaerospermum* and *Aspergillus fumigatus* were also found to produce gibberellin. Recent investigations, such as that of Bader et al. (2020), define not only a single capacity, such as enhanced nutrition absorption or phytohormone synthesis, but also multiple abilities revealed by the same strains. From Argentine Pampas, researchers extracted *Trichoderma* strains from soil and chose four strains with high IAA production and phosphate solubility. After that these strains were inoculated into seeds of tomato and treated plants showed a considerable increase in height of plant, chlorophyll content in leaves, and fresh and dry weight of branches and roots after 45 days. Bader et al. (2020) attribute plant growth promotion by endophytic colonization.

4.6 Indirect Benefits of Interaction with Endophytic Fungi

Interaction with endophytic fungi offers a number of indirect advantages:

4.6.1 Activation of Systemic Resistance

Without providing any advantages to the plant, interactions with endophytic fungi boost the plant's ability to defend itself by triggering the innate immune response (ISR) pathway. This method is similar to the acquired resistance (ASR) method in that it both boosts plant development and protects against diseases and pests (Chadha et al. 2015; Busby et al. 2016). Pathogen-associated molecular patterns (PAMPs) are discovered and activate the immune system (Poveda et al. 2020). MAMP-triggered immunity (MTI) is initiated by plant receptors collecting microbe-associated molecular patterns (MAMPs). MAMPs induced microbe-triggered immunity (MTI) include cell wall of fungal components such as glucans and chitin, as well as secreted enzymes and their products such as chitinases, glucanases, and xylanases (Latz et al. 2018; Yan et al. 2019). Secondary metabolites produced by microorganisms in addition to MAMPs, can induce drug resistance, leading to immunological response (ETI) (Yan et al. 2019; Poveda et al. 2020). Endophytic and pathogenic fungi have similar MTI responses. It has been found, however, that modifying MAMPs in endophytic bacteria may prevent the plant from identifying these diseases. ETI suggests using beneficial microbes to solve specific problems in order to fast up the colonization process (Yan et al. 2019). MAMPs and effectors activate the ASR and ISR pathways, resulting in complicated responses. Protein phosphorylation, dephosphorylation, ion flow, and the synthesis of signaling molecules such as salicylic acid and ethylene are examples of these reactions. It also involves reactive oxygen species (ROS) and increases immune response gene activation. These reactions include phytoalexin and pathogen-associated (PR) protein synthesis, cell wall thickening, and even cell death (Chadha et al. 2015). Endophytic colonization prepares plants for future infection by pathogenic bacteria, nematodes, and herbivores in this environment (Poveda et al. 2020; Latz et al. 2018).

4.6.2 Production of Antibiotics and Secondary Metabolites

In addition to helping their hosts manufacture protective molecules, endophytic fungi provide a rich source of compounds that have other beneficial effects on plants. They are efficient in producing substances that exhibit anti-herbivorous and anti-pathogenic properties including phenols, quinols, flavonoids, polyketones, terpenoids, peptides, steroids, alkaloids, volatile organic compounds (VOCs), and chlorinated compounds (Latz et al. 2018; Lugtenberg et al. 2016; Card et al. 2016). Additionally, research indicates the synthesis of substances have antibacterial, antiviral, antifungal, and insecticidal properties (Card et al. 2016; Latz et al. 2018). Plants produce alkaloids, which are harmful to a variety of pests, insects, and even vertebrates (Johnson et al. 2013; Lugtenberg et al. 2016). It is also stated that *Nodulisporium* sp. produces nodulisporic acid. Because this chemical stimulates glutamate in the chloride channels of nerve cells and insect muscle, it plays a crucial role in regulating insect herbivory as shown in Table 4.2. Endophytic fungi create a vast variety of compounds as secondary metabolites; however, certain routes and chemicals have not yet been thoroughly characterized. For understanding the secondary metabolite chemical structures produced by endophytic fungi, it is recommended that you read the review by Lugtenberg et al. (2016). Endophytic fungus have the ability to develop more than 300 of the diverse secondary metabolites in which volatile organic compounds (VOCs) are also produced (Lugtenberg et al. 2016). For fungal communication with other species including plants, distinct fungal agents are necessary, because of their small molecular composition, high vapor pressure, and ease of diffusibility via soil, atmosphere, and cell membranes. They also have bioactivity against a variety of pathogens (Kaddes et al. 2019). Mycofumigation is a biocontrol method that was first proposed by Strobel et al. (2001) to address postharvest diseases of tubers and fruits.

4.6.2.1 Fungal Endophytes Produced Secondary Metabolites

Fungal endophytes are rich in bioactive natural compounds due to their great biodiversity. Their unique chemical structure has been improved by combining them with higher plants, providing great opportunities for the discovery of new natural products. Recent advances in screening technology have revealed the great potential of endophytic fungi to produce new physiologically active drugs. These chemicals are widely used in agriculture or medicine and especially in plant protection it plays an important role in disease communication, plant adaptation to the environment, circulation, and translocation. Many synthetic farming methods in agriculture have been banned from the market due to safety problems and environmental concerns (Gupta et al. 2023). Endophytic fungi produced secondary metabolite which offer new solutions for the management of different diseases and pests. In this field, there is much to discover about bioactive compounds produced by endophytes. The first step involves identifying plants having such endophytes and their potential to produce new bioactive compounds (Alurappa et al. 2018; Preethi et al. 2021).

4.6.2.2 Production of Anti-microbial Secondary Metabolites

Secondary metabolites produced by the endophytic fungus have a long history of usage in agriculture and horticulture to control pests and diseases. Gunatilaka (2006) examined 230 metabolites produced by plants in association with microbial diseases, including a number of endophytic species. Ambuic acid is an anti-oomycete and antifungal substance which is derived from isolates of the endophytic fungus *Pestalotiopsis microspora*, which have been discovered in numerous rain forests across the world. It is active against several *Pythium ultimum* and *Fusarium* species. A substance called colonolic acid was extracted from *Artemisia annua* from a *Colletotrichum* sp. It demonstrates strong antimicrobial, antifungal, and anti-algal properties. In *Artemisia mongolica, Colletotrichum gloeosporioides* is an endophytic fungus which produces colletotric acid as a byproduct. It exhibits

Table 4.2 Functions of different endophytes

Endophytic fungi	Host plant	Activity	Chemical composition	Reference
Muscodor crispans	Wild pineapple	Antibiotic	Propanoic acid, methyl ester,2- methylbutyl ester Ethanol	Mitchell et al. (2010), Saxena and Strobel (2021)
Ampelomyces sp.	Prickly goldenfleece	Antimicrobial	3-Omethyl alaternin & ltersolanol A	Manjunatha et al. (2020)
Muscodor albus	Dalchini	Antibiotic	E-6 Caryophyllene, phenylethyl alcohol, 2-phenylethyl ester, bulnesene	Roy and Banerjee (2019)
Phomopsis sp.	Frangipani	Antimicrobial	Terpenoid	Terkar and Borde (2021)
Penicillium expansum	Milky Mangrove	Cytotoxic	Polyphenols, expansols A & B	Lu et al. (2021)
Curvularia protuberata	Panic grass	Anti-cancer, anti-inflammatory, phytotoxicity, anti-oxidant, and anti-microbial	Polyketides	Mehta et al. (2022)
Nodulisporium sp.	Thai Bungor	Antibiotic	Alcohols, alkyl alcohols, terpenoid, hydrocarbons	Mends et al. (2013)
Pestalotiopsis microspore	Rain forest	Antimicrobial	(+)-acetylpestalotin, (−)-pestalotin, (6S,7S,8R)-hydroxypestalotin and a lignan, (+)-pinoresinol	Riga et al. (2019)
Gliocladium sp.	Ulmo	Anti-inflammatory, anti-microbial	Chloroformic, ethyl acetate, butanolic, and methanolic	Stinson et al. (2003), Liouane et al. (2010)
Colletotrichum sp.	Wormwood	Antifungal, Antimicrobial and anti-algal	1 (4-hydroxyphenethyl-2′-(R)-hydroxypropanoate), 2 (4-hydroxyphenethyl-2′-(S)-hydroxypropanoate), 3 (1′-(4-hydroxyphenethoxy)-1″-oxopropan-2″-(S)-yl-2′-(S)-hydroxypropanoate), 4 (1′-(4-hydroxyphenethoxy)-1″-oxopropan-2″-(R)-yl-2′-(S)-hydroxypropanoate),	Zou et al. (2000), Feng et al. (2019)

antibacterial efficacy against *Helminthosporium sativum*, a fungus, as well as bacteria. From the fungal endophyte, a secondary metabolite, Phomophomoidin, is obtained and exhibits antibacterial properties against *Bacillus subtilis*, *Staphylococcus aureus,* and *Salmonella enterica* (Horn et al. 1995). Importantly, secondary antibodies produced by these endophytic bacteria have not yet been shown to be produced by bacteria that treat microbial-induced plant diseases (Lugtenberg and Kamilova 2009; Pliego et al. 2011; Ludwig-Muller 2015).

4.6.2.3 Production of Anti-microbial Volatile Compounds

Although antibiotics topic does not appropriate into the previous section, it has been placed in a separate section since special procedures are required for the detection of these compounds. Their resistance is determined by airborne contact between the bacteria producing the bacteria and the bacterial strain. Thereafter, incompatible products can be separated and identified using a chemical mixture and different data generated using Liquid Chromatography-High Resolution Mass Spectrometry (LC-HRMS) or Nuclear Magnetic Resonance (NMR) (Wu et al. 2015). The endophytic bacterium *Muscodor albus*, which was isolated from the cinnamon tree's branches, produces a substance that efficiently inhibit and kill bacteria and viruses as shown in Table 4.2 (Strobel 2018). Several substances have been demonstrated to have some impact on experimental bacteria and fungi, but none of them have proved to be lethal. Nevertheless, a robust correlation was discovered, they generally kill a lot of fungus, bacteria, plants, and humans. There were five categories of chemicals identified: esters, alcohols, acids, ketones, and lipids. Seven other *S. albicans* strains were subsequently identified, and they all produced volatiles in varying amounts. But none of these novel isolates generated esters that are known to originate from primitive mycobacteria. They create fifty distinct volatile compounds in total, each of which has a range of biological functions, such as inhibiting the growth of oomycetes, phytopathogenic fungi, Gram +ve and Gram –ve bacteria (Strobel 2018). In the Bolivian Amazon Basin, *Muscodor crispans* is an endophytic fungus that grows on wild pineapple as shown in Table 4.2. A combination of antifungal and antibacterial volatile organic chemicals were produced by the fungus. It creates a variety of volatile organic compounds that have antifungal and antibacterial properties. Finally, among the fungal phytopathogens evaluated from the Capella volatile mixture, *Pythium* was shown to be the most harmful. Other fungal and oomycete diseases include *Alternaria helianthus*, *Botrytis cinerea*, *Fusarium oxysporum*, *Phytophthora cinnamomi*, *Pseudomonas palmi*, *Rhizoctonia solani, Sclerotinia sclerotiorum*, and *Sclerotinia dahlia*. Interestingly, unlike *M. albus*, the majority of *M. crispans* volatiles are classified as Generally Recognized as Safe (GRAS) by the US Food and Drug Administration (FDA). For example, no naphthalene or azulene compounds were discovered in M. Chips' volatile mixture. When M. crispans' combination of GRAS compounds is tested, the results are remarkably close to natural chemicals. As a result, Mitchell et al. (2010) believe that this mixture has the potential to be used in a wide range of applications, including food preservation, agricultural, domestic, and commercial use. Additionally, *Gliocladium* sp., an endophytic fungus of *Eucryphia cordifolia*, has been identified as an anti-inflammatory drug producer as shown in Table 4.2. Annulene was discovered as the most volatile inhibitor of interest. These chemicals were previously employed as pyrotechnics and have now been discovered to be natural products produced by endophytic bacteria for the first time (Strobel Daisy 2003; Lugtenberg et al. 2016). In Israel, endophytic fungus *Daldinia concentrica*, isolated from olive trees which produces organic compounds (at least 28) that can be exploited for post-harvest management. These volatile chemicals keep dried apricots, prunes, and raisins from deteriorating. They also prevent peanuts against black yeast, oranges against *Penicilliumdigitata*, grapes against *Botrytis cinerea*, and tomatoes against *Penicillium digitatum*. The combination of selected volatiles has significant application potential in the food sector and agriculture (Lugtenberg et al. 2016; Roy and Banerjee 2019).

4.6.2.4 Production of Anti-insect Secondary Metabolites

The combination of synthetic pesticides with environmental harm has led to studies of potency, selectivity, and safety (Strobel and Daisy 2003). Historically, biopesticides represented only a small part of the pesticide industry, but they are becoming increasingly important. Several endophytes have been discovered as producing secondary metabolites having insecticidal capabilities as shown in Fig. 4.4. *Epichlo* endophytes develop a type of alkaloids known as indole diterpenes. Lolitrem B is *Epichlo*-specific and is crucial in generating significant tremor, the mycotoxin that causes ryegrass staggering (Gallagher et al. 1981). *Epichlo coenophialum*, an endophytic fungus found in endophytic tall fescue pastures, is thought to cause prairie poisoning in grazing cattle in the United States. This is because some fungal species produce the alkaloid ergovaline, which causes vasoconstriction and causes serious problems in terms of animal health and welfare (Bacon 1995). Although ergovarine has insecticidal properties (Popay et al. 1981), its association with heat stress and fescue poisoning makes it unsuitable for pest control and inhibits the growth as well as development of ASW larvae.

4.6.2.5 Production of Secondary Metabolites That Have Additional Beneficial Effects on Plants

Although fungi encoding a variety of secondary metabolites, genome sequencing of endophytic fungi has shown that some of these compounds are not produced in laboratory growth models. According to Netzker et al. (2015) and Helaly et al. (2018), this raises the question of what environmental and physical requirements endophytic organisms must meet in order to express secondary metabolites. In particular cases, like *Epichlo*, the expression of significant endophyte secondary metabolic pathways depends on the host genome (Schardl and Panaccione 2005; Wani et al. 2015). The expression of fungal gene clusters that are in control of *Epichlo festucae* var. *Loli's* lolitrem B production is stimulated by plant signaling. This results in high expression in plants but low or undetectable levels in fungal mycelial cultures (Young et al. 2006; Priyadarshini et al. 2022). Secondary metabolite production by endophytes can also be affected by many factors, such as nutrient deficiency, availability of plant material, competition with other organisms, and the presence of other organisms like fungi and their metabolites. These metabolites are also produced when plants are attacked by different bacteria and pests. For example, rhizobial Nod factors are expressed more when flavonoids present in host plant root exudates are present (Van Brussel et al. 1986; Singh and Kumar 2023).

4.6.3 Production of Siderophores

For all living cells, iron is a microelement that is necessary (Rana et al. 2020; Turbat et al. 2020). Some microorganisms, such as endophytic fungi, produce siderophores, which are tiny molecules with iron-chelatin characteristics that are used to bind ferric ions in the rhizosphere (Sr et al. 2020; Chowdappa et al. 2020). Suebrasri et al. (2020) found that endophytic strains of *Macrophomina phaseolina* SS1L10, *Trichoderma koningii* ST-KKU1, and M. phaseolina SS1R10 produced siderophores as shown in Fig. 4.4. The authors of this study proposed that fungal synthesis of siderophores played a significant role in sunchoke plant growth enhancement. Additionally, the synthesis of siderophores is explained for recombinant endophytic strains of *Trichoderma harzianum* that colonize beans (*P. vulgaris*) (Eslahi et al. 2020). It is currently unknown how fungal endophytes' siderophores work and how they are characterized, and there is conjecture regarding how they relate to the ISR (Card et al. 2016).

4.6.4 Protection Against Biotic and Abiotic Stresses

Plant development and maintenance are made more difficult by environmental deterioration caused by agricultural practices and change in global climate. Furthermore, the need for increased yields makes the situation even more challenging for crops. In this case, it is evident that some support is required for healthy plant growth, and endophytic fungi offer a viable substitute for providing biotic and abiotic stress protection for plants. Abiotic stressors that endophytic fungi are resistant to include salt, extremes in temperature, drought, and hazardous heavy metals (Aly et al. 2011; Khan et al. 2015). Fungal endophytes activate ISR and ASR to create compounds that protect against pathogens in response to biotic stress; moreover, competition or parasitism may occur to prevent disease and herbivory (Chadha et al. 2015; Chitnis et al. 2020). Plant morphology and physiology are negatively impacted by abiotic stressors because of genetic regulation of cell pathways that results in a number of dysfunctions (Egamberdieva and Lugtenberg 2014). Through a variety of ways, endophytic fungi support host plants in adapting to stressful environments. This causes the generation of reactive oxygen species (ROS), which causes membrane harm and membrane lipid peroxidation. Endophytic fungi reduce lipid peroxidation by conferring resistance to ROS through processes that are still unknown. Electrolyte leakage is a significant issue that arises from abiotic stresses (such as heat, salinity, and dehydration) on membranes.

4.7 Concerns About the Use of Endophytic Fungi in Agriculture

Some endophytes that promote plant growth are thought to be toxic to humans and other vertebrates. However, it is better to use endophytic fungi for this, because most studies on the toxicity of endophytic microorganisms on humans are associated with diseases. These include organisms from different genera such as Stenotrophomonas, Ralstonia, Staphylococcus, Ochilobacter, Herbispira, Enterobacter, and Burkholderia (Bamisile et al. 2021; Baron and Rigobelo 2022). Fungi can enter tissue and colonize plants through many routes, such as through water, animals, feces, or seeds living in soil. This shows continuity even among the owners of diverse kingdoms (Mendes et al. 2013). In the case of fungi, studies on Epichlo (= Neotyphodium) spp., which colonize endophytically in lawns, have shown their detrimental effects on vertebrate health. These fungi aid in the formation and accumulation of alkaloids in tissues, causing damage to many bone species, especially animals (Sarkar et al. 2021). Only a few studies have been conducted to study the effects of these compounds on vertebrates, and the majority are regarded to be hazardous to the animals. Extensive research is being conducted on the applications of toxic endophytic fungi in agriculture. This research involves examining the outline of alkaloids produced by each species in order to identify those that provide antiherbivore protection in the field against the invertebrates while having no effect on vertebrates (Baron and Rigobelo 2022). Finch et al. (2013) described an experiment in which cattle were fed ryegrass endophytically colonized with two different species of Fescue *Epicchlo festucae* (= Neotyphodium lolii). The authors discovered fungal alkaloids in milk, but their amounts are regarded safe for human consumption. Additionally, these species do not show any negative effects (Finch et al. 2013). It is worth noting that other endophytic fungi, as well as herbaceous endophytes, can also produce mycotoxins harmful to human health (Mousa and Raizada 2013; Chitnis et al. 2020). Caution should be exercised when using biological products, as some bacteria that endophytically infect plants are harmful to humans and other animals. Therefore, comprehensive research is essential to prevent selection of known diseases, whether human or plant diseases (Murphy et al. 2018).

4.8 Endophytes as Bioproducts: Quick, Simple, and Promising

The benefits of using fungus and other compounds in agriculture are well recognized and include the following: (i) improved biosecurity, (ii) reduction of risk to human health and environmental circulation, (iii) particular resistance to pests without affecting the consequences of insects and diseases, (iv) although small productivity, (v) regenerative ability, (vi) absence of drug resistance, (vii) prevention of the development of resistant bacteria, and (viii) relationship to patterns of sexual development (Berg 2009). Mushrooms are often used as anti-bacterial agents, especially anti-bacterial, as a long-term alternative to drug administration. However, their use is limited due to the lack of humidity and UV radiation in agricultural areas and problems arising from field use (Vega 2018). Fungi, as endophytes, can overcome hurdles that have previously prevented their broad usage. Being located within the factory provides significant advantages to those who can carry out vertical transportation. However, this is rarely investigated. This finding means that combinations and methods may not necessarily be used in application (Card et al. 2016). Another broadly accepted theory holds that endophytes modify disease in a manner that is dependent on the environment and impacted by both biotic (pathogens) and abiotic (environment) factors. Changes in soil moisture, pH, and temperature can alter the antagonistic activity of endophytes. At the same time, plant microbiota interacts with endophytes, which may explain why plants respond differently to pathogens (Busby et al. 2016). The severity of the disease response is affected not only by the disease and the host plant, but also by different responses to different diseases in the same plant. Additionally, different plants will respond differently to the same disease. As such, the severity of the disease may vary even after inoculation with the same endophyte. The impacts of these variables must be thoroughly investigated in order to develop methods for the application of endophytic bacteria in agriculture. Information about biological products is spreading around the world, and their use is increasing as people learn more about their ingredients, benefits, and ability to work. Brazil is a world leader in expanding the use of organic products, with products on the market doubling since 2017. In 2019, sales of these products exceeded \$120 million, and the market is expected to grow further in 2020 (Tijani 2020).

4.9 Conclusions

Fungal endophytes are increasingly used as biocontrol agents, a technique discovered many years ago. Due to increasing concerns about the impact of agriculture on the environment and the search for healthy, non-toxic foods, scientific research is choosing to use these organisms. Many species of fungi, including those used for pest control, have been shown to be able to colonize crops through endophytic growth. This approach has proven to be a good strategy for sustainable agriculture. Because they offer numerous direct and indirect advantages to crops, endophytic fungi are advantageous. It can be confirmed that no synthetic drug can provide the same broad range of effective interactions with these organisms. Therefore, the use of endophytic fungi for biocontrol, biofertilization, and biostimulation seems quite promising and shows that these bacteria are good tools for research and business.

References

Abo Nouh FA (2019) Endophytic fungi for sustainable agriculture. Microb Biosyst 4(1):31–44

Afandhi A, Widjayanti T, Emi AAL, Tarno H, Afiyanti M, Handoko RNS (2019) Endophytic fungi Beauveria bassiana Balsamo accelerates growth of common bean (*Phaeseolus vulgaris* L.). Chem Biol Tech Agric 6(1):1–6

Airin AA, Arafat MI, Begum RA, Islam MR, Seraj ZI (2023) Plant growth- promoting endophytic fungi of the wild halophytic rice Oryza coarctata. Ann Microbiol 73(1):36

Alurappa R, Chowdappa S, Narayanaswamy R, Sinniah UR, Mohanty SK, Swamy MK (2018) Endophytic fungi and bioactive metabolites production: an update. In: Microbial biotechnology: Volume 2. Application in food and pharmacology. Springer, pp 455–482

Aly AH, Debbab A, Proksch P (2011) Fungal endophytes: unique plant inhabitants with great promises. Appl Microbiol Biotechnol 90:1829–1845

Anjum R, Afzal M, Baber R, Khan MAJ, Kanwal W, Sajid W, Raheel A (2019) Endophytes: as potential biocontrol agent review and future prospects. J Agric Sci 11(4):113–130

Ansari MW, Trivedi DK, Sahoo RK, Gill SS, Tuteja N (2013) A critical review on fungi mediated plant responses with special emphasis to *Piriformospora indica* on improved production and protection of crops. Plant Physiol Biochem 70:403–410

Bacon CW (1995) Toxic endophyte-infected tall fescue and range grasses: historic perspectives. J Anim Sci 73(3):861–870

Badal S, Brown KJ (2024) Fungal ms: a focus on endophytes. In: Pharmacognosy. Elsevier, pp 455–468

Bader AN, Salerno GL, Covacevich F, Consolo VF (2020) Native *Trichoderma harzianum* strains from Argentina produce indole-3 acetic acid and phosphorus solubilization, promote growth and control wilt disease on tomato (*Solanum lycopersicum* L.). J King Saud Univ-Sci 32(1):867–873

Bahram M, Netherway T (2022) Fungi as mediators linking organisms and ecosystems. FEMS Microbiol Rev 46(2):1–55

Bamisile BS, Dash CK, Akutse KS, Keppanan R, Wang L (2018) Fungal endophytes: beyond herbivore management. Front Microbiol 9:1–49

Bamisile BS, Akutse KS, Siddiqui JA, Xu Y (2021) Model application of entomopathogenic fungi as alternatives to chemical pesticides: prospects, challenges, and insights for next-generation sustainable agriculture. Front Plant Sci 12:1–74

Baron NC, Rigobelo EC (2022) Endophytic fungi: a tool for plant growth promotion and sustainable agriculture. Mycology 13(1):39–55

Baron NC, Rigobelo EC, Zied DC (2019) Filamentous fungi in biological control: current status and future perspectives. Chil J Agric Res 79(2):307–315

Behie SW, Bidochka MJ (2014) Ubiquity of insect-derived Nitrogen transfer to plants by endophytic insect-pathogenic fungi: an additional branch of the soil Nitrogen cycle. Appl Environ Microbiol 1553–1560

Behie SW, Jones SJ, Bidochka MJ (2015) Plant tissue localization of the endophytic insect pathogenic fungi Metarhizium and Beauveria. Fungal Ecol 13:112–119

Behie SW, Zelisko PM, Bidochka MJ (2012) Endophytic insect-parasitic fungi translocate nitrogen directly from insects to plants. Science 336(6088):1576–1577

Berg G (2009) Plant–microbe interactions promoting plant growth and health: perspectives for controlled use of microorganisms in agriculture. Appl Microbiol Biotechnol 84:11–18

Bharadwaj A (2024) Influence of endophytes on plant growth and abiotic stress. In: Plant endophytes and secondary metabolites. Elsevier, pp 159–174

Bhusal KK (2024) Role of bio-fertilizers in organic farming. Asian J Res Rev Agric 2:1–7

Brader G, Compant S, Vescio K, Mitter B, Trognitz F, Ma LJ, Sessitsch A (2017) Ecology and genomic insights into plant-pathogenic and plant-nonpathogenic endophytes. Annu Rev Phytopathol 55:61–83

Briffa J, Sinagra E, Blundell R (2020) Heavy metal pollution in the environment and their toxicological effects on humans. Heliyon 6(9):1–26

Brundrett MC, Tedersoo L (2020) Resolving the mycorrhizal status of important northern hemisphere trees. Plant Soil 454:3–34

Busby PE, Ridout M, Newcombe G (2016) Fungal endophytes: modifiers of plant disease. Plant Mol Biol 90:645–655

Card S, Johnson L, Teasdale S, Caradus J (2016) Deciphering endophyte behaviour: the link between endophyte biology and efficacious biological control agents. FEMS Microbiol Ecol 92(8):1–19

Chadha N, Mishra M, Rajpal K, Bajaj R, Choudhary DK, Varma A (2015) An ecological role of fungal endophytes to ameliorate plants under biotic stress. Arch Microbiol 197(7):869–881

Chagnon M, Kreutzweiser D, Mitchell EA, Morrissey CA, Noome DA, Van der Sluijs JP (2015) Risks of large-scale use of systemic insecticides to ecosystem functioning and services. Environ Sci Pollut Res 22:119–134

Chen J, Akutse KS, Saqib HSA, Wu X, Yang F, Xia X, Wang L, Goettel MS, You M, Gurr GM (2020) Fungal endophyte communities of crucifer crops are seasonally dynamic and structured by plant identity, plant tissue and environmental factors. Front Microbiol 11:1–37

Chitnis VR, Suryanarayanan TS, Nataraja KN, Prasad SR, Oelmüller R, Shaanker RU (2020) Fungal endophyte-mediated crop improvement: the way ahead. Front Plant Sci 11:1–37

Chowdappa S, Jagannath S, Konappa N, Udayashankar AC, Jogaiah S (2020) Detection and characterization of antibacterial siderophores secreted by endophytic fungi from cymbidium aloifolium. Biomolecules 10:1–18

Collinge DB, Jensen DF, Rabiey M, Sarrocco S, Shaw MW, Shaw RH (2022) Biological control of plant diseases–what has been achieved and what is the direction? Plant Pathol 71(5):1024–1047

Das K, Tiwari RKS, Das S (2021) Role and significance of endophytic fungi in crop improvement for sustainable agriculture: current perspective and future trend. Darshan Publishers, Tamil Nadu

Deyett E, Rolshausen PE (2020) Endophytic microbial assemblage in grapevine. FEMS Microbiol Ecol 96(5):1–19

Egamberdieva D, Lugtenberg B (2014) Use of plant growth-promoting rhizobacteria to alleviate salinity stress in plants. In: Use of microbes for the alleviation of soil stresses, vol 1. Springer, pp 73–96

El Mansy SM, Nouh FAA, Mousa MK, Abdel-Azeem AM (2020) Endophytic fungi: diversity, abundance, and plant growth-promoting attributes. In: Agriculturally important fungi for sustainable agriculture: perspective for diversity and crop productivity, vol 1. Springer, pp 21–59

Elnahal AS, El-Saadony MT, Saad AM, Desoky E-SM, El-Tahan AM, Rady MM, AbuQamar SF, El-Tarabily KA (2022) The use of microbial inoculants for biological control, plant growth promotion, and sustainable agriculture: a review. Eur J Plant Pathol 162(4):759–792

Eslahi N, Kowsari M, Motallebi M, Zamani MR, Moghadasi Z (2020) Influence of recombinant Trichoderma strains on growth of bean (*Phaseolus vulgaris* L) by increased root colonization and induction of root growth related genes. Scientia Horticulturae 261:1–6

Fakhro A, Andrade-Linares DR, von Bargen S, Bandte M, Büttner C, Grosch R, Schwarz D, Franken P (2010) Impact of *Piriformospora indica* on tomato growth and on interaction with fungal and viral pathogens. Mycorrhiza 20:191–200

Feng L, Wang J, Liu S et al (2019) Colletopeptides A–D, Anti-inflammatory Cyclic Tridepsipeptides from the Plant Endophytic Fungus *Colletotrichum* sp. S8. J Nat Prod 82:1434–1441

Finch SC, Thom E, Babu J, Hawkes A, Waugh C (2013) The evaluation of fungal endophyte toxin residues in milk. N Z Vet J 61(1):11–17

Franco FP, Túler AC, Gallan DZ, Gonçalves FG, Favaris AP, Penaflor MFG, Leal WS, Moura DS, Bento JMS, Silva-Filho MC (2021) Fungal phytopathogen modulates plant and insect responses to promote its dissemination. ISME J 15(12):3522–3533

Franken P (2012) The plant strengthening root endophyte *Piriformospora indica*: potential application and the biology behind. Appl Microbiol Biotechnol 96:1455–1464

Gallagher R, White E, Mortimer P (1981) Ryegrass staggers: isolation of potent neurotoxins lolitrem A and lolitrem B from staggers-producing pastures. N Z Vet J 29(10):189–190

Garrido-Jurado I, Montes-Moreno D, Sanz-Barrionuevo P, Quesada-Moraga E (2020) Delving into the causes and effects of entomopathogenic endophytic *Metarhizium brunneum* foliar application-related mortality in *Spodoptera littoralis* larvae. Insects 11(7):429

George NM, Mohammed EA, Muhammad G, Bolbol AA (2024) Drought stress impacts and the role of endophytic fungi combating abiotic stress on wheat. Bulletin of Faculty of Science, Zagazig University 2023(4):93–108

Gill SS, Gill R, Trivedi DK, Anjum NA, Sharma KK, Ansari MW, Ansari AA, Johri AK, Prasad R, Pereira E (2016) *Piriformospora indica*: potential and significance in plant stress tolerance. Front Microbiol 7:1–68

Govindasamy V, George P, Raina SK, Kumar M, Rane J, Annapurna K (2018) Plant-associated microbial interactions in the soil environment: role of endophytes in imparting abiotic stress tolerance to crops. In: Advances in crop environment interaction. Springer, pp 245–284

Gunatilaka AL (2006) Natural products from plant-associated microorganisms: distribution, structural diversity, bioactivity, and implications of their occurrence. J Nat Prod 69(3):509–526

Gupta A, Meshram V, Gupta M, Goyal S, Qureshi KA, Jaremko M, Shukla KK (2023) Fungal endophytes: microfactories of novel bioactive compounds with therapeutic interventions; A comprehensive review on the biotechnological developments in the field of fungal endophytic biology over the last decade. Biomol Ther 13(7):1–123

Hardoim PR, Van Overbeek LS, Berg G, Pirttilä AM, Compant S, Campisano A, Döring M, Sessitsch A (2015) The hidden world within plants: ecological and evolutionary considerations for defining functioning of microbial endophytes. Microbiol Mol Biol Rev 79(3):293–320

Hartmann A, Schmid M, van Tuinen D, Berg G (2009) Plant-driven selection of microbes. Springer, pp 235–257

Heimpel GE, Mills NJ (2017) Biological control. Cambridge University Press, pp 48–379

Helaly SE, Thongbai B, Stadler M (2018) Diversity of biologically active secondary metabolites from endophytic and saprotrophic fungi of the ascomycete order Xylariales. Nat Prod Rep 35(9):992–1014

Horn W, Simmonds M, Schwartz R, Blaney W (1995) Phomopsichalasin, a novel antimicrobial agent from an endophytic *Phomopsis* sp. Tetrahedron 51(14):3969–3978

Hossain MM, Sultana F, Islam S (2017) Plant growth-promoting fungi (PGPF): phytostimulation and induced systemic resistance. In: Plant-microbe interactions in agro-ecological perspectives: Volume 2: Microbial interactions and agro-ecological impacts. Springer, pp 135–191

Hussain M, Debnath B, Qasim M, Bamisile BS, Islam W, Hameed MS, Wang L, Qiu D (2019) Role of saponins in plant defense against specialist herbivores. Molecules 24(11):1–83

Hyde KD, Soytong K (2008) The fungal endophyte dilemma. Fungal Divers 33:163–173

Jaroszuk-Ściseł J, Kurek E, Trytek M (2014) Efficiency of indoleacetic acid, gibberellic acid and ethylene synthesized in vitro by *Fusarium culmorum* strains with different effects on cereal growth. Biologia 69:281–292

Johnson LJ, Caradus JR (2019) The science required to deliver Epichloe endophytes to commerce. In: Endophytes for a growing world. Cambridge University Press, pp 343–370

Johnson LJ, de Bonth AC, Briggs LR, Caradus JR, Finch SC, Fleetwood DJ, Fletcher LR, Hume DE, Johnson RD, Popay AJ (2013) The exploitation of epichloae endophytes for agricultural benefit. Fungal Divers 60:171–188

Johnson JM, Alex T, Oelmüller R (2014) *Piriformospora indica*: the versatile and multifunctional root endophytic fungus for enhanced yield and tolerance to biotic and abiotic stress in crop plants. J Trop Agric 52(2):103–122

Johnson LJ, Voisey CR, Faville MJ, Moon CD, Simpson WR, Johnson RD, Stewart AV, Caradus JR, Hume DE (2019) Advances and perspectives in breeding for improved grass-endophyte associations. Grassland Science in Europe 24:351–363

Kaddes A, Fauconnier ML, Sassi K, Nasraoui B, Jijakli MH (2019) Endophytic fungal volatile compounds as solution for sustainable agriculture. Molecules 24:1–16

Kemen E, Jones JD (2012) Obligate biotroph parasitism: can we link genomes to lifestyles? Trends Plant Sci 17(8):448–457

Keswani C, Singh HB, Hermosa R, García-Estrada C, Caradus J, He Y-W, Mezaache-Aichour S, Glare TR, Borriss R, Vinale F (2019) Antimicrobial secondary metabolites from agriculturally important fungi as next biocontrol agents. Appl Microbiol Biotechnol 103:9287–9303

Khan AL, Hussain J, Al-Harrasi A, Al-Rawahi A, Lee I-J (2015) Endophytic fungi: resource for gibberellins and crop abiotic stress resistance. Crit Rev Biotechnol 35(1):62–74

Khan MN, Mobin M, Abbas ZK, Alamri SA (2018) Fertilizers and their contaminants in soils, surface and groundwater. In: Encyclopedia of the Anthropocene, vol 5. Elsevier, pp 225–240

Krings M, Taylor TN, Hass H, Kerp H, Dotzler N, Hermsen EJ (2007) Fungal endophytes in a 400-million-yr-old land plant: infection pathways, spatial distribution, and host responses. New Phytol 174(3):648–657

Kumar P, Devi P, Dey SR (2021) Fungal volatile compounds: a source of novel in plant protection agents. In: Volatiles and metabolites of microbe. Elsevier, pp 83–104

Kumar M, Yadav V, Tuteja N, Johri AK (2009) Antioxidant enzyme activities in maize plants colonized with Piriformospora indica. Microbiology 155:780–790

Latch GCM, Christensen MJ (1985) Artificial infection of grasses with endophytes. Ann Appl Biol 107(1):17–24

Latz MA, Jensen B, Collinge DB, Jørgensen HJ (2018) Endophytic fungi as biocontrol agents: elucidating mechanisms in disease suppression. Plant Ecol Divers 11(5):555–567

Le Cocq K, Gurr SJ, Hirsch PR, Mauchline TH (2017) Exploitation of endophytes for sustainable agricultural intensification. Mol Plant Pathol 18(3):469–473

Liouane K, Saïdana D, Edziri H et al (2010) Chemical composition and antimicrobial activity of extracts from Gliocladium sp. growing wild in Tunisia. Med Chem Res 19:743–756

Lu H, Wei T, Lou H, Shu X, Chen Q (2021) A critical review on communication mechanism within plant-endophytic fungi interactions to cope with biotic and abiotic stresses. J Fungi 1:7–719

Ludlow EJ, Vassiliadis S, Ekanayake PN, Hettiarachchige IK, Reddy P, Sawbridge TI, Rochfort SJ, Spangenberg GC, Guthridge KM (2019) Analysis of the indole diterpene gene cluster for biosynthesis of the epoxy-janthitrems in Epichloë endophytes. Microorganisms 7(11):1–48

Ludwig-Müller J (2015) Plants and endophytes: equal partners in secondary metabolite production? Biotechnol Lett 37:1325–1334

Lugtenberg B, Kamilova F (2009) Plant-growth-promoting rhizobacteria. Ann Rev Microbiol 63:541–556

Lugtenberg BJ, Caradus JR, Johnson LJ (2016) Fungal endophytes for sustainable crop production. FEMS Microbiol Ecol 92(12):1–54

Madouh TA, Quoreshi AM (2023) The function of arbuscular mycorrhizal fungi associated with drought stress resistance in native plants of arid desert ecosystems: a review. Diversity 15(3):1–64

Maji S, Dwivedi DH, Singh N, Kishor S, Gond M (2020) Agricultural waste: its impact on environment and management approaches. In: Emerging eco-friendly green technologies for wastewater treatment. Springer, pp 329–351

Manjunatha L, Singh S, Ravikumara BM et al (2020) Ampelomyces. Beneficial Microbes in Agro-Ecology 833–860

Márquez LM, Redman RS, Rodriguez RJ, Roossinck MJ (2007) A virus in a fungus in a plant: three-way symbiosis required for thermal tolerance. Science 315(11):513–515

Mehta T, Meena M, Nagda A (2022) Bioactive compounds of Curvularia species as a source of various biological activities and biotechnological applications. Front Microbiol 13:1–54

Mejía LC, Rojas EI, Maynard Z, Van Bael S, Arnold AE, Hebbar P, Samuels GJ, Robbins N, Herre EA (2008) Endophytic fungi as biocontrol agents of Theobroma cacao pathogens. Biol Control 46(1):4–14

Mendes R, Garbeva P, Raaijmakers JM (2013) The rhizosphere microbiome: significance of plant beneficial, plant pathogenic, and human pathogenic microorganisms. FEMS Microbiol Rev 37(5):634–663

Miranda-Fuentes P, Quesada-Moraga E, Aldebis HK, Yousef-Naef M (2020) Compatibility between the endoparasitoid Hyposoter didymator and the entomopathogenic fungus *Metarhizium brunneum*: a laboratory simulation for the simultaneous use to control *Spodoptera littoralis*. Pest Manag Sci 76(3):1060–1070

Mitchell AM, Strobel GA, Moore E, Robison R, Sears J (2010) Volatile antimicrobials from Muscodor crispans, a novel endophytic fungus. Microbiology 156(1):270–277

Mousa WK, Raizada MN (2013) The diversity of anti-microbial secondary metabolites produced by fungal endophytes: an interdisciplinary perspective. Front Microbiol 4(65):1–42

Murphy BR, Doohan FM, Hodkinson TR (2015) Persistent fungal root endophytes isolated from a wild barley species suppress seed-borne infections in a barley cultivar. BioControl 60:281–292

Murphy BR, Doohan FM, Hodkinson TR (2018) From concept to commerce: developing a successful fungal endophyte inoculant for agricultural crops. J Fungi 4(1):1–35

Nair DN, Padmavathy S (2014) Impact of endophytic microorganisms on plants, environment and humans. Sci World J 2014:1–22

Netzker T, Fischer J, Weber J, Mattern DJ, König CC, Valiante V, Schroeckh V, Brakhage AA (2015) Microbial communication leading to the activation of silent fungal secondary metabolite gene clusters. Front Microbiol 6(299):1–6

Newsham KK (2011) A meta-analysis of plant responses to dark septate root endophytes. New Phytol 190(3):783–793

Nisa H, Kamili AN, Nawchoo IA, Shafi S, Shameem N, Bandh SA (2015) Fungal endophytes as prolific source of phytochemicals and other bioactive natural products: a review. Microb Pathog 82:50–59

Ortega-Garcia JG, Montes-Belmont R, Rodriguez-Monroy M, Ramirez-Trujillo JA, Suarez-Rodriguez R, Sepulveda Jimenez G (2015) Effect of *Trichoderma asperellum* applications and mineral fertilization on growth promotion and the content of phenolic compounds and flavonoids in onions. Sci Hortic 195:8–16

Ortiz A, Sansinenea E (2022) The role of beneficial microorganisms in soil quality and plant health. Sustain For 14(9):1–34

Philippe G (2016) Lolitrem B and indole diterpene alkaloids produced by endophytic fungi of the genus Epichloë and their toxic effects in livestock. Toxins 8(2):1–52

Pliego C, Kamilova F, Lugtenberg B (2011) Plant growth-promoting bacteria: fundamentals and exploitation. In: Bacteria in agrobiology: crop ecosystems. Springer, pp 295–343

Popay AI (1981) Germination of seeds of five annual species of barley grass. J Appl Ecol 547–558

Poveda J, Abril-Urias P, Escobar C (2020) Biological control of plant-parasitic nematodes by filamentous fungi inducers of resistance: trichoderma, mycorrhizal and endophytic fungi. Front Microbiol 11:992

Preethi K, Manon Mani V, Lavanya N (2021) Endophytic fungi: a potential source of bioactive compounds for commercial and therapeutic applications. In: Endophytes: potential source of compounds of commercial and therapeutic applications. Springer, pp 247–272

Priyadarshini MS, Panigrahi S, Rath CC (2022) Endophytes: novel microorganisms for plant growth promotion. Darshan publishers, pp 1–162

Quesada Moraga E (2020) Entomopathogenic fungi as endophytes: their broader contribution to IPM and crop production. Biocontrol Sci Tech 30(9):864–877

Rabiey M, Ullah I, Shaw M (2015) The endophytic fungus *Piriformospora indica* protects wheat from fusarium crown rot disease in simulated UK autumn conditions. Plant Pathol 64(5):1029–1040

Rai M, Rathod D, Agarkar G, Dar M, Brestic M, Pastore GM, Junior MRM (2014) Fungal growth promotor endophytes: a pragmatic approach towards sustainable food and agriculture. Symbiosis 62:63–79

Rana KL, Kour D, Kaur T, Devi R, Yadav AN, Yadav N et al (2020) Endophytic microbes: biodiversity, plant growth-promoting mechanisms and potential applications for agricultural sustainability. Antonie Van Leeuwenhoek 113:1075–1107

Rashmi M, Kushveer JS, Sarma VV (2019) A worldwide list of endophytic fungi with notes on ecology and diversity. Mycosphere 10(1):798–1079

Rashmi I, Roy T, Kartika KS, Pal R, Coumar V, Kala S, Shinoji KC (2020) Organic and inorganic fertilizer contaminants in agriculture: impact on soil and water resources. In: Contaminants in agriculture: sources, impacts and management. Springer, pp 3–41

Redman RS, Kim YO, Woodward CJ, Greer C, Espino L, Doty SL, Rodriguez RJ (2011) Increased fitness of rice plants to abiotic stress via habitat adapted symbiosis: a strategy for mitigating impacts of climate change. PLoS One 6(7):1–12

Riga R, Happyana N, Hakim EH (2019) Chemical constituents of Pestalotiopsis microspora HF 12440. J Appl Pharm Sci 9:108–124

Rodriguez RJ, Henson J, Van Volkenburgh E, Hoy M, Wright L, Beckwith F, Kim Y-O, Redman RS (2008) Stress tolerance in plants via habitat-adapted symbiosis. ISME J 2(4):404–416

Rodriguez RJ, White JF Jr, Arnold AE, Redman AR (2009) Fungal endophytes: diversity and functional roles. New Phytol 182(2):314–330

Roy S, Banerjee D (2019) Volatile organic compounds from endophytic fungi. In: Recent advancement in white biotechnology through fungi: Volume 2: Perspective for value-added products and environments. Springer, pp 149–175

Roylawar P, Khandagale K, Randive P, Shinde B, Murumkar C, Ade A, Singh M, Gawande S, Morelli M (2021) *Piriformospora indica* primes onion response against stemphylium leaf blight disease. Pathogens 10(9):1085

Sabarwal A, Kumar K, Singh RP (2018) Hazardous effects of chemical pesticides on human health–cancer and other associated disorders. Environ Toxicol Pharmacol 63:103–114

Sánchez-Bayo F (2021) Indirect effect of pesticides on insects and other arthropods. Toxics 9(8):1–180

Sarkar S, Dey A, Kumar V, Batiha GE-S, El-Esawi MA, Tomczyk M, Ray P (2021) Fungal endophyte: an interactive endosymbiont with the capability of modulating host physiology in myriad ways. Front Plant Sci 12:1–62

Saxena S, Strobel GA (2021) Marvellous Muscodor spp.: update on their biology and applications. Microb Ecol 82(1):5–20

Schardl CL (1996) Epichloe species: fungal symbionts of grasses. Annu Rev Phytopathol 34(1):109–130

Schardl CL, Panaccione DG (2005) Biosynthesis of ergot and loline alkaloids. In: Neotyphodium in cool-season grasses. Wiley, pp 73–92

Schardl CL, Leuchtmann A, Spiering MJ (2004) Symbioses of grasses with seedborne fungal endophytes. Annu Rev Plant Biol 55:315–340

Schardl CL, Craven KD, Speakman S, Stromberg A, Lindstrom A, Yoshida R (2008) A novel test for host-symbiont codivergence indicates ancient origin of fungal endophytes in grasses. Syst Biol 57(3):483–498

Sharma S, Gupta S, Dhar MK, Kaul S (2018) Diversity and Bioactive Potential of Culturable Fungal Endophytes of Medicinal Shrub *Berberis aristata* DC.: A First Report. Mycobiology 46:370–381

Shrivastava S, Varma A (2014) From *Piriformospora indica* to rootonic: a review. Afr J Microbiol Res 8(32):2984–2992

Sieber TN (2007) Endophytic fungi in forest trees: are they mutualists? Fungal Biol Rev 21(2):75–89

Singh VK, Kumar A (2023) Secondary metabolites from endophytic fungi: production, methods of analysis, and diverse pharmaceutical potential. Symbiosis 90:1–15

Singh N, Singh A, Dahiya P (2021) Plant growth-promoting endophytic fungi from different habitats and their potential applications in agriculture. In: Recent trends in mycological research: Volume 1: agricultural and medical perspective. Springer, Cham, pp 69–87

Sr I, Sakamoto K, Kuwahara N (2020) Performance of tomato and lettuce to arbuscular mycorrhizal fungi and *penicillium pinophilum* EU0013 inoculation varies with soil, culture media of inoculum, and fungal consortium composition. Rhizosphere 16:1–18

Srivastava S, Rana M, Sarma BK, Meshram S (2024) Endophyte-mediated modulation of secondary metabolism in crops for biotic stress management. In: Plant endophytes and secondary metabolites. Elsevier, pp 205–233

Stinson M, Ezra D, Hess WM, Sears J, Strobel G (2003) An endophytic *Gliocladium* sp. of *Eucryphia cordifolia* producing selective volatile antimicrobial compounds. Plant Sci 165(4):913–922

Strobel G (2018) The emergence of endophytic microbes and their biological promise. J Fungi 4(2):1–46

Strobel G, Daisy B (2003) Bioprospecting for microbial endophytes and their natural products. Microbiol Mol Biol Rev 67(4):491–502

Strobel GA, Dirkse E, Sears J, Markworth C (2001) Volatile antimicrobials from muscodor albus, a novel endophytic fungus. Microbiology 147:2943–2950

Suebrasri T, Harada H, Jogloy S, Ekprasert J, Boonlue S (2020) Auxin-producing fungal endophytes promote growth of sunchoke. Rhizosphere 16:1–28

Tadych M, Bergen MS, White Jr, JF (2014) Epichloë spp. associated with grasses: new insights on life cycles, dissemination and evolution. Mycologia 106:181–201

Takemoto D, Kamakura S, Saikia S, Becker Y, Wrenn R, Tanaka A, Sumimoto H, Scott B (2011) Polarity proteins Bem1 and Cdc24 are components of the filamentous fungal NADPH oxidase complex. Proc Natl Acad Sci 108(7):2861–2866

Terkar A, Borde M (2021) Endophytic fungi: novel source of bioactive fungal metabolites. In: New and future developments in microbial biotechnology and bioengineering. Elsevier, In, pp 95–105

Tidjani C (2020) Readiness to the FinTech industry in developing countries: an overview of prospective factors impacts. In: Handbook on ethics in finance. Springer, pp 1–32

Tiwari P, Bajpai M, Singh LK, Mishra S, Yadav AN (2020) Phytohormones producing fungal communities: metabolic engineering for abiotic stress tolerance in crops. In: Agriculturally important fungi for sustainable agriculture: Volume 1: perspective for diversity and crop productivity. Springer, Cham, pp 171–197

Turbat A, Rakk D, Vigneshwari A et al (2020) Characterization of the plant growth-promoting activities of endophytic fungi isolated from Sophora flavescens. Microorganisms 8:683

Unnikumar K, Sree KS, Varma A (2013) Piriformospora indica: a versatile root endophytic symbiont. Symbiosis 60:107–113

Van Brussel A, Zaat S, Cremers H, Wijffelman C, Pees E, Tak T, Lugtenberg B (1986) Role of plant root exudate and Sym plasmid-localized nodulation genes in the synthesis by Rhizobium leguminosarum of Tsr factor, which causes thick and short roots on common vetch. J Bacteriol 165(2):517–522

Vazquez de Aldana BR, Bills G, Zabalgogeazcoa I (2013) Are endophytes an important link between airborne spores and allergen exposure? Fungal Divers 60:33–42

Vega FE (2018) The use of fungal entomopathogens as endophytes in biological control: a review. Mycologia 110(1):4–30

Vega FE, Goettel MS, Blackwell M, Chandler D, Jackson MA, Keller S, Koike M, Maniania NK, Monzon A, Ownley BH (2009) Fungal entomopathogens: new insights on their ecology. Fungal Ecol 2(4):149–159

Vega FE, Simpkins A, Aime MC, Posada F, Peterson SW, Rehner SA, Infante F, Castillo A, Arnold AE (2010) Fungal endophyte diversity in coffee plants from Colombia, Hawai'i, Mexico and Puerto Rico. Fungal Ecol 3(3):122–138

Verma S, Varma A, Rexer K-H, Hassel A, Kost G, Sarbhoy A, Bisen P, Bütehorn B, Franken P (1998) *Piriformospora indica*, gen. Et sp. Nov., a new root-colonizing fungus. Mycologia 90(5):896–903

Verma A, Shameem N, Jatav HS, Sathyanarayana E, Parray JA, Poczai P, Sayyed RZ (2022) Fungal endophytes to combat biotic and abiotic stresses for climate-smart and sustainable agriculture. Front Plant Sci 13:1–50

Vidal S, Jaber LR (2015) Entomopathogenic fungi as endophytes: plant–endophyte–herbivore interactions and prospects for use in biological control. Curr Sci 109:46–54

Wagner BL, Lewis LC (2000) Colonization of corn, Zea mays, by the entomopathogenic fungus *Beauveria bassiana*. Appl Environ Microbiol 66(8):3468–3473

Waller F, Achatz B, Baltruschat H, Fodor J, Becker K, Fischer M, Heier T, Hückelhoven R, Neumann C, von Wettstein D (2005) The endophytic fungus *Piriformospora indica* reprograms barley to salt-stress tolerance, disease resistance, and higher yield. Proc Natl Acad Sci 102(38):13386–13391

Wani ZA, Ashraf N, Mohiuddin T, Riyaz-Ul-Hassan S (2015) Plant-endophyte symbiosis, an ecological perspective. Appl Microbiol Biotechnol 99:2955–2965

Waqas M, Khan AL, Kamran M, Hamayun M, Kang S-M, Kim Y-H, Lee I-J (2012) Endophytic fungi produce gibberellins and indoleacetic acid and promotes host-plant growth during stress. Molecules 17(9):10754–10773

Weiss M, Selosse M-A, Rexer K-H, Urban A, Oberwinkler F (2004) Sebacinales: A hitherto overlooked cosm of heterobasidiomycetes with a broad mycorrhizal potential. Mycol Res 108(9):1003–1010

Wu C, Kim HK, Van Wezel GP, Choi YH (2015) Metabolomics in the natural products field–a gateway to novel antibiotics. Drug Discov Today: Technol 13:11–17

Yan L, Zhu J, Zhao X, Shi J, Jiang C, Shao D (2019) Beneficial effects of endophytic fungi colonization on plants. Appl Microbiol Biotechnol 103:3327–3340

Young CA, Felitti S, Shields K, Spangenberg G, Johnson RD, Bryan GT, Saikia S, Scott B (2006) A complex gene cluster for indole-diterpene biosynthesis in the grass endophyte *Neotyphodium lolii*. Fungal Genetics Biol 43(10):679–693

Zikankuba VL, Mwanyika G, Ntwenya JE, James A (2019) Pesticide regulations and their malpractice implications on food and environment safety. Cogent Food Agric 5(1):1–40

Zou W, Meng J, Lu H, Chen G, Shi G, Zhang T, Tan R (2000) Metabolites of *Colletotrichum gloeosporioides*, an endophytic fungus in Artemisia mongolica. J Nat Prod 63(11):1529–1530

Fungal Endophytes and Their Role in Bio and Phyto-Remediation Applications

5

Saidul Islam, Sk Hasibul Alam, Arghya Banerjee, Faijuddin Ahammad, Bholanath Mondal, Sahar Murmu, and Maksud Hasan Shah

Abstract

Various endoparasitic microorganisms are being studied extensively around the world in pursuit of sustainable agriculture and the environment. Endophytic microorganisms, such as bacteria, actinobacteria, and fungi, live ubiquitously in the tissues of all plant species without causing damage or disease. Endophytes form symbiotic relationships with various plant species and can modulate numerous host functions, such as tolerance to abiotic and biotic stresses, growth and development, and stimulation of the immune system. Endophytes interact with

S. Islam (✉)
Nadia Krishi Vigyan Kendra, Bidhan Chandra Krishi Viswavidyalaya (BCKV), Gayeshpur, Nadia, West Bengal, India

S. H. Alam · S. Murmu
Department of Plant Pathology, Bidhan Chandra Krishi Viswavidyalaya (BCKV), Mohanpur, Nadia, West Bengal, India

A. Banerjee
Department of Plant Pathology, Bidhan Chandra Krishi Viswavidyalaya (BCKV), Mohanpur, Nadia, West Bengal, India

Department of Plant Pathology, School of Agriculture and Allied Science, The Neotia University, Jhinga, West Bengal, India

F. Ahammad
Department of Plant Pathology, Sam Higgingbottom University of Agriculture, Technology and Science, Prayagraj, Uttar Pradesh, India

B. Mondal
Department of Plant Pathology, Palli Siksha Bhavana, Visva Bharati University, Sriniketan, Bolpur, West Bengal, India

M. H. Shah (✉)
Department of Agronomy, Bidhan Chandra Krishi Viswavidyalaya (BCKV), Mohanpur, Nadia, West Bengal, India

© The Author(s), under exclusive license to Springer Nature Singapore Pte Ltd. 2025
K. A. Abd-Elsalam, A. H. Hashem (eds.), *Fungal Endophytes Volume II*,
https://doi.org/10.1007/978-981-97-8804-0_5

the host through various mechanisms of action. This involves several physiological and mechanical control mechanisms, leading to overall better survivability. Furthermore, the interactions influence the overall productivity, growth promotion, and efficient carbon metabolic processes of the host organism. In some cases, symbiotic co-evolutionary strategies between hosts and endoparasites are used against critical environmental conditions such as heavy metal pollution and soil contamination through various phytoremediation processes. Phytoremediation is an environmentally friendly alternative strategy for removing environmental or toxic pollutants from soil, air, or water using plants. The ability to bind, detoxify, concentrate, or extract environmentally harmful pollutants is primarily due to certain endogenous physicochemical properties of plants. Endophyte-assisted phytoremediation technology eliminates biohazardous pollutants, metal and metalloid pollutants, carcinogens, organic industrial wastes, inorganic pesticides and herbicides, hydrocarbon elements, and naturally occurring chlorinated products, a futuristic form of on-site purification. It serves as a multi-component sustainable pollution control strategy with low environmental impact and low operating costs. This review is an overview approach to highlight the ancillary mechanisms of action involved in endophytic fungi-mediated phytoremediation techniques. Furthermore, we detail the various strategies, their limitations, and the application of transgenic technology in this new field of research.

Keywords

Endophyte fungi · Mechanisms · Bioremediation · Phytoremediation · Sustainable Agriculture

5.1 Introduction

An endophytic fungus is any organism inhabiting plant organs that, at a certain point in its lifetime, can colonize tissues without causing apparent harm (Rozpądek et al. 2018). Endophytic fungi have been a proven source of secondary metabolites with potential uses as anticancer, antibiotics, antivirals, anti-inflammatories, antioxidants, neuroprotective agents, insecticides, and antifungals and have multiple applications in biotechnological developments in pharmaceutical, agriculture, cosmetic, food industry, and environmental processes (Lundberg et al. 2012). Plants engage in interactions with a variety of fungal species that live in the phyllosphere and rhizosphere, which changes essential biological processes and the way they respond to abiotic and biotic challenges (Dastogeer et al. 2020; Glick and Gamalero 2021; Jones et al. 2019; Oukala et al. 2021). Mycorrhizal fungi in the rhizosphere and phyllosphere, along with plant growth-promoting bacteria (PGPB), can either directly stimulate plant growth by increasing essential hormones or indirectly by reducing the detrimental effects of various pathogens (Yu et al. 2020; Vishnupradeep et al. 2022; Nwachukwu and Babalola 2021; Mowafy et al. 2022; Massoni et al. 2020; Chaudhry et al. 2021; Aragón et al. 2017).

Endophytes are fungal species that live inside plant tissues, while epiphytes are those that live on the surfaces of plants (Mengistu 2020; Schlechter et al. 2019; White et al. 2019). De Barry first used the term "endophyte" in 1866 to describe any organism—bacteria, fungi, or their combinations—that multiplies intracellularly or intercellularly into host plants at least once during its lifecycle without exhibiting any overt symptoms of illness. Recent research has demonstrated that these symbiotic microbial species are more important for the growth and development of host plants (Fadiji and Babalola 2020a, b). For instance, the bacterial homologue is said to control and satisfy the host plant's nitrogen needs in the most extensively researched endosymbiotic relationship between rhizobium and legume (Santoyo et al. 2016). By synthesizing and secreting substances that promote plant development and enable host adaptation to specific environmental conditions, endophytes effectively establish symbiotic associations. Several fungal, bacterial, and actinomycetes species are documented to participate in the production and secretion of physiologically active chemicals and secondary metabolites (Anand et al. 2019; Duan et al. 2019; Raimi and Adeleke 2021; Fadiji and Babalola 2020a, b; Figueiredo et al. 2022; Manganyi and Ateba 2020; Techaoei et al. 2020; Zimowska et al. 2020; Adeleke and Babalola 2021; Salazar et al. 2023). Biomolecules produced by bacterial endosymbionts that fall under the classifications of peptides, alkaloids, and so on have a bright future in both medicine and agriculture (Narayanan and Glick 2022; Singh et al. 2022). As an active ingredient in numerous commercial pesticides, such as TreeAzin, AzaMax, BioNEEM, AzaGuard, and AzaSol, azadirachtin has been shown to be effective against over 200 insect species since its first discovery (Gómez and Luiz 2018; Hug et al. 2018; Rojas et al. 2020; Falade et al. 2021; Harman et al. 2021; Farder et al. 2022). Numerous experimental studies have documented the varying effects of variables such as host tissue, weather patterns, and soil properties on bioactive chemicals produced by endophytic microbial species (Sharma et al. 2021). Plant biologists are drawn to unravel the intricacies of endophytic associations in order to enhance crop plants, as evidenced by the significant role that endophytic bacteria play in regulating the composition of metabolic products in host plants. These days, a variety of biotic (such as insects and phytopathogens) and abiotic (such as high or low nutrient levels, drought, floods, extreme temperatures, salinity, and organic/inorganic contamination) stressors brought on by climate change have become significant global crop productivity limiters (Zörb et al. 2019). An estimated 30% of crops' yearly yield is lost due to biotic stress. Specifically, the combined impacts of several abiotic stress elements, including heat and drought, at a specific stage of the plant's growth are more harmful than the effects of any one stressor. In addition to abiotic stressors, pathogenic bacteria, viruses, fungi, insects, and pests present plants with ongoing biological challenges that result in significant losses in food yield globally (Gimenez et al. 2018; Savary et al. 2019; Trebicki 2020; Misiou and Koutsoumanis 2022). To increase crop types' resistance to various stresses, a number of strategies are being employed, including genetic engineering, molecular breeding, and the selection of tolerant varieties.

Nevertheless, most of these techniques are expensive, time-consuming, and unpopular in some places (Ali et al. 2022). Therefore, host plants have evolved numerous biological systems that can work in tandem to counteract the detrimental effects of diverse elements related to abiotic and biotic stress. According to this literature, enhancing plant development and growth through mutualistic associations that result from links between the microbe and the host is both efficient and long-lasting (Rodriguez et al. 2019; Olanrewaju et al. 2019; Lyu et al. 2021). Endophytes are directly related to plants, in contrast to other microbes that stimulate plant growth. Their quick adaptation to biotic and abiotic stress conditions enhances the development and survival of host plants (Aghai et al. 2019; Jones et al. 2019; Mengistu 2020; Fan et al. 2020; Bastías et al. 2021). Moreover, endophytic microorganisms have the ability to produce and secrete enzymes and metabolic products, making them an essential component of the rhizospheric area (Bhadra et al. 2022; Raghav et al. 2022). They aid in reducing the negative effects of plant pathogens. Through the breakdown of pollutants in a way akin to that possessed by plant growth-promoting rhizobacteria (PGPR), they may also enable the host plant to multiply even in contaminated soil (Brader et al. 2014; Cherif-Silini et al. 2021). Utilizing high-throughput contemporary "omics"-based technologies, such as microarrays, metabolomics, and gene sequencing, can help understand the intricate relationships that exist between plants and their endophytes and can be a useful tool for environmentally sustainable development (Crandall et al. 2020; Kaul et al. 2016). Their excellent colonization effectiveness and resilience against abiotic stress make them a suitable candidate for environmental management (Ali et al. 2022; Dwibedi et al. 2022; Kumar et al. 2022; Liu et al. 2022). Endophytes (fungi, bacteria, and actinomycetes) are reported to establish a mutualistic co-existence pattern by precisely balancing secondary metabolite production and signaling pathway regulation. The highest level of endophytic colonization occurs in the root, followed by intercellular space and vascular bundles (Hamilton and Bauerle 2012). Hazardous pollutants and carcinogenic elements originating from industrial and agricultural sources have a great impact on human health and are associated with ecological imbalance and toxicity. The practice of employing a plant's own metabolic system to remove pollutants, heavy metals, hazardous compounds, and xenobiotics from the environment is known as phytoremediation (Khan and Doty 2011), and could be categorized as phytoextraction (removal of heavy metals or toxic compounds from soil or water), phyto-stabilization (immobilization of metals and reduction in bioavailability of contaminants at different trophic levels), bioaccumulation (accumulation or absorption of organic and inorganic contaminants), rhizo-filtration (water filtration via a root system to remove contaminants from the aqueous ecosystem), and evapo-transpiration of volatiles (Glick 2003) (Fig. 5.1).

This review aims to describe the main uses of bioremediation and phytoremediation through endophyte fungi for sustainable agriculture.

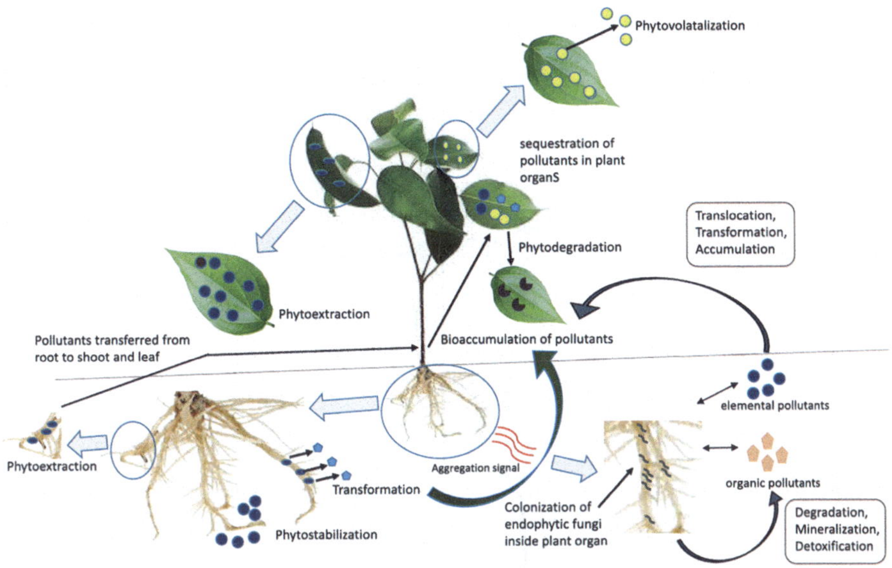

Fig. 5.1 Diagram illustrating the process of phytoremediation and the way in which endophytic fungus sequester organic and elemental contaminants (Nandy et al. 2020)

5.1.1 Plant Disease Resistance Induction

It is well known that endophyte microbiomes have a major impact on host performance, particularly in stressful situations (Hamilton and Bauerle 2012; White et al. 2019). They also play a crucial role in mediating the functioning of the plant microecosystem by drastically modifying how plants respond to environmental changes (Friesen et al. 2011). Due to competition for iron supply between endophytes and plant cells, siderophore production is critical for endophytic growth by boosting mineral availability and iron chelation. It also plays a role in pathogen suppression by inducing the biosynthesis of additional antimicrobial compounds (Sessitsch et al. 2004; Lacava and Azevedo 2013; Patle et al. 2018). A "barrier effect" is produced when endophytes extensively multiply and colonize plant tissues, competing with harmful microbes to keep them from proliferating. Similarly, endophytes are essential for sustaining plant health because they can enhance growth and yield while protecting or preparing the plant against biotic stressors through antibiosis or induced systemic resistance (Lugtenberg et al. 2016). By means of the mechanisms of induced systematic resistance (ISR) and systemic acquired resistance (SAR) (Kumar and Verma 2018), microbes augment disease resistance. Plants have evolved several families of receptor proteins to recognize microbe- or pathogen-associated molecular patterns (MAMPs/PAMPs), which are important features that are conserved and necessary for microbial survival (Plett and Martin 2018) (Fig. 5.2). Pathogen- or microbe-associated molecular patterns (PAMPs or MAMPs) are frequent microbial chemicals that are

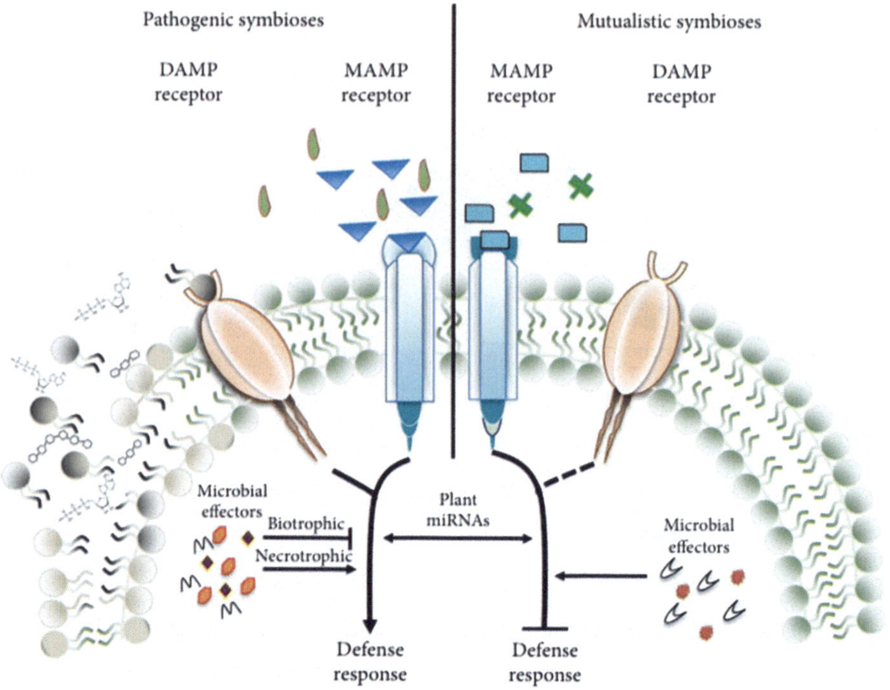

Fig. 5.2 Understanding the difference between the host's pathogenic and mutualistic interactions when the defense response is triggered (Plett and Martin 2018)

recognized by pattern-recognition receptors (PRRs) through evolution. Examples of these molecules include bacterial flagellin and fungal chitin. PAMP-triggered immunity (PTI), a first line of defense that neutralizes most possible invaders, is converted into pattern recognition (Brader et al. 2017). *Beauveria bassiana* treatment of tomato and cotton seeds produced protection against *Rhizoctonia solani* and *Pythium myriotylum.* Analogously, it is known that *Diaphorina citri,* an insect pest, can cause Citrus limon plants to develop resistance to *Beauveria bassiana* (Bamisile et al. 2019). In cucumber plants, the endophytic fungus *Lecanicillium longisporum* reduces aphids and powdery mildew (Bamisile et al. 2018). On *Arabidopsis thaliana*, the endophyte *Bacillus cereus* causes ISR against *Botrytis cinerea* by enhancing PR1 protein expression over time, accumulating hydrogen peroxide, and causing callose deposition. By decoding environmental cues and signal transduction, mitogen-activated protein kinase (MAPK) cascades are essential for both the biotic and abiotic stress responses (Rehman et al. 2020). By activating WRKY53 gene expression and MAPK signaling, endophyte contributes to pathogen-associated molecular pattern-induced (PAMP-) triggered immunity (PTI) through the jasmonic acid (JA)/ethylene (ET) signaling pathway in an NPR1-dependent manner (Nie et al. 2017). In addition to inducing resistance to rice blast, the bacterial endophyte *Azospirillum* sp. also causes systemic disease resistance in rice, and ET signaling is necessary for endophyte-mediated induced systemic resistance (ISR) in rice (Kusajima et al. 2018).

Salicylic acid (SA) signaling, on the other hand, activates resistance against biotrophic and hemi-biotrophic pathogens, while jasmonic acid (JA) and ethylene (ET) signaling together activate resistance against necrotrophic pathogens (Robert-Seilaniantz et al. 2011). At the location of a potential phytopathogen attack, endophytes cause a number of cell wall alterations, including the deposition of callose, pectin, cellulose, and phenolic chemicals, which result in the creation of a structural barrier. Likewise, they trigger the production of defense-associated proteins such as chitinases, peroxidases, and β-1,3-glucanases (Malfanova et al. 2013). *Trichoderma arundinaceum* generates volatile organic compounds (VOCs) such as *trichodiene*, which influence *Botrytis cinerea* by inducing the expression of genes associated with tomato plant defense that encode salicylic acid (SA) and jasmonate (JA) (Malmierca et al. 2012). Barley that has been inoculated with oxo-C14 homoserine lactone (AHL) enhances its resistance to *Puccinia hordei* by generating *Ensifer meliloti* (Wehner et al. 2019), The acyl-homoserine lactone (AHL) biosynthesis genes phzI, csaI, and aurI involved in clear AHL formation were found to be present in the endophytic biocontrol strain of *Pseudomonas chlororaphis* subsp. *aurantiaca* PB-St2 genome study and may have an impact on biocontrol activity (Bauer et al. 2016). Similarly, a significant factor in the decline of the fungus population on wheat field soils is endophytic *Pseudomonas putida* mutated with an antifungal phz gene derived from *Pseudomonas fluorescens* (Glandorf et al. 2001). Abiotic and biotic stress responses in plants are influenced by lipid transfer protein (LTP). Jasmonic acid is bound by LTP1, and when combined, they have a greater affinity for the elicitin binding site and can cause resistance even when applied distantly. The development of the disease was inhibited by the expression of CaLTP-N, an LTP-like protein, indicating that LTP is a functional component of the resistance induced by *Trichoderma* species to *Phytophthora infection* in hot peppers. The endophytic fungi *Aspergillus terreus* LWL5 and *Penicillium citrinum* LWL4 controlled oxidative stress responses in sunflower plants by upregulating catalase and peroxidase and downregulating glutathione and polyphenol oxidase, thereby reducing fungal infection caused by *Alternaria alternata*. Comparably, leaves infected with endophytes had increased amino acid contents, indicating that this alteration slows down cell death and impedes the growth of fungi in plant tissue (Khan et al. 2015).

5.2 Bio-Remediation and Phyto-Remediation Through Endophytes Fungi

The technique known as "bioremediation" uses plants, microbes, or enzymes to clean up pollution found in natural sources. With phytoremediation, hazardous substances can be extracted from water, soil, and air by plants using their own metabolic processes. Metals and metalloid pollutants, carcinogens, inorganic pesticides and herbicides, excess nutrients, chlorinated products, industrial organic waste

Table 5.1 Endophytic fungi-based bio- and phytoremediation techniques. (Ortega et al. 2020)

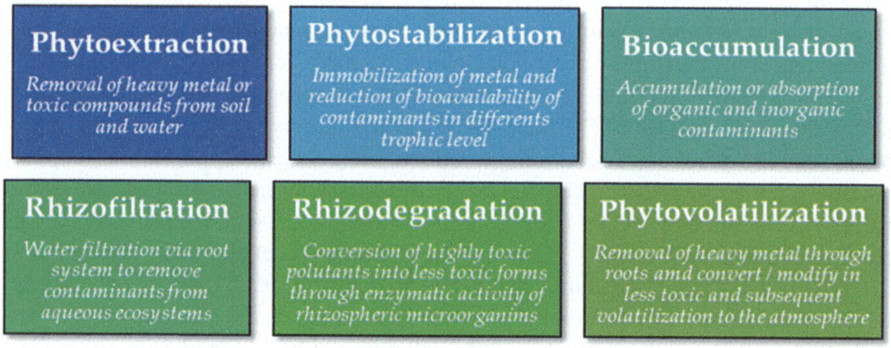

material, and radionuclides are some examples of these compounds or contaminants (Deng and Cao 2017; Nandy et al. 2020; Pietro et al. 2020). Endophytic fungi can break down both big and tiny organic substances through enzymatic processes, break down pollutants from the environment, and enhance the microbiome of the soil (Krishnamurthy and Naik 2017). Additionally, they can modify the morphological and physiological processes of the host plant, enhancing its resistance to metals and offering various detoxification pathways like extracellular scavenging and complexation, compartmentalization, and volatilization (Fomina et al. 2005; He et al. 2020). They can also boost the capacity of host plants to eliminate pollutants from soil, water, sediment, and air (Li et al. 2012). The various bioremediation methods using endophytic fungi are depicted in Table 5.1.

Utilizing ambient microbes like fungi and bacteria, bioremediation is an environmentally benign method of restoring heavy metal-contaminated environments. Through chemical and biological processes, these microorganisms can eliminate and/or deactivate contaminants from contaminated substrates. Additionally, they engage with the native vegetation, safeguarding and promoting plant development under these challenging circumstances. Lead (Pb) has a relative atomic weight of >4 g/cm^3, which makes it a "heavy metal." Lead has no known biological purpose and is hazardous to plants even at low concentrations, whereas other metals like manganese and zinc are necessary for plant growth (Nagajyoti et al. 2010). Because of its great toxicity, it is referred to as a toxic heavy metal (THM) and contributes significantly to environmental pollution (Rahman and Singh 2019; Chibuike and Obiora 2014). Lead does, in fact, naturally occur in rock formations within the crust of the Earth, and owing to human meddling, lead becomes more readily available in the biosphere. The primary sources are mining, ore processing, and recycling of lead-acid batteries. Industrial goods like paints, dyes, and pipes, as well as agricultural operations like pesticides, are next (Nagajyoti et al. 2010; Rahman and Singh 2019). The metal can be found in the atmosphere, hydrosphere, and pedosphere, bound with other chemicals once it has been exposed to the Earth's surface. But because it is not biodegradable, it stays in the environment for a long time and can be absorbed and accumulated by plants and microorganisms, which puts it into the food chain and harms both human and animal health (Rahman and Singh 2019). Because lead is so toxic, plants react to it in different ways. These reactions include: (1) immobilization of the metal in the rhizosphere by release of exudates or biosorption in the walls of the roots (exclusion); (2) assimilation of the metal into the root and activation of detoxification reactions, which involve the sequestration of the metal through the production of metal ligands like phytochelatin and metallothioneins and occasionally accumulation in cellular vacuoles (detoxification); and (3) activation of the NO-dependent stress response and other metabolic pathways to reduce the amount of bioavailable lead (Pb^{2+}) in the cells (non-specific defense mechanism) (Singh et al. 2011; Kaul et al. 2016). Additionally, the production of reactive oxygen species (ROS) aids in the metal's scavenging and causes structural alterations in the root's cell walls, which bind the poisonous element (Singh et al. 2011; Rahman and Singh 2019). However, the Casparian strip of the endodermis functions as a barrier, keeping most of the metal contained at the root level (Kaul et al. 2016). When high levels of lead in plants overcome these defenses, the metal can harm the plant in a variety of ways, such as germination abortion, decreased growth of roots and shoots, inhibition of photosynthesis, disruption of biological reactions where the metal functions as a "molecular mimic," and a decrease in antioxidant enzymes leading to an increase in reactive oxygen species and oxidative cellular damage (Rahman and Singh 2019; Kaul et al. 2016; Asati et al. 2016). Lead poses serious risks to human health and the environment (Sharma et al. 2009; Li et al. 2006). Detoxification of lead-polluted sites is therefore extremely difficult. Microorganisms can be employed in bioremediation, an eco-friendly and novel technique to remove heavy metal pollution from substrates. In fact, a variety of microorganisms have many detoxification processes and are well-suited for harsh environmental conditions (Bishnoi 2005; Leitão 2009; Ahemad and Kibret 2013; Ojuederie and Babalola 2017). By absorbing toxic metals into their membranes (biosorption), converting them chemically into a less harmful form, and actively encouraging plant absorption of the metals (phytoremediation), microorganisms can lower the concentration of toxic metals in the soil while also increasing the amount of organic matter in the soil (Bishnoi 2005; Leitão 2009; Ahemad and Kibret 2013; Ojuederie and Babalola 2017). In the presence of harmful heavy metals, endophytes can also promote plant growth by favorably affecting the contaminant's absorption, breakdown, and dislocation by the plants (Deng and Cao 2017). For instance, foliar endophytes called Neotyphodium have been researched for their advantageous effects on grasses, including Festuca and Lolium species, in soil contaminated with cadmium, zinc, aluminum, and copper (Monnet et al. 2001; Malinowski and Belesky 1999; Malinowski et al. 2004; Ren et al. 2006). In a hydroponic system, Soleimani et al. (Soleimani et al. 2010a, b) showed that *Festuca* sp. treated with *N. coenophialum* and *N. uncinatum* produced more biomass from the leaves and roots than plants that were left untreated, even at high cadmium concentrations. The significance of employing endophytes as a consortium as opposed to a single strain has been highlighted by a recent study. In fact, when co-inoculated with *Rhizopagus irregularis*, an arbuscular mycorrhizal fungus (AMF), *Cochliobolus sativus, Diaporthe* sp., and *Phoma exigua*var. *exigua*, compared to non-inoculated and AMF plants, Wezowicz et al. (2017) found an incremental biomass production of *Verbascum lychnitis* grown in post-mining waste. A detrimental effect was observed in the plants when they were just inoculated with *Diaporthe* sp., but this effect was offset by a positive response when the plants were also inoculated with AMF (Wężowicz et al. 2017). In a related study, Vandegrift et al. (2015), found that co-inoculating an *Epichloë* species with a dark septate endophyte (DSE) decreased its deleterious effect on *Agrostis capillaris*. They postulated that the DSE's excessive nitrogen synthesis was utilized by the *Epichloë* species to produce alkaloids for plant defense, offsetting the costs incurred by the plant in housing the DSE (Vandegrift et al. 2015).

Mercury is a heavy metal that has no biological purpose and is considered an environmental pollutant due to its high mobility, bioaccumulation, prolonged environmental half-life, and significant risk to human health (Dash and Das 2012; Lopez et al. 2014; Cozzolino et al. 2016). Although human activities like mining and the production of metals and fossil fuels produce a significant amount of mercury, the majority of it is released into the environment by natural processes like ocean emissions and biomass burning (Gontia-Mishra et al. 2016; Nelson et al. 2012; Pirrone et al. 2010). According to the Agency for Toxic Compounds and Disease Registry's "priority list of hazardous substances," this heavy metal is ranked third (ATSDR 2015). Mercury emissions into Brazil's natural habitats, especially wetlands such as the Mato Grosso Pantanal, are mostly caused by gold mining (von Tümpling et al. 1995; Leady and Gottgens 2001; Callil and Junk 2011; Ceccatto et al. 2016; Lázaro et al. 2015). Methylmercury (CH_3Hg^+), an organic form of mercury with greater toxicity and potential for bioaccumulation, is produced in such environments by biotic and abiotic processes from inorganic Hg^{+2} (Ceccatto et al. 2016). Hg^{+2} is transported along the trophic chain once accumulation begins, endangering the biodiversity of the Pantanal, particularly the aquatic species (Ceccatto et al. 2016). Although their capacity to bioremediate mercury has not yet been documented, endophytic fungi are frequently isolated from heavy metal-contaminated environments and are widely exploited in the bioremediation of hazardous metals (Xiao et al. 2010; Sim et al. 2016; Khan et al. 2017b; Zahoor et al. 2017). According to some research (Kurniati et al. 2014; Urík et al. 2014), edaphic fungi have the ability to bioremediate mercury through the processes of bioaccumulation and biovolatilization.

5.2.1 Phytoremediation Process

Hyper-accumulator plants' root exudates frequently serve as an "aggregation signal," drawing endophytic microbial colonization (Weyens et al. 2009a, b). After successfully establishing themselves within the plant system, the endophytes change the metabolic pathways of the host, causing disease resistance, stress tolerance, and allelopathic traits. As a result, they enhance growth, maturation of seedlings, germination of seeds, higher biomass production, abscission, senescence, host fitness, and competitive success rate (Glick 2010). As their carbon sources, endophytes receive food from host plants in exchange for improved biotic and abiotic stress tolerance (Hamilton and Bauerle 2012; Hamilton et al. 2012); increased resistance to herbivore or pathogen attack (Guo et al. 2008; Hyde and Soytong 2008); effects on host plant diversity, abundance, and niche utilization (Gazis et al. 2011; Yuan et al. 2011); and decomposition or accumulation of organic and non-organic contaminants (Luo et al. 2011; Shin et al. 2012) at various stages of symbiotic interaction. According to Rajkumar et al. (2009), hazardous environmental conditions frequently change the pH of the soil, the cycling of nutrients, the uptake of minerals, and photosynthesis. In these types of adverse conditions, phytoremediation's capacity is quite helpful. According to Kristin and Miranda (2013), endophytic organisms frequently control the biochemical processes of the plants they inhabit, which may facilitate phytoremediation. We have covered many facets and essential mechanisms of endophytic fungus-mediated phytoremediation in plants in the ensuing paragraphs (Table 5.2).

According to Naranjo-Briceno et al. (Naranjo-Briceño et al. 2013), the powerful remediating endophytic fungus *Pestalotiopsis palmarum* BM-04 could withstand high salinity and contamination from crude oil. It has been observed that *Phomopsis liquidambari* effectively decomposed polycyclic aromatic hydrocarbon (PAH) in *Bischofia polycarpa* (H. Lev.) Airy Shaw, while *Pestalotiopsis microspora* degrades polymeric polyester polyurethane (PUR) in *Dendrobium* sp. (Chen et al. 2013; Dai et al. 2010). *Citrullus lanatus* (Thunb.) Mansf. phenolic acids were broken down by *Ceratobasidium stevensii* (Xiao et al. 2014). Furthermore, compared to the identical uninfected plants, the host plants infected with endophytic fungi may accelerate the breakdown of petroleum hydrocarbons (Xiao et al. 2014). Two such Poaceae species that exhibit increased biomass buildup and accelerated total petroleum hydrocarbon breakdown when infected by endophytic fungi are *Festuca arundinacea* Schreb. and *Festuca pratensis* Huds (Soleimani et al. 2010b). Total petroleum hydrocarbon (TPH) removal in the plant rhizosphere was 64–72% and 80–84%, respectively. TPH removal was higher in endophyte-infected plants than in non-infected plants. A promising green technology that is linked to increased root biomass and the release of water-soluble phenols into the soil, which results in the enzymatic destruction of the C10–C25 chain length of alkanes, is the removal of TPH from oil-contaminated soils by endophytes (Soleimani et al. 2010a, b). Van Hecke et al. (2005) show that endophytic fungi elevate rhizodeposition in plants and impact microbial mineralization in the soil, all while maintaining the diversity of the microbial population. Understanding the potential for mineralization or the biotransformation process is necessary to determine whether this innovative approach will be commercialized. According to reports, chemoheterotrophic metabolism and the filamentous branching hyphal network are actually the superior criteria for metal and metalloid recovery. The "recovery" stage of any complex biochemical process influences its cost-effectiveness (Liang and Gadd 2017). For the metalloid recovery of As, Se, Te, and antimony (Sb), "metalloid oxyanions to elemental metalloids" must be reduced (Gadd 1993; Chasteen and Bentley 2003; Wilson et al. 2010; Pierart et al. 2015). Some essential elements of successful and environmentally friendly phytoremediation via endophytes include the selection and genetic

Table 5.2 Comprehensive details on the method of action, study area, and phytoremediation mediated by endophytic fungus

Plant species	Endophyte fungi	Experiments performed	Aim of the study	Reference
Lolium perenne L. (Poaceae)	*Neotyphodium lolii*	Inoculating seeds with fungi, examining them under a microscope, analyzing their mineral content, determining chlorophyll fluorescence using pulse amplitude modulation, and measuring gas exchange.	Enhances tolerance to excessive zinc levels.	Monnet et al. (2001)
Pinus sabiniana L. (Pinaceae)	*Scleroderma citrinum, Amanita muscaria, Lactarius rufus*	Isolating and characterizing microbes, conducting plant growth promotion assays, and performing chemical analyses.	Shielding against heavy metal exposure.	Krupa and Kozdrój (2007)
Festuca arundinacea Schreb. (Poaceae); *Festuca pratensis* Huds. (Poaceae)	*Neotyphodium coenophialum, Neotyphodium uncinatum,*	Performing a soil assay, establishing microbial cultures, estimating dehydrogenase activity, quantifying water-soluble phenols using colorimetric methods, and analyzing total petroleum hydrocarbons (TPH) and polycyclic aromatic hydrocarbons (PAH).	Removal of TPH from soils contaminated by petroleum.	Soleimani et al. (2010a)

(continued)

Table 5.2 (continued)

Plant species	Endophyte fungi	Experiments performed	Aim of the study	Reference
	Neotyphodium sp.	Establishing hydroponic cultures, measuring chlorophyll fluorescence, conducting soil assays, and estimating cadmium (Cd) levels.	Assessment of Cd accumulation and translocation.	Soleimani et al. (2010b)
Cajanus cajan (L) Huth. (Fabaceae)	*Scleroderma citrinum, Amanita muscaria, Lactarius rufus*	Assessing glutathione levels, analyzing non-thiol proteins, and determining metal concentrations.	Interaction of heavy metals, production of phytochelation and glutathione.	Garg and Aggarwal (2011)
Brassica napus L., *Brassica chinensis* L. (Brassicaceae)	*Mucor* sp. (CBRF59)	Establishing microbial cultures, estimating metal bio-absorption, conducting metal mobility assays, and assessing total biomass production.	Accumulation of lead (Pb) and cadmium (Cd) from soil contaminated with heavy metals.	Deng et al. (2011)
Lolium arundinaceum darbyshire ex. Schreb (Poaceae)	*Neotyphodium coenophialum*	Inoculation and heavy metal analysis.	Phytoextraction (Cd)	Ren et al. (2011)
Arabis hirsuta (L.) Scop. (Poaceae); *Acacia decurrens* Willd. (Fabaceae); *Symplocos paniculata* Miq. (Sym plocaceae); *Rabdosia eriocalyx* L. (Poaceae); *Arenaria serpyllifolia* L. (Caryophyllaceae); *Rosa longicuspis* Bertol. (Rosaceae)	*Phoma* sp., *Alternaria* sp., *Peyronellaea* sp.	Establishing microbial cultures and estimating metal tolerance.	Estimation of Pb and Zn tolerance	Li et al. (2012)

(continued)

Table 5.2 (continued)

Plant species	Endophyte fungi	Experiments performed	Aim of the study	Reference
Brassica napus L. (Brassicaceae)	*Mucor* sp. CBRF59	Isolating, regenerating, and fusing protoplasts, estimating resistance to cadmium (Cd), and conducting soil mobility assays.	Estimating metal resistance through self-fusion of deactivated protoplasts.	Deng et al. (2013)
Zea mays (Poaceae)	Peyronellaea	Identifying, conducting pot experiments, testing chemical components, and determining the dry biomass of shoots and roots.	Absorption and accumulation of heavy metals (lead, zinc, and cadmium).	Shen et al. (2013)
Sorghum bicolor x sudanense (Poaceae)	*Penicillium aculeatum* PDR-4 *Trichoderma* sp. PDR-16	Isolating endophytic fungi, characterizing them, and conducting pot experiments.	Phytostabilization	Babu et al. (2014)
Brassica napus L. (Brassicaceae)	*Mucor* sp. CBRF59 and *Fusarium* sp. CBRF14	Study on inter-species protoplast fusion.	Remediation of multiple metals (Cd, Pb, and Zn) in polluted soil through phytoremediation.	Deng et al. (2014)
Portulaca oleracea L. (Portulacaceae)	*Lasiodiplodia* sp. MXSF31	Isolation, identification, examination of dry biomass and active mycelia, pot assays, and inoculate recovery.	Strong bioabsorption and bioaccumulation abilities for Cd, Pb, and Zn from metal-contaminated solutions and extraction of metals.	Deng et al. (2014)
Festuca arundinacea Schreb. (Poaceae)	*Neotyphodium coenophialum*	Establishing a hydroponic system, analyzing tissue ion concentrations, and conducting growth experiments in saline soil.	Phytoextraction efficiency evaluation	Yin et al. (2014)

(continued)

Table 5.2 (continued)

Plant species	Endophyte fungi	Experiments performed	Aim of the study	Reference
Lolium arundinaceum (Schreb.) Darbysh. (Poaceae)	*Neotyphodium coenophialum*	Study involving inoculation and comparison of hydroponic and soil pots.	Efficiency of phytoextraction in a hydroponic system (accumulating sodium).	Yin et al. (2014)
Salix variegata Franch. (Salicaceae)	*Chrysosporium* sp., *Fusarium* sp. *Gonatobotrysi* sp.	Estimating metal tolerance, conducting inoculation experiments, and molecular identification.	Characterization of Cd-resistant fungi	An et al. (2015)
Solanum lycopersicum Lam. (Solanaceae)	*Penicillium janthinellum* LK5	Isolating endophytes, inoculating them into tolerant plants, conducting ICP-MS analysis, enzymatic assays, and screening.	Improve aluminum phytoextraction	Khan et al. (2015)
Populus L. (Salicaceae)	*Serendipita vermifera* P04	Study on root colonization, fungi identification, metal tolerance testing, and microscopy analysis.	High metal tolerant	Lacercat-Didier et al. (2016)
Dysphania ambrosioides L. (Mosyakin &Clemants.) (Amarantaceae).	*Plectosphaerella* sp., *Cladosporium* sp., and *Verticillium* sp.	Establishment of microbial culture	Assessment of heavy metal tolerance in endophytic fungi.	Li et al. (2016a, b)
Dysphania ambrosioides L. (Amaranthaceae)	FT2G59 (*Penicillium*) and FT2G7 (*P. columnaris*)	Isolation and characterization of fungal endophytes, analysis of heavy metal content, and determination of microbial community.	Protection against heavy metal	Li et al. (2016a, b)

(continued)

Table 5.2 (continued)

Plant species	Endophyte fungi	Experiments performed	Aim of the study	Reference
Imperata cylindrical (L.) P. Beauv.; *Bothriochloa ischaemum* (L.) Keng; *Calamagrostis epigeios* W.D.J. Koch; *Poa annua* Fr. ex Anderson; *Festuca rubra* Krecz. & Bobrov (Poaceae)	Leotiomycetes fungi and Pezizomycetes fungi	Soil examination, analysis of microbial community structure, and estimation of enzyme activity.	Phytoremediation of pb and cd from contaminated soil	Tong et al. (2017)
Solanum nigrum L. (Solanaceae)	*Colletotrichum* sp., *Alternaria* sp., *Fusarium* sp.	Endophytic fungi isolation, measurement of Cd tolerance, and plant inoculation trials.	In-vitro screening of cd tolerance	Khan et al. (2017a)
S. nigrum Korean ecotype (Solanaceae)	*Fusarium tricinctum* RSF-4 L, *Alternaria alternata* RSF 6L	Estimation of bioconcentration factor and tolerance, as well as assessment of Cd tolerance and accumulation.	Phytostabilization of heavy metals	Khan et al. (2017b)
Brassica napus L. (Brassicaceae)	*Fusarium* sp. CBRF44, *Penicillium* sp. CBRF65, *Alternaria* sp. CBSF68	Analysis of IAA and siderophore production, along with phosphate solubilization assay.	Measurement of phosphate-solubilizing activity.	Shi et al. (2017)
Aeschynomene fluminensis Vell. (Fabaceae) and *Zea mays* (poaceae)	*Aspergillus* sp. A31, *C. geniculata* P1, Lindgomycetaceae P87 and Westerdykella sp.	Isolation of endophytic fungi, screening for bioremediation potential, in vitro testing of mercury, and quantification of mercury levels.	Hg bioremediation	Pietro et al. (2020)

engineering of potent endophytic microbes, the creation of transgenic plants, niche division and host specificity, fungal metabolite-mediated reactions, increased pollutant uptake and tolerance, enhanced root and shoot biomass, intact root architecture, maintenance of pH, salinity, and pollutant-induced toxicity in the rhizosphere, management of the microbial population, and balanced symbiotic interaction (Kidd et al. 2009). Yet, transgenic technology is starting to show promise as a novel means of enhancing plants' inherent capacity for phytoremediation (Chatterjee et al. 2013). Since the formation of a characteristic or over-expression of a gene is associated with the sequestration of a particular pollutant, a genetic engineering-based method may improve intrinsic capacities of degradation and detoxification (Schäffner et al. 2002). Nonetheless, safety protocols and reproducibility requirements must be met by these biotechnological methods (Schäffner et al. 2002). One such cutting-edge biotechnological technique used in phytoremediation research is protoplast fusion. The availability of Cd and Pb in the soil was greatly increased by employing the isolated protoplasts of the self-fusing CBRF59T3 strain of Mucor sp. CBRF59 for metal resistance augmentation (Deng et al. 2013).

5.2.2 Problems Associated with Phytoremediation

A futuristic concept called phytoremediation uses processes such as absorption, translocation, transformation or extraction, bioaccumulation, and detoxification to deal with xenobiotics. Effective biotic and abiotic interactions as well as appropriate use of the rhizosphere's physical and chemical characteristics are crucial for phytoremediation (Chatterjee et al. 2013). A tightly packed mycorrhizal sheath and phenolic interhyphal substance shield the contact of plant roots from harmful pollutants (Wenzel 2009). Other significant factors that affect how effective endophytic fungi are as powerful phytoremediation agents include their large surface area, ion exchange capacity, higher rate of adsorption of bioavailable metals and metalloids, strength of the fungal sheath, and metal toxicity tolerance (Wenzel 2009). The sluggish growth rates and modest biomass of hyperaccumulator plants pose a challenge to field testing. Another important component is metal bioavailability, which is caused by a metal pool that exists in *pseudo-equilibrium* between the solid and aqueous soil phases and which indirectly starts the phytoremediation process through complex organic and inorganic colloids (Kidd et al. 2009). The rhizosphere's availability and absorption of organometallic chelating components are limited by the interactions between root exudates and soil constituents, which also lowers the overall amount of metals that are absorbed by plant roots. However, the total metal uptake is strongly influenced by the ion and ligand exchange mediated by microbes, the secretion of metabolite-rich exudates, the release of microbial siderophores, the pH-regulated mobilization and immobilization of chemicals, the high level of tolerance, and the biotransformation capacity promoted by the endogenous enzymatic system (Wenzel 2009; Durán et al. 2011). Sequestration methods such as phyto-stabilization (immobilization within the rhizosphere) and bioaccumulation (translocation, transformation, and storage into aboveground tissue) are generally more common for heavy metals; however, for organic substances, detoxification, or phyto-decontamination is preferred. Mineralization, transformation, enzymatic breakdown, conjugation, and compartmentalization are all aspects of detoxification. The entire phytoremediation process involves several different enzyme classes, including reductases, dehalogenases, glutathione, glycosyl transferase, oxygenases, and cytochrome P450 monohydrolysis (Schäffner et al. 2002).

Coriolopsis polyzona and *Trametes versicolor* both release laccase, which is able to break down aromatic molecules. It has been observed that peroxidase generated by *Phanerochaete laevis* and *Phanerochaete chrysosporidium* causes aliphatic hydrocarbon dehalogenation. Conversely, it has been shown that nitrilase originating from *Aspergillus niger* may cleave cyanide groups off both aromatic and aliphatic nitriles. Despite not being endophytic fungi, all of the enzymes that were extracted from fungal sources helped to sequester pollutants in various ways (Gerhardt et al. 2009).

Although the in situ remediation process is new, there are a number of fundamental obstacles that must be overcome. The primary issue with endophyte-mediated research is its lack of authenticity and reproducibility, which leads to the dearth of extensive field trials. The results from the laboratory must be validated in an uncontrolled, natural setting. If not, the multilayered response mechanism underlying endophytic fungus-assisted phytoremediation will go unstudied. In addition, a detailed understanding of the primary enzymes and released secondary metabolites is crucial for understanding the chemical mechanisms underpinning long-term colonization and remediation. Broad-spectrum understanding in this area is still lacking as of right now. In order to address the fundamental mechanisms of plant-microbe-compound interaction, the first step is to comprehend the binary relationships between the two parties: the plant and the microbe, or pollutant. Although the approach's ecological impact is evident, there are still numerous unidentified interlinking points that require more understanding; therefore, optimizing application methods does not provide comprehensive proof. Transgenic techniques offer benefits, but as sufficient information is still lacking, the main issues are once again repeatability and step-by-step demonstration of mode of action (Redfern and Gunsch 2015). An additional important consideration in this area is cost effectiveness. According to earlier studies, a 21-month field trial evaluating the bioaccumulating capacity of *Lolium annua* L. and *Stenotaphrum secundatum* (Walter) Kuntze revealed a 42% and 50% decrease in the petroleum hydrocarbon content in fields contaminated by oil spills (Gerhardt et al. 2009). However, there are very few statistically validated data in the case of endophytic fungi-mediated phytoremediation, which is detrimental for commercialization. Because it is a multi-factorial approach, "reprogramming" of plant gene expression may produce contradictory results in investigations using transgenics or plant trait enhancement. The absence of cutting-edge technology support in this research, particularly in outdoor trials, is another drawback. In order to separate biologically derived pollutants from "anthropogenic" or "background" hydrocarbons during the study, advanced analytical techniques such as nuclear magnetic resonance (NMR), high-performance liquid chromatography (HPLC), and gas chromatography-mass spectroscopy (GC-MS) must be used. But according to the findings, the majority of field trials are carried out far more haphazardly, meaning that the results of these investigations are not adequately confirmed. The authors have attempted to graphically depict the obstacles that must be overcome in order to apply the phytoremediation technique in Fig. 5.3.

Endophyte-plant interaction offers numerous benefits and drawbacks when used to counteract xenobiotic toxicity. Positive characteristics of endophytic bacteria

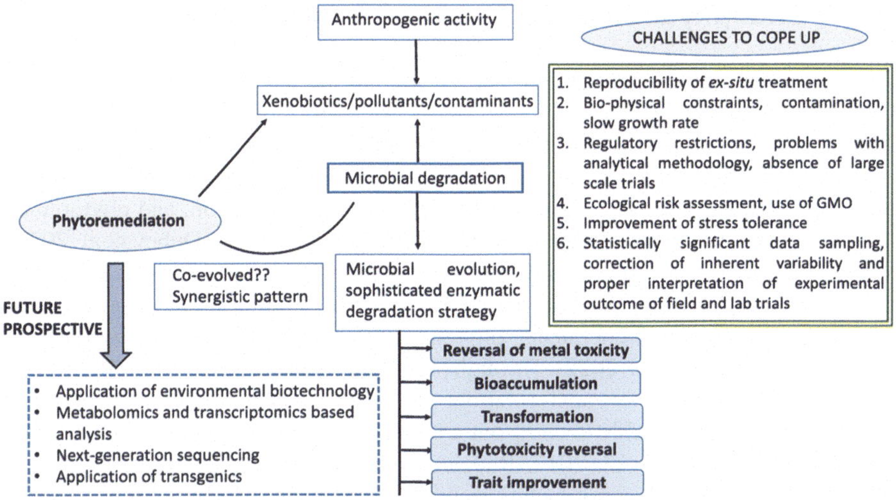

Fig. 5.3 The future problems and prospects of phytoremediation and microbial degradation approaches are interconnected in a synergistic way (Nandy et al. 2020)

include their long-term sustenance, ability to use the host metabolite reservoir, lack of tissue-specific interspecies competition, and ease of genetic alteration (Mercado-Blanco 2014). Nonetheless, communication begins when an "aggregation signal" is received. Only when a delicate balance is established between endophytic virulence factors and plant immune molecules, can colonization be successful. Removal of the pollutants from the environment is the last phase of the plant-microbe interaction in phytoremediation, and it only happens after several biochemical and physiological processes have taken place (Redfern and Gunsch 2016). Accordingly, a three-stage assessment process is needed to evaluate such phytoremediation processes, and "metagenomic, meta-transcriptomic, and metaproteomic" level research is crucial for understanding plant-microbe relationships and their possible impact on the optimization of natural remediation methods (Redfern and Gunsch 2016). Enzyme-linked immunosorbent assay (ELISA), chromatin immunoprecipitation sequencing (ChIPSeq), fluorescent protein tagging, overexpression of pollutant catabolic genes in microbes, and other techniques may help in comprehensive proteomic and genomic analysis, which may be very helpful to validate relevant, substantial data gathered from field trials (Ryan et al. 2007). To comprehend the underlying process of endophytic fungus-assisted phytoremediation, a multitude of characteristics and environmental variables must be evaluated under a comprehensive monitoring system (Redfern and Gunsch 2016). Only with the use of next-generation sequencing and cutting-edge technologies like real-time high-throughput whole plant functional phenotyping could the complex network of microbial communities dispersed throughout the various parts of the plant and their pluralistic interaction (plant-microbiome-soil pollutant) be simultaneously explored (Deng and Cao 2017).

5.3 Endophytic Fungi for Sustainable Agriculture

Broad-spectrum endophytic fungi help plants through either direct or indirect means. Production of phytohormones, nitrogen fixation, phosphate solubilization, siderophore formation, and anti-microbial metabolite formation are all components of the direct beneficial mechanism (El Enshasy et al. 2020). Abiotic and biotic resistance (altering the metabolic process), biocontrol, bioremediation, and phytoremediation are all components of the indirect beneficial mechanism (Dubey et al. 2020; Singh et al. 2021). In order to improve root growth, seed germination, and plant growth promotion, cosmopolitan fungal endophytes function as bio-stimulants to create specific bioactive chemicals, phytohormones, phosphate solubilization factors, etc. (Rustamova et al. 2022). Through phyto-stimulation, phyto-immobilization, phyto-stabilization, phyto-transformation, phytoremediation, biocontrol, etc., fungal endophytes may be able to carry out a variety of widely distributed functions in the host plant (Sahoo et al. 2017; Radziemska et al. 2021; Adeleke et al. 2022). According to reports, they produce bioactive antimicrobial siderophores as secondary metabolites, which may be used as a defense against a variety of infections (Srinivas et al. 2020). The hydroxamate and carboxylate-type siderophores of various classes are produced by the endophytes that are connected with plants, and they are important in the transportation of iron and zinc to the plants. Siderophores have been shown to mobilize and make a variety of metals available to plants by solubilizing them, including actinides, cadmium, molybdenum, manganese, zinc, nickel, and copper (Rana et al. 2020). More than 500 siderophores from different fungi have been identified to date (Ahmed and Holmström 2014; Chowdappa et al. 2020). Numerous fungal endophytes have been discovered recently that solubilize and mobilize potassium (*Aspergillus fumigatus* and *A. niger*), phosphorus (*Penicillium, Aspergillus, Curvularia, Trichoderma, Mesorhizobium,* etc.), and zinc salts (*Alternaria thlaspis* and *Metapochonia rubescens*) (Mehta et al. 2019), increasing plant growth and metabolic activity and leading to increased crop yield. According to Nascimento et al. (2014) and Singh et al. (2015a, b), they carry out phytostimulation by decreasing plant hormone ethylene levels by 1-aminocyclopropane-1-carbo xylate deaminase (ACC), which accelerates plant growth. They can also biodegrade biomass and recycle it in the environment, improving nitrogen availability and zinc and phosphorus uptake for the host, leading to phytoimmobilization. By immobilizing osmolytes and maintaining membrane ion conductivity under stress, phytoimmobilization eventually increases a plant's ability to withstand abiotic stressors. The process of neutralizing soil contaminants, such as pesticides, industrial byproducts, xenobiotics, etc., is known as phytotransformation. Numerous crop species are known to be associated with endophytic fungi such as *Piriformospora* sp., *Exophiala* sp., *Neotyphodium* sp., etc., that are known to remove metal contaminants from soil without causing phytotoxicity (Ningxiao et al. 2016; Sahoo et al. 2017). But in order to fully realize the promise of bioremediation using fungal endophytes for environmental conservation, more research must be done in the near future using endophyte-led nano-sensors (Khanam et al. 2020). Lastly, its two main operations are the use of biocontrol agents to lessen the impact of pests and diseases and biofertilizers to promote plant growth. In this situation, metagenomics and multi-locus sequence typing, two culture-independent molecular approaches, can be used to identify and isolate these endophytes (Gupta et al. 2020; Adeleke et al. 2022; Chouhan et al. 2022). It is strongly anticipated that species-specific fungal endophytes will be identified, isolated, characterized, and studied in relation to inoculation and contact; their molecular signaling mechanisms will be uncovered; and abiotic and biotic stress mitigation will be provided. Furthermore, it is crucial to do investigations on the components that extend the shelf life of these fungal biocontrol agents and biofertilizers, in addition to isolation (Zope et al. 2019).

5.3.1 Abiotic Stress Management

Abiotic stressors significantly inhibit plant growth, and in order for a plant to withstand them, it needs an effective system or mechanism. Numerous microorganisms are typically found obtaining nutrition from different plants; certain relationships are advantageous to the host, while others are detrimental. Ripa et al. (2019) defined "endophytes" as microorganisms that live in plant tissues without causing significant harm or earning a living other than to establish their residence. Every plant has one or a few endophytes that provide nutrients, help the host plant adapt to its surroundings, and provide resistance to both biotic and abiotic stresses. Throughout part or all of its lifecycle, an endophytic fungus resides in intracellular or intercellular gaps in a plant's stem, petiole, and leaves. For the duration of the plant's life, these incredibly effective fungal endophytes can develop, which may also help the plant overcome environmental restrictions (Vandenkoornhuyse et al. 2015; Chouhan et al. 2022). In today's agriculture, fungus endophytes are essential for resilience and sustainability. By working with host plants, they provide a variety of uses in the agricultural system (Redman et al. 2021; Grabka et al. 2022). Their ability to release insoluble phosphate and supply nitrogen to their host plants is crucial, as it promotes successive vegetative development and fortifies related plants to impart resistance (Hassan 2017; Yadav and Yadav 2017). Fungal endophytes emit reduced amounts of compounds that promote plant growth in response to stress. For instance, Penicillium sp., which inhabits *Suaeda japonica,* secretes gibberellins, a phytohormone. Similar to this, the *Aspergillus, Cladosporium, Fusarium, Penicillium, Verticillium*, and *Ascomycete* species that live in *Panax ginseng* secrete ginsenosides and triterpenoid saponins, which promote root growth and the ability to withstand both biotic and abiotic stress (Sahoo et al. 2017). Keswani et al. (2022) examined the biosynthesis of gibberellins, which are advantageous to crop yield and are produced by a variety of fungi. The same is true for *Phoma* sp. that inhabits *Tinospora cordifolia; Trichoderma atroviride, T. polysporum*, and *T. harzianum* that inhabit *Phaseolus vulgaris*; these microorganisms secrete proteolytic enzymes, phosphate solubilizing factors, active volatile, and non-volatile metabolites; and Mucor sp. that inhabits Brassica campestris that increases Indole-3 amino acid (IAA), ACC deaminase, and solubilizes phosphate secretion that promotes plant growth (Sahoo et al. 2017) (Fig. 5.4). In order to promote plant growth under stressful conditions, fungi endophytes carry out phosphate solubilization, siderophore synthesis, antibiotic synthesis, phytohormone synthesis, nutrient mineralization, and other tasks (Yung

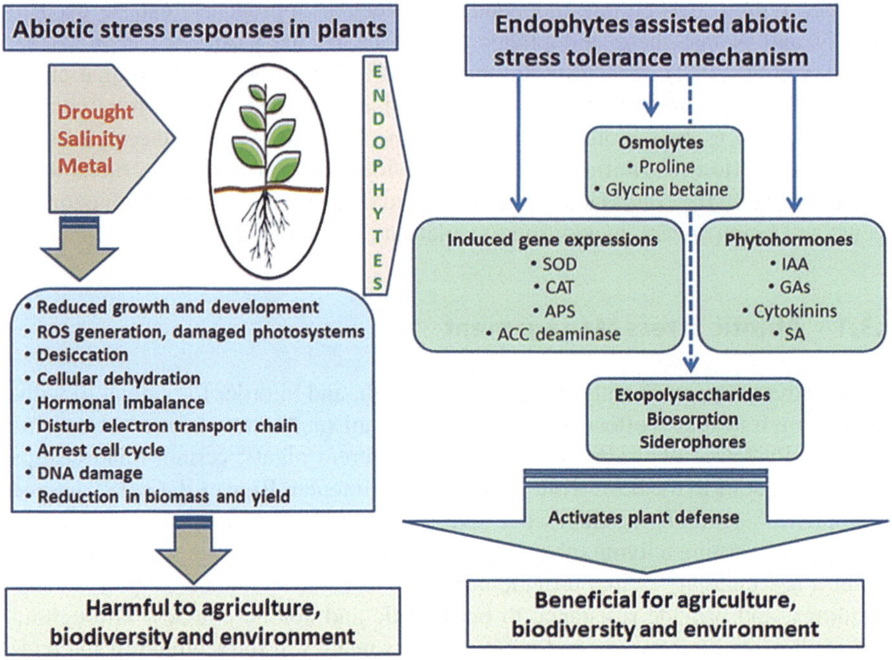

Fig. 5.4 An overview of how plants respond to abiotic stress (left): extended periods of abiotic stress (salinity, heavy metals, and drought) can limit plant growth and productivity by causing desiccation, cellular dehydration, ROS renewal, hormonal imbalances, and other issues. The abiotic stress tolerance system mediated by endophytes (right) involves the induction of gene expressions for plant defense, the generation of osmolytes (proline, glycine betaine, etc.), and the secretion of phytohormones (IAA, cytokinins, gibberellic acid (GAs)) under abiotic conditions (Anand et al. 2023)

et al. 2021). Salicylic acid, jasmonic acid, and abscisic acid are created in response to abiotic stressors and function as chemicals that indicate defense. The majority of soil types have inadequate levels of phosphorus, which is regarded as a necessary component for plant growth. Even when added to the soil as an inorganic fertilizer, they stay insoluble and unusable by plants; for this reason, iron and aluminum hydroxides start to fix in acidic soil. Although iron is an essential element and exhibits redox activity, it is insoluble in soil (ferric hydroxide) and acts as a cofactor for numerous enzymes (Johnson 2008). Fungal endophytes generate siderophores that have a high affinity for iron (Fe_3) and a low molecular weight. This enhances the uptake of iron and phosphates, allowing plants to absorb them and flourish. Endophytes increase biomass output and nitrogen intake while biodegrading the trash (Idbella et al. 2019).

5.3.1.1 High Carbon Di-Oxide

Through enhanced photosynthesis and induced compositional changes at elevated CO_2, endophytic fungi promote internal hyphal development and colonization, carbon sequestration, biomass, and yield (Galani et al. 2022). According to Yadav and Yadav (2017), a number of researchers have discovered the benefits of symbiotic relationships in this context. These relationships have been shown to increase plant growth in crops such as potatoes, rice, beans, and tomatoes, as well as to secrete ascorbate and glutathione, which reduce the production of reactive oxygen species (ROS) and antioxidants. In order to improve survival without compromising yield, this results in efficient nutrient uptake, photosynthetic rate, and water use efficiency (Elhindi et al. 2016), as well as increased production of phytohormones, amplified gene expression, supplemented protein function, metabolite synthesis, secretion of signaling molecules, and stress-responsive pathways (Ma et al. 2019). On the other hand, some contrast data that showed reduced growth of roots and shoots overstated the fact that endophytes have different impacts on the same host plants when they are exposed to other niches, preferentially in certain growing conditions (Gavito et al. 2000). Additionally, the type of plant has an impact on colonization. For instance, C3 plants are less likely to benefit from colonization because they require more carbon to produce biomass, whereas C4 plants allocate more carbohydrates and benefit from endophytic fungi associations (Tang et al. 2009). As a result, current research indicates that the role of microbiota in biological and ecological processes is still unclear (Rehman et al. 2016). This is particularly true for potential agricultural applications, as endophytes use molecular mechanisms to induce the synthesis of reactive oxygen species (ROS) scavenger molecules, anti-stress metabolites, and genes responsive to stress.

5.3.1.2 Drought

One of the major factors lowering agricultural output and directly impeding developing nations' efforts to achieve food security is drought. In response to dry circumstances, plants undergo a variety of physiological and biochemical reactions, including stomatal closure, increased respiration, growth suppression, and photosynthesis. Two categories of genes that are induced by drought stress are said to give plant tolerance: the first group consists of proteins, and the second group consists of regulatory proteins that are involved in signal transduction (Shinozaki and Yamaguchi-Shinozaki 2007). The hyphae of the fungal endophyte *Piriformospora indica* (family: Sebacinaceae) are found in the outer layer of the host's roots and aid in plant growth, higher chlorophyll content, and elevated yield under drought stress conditions. Ripa et al. (2019) studied how these fungal endophytes led to drought stress tolerance in wheat at diverse temperature regimes. The fungus genera *Fusarium* sp., *Trichoderma* sp., *Penicillium* sp., *Aspergillus* sp., *Alternaria* sp., etc., were examined for plant growth metrics and stress tolerance. It was discovered that the majority of them gave wheat tolerance to drought. Rice, tomatoes, and dune grass—plants associated with endophytic fungi—have utilized a notably smaller amount of water and produced more biomass than plants that are not symbiotic. The phenomenon of drought tolerance could be explained by reduced leaf conductance and a slowing of the transpiration stream, by an increased solute buildup in the tissues of endophyte-infected plants compared to non-infected plants, or by the creation of thicker cuticles (Ripa et al. 2019). Changes in gene expression patterns were found during the characterization of *Trichoderma* sp., which inhabits *Theobroma cacao*. These findings suggest that *Trichoderma* sp. may induce tolerance to abiotic conditions, perhaps including drought (Bailey et al. 2006). The plant–water connection is impacted by heat, salt stress, and drought, which cause complicated plant reactions such as increased osmolyte synthesis. Marquez et al. (Márquez et al. 2007) state that the solute potential and matrix potential are the main factors determining the osmotic potential; symbiotic fungi are likely to contribute to the matrix potential, which is crucial for improving plant drought tolerance and water retention. As a result, plants associated with symbiotic fungi use a lot less water than plants that are not symbiotic (Márquez et al. 2007).

5.3.1.3 Salinity

Another important abiotic stressor that restricts plant development and productivity globally is salinity. The reduction in antioxidant enzymes, such as superoxide dismutases (SOD), catalases (CAT), ascorbate- or thiol-dependent peroxidases (APX), glutathione reductases (GR), dehydro ascorbate reductases (DHAR), and monodehydro ascorbate reductases (MDHAR), is linked to plants' ability to withstand salt stress (Rouhier et al. 2008). They have a direct role in eliminating ROS (SOD, CAT, and APX) or a secondary role in restoring glutathione and ascorbate within the cell. By enhancing antioxidative capacity, fatty acid composition, lipid peroxidase, dehydroascorbate reductase, catalase, and glutathione reductase enzyme accumulation in barley, for example, *Piriformospora indica*, a root fungal endophyte, increased biomass production under salt stress conditions (Baltruschat et al. 2008). In a different case, rice seemed to be able to withstand drought and salinity stress thanks to the endophytic fungus *Exopiala* sp. LHL08 that was isolated from cucumber roots (Murphy et al. 2018). Comparably, *S. vermifera* and *P. indica* demonstrated salt tolerance along with higher growth and yield, and the combination of biological agents had significant benefits (Franken 2012). Under salinity and drought, rice seedlings and the endophytic fungus *Fusarium culmorum* (FcRed1) were discovered to have a symbiotic relationship that produced higher plant height, root length, biomass, and osmolyte accumulation than the non-symbiotic rice seedlings. Uncovering how fungal endophytes enhance plants' defense mechanisms against abiotic stress is still crucial (Egamberdieva et al. 2017; Ma et al. 2019).

5.3.1.4 Temperature

Depending on their biology, the developmental stage, and the cultivar of the host plant, endophytic fungi may alter, regulate, or modify plant physiological, biochemical, and metabolic activities to mitigate the effects of high temperatures. Increases in the quantity and diversity of endophytic populations impacting metabolite synthesis can also be attributed to environmental factors that impact the endophytic community. Numerous scientists have examined a variety of isolated fungal endophytes, their complex relationships that protect the host from high-temperature stress situations, and their specific functions in acclimating to the environment and granting tolerance (Lata et al. 2018). For example, a study discovered that the fungal endophyte *Curvularia protuberata* gives tomatoes and panic grass thermotolerance. *Arabidopsis* sp. is able to withstand heat thanks to the heat shock protein 90 (HSP90) inhibitor produced by *Paraphaeosphaeria quadriseptata*. When a virus infects a plant, as demonstrated by research on *Stenotrophomonas maltophilia* SBP-9's ability to stimulate wheat growth in salt stress conditions, the plant may impart heat tolerance (Márquez et al. 2007; Yadav and Yadav 2017). It was therefore necessary to critically assess a more exciting and puzzling problem involving numerous fungal endophytes in order to manufacture host metabolites in order to potentially utilize them in a variety of environments (Khare et al. 2018; Galani et al. 2022).

5.3.2 Biotic Stress Management

More sustainable solutions are required as a result of pesticide and agrochemical resistance developing in pests and pathogens, ideally pesticide environmental stewardship under biocontrol agents (Baron and Rigobelo 2022). For this purpose, isolates of non-pathogenic fungal endophytes have shown promise in strengthening other plants' resistance against future pathogen assaults (Mane et al. 2018; Grabka et al. 2022). Bioactive chemicals, phenolic compounds, antioxidants, phytoalexins, phytohormone synthesis, ecological occupancy, and other processes are all upregulated as a result of them. It is their duty to protect crops from biotic stress (Singh et al. 2021). Some of the main endophytes that initiate biotic stress resistance and simultaneously improve the growth and yield components of the host plant are *Aspergillus* sp., *Colletotrichum* sp., *Gliocladium* sp., *Fusarium* sp., *Petriella* sp., *Piriformospora* sp., *Trichoderma* sp., *Epicoccum* sp., *Epichloe* sp., etc. According to Adeleke et al. (2022), these important fungal endophytes have the ability to coexist with crops in an environmentally beneficial way, making them suitable for use as biocontrol agents. In addition to controlling herbivorous insects, *Epichloe* sp. provides biotic resistance to its host plant (Laihonen et al. 2022). The filamentous fungus *Piriformospora indica*, which belongs to the Basidiomycota class and colonizes roots, has been discovered to be associated with xerophytic plants. Its host range is vast and includes most angiosperms, such as wheat, rice, maize, barley, tomatoes, and chilies (Arora et al. 2022). By speeding up the synthesis of phytoalexins, antimicrobial substances, the activation of the antioxidant defense enzyme system (glutathione-ascorbate), phytohormone synthesis, etc., resistance to pests and pathogens is achieved (Waller et al. 2005). There are primarily three countermeasures against pests and diseases that have been documented: (a) mycoparasitism; (b) competition for nutrients; and (c) antibiosis (synthesis of antibiotics) (Islam et al. 2023). *P. indica* is a viable candidate for biocontrol agent status because of its ability to effectively control a variety of pests and diseases, including rots, mildews, blights, nematodes, leaf mosaics, and wilts (Waller et al. 2005; Ali et al. 2019). *Trichoderma* sp. is an anamorphic filamentous saprophytic fungus that grows on the roots, twigs, and stems of host plants. It can operate as a biocontrol agent by causing cytotoxic, antifungal, and antibacterial effects. Additionally, they readily adapt to edaphic environments and thrive naturally in a variety of habitats under a range of climatic circumstances. *Trichoderma* sp. is a ubiquitous soil resident that lives in the roots

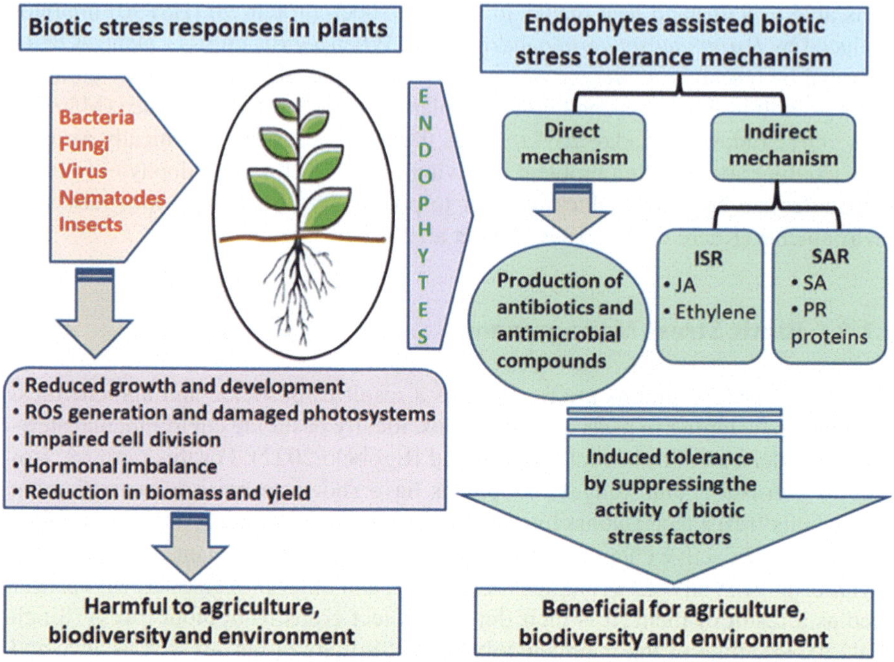

Fig. 5.5 A summary of how plants react to biotic stress (left): infection by pathogens damages photosystems, inhibits cell division, and produces reactive oxygen species (ROS), all of which result in decreased plant growth and development. Endophytic mediated biotic stress tolerance mechanism (right): endophytes activate systemic defense responses involving salicylic acid (SA), oligogalacturonoids (OGAs), and jasmonic acid (JA) signaling pathways, as well as indirectly through the production of lytic enzymes. They also directly trigger defense mechanisms through the production of antimicrobial compounds (Anand et al. 2023).

and root hairs of different plants. It secretes chemicals both inside and outside the cell that have cytotoxic effects on fungal pathogens (Srivastava et al. 2012, 2014). It has been shown that a number of fungal isolates, including *T. harzianum, T. koningii, T. atroviride, T. virens, T. reesei*, and *T. citrinoviride*, have a pathogenic potential against various pathogens. Its effective defense against powerful pests demonstrates its suitability as a candidate for bio-control agents (Singh et al. 2021). A few of its strains are currently being utilized on farms as biofertilizers, which lowers the need for artificial fertilizers. *Trichoderma* sp. also produces siderophores, phytohormones (indole-3-acetic acid (IAA) or auxin analogs, gibberellin, and abscisic acid), and other substances that promote plant growth (e.g., 6-pentyl pyrone, cyclonerodiol, harzianolide, and koninginins), which enrich soil microflora and also help to improve soil fertility (El Enshasy et al. 2020; Keswani et al. 2020) (Fig. 5.5). Recently, a novel strain of *Trichoderma phayaoense* was identified that exhibits a 91% mycelial reduction in its ability to inhibit fungal diseases through parasitic and anti-fungal capabilities (Nuangmek et al. 2021). It has been reported that *T. botryosum, T. caeruloviride, T. lentissimum*, and *T. pseudopyramidale*, which were identified from wild coffee in Africa, use mycoparasitism to provide biological control

(Rodríguez et al. 2021). Recently, a variety of techniques have been used to culture highly efficient *Trichoderma* sp. that yield significant levels of xylanase (Ambehabati et al. 2022). *Myrothecium verrucaria* is another soil-borne ascomycete fungal endophyte that has antagonistic properties and lives in a variety of host plants. Because of its high production of lytic enzymes and phytotoxins, it is regarded as a natural mycoherbicide (Clarke et al. 2007). *Meloidogyne incognita* is one root knot nematode species that can be colonized by the fungus *Purpureocillium lilacinum* (previously *Paecilomyces lilacinus*), which then eliminates the females, cysts, and eggs (Sharma et al. 2020). Comparably, *Alternaria* sp., which causes post-harvest illness in citrus, can be combated by *Aureobasidium pullulans* and *Candida orthopsilosis,* which were identified in teak plants (Sukmawati et al. 2021). Fungal endophytes are known to produce bioactive defense-related compounds and bioactive natural substances (Rustamova et al. 2022). These substances include bioactive metabolites that are antimicrobial, antifungal, and antiviral, such as perfumoid, phomoenamide, dimethyl disulfide, dibenzofuran, methanethiol, ketones, enzymes that degrade b-1,3-glucan, b-1,4-glucan, and b-glucosides; chitin and glucans; cellulose, protease, hydrolyzing enzymes, fumonisin, and beauvericin, among others, which aid in the control of pathogens. Numerous pieces of compelling data have previously been published, demonstrating the potential of these fungal endophytes to mitigate biotic stress in a variety of agriculturally significant crops through host adaptation and broad-spectrum pathogenesis. For various crop species, many of them can be used as safe, environmentally acceptable, and efficient biocontrol agents.

5.4 Conclusion

The conclusion is that a variety of symbiotic fungal endophyte communities are naturally found in crops. These communities offer a number of advantages, including high food production through biotic and abiotic resistance. With considerable potential to combat crop infections that are growing more and more resistant to pesticides, fungal endophytes are a potentially essential supply of metabolites, or bioactive chemicals, for sustained agriculture. Since some of these endophytes cannot be cultivated, using artificial inoculation or bioagents in cultivable crops to harvest their benefits requires a more sophisticated strategy. To further understand their ecological roles and microbial interactions, we must find effective endophytes for climate-resilient and sustainable agriculture. In general, fungal endophytes offer a great deal of promise to address climate change and plant protection in an environmentally responsible way.

References

Adeleke BS, Babalola OO (2021) Pharmacological potential of fungal endophytes associated with medicinal plants: a review. J Fungi 7(2):147

Adeleke BS, Ayilara MS, Akinola SA, Babalola OO (2022) Biocontrol mechanisms of endophytic fungi. Egypt J Biol Pest Control 32(1):1–17

Agency for Toxic Substances and Disease Registry. (2015). Priority list of hazardous substances
Aghai MM, Khan Z, Joseph MR, Stoda AM, Sher AW, Ettl GJ, Doty SL (2019) The effect of microbial endophyte consortia on Pseudotsuga menziesii and Thuja plicata survival, growth, and physiology across edaphic gradients. Front Microbiol 10:1353
Ahemad M, Kibret M (2013) Recent trends in microbial biosorption of heavy metals: a review. Biochem Mol Biol 1(1):19–26
Ahmed E, Holmström SJ (2014) Siderophores in environmental research: roles and applications. Microb Biotechnol 7(3):196–208
Ali A, Bilal S, Khan AL, Mabood F, Al-Harrasi A, Lee IJ (2019) Endophytic Aureobasidium pullulans BSS6 assisted developments in phytoremediation potentials of Cucumis sativus under Cd and Pb stress. J Plant Interact 14(1):303–313
Ali B, Hafeez A, Javed MA, Afridi MS, Abbasi HA, Qayyum A et al (2022) Role of endophytic bacteria in salinity stress amelioration by physiological and molecular mechanisms of defense: a comprehensive review. S Afr J Bot 151:33–46
Ambehabati KK, Hanapi SZ, El Baz AF, Sayyed RZ, Dailin DJ, El Enshasy HA (2022) Isolation and identification studies on potential xylanase producing strain Trichoderma sp. WICC F46 isolated from tropical soil. J Sci Ind Res 79(2):153–159
An H, Liu Y, Zhao X, Huang Q, Yuan S, Yang X, Dong J (2015) Characterization of cadmium-resistant endophytic fungi from Salix variegata Franch. in three gorges reservoir region, China. Microbiol Res 176:29–37
Anand U, Jacobo-Herrera N, Altemimi A, Lakhssassi N (2019) A comprehensive review on medicinal plants as antimicrobial therapeutics: potential avenues of biocompatible drug discovery. Meta 9(11):258
Anand U, Pal T, Yadav N, Singh VK, Tripathi V, Choudhary KK et al (2023) Current scenario and future prospects of endophytic microbes: promising candidates for abiotic and biotic stress management for agricultural and environmental sustainability. Microb Ecol:1–32
Aragón W, Reina-Pinto JJ, Serrano M (2017) The intimate talk between plants and microorganisms at the leaf surface. J Exp Bot 68(19):5339–5350
Arora H, Sharma A, Poczai P, Sharma S, Haron FF, Gafur A, Sayyed RZ (2022) Plant-derived protectants in combating soil-borne fungal infections in tomato and chilli. J Fungi 8(2):213
Asati A, Pichhode M, Nikhil K (2016) Effect of heavy metals on plants: an overview. Int J Appl Innov Eng Manage 5(3):56–66
Babu AG, Shim J, Bang KS, Shea PJ, Oh BT (2014) Trichoderma virens PDR-28: a heavy metal-tolerant and plant growth-promoting fungus for remediation and bioenergy crop production on mine tailing soil. J Environ Manag 132:129–134
Bailey BA, Bae H, Strem MD, Roberts DP, Thomas SE, Crozier J et al (2006) Fungal and plant gene expression during the colonization of cacao seedlings by endophytic isolates of four Trichoderma species. Planta 224:1449–1464. https://doi.org/10.2135/cropsci2000.404923x
Baltruschat H, Fodor J, Harrach BD, Niemczyk E, Barna B, Gullner G et al (2008) Salt tolerance of barley induced by the root endophyte Piriformospora indica is associated with a strong increase in antioxidants. New Phytol 180(2):501–510. https://doi.org/10.1111/j.1469-8137.2008.02583.x
Bamisile BS, Dash CK, Akutse KS, Keppanan R, Afolabi OG, Hussain M et al (2018) Prospects of endophytic fungal entomopathogens as biocontrol and plant growth promoting agents: an insight on how artificial inoculation methods affect endophytic colonization of host plants. Microbiol Res 217:34–50
Bamisile BS, Dash CK, Akutse KS, Qasim M, Ramos Aguila LC, Wang F et al (2019) Endophytic Beauveria bassiana in foliar-treated Citrus limon plants acting as a growth suppressor to three successive generations of Diaphorina citri Kuwayama (Hemiptera: Liviidae). Insects 10(6):176
Baron NC, Rigobelo EC (2022) Endophytic fungi: a tool for plant growth promotion and sustainable agriculture. Mycology 13(1):39–55. https://doi.org/10.1080/21501203.2021.1945699
Bastías DA, Gianoli E, Gundel PE (2021) Fungal endophytes can eliminate the plant growth–defence trade-off. New Phytol 230(6):2105–2113
Bhadra F, Gupta A, Vasundhara M, Reddy MS (2022) Endophytic fungi: a potential source of industrial enzyme producers. 3 Biotech 12(4):86

Bishnoi NR (2005) Fungus-an alternative for bioremediation of heavy metal containing wastewater: a review. J Sci Ind Res 64(2)

Brader G, Compant S, Mitter B, Trognitz F, Sessitsch A (2014) Metabolic potential of endophytic bacteria. Curr Opin Biotechnol 27:30–37

Brader G, Compant S, Vescio K, Mitter B, Trognitz F, Ma LJ, Sessitsch A (2017) Ecology and genomic insights into plant-pathogenic and plant-nonpathogenic endophytes. Annu Rev Phytopathol 55:61–83

Bauer JS, Hauck N, Christof L, Mehnaz S, Gust B (2016) Gross H e systemic investigation of the quorum sensing system of the biocontrol strain Pseudomonas chlororaphis subsp. aurantiaca PB-St2 unveils aurl to Be a biosynthetic origin for 3-oxo-homoserine lactones Plos One, 11.

Callil CT, Junk WJ (2011) Gold-mining near Pocone: environmental, social, and economic impact. In: The Pantanal: ecology, biodiversity and sustainable management of a large neotropical seasonal wetland. Pensoft, Moscow, pp 695–717

Ceccatto AP, Testoni MC, Ignácio AR, Santos-Filho M, Malm O, Díez S (2016) Mercury distribution in organs of fish species and the associated risk in traditional subsistence villagers of the Pantanal wetland. Environ Geochem Health 38:713–722. https://doi.org/10.1007/s10653-015-9754-4

Chasteen TG, Bentley R (2003) Biomethylation of selenium and tellurium: microorganisms and plants. Chem Rev 103(1):1–26

Chatterjee S, Mitra A, Datta S, Veer V (2013) Phytoremediation protocols: an overview. In: Plant-based remediation processes, vol 35. Springer, Heidelberg, pp 1–18

Chaudhry V, Runge P, Sengupta P, Doehlemann G, Parker JE, Kemen E (2021) Shaping the leaf microbiota: plant–microbe–microbe interactions. J Exp Bot 72(1):36–56

Chen Y, Wang HW, Li L, Dai CC (2013) The potential application of the endophyte Phomopsis liquidambari to the ecological remediation of long-term cropping soil. Appl Soil Ecol 67:20–26

Cherif-Silini H, Silini A, Chenari Bouket A, Alenezi FN, Luptakova L, Bouremani N et al (2021) Tailoring next generation plant growth promoting microorganisms as versatile tools beyond soil desalinization: a road map towards field application. Sustain For 13(8):4422

Chibuike GU, Obiora SC (2014) Heavy metal polluted soils: effect on plants and bioremediation methods. Appl Environ Soil Sci 2014:752708

Chouhan S, Agrawal L, Prakash A (2022) Amelioration in traditional farming system by exploring the different plant growth-promoting attributes of endophytes for sustainable agriculture. Arch Microbiol 204(2):151. https://doi.org/10.1007/s00203-021-02637-4

Chowdappa S, Jagannath S, Konappa N, Udayashankar AC, Jogaiah S (2020) Detection and characterization of antibacterial siderophores secreted by endophytic fungi from Cymbidium aloifolium. Biomol Ther 10(10):1412. https://doi.org/10.3390/biom10101412

Clarke TC, Shetty KG, Jayachandran K, Norland MR (2007) Myrothecium verrucaria–a potential biological control agent for the invasive 'old world climbing fern' (Lygodium microphyllum). BioControl 52:399–411

Cozzolino V, De Martino A, Nebbioso A, Di Meo V, Salluzzo A, Piccolo A (2016) Plant tolerance to mercury in a contaminated soil is enhanced by the combined effects of humic matter addition and inoculation with arbuscular mycorrhizal fungi. Environ Sci Pollut Res 23:11312–11322. https://doi.org/10.1007/s11356-016-6337-6

Crandall SG, Gold KM, Jiménez-Gasco MDM, Filgueiras CC, Willett DS (2020) A multi-omics approach to solving problems in plant disease ecology. PLoS One 15(9):e0237975

Dai CC, Tian LS, Zhao YT, Chen Y, Xie H (2010) Degradation of phenanthrene by the endophytic fungus Ceratobasidum stevensii found in Bischofia polycarpa. Biodegradation 21:245–255

Dash HR, Das S (2012) Bioremediation of mercury and the importance of bacterial mer genes. Int Biodeterior Biodegradation 75:207–213. https://doi.org/10.1016/j.ibiod.2012.07.023

Dastogeer KM, Tumpa FH, Sultana A, Akter MA, Chakraborty A (2020) Plant microbiome–an account of the factors that shape community composition and diversity. Curr Plant Biol 23:100161

Deng Z, Cao L (2017) Fungal endophytes and their interactions with plants in phytoremediation: a review. Chemosphere 168:1100–1106

Deng Z, Cao L, Huang H, Jiang X, Wang W, Shi Y, Zhang R (2011) Characterization of Cd-and Pb-resistant fungal endophyte Mucor sp. CBRF59 isolated from rapes (Brassica chinensis) in a metal-contaminated soil. J Hazard Mater 185(2–3):717–724

Deng Z, Zhang R, Shi Y, Tan H, Cao L (2013) Enhancement of phytoremediation of Cd-and Pb-contaminated soils by self-fusion of protoplasts from endophytic fungus Mucor sp. CBRF59. Chemosphere 91(1):41–47

Deng Z, Cao L, Zhang R, Wang W, Shi Y, Tan H et al (2014) Enhanced phytoremediation of multi-metal contaminated soils by interspecific fusion between the protoplasts of endophytic Mucor sp. CBRF59 and Fusarium sp. CBRF14. Soil Biol Biochem 77:31–40

Duan X, Xu F, Qin D, Gao T, Shen W, Zuo S et al (2019) Diversity and bioactivities of fungal endophytes from Distylium chinense, a rare waterlogging tolerant plant endemic to the Three Gorges Reservoir. BMC Microbiol 19:1–14

Dubey A, Malla MA, Kumar A, Dayanandan S, Khan ML (2020) Plants endophytes: unveiling hidden agenda for bioprospecting toward sustainable agriculture. Crit Rev Biotechnol 40(8):1210–1231

Durán N, Marcato PD, Durán M, Yadav A, Gade A, Rai M (2011) Mechanistic aspects in the biogenic synthesis of extracellular metal nanoparticles by peptides, bacteria, fungi, and plants. Appl Microbiol Biotechnol 90:1609–1624

Dwibedi V, Rath SK, Joshi M, Kaur R, Kaur G, Singh D et al (2022) Microbial endophytes: application towards sustainable agriculture and food security. Appl Microbiol Biotechnol 106(17):5359–5384

Egamberdieva D, Wirth SJ, Alqarawi AA, Abd_Allah EF, Hashem A (2017) Phytohormones and beneficial microbes: essential components for plants to balance stress and fitness. Front Microbiol 8:2104. https://doi.org/10.3389/fmicb.2017.02104

El Enshasy HA, Ambehabati KK, El Baz AF, Ramchuran S, Sayyed RZ, Amalin D et al (2020) Trichoderma: biocontrol agents for promoting plant growth and soil health. In: Agriculturally important fungi for sustainable agriculture: Volume 2: Functional annotation for crop protection, pp 239–259. https://doi.org/10.1007/978-3-03048474-3_8

Elhindi KM, El-Hendawy S, Abdel-Salam E, Schmidhalter U, Rehman SU (2016) Foliar application of potassium nitrate affects the growth and photosynthesis in coriander (Coriander sativum L.) plants under salinity. Prog Nutr 18(1):63–73. https://mattiolihealth.com/progress-in-nutrition/

Fadiji AE, Babalola OO (2020a) Elucidating mechanisms of endophytes used in plant protection and other bioactivities with multifunctional prospects. Front Bioeng Biotechnol 8:467

Fadiji AE, Babalola OO (2020b) Exploring the potentialities of beneficial endophytes for improved plant growth. Saudi J Biol Sci 27(12):3622–3633

Falade AO, Adewole KE, Ekundayo TC (2021) Therapeutic potentials of endophytes for healthcare sustainability. Egypt J Basic Appl Sci 8(1):117–135

Fan D, Subramanian S, Smith DL (2020) Plant endophytes promote growth and alleviate salt stress in Arabidopsis thaliana. Sci Rep 10(1):12740

Farder-Gomes CF, Saravanan M, Martinez LC, Plata-Rueda A, Zanuncio JC, Serrao JE (2022) Azadirachtin-based biopesticide affects the respiration and digestion in Anticarsia gemmatalis caterpillars. Toxin Rev 41(2):466–475

Figueiredo G, Gomes M, Covas C, Mendo S, Caetano T (2022) The unexplored wealth of microbial secondary metabolites: the sphingobacteriaceae case study. Microb Ecol 83(2):470–481

Fomina MA, Alexander IJ, Colpaert JV, Gadd GM (2005) Solubilization of toxic metal minerals and metal tolerance of mycorrhizal fungi. Soil Biol Biochem 37(5):851–866. https://doi.org/10.1016/j.soilbio.2004.10.013

Franken P (2012) The plant strengthening root endophyte Piriformospora indica: potential application and the biology behind. Appl Microbiol Biotechnol 96:1455–1464. https://doi.org/10.1007/s00253-012-4506-1

Friesen ML, Porter SS, Stark SC, von Wettberg EJ, Sachs JL, Martinez-Romero E (2011) Microbially mediated plant functional traits. Annu Rev Ecol Evol Syst 42:23–46

Gadd GM (1993) Microbial formation and transformation of organometallic and organometalloid compounds. FEMS Microbiol Rev 11(4):297–316

Galani YJH, Hansen EMØ, Droutsas I, Holmes M, Challinor AJ, Mikkelsen TN, Orfila C (2022) Effects of combined abiotic stresses on nutrient content of European wheat and implications for nutritional security under climate change. Sci Rep 12(1):5700. https://doi.org/10.1038/s41598-022-09538-6
Garg N, Aggarwal N (2011) Effects of interactions between cadmium and lead on growth, nitrogen fixation, phytochelatin, and glutathione production in mycorrhizal Cajanus cajan (L.) Millsp. J Plant Growth Regul 30:286–300
Gavito ME, Curtis PS, Mikkelsen TN, Jakobsen I (2000) Atmospheric CO2 and mycorrhiza effects on biomass allocation and nutrient uptake of nodulated pea (Pisum sativum L.) plants. J Exp Botany 51(352):1931–1938. https://doi.org/10.1093/jexbot/51.352.1931
Gazis R, Rehner S, Chaverri P (2011) Species delimitation in fungal endophyte diversity studies and its implications in ecological and biogeographic inferences. Mol Ecol 20(14):3001–3013
Gerhardt KE, Huang XD, Glick BR, Greenberg BM (2009) Phytoremediation and rhizoremediation of organic soil contaminants: potential and challenges. Plant Sci 176(1):20–30
Gimenez E, Salinas M, Manzano-Agugliaro F (2018) Worldwide research on plant defense against biotic stresses as improvement for sustainable agriculture. Sustain For 10(2):391
Glandorf DC, Verheggen P, Jansen T, Jorritsma JW, Smit E, Leeflang P et al (2001) Effect of genetically modified Pseudomonas putida WCS358r on the fungal rhizosphere microflora of field-grown wheat. Appl Environ Microbiol 67(8):3371–3378
Glick BR (2003) Phytoremediation: synergistic use of plants and bacteria to clean up the environment. Biotechnol Adv 21(5):383–393
Glick BR (2010) Using soil bacteria to facilitate phytoremediation. Biotechnol Adv 28(3):367–374
Glick BR, Gamalero E (2021) Recent developments in the study of plant microbiomes. Microorganisms 9(7):1533
Gómez OC, Luiz JHH (2018) Endophytic fungi isolated from medicinal plants: future prospects of bioactive natural products from Tabebuia/Handroanthus endophytes. Appl Microbiol Biotechnol 102(21):9105–9119
Gontia-Mishra I, Sapre S, Sharma A, Tiwari S (2016) Alleviation of mercury toxicity in wheat by the interaction of mercury-tolerant plant growth-promoting rhizobacteria. J Plant Growth Regul 35:1000–1012. https://doi.org/10.1007/s00344-016-9598-x
Grabka R, d'Entremont TW, Adams SJ, Walker AK, Tanney JB, Abbasi PA, Ali S (2022) Fungal endophytes and their role in agricultural plant protection against pests and pathogens. Plan Theory 11(3):384. https://doi.org/10.3390/plants11030384
Guo B, Wang Y, Sun X, Tang K (2008) Bioactive natural products from endophytes: a review. Appl Biochem Microbiol 44:136–142
Gupta S, Chaturvedi P, Kulkarni MG, Van Staden J (2020) A critical review on exploiting the pharmaceutical potential of plant endophytic fungi. Biotechnol Adv 39:107462. https://doi.org/10.1016/j.biotechadv.2019.107462
Hamilton CE, Bauerle TL (2012) A new currency for mutualism? Fungal endophytes alter antioxidant activity in hosts responding to drought. Fungal Divers 54:39–49
Hamilton CE, Gundel PE, Helander M, Saikkonen K (2012) Endophytic mediation of reactive oxygen species and antioxidant activity in plants: a review. Fungal Divers 54:1–10
Harman G, Khadka R, Doni F, Uphoff N (2021) Benefits to plant health and productivity from enhancing plant microbial symbionts. Front Plant Sci 11:610065
Hassan SED (2017) Plant growth-promoting activities for bacterial and fungal endophytes isolated from medicinal plant of Teucrium polium L. J Adv Res 8(6):687–695. https://doi.org/10.1016/j.jare.2017.09.001
He W, Megharaj M, Wu CY, Subash Chandrabose SR, Dai CC (2020) Endophyte-assisted phytoremediation: mechanisms and current application strategies for soil mixed pollutants. Crit Rev Biotechnol 40(1):31–45
Hug JJ, Bader CD, Remškar M, Cirnski K, Müller R (2018) Concepts and methods to access novel antibiotics from actinomycetes. Antibiotics 7(2):44
Hyde KD, Soytong K (2008) The fungal endophyte dilemma. Fungal Divers 33(163):e173

Idbella M, Zotti M, Cesarano G, Fechtali T, Mazzoleni S, Bonanomi G (2019) Fungal endophytes affect plant response to leaf litter with contrasting chemical traits. Community Ecol 20:205–213. https://doi.org/10.1556/168.2019.20.2.10

Islam S, Banerjee A, Shah MH, Ahammad F, Rahman SS, Hossain A (2023) Beneficial fungi as a biocontrol agent against fungi. In: Biofungicides: eco-safety and future trends. CRC Press, pp 65–93

Johnson L (2008) Iron and siderophores in fungal–host interactions. Mycol Res 112(2):170–183. https://doi.org/10.1016/j.mycres.2007.11.012

Jones P, Garcia BJ, Furches A, Tuskan GA, Jacobson D (2019) Plant host-associated mechanisms for microbial selection. Front Plant Sci 10:862

Kaul S, Sharma T, Dhar KM (2016) "Omics" tools for better understanding the plant–endophyte interactions. Front Plant Sci 7:955

Keswani C, Singh SP, Cueto L, García-Estrada C, Mezaache-Aichour S, Glare TR et al (2020) Auxins of microbial origin and their use in agriculture. Appl Microbiol Biotechnol 104:8549–8565. https://doi.org/10.1007/s00253-020-10890-8

Keswani C, Singh SP, García-Estrada C, Mezaache-Aichour S, Glare TR, Borriss R et al (2022) Biosynthesis and beneficial effects of microbial gibberellins on crops for sustainable agriculture. J Appl Microbiol 132(3):1597–1615. https://doi.org/10.1111/jam.15348

Khan Z, Doty S (2011) Endophyte-assisted phytoremediation. Curr Topics Plant Biol 11:97–105

Khan AL, Waqas M, Hussain J, Al-Harrasi A, Hamayun M, Lee IJ (2015) Phytohormones enabled endophytic fungal symbiosis improve aluminum phytoextraction in tolerant Solanum lycopersicum: an example of Penicillium janthinellum LK5 and comparison with exogenous GA3. J Hazard Mater 295:70–78

Khan AR, Ullah I, Waqas M, Park GS, Khan AL, Hong SJ et al (2017a) Host plant growth promotion and cadmium detoxification in Solanum nigrum, mediated by endophytic fungi. Ecotoxicol Environ Saf 136:180–188

Khan AR, Waqas M, Ullah I, Khan AL, Khan MA, Lee IJ, Shin JH (2017b) Culturable endophytic fungal diversity in the cadmium hyperaccumulator Solanum nigrum L. and their role in enhancing phytoremediation. Environ Exp Bot 135:126–135. https://doi.org/10.1016/j.envexpbot.2016.03.005

Khanam Z, Gupta S, Verma A (2020) Endophytic fungi-based biosensors for environmental contaminants-A perspective. S Afr J Bot 134:401–406. https://doi.org/10.1016/j.sajb.2020.08.007

Khare E, Mishra J, Arora NK (2018) Multifaceted interactions between endophytes and plant: developments and prospects. Front Microbiol 9:2732. https://doi.org/10.3389/fmicb.2018.02732

Kidd P, Barceló J, Bernal MP, Navari-Izzo F, Poschenrieder C, Shilev S et al (2009) Trace element behaviour at the root–soil interface: implications in phytoremediation. Environ Exp Bot 67(1):243–259

Krishnamurthy YL, Naik BS (2017) Endophytic fungi bioremediation. In: Endophytes: crop productivity and protection: Volume 2. Springer, Cham, pp 47–60

Kristin A, Miranda H (2013) The root microbiota—a fingerprint in the soil? Plant Soil 370:671–686

Krupa P, Kozdrój J (2007) Ectomycorrhizal fungi and associated bacteria provide protection against heavy metals in inoculated pine (Pinus sylvestris L.) seedlings. Water Air Soil Pollut 182:83–90

Kumar A, Verma JP (2018) Does plant—microbe interaction confers stress tolerance in plants: a review. Microbiol Res 207:41–52

Kumar S, Thakur N, Singh AK, Gudade BA, Ghimire D, Das S (2022) Microbes-assisted phytoremediation of contaminated environment: global status, progress, challenges, and future prospects. In: Phytoremediation technology for the removal of heavy metals and other contaminants from soil and water. Elsevier, pp 555–570

Kurniati E, Arfarita N, Imai T, Higuchi T, Kanno A, Yamamoto K, Sekine M (2014) Potential bioremediation of mercury-contaminated substrate using filamentous fungi isolated from forest soil. J Environ Sci 26(6):1223–1231. https://doi.org/10.1016/S10010742(13)60592-6

Kusajima M, Shima S, Fujita M, Minamisawa K, Che FS, Yamakawa H, Nakashita H (2018) Involvement of ethylene signaling in Azospirillum sp. B510-induced disease resistance in rice. Biosci Biotechnol Biochem 82(9):1522–1526

Lacava PT, Azevedo JL (2013) Endophytic bacteria: a biotechnological potential in agrobiology system. In: Bacteria in agrobiology: crop productivity. Springer, Berlin Heidelberg, pp 1–44

Lacercat-Didier L, Berthelot C, Foulon J, Errard A, Martino E, Chalot M, Blaudez D (2016) New mutualistic fungal endophytes isolated from poplar roots display high metal tolerance. Mycorrhiza 26:657–671

Laihonen M, Saikkonen K, Helander M, Vázquez de Aldana BR, Zabalgogeazcoa I, Fuchs B (2022) Epichloë endophyte-promoted seed pathogen increases host grass resistance against insect herbivory. Front Microbiol 12:786619. https://doi.org/10.3389/fmicb.2021.786619

Lata R, Chowdhury S, Gond SK, White JF Jr (2018) Induction of abiotic stress tolerance in plants by endophytic microbes. Lett Appl Microbiol 66(4):268–276. https://doi.org/10.1111/lam.12855

Lázaro WL, de Oliveira RF, dos Santos-Filho M, da Silva CJ, Malm O, Ignácio ÁR, Díez S (2015) Non-lethal sampling for mercury evaluation in crocodilians. Chemosphere 138:25–32. https://doi.org/10.1016/j.chemosphere.2015.05.007

Leady BS, Gottgens JF (2001) Mercury accumulation in sediment cores and along food chains in two regions of the Brazilian Pantanal. Wetl Ecol Manag 9:349–361

Leitão AL (2009) Potential of Penicillium species in the bioremediation field. Int J Environ Res Public Health 6(4):1393–1417

Li J, Xie ZM, Xu JM, Sun YF (2006) Risk assessment for safety of soils and vegetables around a lead/zinc mine. Environ Geochem Health 28:37–44

Li HY, Wei DQ, Shen M, Zhou ZP (2012) Endophytes and their role in phytoremediation. Fungal Divers 54:11–18

Li X, Li W, Chu L, White JF Jr, Xiong Z, Li H (2016a) Diversity and heavy metal tolerance of endophytic fungi from Dysphania ambrosioides, a hyperaccumulator from Pb–Zn contaminated soils. J Plant Interact 11(1):186–192

Li N, Alfiky A, Vaughan MM, Kang S (2016b) Stop and smell the fungi: fungal volatile metabolites are overlooked signals involved in fungal interaction with plants. Fungal Biol Rev 30(3):134–144. https://doi.org/10.1016/j.fbr.2016.06.004

Liu C, Lin H, Li B, Dong Y, Qiu Y (2022) Screening endophyte with capability to improve phytoremediation efficiency from hyperaccumulators: a novel and efficient microfluidic method. Chemosphere 286:131723

Lugtenberg BJ, Caradus JR, Johnson LJ (2016) Fungal endophytes for sustainable crop production. FEMS Microbiol Ecol 92(12):fiw194

Lundberg DS, Lebeis SL, Paredes SH, Yourstone S, Gehring J, Malfatti S et al (2012) Defining the core Arabidopsis thaliana root microbiome. Nature 488(7409):86–90

Luo SL, Chen L, Chen JL, Xiao X, Xu TY, Wan Y et al (2011) Analysis and characterization of cultivable heavy metal-resistant bacterial endophytes isolated from Cd-hyperaccumulator Solanum nigrum L. and their potential use for phytoremediation. Chemosphere 85(7):1130–1138

Lyu D, Msimbira LA, Nazari M, Antar M, Pagé A, Shah A et al (2021) The coevolution of plants and microbes underpins sustainable agriculture. Microorganisms 9(5):1036

Lopez DC, Zhu-salzman K, Ek-ramos MJ, Sword GA, (2014) The entomopathogenic fungal endophytes Purpureocillium lilacinum (formerly Paecilomyces lilacinus) and Beauveria bassiana negatively affect cotton aphid reproduction under both greenhouse and field conditions 9. https://doi.org/10.1371/journal.pone.0103891

Liang X, Gadd GM (2017) Metal and metalloid biorecovery using fungi. Microbial bio technology 10:1199–1205.

Ma Y, Vosátka M, Freitas H (2019) Beneficial microbes alleviate climatic stresses in plants. Front Plant Sci 10:595. https://doi.org/10.3389/fpls.2019.00595

Malfanova N, Lugtenberg BJ, Berg G (2013) Bacterial endophytes: who and where, and what are they doing there? Mol Microb Ecol Rhizosphere 1:391–403

Malinowski DP, Belesky DP (1999) Tall fescue aluminum tolerance is affected by Neotyphodium coenophialum endophyte. J Plant Nutr 22(8):1335–1349
Malinowski DP, Zuo H, Belesky DP, Alloush GA (2004) Evidence for copper binding by extracellular root exudates of tall fescue but not perennial ryegrass infected with Neotyphodium spp. endophytes. Plant Soil 267:1–12
Malmierca MG, Cardoza RE, Alexander NJ, McCormick SP, Hermosa R, Monte E, Gutiérrez S (2012) Involvement of Trichoderma trichothecenes in the biocontrol activity and induction of plant defense-related genes. Appl Environ Microbiol 78(14):4856–4868
Mane RS, Paarakh PM, Vedamurthy AB (2018) Brief review on fungal endophytes. Int J Second Metab 5(4):288–303
Manganyi MC, Ateba CN (2020) Untapped potentials of endophytic fungi: a review of novel bioactive compounds with biological applications. Microorganisms 8(12):1934
Márquez LM, Redman RS, Rodriguez RJ, Roossinck MJ (2007) A virus in a fungus in a plant: three-way symbiosis required for thermal tolerance. Science 315(5811):513–515
Massoni J, Bortfeld-Miller M, Jardillier L, Salazar G, Sunagawa S, Vorholt JA (2020) Consistent host and organ occupancy of phyllosphere bacteria in a community of wild herbaceous plant species. ISME J 14(1):245–258
Mehta P, Sharma R, Putatunda C, Walia A (2019) Endophytic fungi: role in phosphate solubilization. In: Advances in endophytic fungal research: present status and future challenges. Springer, pp 183–209. https://doi.org/10.1007/978-3-030-03589-1_9
Mengistu AA (2020) Endophytes: colonization, behaviour, and their role in defense mechanism. Int J Microbiol 2020:6927219
Mercado-Blanco J (2014) Life of microbes inside the plant. In: Principles of plant-microbe interactions: microbes for sustainable agriculture. Springer International Publishing, Cham, pp 25–32
Misiou O, Koutsoumanis K (2022) Climate change and its implications for food safety and spoilage. Trends Food Sci Technol 126:142–152
Monnet F, Vaillant N, Hitmi A, Coudret A, Sallanon H (2001) Endophytic Neotyphodium lolii induced tolerance to Zn stress in Lolium perenne. Physiol Plant 113(4):557–563
Mowafy AM, Agha MS, Haroun SA, Abbas M, Elbalkini M (2022) Insights in nodule-inhabiting plant growth promoting bacteria and their ability to stimulate Vicia faba growth. Egypt J Basic Appl Sci 9(1):51–64. https://doi.org/10.1080/2314808X.2021.2019418
Murphy BR, Doohan FM, Hodkinson TR (2018) From concept to commerce: developing a successful fungal endophyte inoculant for agricultural crops. J Fungi 4(1):24. https://doi.org/10.3390/jof4010024
Nagajyoti PC, Lee KD, Sreekanth TVM (2010) Heavy metals, occurrence and toxicity for plants: a review. Environ Chem Lett 8:199–216
Nandy S, Das T, Tudu CK, Pandey DK, Dey A, Ray P (2020) Fungal endophytes: futuristic tool in recent research area of phytoremediation. S Afr J Bot 134:285–295. Available from: https://doi.org/10.1016/j.sajb.2020.02.015
Naranjo-Briceño L, Pernía B, Guerra M, Demey JR, De Sisto Á, Inojosa Y et al (2013) Potential role of oxidative exoenzymes of the extremophilic fungus P estalotiopsis palmarum BM-04 in biotransformation of extra-heavy crude oil. Microb Biotechnol 6(6):720–730
Narayanan Z, Glick BR (2022) Secondary metabolites produced by plant growth-promoting bacterial endophytes. Microorganisms 10(10):2008
Nascimento FX, Rossi MJ, Soares CR, McConkey BJ, Glick BR (2014) New insights into 1-aminocyclopropane-1-carboxylate (ACC) deaminase phylogeny, evolution and ecological significance. PLoS One 9(6):e99168. https://doi.org/10.1371/journal.pone.0099168
Nelson PF, Morrison AL, Malfroy HJ, Cope M, Lee S, Hibberd ML et al (2012) Atmospheric mercury emissions in Australia from anthropogenic, natural and recycled sources. Atmos Environ 62:291–302. https://doi.org/10.1016/j.atmosenv.2012.07.067
Nie P, Li X, Wang S, Guo J, Zhao H, Niu D (2017) Induced systemic resistance against Botrytis cinerea by Bacillus cereus AR156 through a JA/ET-and NPR1-dependent signaling pathway and activates PAMP-triggered immunity in Arabidopsis. Front Plant Sci 8:238

Nuangmek W, Aiduang W, Kumla J, Lumyong S, Suwannarach N (2021) Evaluation of a newly identified endophytic fungus, Trichoderma phayaoense for plant growth promotion and biological control of gummy stem blight and wilt of muskmelon. Front Microbiol 12:634772. https://doi.org/10.3389/fmicb.2021.634772

Nwachukwu BC, Babalola OO (2021) Perspectives for sustainable agriculture from the microbiome in plant rhizosphere. Plant Biotechnol Reports 15(3):259–278

Ningxiao LI, Alfiky A, Vaughan MM, Kang S (2016) Stop and smell the fungi: Fungal volatile metabolites are overlooked signals involved in fungal interaction with plants. Fungal Biol Rev 30:134–144 https://doi.org/10.1016/j.fbr.2016.06.004

Ojuederie OB, Babalola OO (2017) Microbial and plant-assisted bioremediation of heavy metal polluted environments: a review. Int J Environ Res Public Health 14(12):1504

Olanrewaju OS, Ayangbenro AS, Glick BR, Babalola OO (2019) Plant health: feedback effect of root exudates-rhizobiome interactions. Appl Microbiol Biotechnol 103:1155–1166

Ortega HE, Torres-Mendoza D, Cubilla-Rios L (2020) Patents on endophytic fungi for agriculture and bio-and phytoremediation applications. Microorganisms 8(8):1237

Oukala N, Aissat K, Pastor V (2021) Bacterial endophytes: the hidden actor in plant immune responses against biotic stress. Plan Theory 10(5):1012

Patle PN, Navnage NP, Ramteke PR (2018) Endophytes in plant system: roles in growth promotion, mechanism and their potentiality in achieving agriculture sustainability. Int J Chem Stud 6(1):270–274

Pierart A, Shahid M, Séjalon-Delmas N, Dumat C (2015) Antimony bioavailability: knowledge and research perspectives for sustainable agricultures. J Hazard Mater 289:219–234

Pietro-Souza W, de Campos Pereira F, Mello IS, Stachack FFF, Terezo AJ, da Cunha CN et al (2020) Mercury resistance and bioremediation mediated by endophytic fungi. Chemosphere 240:124874

Pirrone N, Cinnirella S, Feng X, Finkelman RB, Friedli HR, Leaner J et al (2010) Global mercury emissions to the atmosphere from anthropogenic and natural sources. Atmos Chem Phys 10(13):5951–5964. https://doi.org/10.5194/acp-10-5951-2010

Plett JM, Martin FM (2018) Know your enemy, embrace your friend: using omics to understand how plants respond differently to pathogenic and mutualistic microorganisms. Plant J 93(4):729–746

Radziemska M, Gusiatin ZM, Cydzik-Kwiatkowska A, Cerdà A, Pecina V, Bęś A et al (2021) Insight into metal immobilization and microbial community structure in soil from a steel disposal dump phytostabilized with composted, pyrolyzed or gasified wastes. Chemosphere 272:129576. https://doi.org/10.1016/j.chemosphere.2021.129576

Raghav D, Jyoti A, Siddiqui AJ, Saxena J (2022) Plant-associated endophytic fungi as potential bio-factories for extracellular enzymes: progress, challenges and strain improvement with precision approaches. J Appl Microbiol 133(2):287–310. https://doi.org/10.1111/jam.15574

Rahman Z, Singh VP (2019) The relative impact of toxic heavy metals (THMs)(arsenic (As), cadmium (Cd), chromium (Cr)(VI), mercury (Hg), and lead (Pb)) on the total environment: an overview. Environ Monit Assess 191:1–21

Raimi A, Adeleke R (2021) Bioprospecting of endophytic microorganisms for bioactive compounds of therapeutic importance. Arch Microbiol 203(5):1917–1942

Rajkumar M, Ae N, Freitas H (2009) Endophytic bacteria and their potential to enhance heavy metal phytoextraction. Chemosphere 77(2):153–160

Rana KL, Kour D, Kaur T, Devi R, Yadav AN, Yadav N et al (2020) Endophytic microbes: biodiversity, plant growth-promoting mechanisms and potential applications for agricultural sustainability. Antonie Van Leeuwenhoek 113:1075–1107. https://doi.org/10.1007/s10482-020-01429-y

Redfern LK, Gunsch CK (2015) Endophytic phytoaugmentation: treating wastewater and runoff through augmented phytoremediation. Indus Biotechnol 12(2):83–90. https://doi.org/10.3390/microorganisms9050920

Rehman N, Khan MR, Abbas Z, Rafique RS, Zaynab M, Qasim M et al (2020) Functional characterization of mitogen-activated protein kinase kinase (MAPKK) gene in halophytic Salicornia europaea against salt stress. Environ Exp Bot 171:103934

Ren A, Gao Y, Zhang L, Xie F (2006) Effects of cadmium on growth parameters of endophyte-infected endophyte-free ryegrass. J Plant Nutr Soil Sci 169(6):857–860
Ren A, Li C, Gao Y (2011) Endophytic fungus improves growth and metal uptake of Lolium arundinaceum Darbyshire ex. Schreb. Int J Phytoremed 13(3):233–243
Ripa FA, Cao WD, Tong S, Sun JG (2019) Assessment of plant growth promoting and abiotic stress tolerance properties of wheat endophytic fungi. Biomed Res Int 2019. https://doi.org/10.1155/2019/6105865
Robert-Seilaniantz A, Grant M, Jones JD (2011) Hormone crosstalk in plant disease and defense: more than just jasmonate-salicylate antagonism. Annu Rev Phytopathol 49:317–343
Rodriguez PA, Rothballer M, Chowdhury SP, Nussbaumer T, Gutjahr C, Falter-Braun P (2019) Systems biology of plant-microbiome interactions. Mol Plant 12(6):804–821
Rojas V, Rivas L, Cárdenas C, Guzmán F (2020) Cyanobacteria and eukaryotic microalgae as emerging sources of antibacterial peptides. Molecules 25(24):5804
Rouhier N, San Koh C, Gelhaye E, Corbier C, Favier F, Didierjean C, Jacquot JP (2008) Redox based anti-oxidant systems in plants: biochemical and structural analyses. Biochimica et Biophysica Acta (BBA)-General Subjects 1780(11):1249–1260
Rozpądek P, Domka A, Ważny R, Nosek M, Jędrzejczyk R, Tokarz K, Turnau K (2018) How does the endophytic fungus Mucor sp. improve Arabidopsis arenosa vegetation in the degraded environment of a mine dump? Environ Exp Bot 147:31–42. https://doi.org/10.1016/j.envexpbot.2017.11.009
Rustamova NIGORA, Litao N, Bozorov KHURSHED, Sayyed RIYAZ, Aisa HA, Yili ABULIMITI (2022) Plant-associated endophytic fungi: a source of structurally diverse and bioactive natural products. Plant Cell Biotechnol Mol Biol 23:1–19. https://www.ikppress.org/index.php/PCBMB /article/view/7454
Ryan RP, Ryan D, Dowling DN (2007) Plant protection by the recombinant, root-colonizing Pseudomonas fluorescens F113rifPCB strain expressing arsenic resistance: improving rhizoremediation. Lett Appl Microbiol 45(6):668–674
Redfern LK, Gunsch CK (2016) Endophytic phytoaugmentation: treating wastewater and runoff through augmented phytoremediation. Industrial Biotechnology 12:83–90.
Redman RS, Kim YO, Cho S, Mercer M, Rienstra M, Manglona R et al. (2021) A Symbiotic Approach to Generating Stress Tolerant Crops. Microorganisms Microorg. 9:920. https://doi.org/10.3390/microorganisms9050920
Rehman S, Akhtar M, Abdullah SNA (2016) Plant, Soil and Microbes. Cham:Springer, doi: 978-3-319-29573-2
Rodríguez MCH, Evans HC, de Abreu, LM, Macedo M, Miraine K, Bekele KB, et al (2021) New species and records of Trichoderma isolated as mycoparasites and endophytes from cultivated and wild coffee in Africa. Sci Rep 11:5671. https://doi.org/10.1038/s41598-021-84111-1
Sahoo S, Sarangi S, Kerry RG (2017) Bioprospecting of endophytes for agricultural and environmental sustainability. In: Microbial biotechnology: Volume 1. Applications in agriculture and environment. Springer, pp 429–458
Salazar B, Ortiz A, Keswani C, Minkina T, Mandzhieva S, Pratap Singh S et al (2023) Bacillus spp. as bio-factories for antifungal secondary metabolites: innovation beyond whole organism formulations. Microb Ecol 86(1):1–24. https://doi.org/10.1007/s00248-022-02044-2
Santoyo G, Moreno-Hagelsieb G, del Carmen Orozco-Mosqueda M, Glick BR (2016) Plant growth-promoting bacterial endophytes. Microbiol Res 183:92–99
Savary S, Willocquet L, Pethybridge SJ, Esker P, McRoberts N, Nelson A (2019) The global burden of pathogens and pests on major food crops. Nat Ecol Evol 3(3):430–439
Schäffner A, Messner B, Langebartels C, Sandermann H (2002) Genes and enzymes for in-planta phytoremediation of air, water and soil. Acta Biotechnol 22(1–2):141–151
Schlechter RO, Miebach M, Remus-Emsermann MN (2019) Driving factors of epiphytic bacterial communities: a review. J Adv Res 19:57–65
Sessitsch A, Reiter B, Berg G (2004) Endophytic bacterial communities of field-grown potato plants and their plant-growth-promoting and antagonistic abilities. Can J Microbiol 50(4):239–249

Sharma RK, Agrawal M, Marshall FM (2009) Heavy metals in vegetables collected from production and market sites of a tropical urban area of India. Food Chem Toxicol 47(3):583–591

Sharma A, Gupta A, Dalela M, Sharma S, Sayyed RZ, Enshasy HAE, Elsayed EA (2020) Linking organic metabolites as produced by Purpureocillium lilacinum 6029 cultured on Karanja deoiled cake medium for the sustainable management of root-knot nematodes. Sustain For 12(19):8276. https://doi.org/10.3390/su12198276

Sharma H, Rai AK, Dahiya D, Chettri R, Nigam PS (2021) Exploring endophytes for in vitro synthesis of bioactive compounds similar to metabolites produced in vivo by host plants. AIMS Microbiol 7(2):175

Shen M, Liu L, Li DW, Zhou WN, Zhou ZP, Zhang CF et al (2013) The effect of endophytic Peyronellaea from heavy metal-contaminated and uncontaminated sites on maize growth, heavy metal absorption and accumulation. Fungal Ecol 6(6):539–545

Shi Y, Xie H, Cao L, Zhang R, Xu Z, Wang Z, Deng Z (2017) Effects of Cd-and Pb-resistant endophytic fungi on growth and phytoextraction of Brassica napus in metal-contaminated soils. Environ Sci Pollut Res 24:417–426

Shin MN, Shim J, You Y, Myung H, Bang KS, Cho M et al (2012) Characterization of lead resistant endophytic Bacillus sp. MN3-4 and its potential for promoting lead accumulation in metal hyperaccumulator Alnus firma. J Hazard Mater 199:314–320

Shinozaki K, Yamaguchi-Shinozaki K (2007) Gene networks involved in drought stress response and tolerance. J Exp Bot 58(2):221–227. https://doi.org/10.1093/jxb/erl164

Sim CSF, Tan WS, Ting ASY (2016) Endophytes from Phragmites for metal removal: evaluating their metal tolerance, adaptive tolerance behaviour and biosorption efficacy. Desalin Water Treat 57(15):6959–6966. https://doi.org/10.1080/19443994.2015.1013507

Singh R, Gautam N, Mishra A, Gupta R (2011) Heavy metals and living systems: an overview. Indian J Pharmacol 43(3):246

Singh RP, Shelke GM, Kumar A, Jha PN (2015a) Biochemistry and genetics of ACC deaminase: a weapon to "stress ethylene" produced in plants. Front Microbiol 6:937. https://doi.org/10.3389/fmicb.2015.00937

Singh VP, Singh S, Kumar J, Prasad SM (2015b) Investigating the roles of ascorbate-glutathione cycle and thiol metabolism in arsenate tolerance in ridged Luffa seedlings. Protoplasma 252:1217–1229

Singh N, Singh A, Dahiya P (2021) Plant growth-promoting endophytic fungi from different habitats and their potential applications in agriculture. In: Recent trends in mycological research: Volume 1: Agricultural and medical perspective, pp 69–87. https://doi.org/10.1007/978-3-030-60659-6_3

Singh N, Mishra SK, Kumar PR, Kumar N, Kumar D (2022) Role of endophyte metabolites in plant protection and other metabolic activities. In: Bacterial endophytes for sustainable agriculture and environmental management. Springer Singapore, Singapore, pp 213–233

Soleimani M, Afyuni M, Hajabbasi MA, Nourbakhsh F, Sabzalian MR, Christensen JH (2010a) Phytoremediation of an aged petroleum contaminated soil using endophyte infected and non-infected grasses. Chemosphere 81(9):1084–1090

Soleimani M, Hajabbasi MA, Afyuni M, Mirlohi A, Borggaard OK, Holm PE (2010b) Effect of endophytic fungi on cadmium tolerance and bioaccumulation by Festuca arundinacea and Festuca pratensis. Int J Phytoremediation 12(6):535–549

Sukmawati D, Family N, Hidayat I, Sayyed RZ, Elsayed EA, Dailin DJ et al (2021) Biocontrol activity of Aureubasidium pullulans and Candida orthopsilosis isolated from Tectona grandis L. phylloplane against Aspergillus sp. in post-harvested citrus fruit. Sustain For 13(13):7479. https://doi.org/10.3390/su13137479

Srinivas C, Shubha S, Narasimhamurthy K, Arakere U, Sudisha J (2020) Detection and Characterization of Antibacterial Siderophores Secreted by Endophytic Fungi from Cymbidium aloifolium. Biomolecules 10:1412.https://doi.org/10.3390/biom10101412

Srivastava A, Srivastava N, Sayyed RZ (2012) In vitro bio-control activity of Trichoderma species against phytopathogenic A. brassicae. Int J Biotech BioSci 2:6–9.

Srivastava A, Srivastava N, Sayyed RZ (2014) Bio-control Potential of Trichoderma Species Against Alternaria brassicae. Recent advances in biofertilizers & biofungicides (PGPR) for sustainableAgriculture. London, UK: Cambridge Scholars Press, 393–400.

Tang J, Xu L, Chen X, Hu S (2009) Interaction between C4 barnyard grass and C3 upland rice under elevated CO2: impact of mycorrhizae. Acta Oecol 35(2):227–235. https://doi.org/10.1016/j.actao.2008.10.005

Techaoei S, Jirayuthcharoenkul C, Jarmkom K, Dumrongphuttidecha T, Khobjai W (2020) Chemical evaluation and antibacterial activity of novel bioactive compounds from endophytic fungi in Nelumbo nucifera. Saudi J Biol Sci 27(11):2883–2889

Tong J, Miaowen C, Juhui J, Jinxian L, Baofeng C (2017) Endophytic fungi and soil microbial community characteristics over different years of phytoremediation in a copper tailings dam of Shanxi, China. Sci Total Environ 574:881–888

Trebicki P (2020) Climate change and plant virus epidemiology. Virus Res 286:198059

Urík M, Hlodák M, Mikušová P, Matúš P (2014) Potential of microscopic fungi isolated from mercury contaminated soils to accumulate and volatilize mercury (II). Water Air Soil Pollut 225:1–11. https://doi.org/10.1007/s11270-014-2219-z

von Tümpling W, Zeilhofer P, Ammer U, Einax J, Wilken RD (1995) Estimation of mercury content in tailings of the gold mine area of Poconé, Mato Grosso, Brazil. Environ Sci Pollut Res 2:225–228

Van Hecke MM, Treonis AM, Kaufman JR (2005) How does the fungal endophyte Neotyphodium coenophialum affect tall fescue (Festuca arundinacea) rhizodeposition and soil microorganisms? Plant Soil 275:101–109

Vandegrift R, Roy BA, Pfeifer-Meister L, Johnson BR, Bridgham SD (2015) The herbaceous landlord: integrating the effects of symbiont consortia within a single host. PeerJ 3:e1379

Vandenkoornhuyse P, Quaiser A, Duhamel M, Le Van A, Dufresne A (2015) The importance of the microbiome of the plant holobiont. New Phytol 206(4):1196–1206. https://doi.org/10.1111/nph.13312

Vishnupradeep R, Bruno LB, Taj Z, Karthik C, Challabathula D, Kumar A et al (2022) Plant growth promoting bacteria improve growth and phytostabilization potential of Zea mays under chromium and drought stress by altering photosynthetic and antioxidant responses. Environ Technol Innov 25:102154

Waller F, Achatz B, Baltruschat H, Fodor J, Becker K, Fischer M et al (2005) The endophytic fungus Piriformospora indica reprograms barley to salt-stress tolerance, disease resistance, and higher yield. Proc Natl Acad Sci 102(38):13386–13391

Wehner G, Kopahnke D, Richter K, Kecke S, Schikora A, Ordon F (2019) Priming is a suitable strategy to enhance resistance towards leaf rust in barley. Phytobiomes J 3(1):46–51

Wenzel WW (2009) Rhizosphere processes and management in plant-assisted bioremediation (phytoremediation) of soils. Plant Soil 321:385–408

Weyens N, Van Der Lelie D, Artois T, Smeets K, Taghavi S, Newman L et al (2009a) Bioaugmentation with engineered endophytic bacteria improves contaminant fate in phytoremediation. Environ Sci Technol 43(24):9413–9418

Weyens N, van der Lelie D, Taghavi S, Newman L, Vangronsveld J (2009b) Exploiting plant–microbe partnerships to improve biomass production and remediation. Trends Biotechnol 27(10):591–598. https://doi.org/10.1016/j.tibtech.2009.07.006

Wężowicz K, Rozpądek P, Turnau K (2017) Interactions of arbuscular mycorrhizal and endophytic fungi improve seedling survival and growth in post-mining waste. Mycorrhiza 27:499–511

White JF, Kingsley KL, Zhang Q, Verma R, Obi N, Dvinskikh S et al (2019) Endophytic microbes and their potential applications in crop management. Pest Manag Sci 75(10):2558–2565

Wilson SC, Lockwood PV, Ashley PM, Tighe M (2010) The chemistry and behaviour of antimony in the soil environment with comparisons to arsenic: a critical review. Environ Pollut 158(5):1169–1181

Xiao X, Luo S, Zeng G, Wei W, Wan Y, Chen L et al (2010) Biosorption of cadmium by endophytic fungus (EF) Microsphaeropsis sp. LSE10 isolated from cadmium hyperaccumulator Solanum nigrum L. Bioresour Technol 101(6):1668–1674. https://doi.org/10.1016/j.biortech.2009.09.083

Xiao Y, Dai CC, Wang XX, Liu FY, Wang HW, Li XG (2014) Effect of the endophyte Ceratobasidium stevensii on 4-HBA degradation and watermelon seed germination. Afr J Microbiol Res 8:1535–1543

Yadav A, Yadav K (2017) Exploring the potential of endophytes in agriculture: a minireview. Adv Plants Agric Res 6(4):00221. https://doi.org/10.15406/apar.2017.06.00221

Yin L, Ren A, Wei M, Wu L, Zhou Y, Li X, Gao Y (2014) Neotyphodium coenophialum-infected tall fescue and its potential application in the phytoremediation of saline soils. Int J Phytoremediation 16(3):235–246

Yu Z, Duan X, Luo L, Dai S, Ding Z, Xia G (2020) How plant hormones mediate salt stress responses. Trends Plant Sci 25(11):1117–1130

Yuan ZL, Rao LB, Chen YC, Zhang CL, Wu YG (2011) From pattern to process: species and functional diversity in fungal endophytes of Abies beshanzuensis. Fungal Biol 115(3):197–213

Yung L, Sirguey C, Azou-Barré A, Blaudez D (2021) Natural fungal endophytes from Noccaea caerulescens mediate neutral to positive effects on plant biomass, mineral nutrition and Zn Phytoextraction. Front Microbiol 12:689367. https://doi.org/10.3389/fmicb.2021.689367

Zahoor M, Irshad M, Rahman H, Qasim M, Afridi SG, Qadir M, Hussain A (2017) Alleviation of heavy metal toxicity and phytostimulation of Brassica campestris L. by endophytic Mucor sp. MHR-7. Ecotoxicol Environ Saf 142:139–149

Zimowska B, Bielecka M, Abramczyk B, Nicoletti R (2020) Bioactive products from endophytic fungi of sages (Salvia spp.). Agriculture 10(11):543

Zope VP, Jadhav HP, Sayyed RZ (2019) Neem cake carrier prolongs shelf life of biocontrol fungus Trichoderma viridae. Indian J Exp Biol 57(4):372–375

Zörb C, Geilfus CM, Dietz KJ (2019) Salinity and crop yield. Plant Biol 21:31–38

The Potential of Fungal Endophytes in Wheat

6

J. M. Rajwade and M. D. Oak

Abstract

In 1866, De Bary proposed the concept of "endophytes" which are microbes exhibiting symbiotic, probably mutualistic relationships with several wild and cultivated plant species. It is established that factors such as geographical location, agricultural practices, environmental conditions, plant genotype, leaf position, growth stage, cultivars, etc. contribute to the fungal and bacterial endophyte community. The focus of this chapter remains on studies on fungal endophytes in wheat with reference to the diversity, ecological roles, and potential applications, particularly for maintaining plant health and productivity. Studies on plant growth-promoting, biocontrol, stress alleviating, and micronutrient mobilizing activities are valuable for sustainable agricultural practices aimed at reducing the reliance on synthetic growth regulators, chemical fertilizers, insecticides, and pesticides and improving overall plant health. Particularly, the seed-associated microbiota shows vertical transfer, and coadaptation with the host over generations and establishes a healthy crop stand. With the advances in sequencing technologies and microscopic techniques, in-depth studies on fungal endophytes can be undertaken for ensuring sustainable agricultural practices and food security for the future.

Keywords

Fungal endophytes · Vascular arbuscular mycorrhiza · Synthetic growth regulators · Chemical fertilizers · Plant health

J. M. Rajwade (✉)
Nanobioscience group, Agharkar Research Institute, Pune, India
e-mail: jrajwade@aripune.org

M. D. Oak
Genetics and Plant breeding group, Agharkar Research Institute, Pune, India

© The Author(s), under exclusive license to Springer Nature Singapore Pte Ltd. 2025

K. A. Abd-Elsalam, A. H. Hashem (eds.), *Fungal Endophytes Volume II*,
https://doi.org/10.1007/978-981-97-8804-0_6

6.1 Introduction

Wheat (*Triticum aestivum* L.) is a key cereal crop for the human food supply and one of the most widely cultivated staple food crops worldwide (Shiferaw et al. 2013; Singh et al. 2016; Erenstein et al. 2022) which has helped overcome famines and saved millions of lives. Wheat has played a vital role in global food security since it was first domesticated around 10,000 years ago and is often referred to as the "miracle crop" of the last century (Erenstein et al. 2022). Nowadays, it provides one-fifth (20%) of the world's food calories and protein, making it the most widely cultivated crop on the planet, grown on 217 million hectares annually. The adaptability of wheat for cultivation in various climatic regions, ease of storage compared to most other commercial crops, and the ability to convert the grain to flour for making various foods make wheat one of the most popular crops cultivated across the globe. Currently, China and India are the top two wheat-producing countries because wheat requires less water for cultivation than other comparable crops and is the main ingredient of a variety of processed foods valued in modern, mainly urban life. However, there is an urgent need to enhance wheat productivity to ensure global food security, given the continued global population growth and the growing popularity of wheat-based processed foods. This will require substantive investments in research and development (Igrejas and Branlard 2020; Erenstein et al. 2022).

Although global wheat production has reached approximately 750 million metric tons, it still faces big challenges in supplying the human population, which will reach 9 billion by 2050 (Boyd et al. 2013), plant disease being one of them (Bebber et al. 2013). Diseases and ensuing loss of crops remain a major threat to food security in developing countries (Ghini et al. 2011; Chakraborty and Newton 2011). Historically, infections due to fungal pathogens have led to crisis situations. For example, in 1931 Karnal bunt was reported in South Asia and Iraq which subsequently spread to Mexico and USA (Anderson et al. 2004; Chakraborty and Newton 2011). The fungus *Tilletia* (Neovossia) *indica* is the causative agent for the disease which occurs in wheat, durum, rye, and triticale. Wheat rust caused by another fungus, *Puccinia graminis*, maize smut caused by *Ustilago maydis,* and soybean rust caused by *Phakospora pachyrizi* are some diseases that were reported for huge crop losses (Fisher 2013). In 1970, other diseases in rice and wheat crops by Rice blast (caused by *Pyricularia grisea*) and Wheat scab (caused by *Fusarium graminearum*) were reported and continue to occur. The incidence of the disease was linked to the increased temperature during the growth period (Coakley et al. 1999).

To ensure food security for the ever-increasing population, more crops need to be cultivated and measures to control plant pathogens and pests in agriculture (Mousa and Raizada 2013). Chemical pesticides are useful in controlling pathogens but have adverse effects on the consumers and the environment which has prompted governments to restrict their usage (Ott et al. 1991; Magnusson and Cranfield 2005; Bailey 2010; Glare et al. 2012). The use of clean technologies for crop production needs to be implemented. Resistance to disease control products is growing and constantly the industries are required to develop new products to meet the farmer's requirements. Commercially feasible products and technologies for disease management are the demand from the industry and the researchers. Products which can be an alternative to chemical pesticides, products which can have fertilizer characteristics, which are economical and environment friendly are a solution to the problem. To this end, biopesticides, integrated farm management practices, biofertilizers, phage therapy, bioactive botanicals, etc. are available and are being bettered for wider applications. To complement this list, beneficial endophytes are emerging as biological control agents for crop protection.

6.2 What Are Endophytes?

The discovery of endophytes, which are microorganisms that live inside plant tissues without causing any harm, can be traced back to the late nineteenth century. However, their significance and widespread occurrence were not fully recognized until the latter half of the twentieth century. One of the earliest mentions of endophytes can be found in a scientific publication from 1892 by German mycologist Heinrich Anton de Bary. In his work, de Bary described a fungus that lived within the leaves of certain plants and proposed the term "endophyte" to describe this internal fungal association (Segaran and Sathiavelu 2019). Endophytic microorganisms inhabit inside different tissues of the host plant, without inducing any apparent expression of any disease (Bacon and White 2000; Currie et al. 2014). Instead, endophytes can have neutral or beneficial interactions with the host plant. Several wild and cultivated plant species exhibit symbiotic, and possibly mutualistic, interactions with endophytes (Rana et al. 2019). Phytopathogenic and mycorrhizal fungi have traditionally dominated plant-fungal studies, which have increasingly focused on endophytes. More than one million endophytic species are estimated to exist in 3,00,000 plant species, yet only a small fraction have been isolated and studied for their functions. A timeline providing the description of endophytes is presented in Fig. 6.1.

Koch's postulates, which are a set of criteria used to establish a causal relationship between a microorganism and a disease, were originally formulated for studying pathogenic microorganisms, and may not be directly applicable to endophytes. Therefore, the application of Koch's postulates in the context of endophytes is limited.

Koch's postulates are typically used to demonstrate that a specific microorganism is the causative agent of a particular disease by fulfilling four criteria:

- *The microorganism must be consistently present in individuals suffering from the disease.*
- *The microorganism must be isolated and grown in pure culture.*
- *Inoculation of a healthy individual with the cultured microorganism should result in the development of the same disease.*
- *The microorganism must be re-isolated from the inoculated individual and be identical to the original microorganism.*

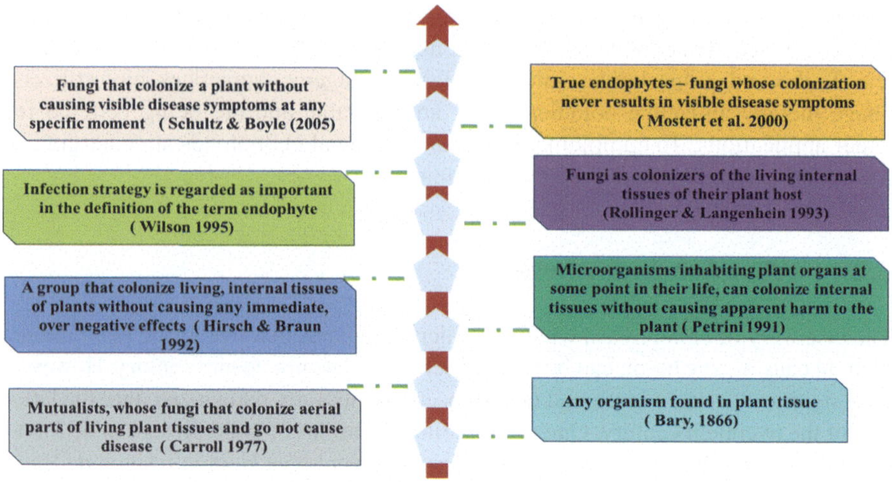

Fig. 6.1 Historical description of endophytes

Instead, researchers employ a combination of different techniques, such as surface sterilization, isolation, molecular identification, and functional assays, to study the presence, diversity, and ecological roles of endophytic microorganisms in plants. These methods focus on understanding the association between endophytes and host plants and their potential benefits, rather than proving a causal relationship between the microorganism and a disease.

Bacterial endophytes are microorganisms that reside within the tissues of plants without causing any harm. In the case of wheat (*Triticum* spp.), several bacterial species have been identified as endophytes (Pang et al. 2022). These bacterial endophytes play important roles in promoting plant growth, nutrient uptake, disease resistance, and overall plant health (Pang et al. 2022; Ahlawat et al. 2022; Makar et al. 2021; Herrrera et al. 2016). Various species of *Pseudomonas*, such as *Pseudomonas fluorescens* and *Pseudomonas putida*, have been isolated from wheat plants. These bacteria have been shown to enhance plant growth by producing plant growth-promoting substances, including phytohormones like auxins and cytokinins. They also contribute to nutrient acquisition by fixing atmospheric nitrogen or solubilizing phosphates, making these nutrients more available to the plant. *Bacillus subtilis* and *Bacillus amyloliquefaciens* have been isolated from wheat tissues. These bacteria have multiple beneficial effects on the host plant, including the production of antimicrobial compounds that help protect the plant against pathogens (Kaul et al. 2022). They also assist in nutrient uptake, improve plant tolerance to abiotic stresses, and stimulate the plant's defense mechanisms (Etesami and Alikhani 2018). *Enterobacter*, *Pantoea*, and *Stenotrophomonas* have also been reported in wheat (Rana et al. 2019). Each of these bacterial species may contribute to the overall health and growth of wheat plants in different ways. It is important to note that the composition and abundance of bacterial endophytes in wheat can vary depending on several factors, including plant genotype, environmental conditions, and agricultural practices (Robinson et al. 2016). Agnolucci et al. (2019) have reviewed the effect of the beneficial microorganisms on the bacterial endophyte communities in the durum wheat roots. They used different molecular approaches in their studies. Ongoing research continues to explore the diversity and functional roles of bacterial endophytes in wheat and their potential applications in sustainable agriculture (Latz et al. 2021). Recent studies with endophytic bacteria in wheat elucidated the mechanisms underlying plant growth promotion and protective effects under drought conditions (Lastochkina and Allagulova 2023a, b). Makar and Romanyuk (2022) studied bacterial endophytes in wheat for their effect on the improvement of micronutrients in grain.

Studies have shown that the diversity of fungal endophytes inside plants is underestimated, and some endophytes are host and/or environment-specific (Hardoim et al. 2015; Grabka et al. 2022). Fungal endophytes are asymptomatic inhabitants of plant tissue and have been found in all parts of plants (Caroll 1988; Dutta et al. 2014; Schouten 2019). Endophytic species may remain localized in a plant and provide tissue-specific protection from disease (Witzel et al. 2014; Kumar et al. 2023), or they may spread systemically to herbaceous plants (Rodriguez et al. 2009; Kumar et al. 2023). Several wild and cultivated plant species exhibit symbiotic, and possibly mutualistic, interactions with endophytes (Schulz 2006; Khare et al. 2018; Mishra et al. 2021; Anand et al. 2023). Fungal endophytes belong to ascomycetous fungi and occur within the asymptomatic photosynthetic tissues of plants. According to Arnold (2007), the interactions between the plant hosts and other co-occurring endophytes are quite difficult to understand, however, with the advances in various molecular techniques, this notion is changing (Busby et al. 2016).

6.3 Differences in the Endophytic Fungi and Vascular Arbuscular Mycorrhiza

Endophytic fungi and arbuscular mycorrhizal (AM) fungi, specifically vascular arbuscular mycorrhizae (VAM), are two distinct types of fungi that interact with plants in different ways. Here are the key differences between endophytic fungi and vascular arbuscular mycorrhizae:

Location and Colonization: Endophytic fungi reside within the internal tissues of plants, including the roots, stems, leaves, and other plant parts, without causing visible symptoms of disease. In contrast, vascular arbuscular mycorrhizae (VAM) colonize the roots of plants. They form a mutualistic association where fungal structures, such as arbuscules and vesicles, penetrate the plant root cells, establishing a specialized symbiotic interface for nutrient exchange.

Plant Interactions: Endophytic fungi generally have neutral or beneficial interactions with plants. They can promote plant growth, enhance stress tolerance, and provide protection against pathogens and herbivores. On the other hand, VAM fungi have a mutualistic relationship with plants. They form extensive networks of hyphae that extend into the surrounding soil, facilitating the uptake of nutrients, especially phosphorus, from the soil. In return, the host plant provides the fungi with carbohydrates produced through photosynthesis.

Fungal Diversity and Taxonomy: Endophytic fungi comprise a diverse group of fungi from various taxonomic orders and families, including *Ascomycota*, *Basidiomycota*, and others. They can include both beneficial and potentially harmful species. Vascular arbuscular mycorrhizae, on the other hand, specifically refer to a type of AM fungi that belong to the *Glomeromycota* phylum. VAM fungi are a monophyletic group and share a similar morphological and ecological adaptation for symbiosis with plants.

Nutrient Exchange: Endophytic fungi can facilitate nutrient uptake for the host plant by solubilizing nutrients, fixing nitrogen, or promoting nutrient mobilization in the soil. However, the specific mechanisms and nutrient preferences may vary depending on the fungal species. Vascular arbuscular mycorrhizae have a well-established role in enhancing phosphorus uptake in plants. The fungal hyphae extend into the soil, effectively increasing the surface area for nutrient absorption and transporting phosphorus and other nutrients to the host plant through the arbuscules.

Research Focus: Endophytic fungi have been studied for their potential applications in agriculture, bioremediation, and plant protection. Their beneficial effects on plant growth and stress tolerance have garnered interest in harnessing their potential for sustainable crop production. Vascular arbuscular mycorrhizae have been extensively studied for their roles in nutrient cycling, ecosystem functioning, and promoting plant diversity and productivity in natural and agricultural ecosystems.

6.4 Methods to Isolate and Confirm the Fungal Endophytes

Confirming the presence of fungal endophytes in plant tissues involves a series of procedures to ensure accurate identification and confirmation. The specific steps commonly followed to confirm the presence of fungal endophytes are

Surface Sterilization: The first step is to perform surface sterilization of the plant tissues to remove any externally associated fungi. This is done by treating the plant samples with disinfectants or sterilizing agents, such as sodium hypochlorite or ethanol. Surface sterilization is important to ensure that the fungi isolated afterward are truly residing inside the plant tissues and not contaminants from the surface.

Tissue Disruption and Homogenization: After surface sterilization, the plant tissues are disrupted, crushed, ground, or homogenized to release the fungal endophytes present within. This step helps in liberating the endophytic fungi from the plant cells and allows for their extraction.

Fungal Isolation: The homogenized plant tissues are then plated onto selective media to isolate the fungal endophytes. Different media can be used based on the targeted fungal groups or specific growth requirements. It is important to use media that support the growth of both fast-growing and slow-growing fungi to capture a wide range of endophytic fungi.

Colony Purification: Once fungal colonies start to emerge on the plates, individual colonies can be selected for purification. This involves transferring a small portion of the fungal colony to fresh media plates to obtain pure cultures of the fungal endophytes. Purification is important to eliminate potential contamination and ensure that the isolated fungi are indeed endophytes.

Microscopic Observation: Microscopic examination of the isolated fungal cultures can help confirm the presence of endophytic fungi. Morphological features, such as hyphal characteristics, spore formation, and reproductive structures, are examined using light microscopy or other relevant techniques. These observations can provide initial identification or characterization of the fungal endophytes.

Molecular Identification: To obtain accurate identification and confirmation of the fungal endophytes, molecular techniques are often employed. PCR amplification of specific DNA regions, such as the internal transcribed spacer (ITS) region, is commonly used. The amplified DNA can be sequenced, and the sequences are compared to databases to determine the fungal species or genus. DNA sequencing can provide a more precise identification of the fungal endophytes.

By following these specific procedures, researchers can accurately confirm the presence of fungal endophytes in plant tissues and obtain valuable information about their identity and diversity. Figure 6.2 is a schematic providing the methods for isolation and characterization of endophytes. These steps help in understanding the ecological roles, interactions, and potential applications of fungal endophytes in various plant species, including wheat.

In a detailed study on the species composition of endophytes from wheat-healthy plants, plant parts such as leaves, stems, glumes, and grains from five wheat cultivars were processed. Parts were collected at five growth stages. The analysis of 1750 plant segments showed 722 isolates representing 30 fungal genera. Highest colonization frequency (in all the plant tissues) was due to *Alternaria alternata, Cladosporium herbarum*, *Epicoccum nigrum*, *Cryptococcus* sp., *Rhodotorula rubra*, *Penicillium* sp., and *Fusarium graminearum*. Leaves showed abundant taxa in comparison to other parts (Larrán et al. 2002, Larran et al. 2007).

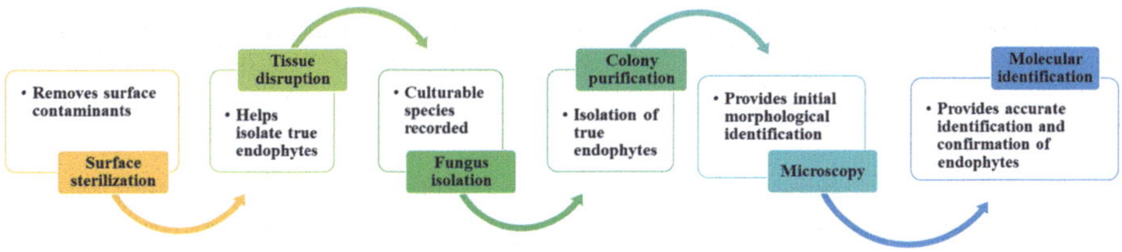

Fig. 6.2 Typical workflow in isolation and identification of endophytes

6.5 Techniques Used to Study Fungal Endophytes

There are several challenges in the studies on endophytes. Some of them are (a) the microbe cannot be isolated and cultured, (b) difficulties in conducting experiments because plants free of endophytes cannot be obtained, (c) new techniques have to be developed and methods have to be established to understand their potential applications (Gamboa et al. 2002; Rodriguez et al. 2009; Varma et al. 2019). The study of fungal endophytes involves a range of techniques to explore their diversity, identify species, and understand their interactions with host plants (Agnolucci et al. 2019). Here are some commonly used techniques:

6.5.1 Culture-Dependent Methods

Culturing fungi is a traditional approach to study fungal endophytes. The disrupted plant tissues or homogenates are plated onto various nutrient media that support fungal growth. Colonies that emerge on the plates can be isolated, purified, and characterized based on their morphological features, such as colony appearance, spore characteristics, and other microscopic observations.

In a study, two wheat cultivars, viz., Apache and Caphorn, showing differential susceptibility toward Fusarium head blight were analyzed for variation in microbial diversity and assemblages depending on the host cultivar, organ (aerial organs vs. roots), or maturity. Bacterial and fungal communities were isolated using culture-based methods and the composition of the community was analyzed over space and time. A rich species diversity was observed along with the co-occurrence of endophytes and pathogens inside plant tissues (Comby et al. 2016). The study was carried out to find potential biocontrol agents for wheat.

6.5.2 Culture-Independent Methods

These methods focus on studying fungal endophytes directly from plant tissues without relying on traditional culturing techniques because many fungal endophytes cannot be readily cultured in the laboratory.

(a) Polymerase Chain Reaction (PCR) and DNA Sequencing: PCR amplification of specific fungal DNA regions, such as the internal transcribed spacer (ITS) region, allows for the identification and classification of fungal endophytes. The amplified DNA can be sequenced to determine the fungal species present in the plant tissues.
(b) Next-Generation Sequencing (NGS): NGS technologies, such as metagenomics and amplicon sequencing, provide a high-throughput approach to study the entire fungal community within plant tissues. These methods can provide a comprehensive understanding of fungal diversity, abundance, and functional potential in the endophytic community.

Following paragraphs provide the efforts at identifying the fungal endophytes in wheat. Clapa et al. (2020) described a rapid, simple, and low-cost method based on high-resolution melting PCR (HRM-PCR) for the identification and differentiation of wheat endogenous fungal isolates. The researchers employed the HRM-PCR and differentiated 28 fungal isolates. These belonged to different genera, viz., *Alternaria*, *Penicillium*, *Epicoccum*, *Fusarium*, and *Trichoderma*. The results indicate the utility of the method for large-scale screening of cultured samples.

Kuzniar et al. (2020a, b, c) adopted the culture-independent technique to determine the microbiome of the embryo and endosperm of wheat seed of eight cultivars and proved that it is independent of cultivars and differs between the cultivars *Triticum aestivum* L.(cv.'Hondia') and *Triticum spelta* L.(cv.'Rokosz').

The data by Ofek-Lalzar et al. (2016) show that wild grass relatives of wheat (*Triticum dicoccoides* and *Aegilops sharonensis*) contained taxonomically diverse fungal endophytes belonging to taxa with known beneficial effects; which, were not found in modern day *T. aestivum* wheat. The work was carried out using both cultivation and cultivation-independent methods. In the former, the ITS sequences from 512 cultures resulted in 67 operational taxonomic units (OTUs) which were evenly distributed between the different plant species. Each wheat type had unique endophyte taxa, with a narrow core community of *Alternaria* spp. present in all samples. As expected, a larger number of OTUs were obtained in the cultivation-independent analysis. Both methods of analysis yielded a similar number of taxa for OTUs that had >0.5% relative abundance.

Sapkota et al. (2017) studied the wheat mycobiome by using metabarcoding of the fungal ITS1 region as a technique for analysis and the study involved leaf samples from four varieties collected from two locations. To understand the spatiotemporal variations of the mycobiome, samples were collected at three growth stages in triplicates. Results showed that several factors such as geographical location, leaf position, growth stage, and cultivar were important factors contributing to the fungal community. A network analysis was performed revealing the negative coexistence between *Zymoseptoria tritici* and the yeasts *Sporobolomyces*, *Dioszegia*, and *Cystofilobasidiaceae*. It was observed that the fast-growing fungi occupied the empty spaces in the phyllosphere rapidly.

Ren et al. (2023) prove the utility of a combined approach of amplicon sequencing with a based method for determining the seed microbiome endophytes in wheat infected by *Tilletia controversa*. Fragments plating method and the high-throughput culturing method were reported by Comby et al. (2016) for the isolation of endophytes from wheat. The endophytes were tested for antagonistic effects against *Fusarium* sp. by dual culture assay. Another assay was performed on detached spikelets for assessment of antagonistic effects against *F. graminareum*.

6.5.3 Fluorescent In Situ Hybridization (FISH)

Similar to bacterial endophytes, FISH can be utilized to visualize and localize fungal endophytes within plant tissues. Fluorescently labeled DNA probes specific to target fungal groups can bind to the fungal DNA, enabling their detection and identification using fluorescence microscopy. Fluorescent dyes such as fluorescein diacetate (FDA), carboxyfluorescein diacetate (CFDA), chloromethylfluorescein diacetate (CMFDA), SYTO9 (S9), propidium iodide (PI), diamidino-2-phenylindole (DAPI), etc. allow for the microscopic observation of endophytic bacteria and fungi that colonize the plant (Varma et al. 2019).

6.5.4 Microscopic Observations

Microscopic techniques, such as light microscopy and electron microscopy, allow for the examination of fungal structures, spores, and hyphae within the plant tissues. These observations provide valuable information for species identification and characterization.

6.5.5 Functional Assays

Functional assays can be performed to investigate the roles of fungal endophytes in plant health and interactions. These assays may include tests for antifungal activity, plant growth promotion, stress tolerance enhancement, and production of bioactive compounds. Some examples specific for fungal endophytes in wheat are provided in the subsequent paragraphs.

Fungal endophyte communities (FECs) from domesticated bread wheat, *Aegilops sharonensis,* and *Triticum dicoccoides* were characterized and compared to identify taxa that exhibit potential for crop improvement. With NGS, 1666 taxa were identified and the presence of random taxa with low abundance was noted. Interestingly, the collective diversity of FECs of the three species was similar. In comparison to domesticated wheat, the two wheat ancestors *Aegilops sharonensis* and *Triticum dicoccoides* had higher numbers of differentially abundant fungal taxa. It was observed that *Alternaria infectoria* was the most abundant species in the two wheat ancestors while bread wheat plants had *Candida sake*. Presence of a small number of taxa with a relatively high number of co-occurrence associations was observed in network analysis. Thus, the authors reported limited types of endophytes in domesticated wheat in comparison to wild plants, indicating a potential for improvement using the functional taxa from the latter (Sun et al. 2020).

Ofek-Lalzar et al. (2016) state that the endophytes exert a great impact on the agriculturally important traits in crop plants. Furthermore, the authors hypothesized that endophytes present in wild plants are indeed beneficial endophytes and were absent or underrepresented in domesticated crops. By employing these techniques, researchers can gain insights into the diversity, ecology, functional attributes, and beneficial effects of fungal endophytes in plants. Based on a synergistic culture-dependent and independent approach, Becker et al. (2024) reported that the wheat seed mycobiome is conserved.

6.6 Analysis of Endophyte Diversity and Interactions with Host Plant

Some commonly reported fungal genera as endophytes in wheat
Alternaria, Aspergillus, Cladosporium, Colletotrichum, Curvularia Fusarium, Penicillium, Phoma, Phomopsis, Pyrenophora, Trichoderma Xylaria

Several fungal genera have been reported as endophytes in wheat. These endophytic fungi can colonize different plant tissues, including the roots, stems, leaves, and seeds of wheat plants.

Culture-based methods were used to understand the diversity of endophytic microbes cultivated on different culture media and colonization frequency, isolation frequency, alpha diversity indices (Simpson's, Shannon–Wiener diversity indices), species richness-evenness, etc. were calculated (Hata and Futai 1995; Mishra et al. 2012; Verma et al. 2014).

The colonization frequency (%CF) can be calculated using the formula given by Hata and Futai (1995).

$$\%\mathrm{CF} = N_{\mathrm{col}} / N_t \times 100$$

where,

N_{col} = number of segments colonized by each fungus,
N_t = total number of segments studied.

Diversity indices:

$$\text{Simpson's diversity index} = 1 - \varsigma\left(p_i\right)2$$

$$\text{Shannon} - \text{Wiener diversity} = -\varsigma\, s\left(p_i \log p_i\right),$$

where p_i = proportion of frequency of the ith species in a sample.

Species evenness

$$\text{Evenness}(E) = H / \log(S),$$

where

H = Shannon Wiener diversity and
S = species richness (i.e., total number of species).

Shannon–Wiener diversity indices, and species richness can be calculated manually or by software such as PAST, BioDiversity Pro, and Origin software (Verma et al. 2014; Orduna et al. 2011). The authors suggest the use of Principal Component

Analysis (PCA) to further determine the interrelationship and tissue and site specificity of endophytic microbial populations among different tissue samples and sites, ecological associations among endophytes with the different tissues, and different localities of the host.

One of the most well-known fungal endophytes in wheat is the genus *Epichloë*. *Epichloë coenophiala* and *Epichloë typhina* form mutualistic associations with certain grasses, including some wild and cultivated wheat varieties. These endophytes produce alkaloids that can provide protection against herbivores and pathogens. However, it is important to note that some *Epichloë* strains can also produce toxic alkaloids that may be harmful to livestock if consumed in large quantities. Apart from *Epichloë*, other fungal endophytes found in wheat include species from the genera *Cladosporium*, *Alternaria*, *Penicillium*, and *Fusarium*, among others. These fungal endophytes can have both beneficial and detrimental effects on the host plant. Some endophytic fungi may enhance plant growth and nutrient acquisition by producing plant growth-promoting substances or facilitating nutrient uptake. However, certain fungal endophytes can also produce mycotoxins, which are toxic compounds that can negatively affect plant health and can pose risks to human and animal health if present in high concentrations.

In 2002, Larran et al. carried out surface-sterilization of three wheat cultivars (Klein centauro, Klein dragon, and buck ombu) to eliminate the epiphytes and test the infection frequencies of endophytes at three growth stages. In this study, 130 fungal isolates were isolated from 450 leaf segments, and 19 species were identified. Amongst the fungal endophytes, *Rhodotorula rubra*, *Alternaria alternata*, *Cladosporium herbarum*, and *Epicoccum nigrum* were isolated in the highest frequency. The endophytes could be grouped as (a) well-known and economically important pathogens of wheat, (b) commonly abundant phylloplane fungi considered to be primary saprobic and minor pathogens, and (c) species occasionally present in wheat.

Sun et al. (2020) carried out studies to identify the endophytic fungal taxa that possess a potential for crop improvement. In this study, the fungal endophyte communities (FECs) from domesticated bread wheat and two wheat ancestors, viz., *Aegilops sharonensis* and *Triticum dicoccoides* were characterized and compared using NGS. The sequencing revealed the presence of a total of 1666 taxa. The FECs in the three plant species contained high proportions of random taxa with low abundance and abundant taxa shared a commonality with all host plants. Further, the collective FECs of each of the three plant species had similar diversity. The study revealed that the FECs from the wild plants in specific sites were more diverse and had greater richness than the specific fields. The wild plants had higher numbers of differentially abundant fungal taxa than domesticated bread wheat. *Alternaria infectoria* and *Candida sake* were the most abundant species in wild plants and domesticated wheat respectively. The authors carried out network analysis on co-occurrence association which indicated a small number of taxa with a relatively high number of co-occurrence associations, which might be important in community assembly. The study proved that the number and types of endophytes in domesticated wheat plants were smaller relative to wild plants. The study highlighted both (putative functional and hub) fungal taxa with potential for wheat improvement.

In wheat, while the rhizospheric communities and root endophytes are the focus of research, studies on grain endophytes are limited (Makar et al. 2021). Considering the ability of vertical transmission between the generations, the agrobiotechnological potential of seed endophytes has to be explored (Compant et al. 2021; Kuzniar et al. 2020a, b, c; Pastoshchuk et al. 2021a, b). Seed endophytes have unique characteristics such as adaptation to unfavorable conditions, dormancy during grain/seed storage, etc. The bacterial endophytes residing in wheat seeds were reported to confer resistance against phytopathogens (Pastoshchuk et al. 2018). Studies by Herrera et al. (2016), Geisen et al. (2017), Ridout et al. (2019), and Kuzniar et al. (2020a, b, c) suggest that the seed endophytes help in establishing the endophytes in the whole plant, support plant growth and productivity.

All over the world, wheat dwarf bunt caused by teliospores of *Tilletia controversa* Kühn. is regarded as a devastating disease. Ren et al. (2023) analyzed the seed microbiome of endophytes between resistant and susceptible cultivars after infection with *T. controversa*. The study employed a combination of amplicon sequencing and isolation approach, the results of which suggested that the prevalence *of T. controversa* was lower in resistant cultivars. The latter had higher microbial diversity and specific promising antagonists (viz., such as *Phoma* sp. 4 BRO-2013, *Sarocladium* sp., *P. lurida*, *P. brenneri*, *C. flaccumfaciens*, *F. fujikuroi*, *F. proliferatum*, *L. chromiiresistens*, *B. nealsonii*, *F. avenaceum*, *A. penicillioides*, and *Alternaria* sp.) contributed to the ability of pathogen suppression. The germination of *Tilletia controversa* teliospores was significantly suppressed by the seed endophyte communities, especially *Sarocladium* sp., which provides clues for future studies against wheat dwarf bunt.

It is important to note that the composition and abundance of fungal endophytes in wheat can vary depending on factors such as plant genotype, geographic location, agricultural practices, and environmental conditions. Ongoing research continues to explore the diversity, ecological roles, and potential applications of fungal endophytes in wheat and their contribution to plant health and productivity.

6.7 Bioactivities of Fungal Endophytes in Wheat

Through various mechanisms, endophytic microbes contribute to plant growth and protection against pests and pathogens. Secondary metabolites and biochemicals produced by them suppress or reduce the negative effects of pathogens. In addition, they induce plant defense mechanisms, such as Systemic acquired resistance (SAR) or Induced systemic resistance (ISR), and combat against abiotic stress. Figure 6.3 is a schematic showing all such roles of endophytes.

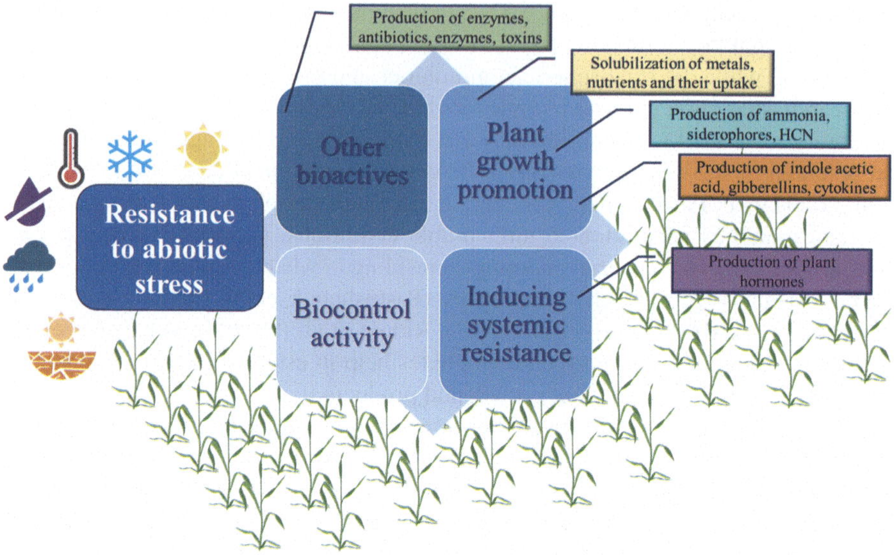

Fig. 6.3 Roles of endophytic fungi (schematic prepared using BioRender.com)

6.7.1 Endophytes Promote Plant Growth

Fungal endophytes present in wheat can produce various plant growth-promoting substances that positively influence the growth and development of the host plant. Here are some examples of plant growth-promoting substances produced by fungal endophytes in wheat:

Indole-3-Acetic Acid (IAA): Fungal endophytes can produce IAA, which is a type of auxin hormone. IAA plays a crucial role in regulating plant growth and development, including cell elongation, root development, and lateral root formation. By producing IAA, fungal endophytes can promote root growth and overall plant growth in wheat.

Gibberellins (GA): Some fungal endophytes are capable of producing gibberellins, a group of plant hormones that stimulate stem elongation, leaf expansion, and flowering. Gibberellins can enhance wheat plant growth by promoting cell division and elongation, leading to increased biomass and overall plant vigor.

Cytokinins: Fungal endophytes can also produce cytokinins, which are plant hormones that regulate cell division, shoot development, and nutrient allocation. Cytokinins can influence wheat plant growth by promoting shoot and root growth, enhancing nutrient uptake, and improving stress tolerance.

Abscisic Acid (ABA) Modulation: Fungal endophytes can influence the levels and response of abscisic acid in plants. ABA is a plant hormone involved in stress responses, dormancy, and stomatal regulation. By modulating ABA levels and response, fungal endophytes can help wheat plants better cope with abiotic stresses and maintain water balance.

Volatile Organic Compounds (VOCs): Some fungal endophytes release volatile organic compounds that can positively affect plant growth. These compounds can stimulate root growth, enhance nutrient uptake, and improve plant resilience to abiotic stresses. VOCs emitted by endophytic fungi can also induce systemic responses in the host plant, resulting in enhanced growth and stress tolerance.

Enzyme Production: Fungal endophytes can produce enzymes, such as cellulases, chitinases, and proteases, which can break down complex organic compounds in the soil and release nutrients for the plant's uptake. These enzymes contribute to nutrient mobilization and availability, promoting plant growth in wheat.

It is known that plant endophytes exhibit a potential for promoting plant growth and controlling diseases. It is likely that long-term fertilization regimes would impact the diversity and community of plant endophytes. With newer and more efficient techniques of genome analysis in place, studies on understanding the diversity and community of wheat leaf endophytes under different fertilization regimes need to be pursued (Sun et al. 2020).

The production of these plant growth-promoting substances by fungal endophytes can enhance wheat plant growth, nutrient uptake, stress tolerance, and overall crop productivity. These beneficial effects make fungal endophytes valuable for sustainable agricultural practices aimed at reducing the reliance on synthetic growth regulators and improving plant health (Le Cocq et al. 2017; Devi et al. 2023).

6.7.2 Endophytes Affect Plant Pathogens

Endophytic fungi can produce a wide array of antimicrobial compounds that inhibit phytopathogens (Aamir et al. 2020). Some examples of antimicrobial compounds produced by endophytes known to have inhibitory effects on phytopathogens are enlisted.

Alkaloids: Alkaloids are a diverse group of compounds that exhibit antimicrobial activity. For example, endophytic fungi of the genus *Epichloë* produce alkaloids such as lolines, ergot alkaloids, and peramine, which can inhibit a range of phytopathogens.

Polyketides: Polyketides are complex secondary metabolites with diverse chemical structures and antimicrobial properties. Many endophytic fungi produce polyketides that exhibit inhibitory effects on phytopathogens. Examples include alternariol, beauvericin, fusaric acid, and cercosporin.

Terpenoids: Endophytic fungi are known to produce various terpenoid compounds that possess antimicrobial activity. For instance, the endophytic fungus *Pestalotiopsis microspora* produces pestalotiopenes, which show antifungal properties against phytopathogens.

Peptides: Some endophytic fungi produce antimicrobial peptides that can effectively inhibit phytopathogens. For instance, the endophyte *Trichoderma* spp. produces antifungal peptides such as peptaibols and chitinases that help suppress fungal pathogens.

Volatile Organic Compounds (VOCs): Some endophytic fungi produce volatile organic compounds that exhibit antimicrobial properties. These compounds can be released into the air and inhibit the growth and activity of phytopathogens. Examples of such VOCs include 2,4-diacetylphloroglucinol (DAPG) produced by certain strains of *Pseudomonas* spp.

Phenolic Compounds: Endophytic fungi can also produce phenolic compounds with antimicrobial activity. These compounds can disrupt the growth and development of phytopathogens. For example, endophytic fungi of the genus *Colletotrichum* produce phenolic compounds that inhibit other fungi.

Endophytes have antagonistic activity against disease-causing phytopathogens and are also capable of producing antimicrobial, insecticidal, antioxidant, antitumor, and antiviral metabolites. Alkaloids such as perfumoid, phomoenamide, joxysporidinone, alantrypinene, alantryleunone, (−)-4,6′-anhydrooxysporidinone, and (−)-6-deoxyoxysporidinone were reported in certain endophytic fungi. Nidurufin, sterigmatocystin, averantin, 11a-methoxycurvularin 4, 11b-methoxycurvularin, tenellone H, phomopene, and 1-chloro-2,4-dihydroxy-5-methoxy-7-methylanthraquinone produced by the endophytic fungi were reported to have cytotoxic effects (Segaran and Sathiavelu 2019).

Two endophytic isolates of *Epicoccum nigrum* were isolated from wheat grains and were tested for their competitive effects against two *Fusarium graminearum* isolates in in vitro experiments. *Epicoccum nigrum* reduced the mycelial growth of *Fusarium graminearum*. Further, the crude cultural filtrate of *E. nigrum* inhibited *F. graminarium* but the spores had nosuch effects. Apical hyphal growth was significantly reduced and application of *E. nigrum* to seeds prior to *F. graminareum* resulted in inhibition of the latter. Although *planta* experiments are required, the in vitro data suggests the potential of the endophyte *E. nigrum* in reducing *F. graminareum* infections in wheat (Jensen et al. 2016).

The winter wheat "ac Morley" and "25r34" and spring wheat "Sumai" and "Scotia" are known to be moderately resistant to fusarium head blight. To find if the endophytes contributed to resistance, 90 endophytic fungal isolates with 54 distinct morphotypes were obtained from tissues of the winter and spring wheat growing in Ariss, Ontario in 2016 and screened for antagonistic effects against *Fusarium graminearum*, *Waitea circinata*, and *Microdochium majus*. Further, 33 isolates were used in specific assays for pathogenicity; from which 16 were observed to be non-phytopathogenic. Amongst 16, *Valsa friesii*, *Simplicillium lamellicola*, and *Cladorrhinum flexuosum* were identified by ITS-PCR. The hyphal plugs of these fungi were observed to inhibit *F. graminearum*, and *W. circinata* on leaves of wheat seedlings, and reduce the lesion lengths, proving them to be promising "biocontrol" agents (Abaya et al. 2021).

The presence of endophytic fungi was correlated with a decrease in the density of pustules of *Puccinia recondita* f. sp. *tritici* (Dingle and Mcgee 2003). *Alternaria* spp., *Cladosporium* spp., and *Leptosphaeria* spp. isolated from wheat (Huang et al. 2016) showed good biocontrol activity and were demonstrated to be transmitted from seed to sprout. Endophytic *Chaetomium* spp. (from healthy wheat leaves) reduced the number of pustules caused by *Puccinia recondita*. This reduction in the number of pustules resulted from the use of some endophyte culture washings (i.e., cell-free washings from endophyte culture plates were prepared by passing the solutions of spores and hyphal fragments through a sterile 0.2 mm filter) through activating plant defense mechanisms or due to direct inhibitory substances (Dingle and Mcgee 2003; O'Hanlon et al. 2012; Larran et al. 2016). Similarly, endophytic *Chaetomium globosum* and its culture washings suppressed the development of the tan spot pathogen, *Pyrenophora tritici-repentis* in wheat leaf segments (O'Hanlon et al. 2012). An experiment confirmed the accumulation of proteins in the presence of endophytes in leaves and no effect on pathogen growth in vitro due to culture washings. Therefore, it was suggested that induction of plant resistance is the mechanism responsible for suppressing pathogens (O'Hanlon et al. 2012).

Trichoderma hamatum, *Chaetomium globosum,* and *Fusarium graminearum* significantly reduced tan spot disease severity as compared to the control. Larran et al. (2016) concluded, that *Bacillus* sp. and *Trichoderma hamatum* are good candidates as biocontrol agents against *Drechslera tritici-repentis*. Perello et al. (2009) demonstrated different antagonistic activities of nine leaf endophyte species against four different necrotrophic wheat pathogens, including *Alternaria triticimaculans*, *Bipolaris sorokiniana*, *Drechslera tritici-repentis,* and *Zymoseptoria tritici*, the most common phytopathogens in wheat. The study by Gdanetz and Trail (2017) shows that the endophytes in wheat are affected by crop management strategies and potentially useful endophytes that offer disease protection against *Fusarium graminerium* were recorded. In this study, the fungal and bacterial communities of leaves, stems, and roots of wheat throughout the growing season were recorded using 16S and fungal internal transcribed spacer 2 rRNA gene amplicon sequencing. Seedling assays were performed to assess antagonistic strains. Salamon et al. (2023) also reported on the variations in endophytic mycobiome in spring and winter wheat cultivars grown under various conditions.

Products containing metabolites of endophytes or active cultures were introduced in agriculture almost two decades ago. *Ampelomyces quisqualis* is an ingredient in AQ10, a product that is targeted against powdery mildew in wheat and other crops (Kiss et al. 2004). Biomal, another product containing *Colletotrichum gloeosporioides* f. sp. *malvae* is commercialized in Canada for the control of wheat and lentils pathogens (Butt and Copping 2000). In the USA, product containing *Pseudozyma flocculosa* (Sporodex) is manufactured by Ecogen Inc., USA for control of powdery mildew in wheat and other crops.

It is important to note that the antimicrobial compounds produced by endophytic fungi can vary depending on the fungal species, environmental conditions, and the specific phytopathogens involved. Additionally, the efficacy of these compounds may vary against different pathogens. Therefore, it is crucial to study the specific interactions between endophytes and phytopathogens to understand the antimicrobial mechanisms and optimize their application for disease management in agriculture.

Fungal endophytes can prevent phytopathogens in wheat through various mechanisms which include:

Competition for Space and Nutrients: Fungal endophytes establish a symbiotic relationship with the host plant, occupying niches within the plant tissues. By occupying these spaces, they limit the colonization and establishment of phytopathogens. Fungal endophytes compete with pathogens for resources such as nutrients, space, and other essential factors, thereby reducing the availability of resources for the pathogens to thrive.

Antibiosis and Production of Antimicrobial Compounds: Many fungal endophytes produce antimicrobial compounds that can directly inhibit the growth and activity of phytopathogens. These compounds can include antibiotics, antifungal metabolites, enzymes, and toxins. By secreting these antimicrobial substances, fungal endophytes can directly suppress the growth and development of pathogens in the wheat plant tissues.

Induced Systemic Resistance (ISR): Fungal endophytes can trigger the plant's defense mechanisms and induce systemic resistance against pathogens. They can stimulate the production of plant defense-related compounds, such as phytohormones (e.g., salicylic acid and jasmonic acid), and pathogenesis-related proteins. This systemic response prepares the plant to defend against potential pathogen attacks and provides a long-lasting defense mechanism.

Enhanced Plant Immune Response: Fungal endophytes can enhance the plant's immune response by activating specific defense pathways. They can trigger the expression of genes involved in defense responses, such as the production of cell wall reinforcement compounds, antimicrobial proteins, and enzymes that degrade pathogen cell walls. These immune responses strengthen the plant's defense system, making it more resistant to pathogen attacks.

Indirect Effects on Phytopathogens: Fungal endophytes can indirectly impact phytopathogens through modifications in the plant's physiology or environment. They can modulate plant growth, alter root exudates, and influence the chemical composition of the plant tissues. These changes can make the plant less favorable or more resistant to pathogen colonization and infection.

The effectiveness of fungal endophytes in pathogen control may vary depending on the specific endophyte species, the pathogen species, and the environmental conditions. Additionally, not all fungal endophytes exhibit strong antagonistic activity against pathogens. Therefore, careful selection and screening of beneficial endophyte strains is crucial for their effective application in disease management strategies for wheat crops. *Trichoderma* sp. is an endophyte which is widely used as a biocontrol agent against several plant diseases. However, the development of endophytes as putative biocontrol agents on a commercial scale would require several experiments to be conducted under standardized conditions and also in fields.

To prove that endophytes are an attractive alternative to chemical pesticides, the physiological relationships between host plants and their environmental effects have to be studied by employing techniques such as metagenomics, advanced statistics, and ecological dynamics. Extensive screening, and testing antagonistic ability in greenhouse and field trials, coupled with biomass production, will help to ensure global agricultural security (De Silva et al. 2019).

6.7.3 Fungal Endophytes Have Potential in Managing Abiotic Stresses in Wheat

Abiotic stresses, such as drought, salinity, high temperatures, and heavy metal toxicity, can significantly impact crop productivity. These are the mechanisms by which fungal endophytes mitigate the abiotic stresses in wheat.

Drought Tolerance: Fungal endophytes can enhance the drought tolerance of wheat plants. They promote plant water status regulation by producing osmolytes and substances that improve water-use efficiency. These endophytes also help in maintaining stomatal closure, reducing water loss through transpiration. By improving the water status of the plants, fungal endophytes enable wheat crops to better withstand drought conditions and maintain productivity even during water scarcity.

Salt Stress Management: Fungal endophytes play a role in alleviating salt stress in wheat crops. They facilitate ion homeostasis and regulate sodium uptake, preventing the toxic buildup of salts in plant tissues. Endophytes can enhance the production of osmolytes, such as proline, that protect cells from osmotic stress caused by high salt concentrations. By improving salt tolerance, fungal endophytes help wheat plants maintain growth and productivity in saline environments.

Thermotolerance: High temperatures can negatively impact wheat growth and development. Fungal endophytes can help mitigate the adverse effects of heat stress. They produce heat-shock proteins and antioxidants that protect plant cells from heat-induced damage. These endophytes also enhance the expression of genes involved in heat stress response pathways. By improving thermotolerance, fungal endophytes assist wheat crops in coping with high-temperature conditions and reducing yield losses.

Heavy Metal Detoxification: Fungal endophytes are known for their ability to tolerate and detoxify heavy metals in contaminated soils. They can accumulate heavy metals within their structures, preventing their translocation into the plant tissues. This capability helps in reducing heavy metal toxicity in wheat crops grown in contaminated environments. Fungal endophytes can also enhance plant growth by promoting nutrient uptake and stimulating antioxidant systems that counteract heavy metal-induced oxidative stress.

Hormonal Regulation: Fungal endophytes produce plant growth-promoting substances such as auxins, cytokinins, and gibberellins. These hormones play a crucial role in regulating plant growth, development, and stress responses. By influencing hormone balance, fungal endophytes can enhance the adaptive capacity of wheat plants to various abiotic stresses.

The response of *Alternaria alternata* (Lq1230) (endophyte) inoculated in wheat to drought stress was evaluated by Qiang et al. (2019). The endophyte inoculated in the plants secreted indole acetic acid by tryptophan-dependent as well as tryptophan-independent pathways. The endophyte regulated the antioxidant enzyme activities and compatible solutes (such as soluble sugars and proline) content, enhanced growth, and drought tolerance. Additionally, improvement in photosynthesis and better C:N content leading to better dry biomass were recorded during water deficit and rewatering. Thus, the fungal endophyte *Alternaria alternata* (Lq1230) was responsible for reprogramming the metabolism in wheat plants and alleviating drought stress (Qiang et al. 2019).

Hubbard et al. (2014a, b) studied the effect of fungal endophytes on the growth, ecophysiological, and reproductive success of wheat exposed to heat and drought. Several parameters such as [a] efficiency of photosystem ii (f-v/f-m), [b] plant height, [c] average seed weight (asw), [d] total seed weight (tsw), [e] water-use efficiency (wue), [f] time to 50% germination, and [g] percentage germination of second-generation seeds were measured under heat-, drought-stress, and well-watered conditions. Seeds containing endophytes germinated rapidly. A consortium of endophytes was characterized and demonstrated to protect the wheat from abiotic stress. Intergenerational endophyte effects were also observed, but elucidation of underlying mechanisms would require in-depth research.

It was demonstrated that colonization by endophytic fungi is an eco-friendly strategy which shows a response to environmental stresses while inducing tolerance in the host plant (Qiang et al. 2019). Thus, by harnessing the beneficial effects of fungal endophytes, wheat crops can exhibit improved tolerance to abiotic stresses, leading to enhanced yield and quality. Incorporating these endophytes in sustainable agricultural practices can help mitigate the negative impacts of abiotic stresses, reduce the need for external inputs like irrigation water and chemical treatments, and contribute to more resilient and productive wheat cultivation systems.

The role of endophytic fungi living inside host plants has been demonstrated to affect plant health positively, but knowledge on the effect of plant genotype on the fungal endophytic microbiome is lacking. For modern agriculture, a thorough understanding of how the plant genotype and its interaction with abiotic and biotic environments affect the endophytic community is of utmost importance. This would have direct implications on crop productivity, disease management, and development of biocontrol agents (Latz et al. 2021). Thus, from various reports, it is clear that the microbiome that resides in various plant tissues can be considered a major deciding factor in plant fitness by resisting biotic and abiotic factors (Saijo and Loo 2020).

6.7.4 Nutritional Enhancement by Presence of Endophytes

Earlier paragraphs have elucidated studies that show that specific endophytes help to acquire resistance to host plants against bacterial/fungal diseases and that the endophytes can be used as biocontrol agents in addition to their contributions toward susceptibility to stress such as drought, salt, cold, heat, etc. Another activity is that they colonize the roots contributing to the health of the plant by releasing phytosiderophores that help in metal and micronutrient absorption.

The wheat crop can be biofortified with iron to cope with iron deficiency which is the root cause of hidden hunger. Foliar fertilization, nutritional breeding, and selection are the strategies to enhance grain iron content. Wheat-associated microbes (rhizobacteria and endophytes) can be useful in aiding iron enrichment because of their ability to solubilize complex metals, transport them to tissues, etc. Shi et al. suggest that a combination of plant and microbial genetics (especially engineering of the plant with genes from the endophytes that aid in iron mobilization) will help in achieving iron biofortification in wheat (Shi et al. 2020).

6.8 Endophytes: Host Colonization and Transmission

Seed germination represents the most crucial stage in the life cycle of a plant. It influences plant survival and initial establishment in stressful environments (Hubbard et al. 2012). According to research findings, right through all stages of plant development, endophytes are associated with host plants. The seed is also colonized by endophytic microbes which could be transmitted vertically from parent plants to the next generation. This vertical transmission selects against pathogenicity and favors mutualism as these endophytes depend entirely on their host for survival and reproduction (Huang et al. 2016). The works of Shahzad et al. (2018), Walitang et al. (2019), and Kim et al. (2020) indicate that the seeds provide a stable microenvironment for microbial communities, which drive the root microbiome after germination.

Illumina-based sequencing was performed for determining the vertically transmitted seed microbiome from sprouts from sterilized seeds and seeds of healthy wheat plants. More fungal diversity was recorded from sprouts of sterilized seed, with *Emericella* being the most prominent fungal taxon. According to the authors, core endophyte microbiome illustrates an intimate relation with plants and would be helpful to study plant-microbe interaction and also in the selection of appropriate and effective biocontrol agents during wheat production (Huang et al. 2016).

Attempts at understanding the host-endophyte interaction and association of epigenetic mechanisms were reported (Hubbard et al. 2014a, b). Drought-stressed wheat seedlings cocultured with the endophyte SMCD 2206 showed DNA methylation patterns which were similar to unstressed controls. The DNA sequences isolated from prominent polymorphic MSAP bands were like a cacta-type transposon and two retrotransposons of gypsy and copia types. Another polymorphic band was observed to be similar to a wheat cytochrome p450 expressed sequence tag. These observations suggested that inoculation may be linked to altered plant responses to oxidative stress in drought conditions. Thus, SMCD 2206 colonization coincided with epigenetic differences in wheat.

A conceptual model viz., the hydrothermal time (HTT) model of germination was applied for the first time to predict the timing and energy of germination (EG) to find out if one (or more) endophytic fungus can enhance the heat and drought tolerance in wheat. The fungal endophytes had positive effects and indeed conferred increased percentage germination and relieved stress. This was evidenced by monitoring seed parameters such as fresh seedling weight (Hubbard et al. 2012).

The phyllosphere is an important entry point for many pathogens and forms an important habitat for a diverse microbiome. Currently, the understanding of the factors shaping the phyllosphere microbiome, the coexistence among members, and their effects on disease development is very limited (Sapkota et al. 2017). It is likely that the seed microbiota undergoes a strong directional selection and the assembly and transmission of microbiota to next generations change with the genetic composition of plants during domestication and cultivation (Ozkurt et al. 2020). It is accepted that the composition of the plant microbiota may be altered by ecological and evolutionary changes in the host population. The study by Ozkurt et al. (2020) provides insight into the diversity of vertically transmitted microbiota of wheat. The work contributes to our understanding of wheat as a "metaorganism." This insight into the wheat microbiota is of fundamental importance for the development of improved crops. The authors reported significant differences in the seed-associated bacterial communities of the wheat species, but less effects on the fungal communities.

In 2020, Latz et al. carried out a comprehensive study of the fungal microbiome from seed to plant to seed in wheat. The wheat cultivar was cultivated over one growth season under two environments. In this study, the effects of host genotype, temperature, humidity, precipitation (abiotic factors), and fungi present in the seed stock, air, and soil on the succession of the endophytic fungal communities were studied. The endophytes from roots, flag leaves, and seeds at harvest were analyzed. Data revealed that the effect of host genotype was location-specific and limited to leaves and roots. Abiotic factors did not affect the root microbiome but the phyllosphere microbiome showed differences with respect to abundance, diversity, and pathogenic genera. In aerial plant parts, air-borne fungi were the source of inoculum, with the host genotype and the environment contributing to the fungal endophyte diversity. Studies by Sharon et al. (2023) discuss the transmission mode and assembly of seed fungal endophyte communities in wheat and wheat wild relatives and the recent study by Sharon et al. (2024) proves the inseparable nature of wheat and the fungal endophyte communities which further has an influence on plant development.

6.9 Newer Technologies to Study Endophytes

It is important to note that the field of endophyte research is continuously evolving, and new guidelines and criteria specific to the study of endophytes may be developed in the future to further understand their interactions with host plants. In recent years, several studies have focused on endophytes for the control of plant diseases to ensure food security. The advances in molecular techniques have played a very important role in understanding the interaction between plants, endophytes, and pathogens. Genomics, proteomics, transcriptomics, and metabolomics are some of the powerful techniques that can be used to elucidate pathways that increase plant tolerance, disease resistance, metabolite production, pathogen response, etc. These techniques are referred to as the "omics" (Kaul et al. 2014; Laur et al. 2018). Recently, Laur et al. (2018) applied RNAseq as a tool to gain insight into the interaction of the yeast BCA *Pseudozyma flocculosa*, powdery mildew, and barley. The responses of all three organisms at different time points were simultaneously recorded. These studies were supported by light microscopy, scanning, and transmission electron microscopy observations of the various phenotypes.

Gene expression studies, enzyme assays, and the use of knockout mutants or RNA interference are some of the other tools to study the antagonistic behavior of endophytes. The investigations of the tripartite interactions between the pathogen, plant, and endophyte can be visualized by confocal microscopy and light microscopy using specific fluorescent dyes (Karlsson et al. 2015). Mass Spectrometry-Based Molecular 3DCartography is a recent development which is a method for determining the distribution of specific metabolites in relation to microorganisms in situ (Floros et al. 2017). If this method is linked with molecular genetic studies, it can be extremely useful for understanding the role of specific metabolites.

Evolution of newer methods which are ecologically safe and can control disease whilst promoting crop yields is required. Unsustainable methods of fertilizer use can be replaced after a good understanding of the plant microbiome and these would truly represent the methods of the future (Prudence et al. 2021). According to Latz et al. (2018), the analysis and in-depth knowledge about the mutualistic relationships between plants and the endophytes will be helpful in the efficient control of pathogens and would also help in registering these as products for plant protection (Latz et al. 2018).

6.10 Conclusion

In-depth studies on the nature of wheat-endophyte interactions are required for understanding the requirements for successful establishment. Such studies will not only be locale-specific, but also will depend on the wheat species and soil type. The role of fungal compounds in disease reduction as well as those inducing responses in host plants will have to be investigated.

With this, markers for specific biocontrol agent(s) will be identified which will help boosting productivity and inducing disease resistance. The schematic (Fig. 6.4) clearly indicates that our current understanding of wheat endophytes is rather limited and several aspects need further exploration.

Fungal endophytes would thus contribute to the following aspects.

- *Enhanced Nutrient Uptake: Fungal endophytes, including certain species of the genus Epichloë, can improve nutrient uptake in wheat plants. They can enhance the plant's ability to acquire nutrients such as nitrogen and phosphorus from the*

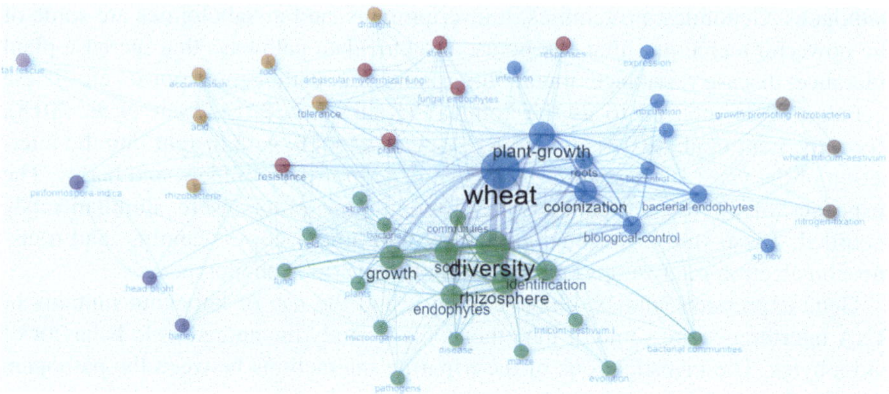

Fig. 6.4 Co-occurrence network

soil. This increased nutrient availability reduces the need for synthetic fertilizers, thus promoting sustainable agricultural practices.

- *Disease Resistance: Many fungal endophytes possess biocontrol properties, meaning they can protect plants from pathogens and pests. These endophytes produce antimicrobial compounds and induce systemic resistance in the host plant, reducing the dependence on chemical pesticides. By harnessing the natural defenses provided by endophytic fungi, farmers can adopt sustainable disease management strategies.*
- *Drought and Stress Tolerance: Fungal endophytes can improve the drought and stress tolerance of wheat plants. They produce substances that help regulate plant water status and protect against oxidative stress caused by environmental factors. This increased resilience to abiotic stresses reduces the impact of adverse conditions on crop productivity, contributing to sustainable agricultural practices.*
- *Reduced Environmental Impact: Sustainable agriculture aims to minimize the negative environmental impact of farming practices. Fungal endophytes can play a role in reducing chemical inputs, such as fertilizers and pesticides, thus decreasing pollution and ecological harm. By relying on the natural symbiotic interactions between plants and endophytic fungi, sustainable agriculture practices can be implemented with a reduced environmental footprint.*
- *Biodiversity Conservation: Studies on fungal endophytes contribute to the preservation of biodiversity in agricultural ecosystems. By exploring the diversity and ecological roles of these fungi, researchers can better understand the complex interactions within plant communities. This knowledge can help design agricultural systems that promote biodiversity conservation and ecological balance.*

Overall, research on fungal endophytes in wheat aligns with the principles of sustainable agriculture by enhancing nutrient uptake, disease resistance, stress tolerance, reducing chemical inputs, and promoting biodiversity conservation. By harnessing the beneficial properties of these endophytic fungi, sustainable agricultural practices can be developed and implemented, leading to more environmentally friendly and economically viable farming systems.

Acknowledgments The authors thank the Director, Agharkar Research Institute, Pune for providing the infrastructural facilities.

References

Aamir M, Rai KK, Zehra A, Kumar S, Yadav M, Shukla V, Upadhyay RS (2020) Fungal endophytes: classification, diversity, ecological role, and their relevance in sustainable agriculture. In: Microbial endophytes. Woodhead Publishing, pp 291–323

Abaya A, Xue A, Hsiang T (2021) Selection and screening of fungal endophytes against wheat pathogens. Biol Control 154:104511

Agnolucci M, Palla M, Cristani C, Cavallo N, Giovannetti M, De Angelis M et al (2019) Beneficial plant microorganisms affect the endophytic bacterial communities of durum wheat roots as detected by different molecular approaches. Front Microbiol 10:482306

Ahlawat OP, Yadav D, Kashyap PL, Khippal A, Singh G (2022) Wheat endophytes and their potential role in managing abiotic stress under changing climate. J Appl Microbiol 132(4):2501–2520

Anand U, Pal T, Yadav N, Singh VK, Tripathi V, Choudhary KK et al (2023) Current scenario and future prospects of endophytic microbes: promising candidates for abiotic and biotic stress management for agricultural and environmental sustainability. Microb Ecol 86(3):1455–1486

Anderson PK, Cunningham AA, Patel NG, Morales FJ, Epstein PR, Daszak P (2004) Emerging infectious diseases of plants: pathogen pollution, climate change and agrotechnology drivers. Trends Ecol Evol 19(10):535–544

Arnold AE (2007) Understanding the diversity of foliar endophytic fungi: progress, challenges, and frontiers. Fungal Biol Rev 21(2–3):51–66

Bacon CW, White JF (2000) Microbial endophytes. Marcel Dekker Inc, New York, pp 3–29

Bailey KL (2010) Canadian innovations in microbial biopesticides. Can J Plant Pathol 32:113–121

Bary A (1866) Morphologie und Physiologie der Pilze, Flechten, und Myxomyceten. In: Hofmeister's handbook of physiological botany, vol II. Verlag von Wilhelm Engelmann, Leipzig

Bebber DP, Ramotowski MAT, Gurr SJ (2013) Crop pests and pathogens move polewards in a warming world. Nat Clim Chang 3:985–988. https://doi.org/10.1038/nclimate1990

Becker LE, Marshall D, Cubeta MA (2024) A synergistic culture dependent and independent approach reveals a conserved wheat seed mycobiome. bioRxiv:2024–2002

Boyd LA, Ridout C, O'Sullivan DM, Leach JE, Leung H (2013) Plant–pathogen interactions: disease resistance in modern agriculture. Trends Genet 29:233–240. https://doi.org/10.1016/j.tig.2012.10.011

Busby PE, Peay KG, Newcombe G (2016) Common foliar fungi of Populus trichocarpa modify Melampsora rust disease severity. New Phytol 209(4):1681–1692

Butt TM, Copping LG (2000) Fungal biological control agents. Pesticide Outlook 11(5):186–191

Carroll F (1977) Preliminary studies on the incidence of needle endophytes in some European conifers. Sydowia 29:87–103

Carroll G (1988) Fungal endophytes in stems and leaves: from latent pathogen to mutualistic symbiont. Ecology 69(1):2–9

Chakraborty S, Newton AC (2011) Climate change, plant diseases and food security: an overview. Plant Pathol 60(1):2–14

Cłapa T, Mikołajczak K, Błaszczyk L, Narożna D (2020) Development of high-resolution melting PCR (HRM-PCR) assay to identify native fungal species associated with the wheat endosphere. J Appl Genet 61:629–635

Coakley SM, Scherm H, Chakraborty S (1999) Climate change and plant disease management. Annu Rev Phytopathol 37(1):399–426

Comby M, Lacoste S, Baillieul F, Profizi C, Dupont J (2016) Spatial and temporal variation of cultivable communities of co-occurring endophytes and pathogens in wheat. Front Microbiol 7:190376

Compant S, Cambon MC, Vacher C, Mitter B, Samad A, Sessitsch A (2021) The plant endosphere world–bacterial life within plants. Environ Microbiol 23(4):1812–1829

Currie AF, Wearn J, Hodgson S, Wendt H, Broughton SJ, Jin L (2014) Foliar fungal endophytes in herbaceous plants: a marriage of convenience. In: Verma VC, Gange AC (eds) Advances in endophytic research. Springer, New Delhi, pp 61–81

De Silva NI, Brooks S, Lumyong S, Hyde KD (2019) Use of endophytes as biocontrol agents. Fungal Biol Rev 33:133–148

Devi R, Verma R, Dhalaria R, Kumar A, Kumar D, Puri S et al (2023) A systematic review on endophytic fungi and its role in the commercial applications. Planta 257(4):70

Dingle J, Mcgee PA (2003) Some endophytic fungi reduce the density of pustules of *Puccinia recondita* f. sp. *tritici* in wheat. Mycol Res 107(3):310e316

Dutta D, Puzari KC, Gogoi R, Dutta P (2014) Endophytes: exploitation as a tool in plant protection. Braz Arch Biol Technol 57:621–629

Erenstein O, Jaleta M, Mottaleb KA, Sonder K, Donovan J, Braun HJ (2022) Global trends in wheat production, consumption and trade. In: Reynolds MP, Braun HJ (eds) Wheat improvement. Springer, Cham. https://doi.org/10.1007/978-3-030-90673-3_4

Etesami H, Alikhani HA (2018) *Bacillus* species as the most promising bacterial biocontrol agents in rhizosphere and endorhiza of plants grown in rotation with each other. Eur J Plant Pathol 150:497–506

Fisher MX (2013) Emerging fungal threats to animal, plant and ecosystem health. In: MYCOSES, vol 56. Wiley-Blackwell, Hoboken, p 13

Floros DJ, Petras D, Kapono CA, Melnik AV, Ling TJ, Knight R, Dorrestein PC (2017) Mass spectrometry-based molecular 3D-cartography of plant metabolites. Front Plant Sci 8:429

Gamboa MA, Laureano S, Bayman P (2002) Measuring diversity of endophytic fungi in leaf 642 fragments: dose sizematter? Mycopathologia 156:41–45

Gdanetz K, Trail F (2017) The wheat microbiome under four management strategies, and potential for endophytes in disease protection. Phytobiomes 1(3):158–168

Geisen S, Kostenko O, Cnossen MC, Ten Hooven FC, Vreš B, van Der Putten WH (2017) Seed and root endophytic fungi in a range expanding and a related plant species. Front Microbiol 8:1645

Ghini R, Bettiol W, Hamada E (2011) Diseases in tropical and plantation crops as affected by climate changes: current knowledge and perspectives. Plant Pathol 60(1):122–132

Glare T, Caradus J, Gelernter W, Jackson T, Keyhani N, Köhl J, Marrone P, Morin L, Stewart A (2012) Have biopesticides come of age? Trends Biotechnol 30:250–258

Grabka R, d'Entremont TW, Adams SJ, Walker AK, Tanney JB, Abbasi PA, Ali S (2022) Fungal endophytes and their role in agricultural plant protection against pests and pathogens. Plan Theory 11(3):384

Hallmann J, Sikora RA (1996) Toxicity of fungal endophyte secondary metabolites to plant parasitic nematodes and soil-borne plant pathogenic fungi. Eur J Plant Pathol 102:155–162

Hardoim PR, Van Overbeek LS, Berg G, Pirttilä AM, Compant S, Campisano A et al (2015) The hidden world within plants: ecological and evolutionary considerations for defining functioning of microbial endophytes. Microbiol Mol Biol Rev 79(3):293–320

Hata F, Futai K (1995) Endophytic fungi associated with healthy pine needles and needles infested by pine needle gall midge *Thecodiplosis japonensis*. Can J Bot 73:384–390

Herrera SD, Grossi C, Zawoznik M, Groppa MD (2016) Wheat seeds harbour bacterial endophytes with potential as plant growth promoters and biocontrol agents of *Fusarium graminearum*. Microbiol Res 186:37–43

Hirsch G, Braun U (1992) Communities of parasitic microfungi. In: Fungi in vegetation science. Springer, pp 225–250

Huang Y, Kuang Z, Wang W, Cao L (2016) Exploring potential bacterial and fungal biocontrol agents transmitted from seeds to sprouts of wheat. Biol Control 98:27e33

Hubbard M, Germida J, Vujanovic V (2012) Fungal endophytes improve wheat seed germination under heat and drought stress. Botany 90(2):137–149

Hubbard M, Germida JJ, Vujanovic V (2014a) Fungal endophyte colonization coincides with altered DNA methylation in drought-stressed wheat seedlings. Can J Plant Sci 94(2):223–234

Hubbard M, Germida JJ, Vujanovic V (2014b) Fungal endophytes enhance wheat heat and drought tolerance in terms of grain yield and second-generation seed viability. J Appl Microbiol 116(1):109–122

Igrejas G, Branlard G (2020) The importance of wheat. In: Igrejas G, Ikeda T, Guzmán C (eds) Wheat quality for improving processing and human health. Springer, Cham. https://doi.org/10.1007/978-3-030-34163-3_1

Jensen BD, Knorr K, Nicolaisen M (2016) *In vitro* competition between *Fusarium graminearum* and *Epicoccum nigrum* on media and wheat grains. Eur J Plant Pathol 146:657–670

Karlsson M, Durling M B, Choi J, Kosawang C, Lackner G, Tzelepis G D, et al (2015) Insights on the evolution of mycoparasitism from the genome o Cf lonostachys rosea. Genome Biol Evol 7:465–480

Kaul S, Sharma T, Dhar MK (2014) Omics tools for better understanding the plant 704 endophyte interactions. Front Plant Sci. https://doi.org/10.3389/fpls.2016.00955

Kaul N, Kashyap PL, Kumar S, Singh D, Singh GP (2022) Diversity and exploration of endophytic bacilli for the management of head scab (*Fusarium graminearum*) of wheat. Pathogens 11(10):1088

Khare E, Mishra J, Arora NK (2018) Multifaceted interactions between endophytes and plant: developments and prospects. Front Microbiol 9:411604

Kim H, Lee KK, Jeon J, Harris WA, Lee YH (2020) Domestication of Oryza species eco-evolutionarily shapes bacterial and fungal communities in rice seed. Microbiome 8(1):20

Kiss L, Russell JC, Szentiványi O, Xu X, Jeffries P (2004) Biology and biocontrol potential of *Ampelomyces* mycoparasites, natural antagonists of powdery mildew fungi. Biocontrol Sci Tech 14(7):635–651

Kumar AP, Murali V, Nagaraju K, Srinivas M (2023) Fungal endophytes: a potential application in integrated plant health management. Int J Plant Soil Sci 35(18):1570–1578

Kuźniar A, Włodarczyk K, Grządziel J, Goraj W, Gałązka A, Wolińska A (2020a) Culture-independent analysis of an endophytic core microbiome in two species of wheat: *Triticum aestivum* L.(cv.'Hondia') and the first report of microbiota in *Triticum spelta* L.(cv.'Rokosz'). Syst Appl Microbiol 43(1):126025

Kuźniar A, Włodarczyk K, Grządziel J, Woźniak M, Furtak K, Gałązka A, Dziadczyk E, Skórzyńska-Polit E, Wolińska A (2020b) New insight into the composition of wheat seed microbiota. Int J Mol Sci 21(13):4634

Kuźniar A, Włodarczyk K, Grządziel J, Woźniak M, Furtak K, Gałązka A et al (2020c) New insight into the composition of wheat seed microbiota. Int J Mol Sci 21(13):4634

Larrán S, Perello A, Simón MR, Moreno V (2002) Isolation and analysis of endophytic microorganisms in wheat (Triticum aestivum L.) leaves. World J Microbiol Biotechnol 18:683–686

Larran S, Perelló A, Simón MR, Moreno V (2007) The endophytic fungi from wheat (*Triticum aestivum* L.). World J Microbiol Biotechnol 23:565–572

Larran S, Simon, MR, Moreno MV, Siurana MS, Perelló A (2016) Endophytes from wheat as biocontrol agents against tan spot disease. Biological Control 92:17–23

Lastochkina OV, Allagulova CR (2023a) The mechanisms of the growth promotion and protective effects of endophytic PGP bacteria in wheat plants under the impact of drought. Appl Biochem Microbiol 59(1):14–32

Lastochkina OV, Allagulova CR (2023b) Mechanisms of growth promotional and protective effects of endophytic PGP-bacteria in wheat plants under the impact of drought. Prikladnaâ Biohimiâ i Mikrobiologiâ 59(1):17–37

Latz MA, Jensen B, Collinge DB, Jørgensen HJ (2018) Endophytic fungi as biocontrol agents: elucidating mechanisms in disease suppression. Plant Ecol Divers 11(5–6):555–567

Latz MA, Kerrn MH, Sørensen H, Collinge DB, Jensen B, Brown JK et al (2021) Succession of the fungal endophytic microbiome of wheat is dependent on tissue-specific interactions between host genotype and environment. Sci Total Environ 759:143804

Laur J, Ramakrishnan GB, LabbÚ C, Lefebvre F, Spanu PD, BÚlanger RR (2018) Effectors involved in fungal–fungal interaction lead to a rare phenomenon of hyperbiotrophy in the tritrophic system biocontrol agent–powdery mildew–plant. New Phytol 217(2):713–725

Le Cocq K, Gurr SJ, Hirsch PR, Mauchline TH (2017) Exploitation of endophytes for sustainable agricultural intensification. Mol Plant Pathol 18(3):469–473

Magnusson E, Cranfield JAL (2005) Consumer demand for pesticide free food products in Canada: a probit analysis. Can J Agric Econ Rev Can D'Agroecon 53:67–81

Makar OO, Romanyuk ND (2022) Endophytic bacteria of wheat and the potential to improve microelement composition of grain. Біологічні студії/Studia Biologica 16(3):101–128

Makar O, Kuźniar A, Patsula O, Kavulych Y, Kozlovskyy V, Wolińska A et al (2021) Bacterial endophytes of spring wheat grains and the potential to acquire Fe, Cu, and Zn under their low soil bioavailability. Biology 10(5):409

Mishra A, Gond SK, Kumar A, Sharma VK, Verma SK, Kharwar RN, Sieber TN (2012) Season and tissue type affect fungal endophyte communities of the Indian medicinal plant *Tinospora cordifolia* more strongly than geographic location. Microb Ecol 64:3288–3398

Mishra S, Bhattacharjee A, Sharma S (2021) An ecological insight into the multifaceted world of plant-endophyte association. Crit Rev Plant Sci 40(2):127–146

Mostert L, Crous P, Petrini O (2000) Endophytic fungi associated with shoots and leaves of Vitis vinifera, with specific reference to the Phomopsis viticola complex. Sydowia 52:46–58

Mousa WK, Raizada MN (2013) The diversity of anti-microbial secondary metabolites produced by fungal endophytes: an interdisciplinary perspective. Front Microbiol 4:65

O'Hanlon KA, Knorr K, Jørgensen LN, Nicolaisen M, Boelt B (2012) Exploring the potential of symbiotic fungal endophytes in cereal disease suppression. Biol Control 63(2):69–78

Ofek-Lalzar M, Gur Y, Ben-Moshe S, Sharon O, Kosman E, Mochli E, Sharon A (2016) Diversity of fungal endophytes in recent and ancient wheat ancestors *Triticum dicoccoides* and *Aegilops sharonensis*. FEMS Microbiol Ecol 92(10):152

Orduña FNR, Sanchez RAS, Bustamante ZRF, Rodriguez JNG, Cotera LBF (2011) Diversity of endophytic fungi of *Taxus globosa* (Mexican yew). Fungal Divers 47:65–74

Ott SL, Huang CL, Misra SK (1991) Consumers' perceptions of risks from pesticide residues and demand for certification of residue-free produce. In: Caswell JA (ed) Economics of food safety. Springer, Dordrecht, pp 175–188. ISBN 9789401170765

Özkurt E, Hassani MA, Sesiz U, Künzel S, Dagan T, Özkan H, Stukenbrock EH (2020) Seed-derived microbial colonization of wild emmer and domesticated bread wheat (*Triticum dicoccoides* and *T. aestivum*) seedlings shows pronounced differences in overall diversity and composition. MBio 11(6):1110–1128

Pang F, Tao A, Ayra-Pardo C, Wang T, Yu Z, Huang S (2022) Plant organ-and growth stage-diversity of endophytic bacteria with potential as biofertilisers isolated from wheat (Triticum aestivum L.). BMC Plant Biol 22(1):276

Pastoshchuk AY, Skivka LM, Butsenko LM, Patyka VP (2018) Vplyvzbudnyka bazalnoho bakteriozu na prorostannia zeren ta rist parostkiv pshenytsi riznykh sortiv [Effect of causal agent ofbasal bacteriosis on seed germination and root growth of different wheat varieties]. Mikrobiolohiia i Biotekhnolohiia 42:39–48 (in Ukrainian)

Pastoshchuk A, Yumyna Y, Zelena P, Nudha V, Yanovska V, Kovalenko M et al (2021a) Beneficial traits of grain-residing endophytic communities in wheat with different sensitivity to Pseudomonas syringae. Regul Mech Biosyst 12(3):498–505

Pastoshchuk A, Yumyna Y, Zelena P, Skivka L (2021b) Survey of plant growth promoting and antagonistic traits in winter wheat grain endophytic bacteria. EUREKA: Life Sci 4:66–72

Perelló AE, Moreno MV, Mónaco C, Simón MR, Cordo C (2009) Biological control of Septoria tritici blotch on wheat by Trichoderma spp. under field conditions in Argentina. BioControl 54:113–122

Petrini O (1991) Fungal endophytes of tree leaves. In: Andrews JH, Hirano SS (eds) Microbial ecology of leaves. Springer, New York, pp 179–197

Prudence SM, Newitt JT, Worsley SF, Macey MC, Murrell JC, Lehtovirta-Morley LE, Hutchings MI (2021) Soil, senescence and exudate utilisation: characterisation of the Paragon var. spring bread wheat root microbiome. Environ Microbiome 16(1):12

Qiang X, Ding J, Lin W, Li Q, Xu C, Zheng Q, Li Y (2019) Alleviation of the detrimental effect of water deficit on wheat (*Triticum aestivum* L.) growth by an indole acetic acid-producing endophytic fungus. Plant Soil 439:373–391

Rana KL, Kour D, Sheikh I, Yadav N, Yadav AN, Kumar V et al (2019) Biodiversity of endophytic fungi from diverse niches and their biotechnological applications. In: Advances in endophytic fungal research: present status and future challenges. Springer, pp 105–144

Ren Z, Chen AJ, Zong Q, Du Z, Guo Q, Liu T et al (2023) Microbiome signature of endophytes in wheat seed response to wheat dwarf bunt caused by *Tilletia controversa* Kühn. Microbiol Spect 11(1):e00390–e00322

Ridout ME, Schroeder KL, Hunter SS, Styer J, Newcombe G (2019) Priority effects of wheat seed endophytes on a rhizosphere symbiosis. Symbiosis 78:19–31

Robinson RJ, Fraaije BA, Clark IM, Jackson RW, Hirsch PR, Mauchline TH (2016) Endophytic bacterial community composition in wheat (Triticum aestivum) is determined by plant tissue type, developmental stage and soil nutrient availability. Plant Soil 405:381–396

Rodriguez RJ, White JF, Arnold JAE, Redman RS (2009) Fungal endophytes: diversity and 801 functional roles. New Phytol 182:314–330

Rollinger JL, Langenheim JH (2018) Geographic survey of fungal endophyte community composition in leaves of coastal redwood. Mycologia 85(2):149–156

Saijo Y, Loo EP (2020) Plant immunity in signal integration between biotic and abiotic stress responses. New Phytol 225:87–104

Salamon S, Mikołajczak K, Błaszczyk L (2023) Constellation of the endophytic mycobiome in spring and winter wheat cultivars grown under various conditions. Sci Rep 13(1):6089. (to add)

Sapkota R, Jørgensen LN, Nicolaisen M (2017) Spatiotemporal variation and networks in the mycobiome of the wheat canopy. Front Plant Sci 8:287144

Schouten A (2019) Endophytic fungi: definitions, diversity, distribution and their significance in plant life. In: Endophyte biotechnology: potential for agriculture and pharmacology. CAB International, pp 6–31

Schulz B (2006) Mutualistic interactions with fungal root endophytes. In: Microbial root endophytes. Springer, pp 261–279

Schulz B, Boyle C (2005) The endophytic continuum. Mycol Res 109:661–686

Segaran G, Sathiavelu M (2019) Fungal endophytes: a potent biocontrol agent and a bioactive metabolites reservoir. Biocatal Agric Biotechnol 21:101284

Shahzad R, Khan A, Bilal S, Asaf S, Lee IJ (2018) What is there in seeds? Vertically transmitted endophytic resources for sustainable improvement in plant growth. Front Plant Sci 9:24

Sharon O, Sun X, Ezrati S, Kagan-Trushina N, Sharon A (2023) Transmission mode and assembly of seed fungal endophyte communities in wheat and wheat wild relatives. Phytobiomes J 7(1):113–124

Sharon O, Kagan-Trushina N, Sharon A (2024) Wheat fungal endophyte communities are inseparable from the host and influence plant development. MBio 15(2):e02533–e02523

Shi Y, Li J, Sun Z (2020) Success to iron biofortification of wheat grain by combining both plant and microbial genetics. Rhizosphere 15:100218

Shiferaw B, Smale M, Braun HJ et al (2013) Crops that feed the world 10. Past successes and future challenges to the role played by wheat in global food security. Food Sec 5:291–317. https://doi.org/10.1007/s12571-013-0263-y

Singh A, Knox RE, DePauw RM, Singh AK, Cuthbert RD, Kumar S, Campbell HL (2016) Genetic mapping of common bunt resistance and plant height QTL in wheat. Theor Appl Genet 129:243–256. https://doi.org/10.1007/s00122-015-2624-8

Sun X, Kosman E, Sharon A (2020) Stem endophytic mycobiota in wild and domesticated wheat: structural differences and hidden resources for wheat improvement. J Fungi 6(3):180

Verma SK, Gond SK, Mishra A, Sharma VK, Kumar J, Singh DK, Kumar A, Goutam J, Kharwar RN (2014) Impact of environmental variables on the isolation, diversity and antibacterial activity of endophytic fungal communities from *Madhuca indica* Gmel. at different locations in India. Ann Microbiol 64(2):721–734

Verma SK, Kharwar RN, Gond SK, Kingsley KL, White JF (2019) Exploring endophytic communities of plants: methods for assessing diversity, effects on host development and potential biotechnological applications. In: Seed endophytes: biology and biotechnology. Springer, pp 55–82

Walitang DI, Kim CG, Jeon S, Kang Y, Sa T (2019) Conservation and transmission of seed bacterial endophytes across generations following crossbreeding and repeated inbreeding of rice at different geographic locations. Microbiol Open 8(3):e00662

Wilson D (1995) Endophyte: the evolution of a term, and clarification of its use and definition. Oikos 73(2):274

Witzell J, Martín JA, Blumenstein K (2014) Ecological aspects of endophyte-based biocontrol of forest diseases. Adv Endophytic Res:321–333

Part II

Plant Protection Application

Endophytic Fungi for Reduction of Mycotoxin Contamination in Crops

7

Heba Yousef, Kamel A. Abd-Elsalam,
and Salah M. Abdel-Momen

Abstract

Common pollutants in agricultural commodities include mycotoxins, produced by toxigenic fungi such as *Fusarium* and *Aspergillus*. Conventional techniques using chemical fungicides can pollute the environment and lead to fungicide resistance. Eco-friendly and sustainable alternatives are provided by endophytic fungi like *Penicillium* and *Trichoderma*. These fungi produce antifungal chemicals, induce systemic resistance against infections, and compete for resources. By utilizing the biocontrol capabilities of endophytic fungi, farmers can protect their crops from mycotoxin contamination and enhance food safety. To fully investigate the advantages of these helpful fungi and create useful agricultural applications, further study must be conducted.

Keywords

Endophytic fungi · Mycotoxin contamination · Crops · Biological control · Food safety

7.1 Introduction

Secondary metabolites known as mycotoxins are produced by toxigenic fungi, such as *Aspergillus* and *Fusarium* species, which often contaminate various agricultural products. When inhaled, these mycotoxins can cause detrimental effects on both humans and animals, leading to a range of acute and long-term health issues (Yousef et al. 2022; Ekwomadu et al. 2021). Chemical fungicides are commonly used to

H. Yousef (✉) · K. A. Abd-Elsalam · S. M. Abdel-Momen
Plant Pathology Research Institute, Agricultural Research Center (ARC), Giza, Egypt
e-mail: heba.yousef@arc.sci.eg

© The Author(s), under exclusive license to Springer Nature Singapore Pte Ltd. 2025

K. A. Abd-Elsalam, A. H. Hashem (eds.), *Fungal Endophytes Volume II*,
https://doi.org/10.1007/978-981-97-8804-0_7

prevent mycotoxin contamination and control toxigenic fungi in conventional agriculture; however, they pose their own risks, including fungicide resistance and environmental contamination (Martínez-Culebras et al. 2021). Toxigenic fungus and mycotoxin contamination can be reduced by using endophytic fungi, a potentially effective alternative to chemical fungicides. Endophytic fungi can produce antifungal chemicals, stimulate systemic resistance to infections, and compete for resources through their interactions with the host plant (Wen et al. 2022; Terkar and Borde 2021). There is evidence that certain endophytic fungi can inhibit the growth or toxin production of toxigenic fungi. This natural method of biocontrol provides an eco-friendly and sustainable alternative to chemical fungicides (Baron and Rigobelo 2022; Adeleke et al. 2022). Studies have demonstrated that mycotoxigenic fungi, such as *Aspergillus* and *Fusarium*, are significantly inhibited by specific endophytic fungi, including *Trichoderma* and *Penicillium* species. These beneficial endophytes can colonize plant tissues, compete with pathogenic fungi for resources and space, and produce antifungal substances that reduce the formation of mycotoxins (Tyśkiewicz et al. 2022; Harman et al. 2021). By utilizing the biocontrol capabilities of endophytic fungi, farmers can enhance the safety of food and feed products and protect their crops from mycotoxin contamination (Bhardwaj et al. 2023; Fite et al. 2023). To investigate the advantages of helpful fungi and create useful agricultural applications, further study is required. Using endophytic fungi to their full potential could transform our understanding of mycotoxin and fungus-related diseases (Devi et al. 2023). This chapter focuses on how endophytic fungi can be used as biocontrol agents to minimize crop contamination with mycotoxins. We will review the processes by which endophytic fungi combat mycotoxigenic fungi, the factors that affect their efficacy, and the opportunities and challenges associated with using them in agricultural practices. Improving our understanding of the role of endophytic fungi in mycotoxin management will enable us to develop more effective strategies for ensuring the quality and safety of crops intended for human and animal consumption.

7.2 Mycotoxins: Definition and Significance in Crop Contamination

7.2.1 What Are Mycotoxins?

Natural poisons are the main sources of contamination for foodstuffs and agricultural commodities. As a defense strategy against predators, insects, or microbes, living organisms such as algae, bacteria, fungi, and plants produce these chemicals, which are toxic secondary metabolites. Toxins can cause severe health issues like allergic reactions, excruciating stomachaches, and even death. They also result in financial losses for the food industry and manufacturers (Marins-Gonçalves et al. 2023). Since 1960, more than 400 different types of mycotoxins have been discovered and documented. The most prevalent and dangerous mycotoxins found in food include aflatoxins (AFs), trichothecenes (TCNs), ochratoxins (OTs), zearalenone (ZEA), fumonisins (FMs), and patulins (PATs). These significant pollutants can contaminate raw agricultural commodities and may be produced by various toxigenic fungi due to improper harvesting, storage, or transportation conditions (Liu et al. 2022). Most of them have complex chemical structures with specific groups that confer toxic effects.

7.2.1.1 Aflatoxins

The most researched mycotoxin found in agricultural commodities is aflatoxin (AF). The four principal Aspergillus species that produce them are *A. flavus, A. parasiticus, A. nomius,* and *A. tamarii*. AFB1, the most dangerous of the AFs, is categorized as a Group 1 naturally occurring carcinogen. AFs are furanocoumarins linked to a coumarin that is related to a bisdihydrodifuran or tetrahydrobisfuran and is substituted by a lactone or cyclopentenone (Afshar et al. 2020). The cytotoxicity and carcinogenic potential of AFs are caused by the lactone ring and the double bond in the terminal furan ring (Nazhand et al. 2020; Liu et al. 2022).

7.2.1.2 Trichothecenes

Trichothecenes (TCNs) are a significant class of mycotoxins with a considerable economic impact on the food chain annually. *Aspergillus, Penicillium, Fusarium, Spicellum,* and *Stachybotrys* are some of the filamentous fungi that produce TCNs in nature; *Fusarium* is the most common toxin source in cereals, vegetables, and feedstock (Polak-Śliwińska and Paszczyk 2021). TCNs are tetracyclic sesquiterpenoid epoxides that constitute a family of over 200 toxins sharing a common core structure: an atricyclic 12,13-epoxytrichothec-9-ene (EPT). Toxins in this family are categorized into four classes (types A, B, C, and D) based on their structural makeup and the EPT substitution pattern. Types A and B are particularly interesting due to their high levels of toxicity (Khaneghah et al. 2020). Representative type A trichothecene mycotoxins (ATCNs) include T-2 toxin (T-2), HT-2 toxin (HT-2), and neosolaniol (NEO). Type B trichothecene mycotoxins (BTCNs) are represented by deoxynivalenol (DON) and its acetyl derivatives (Liu et al. 2022). Trichothecenes can be further divided into four classes: Type A–D. Type B trichothecenes have a ketone group on C8 of the EPT nucleus, differentiating them from Type A trichothecenes (Abraham et al. 2022). Trichothecenes of Type D have a di- or tri-ester connection across the C4 and C15 regions, while Type C has an epoxide ring at C8 (Fig. 7.1).

7.2.1.3 Ochratoxins

Ochratoxins (OTs) are a family of mycotoxins commonly found in various types of food and feed. Penicillium and Aspergillus species, such as *P. verrucosum, P. nordicum, A. ochraceus, A. carbonarius,* and *A. niger*, are the primary producers of OTs (Kumar et al. 2020). They belong to a class of coumarin compounds which includes ochratoxin A (OTA), ochratoxin B (OTB), and ochratoxin C (OTC), with OTA being the most common and toxic (Liu et al. 2022).

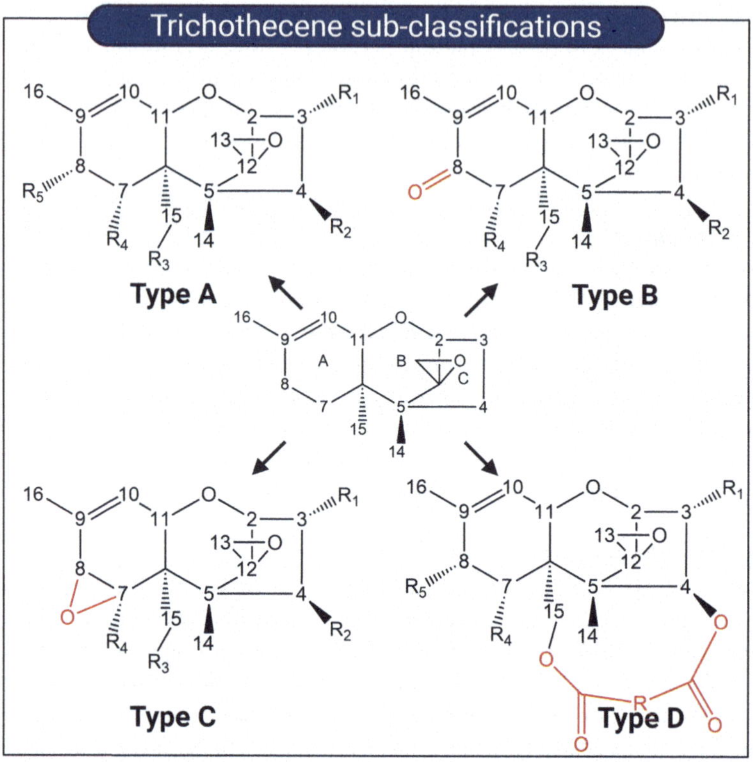

Fig. 7.1 Subdivision of trichothecenes. Based on substitution patterns surrounding the EPT nucleus, trichothecenes can be further divided into four classes. Red highlights indicate important functional groupings that set each category to separate from the others. (Reprinted from Abraham et al. 2022)

7.2.1.4 Zearalenone

Numerous grains are contaminated with zearalenone (ZEA), a nonsteroidal mycotoxin produced by several species of mold fungi in the Fusarium family (Rai et al. 2018). The lactone group in ZEA's chemical structure is essential to its toxicity. Reports indicate that cleaving the lactone ring can drastically reduce the harmful effects of ZEA (Al-Jaal et al. 2019).

7.2.1.5 Fumonisins

Fumonisins (FBs), mycotoxins originating from polyketides, are predominantly produced by *Fusarium* mold species. Four primary groups of fumonisin analogs have been identified, each with unique structures. Among these, FB1 is the most widespread and toxic. The toxicity mechanism of FBs is largely determined by the tricarballylic acid (TCA) side chains and the free amino group (Li et al. 2012).

7.2.1.6 Patulin

The most common source of patulin (PAT), a tetraketide mycotoxin, is apples and products made from them (Diao EnJie et al. 2018). Many filamentous fungi, including *Penicillium, Aspergillus,* and *Byssochlamys*, are responsible for its production. Since this mycotoxin can readily form covalent bonds with sulfhydryl-containing compounds, which is the principal mechanism associated with PAT toxicity, it has also been demonstrated that PAT affects the activity of thiol-containing enzymes that play significant roles in glycolysis (Awuchi et al. 2022). Mycotoxins, being poisonous secondary metabolites, do not directly impact the growth and reproduction of mold. Instead, they are believed to enhance the competitiveness of molds against other microbes (Bulgaru et al. 2021). Mycotoxin contamination of crops, both before and after harvest, leads to the emergence of mycotoxins in the food chain. Exposure can occur either directly through the consumption of contaminated food or indirectly through feeding animals contaminated feed (Awuchi et al. 2022). Mycotoxins pose a serious concern in agriculture due to their ability to infect crops and subsequently infiltrate the food chain, presenting considerable health risks to both humans and animals. Therefore, it is imperative to implement appropriate measures to reduce mycotoxin contamination in crops (Awuchi et al. 2022).

7.3 Significance of Mycotoxins in Crop Contamination

Mycotoxins are harmful substances found in contaminated crops or their derived products. When consumed, they can lead to various health issues, ranging from acute poisoning to long-term effects such as immune deficiency and cancer. Different mycotoxins target specific organs and systems, causing problems such as liver damage, immune suppression, and neurological disorders (Awuchi et al. 2021). Crop contamination by mycotoxins results in economic losses for farmers and the food industry. Infected crops are often downgraded or rejected in international markets, leading to reduced profits and trade restrictions. Stringent testing increases production costs. Mycotoxins threaten food security, especially in regions where staple crops such as maize, peanuts, wheat, and rice are susceptible to contamination. Climate change exacerbates mycotoxin risks by creating favorable conditions for mold growth (Zingales et al. 2022). Regulations and guidelines have been established to limit mycotoxin levels in food and promote safe agricultural practices (López-García 2022). Scientific research greatly aids the development of novel strategies to control and avoid mycotoxin contamination. This includes the creation of biological control agents, resistant crop varieties, and improved storage methods. Mycotoxins are hazardous substances produced by molds and pose a significant challenge to the food and agricultural industries. They are influenced by climate change, lead to economic losses, threaten food security, and endanger the health of humans and animals. To address this issue, continued research and regulatory actions are essential. Considering the role of mycotoxins in agricultural contamination is critical for protecting public health, ensuring food security, and preserving the sustainability of agriculture in a rapidly changing world.

7.4 Current Challenges and Approaches for Mycotoxins Reduction

Mycotoxins pose a significant concern in food safety, creating issues in food toxicology due to naturally occurring pollutants. These low-molecular-weight compounds are produced by various fungi as secondary metabolites. Mycotoxins enter the global food chain as contaminants when crops intended for human and animal consumption, or raw materials, become infected prior to harvest or are stored improperly (Karlovsky et al. 2016; Patel et al. 2021). They represent a serious health risk as they can cause a variety of adverse symptoms and are often found in food and feed. Fungal infections and the production of mycotoxins are extremely difficult to prevent, control, and avoid. In contrast, contamination by man-made toxicants typically ceases once the sources have been addressed. Consequently, mycotoxins are common in a wide range of food products and can serve as a persistent source of contamination. Moreover, as a result of the carryover phenomenon that occurs when animals are fed contaminated feedstuffs, mycotoxins can be detected in animal-derived products as well as grains and grain-based diets. Therefore, mycotoxins may contaminate the diets of various demographic groups, including newborns and the elderly. Reducing mycotoxin contamination in our food chain is a critical challenge that affects the entire world. This chapter examines the difficulties posed by mycotoxins and the various strategies used to reduce their incidence (Awuchi et al. 2022).

7.4.1 The Current Challenges for Mycotoxins Reduction

7.4.1.1 Environmental Factors

The environmental factors affecting mycotoxin contamination include temperature, humidity, and rainfall. These variables can change due to climate change, making it challenging to predict and mitigate mycotoxin contamination in a changing global environment. The absence of practical and economically viable techniques to prevent and control the growth of fungi and the formation of mycotoxins in storage and in the field remains a concern (Zingales et al. 2022).

7.4.1.2 Globalization of Food Trade

Considering the rising globalization of the economy, there is a greater chance of consumer exposure as mycotoxin-contaminated products can easily cross international boundaries. Additionally, farmers, consumers, and policymakers lack adequate awareness and knowledge regarding the health dangers and economic effects of mycotoxins. International collaboration and standardized mycotoxin management standards are necessary to address this issue (Agriopoulou et al. 2020).

7.4.1.3 Detection and Monitoring

The variety, complexity, and origins of mycotoxins make their identification and detection in food and feed challenging. The procedures used to detect mycotoxins are often expensive, time-consuming, and sophisticated. Developing quick and precise detection techniques to ensure the safety of food and feed is challenging (Janik et al. 2021).

7.4.1.4 Regulatory Challenges

The variations between national and regional mycotoxin limits, regulations, and enforcement methods present challenges. Mycotoxin restrictions are regulated differently across nations, complicating adherence to uniform guidelines for the international food sector. Integrating regulations is essential for improving food safety and facilitating global trade (Imade et al. 2021).

Mycotoxins pose a worldwide threat to food safety and security, necessitating a multidisciplinary response. No single method can completely remove toxins such as mycotoxins from food or feed. To minimize mycotoxin contamination, various techniques should be employed at different stages of the food chain.

7.4.2 The Current Approaches for Mycotoxins Reduction

7.4.2.1 Pre-harvest Management

Crop rotation, which involves alternating crops, can disrupt the life cycles of mycotoxin-producing fungi. Additionally, developing and utilizing genetically resistant crop varieties helps reduce mycotoxin contamination. Implementing Good Agricultural Practices, such as proper irrigation and crop sanitation, further contributes to mitigating mycotoxin contamination before harvest.

7.4.2.2 Post-harvest Management

In grain production, a comprehensive approach ensures safety and quality. Advanced sorting and cleaning technologies remove contaminants from the production line, while efficient drying methods reduce moisture content to prevent fungal growth. Proper storage conditions, including temperature and humidity control, are crucial in minimizing mycotoxin development during storage. Preventive measures based on good agricultural practices are designed to reduce the growth of fungi and the production of mycotoxins in the field and during storage. These methods, which include selecting resistant crop varieties, pest management, irrigation control, fertilization, drying, cleaning, sorting, and storing under optimal temperature, humidity, aeration, and sanitation conditions, can improve crop quality and yield.

7.4.2.3 Chemical Control

Chemical techniques can destroy the structure of mycotoxins, generating mildly toxic or nontoxic products. Chemical decontamination involves reacting with or neutralizing mycotoxins in food and feed using acids, bases, oxidizing agents, reducing agents, or other substances. For example, ammonia treatment disrupts the aflatoxin's lactone ring, lowering aflatoxin levels in corn. Similarly, sodium bisulfite breaks the toxin's amide link, reducing the amount of ochratoxin A in wine. However, these techniques might also impact the sensory qualities and nutritional value of the food products. Feed additives such as clay minerals and activated carbon can serve as mycotoxin binders, reducing their bioavailability by binding to them. In the field and during storage, fungicides can prevent mold growth (Liu et al. 2022). The commonly used methods for the chemical detoxification of mycotoxins are described below.

7.4.2.3.1 Alkaline Treatment

Various mycotoxins found in moldy feedstocks can be destroyed using alkaline chemicals such as sodium hydroxide, potassium hydroxide, ammonia, and sodium carbonate (Jalili et al. 2011). By producing coumarin sodium salt through base hydrolysis, the lactone ring structure of AFB1 may be opened, which can then be further removed by washing with water (Javanmardi et al. 2022). AFB1 is commonly removed from feed ingredients using ammoniation and hydroxide salt treatments, which have a clearance rate of more than 95% in various cereals. Alkaline conditions can destroy an epoxide at C-12 and C-13, which is crucial for the toxicity of DON. In various feedstocks, treatments with sodium carbonate and hydroxide salts can reduce DON levels by 83.9–100% (Bretz et al. 2006).

7.4.2.3.2 Ozone Treatment

Treatment with mycotoxin oxidizing agents is a useful detoxifying technique that modifies the molecular structure of mycotoxins. Ozone, hydrogen peroxide, sodium hypochlorite, calcium hypochlorite, chlorine, and other oxidizers are frequently utilized (Santos Alexandre et al. 2018). Ozone has been demonstrated to effectively degrade AFs, DON, ZEN, and FB1. Agriopoulou et al. (2020) showed that ozone is capable of breaking down AFs (AFB1, AFB2, AFG1, and AFG2). According to Jr. and King (2002), ozone can reduce AFs by 91% in cottonseed or 78% in peanut meal and by 92–95% in corn. Mycotoxins can be efficiently broken down by oxidizing agents such as hydrogen peroxide and sodium hypochlorite (Altugˇ et al. 1990).

7.4.2.4 Physical Control

Physical procedures for decontaminating mycotoxins include sorting and separation, washing, solvent extraction, heating, irradiation, and adsorption. The efficiency of these techniques is affected by the type and amount of mycotoxin, the temperature and duration of exposure, and the pH and moisture content of the food matrix, among other variables (Matumba et al. 2015).

7.4.2.4.1 Sorting and Separation

Cereals contaminated with mycotoxins have a specific gravity that is lower than usual. These properties allow for the separation of mycotoxin-contaminated feedstocks using techniques such as sieving, aspiration, gravity separation, photoelectric separation, and image processing (Tibola et al. 2016). Matumba et al. state that flotation, dehulling, and manual sorting can eliminate at least 51%, 63%, and 93% of aflatoxins (AFs), trichothecenes, and fumonisins from shelled white maize, respectively. Aspiration and gravity separation techniques can reduce the deoxynivalenol (DON) levels in wheat, albeit at the cost of a lower yield (Salgado et al. 2011).

7.4.2.4.2 Washing and Solvent Extraction

Mycotoxins can be decontaminated by washing with water or by extraction using an organic solvent, depending on whether they are fat- or water-soluble. Aflatoxins (AFs), trichothecenes, zearalenone (ZEN), and fumonisins can be eliminated from grains by 51–72%, 64–69%, 2–61%, and 73–74%, respectively, through floating and washing with water. Furthermore, the removal of fumonisins from corn and wheat can be enhanced by floating the grains and washing them with a water solution containing 10–30% NaCl, 30% sucrose, or 1 mol/L sodium carbonate (Karlovsky et al. 2016). Up to 84% of fumonisins can be reduced by combining washing with hand sorting technologies. The most commonly used solvents for mycotoxin extraction are methanol, ethanol, hexane, acetonitrile, isopropanol, and aqueous acetone. Prior research revealed that dimethyl ether and a hexane-aqueous acetone mixture (56%:42%:2%) could remove more than 98% of aflatoxin fumarates from oil crops (Van der Westhuizen et al. 2011).

7.4.2.4.3 Heating

For many years, mycotoxins in feed have been decontaminated via thermal treatment. The effectiveness of this approach is affected by the chemical makeup and concentration of mycotoxins, as well as the temperature, time, moisture content, pH, and ionic concentration during heat treatment. Kabak et al. (2006) state that AFB1, DON, ZEN, and FB1 are heat-stable compounds with decomposition temperatures greater than 237 °C, 175 °C, 220 °C, and 150 °C respectively. AFB1 can be degraded by 78–88% in rice through conventional hydrothermal treatment under pressure at 160 °C for 20 min (Park and Kim 2006), and by 95% in moist peanut powder by pressure heating at 120 °C for 4 h (Fan et al. 2013). According to Yumbe-Guevara et al., barley powder can lose 90% of its DON or ZEN after being heated at 220 °C for either 11 or 85 min.

7.4.2.4.4 Irradiation

Mycotoxins can be removed by radiation, which is divided into non-ionizing (ultraviolet rays, infrared, and microwave) and ionizing (X-rays, γ-rays, and electron beams). In particular, AFB1 levels in various cereals can be reduced by 43.0–87.8% with γ-rays, 65.7–71.5% with electron beams, 22.0–100% with ultraviolet light, and 90.7% with microwave radiation (Pankaj et al. 2018).

7.4.2.4.5 Adsorption

A highly effective mycotoxin absorbent should possess low nonspecific binding to nutrients, high safety, stability, and palatability, along with a high adsorption capacity for both types of mycotoxins. Adsorption binders may also form complexes with mycotoxins (Kabak et al. 2006).

7.4.2.5 Biological Control

Biological detoxification involves using microorganisms, their biofilms, or enzymes to break down or bind mycotoxins in food and feed. Certain bacteria, yeasts, and fungi can convert aflatoxins, ochratoxin A, patulin, zearalenone, and trichothecenes into less hazardous or nontoxic substances. Some microbes can create biofilms that adsorb mycotoxins on their surfaces. Additionally, enzymes such as lactonases, oxidoreductases, and esterases can hydrolyze or alter the chemical structure of mycotoxins.

7.4.2.6 Microorganisms with Detoxification Activities

Numerous fungi have been shown to detoxify AFB1. Strains such as S. cerevisiae have demonstrated a 69.0% degradation of AFB1 (Chlebicz and Śliżewska 2020). Similarly, research has indicated that different strains of Aspergillus, including A. niger, can degrade AFB1 to levels ranging from 88.6% to 98.7% (Qiu et al. 2021). Bacteria primarily break down AFs through extracellular enzymes. Certain strains of *Flavobacterium aurantiacum, Bacillus,* and *Nocardia corynebacterioides* have also been shown to degrade AFB1. Smiley and Draughon discovered that *Nocardia corynebacterioides* could degrade AFB1 with 74.5% efficiency within a 24-h period (Smiley and Draughon 2000). AFB1 has also been shown to be degraded to levels ranging from 67.2% to 94.7% by various *Bacillus* species, including *Bacillus licheniformis, Bacillus velezensis, Bacillus subtilis,* and *Bacillus shackletonii* (Shu et al. 2018). *Pseudomonas putida, Escherichia coli,* and *Stenotrophomonas* sp. demonstrated highly effective biodegradation rates of up to 90% or more for AFB1, as reported by Cai et al. (2020). Similarly, certain bacterial and fungal microbes have been reported to break down fumonisins (Štyriak et al. 2001). The toxicity of mycotoxins is due to specific chemical groups that microbial enzymes can selectively modify or remove, as noted by Abraham et al. (2022). These enzymes are beneficial for the detoxification of mycotoxins in food and feed because they catalyze the biotransformation of mycotoxins under benign conditions. For successful implementation, cofactors are essential (Fig. 7.2).

7.4.2.7 The Usage of Catabolizing Enzymes

Certain bacteria exhibit significant activity in the biodegradation of mycotoxins, while others may emit toxic compounds or be unable to survive in the animal nervous system. The two primary fungal enzymes known to be active in AFB1 breakdown are laccase and oxidase. The enzyme known as aflatoxin detoxifizyme was responsible for AFB1 detoxification (Cao et al. 2011). Peroxidases capable of degrading DON include lignin peroxidase and manganese peroxidase (Tso et al. 2021). Copper-containing oxidases, known as laccases, show great promise for degrading the heat-stable mycotoxin zearalenone, which has numerous industrial applications. Reducing mycotoxin contamination is a complex issue influenced by globalization, environmental factors, and regulatory disparities. To mitigate mycotoxin levels in food and feed, various strategies are employed, such as analytical methods, pre- and post-harvest management, biological and chemical control, and international cooperation. It is essential for governments, businesses, and researchers to collaborate in tackling these challenges to enhance the safety and quality of our food supply. A comprehensive and scientifically grounded approach can reduce the hazards associated with mycotoxins, protect public health, and preserve the financial interests of the food and agriculture sectors. To enhance their ability to

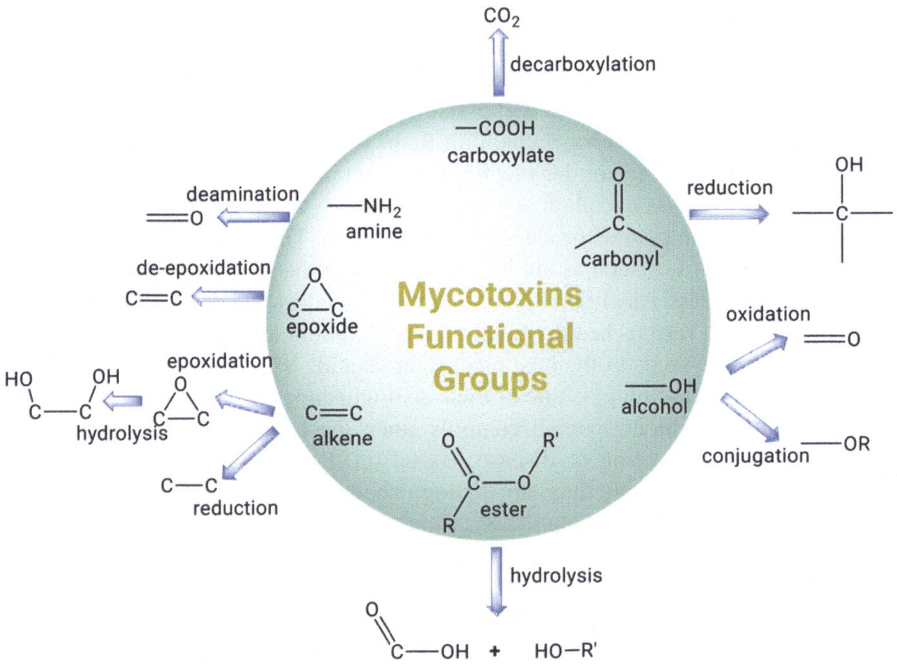

Fig. 7.2 Explanation of the common functional groups found in mycotoxins (circle) and how microbial enzymes may be able to change them. (Reprinted from Abraham et al. 2022)

manage this challenge, all parties involved in the food chain must communicate, educate, and increase their awareness of one another.

7.5 Definition and Characteristics of Endophytic Fungi

Over time, the definition of the term endophyte—endo meaning within, and phyte meaning plant—has evolved. De Bary (1866) originally coined the term to describe any organism capable of living inside plant tissues. A more precise definition was provided by Petrini et al. (1991), who stated that endophytes are organisms that reside within plant organs without harming the host. This encompasses organisms that may at some point in their life cycle reside on the plant's surface or cause infection. Examples of endophytes include protists, bacteria, and fungi (Kobayashi and Palumbo 2000; Stone et al. 2000; Carvalho et al. 2012). The most prevalent kind of endophytes is fungi (Strobel and Daisy 2003; Strobel et al. 2004; Tan and Zou 2001). In natural ecosystems, it appears that all plants interact in some way with endophytic or other related fungi (Rodriguez et al. 2009). Eukaryotic microorganisms known as fungi are crucial to agriculture, biotechnology, the natural world, and other living organisms. Fungi can be isolated from both terrestrial and marine plant tissues, as well as from various environments. Fungal partners inside the tissues of higher plants, known as endophytic fungi, do not harm their hosts. Unlike pathogenic fungi that cause harm to the host, endophytes typically provide certain benefits to the plant. Endophytic fungi have the ability to protect their host plants from a variety of environmental stresses and assist in their growth. Endophytic fungi possess unique metabolic pathways enabling them to produce physiologically active substances, in addition to similar metabolites generated by their hosts. Endophytic fungi are recognized as excellent sources of bioactive compounds, serving as substitutes for phytocompounds produced by most plants. These fungi produce metabolites with bioactivities such as antibiotics, phytohormones, antifungals, antivirals, and anticancer agents. Any higher plant may be invaded by one or more endophytes. They can colonize various healthy plant tissues including leaves, stems, bark, roots, fruits, flowers, and seeds of the host plant (Gupta et al. 2023). The secret or hidden lifestyle of endophytic fungi is one of their distinguishing traits. They are often difficult to identify without specialized tools since the host plant does not exhibit apparent signs of infection or disease. Based on their association with grasses, *clavicipitaceous* endophytic fungi are distinguished from nonclavicipitaceous endophytic fungi (Hyde and Soytong 2008). *Clavicipitaceous* and *nonclavicipitaceous* fungi were divided into four classes by Rodriguez et al. (2009). As a result of their ability to thrive as symptomless endophytes in various plant species and their unique capacity to infect and colonize insects, Lugtenberg et al. (2016) have proposed the addition of a new class for endophytic entomopathogenic fungi. Bamisile et al. (2018) support this classification. Endophytic fungi can be categorized based on their function, diversity, and ecology. They can be classified as either sexual or asexual depending on whether they reproduce through sexual or asexual spores. Furthermore, their mode of transmission categorizes them as either vertically or horizontally transmitted. Vertical transmission is the transfer of a fungus to plant offspring through seeds, while horizontal transmission occurs when the endophyte produces vegetative propagules or spores that are dispersed to the plant population through air or by other vectors (Lugtenberg et al. 2016). Endophytic fungi can be classified as either necrotrophic or biotrophic based on their feeding habits (Kemen and Jones 2012). Necrotrophic endophytic fungi induce necrosis to derive nutrients from dead tissues, whereas biotrophic endophytic fungi extract nutrients from living tissues. Endophytic fungi are an excellent source of bioactive secondary metabolites, offering numerous advantages to host plants (Table 7.1).

7.5.1 Enhanced Growth

Certain endophytes produce nutrients and phytohormones, among other growth-promoting substances, that can accelerate the plant's overall growth and development (Santoyo et al. 2016).

Table 7.1 Fungal endophyte metabolites by type and potential applications

Metabolite type	Examples	Potential applications
Antibiotics	Lolitrem B, Sphaeropsidin A	Control of plant and animal pathogens
Enzymes	Cellulases, Lacases, Amylases	Biodegradation of pollutants, biomass conversion, food processing
Antioxidants	Resveratrol, Curcumin	Protection against oxidative stress, reducing risk of chronic diseases
Antifungals	Ergosterol, Paclitaxel	Treatment of fungal infections
Anticancer	Artemisinin, Onnamide A	Cancer treatment and prevention
Anti-inflammatories	Beta-sitosterol, Quercetin	Reducing inflammation and pain
Immunomodulators	Cyclosporine A, Ergosterol	Modulating the immune system
Plant growth promoters	Auxins, Gibberellins	Enhancing plant growth and yield
Insecticides	Lolitrem B, Beauvericin	Controlling insect pests

7.5.2 Increased Tolerance

Various endophytes produce chemicals that help the host plant resist environmental stressors or repel herbivores, thereby increasing the plant's tolerance to a variety of conditions, such as drought, salt, and herbivore attacks (Kamran et al. 2022).

7.5.3 Disease Resistance

Certain endophytic fungi can protect the host plant from pathogenic fungi through the production of antimicrobial chemicals or resource competition (Akram et al. 2023).

7.5.4 Improved Nutrient Uptake

Endophytes enhance a plant's ability to absorb essential nutrients from the soil, aiding in nutrient uptake (García-Latorre et al. 2021).

7.5.5 Biodiversity

Numerous species and strains of endophytic fungi have evolved to live in specific host plants, contributing to their immense diversity. This variety enhances the overall biodiversity of ecosystems (Eo et al. 2022). Utilizing endophytic fungi as biological control agents against mycotoxin-producing fungi is another viable strategy.

This approach is based on the theory that endophytic fungi can establish a symbiotic relationship with their host plants, protecting them from both abiotic and biotic stresses by producing secondary metabolites that inhibit the growth or activity of pathogenic organisms. Several reports indicate that endophytic fungi can mitigate mycotoxin-producing fungal infections or colonization in various crops. Additionally, some endophytic fungi can produce secondary metabolites that prevent the biosynthesis of mycotoxins or their accumulation in plant tissues. These secondary metabolites belong to various chemical groups, such as polyketides, peptides, lipopeptides, terpenoids, and alkaloids (Billar de Almeida et al. 2020). To date, culture-based methods have been employed to assess the diversity and composition of endophytic fungal communities in all investigated plant species. Morphological characteristics, along with molecular analysis techniques, have been employed to identify non-culturable endophytic fungi (Premalatha and Kalra 2013). The identification of fungal DNA in plant tissues using molecular techniques has revealed various aspects of fungal endophytic populations, rendering cultivation methods unfeasible (Dissanayake et al. 2018). Through the use of fingerprinting and ITS amplicon amplification from collected DNA, several endophytic populations have been identified. The predominant fungal communities differed significantly between culture-dependent and culture-independent approaches (Akram et al. 2023). The detection method determines which taxa can be detected. Fungi represent numerous ecological roles, including endophytes, pathogens, and saprobes. The species of fungi detected by the DNA detection method are comparable to those identified using a combination of culture and direct observation techniques (Franco-Duarte et al. 2019). The ecology and distribution of some endophytes cannot be fully understood through culture-independent methods such as next-generation sequencing. This technique has identified all commonly found fungal genera from a culture-dependent method with a relative abundance greater than 5%. Numerous researchers, including Moncada-Torres et al. (2014), have discovered that up to 99% of prokaryotic microbes may not be cultivable. Endophytic fungi are a diverse group of microorganisms that form symbiotic relationships with plants, offering a variety of benefits to their hosts. They are of great scientific interest and have real-world applications in environmental management, agriculture, and medicine due to their modest and often undetectable presence in plant tissues.

7.6 Mechanisms of Endophyte Fungal Colonization of Plant Tissue

A broad class of microorganisms known as endophytic fungi lives inside the tissues of plants without causing noticeable harm to their hosts. These fungi engage in complex mechanisms that facilitate their penetration, propagation, and survival within plant tissues, contributing to the formation and maintenance of endophytic relationships. Endophytes are found in all plant species, regardless of origin. They are unique in their ability to infiltrate and thrive within host tissues, exhibiting intricate interactions with the host plant. The presence of endophytes is known to influence a number of host functions. According to Khare et al. (2018), they can stimulate plant growth, trigger defense mechanisms against pathogen invasion, and mitigate the effects of abiotic stressors. They also act as remediators of abiotic stress, stimulating defense responses against pathogen attack and boosting plant development. Various mechanisms are involved in the colonization process of endophytic fungi. For example, endophytes release a variety of enzymes to help hydrolyze the plant cell wall in order to colonize the plant surface. They might also use strategies such as altering plant elicitors, producing harmful metabolites, and suppressing the plant immune system.

Numerous openings, such as tissue wounds, stomata, lenticels, root fissures, and germination radicles, allow fungi to penetrate plant tissues. Numerous modified secretory trichomes, originating from the lower epidermis, are found at the terminals of plant shoots in nodulated dicot species. This indicates a significant function of trichomes in host-endophyte interactions (Lu et al. 2021). Certain endophytic fungi can infiltrate plants through naturally occurring apertures, including lenticels on stems or stomata on leaves. The internal tissues of the plant can be directly accessed through these entry sites (Averill et al. 2014). An endophytic fungus must penetrate the physical barrier of the plant cell walls to colonize plant tissues. The plant's cell wall becomes less rigid upon contact with a microbe, allowing the microbe to be absorbed and its plasma membrane to invade. Consequently, endophytic fungi do not come into direct contact with the plant's cytoplasm; instead, they remain trapped in membrane compartments. The plasma membrane and cell wall of endophytes facilitate molecular exchange with plant cytoplasm, creating a functional area known as the symbiotic interface. The majority of interactions between plants and microbes are mediated by the cell wall, which plays a significant role in their molecular conversation (Geurts et al. 2016). Cell wall-degrading enzymes such as cellulase, laccase, pectinase, and xylanase are secreted by endophytic fungi, enabling them to penetrate, colonize, and proliferate within plant tissues. This interaction can break down the plant cell wall and modify its structure. Fourier transform infrared spectroscopy has shown that plant cell walls respond to infections in a similar manner. Consequently, most endophytes react to infected hosts nonpathogenically and asymptomatically, without degrading lignin or carbohydrates. Cell wall-degrading enzymes such as cellulase, laccase, pectinase, and xylanase are secreted by endophytic fungi, which penetrate, colonize, and multiply within plant tissues. These enzymes can break down the plant cell wall and alter its structure. The response of plant cell walls to these enzymes was similar to that observed in infections, as indicated by Fourier transform infrared spectroscopy detection results. The majority of endophytes interact with their hosts in a non-pathogenic and asymptomatic manner, not breaking down lignin or carbohydrates. Endophytes are also more likely to infect the host under the same conditions than pathogens (Chow and Ting 2019). Some endophytes form specialized structures known as vesicles that serve as nutrient storage organelles, helping to maintain the symbiotic relationship. Endophytic fungi may use chemotactic signals to locate host tissues and navigate through a plant. Fungal hyphae are directed toward the most advantageous colonization locations by chemical signals from the host, such as root exudates. Establishing a suitable relationship and recognizing host-specific signals are critical for successful colonization. Fungal hyphae can respond to signals released by plants by modifying their metabolic processes and gene expression (Mengistu 2020). The "barrier effect" occurs when endophytes proliferate and colonize plant tissues extensively, competing with harmful microbes to prevent their proliferation. Endophytes are crucial for maintaining plant health as they can boost growth and yields while protecting or priming the plant against biotic challenges through antibiosis or induced systemic resistance (Lugtenberg et al. 2016). Microbes enhance disease resistance through the mechanisms of induced systemic resistance (ISR) and systemic acquired resistance (SAR). Plants have evolved various families of receptor proteins to recognize microbe- or pathogen-associated molecular patterns (MAMPs/PAMPs), which are conserved structures vital for microbial survival. This recognition triggers the plant's immune system (Plett and Martin 2018). In cucumber plants, the endophytic fungus *Lecanicillium longisporum* reduces aphids and powdery mildew. Bacillus cereus, another endophyte, induces systemic resistance (ISR) against Botrytis cinerea in Arabidopsis thaliana by increasing PR1 protein expression over time, accumulating hydrogen peroxide, and causing callose deposition. This endophyte stimulates WRKY53 gene expression and MAPK signaling, both of which are implicated in pathogen-associated molecular pattern-triggered immunity (PTI) through the JA/ET signaling pathway in an NPR1-dependent manner (Nie et al. 2017). At the site of a potential phytopathogen attack, endophytes induce several cell wall modifications, such as the deposition of callose, pectin, cellulose, and phenolic compounds, resulting in the formation of a structural barrier. Similarly, they activate defense-related proteins, including chitinases, peroxidases, and β-1,3-glucanases (Malfanova et al. 2013). *Aspergillus terreus* and *Penicillium citrinum*, two endophytic fungi, inhibited fungal infection in sunflower plants caused by *Alternaria alternata* by controlling oxidative stress responses. This was achieved by upregulating catalase and peroxidase activities and activating glutathione and polyphenol oxidase. Similarly, leaves injected with endophytes showed increased amino acid contents, suggesting that this alteration slows down cell death and impedes the growth of fungi in plant tissue (Waqas et al. 2015). The tomato plants treated with bacterial endophytes exhibited elevated levels of defense-related oxidative enzymes, including phenyl ammonia lyase (PAL), polyphenol oxidase (PPO), and peroxidase (PO), leading to the induction of systemic resistance against Fusarium wilt of tomatoes (Shafi et al. 2017).

7.7 Diversity and Distribution of Endophytic Fungi in Various Crops

Endophytic fungi are characterized by a continuous gradient distribution of features in their geographical spread. According to Violle et al. (2014), the foundation of this continuous distribution is biodiversity at multiple levels, including species, function, and phylogenesis. Fungal strains mediate various activities and can be important in their habitats, although the mechanisms of biodiversity development in a given geographical area are not well understood (Alam et al. 2021). A diverse class of microorganisms known as endophytic fungi lives within the interior tissues of plants in mutualistic or commensal interactions. These fungi are quite adaptable and have been isolated from a variety of plant species in different environments. Researchers are increasingly interested in the relationships that endophytic fungi have with their host plants, as these interactions can influence the health, growth, and susceptibility of the plants to environmental stresses (Kashyap et al. 2023). Thousands of distinct species of endophytic fungi have been identified to date, demonstrating their enormous diversity. These fungi may fall into a number of taxonomic categories. Seasons, environmental factors, plant genotypes, and plant organs all influence the endophytic microbial communities. However, it is frequently reported that the Proteobacteria phylum is the most prevalent type of bacteria, followed by Actinobacteria, Firmicutes, and Pseudomonas. According to reports, the three main endophytic bacterial genera found in most plants and plant organs are Pantoea, Acinetobacter, and Enterobacter. Arbuscular mycorrhizal fungi are the most common endophytic fungal genera in various host plants, with Glomeromycota being the most prevalent fungal phylum, followed by Ascomycota and Basidiomycota (Agrawal and Bhatt 2023). The host plant's organs had different endophytic fungal communities. However, as most fungi can move through the air and opportunistically infect plants through mechanisms such as spores dropping on the soil or leaves, it is not possible to conclude that a certain taxon is found only in specific organ tissues. Precisely determining the spread of endophytic fungi is challenging since their degree of isolation can vary based on time and medium (Eo et al. 2022). Endophytic fungi are found in various types of crops, including cereals, fruits, vegetables, legumes, and trees. Endophytic fungi, found in various crops, serve multiple purposes. They are essential for nutrient uptake, stress tolerance, and disease resistance in cereal crops such as wheat, rice, and maize. Instances of their dissemination across different crops have been observed. They can help improve the overall well-being and productivity of these staple crops (Jana et al. 2022). Fruit-bearing plants, including tomato plants, apple trees, and grapevines, often harbor endophytic fungi. These fungi are beneficial for both commercial and subsistence farming as they can enhance fruit quality and shelf life. Additionally, endophytic fungi aid in nitrogen fixation and promote the growth of leguminous crops such as peanuts and soybeans. This relationship is mutually beneficial to both the fungi and the plants. Many tree species contain endophytic fungi, which are plentiful and contribute to the dynamics of forests and ecosystems. Enabling plants to survive heavy metals or other contaminants in disturbed soils supports phytoremediation (Zheng et al. 2016). Comprehending the variety and distribution of endophytic fungi in different crops is essential to promote sustainable agricultural operations. Utilizing the potential of these advantageous microbes can increase crop yields, reduce reliance on chemical inputs, and enhance plant resistance to biotic and abiotic stresses. Furthermore, endophytic fungi offer new insights into developing biopesticides, biofertilizers, and bioremediation techniques. In modern agriculture, endophytic fungi are becoming increasingly significant in promoting economically and environmentally sustainable farming practices (Andrés et al. 2017).

7.8 Role of Endophytic Fungi in Mycotoxin Reduction

Endophytic fungi represent a fascinating group of fungal communities that associate with host plants, residing within the intercellular or intracellular spaces of their tissues. They provide benefits and impart positive effects on their hosts (Jha et al. 2023). The production of secondary metabolites, especially alkaloids, by these fungi is one of the main advantages for plants. These compounds accumulate in plant tissues and exhibit bioactivity against a range of vertebrates, invertebrates, and diseases such as fungi, as well as confer tolerance to abiotic stresses. The majority of endophytes have been shown to produce secondary metabolites, some of which possess antibacterial and antifungal properties that help prevent the establishment of phytopathogenic microbes. The biosynthesis of these secondary metabolites in symbiosis occurs through intricate gene expression regulation mechanisms within the symbiotic continuum and by stimulating or altering the production of phytochemicals in host plants. These secondary metabolites act as effectors in these multidimensional interactions (Singh and Kumar 2023). The concept of long-term coevolution within biological communities suggests that the reciprocal relationship between endophytic fungi and their hosts leads to each fungus developing a specific range of host species (Ji et al. 2010). This enables them to accumulate within a particular eukaryotic host group (Fig. 7.3).

Endophytic fungi play a crucial role in reducing mycotoxins, which are harmful substances produced by certain fungi that pose significant risks to human and animal health when present in food and feed. These fungi can decrease mycotoxin levels by inhibiting the growth of mycotoxin-producing fungi through the production of antifungal secondary compounds. Additionally, some endophytic fungi may have the ability to degrade mycotoxins, providing an alternative method for reducing their levels. It has been established that certain endophytic fungi, such as those that produce mycotoxins, can function as natural biocontrol agents against phytopathogenic fungi. These endophytes affect the development and activity of pathogenic fungi through nutritional competition, synthesis of antimicrobial chemicals, and stimulation of host defensive responses (Devi et al. 2023).

7.9 Production of Antifungal Metabolites by Endophytic Fungi

Fungal endophytes are microorganisms found in the newly formed tissues of living plants and are considered untapped resources for novel natural products useful in agriculture. Chemical analysis of endophytic fungi has yielded a vast array of antifungal compounds with potential use in developing biopesticides. Elicitors, which are biomolecules such as polysaccharides, polypeptides, unsaturated fatty acids, and glycoproteins, act as signaling molecules triggered by the accumulation of bioactive phytochemicals in plants and are produced by endophytic fungi (Table 7.2 and Fig. 7.4). Some of them protect plants from organisms that cause disease. Additionally, they promote the synthesis of various phytochemicals, such as

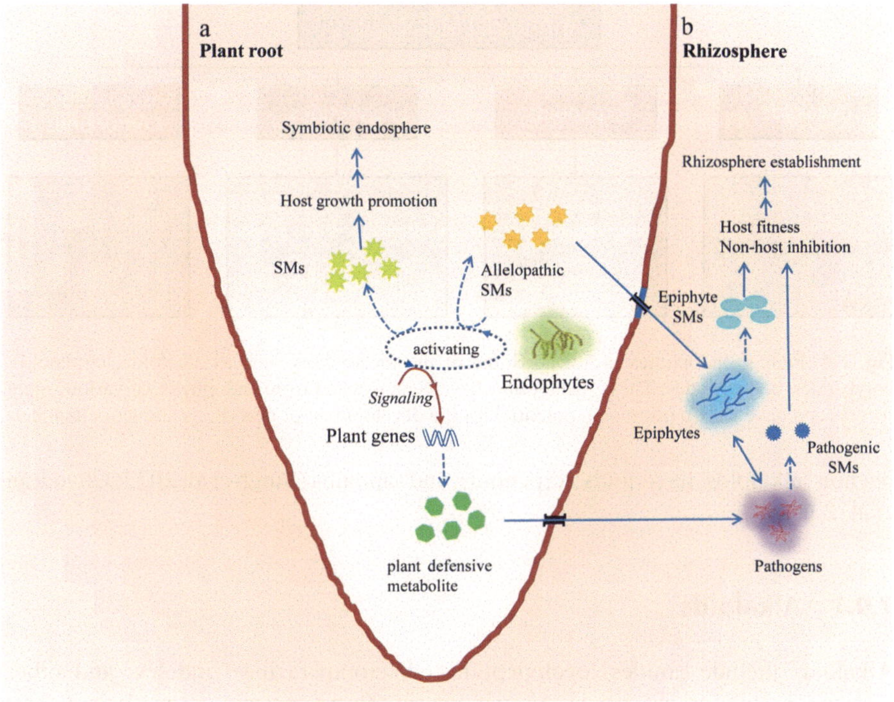

Fig. 7.3 Plant-microbiome mutual orientation (**a**) and multidirectional chemical communications or crosstalks among endophytes, epiphytes, and pathogens in natural ecological settings (**b**). Pathogens, pathogenic SM, endophytic fungus (EF), EF secondary metabolite (SM), allelopathic SMs, epiphytes, epiphytic SM, and plant defensive metabolite. (Reprinted from Alam et al. 2021)

Table 7.2 Types of endophytic fungi metabolites and their properties

Class of metabolite	Examples	Potential properties
Alkaloids	Ergonovine, Lolitrem B	Antifungal, anti-insect, antitumor
Terpenoids	Paclitaxel, Artemisinin	Anticancer, antimalarial
Steroids	Ergosterol, Beta-sitosterol	Anti-inflammatory, immunomodulatory
Phenolics	Resveratrol, Curcumin	Antioxidant, anti-inflammatory
Polyketides	Lovastatin, Tetracycline	Cholesterol-lowering, antibiotic
Quinones	Lapachol, Hypericin	Anticancer, antiviral
Flavonoids	Quercetin, Rutin	Antioxidant, anti-inflammatory
Peptides	Cyclosporine A, Onnamide A	Immunosuppressant, anticancer

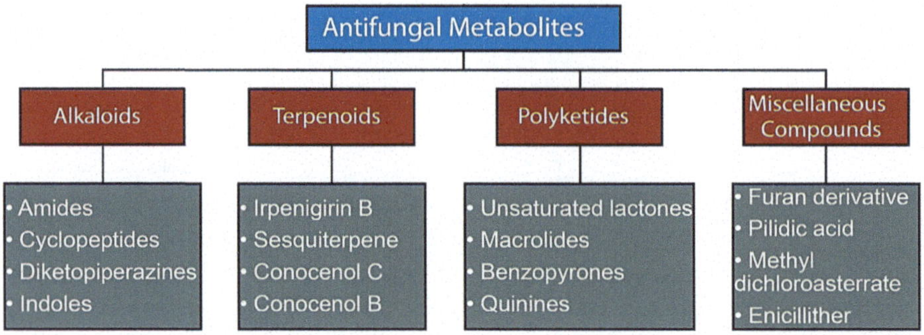

Fig. 7.4 Fungal endophytes produced Some of the metabolites such as alkaloids, terpenoids, polyketides, and peptides. These metabolites have been isolated from endophytes of various plant species and have shown promising potential for the development of new drugs and agrochemicals

phenols, alkaloids, flavonoids, terpenoids, and saponins (Singh et al. 2023; Chandran et al. 2020).

7.9.1 Alkaloids

Alkaloids include amides, cyclopeptides, diketopiperazines, indoles, and other nitrogen-containing chemicals. An endophytic *Mucor* sp. from the medicinal plant *Centaurea stoebe* yielded two prenylated tryptophan analogs (Abdou et al. 2020). Chemical analysis of the endophytic fungus *C. gloeosporioides* from *Michelia champaca* leaves led to the isolation of a novel indole alkaloid, 2-phenylethyl 1H-indol-3-yl acetate. This alkaloid exhibited moderate activity against both *C. sphaerospermum* and *C. cladosporioides* (Chapla et al. 2014). *Aspergillus fumigatus*, an endophytic fungus isolated from the stem bark of *Melia azedarach*, produced three prenylated indole diketopiperazine alkaloids: 12b-hydroxy-13a-methoxyverruculogen, fumitremorgin B, and verruculogen. These alkaloids exhibit broad-spectrum antiphytopathogenic activities against eight fungi: *B. cinerea, A. solani, A. alternata, C. gloeosporioides, Fusarium solani, F. oxysporum f.* sp. *niveum, F. oxysporum f.* sp. *vasinfectum,* and *G. saubinettii* (Li et al. 2012). *Ginkgo biloba L.* yielded an endophytic Xylaria sp., which demonstrated wide antibacterial activity. Through bioactivity-guided fractionation, 7-amino-4-methylcoumarin was identified as the active compound. Among the 13 pathogens evaluated were *Penicillium expansum* and *A. niger*; antimicrobial analysis revealed that 35 of them suppressed their growth (Liu et al. 2008).

7.9.2 Terpenoids

Fungi produce terpenoids, which are among the most abundant and structurally diverse secondary metabolites with a wide range of pharmacological effects. The coculture of the endophytic fungus *Irpex lacteus* and the phytopathogenic fungus *Nigrospora oryzae*, both from the same host, *Dendrobium officinale*, yielded a new squalene derivative named irpenigirin B, as well as new tremulane sesquiterpenes 5-demethyl conocenol C and B. These novel compounds exhibited antifungal activity against *Didymella glomerata* and were also active against *C. gloeosporioides*. Additionally, it has been shown that the antagonistic relationship between endophytes and phytopathogens can result in the synthesis of antibiotics that impede the development of specific phytotoxins and prevent the proliferation of phytopathogens. Interestingly, all identified sesquiterpenes exhibited strong antifungal activity against *C. gloeosporioides*, indicating they are metabolites of the endophytic fungus from *L. chinensis*. The phytopathogenic fungus *C. gloeosporioides* can induce anthracnose in *L. chinensis*, suggesting that metabolites from endophytes may provide protection against invasive phytopathogens (Wu et al. 2019).

7.9.3 Polyketides

Polyketides are a diverse class of naturally occurring compounds with various structures and bioactivities. Polyketide synthases are responsible for the biosynthesis of these metabolites, and several have hybrid chemotypes derived from multiple biosynthetic pathways. Polyketides encompass phenols, unsaturated lactones, macrolides, benzopyrones, quinones, and chromones. A perylenequinone derivative named stemphyperylenol was produced by the fungus *Botryosphaeria dothidea*, isolated from the stems of white cedar (*Melia azedarach L.*). This derivative exhibited strong antifungal activity against A. solani, surpassing the commercially sold fungicide carbendazim (Xiao et al. 2014b). Three novel naphthoquinone spiroketals were produced through the cultivation of the recently identified fungal strain *Edenia gomezpompae*, an endophyte isolated from the leaves of *Callicarpa acuminata* (Verbenaceae). These spiroketals showed significant growth suppression against four economically relevant phytopathogens: *Phytophthora capsici, P. parasitica, F. oxysporum,* and *A. solani* (Macías-Rubalcava et al. 2008). Additionally, the endophytic fungus *Aspergillus* sp., found in the stem bark of *Melia azedarach*, was used to characterize Fonsecinone A, a dimeric naphtho-γ-pyrone. The substance exhibited significant suppression of *G. saubinetti, Magnaporthe grisea, B. cinerea,* and *A. solani,* with effects that were either superior to or comparable with those of hymexazol (Xiao et al. 2014a). The mangrove endophytic fungus *Phoma* sp. produced macrosporin, an anthraquinone with broad-spectrum antifungal activity against pathogens such as *F. oxysporum, F. graminearum, C. musae, P. italicum, R. solani,* and *C. gloeosporioides* (Huang et al. 2018). *Trichoderma koningiopsis*, isolated from the medicinal plant *Panax notoginseng*, was the source of a novel fungal polyketide named *koningiopisin C*. This compound was extracted from the culture broth and exhibited antifungal activity against *Plectosphaerella cucumerina* (Liu et al. 2016). A native Australian grass, *Neurachne alopecuroidea*, was found to contain the endophyte *Drechslera* sp., which showed promise in combating various plant diseases. Metabolomic analysis identified two primary bioactive metabolites:

monocerin, effective against *B. cinerea and S. sclerotiorum*, and an alkynyl-substituted epoxycyclohexenone derivative (5-hydroxy-1-(3-oxo-but-1-ynyl)-7-oxa bicyclo[3.2.1]hept-3-en-2-one) (d'Errico et al. 2021).

7.9.4 Miscellaneous Compounds

An endophytic *Aspergillus tubingensis* from *Decaisnea* insignis yielded a novel furan derivative known as 3-(5-oxo-2,5-dihydrofuran-3-yl) propanoic acid, which demonstrated strong antifungal activity against *F. graminearum* (Yang et al. 2019). When endophytic fungi *Xylaria* sp. were isolated from guarana plant leaves, they produced pilidic acid. Pilidic acid is derived from octanoate, which is produced by fatty acid synthase and exhibits antifungal action against *C. gloeosporioides* (Elias et al. 2018). Two chlorinated diphenyl ethers, methyl dichloroasterrate and enicillither, were isolated from *Aspergillus* on a healthy oilseed rape plant (*Brassica napus*). Both compounds suppressed four plant pathogenic fungi: *B. cinerea, M. fructicola, S. sclerotiorum, and S. trifoliorum* (Qin et al. 2019). Endophytic fungi have the potential to produce antifungal compounds with numerous applications. By increasing crop resistance to fungal infections, endophytic fungi can help minimize the need for chemical fungicides and advance sustainable agriculture. These fungi generate bioactive chemicals that have applications in biotechnology, such as the creation of innovative biopesticides and bioremediation techniques.

7.10 Competition and Antagonism Against Mycotoxin-Producing Fungi

Microorganisms often become antagonistic to one another. Fungal and bacterial antagonists can impact harmful bacteria and fungi in plants. This natural conflict between plant diseases and beneficial bacteria contributes to the natural buffering of cropping systems, which prevents or slows disease spread. This is particularly evident in suppressive soils where a pathogen population is present, and an antagonistic potential has developed (Weller et al. 2002). For many years, the use of BCAs to manage fungal pathogens in a variety of durable and horticultural crops has received significant attention. This includes treatment for both soil-borne and foliar diseases, as well as weed control. A common mode of action is the direct antagonistic relationship between the fungal pathogen and the BCA. This is followed by competitive exclusion of the pathogen through niche occupation, generation of secondary metabolites, hyperparasitism, and production of volatile organic compounds. The goal in nearly all cases has been to reduce the symptoms of plant diseases, as well as minimize yield loss and quality degradation. Both plant disease management and mycotoxin contamination reduction techniques share parallels and contrasts. The application of BCAs to the management of mycotoxigenic fungi has centered on two main goals: (a) reducing the production of the associated mycotoxin by targeting the biosynthetic genes involved in toxin biosynthesis, and (b) increasing the efficacy of BCAs in controlling the germination, growth, and colonization of mycotoxigenic fungi in raw or processed food commodities (Medina et al. 2017). The screening process has been used for many years to identify biological control agents (BCAs). To commercialize both bacterial and fungal BCAs, systematic procedures are required. A different strategy may be necessary to develop BCAs for managing mycotoxigenic molds and reducing mycotoxin levels. This is because many mycotoxigenic species can produce additional mycotoxins when they cannot properly colonize a host due to environmental stress (Magan and Medina 2016). When developing biological control agents (BCAs) to manage mycotoxins, it is important to consider: (a) the life cycle of the mycotoxin-producing species; (b) fluctuations in environmental conditions; (c) plant agronomy; and (d) interactions with pests, which often act as entry points for mycotoxin-producing fungi, especially during post-harvest processing of coffee, cocoa, and nuts or the ripening stages of cereals. In liquid fermentation systems and on cotton balls, an atoxigenic strain was able to compete effectively with the same inoculum ratio or even less atoxigenic inoculum. It was claimed that the atoxigenic strains employed two modes of action: exclusion of the toxigenic strain from the niche and competition for resources needed for aflatoxin production. The ability of a biocontrol agent (BCA) to effectively compete with a mycotoxigenic species for niche exclusion or replacement, as well as its capacity to reduce the biosynthetic pathways involved in toxin metabolite formation, is likely the mechanism of action (Bayman 1993). The competition between mycotoxigenic species and biological control agents (BCAs) for primary resources must be considered. True mycotoxigenic pathogens, such as *Fusarium graminearum* and *F. verticillioides*, are involved in acquiring primary resources; their successful establishment is crucial for infection (Abbas et al. 2011). Species from *Aspergillus* section *Flavi*, Section *Circumdati*, and Section *Nigri* are examples of secondary pathogens and saprophytes that can colonize niches with less competition due to their xerophilic nature. Often, pest damage or drought-stressed hosts facilitate this colonization process. The BCA must effectively contend in such environmental regimes. The reductions in AFB1 production by toxic strains of *A. flavus* cannot be explained solely by exploitative competition (Abbas et al. 2011). Two key mechanisms of action have been identified: niche exclusion and carbon source usage patterns. The temporal rate of carbon source utilization patterns (Temporal Carbon Utilization Sequence) and the Niche Overlap Index were recently studied for *A. flavus* and other mycotoxigenic species, between toxigenic and atoxigenic strains of *A. flavus,* and between *F. verticillioides* and the biocontrol agent *Clonostachys rosea* (Giorni et al. 2009). The carbon sources used by the pathogen *F. verticillioides* and the antagonist *C. rosea* differed slightly when analyzing the effects on carbon source utilization patterns at 25 and 30 °C. Various species appear to have utilized the niche differently, suggesting that under certain circumstances, the niche may be exclusive. According to the Temporal Carbon Utilization Sequence, the *F. verticillioides* strain rapidly utilized carbohydrates, followed by amino acids, and ultimately palmitic acid, a fatty acid. Similarly, although at a slower pace, the antagonist *C. rosea* consumed both carbohydrates and amino acids. There were also variations in the utilization of specific amino acids and carbohydrates. These differences likely influenced the occupation of various niches within the context of interacting environmental factors. By utilizing specific nutrients more rapidly than the pathogen, *C. rosea* may prevent *F. verticillioides* from occupying microniches. Thus, the mechanism of action may involve niche exclusion or the rate at which nutritional sources are utilized, affecting the mycotoxigenic species' capacity to occupy the niche successfully. This could be related to the ability to produce relevant hydrolytic enzymes that enable occupation in diverse environmental conditions. Temperature has been shown to affect the hydrolytic enzymes produced by toxic strains of *A. flavus* and *F. verticillioides* (Al-Saad et al. 2016). Mohale et al. (2013) studied the production of these enzymes, which are crucial for niche occupation, comparing atoxigenic and toxic strains of *A. flavus*. The hydrolytic enzyme profiles of toxigenic and atoxigenic strains of *A. flavus* were compared. The atoxigenic strain produced a somewhat smaller range of enzymes than the toxigenic strain. When a toxigenic *A. flavus* strain was grown in vitro and in situ in peanut storage, resulting in the production of AFB1, secondary metabolites from *Streptomyces* isolated from peanuts effectively prevented its growth. In fact, regardless of the biocontrol agent (BCA) concentration or environmental conditions, *Streptomyces* metabolites were more successful in inhibiting growth than using bacterial cells as an inoculum (Sultan and Magan 2011). Verheecke et al. (2015) investigated the relationship between strains of *Streptomyces* and *A. flavus* and *A. parasiticus* in a recent study. To investigate the mechanism of action and measure the effects on five biosynthetic genes (aflD, aflM, aflP, aflR, and aflS), RT-qPCR was employed. This demonstrated that compared to *A. flavus, A. parasiticus* exhibited higher levels of gene expression repression caused by Streptomyces strains. In both species, the expression of the regulatory genes aflR and aflS was significantly repressed. The suppression of aflM gene expression correlated with the inhibition of AFB1 production. These findings suggest that aflM expression could serve as a potential marker for the inhibition or regulation of AFB1 production. Biocontrol research has focused on controlling patulin in pome fruits, particularly through the use of various yeasts. Yang et al. (2015) added phytic acid to treatments to enhance the biocontrol efficacy of the yeast *Rhodotorula mucilaginosa*, reducing patulin synthesis and *Penicillium expansum* infection. They also suggested that phytic acid might degrade patulin in vitro.

For instance, *T. afroharzianum*, one of the 15 *Trichoderma* spp. isolated for this investigation, demonstrated encouraging outcomes for managing *fusarium* head blight and crown rot in wheat in Algeria (Bouanaka et al. 2021). There are significant differences in the antagonistic impact of *Trichoderma* isolates in direct confrontation (DC) against *F. culmorum* (Fig. 7.5). It is advised to use *Trichoderma afroharzianum*, recently assessed in Algeria as a biocontrol agent, as a preventive measure that won't affect grain physical characteristics or yield. BCAs are applied in conjunction with fewer fungicide applications. Especially in sustainable production systems, integrated pest management can be envisioned by combining BCAs for mycotoxin control with entomogenous fungi for pest control and minimal pesticide/fungicide inputs. Pests, a significant source of harm to ripening crops, facilitate the spread of mycotoxigenic fungi. Therefore, a synergistic effect on toxin control levels may result from combining BCAs based on entomogenous fungi with those

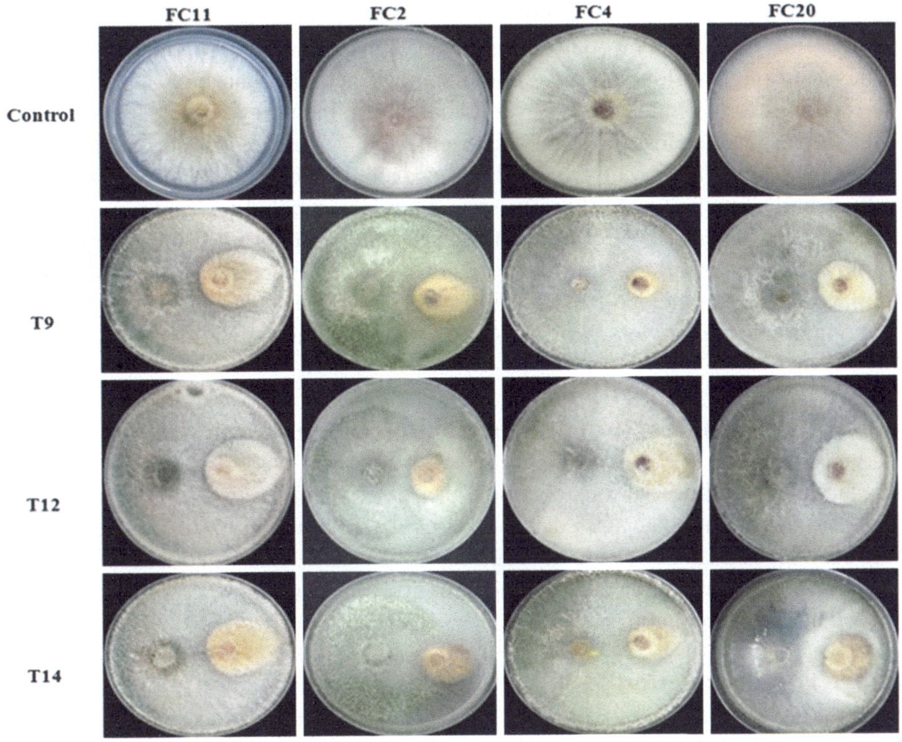

Fig. 7.5 *F. culmorum* strains FC11, FC2, FC4, and FC20 are directly confronted in vitro by *Trichoderma* isolates T9, T12, and T14. (Reprinted from Bouanaka et al. 2021)

that control toxigenic fungi. Certain mycotoxigenic fungi may produce higher levels of mycotoxins under conditions influenced by climate change. An integrated strategy that utilizes biocontrol agents (BCAs) to reduce chemical inputs while controlling toxic fungi and related pests is advantageous for sustainable crop protection. The use of competition and antagonism as techniques to reduce mycotoxin contamination in agriculture is important. Promising approaches to lowering hazards related to mycotoxins in food and feed production involve understanding these ecological relationships and harnessing the potential of beneficial microorganisms as biocontrol agents.

7.11 Detoxification and Biotransformation of Mycotoxins by Endophytic Fungi

Several detoxification techniques have been reported, including chemical, physical, and biotransformation methods. The structure of mycotoxins can be chemically altered using acids, bases, oxidizing agents, and aldehydes. This has raised public concerns about the presence of chemical residues in food and feed (Zhu et al. 2016).

However, certain physical techniques, such as adsorption agents, have proven to be insufficiently effective in controlling mycotoxins. Mycotoxin reduction can be achieved through various physical and chemical methods, but their application in the food industry is hindered by several challenges, including high costs, potential safety issues, and the loss of essential nutrients. Conversely, biotransformation techniques have become a promising method for decontaminating mycotoxins because of their high specificity, ability to produce harmless metabolites, and potential for complete detoxification under mild and eco-friendly conditions. Biotransformation employs isolated enzymes and living organisms such as bacteria, fungi, plants, and animals. These can metabolize, destroy, or render mycotoxins inert, leading to stable compounds that are less harmful or even benign (Wang et al. 2019).

7.11.1 Types of Reaction of Mycotoxin Biotransformation

The summary outlines various chemical processes involved in the biotransformation of mycotoxins, categorized into classes: hydroxylation, oxidation-reduction between alcohols and ketones, hydrogenation of carbon-carbon double bonds, de-epoxidation, methylation, esterification, hydrolysis, sulfation, deamination, and miscellaneous reactions.

7.11.1.1 Hydroxylation

Mycotoxins undergo hydroxylation, a biotransformation process that adds a hydroxyl group (-OH) to the molecule. The enzymes known as hydroxylases often assist in introducing hydroxyl groups at various locations on the molecule. Hydroxylation typically makes mycotoxins more polar and reduces their toxicity. In channel catfish liver, aflatoxin B1 (AFB1) is converted into aflatoxin M1 (AFM1). Similarly, animal liver microsomes convert aflatoxin B2 (AFB2) into aflatoxin M2 (AFM2) (Wu et al. 2009). Fusaric acid (FA), also known as 5-butylpicolinic acid, is a phytotoxin that is not specific to any one host and is produced by fungi belonging to the Fusarium genus. The fungus *Mucor rouxii* hydroxylates FA to produce 8-hydroxyfusaric acid (Crutcher et al. 2017). Ochratoxin A (OTA) consists of an amide bond that connects a chlorinated dihydroisocoumarin to L-phenylalanine via a 7-carboxyl group. Seven-carboxy-(20-hydroxy-1-phenylalanine-amide) is formed by hydroxylating OTA, resulting in a dihydrodiol derivative and 5-chloro-8-hydroxy-3,4-dihydro-3R-methylisocoumarin, produced by the Gram-negative bacterium *Phenylobacterium immobile*. Rabbit liver microsomes hydroxylate OTA to produce (4R)-4-hydroxyochratoxin A and (4S)-4-hydroxyochratoxin A (Ruhland et al. 1995). The fungus *Cunninghamella bainieri* hydroxylates the aliphatic ring of zearalenone (ZEN), converting it into (5S)-5-hydroxy ZEN (El-Sharkawy and Abul-Hajj 1988).

7.11.1.2 Oxidoreduction Between Alcohols and Ketones

Two types of oxidation and reduction occur between the alcohols and ketones of mycotoxins: (i) oxidation of the hydroxyl group and (ii) reduction of the carbonyl group. Numerous fungi, including Eurotium herbariorum, Rhizopus sp., and Aspergillus flavus, which do not produce aflatoxin, reduce the keto group of aflatoxin B1 (AFB1) to a hydroxyl group, converting it to aflatoxicol. During the infection process, the forest pathogen Heterobasidion annosum produces a phytotoxin called fomannoxin, which is a benzohydrofuran. The phytotoxin transformed into the nontoxic formannoxin alcohol when fomannoxin was administered to *Pinus sylvestris* cell cultures (Zweimüller et al. 1997). The fungi *Candida tropicalis, Saccharomyces cerevisiae, Torulaspora delbrueckii,* and *Zygosaccharomyces rouxii* reduced zearalenone (ZEN) to both α- and β-zearalenol (α-ZEL). Conversely, Pichia fermentans and several yeast strains from the genera *Candida, Hansenula, Brettanomyces, Schizosaccharomyces,* and *Saccharomycopsis*, as well as *Rhizopus* sp. and *Aspergillus* sp., could only transform ZEN into α-ZEL (Brodehl et al. 2014).

7.11.1.3 Hydrogenation of the Carbon–Carbon Double Bond

The reduction of a C-C double bond through biotransformation is known as the hydrogenation of the carbon–carbon double bond in mycotoxins. *Penicillium raistrickii*, a fungus, hydrogenated aflatoxin B1 (AFB1) to produce aflatoxin B2 (Wu et al. 2009). As the carbon–carbon bond was reduced in zearalenone (ZEN), it was transformed into zearalanone (ZAN) (Miles et al. 1996).

7.11.1.4 De-epoxidation

Mycotoxin de-epoxidation is frequently observed in trichothecene analogs. The epoxide moiety of the molecule plays a major role in the toxicity of trichothecenes, and opening the epoxide ring significantly lowers toxicity. Eubacterium sp. conducted de-epoxidation to transform deoxynivalenol (DON) into deepoxydeoxynivalenol (DOM) (Fuchs et al. 2000). Group B trichothecenes include nivalenol (NIV). The bacteria Eubacterium sp. used de-epoxidation to transform NIV into de-epoxynivalenol (de-epoxy-NIV) (Zhou et al. 2008).

7.11.1.5 Methylation

Mycotoxins have been shown to undergo methylation on their hydroxyl groups. O-methyltransferase catalyzes this transformation. The fungus Cunninghamella bainieri converts zearalenone (ZEN) into zearalenone 16-methyl ether and zearalenone 14,16-bis(methyl ether).

7.11.1.6 Glycosylation and Glucuronidation

The process by which glucose or glucuronic acid is covalently bonded to a hydroxyl group is known as glycosylation or glucuronidation of mycotoxins. Mycotoxins are often made more polar by glycosylation or glucuronidation, which also reduces their toxicity (Wetterhorn et al. 2016). Fungal species *Blastobotrys muscicola* and *B. robertii* glycosylate T-2 toxin to produce T-2 toxin 3-O-α-D-glucoside (McCormick et al. 2012). Zearalenone (ZEN) was converted into zearalenone 14-O-glucoside through the glycosylation process by Arabidopsis UDP-glucosyltransferases produced in *Saccharomyces cerevisiae*. Similarly, the fungi

Mucor vainieri and *Thamnidium elegans* glycosylated ZEN to produce ZEN 14-O-Glc (Li et al. 2020).

7.11.1.7 Esterification

Mycotoxin esterification predominantly forms esters from alcohols and carboxylic acids. Wheat and maize cell cultures transformed ochratoxin A (OTA) into OTA methyl ester (Ruhland et al. 1995). In vitro, bovine rumen fluid acetylated the T-2 toxin, converting it to 3-acetyl T-2 toxin (Li et al. 2020).

7.11.1.8 Hydrolysis

Mycotoxin amide and ester linkages have been the subject of extensive hydrolysis research (Kim SeonHwa and Vujanovic 2018). Significant esterase activity was detected in a crude cell-free extract obtained from Fusarium sp. cultures, which degraded 3-acetyl deoxynivalenol (3-acetyl DON) to DON. The fungus *Sphaerodes mycoparasitica*, cultured with 3-acetyl DON, also exhibited this deacetylation (Kim SeonHwa and Vujanovic 2018).

When the fungus *Exophiala spinifera* degraded fumonisin B1 (FB1), it produced aminopentol 1 (AP1), the hydrolyzed form of FB1. Additionally, FB1 was converted into AP1 using an enzyme derived from *Sphingopyxis* sp. (Heinl et al. 2010). The crude lipase from Aspergillus niger was tested for its ability to degrade ochratoxin A (OTA) into less harmful compounds, OTα and L-β-phenylalanine (Stander et al. 2000).

7.11.1.9 Sulfation

Sulfotransferase catalyzes the conjugation of a sulfo group to a hydroxyl group, resulting in the sulfation of mycotoxins. The fungus *Sphaerodes mycoparasitica* transformed deoxynivalenol (DON) and zearalenone (ZEN) into their respective sulfates, DON 3-sulfate and ZEN 14-sulfate. Both DON 3-sulfate and ZEN 14-sulfate are less hazardous than DON and ZEN, respectively (Kim SeonHwa and Vujanovic 2018).

7.11.1.10 Demethylation

The process of removing a methyl group from a molecule is known as demethylation of mycotoxins. A common method of mycotoxin demethylation is the substitution of a hydroxyl group for a methoxyl group. The enzyme CYP321A1 from the corn earworm *Helicoverpa zea* uses demethylation to transform aflatoxin B1 (AFB1) into aflatoxin P1 (AFP1). AFP1 is more polar and less toxic than AFB1 (Niu et al. 2008).

7.11.1.11 Deamination

The process by which deaminase removes an amino group from a molecule is known as deamination. Aspergillus sp. uses oxidative deamination to convert fumonisin B4 into fumonisins La4 and Py4 (Burgess et al. 2016). Compared to fumonisin B mycotoxins, both Fla4 and FPy4 were significantly less toxic. This indicates that *Aspergillus fungi* can produce enzymes that may be used to detoxify fumonisins (Burgess et al. 2016). Fumonisin B1, also known as aminopentol 1, is hydrolyzed and transformed into 2-keto aminopentol 1, or hydrolyzed fumonisin B1. This compound is produced by the fungus *Exophiala spinifera* through oxidative deamination (Burgess et al. 2016).

7.11.1.12 Miscellaneous Reactions

Numerous other mycotoxin biotransformation processes, such as multi-step conversions, epoxidation, dehydrogenation, and dechlorination, have also been reported (Li et al. 2020).

7.11.2 Detoxification of Different Types of Mycotoxins

Mycotoxin toxicity is often associated with specific functional groups in their chemical structure. Thus, the toxicity of mycotoxins can be reduced or eliminated by selectively modifying these functional groups. Microorganisms and their associated enzymes are specialized and can catalyze the biotransformation of mycotoxins into benign substances, making them useful for detoxifying mycotoxins in food (Awuchi et al. 2022).

7.11.2.1 Detoxification of Aflatoxins

Furanocoumarin mycotoxins, known as aflatoxins (AFs), are predominantly produced by Aspergillus species. The International Agency for Research on Cancer (IARC) has categorized them as Group 1 carcinogens due to their risk to both human and animal health (Loi et al. 2017). The furofuran ring of AFs has been identified as the source of their toxic and carcinogenic properties, which result from the metabolic activation of the C8–C9 double bond to form an 8 to 9 epoxide (Loi et al. 2017). This epoxidation process allows binding to N7-guanine, leading to G to T transversions in the DNA molecule, which is critical to the carcinogenicity of AFs. Acute toxicity may also arise from activated aflatoxins' capacity to form Schiff bases with cellular and microsomal proteins, involving methionine, histidine, and lysine. The toxicity and carcinogenicity of aflatoxins are influenced by the lactone ring. DNA alkylation depends on the lactone and difuranocoumarin moieties of aflatoxins (Li et al. 2020). Research has examined two main detoxification pathways: modifications to the difuran ring and coumarin structure of aflatoxins (AFs). Assessments of biotransformation-based detoxification of aflatoxins have been conducted (Rushing and Selim 2019). Aflatoxin B1 (AFB1) is the most potent and prevalent among the isolated aflatoxin-forming bacteria. The primary biotransformation processes for AFB1 include hydroxylation of the difuran ring, epoxidation of the furan ring's carbon–carbon double bond, demethylation of the aromatic ring's methoxy group, reduction of the cyclopentenone's carbonyl group, and C3 hydroxylation of the cyclopentenone (Rushing and Selim 2019). Several fungal species reduce the carbonyl group in the pentan cycle to a hydroxyl group, while *Pseudomonas putida* multistep oxidation–reduction processes convert AFB1 to AFD1, AFD2, and AFD3 as successful detoxification reactions (Eshelli et al. 2015).

7.11.2.2 Detoxification of Alternaria Toxins

A number of mycotoxin families produced by *Alternaria* species can harm plants, humans, and animals. The primary Alternaria mycotoxins include Altertoxin II, Tentoxin, Altertoxin III, Altertoxin I (AOH), Altertoxin 9-methyl ether (AME), Altenuisol, Alteruene, and Tenuazonic Acid. AME and AOH are classified as dibenzo-α-pyranones and are the primary mycotoxins. They are mutagenic and genotoxic and have been linked to esophageal cancer (Lou et al. 2013). Plants use glycosylation and glucuronidation as their primary detoxifying processes for Alternaria toxins.

7.11.2.3 Detoxification of Fomannoxin

The fungal phytotoxin known as fomannoxin is a benzohydrofuran aldehyde produced by the forest pathogenic basidiomycete Heterobasidion annosum during the infection process. Fomannoxin exhibited growth-inhibiting properties on callus and suspension cultures of conifer cells. The most efficient detoxification method was the reduction of fomannoxin by cultivated cells of Pinus sylvestris (Zweimüller et al. 1997).

7.11.2.4 Detoxification of Fumonisins

Fumonisins are primarily produced by *F. proliferatum* and *F. verticillioides.* They share structural similarities with long-chain sphingolipid bases such as sphingosine and sphinganine. The most common fumonisin, fumonisin B1, poses the greatest risk to animal and human nutrition among the fumonisins. They contaminate processed foods containing corn. Their toxicity mechanism is closely associated with their ability to inhibit the production of sphingolipids in plants, animals, and yeasts. Animal disorders linked to fumonisins include pulmonary edema in pigs and liver cancer in rats (Lumsangkul et al. 2019). In some populations that consume maize, fumonisins may cause neural tube defects. The two main biotransformation pathways involved in fumonisin detoxification are aminotransferase-mediated deamination and carboxylesterase-mediated hydrolysis, resulting in the elimination of two tricarballylic side-chains (Heinl et al. 2010).

7.11.2.5 Detoxification of Ochratoxins

Ochratoxins (OTs) are a class of mycotoxins that are carcinogenic to humans and share an isocoumarin moiety substituted with: (i) a phenylalanine group, which includes ochratoxin A, 10-hydroxy-ochratoxin A, and ochratoxin B; (ii) a phenylalanine ester group, which includes ochratoxin C, ochratoxin A methyl ester, ochratoxin B methyl ester, and ochratoxin B ethyl ester; and (iii) a hydroxyl group, which includes ochratoxin α and OTβ. Regarding Group 2B carcinogenicity to humans, OTA is the most dangerous mycotoxin among them. Additionally, OTA has demonstrated several harmful effects on animals, the most notable of which is nephrotoxicity. Many species of *Penicillium* and *Aspergillus* can produce OTA. Hydrolysis of OTA by various bacterial and fungal species yields altered compounds such as OTα and phenylalanine, which has proven to be the most efficient detoxification method (Cho et al. 2016).

7.11.2.6 Detoxification of Patulin

The polyketide lactone patulin (PAT) exhibits cytotoxic, mutagenic, carcinogenic, teratogenic, and genotoxic effects on people, animals, and plants. Several fungal species from the genera Penicillium, *Aspergillus*, and *Byssochlamys* can produce PAT. However, *Penicillium expansum* is considered the main producer of PAT and is the primary cause of blue mold disease in stored apples. PAT can be effectively detoxified through biotransformation using yeast, bacteria, and fungi. This method is appealing as it operates in a moderate and environmentally benign environment. The modified products become less toxic or nontoxic if PAT's furan or pyran ring is destroyed (Saleh and Goktepe 2019).

7.11.2.7 Detoxification of Trichothecenes

The majority of trichothecenes are produced by fungi of the Fusarium genus. Other genera such as *Cephalosporium, Myrothecium, Stachybotrys, Trichoderma,* and *Trichothecium* also have species that produce trichothecenes, albeit to a lesser extent. Trichothecenes are often found in contaminated grains, particularly maize, wheat, barley, and oats. They can cause developmental growth retardation, immunosuppression, reproductive issues, and even mortality in both humans and animals. Trichothecenes attach to the ribosome and cause ribotoxic stress at the molecular level, activating MAP kinases, disrupting the cell cycle, and leading to cell death (Payros et al. 2016). These toxins can be detoxified through various biotransformation pathways, including oxygenation, de-epoxidation at C-12 and C-13 positions, epimerization of hydroxyl groups at C-3 position, glycosylation of hydroxyl groups, and hydrolysis of acetoxy groups (McCormick et al. 2012). Numerous bacteria, fungi, and their respective enzymes are components of biotransformation systems.

7.11.2.8 Detoxification of Zearalenone Analogues

Zearalenone (ZEN) is a β-resorcylic acid lactone that is macrocyclic and phenolic. It is primarily produced by fungi in the Fusarium genus, namely *F. graminearum* and *F. culmorum.* ZEN exhibits estrogenic activity in sheep, cattle, and pigs. Furthermore, ZEN and its metabolites, including α- and β-zearalenol, have been shown to influence animal reproductive function (Dai MeiLing et al. 2016). Additionally, ZEN impairs immunological functions, damages the liver and kidneys, and causes cytotoxicity and immunotoxicity (Wang et al. 2019). Possible mechanisms for ZEN biotransformation include modification of the hydroxyl group (glycosylation and sulfation), reduction of the carbon–carbon double bond, hydrolysis of the lactone ring, and reduction of the ketonic carbonyl group. Biotransformation-based detoxification of mycotoxins provides a reliable benchmark approach for managing mycotoxins in food and feed. Offering fundamental knowledge for the risk assessment of mycotoxins in food and feed safety, this research will also aid in understanding the fate of mycotoxins in humans and animals. Further research is required for mycotoxin detoxification by biotransformation, particularly in developing techniques for multi-reaction processes and cloning and expressing detoxification enzyme genes in organisms.

Table 7.3 Detoxification and biotransformation of mycotoxins by endophytic fungi

Mechanism	Description	Examples	Benefits
Hydrolysis	Breaking down mycotoxin molecules by adding water molecules	Degradation of aflatoxins, fumonisins, and ochratoxin A	Reduced toxicity and easier elimination
Oxidation	Introducing oxygen atoms into the mycotoxin structure	Conversion of zearalenone to less toxic forms by Penicillium sp.	Decreased harmful effects on animals
Reduction	Removing oxygen atoms from the mycotoxin structure	Detoxification of fumonisin B1 by Fusarium equiseti	Lower risk of health problems in livestock and humans
Conjugation	Attaching a harmless molecule to the mycotoxin, making it more water-soluble for easier excretion	Detoxification of deoxynivalenol by Acremonium coenophialum	Reduced accumulation of mycotoxins in plant tissues
Dehalogenation	Removing halogen atoms (like chlorine) from the mycotoxin structure	Detoxification of trichothecenes by Epicoccum nigrum	Decreased potential for immune system suppression

In summary, different endophytic fungi may have varying abilities to detoxify specific mycotoxins (Table 7.3). The effectiveness of biotransformation can be influenced by factors like the type of mycotoxin, plant host, and environmental conditions. Combining endophytes with other strategies, like good agricultural practices and proper storage conditions, can further enhance food and feed safety against mycotoxins.

7.12 Limitations of Endophytic Fungi

When endophytic fungi interact with plants and other microbes, they encounter several restrictions and challenges. Various factors can influence the colonization rate and dispersion of these fungi in plant tissues, such as the type of plant, tissue, environment, and specific fungal strain (Xin-Chen et al. 2023). These factors might affect their ability to confer benefits or produce metabolites. Additionally, they can engage in antagonistic or cooperative interactions with other endophytes or pathogens for nutrients and space within the plant, influencing the diversity and structure of the fungal community. Fungal signals or chemicals may be recognized by the plant's immune system, which can react by initiating defense mechanisms to prevent or limit fungal colonization. Endophytic fungi's capacity for evolution and their ability to withstand environmental stresses may be hampered by decreased genetic diversity or functional loss due to adaptation to the endophytic lifestyle (Beena et al. 2021). Lastly, they might grow slowly or exhibit poor culturability in artificial media, making it difficult to isolate, identify, and search for new bioactive compounds or biotechnological applications.

7.13 Conclusion

Mycotoxins are harmful secondary metabolites produced by various organisms, posing significant challenges for food safety and agriculture. Aflatoxin, trichothecenes, ochratoxins, zearalenone, fumonisins, and patulin are among the more than 400 known mycotoxins that underscore the extent of contamination. They cause financial losses and major health issues that threaten food security and necessitate regulatory actions. Climate change exacerbates these problems by creating conditions favorable to contamination. Governments and international organizations have enacted laws emphasizing sustainable agricultural practices and establishing maximum allowable levels of contaminants. Innovative approaches to contamination, including the development of biological control agents and resistant crop varieties, have also been explored through scientific research. Potential biocontrol agents include endophytic fungi, which live inside plants without causing harm. Their diverse advantages, such as chemotactic signaling and enzymatic hydrolysis, support plant health and defense. Studies on crop diversity have demonstrated its impact, creating opportunities for sustainable farming methods. Endophytic fungi produce antifungal metabolites that limit mycotoxin production and may even break down mycotoxins, helping to reduce their levels. Promising strategies can be achieved by combining these methods with other biological control agents like fungi and bacteria. Issues such as low rates of colonization and plant immunological responses need to be addressed. In summary, a comprehensive strategy that integrates biological controls with reduced chemical inputs is essential for long-term crop protection and efficient, environmentally safe mycotoxin mitigation.

References

Abbas HK, Weaver MA, Horn BW, Carbone I, Monacell JT, Shier WT (2011) Selection of Aspergillus flavus isolates for biological control of aflatoxins in corn. Toxin Rev 30(2–3):59–70

Abdou R, Shabana S, Rateb ME (2020) Terezine E, bioactive prenylated tryptophan analogue from an endophyte of Centaurea stoebe. Nat Prod Res 34(4):503–510

Abraham N, Chan ETS, Zhou T, Seah SY (2022) Microbial detoxification of mycotoxins in food. Front Microbiol 13:957148

Adeleke BS, Ayilara MS, Akinola SA, Babalola OO (2022) Biocontrol mechanisms of endophytic fungi. Egypt J Biol Pest Control 32(1):1–17

Afshar P, Shokrzadeh M, Raeisi SN, Ghorbani-HasanSaraei A, Nasiraii LR (2020) Aflatoxins biodetoxification strategies based on probiotic bacteria. Toxicon 178:50–58

Agrawal S, Bhatt A (2023) Microbial endophytes: emerging trends and biotechnological applications. Curr Microbiol 80(8):249

Agriopoulou S, Stamatelopoulou E, Varzakas T (2020) Advances in occurrence, importance, and mycotoxin control strategies: prevention and detoxification in foods. Food Secur 9(2):137

Akram S, Ahmed A, He P, He P, Liu Y, Wu Y et al (2023) Uniting the role of endophytic fungi against plant pathogens and their interaction. J Fungi 9(1):72

Alam B, Lǐ J, Gě Q, Khan MA, Gōng J, Mehmood S et al (2021) Endophytic fungi: from symbiosis to secondary metabolite communications or vice versa? Front Plant Sci 12:3060

Al-Jaal BA, Jaganjac M, Barcaru A, Horvatovich P, Latiff A (2019) Aflatoxin, fumonisin, ochratoxin, zearalenone and deoxynivalenol biomarkers in human biological fluids: a systematic literature review, 2001–2018. Food Chem Toxicol 129:211–228

Al-Saad LA, Al-Badran AI, Al-Jumayli SA, Magan N, Rodríguez A (2016) Impact of bacterial biocontrol agents on aflatoxin biosynthetic genes, aflD and aflR expression, and phenotypic aflatoxin B1 production by Aspergillus flavus under different environmental and nutritional regimes. Int J Food Microbiol 217:123–129

Altugˇ T, Yousef AE, Marth EH (1990) Degradation of aflatoxin B1 in dried figs by sodium bisulfite with or without heat, ultraviolet energy or hydrogen peroxide. J Food Prot 53(7):581–583

Andrés MF, Diaz CE, Giménez C, Cabrera R, González-Coloma A (2017) Endophytic fungi as novel sources of biopesticides: the Macaronesian Laurel forest, a case study. Phytochem Rev 16:1009–1022

Averill C, Turner BL, Finzi AC (2014) Mycorrhiza-mediated competition between plants and decomposers drives soil carbon storage. Nature 505(7484):543–545

Awuchi CG, Ondari EN, Ogbonna CU, Upadhyay AK, Baran K, Okpala COR et al (2021) Mycotoxins affecting animals, foods, humans, and plants: types, occurrence, toxicities, action mechanisms, prevention, and detoxification strategies—A revisit. Food Secur 10(6):1279

Awuchi CG, Ondari EN, Nwozo S, Odongo GA, Eseoghene IJ, Twinomuhwezi H et al (2022) Mycotoxins' toxicological mechanisms involving humans, livestock and their associated health concerns: a review. Toxins 14(3):167

Bamisile BS, Dash CK, Akutse KS, Keppanan R, Wang L (2018) Fungal endophytes: beyond herbivore management. Front Microbiol 9:329443

Baron NC, Rigobelo EC (2022) Endophytic fungi: a tool for plant growth promotion and sustainable agriculture. Mycology 13(1):39–55

Bary A (1866) Morphologie und physiologie der pilze, flechten und myxomyceten, (vol. 1). Leipzig: Engelmann

Beena R, Veena V, Jaslam MPK, Nithya N, Adarsh VS (2021) Germplasm innovation for high-temperature tolerance from traditional rice accessions of Kerala using genetic variability, genetic advance, path coefficient analysis and principal component analysis. J Crop Sci Biotechnol 24(5):555–566

Bhardwaj M, Kailoo S, Khan SS (2023) Harnessing fungal endophytes for natural management: a biocontrol perspective. Front Microbiol 14:1280258

Billar de Almeida A, Concas J, Campos MD, Materatski P, Varanda C, Patanita M et al (2020) Endophytic fungi as potential biological control agents against grapevine trunk diseases in Alentejo region. Biology 9(12):420

Bouanaka H, Bellil I, Harrat W, Boussaha S, Benbelkacem A, Khelifi D (2021) On the biocontrol by Trichoderma afroharzianum against Fusarium culmorum responsible of Fusarium head blight and crown rot of wheat in Algeria. Egypt J Biol Pest Control 31:1–13

Bretz M, Beyer M, Cramer B, Knecht A, Humpf HU (2006) Thermal degradation of the Fusarium mycotoxin deoxynivalenol. J Agric Food Chem 54(17):6445–6451

Brodehl A, Möller A, Kunte HJ, Koch M, Maul R (2014) Biotransformation of the mycotoxin zearalenone by fungi of the genera Rhizopus and Aspergillus. FEMS Microbiol Lett 359(1):124–130

Bulgaru CV, Marin DE, Pistol GC, Taranu I (2021) Zearalenone and the immune response. Toxins 13(4):248

Burgess KM, Renaud JB, McDowell T, Sumarah MW (2016) Mechanistic insight into the biosynthesis and detoxification of fumonisin mycotoxins. ACS Chem Biol 11(9):2618–2625

Cai M, Qian Y, Chen N, Ling T, Wang J, Jiang H et al (2020) Detoxification of aflatoxin B1 by Stenotrophomonas sp. CW117 and characterization the thermophilic degradation process. Environ Pollut 261:114178

Cao H, Liu D, Mo X, Xie C, Yao D (2011) A fungal enzyme with the ability of aflatoxin B1 conversion: purification and ESI-MS/MS identification. Microbiol Res 166(6):475–483

Carvalho CR, Gonçalves VN, Pereira CB, Johann S, Galliza IV, Alves TM et al (2012) The diversity, antimicrobial and anticancer activity of endophytic fungi associated with the medicinal plant Stryphnodendron adstringens (Mart.) Coville (Fabaceae) from the Brazilian savannah. Symbiosis 57:95–107

Chandran H, Meena M, Barupal T, Sharma K (2020) Plant tissue culture as a perpetual source for production of industrially important bioactive compounds. Biotechnol Rep 26:e00450

Chapla VM, Zeraik ML, Leptokarydis IH, Silva GH, Bolzani VS, Young MCM et al (2014) Antifungal compounds produced by Colletotrichum gloeosporioides, an endophytic fungus from Michelia champaca. Molecules 19(11):19243–19252
Chlebicz A, Śliżewska K (2020) In vitro detoxification of aflatoxin B 1, deoxynivalenol, fumonisins, T-2 toxin and zearalenone by probiotic bacteria from genus Lactobacillus and Saccharomyces cerevisiae yeast. Probiotics Antimicro Proteins 12:289–301
Cho SM, Jeong SE, Lee KR, Sudhani HP, Kim M, Hong SY, Chung SH (2016) Biodegradation of ochratoxin A by Aspergillus tubingensis isolated from meju. J Microbiol Biotechnol 26(10):1687–1695
Chow YY, Ting ASY (2019) Influence of fungal infection on plant tissues: FTIR detects compositional changes to plant cell walls. Fungal Ecol 37:38–47
Cotty PJ, Bayman P (1993) Competitive exclusion of a toxigenic strain of Aspergillus flavus by an atoxigenic strain. Phytopathology 83(12):1283–1287
Crutcher FK, Puckhaber LS, Bell AA, Liu J, Duke SE, Stipanovic RD, Nichols RL (2017) Detoxification of fusaric acid by the soil microbe Mucor rouxii. J Agric Food Chem 65(24):4989–4992
d'Errico G, Aloj V, Flematti GR, Sivasithamparam K, Worth CM, Lombardi N et al (2021) Metabolites of a Drechslera sp. endophyte with potential as biocontrol and bioremediation agent. Nat Prod Res 35(22):4508–4516
Dai MeiLing DM, Jiang ShuZhen JS, Yuan XueJun YX, Yang WeiRen YW, Yang ZaiBin YZ, Huang LiBo HL (2016) Effects of zearalenone-diet on expression of ghrelin and PCNA genes in ovaries of post-weaning piglets. Anim Reprod Sci 168:126–137
Devi R, Verma R, Dhalaria R, Kumar A, Kumar D, Puri S et al (2023) A systematic review on endophytic fungi and its role in the commercial applications. Planta 257(4):70
Diao EnJie DE, Hou HanXue HH, Hu WeiCheng HW, Dong HaiZhou DH, Li XiangYang LX (2018) Removing and detoxifying methods of patulin: a review. Trends Food Sci Technol 81:139–145
Dissanayake AJ, Purahong W, Wubet T, Hyde KD, Zhang W, Xu H et al (2018) Direct comparison of culture-dependent and culture-independent molecular approaches reveal the diversity of fungal endophytic communities in stems of grapevine (Vitis vinifera). Fungal Divers 90:85–107
Ekwomadu TI, Akinola SA, Mwanza M (2021) Fusarium mycotoxins, their metabolites (free, emerging, and masked), food safety concerns, and health impacts. Int J Environ Res Public Health 18(22):11741. https://doi.org/10.3390/ijerph182211741
Elias LM, Fortkamp D, Sartori SB, Ferreira MC, Gomes LH, Azevedo JL et al (2018) The potential of compounds isolated from Xylaria spp. as antifungal agents against anthracnose. Braz J Microbiol 49:840–847
El-Sharkawy SH, Abul-Hajj YJ (1988) Microbial transformation of zearalenone. 2. Reduction, hydroxylation, and methylation products. J Org Chem 53(3):515–519
Eo JK, Choi JW, Eom AH (2022) Diversity, distribution, and host plant of endophytic fungi: a focus on Korea. Mycobiology 50(6):399–407
Eshelli M, Harvey L, Edrada-Ebel R, McNeil B (2015) Metabolomics of the bio-degradation process of aflatoxin B1 by actinomycetes at an initial pH of 6.0. Toxins 7(2):439–456
Fan Y, Zhao L, Ma Q, Li X, Shi H, Zhou T, Ji C (2013) Effects of Bacillus subtilis ANSB060 on growth performance, meat quality and aflatoxin residues in broilers fed moldy peanut meal naturally contaminated with aflatoxins. Food Chem Toxicol 59:748–753
Fite T, Kebede E, Tefera T, Bekeko Z (2023) Endophytic fungi: versatile partners for pest biocontrol, growth promotion, and climate change resilience in plants. Front Sustain Food Syst 7:1322861
Franco-Duarte R, Černáková L, Kadam S, Kaushik KS, Salehi B, Bevilacqua A et al (2019) Advances in chemical and biological methods to identify microorganisms—from past to present. Microorganisms 7(5):130
Fuchs E, Binder EM, Heidler D, Krska R (2000) Characterisation of metabolites after the microbial degradation of A-and B-trichothecenes by BBSH 797. Mycotoxin Res 16(Suppl 1):66–69

García-Latorre C, Rodrigo S, Santamaría O (2021) Endophytes as plant nutrient uptake-promoter in plants. In: Endophytes: Mineral Nutrient Management, vol 3. CRC Press, pp 247–265

Geurts R, Xiao TT, Reinhold-Hurek B (2016) What does it take to evolve a nitrogen-fixing endosymbiosis? Trends Plant Sci 21(3):199–208

Giorni P, Magan N, Battilani P (2009) Environmental factors modify carbon nutritional patterns and niche overlap between Aspergillus flavus and Fusarium verticillioides strains from maize. Int J Food Microbiol 130(3):213–218

Gupta A, Meshram V, Gupta M, Goyal S, Qureshi KA, Jaremko M, Shukla KK (2023) Fungal endophytes: microfactories of novel bioactive compounds with therapeutic interventions; A comprehensive review on the biotechnological developments in the field of fungal endophytic biology over the last decade. Biomol Ther 13(7):1038

Harman GE, Doni F, Khadka RB, Uphoff N (2021) Endophytic strains of Trichoderma increase plants' photosynthetic capability. J Appl Microbiol 130(2):529–546

Heinl S, Hartinger D, Thamhesl M, Vekiru E, Krska R, Schatzmayr G, Grabherr R (2010) Degradation of fumonisin B1 by the consecutive action of two bacterial enzymes. J Biotechnol 145(2):120–129

Huang R, Jiang BG, Li XN, Wang YT, Liu SS, Zheng KX et al (2018) Polyoxygenated cyclohexenoids with promising α-glycosidase inhibitory activity produced by Phomopsis sp. YE3250, an endophytic fungus derived from Paeonia delavayi. J Agric Food Chem 66(5):1140–1146

Hyde KD, Soytong K (2008) The fungal endophyte dilemma. Fungal Divers 33(163):e173

Imade F, Ankwasa EM, Geng H, Ullah S, Ahmad T, Wang G et al (2021) Updates on food and feed mycotoxin contamination and safety in Africa with special reference to Nigeria. Mycology 12(4):245–260

Jalili M, Jinap S, Son R (2011) The effect of chemical treatment on reduction of aflatoxins and ochratoxin A in black and white pepper during washing. Food Addit Contam 28(4):485–493

Jana SK, Islam MM, Mandal S (2022) Endophytic microbiota of rice and their collective impact on host fitness. Curr Microbiol 79(2):37

Janik E, Niemcewicz M, Podogrocki M, Ceremuga M, Gorniak L, Stela M, Bijak M (2021) The existing methods and novel approaches in mycotoxins' detection. Molecules 26(13):3981

Javanmardi F, Khodaei D, Sheidaei Z, Bashiry M, Nayebzadeh K, Vasseghian Y, Mousavi Khaneghah A (2022) Decontamination of aflatoxins in edible oils: a comprehensive review. Food Rev Intl 38(7):1410–1426

Jha P, Kaur T, Panja A, Paul S, Kumar V, Malik T (2023) Endophytic fungi: hidden treasure chest of antimicrobial metabolites interrelationship of endophytes and metabolites. Front Microbiol 14:1227830

Ji KX, Chi F, Yang MF, Shen SH, Jing YX, Dazzo FB, Cheng HP (2010) Movement of rhizobia inside tobacco and lifestyle alternation from endophytes to free-living rhizobia on leaves. J Microbiol Biotechnol 20(2):238–244

Jr ADP, King JM (2002) Efficacy and Safety Evaluation of Ozonation to Degrade Aflatoxin in Corn. J Food Sci 67:2866–2872

Kabak B, Dobson AD, Var I (2006) Strategies to prevent mycotoxin. Crit Rev Food Sci Nutr 46:8

Kamran M, Imran QM, Ahmed MB, Falak N, Khatoon A, Yun BW (2022) Endophyte-mediated stress tolerance in plants: a sustainable strategy to enhance resilience and assist crop improvement. Cells 11(20):3292

Karlovsky P, Suman M, Berthiller F, De Meester J, Eisenbrand G, Perrin I et al (2016) Impact of food processing and detoxification treatments on mycotoxin contamination. Mycotoxin Res 32:179–205

Kashyap N, Singh SK, Yadav N, Singh VK, Kumari M, Kumar D et al (2023) Biocontrol screening of endophytes: applications and limitations. Plan Theory 12(13):2480

Kemen E, Jones JD (2012) Obligate biotroph parasitism: can we link genomes to lifestyles? Trends Plant Sci 17(8):448–457

Khaneghah AM, Farhadi A, Nematollahi A, Vasseghian Y, Fakhri Y (2020) A systematic review and meta-analysis to investigate the concentration and prevalence of trichothecenes in the cereal-based food. Trends Food Sci Technol 102:193–202

Khare E, Mishra J, Arora NK (2018) Multifaceted interactions between endophytes and plant: developments and prospects. Front Microbiol 9:2732
Kim SeonHwa KS, Vujanovic V (2018) Biodegradation and biodetoxification of Fusarium mycotoxins by Sphaerodes mycoparasitica. AMB Expr 7:145
Kobayashi DY, Palumbo JD (2000) Bacterial endophytes and their effects on plants and uses in agriculture. In: Microbial Endophytes. CRC Press., pp 213–250
Kumar P, Mahato DK, Sharma B, Borah R, Haque S, Mahmud MC, ..., Bui S (2020) Ochratoxins in food and feed: Occurrence and its impact on human health and management strategies. Toxicon 187:151–162
Li XJ, Zhang Q, Zhang AL, Gao JM (2012) Metabolites from *Aspergillus fumigatus*, an endophytic fungus associated with Melia azedarach, and their antifungal, antifeedant, and toxic activities. J Agric Food Chem 60(13):3424–3431
Li P, Su R, Yin R, Lai D, Wang M, Liu Y, Zhou L (2020) Detoxification of mycotoxins through biotransformation. Toxins 12(2):121
Liu X, Dong M, Chen X, Jiang M, Lv X, Zhou J (2008) Antimicrobial activity of an endophytic Xylaria sp. YX-28 and identification of its antimicrobial compound 7-amino-4-methylcoumarin. Appl Microbiol Biotechnol 78:241–247
Liu K, Yang Y, Miao CP, Zheng YK, Chen JL, Chen YW et al (2016) Koningiopisins A–H, polyketides with synergistic antifungal activities from the endophytic fungus Trichoderma koningiopsis. Planta Med 82(04):371–376
Liu L, Xie M, Wei D (2022) Biological detoxification of mycotoxins: current status and future advances. Int J Mol Sci 23(3):1064
Loi M, Fanelli F, Liuzzi VC, Logrieco AF, Mulè G (2017) Mycotoxin biotransformation by native and commercial enzymes: present and future perspectives. Toxins 9(4):111
López-García R (2022) Mycotoxin management: an international challenge. In: Ensuring Global Food Safety. Academic Press, pp 213–219
Lou F, Sima CS, Adusumilli PS, Bains MS, Sarkaria IS, Rusch VW, Rizk NP (2013) Esophageal cancer recurrence patterns and implications for surveillance. J Thorac Oncol 8(12):1558–1562
Lu H, Wei T, Lou H, Shu X, Chen Q (2021) A critical review on communication mechanism within plant-endophytic fungi interactions to cope with biotic and abiotic stresses. J Fungi 7(9):719
Lugtenberg BJ, Caradus JR, Johnson LJ (2016) Fungal endophytes for sustainable crop production. FEMS Microbiol Ecol 92(12):fiw194
Lumsangkul C, Chiang HI, Lo NW, Fan YK, Ju JC (2019) Developmental toxicity of mycotoxin fumonisin B1 in animal embryogenesis: An overview. Toxins 11(2):114
Macías-Rubalcava ML, Hernández-Bautista BE, Jiménez-Estrada M, González MC, Glenn AE, Hanlin RT et al (2008) Naphthoquinone spiroketal with allelochemical activity from the newly discovered endophytic fungus Edenia gomezpompae. Phytochemistry 69(5):1185–1196
Magan N, Medina A (2016) Integrating gene expression, ecology and mycotoxin production by Fusarium and Aspergillus species in relation to interacting environmental factors. World Mycotoxin J 9(5):673–684
Malfanova N, Lugtenberg BJ, Berg G (2013) Bacterial endophytes: who and where, and what are they doing there? In: Molecular microbial ecology of the rhizosphere, vol 1. CRC Press, pp 391–403
Marins-Gonçalves L, Ferreira MM, Guidi LR, De Souza D (2023) Is chemical analysis suitable for detecting mycotoxins in agricultural commodities and foodstuffs? Talanta 265:124782
Martínez-Culebras PV, Gandía M, Garrigues S, Marcos JF, Manzanares P (2021) Antifungal peptides and proteins to control toxigenic fungi and mycotoxin biosynthesis. Int J Mol Sci 22(24):13261
Matumba L, Van Poucke C, Njumbe Ediage E, Jacobs B, De Saeger S (2015) Effectiveness of hand sorting, flotation/washing, dehulling and combinations thereof on the decontamination of mycotoxin-contaminated white maize. Food Addit Contam: Part A 32(6):960–969
McCormick SP, Price NP, Kurtzman CP (2012) Glucosylation and other biotransformations of T-2 toxin by yeasts of the Trichomonascus clade. Appl Environ Microbiol 78(24):8694–8702

Medina A, Mohale S, Samsudin NIP, Rodriguez-Sixtos A, Rodriguez A, Magan N (2017) Biocontrol of mycotoxins: dynamics and mechanisms of action. Curr Opin in Food Sci 17:41–48

Mengistu AA (2020) Endophytes: colonization, behaviour, and their role in defense mechanism. Int J Microbiol 2020:6927219

Miles CO, Erasmuson AF, Wilkins AL, Towers NR, Smith BL, Garthwaite I et al (1996) Ovine metabolism of zearalenone to α-zearalanol (zeranol). J Agric Food Chem 44(10):3244–3250

Mohale S, Medina A, Magan N (2013) Effect of environmental factors on in vitro and in situ interactions between atoxigenic and toxigenic A. flavus strains and control of aflatoxin contamination of maize. Biocontrol Sci Tech 23(7):776–793

Moncada-Torres A, Leuenberger K, Gonzenbach R, Luft A, Gassert R (2014) Activity classification based on inertial and barometric pressure sensors at different anatomical locations. Physiol Meas 35(7):1245

Nazhand A, Durazzo A, Lucarini M, Souto EB, Santini A (2020) Characteristics, occurrence, detection and detoxification of aflatoxins in foods and feeds. Food Secur 9(5):644

Nie P, Li X, Wang S, Guo J, Zhao H, Niu D (2017) Induced systemic resistance against Botrytis cinerea by Bacillus cereus AR156 through a JA/ET-and NPR1-dependent signaling pathway and activates PAMP-triggered immunity in Arabidopsis. Front Plant Sci 8:238

Niu G, Wen Z, Rupasinghe SG, Zeng RS, Berenbaum MR, Schuler MA (2008) Aflatoxin B1 detoxification by CYP321A1 in Helicoverpa zea. Arch Insect Biochem Physiol 69(1):32–45

Pankaj SK, Shi H, Keener KM (2018) A review of novel physical and chemical decontamination technologies for aflatoxin in food. Trends Food Sci Technol 71:73–83

Park JW, Kim YB (2006) Effect of pressure cooking on aflatoxin B1 in rice. J Agric Food Chem 54(6):2431–2435

Patel HK, Kalaria RK, Kahimani MR, Shah GS, Dholakiya BZ (2021) Prevention and control of mycotoxins for food safety and security of human and animal feed. In: Fungi bio-prospects in sustainable agriculture, environment and nano-technology. Academic Press. pp 315–345

Payros D, Alassane-Kpembi I, Pierron A, Loiseau N, Pinton P, Oswald IP (2016) Toxicology of deoxynivalenol and its acetylated and modified forms. Arch Toxicol 90:2931–2957

Petrini G, Cesana A, De Alberti G, Genoni F, Leofanti G, Padovan M, ... Roffia P (1991) Deactivation phenomena on Ti-silicalite. In: Studies in surface science and catalysis, vol 68. Elsevier, pp 761–766

Plett JM, Martin FM (2018) Know your enemy, embrace your friend: using omics to understand how plants respond differently to pathogenic and mutualistic microorganisms. Plant J 93(4):729–746

Polak-Śliwińska M, Paszczyk B (2021) Trichothecenes in food and feed, relevance to human and animal health and methods of detection: a systematic review. Molecules 26(2):454

Premalatha K, Kalra AJFE (2013) Molecular phylogenetic identification of endophytic fungi isolated from resinous and healthy wood of Aquilaria malaccensis, a red listed and highly exploited medicinal tree. Fungal Ecol 6(3):205–211

Qin J, Lyu A, Zhang QH, Yang L, Zhang J, Wu MD, Li GQ (2019) Strain identification and metabolites isolation of Aspergillus capensis CanS-34A from Brassica napus. Mol Biol Rep 46:3451–3460

Qiu T, Wang H, Yang Y, Yu J, Ji J, Sun J et al (2021) Exploration of biodegradation mechanism by AFB1-degrading strain Aspergillus niger FS10 and its metabolic feedback. Food Control 121:107609

Rai A, Dixit S, Singh SP, Gautam NK, Das M, Tripathi A (2018) Presence of zearalenone in cereal grains and its exposure risk assessment in Indian population. J Food Sci 83(12):3126–3133

Rodriguez RJ, White JF Jr, Arnold AE, Redman ARA (2009) Fungal endophytes: diversity and functional roles. New phytol 182(2):314–330

Ruhland M, Engelhardt G, Wallnöfer PR, Schäfer W (1995) Transformation of the mycotoxin ochratoxin A in wheat and maize cell suspension cultures. Naturwissenschaften 81(10):453–454

Rushing BR, Selim MI (2019) Aflatoxin B1: a review on metabolism, toxicity, occurrence in food, occupational exposure, and detoxification methods. Food Chem Toxicol 124:81–100

Saleh I, Goktepe I (2019) The characteristics, occurrence, and toxicological effects of patulin. Food Chem Toxicol 129:301–311

Salgado JD, Wallhead M, Madden LV, Paul PA (2011) Grain harvesting strategies to minimize grain quality losses due to Fusarium head blight in wheat. Plant Dis 95(11):1448–1457

Santos Alexandre AP, Vela-Paredes RS, Santos AS, Costa NS, Canniatti-Brazaca SG, Calori-Domingues MA, Augusto PED (2018) Ozone treatment to reduce deoxynivalenol (DON) and zearalenone (ZEN) contamination in wheat bran and its impact on nutritional quality. Food Addit Contam: Part A 35(6):1189–1199

Santoyo G, Moreno-Hagelsieb G, del Carmen Orozco-Mosqueda M, Glick BR (2016) Plant growth-promoting bacterial endophytes. Microbiol Res 183:92–99

Shafi J, Tian H, Ji M (2017) Bacillus species as versatile weapons for plant pathogens: a review. Biotechnol Biotechnol Equip 31(3):446–459

Shu X, Wang Y, Zhou Q, Li M, Hu H, Ma Y et al (2018) Biological degradation of aflatoxin B1 by cell-free extracts of Bacillus velezensis DY3108 with broad pH stability and excellent thermostability. Toxins 10(8):330

Singh VK, Kumar A (2023) Secondary metabolites from endophytic fungi: production, methods of analysis, and diverse pharmaceutical potential. Symbiosis 90(2):111–125

Singh M, Naskar A, Rupashree A, Rajput M, Singh VK (2023) Analysis of endophytic microbes harboring in medicinal plants of Himalayan region with their medicinal properties. Biocatal Agric Biotechnol 102857

Smiley RD, Draughon FA (2000) Preliminary evidence that degradation of aflatoxin B1 by Flavobacterium aurantiacum is enzymatic. J Food Prot 63(3):415–418

Stander MA, Bornscheuer UT, Henke E, Steyn PS (2000) Screening of commercial hydrolases for the degradation of ochratoxin A. J Agric Food Chem 48(11):5736–5739

Stone JK, Bacon CW, White JF Jr (2000) An overview of endophytic microbes: endophytism defined. Microbial Endophytes 1:17–44

Strobel G, Daisy B (2003) Bioprospecting for microbial endophytes and their natural products. Microbiol Mol Biol Rev 67(4):491–502

Strobel G, Daisy B, Castillo U, Harper J (2004) Natural products from endophytic microorganisms. J Nat Prod 67(2):257–268

Štyriak I, Conková E, Kmec V, Böhm J, Razzazi E (2001) The use of yeast for microbial degradation of some selected mycotoxins. Mycotoxin Res 17:24–27

Sultan Y, Magan N (2011) Impact of a Streptomyces (AS1) strain and its metabolites on control of Aspergillus flavus and aflatoxin B1 contamination in vitro and in stored peanuts. Biocontrol Sci Tech 21(12):1437–1455

Tan RX, Zou WX (2001) Endophytes: a rich source of functional metabolites. Nat Prod Rep 18(4):448–459

Terkar A, Borde M (2021) Endophytic fungi: novel source of bioactive fungal metabolites. In: New and future developments in microbial biotechnology and bioengineering. Elsevier, pp 95–105

Tibola CS, Fernandes JMC, Guarienti EM (2016) Effect of cleaning, sorting and milling processes in wheat mycotoxin content. Food Control 60:174–179

Tso KH, Lumsangkul C, Ju JC, Fan YK, Chiang HI (2021) The potential of peroxidases extracted from the spent mushroom (Flammulina velutipes) substrate significantly degrade mycotoxin deoxynivalenol. Toxins 13(1):72

Tyśkiewicz R, Nowak A, Ozimek E, Jaroszuk-Ściseł J (2022) Trichoderma: the current status of its application in agriculture for the biocontrol of fungal phytopathogens and stimulation of plant growth. Int J Mol Sci 23(4):2329

Van der Westhuizen L, Shephard GS, Rheeder JP, Burger HM, Gelderblom WCA, Wild CP, Gong YY (2011) Optimising sorting and washing of home-grown maize to reduce fumonisin contamination under laboratory-controlled conditions. Food Control 22(3–4):396–400

Verheecke C, Liboz T, Anson P, Zhu Y, Mathieu F (2015) Streptomyces–Aspergillus flavus interactions: impact on aflatoxin B accumulation. Food Addit Contam Part A Chem Anal Control Expo Risk Assess 32(4):572–576

Violle C, Reich PB, Pacala SW, Enquist BJ, Kattge J (2014) The emergence and promise of functional biogeography. Proc Natl Acad Sci 111(38):13690–13696

Wang N, Wu W, Pan J, Long M (2019) Detoxification strategies for zearalenone using microorganisms: a review. Microorganisms 7(7):208

Waqas M, Khan AL, Hamayun M, Shahzad R, Kim YH, Choi KS, Lee IJ (2015) Endophytic infection alleviates biotic stress in sunflower through regulation of defence hormones, antioxidants and functional amino acids. Eur J Plant Pathol 141:803–824

Weller DM, Raaijmakers JM, Gardener BBM, Thomashow LS (2002) Microbial populations responsible for specific soil suppressiveness to plant pathogens. Annu Rev Phytopathol 40(1):309–348

Wen J, Okyere SK, Wang S, Wang J, Xie L, Ran Y, Hu Y (2022) Endophytic fungi: an effective alternative source of plant-derived bioactive compounds for pharmacological studies. J Fungi 8(2):205

Wetterhorn KM, Newmister SA, Caniza RK, Busman M, McCormick SP, Berthiller F et al (2016) Crystal structure of Os79 (Os04g0206600) from Oryza sativa: a UDP-glucosyltransferase involved in the detoxification of deoxynivalenol. Biochemistry 55(44):6175–6186

Wu Q, Jezkova A, Yuan Z, Pavlikova L, Dohnal V, Kuca K (2009) Biological degradation of aflatoxins. Drug Metab Rev 41(1):1–7

Wu YM, Zhou QY, Yang XQ, Luo YJ, Qian JJ, Liu SX et al (2019) Induction of antiphytopathogenic metabolite and squalene production and phytotoxin elimination by adjustment of the mode of fermentation in cocultures of phytopathogenic Nigrospora oryzae and Irpex lacteus. J Agric Food Chem 67(43):11877–11882

Xiao J, Zhang Q, Gao YQ, Shi XW, Gao JM (2014a) Antifungal and antibacterial metabolites from an endophytic Aspergillus sp. associated with Melia azedarach. Nat Prod Res 28(17):1388–1392

Xiao J, Zhang Q, Gao YQ, Tang JJ, Zhang AL, Gao JM (2014b) Secondary metabolites from the endophytic Botryosphaeria dothidea of Melia azedarach and their antifungal, antibacterial, antioxidant, and cytotoxic activities. J Agric Food Chem 62(16):3584–3590

Xin-Chen L, Xi C, Xue-Jing D, Lai-Le W, Li YC, Zhi-Hua S, Xiao-Ling S, Sheng Q, Wei-Men D, Dai WM (2023) Test of the colonization ability of the invasive alien plant Flaveria bidentis (Asteraceae) in southern China. Biol Inv 25(11):3409–3424

Yang Q, Zhang H, Zhang X, Zheng X, Qian J (2015) Phytic acid enhances biocontrol activity of Rhodotorula mucilaginosa against Penicillium expansum contamination and patulin production in apples. Front Microbiol 6:1296

Yang XF, Wang NN, Kang YF, Ma YM (2019) A new furan derivative from an endophytic Aspergillus tubingensis of Decaisnea insignis (Griff.) Hook. f. & Thomson. Nat Prod Res 33(19):2777–2783

Yousef H, Metwaly HA, Hassanin MM (2022) Effect of plant extracts on suppression of Aspergillus flavus growth and aflatoxins production in peanuts. Egypt J Phytopathol 50(2):25–32

Zheng YK, Qiao XG, Miao CP, Liu K, Chen YW, Xu LH, Zhao LX (2016) Diversity, distribution and biotechnological potential of endophytic fungi. Ann Microbiol 66:529–542

Zhou T, He J, Gong J (2008) Microbial transformation of trichothecene mycotoxins. World Mycotoxin J 1(1):23–30

Zhu Y, Hassan YI, Watts C, Zhou T (2016) Innovative technologies for the mitigation of mycotoxins in animal feed and ingredients—A review of recent patents. Anim Feed Sci Technol 216:19–29

Zingales V, Taroncher M, Martino PA, Ruiz MJ, Caloni F (2022) Climate change and effects on molds and mycotoxins. Toxins 14(7):445

Zweimüller M, Antus S, Kovács T, Sonnenbichler J (1997) Biotransformation of the fungal toxin fomannoxin by conifer cell cultures. Biol Chem 378(8):915–921

Evaluating the Efficacy of Endophytic Fungal Compounds as Biopesticides in Agriculture

8

Debora Luiza Costa Barreto, Camila Rodrigues de Carvalho, Victor Pena Ribeiro, Láuren Machado Drumond de Souza, Elisa Amorim Amâncio Teixeira, Henrique Oliveira Chuma n Borges, Sonia Claudia do Nascimento de Queiroz, Kumudini Mangalika Meepagala, Stephen Oscar Duke, Charles Lowell Cantrell, and Luiz Henrique Rosa

Abstract

Substantial losses in agriculture due to plant diseases, pests, and weeds have made the use of pesticides indispensable. Chemical interventions raise environmental concerns, and the increasing evolution and spread of resistance of these agricultural pests to commercially available products highlight the urgency of finding new alternatives. Endophytic fungi from different environments have

D. L. C. Barreto · L. M. D. de Souza · E. A. A. Teixeira · H. O. C. Borges · L. H. Rosa (✉)
Departamento de Microbiologia, Instituto de Ciências Biológicas, Universidade Federal de Minas Gerais, Belo Horizonte, Minas Gerais, Brazil
e-mail: lhrosa@icb.ufmg.br

C. R. de Carvalho
René Rachou Institute, FIOCRUZ-MG, Belo Horizonte, Minas Gerais, Brazil

V. P. Ribeiro · C. L. Cantrell
Natural Products Utilization Research Unit, Agricultural Research Service, United States Department of Agriculture, Oxford, MS, USA

S. C. do Nascimento de Queiroz
Laboratório de Resíduos e Contaminantes, Embrapa Meio Ambiente, Jaguariúna, São Paulo, Brazil

K. M. Meepagala
The National Center for Agricultural Utilization Research, Crop Bioprotection Research Unit, Peoria, IL, USA

S. O. Duke
National Center for Natural Products Research, School of Pharmacy, Oxford, MS, USA

© The Author(s), under exclusive license to Springer Nature Singapore Pte Ltd. 2025

K. A. Abd-Elsalam, A. H. Hashem (eds.), *Fungal Endophytes Volume II*,
https://doi.org/10.1007/978-981-97-8804-0_8

emerged as a promising source of bioactive compounds with pesticidal properties. Their potential to protect crops and livestock makes them greener and safer allies than many currently used synthetic pesticides in controlling pests. In this chapter, we discuss compounds with fungicidal, insecticidal, nematicidal, acaricidal, and herbicidal activity produced by endophytic fungi, revealing vast pesticide potential to be explored.

Keywords

Endophytic fungi · Natural compounds · Agriculture · Pesticides

8.1 Introduction

Pesticides are compounds or mixtures of compounds used to prevent, kill, or control pests, including fungi, insects, nematodes, mites, and weeds. These unwanted species may cause damage during the production and storage of crops (Zacharia 2011). According to the FAO (2019), biotic stress caused by phytopathogens, weeds, and herbivores (including arthropods) is responsible for an estimated annual loss of 20–40% of global agricultural production. This makes the use of pesticides indispensable for maintaining crop productivity and food security. The escalating concerns over the accumulation of pesticides in the environment, due to insufficient decomposition, and the toxic effects on human and animal health associated with synthetic pesticides have spurred the search for safer and more sustainable alternatives. Endophytic fungi have emerged as promising sources in this quest (Dufossé et al. 2021).

Endophytic fungi represent a diverse group of microorganisms that inhabit the internal tissues of plants without causing any apparent harm to their hosts (Wani et al. 2015). In plant-endophyte symbiosis, plants provide carbohydrates and other nutrients to endophytes, and in return, endophytes aid in resistance development against biotic and abiotic stresses and support plant growth (Hamilton and Bauerle 2012). Over the past few decades, endophytic fungi have been extensively studied due to their production of bioactive natural compounds, making them suitable for research on new compounds that could be utilized in controlling agricultural pests (Barreto et al. 2021). Bioactive metabolites exhibit a wide range of biological activities, including fungicidal, insecticidal, nematicidal, acaricidal, and herbicidal effects. Their potential to regulate and suppress pests and pathogens holds immense promise for integrated pest management strategies that promote reduced chemical dependency and ecological balance (Abo Nouh 2019). This review discusses endophytic fungi as potential sources of pesticidal agents and explores isolated compounds from these fungi originating from various plant sources, examining their effectiveness in pest control. Emphasizing the importance of continued research in this area, we envision a future where endophytic fungi play a pivotal role in reducing the environmental impact of conventional pesticides and contribute to the development of greener, more environmentally friendly pest management strategies.

8.2 Antifungal Compounds Isolated from Endophytic Fungi

Plant diseases result in estimated annual losses of 10–15% of major crops worldwide (Chatterjee et al. 2016). Of these diseases, 70–80% are caused by pathogenic fungi (Li et al. 2017). Phytopathogenic fungi produce phytotoxins that play a crucial role in the development of plant diseases, negatively impacting host plants (Soyer et al. 2015). These toxins are primarily low-molecular weight secondary metabolites capable of inducing specific symptoms such as wilting, growth inhibition, chlorosis, necrosis, and leaf spots (Yin et al. 2016). After infection, phytopathogenic fungi disrupt plant metabolism and impact normal functions, leading to disease symptoms. They do this by producing enzymes, toxins, and other metabolic inhibitors, such as hormones, or by absorbing nutrients from the host plants. Fungi can grow internally or externally on the host plant (Coque et al. 2020; Peng et al. 2021). In response, plants activate defense mechanisms to neutralize the virulence factors of fungi. If successful, plants may prevent the survival of pathogenic fungi (de León and Montesano 2013; Westrick et al. 2021). Synthetic chemical fungicides are extensively employed to address this issue; however, excessive and prolonged use of these products raises serious concerns regarding human, animal, and environmental health. Additionally, the growing resistance of fungal pathogens to available antifungal agents poses a significant threat to agriculture (Khan et al. 2020; Corkley et al. 2022). Suitable solutions may be found in untapped natural sources, with endophytic metabolites increasingly recognized as potential contributors in this domain. Many articles have been published on endophytic fungi producing fungicidal compounds. This review will focus on the most recently published research, primarily from the years 2022 to 2024.

The endophytic fungus *Fusarium decemcellulare* was isolated from the stem of the Chinese medicinal plant *Mahonia fortunei* (Lindl.) Fedde (Song et al. 2023). Seven new compounds, along with 13 known compounds, were isolated from the rice culture of *F. decemcellulare*. The antifungal evaluation of pure compounds against *Colletotrichum musae* was determined using the broth dilution method and provided minimum inhibitory concentration (MIC) values. *C. musae* is the causative agent of anthracnose disease in guava and banana, responsible for significant postharvest losses (Murmu and Mishra 2018). Cycloheximide was used as a positive control in parallel to reveal the comparative antifungal efficacy of compounds. Enniatin H (**1**), enniatin I (**2**), and nectriacid B (**3**) exhibited inhibitory activities against *C. musae* with MIC values of 256, 64, and 128 μg mL^{-1}, respectively. The MIC of the positive control cycloheximide was 32 μg mL^{-1}. Siebatcheu et al. (2023) isolated *Trichoderma erinaceum* from a healthy and asymptomatic common bean field crop. Chemical investigation of the endophytic fungus resulted in the isolation of a new alkene, (Z)-5-amino-5-(1,1,2-trihydroxybuta-1,3-dienyloxy)pentane-6,7,8,9-tetraol (**4**), along with five known compounds. Compound (**4**) inhibited the mycelial growth of *Pythium ultimum* by 60% at 200 μg mL^{-1}.

Hypoxylon vinosopulvinatum is an endophytic fungus isolated from *Cinnamomum cassia* Presl and has an inhibitory effect on *Lasiodiplodia pseudotheobromae*, an important plant pathogen that causes die-back, canker, and blight in many crops worldwide (Pipattanapuckdee et al. 2019; Lv et al. 2023). Three new furanones, hypoxylonone A-C, as well as three known compounds, were isolated from an EtOAc extract of *H. vinosopulvinatum*. In the antifungal bioassay, hypoxylonone B (**5**) and C (**6**) exhibited strong inhibitory effects on *L. pseudotheobromae* with IC_{50} values of 1.01 and 2.40 μg mL^{-1}, respectively. Wu et al. (2023) isolated the endophytic fungus *Trichoderma* sp. from the medicinal plant *Uncaria rhynchophylla* (Miq.) Jacks. This fungus produced eight octahydronaphthalene derivatives, including four new natural products identified as trichodermic acids E-H (**7–10**). The new compounds exhibited moderate antifungal activity against *Botrytis cinerea*, an important phytopathogen that causes gray mold in fruits and vegetables, and *Alternaria alternata*, which causes leaf spots on vegetables, with an MIC of 64 μg mL^{-1}. Notably, the known compound trichodermic acid (**11**) demonstrated potent antifungal activity against *B. cinerea* with an MIC of 4 μg mL^{-1} and against *A. alternata* with an MIC of 16 μg mL^{-1}. Furthermore, compound (**11**) also exhibited moderate antifungal activity against *Fusarium moniliforme* with an MIC of 32 μg mL^{-1}. The MIC for the positive control amphotericin B was 0.5 μg mL^{-1}. These findings suggest that trichoderma acids could serve as natural fungicides in agriculture. *Fusarium moniliforme* is one of the most prevalent fungal infections in corn, livestock, and humans (Nelson 1992).

Zhang et al. (2023) isolated the endophytic fungus *Aspergillus ochraceus* from *Lagopsis supina* (Steph. ex Willd.) I.K.-Gal. ex Knorr. The antifungal activity of the ethyl acetate extract derived from *A. ochraceus* malt extract culture broth was tested against the phytopathogenic fungus *Fusarium graminearum*, commonly found in maize and other grains, using the cup-plate method. It demonstrated effective inhibition of pathogen growth. Consequently, the endophytic fungus was cultivated on a large scale for active compound isolation. Four compounds, including 2-methoxy-6-methyl-1,4-benzoquinone (**12**) and penicillic acid (**13**), were discovered. Compounds **12** and **13** exhibited antifungal activity against several agriculturally important plant pathogenic fungi: *F. graminearum*, *F. oxysporum*, *F. moniliforme*, *F. solani*, *B. cinerea*, *Magnaporthe oryzae*, and *Verticillium dahliae*. They showed a range of MIC values (128–512 μg mL^{-1}), which were close to those of the positive control actidione (64–128 μg mL^{-1}). Lasiodiplodin (**14**) was isolated from the endophytic fungus *Lasiodiplodia pseudotheobromae* associated with the medicinal plant *Sarcandra glabra* (Thunb.) Nakai (Luo et al. 2023). An in vitro antifungal assay demonstrated the inhibitory activity of lasiodiplodin against the growth of six phytopathogenic fungi, with IC_{50} values ranging between 15.5 and 52.3 μg mL^{-1}. The highest antifungal activities were observed against *Exserohilum turcicum* (pathogen infecting maize), *Colletotrichum capsici* (pathogen infecting peppers), and *Pestalotiopsis theae* (pathogen infecting tea), with IC_{50} values of 15.5, 15.9, and 17.6 μg mL^{-1}, respectively. The antifungal activity of lasiodiplodin against *E. turcicum* caused an alteration of its colony morphology and disturbance of its cell membrane integrity. Additionally, optimizing the culture conditions of *L. pseudotheobromae*—including pH, temperature, liquid volume, carbon-to-nitrogen ratio, and culture duration—increased the yield of lasiodiplodin from

0.6 mg L^{-1} to 52.3 mg L^{-1}. The results of this study suggest that lasiodiplodin is a natural compound with significant potential as a biopesticide against phytopathogens.

Two compounds, glaucanic acid (**15**) and dihydrocompactin acid (**16**), were isolated from *Penicillium polonicum*, an endophyte of *Zingiber officinale* Roscoe. These active constituents were obtained from the ethyl acetate extract of *P. polonicum* culture broth (Gupta et al. 2023). Compounds **15** and **16** reduced the growth of *Colletotrichum gloeosporioides* by more than 50%. This marks the first report of the isolation of glaucanic acid (**15**) and dihydrocompactin acid (**16**) from an endophytic fungus. The endophytic fungus *Cladosporium sphaerospermum*, isolated from the plant *Fritillaria unibracteata* var. wabuensis (S.Y. Tang & S.C. Yueh) Z.D. Liu, S. Wang & S.C. Chen, has proven highly effective in combating plant pathogens, particularly *B. cinerea* (Pan et al. 2023). The hybrid polyketide compound known as cladodionen (CLD) (**17**) has emerged as the primary active ingredient, demonstrating remarkable inhibitory effects on the mycelial growth and spore germination of *B. cinerea*. It has also been shown to be a quorum-sensing inhibitor against *Pseudomonas aeruginosa*. With IC_{50} values of 1.13 mM for the inhibition of mycelial growth and 0.09 mM for spore germination, respectively, CLD exhibits significant antifungal effects and prevents fungal growth on harvested strawberries and tomatoes. In specific tests, CLD did not show significant toxicity in zebrafish embryos, even at concentrations of up to 0.23 mM, highlighting its potential as a safe biological control agent. Thus, CLD emerges as a promising candidate for plant pathogen control, serving as an effective fungicide for the management of *B. cinerea*.

The chemical investigation of the endophyte *Pseudofusicoccum stromaticum*, isolated from the medicinal plant *Stryphnodendron adstringens* (Mart.) Coville, resulted in the identification of ten compounds (Iantas et al. 2023). The assessment of the antifungal activity of these isolated compounds was directed toward an important citrus pathogen, *Phyllosticta citricarpa*. Among them, cytochalasin H (**18**) (78.3%), fomoxanthone A (**19**) (70.2%), fomoxanthone B (**20**) (63.1%), and paecilomycesin Q (**21**) (50.5%) demonstrated the ability to reduce the number of pycnidia produced by *P. citricarpa* in vitro. Additionally, compounds (**18**) and (**19**) inhibited the development of symptoms of citrus black spots on citrus fruits. Cytochalasin H and paecilin Q stand out as particularly promising, exhibiting strong activity against the citrus pathogen with low or no cytotoxic activity. These results suggest that these compounds may represent valuable tools in controlling disease spread, indicating potential applications in agriculture.

Bipolarithizole A (**22**), a newly discovered phenylthiazole-sativene sesquiterpenoid hybrid, has been isolated from cultures of the potato endophytic fungus *Bipolaris eleusines* (Ai et al. 2022). The compound exhibits antifungal activity against the soil-borne fungus *Rhizoctonia solani*, with an MIC value of 16 μg mL^{-1}. Additionally, a fungus belonging to *Emericella* sp. isolated from *Panax notoginseng* (Burk.) F.H. Chen leaves was investigated for potential antifungal agents (Xian et al. 2022). Fifteen compounds, including seven previously undescribed structures, were obtained from this species. Emerlactam A (**23**) was the most active against

Helminthosporium maydis, a pathogenic fungus on maize, with an MIC value of 12.5 μg mL^{-1}. 5-Alkenylresorcinol (**24**) showed significant antifungal activity against all tested phytopathogenic fungi (*Verticillium dahliae*, *Rhizoctonia solani*, *Sclerotinia sclerotiorum*, *Fusarium oxysporum*, *Alternaria fragriae*, *Helminthosporium maydis*, and *Botryosphaeria berengeriana*), with MIC values ranging from 6.3 to 12.5 μg mL^{-1}. Emerlactam B (**25**) exhibited moderate antifungal activity against *H. maydis* and *S. sclerotiorum*, with MIC values of 25 μg mL^{-1}.

During a screening for antifungal secondary metabolites, six new derivatives of mono−/bis-alkenoic acid and a known derivative of alkenoic acid, hymeglusin (**26**), were isolated from an endophytic fungus, *Scopulariopsis candelabrum*, obtained from the stems of tea trees *Camellia sinensis* (L.) O. Ktze (Tang et al. 2022). All compounds were tested for antifungal activity against the maize pathogen *Exserohilum turcicum* using the agar diffusion method on filter paper. Compound (**26**) exhibited significant inhibitory activity against *E. turcicum*, with an IC_{50} value of 21.2 μg mL^{-1}, which was markedly lower than that of the positive control, cycloheximide (IC_{50} 46.7 μg mL^{-1}). Therefore, it can be concluded that the β-lactone ring is a crucial component for antifungal activity. Consistent with previous reports, the antifungal activity of alkenoic acid derivatives is consistently associated with the presence of the β-lactone ring (Liu et al. 2018; Niu et al. 2019; Tang et al. 2019). The β-lactone ring of some compounds plays a key role in inhibiting fungal HMG-CoA (3-hydroxy-3-methylglutaryl-CoA) synthase activity (Greenspan et al. 1987; Tomoda et al. 2004; Skaff et al. 2012).

Three new highly oxygenated azaphilones, named penicilones A-C, were identified along with four previously known analogs from the culture of a *Penicillium* sp., an endophyte isolated from the plant *Euphorbia pekinensis* Rupr (Feng et al. 2022). The isolated compounds were subjected to in vitro evaluation for their antifungal activity against four phytopathogenic fungi: *Rhizoctonia solani*, *Fusarium oxysporum*, *Penicillium citrinum*, and *Phytophthora melonis*. For comparison, carbendazim, a commercial antifungal agent, was used as a positive control. Compounds penicilphilone A (**27**), penicilphilone C (**28**), and 7-epi-WB (**29**) exhibited notable activities against the tested phytopathogenic fungi, with inhibitory rates ranging from 49.8% to 78.6% at a concentration of 50 μg mL^{-1}. Notably, these three compounds showed higher inhibitory effects against *F. oxysporum* compared to carbendazim at the same concentration. These results underscore the promising efficacy of these compounds as antifungal agents, particularly against *F. oxysporum*.

The analysis of metabolites from an endophytic fungus, *Xylaria* sp., isolated from *Azadirachta indica* A. Juss, led to the isolation of six previously undescribed compounds, namely xylarioxides A–F (Wang et al. 2022a). All isolated compounds were tested for antifungal activities against seven phytopathogenic fungi. Remarkably, xylarioxides A (**30**), E (**31**), and F (**32**) exhibited potent activities against the tested phytopathogens. Specifically, xylarioxide E (**31**) showed the highest activity against *Gibberella saubinetii*, *Curvularia lunata*, and *Colletotrichum gloeosporioides*, with MIC values of 4, 4, and 8 μg mL^{-1}, respectively. The results were comparable to those obtained with the positive control, nystatin. Liu et al. (2022) isolated the endophytic fungus *Aspergillus polyporicola* from the roots of *Synsepalum dulcificum* Daniell. Four new compounds, along with 11 known compounds, were isolated and evaluated for their antifungal activity. The compound 11-dehydrocurvularin (**33**) demonstrated significant inhibitory activity against *B. cinerea* and *F. graminearum*, with MIC values of 32 and 16 μg mL^{-1}, respectively. The compound (+)-2,5-dimethyl-3(2H)-benzofuranone (**34**) exhibited antifungal activity against *F. graminearum* with an MIC value of 32 μg mL^{-1}.

Two new compounds, aflaxanthone A (**35**) and aflaxanthone B (**36**), were isolated from the endophytic fungus *Aspergillus flavus* obtained from a freshly collected leaf of the *Kandelia candel* (L.), also known as the Druce mangrove plant (Zang et al. 2022). These compounds exhibited promising broad-spectrum antifungal activity against *C. gloeosporioides*, *F. oxysporum*, and *C. musae* in vitro bioassays, with MIC values ranging from 3.1 to 25 μM. This work provides further evidence supporting mangrove endophytic fungi as a sustainable source of chemical diversity. Wei et al. (2023) reported Equisetin (**37**), a well-known fungal metabolite, is produced by the endophytic fungus *Fusarium* sp., which was obtained from *Mirabilis jalapa.* They reported the EC_{50} of equisetin against *B. cinerea*, *F. graminearum*, *S. sclerotiorum*, and *R. solani* to be 10.7, 12.9, 17.1, and 21.0 μg mL^{-1}, respectively. The control fungicide boscalid was more active against all these plant pathogens than equisetin. In summary, fungicidal compounds produced by endophytic fungi represent a promising area in biotechnological and agricultural research. The ability of these microorganisms to synthesize bioactive substances with antifungal properties offers a sustainable and effective alternative for controlling plant pathogens. Increasing understanding of the diversity of these compounds and their modes of action will enhance their potential in developing new plant disease control agents.

The endophytic fungus *Phoma* sp., isolated from the plant *Saurauia scaberrinae*, produces the compound cercosporamide (**38**), known for its antifungal properties (Liu et al. 2024). This study cultivated *Phoma* sp. in seven growth media to identify the optimal culture conditions for cercosporamide production. The maximum production of cercosporamide was recorded on the eighteenth day of incubation in bovine peptone dextrose broth (BPD), reaching an average concentration of 77.5 μg mL^{-1} in BPD. Paper disc diffusion assays revealed that the culture filtrate containing cercosporamide exhibited significant antifungal activities against two pepper anthracnose pathogens, *Colletotrichum gloeosporioides* and *C. scovillei*, with EC_{50} values of 3.8 and 7.0 μg mL^{-1}, respectively. This study suggests the potential application of cercosporamide as an effective antifungal agent for controlling pepper anthracnose. Dibutyl phthalate (**39**) was initially extracted from the endophytic fungus *Gymnoascus thermotolerans*, sourced from the roots, stems, and leaves of *Euphorbia geniculate* (Kamel et al. 2024). It displayed notable inhibition of *Fusarium solani* spore germination, achieving 90.8% inhibition at the highest concentration tested, which was 2.5 mg mL^{-1} (Fig. 8.1).

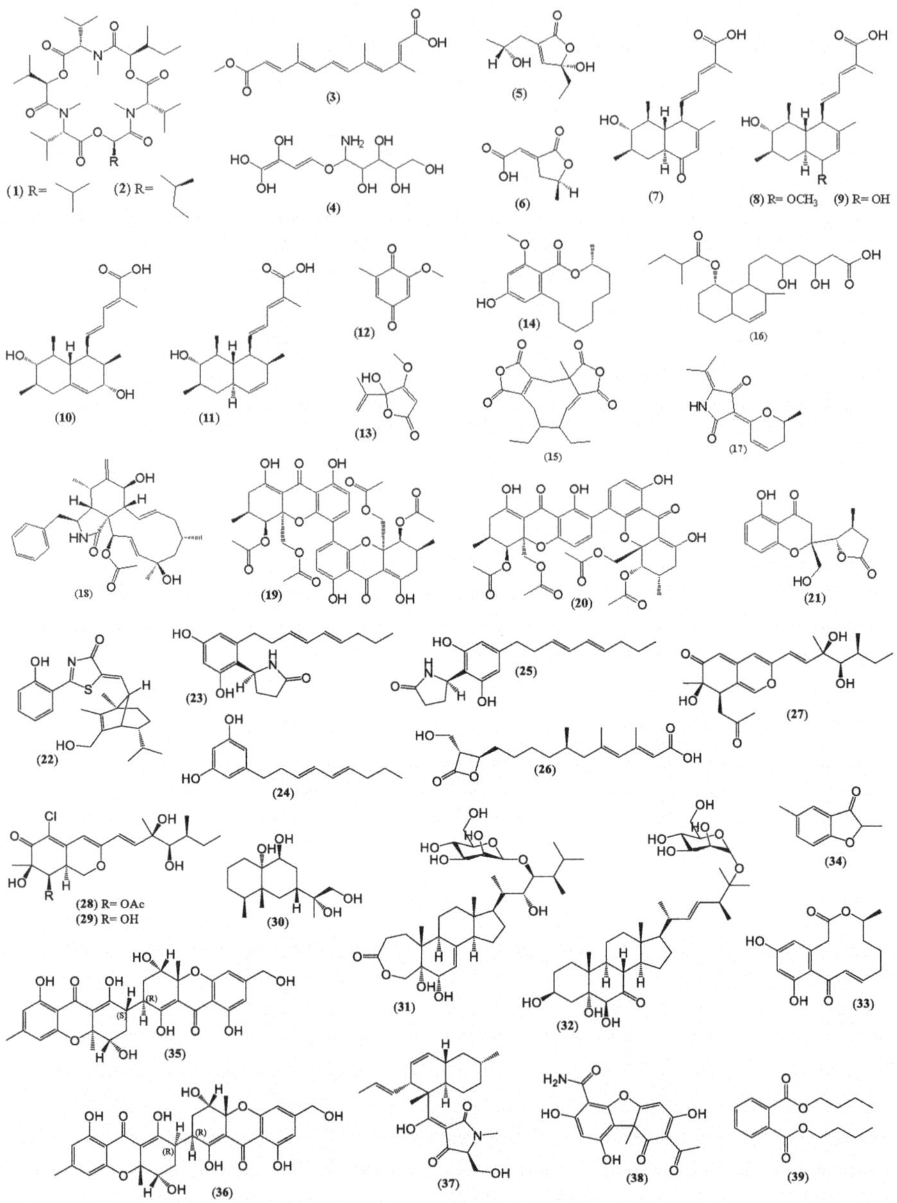

Fig. 8.1 Compounds with antifungal activity isolated from endophytic fungi. Enniatin H (**1**), enniatin I (**2**), nectriacid B (**3**), (Z)-5-amino-5-(1,1,2-trihydroxybuta-1,3-dienyloxy)pentane-6,7,8,9-tetraol (**4**), hypoxylonone B (**5**), hypoxylonone C (**6**), trichodermic acid E (**7**), trichodermic acid F (**8**), trichodermic acid G (**9**), trichodermic acid H (**10**), trichodermic acid (**11**), 2-methoxy-6-methyl-1,4-benzoquinone (**12**), penicillic acid (**13**), lasiodiplodin (**14**), glaucanic acid (**15**), dihydrocompactin acid (**16**), cladodionen (**17**), cytochalasin H (**18**), fomoxanthone A (**19**), fomoxanthone B (**20**), paecilin Q (**21**), Bipolarithizole A (**22**), emerlactam A (**23**), 5-alkenylresorcinol (**24**), emerlactam B (**25**), hymeglusin (**26**), penicilphilone A (**27**), penicilphilone C (**28**), 7-epi-WB (**29**), xilarioxides A (**30**), xilarioxides E (**31**), xilarioxides F (**32**), 11-dehydrocurvularin (**33**), (+)-2,5-dimethyl-3(2H)-benzofuranone (**34**), aflaxanthone A (**35**), aflaxanthone B (**36**), equisetin (**37**), cercosporamide (**38**), dibutyl phthalate (**39**)

8.3 Insecticidal Compounds Isolated from Endophytic Fungi

Insecticides, synthetic or natural chemical compounds, are used in insect control, leading to the death of insects or the prevention of behaviors considered destructive. Despite the significant discovery and development of synthetic insecticides to increase agricultural production, many problems have resulted from their intensive, incorrect, and indiscriminate use over decades. Among these undesired effects are toxic residues in food, poisoning of farmers and consumers, water and soil contamination, the emergence of resistant insect pests, and damage to nontarget organism populations (Kim et al. 2003; Aguiar-Menezes 2005; Hassan and Prijono 2011; Morais and Marinho-Prado 2016). As the world becomes increasingly cautious about the ecological damage caused by synthetic insecticides, research into endophytic microbes intensifies in the pursuit of potent, selective, and safe alternatives. Several endophytes are known to possess insecticidal properties, and the market for bioinsecticides is growing significantly (Tenguria et al. 2011).

Webber (1981) was the first to report plant protection provided by an endophytic fungus, *Phomopsis oblonga*, in elms against the beetle *Physocnemum brevilineum*. The endophytic fungus was suggested to play a role in reducing the spread of Dutch elm disease, caused by *Ceratocystis ulmi*, by controlling its insect vector, *P. brevilineum* (Dutta et al. 2014). Several alkaloids discovered in endophytic fungi have useful biological properties, such as insecticidal, antimicrobial, cytotoxic, and anticancer activities (Zhang et al. 2012). Bioactive alkaloids such as ergot, peramine, lolines, and pyrrolizidine, found in endophytes, are potential broad-spectrum insecticides and also play significant roles in agriculture, medicine, and the food industry (Zhang et al. 2012). Many of these endophytic fungi produce alkaloids that enhance their host's resistance to insects and are toxic to herbivorous mammals. Examples include ergot alkaloids and lolitrems (indole-diterpenes) that cause neurotoxic effects in herbivorous or granivorous vertebrates, peramine (a pyrrolopyrazine), which is an insect feeding inhibitor, and lolines (1-aminopyrrolizidines) that act as insecticides (Bush et al. 1997; Wilkinson et al. 2000; Zhang et al. 2012).

Toxic metabolites, such as loline alkaloids (**40–42**) and peramine (**43**), produced by endophytic fungi *Acremonium* sp., *Epichloë* sp., and *Neotyphodium* sp., have been found in fescue grasses. These compounds reduce populations of herbivorous insects (Rudgers and Clay 2008; Rohlfs and Churchill 2011; Fe Andres et al. 2017). Conversely, other studies suggest that alkaloids produced by endophytes may increase herbivore abundance and species richness, potentially negatively impacting the natural enemies of herbivores (Jani et al. 2010; Faeth and Shochat 2010; Fe Andres et al. 2017).

The first nodulisporic compounds were isolated from the endophytic fungus *Nodulisporium* sp., found in the plant *Bontia daphnoides* L. Nodulisporic acids (**44**) are indole diterpenes with potent insecticidal properties against blowfly larvae and other blood-sucking insects, functioning by activating glutamate-gated chloride channels in insects (Ondeyka et al. 1997; Fe Andres et al. 2017). Another endophytic fungus, *Muscodor vitigenus*, isolated from the liana *Paullinia paullinioides*, produces naphthalene (**45**), which effectively repels the adult stage of the wheat stem sawfly *Cephus cinctus*. Naphthalene, at concentrations comparable to those used in the tests, mimicked the insect-repellent activity observed for the fungus (Daisy et al. 2002). This compound, known for over two centuries and once the primary ingredient in mothballs, has also been used as a soil fumigant for insect management. Its potential adverse effects on human health led to the banning of its use in mothballs in the European Union in 2008. Therefore, being a natural compound does not always guarantee safety.

Rugulosin (**46**), an insect antifeedant compound, is produced by the fungal endophyte *Phialocephala scopiformis* from spruce needles attacked by the forest pest *Choristoneura fumiferana* (Miller et al. 2002, 2008; Sumarah et al. 2008; Sumarah and Miller 2009; Fe Andres et al. 2017). Several insecticidal metabolites active against *C. fumiferana* have been isolated, including two compounds: 5-hydroxy-2-(1′-oxo-5′-methyl-4′-hexenyl) benzofuran (**47**) and 5-hydroxy-2-(1′-hydroxy-5′-methyl-4′-hexenyl) benzofuran (**48**), which were isolated from an unidentified endophytic fungus of wintergreen, *Gaultheria procumbens* L. (Findlay et al. 1997), and nine compounds were isolated from the endophytic fungi of red spruce needles (*Picea rubens*), including vermiculine (**49**), 4-oxo-5-phenylpentanoic acid (**50**), 5-benzyl-dihydrofuran-2-one (**51**), 3-epi-isopetasol (**52**), 15-hydroxy-3-epi-isopetasol (**53**), two new isomeric compounds related to cordianhydrides A and B (**54, 55**), and cordianhydrides A and B themselves (**56, 57**) (Sumarah et al. 2010; Fe Andres et al. 2017).

Alternaria alternata, an endophyte of *Catharanthus roseus* (L.) G. Don, produced an acetylcholinesterase (AChE) inhibitor identified as altenuene (**58**), which exhibited 78% inhibition of growth and significant larval mortality in *Spodoptera litura* (Bhagat et al. 2016). The inhibitory activity of α-glucosidase against *S. litura* was demonstrated by partially purified fractions obtained from the endophytic *Cladosporium velox*, isolated from *Tinospora cordifolia* (Willd.) Miers. Considerable mortality, a significant extension of the overall developmental period, an influence on female longevity, and reproductive potential were observed (Singh et al. 2016). The study demonstrated the presence of phenolic compounds in the active fraction, with chlorogenic acid (**59**) identified as the compound responsible for bioactivity. This suggests that the resistance conferred by the endophyte to insects in plants may be mediated through chlorogenic acid targeting the α-glucosidases present in the insect's gut. Chlorogenic acid is an almost ubiquitous phytochemical and has been known to be toxic to insects for many years. However, the search for new insecticides has stimulated a reexamination of this compound for this use (Lin et al. 2023).

Aspergillus sp., an endophyte isolated from the almonds of *Bertholletia excelsa* Humb. & Bonpl., collected in the Brazilian Amazon, exhibited larvicidal activity against *Aedes aegypti* with an LC_{50} of 26.86 μg mL^{-1} at 24 h and 18.75 μg mL^{-1} at 48 h. Molecular docking studies indicated that the compound aspergillol B (**60**) is a potent larvicide by inhibiting the acetylcholinesterase enzyme (Araújo et al. 2022). The cited works suggest that endophytic fungi could be a promising source for the production of green insecticides, warranting increased research investment in this field (Wang et al. 2023) (Fig. 8.2).

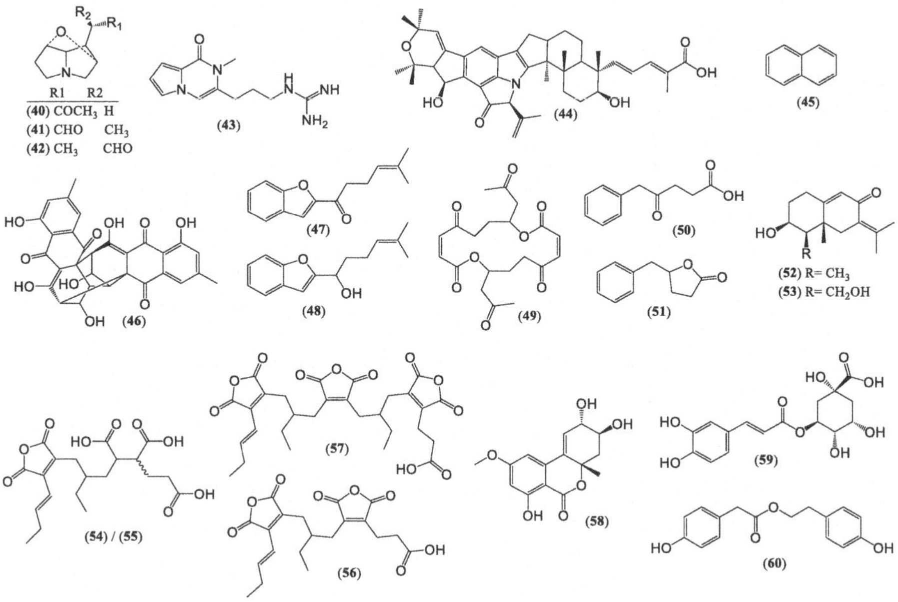

Fig. 8.2 Compounds with insecticidal properties isolated from endophytic fungi. Loline alkaloids (**40**, **41**, **42**), peramine (**43**), nodulisporic acid (**44**), naphthalene (**45**), rugulosin (**46**), 5-hydroxy-2-(1′-oxo-5′-methyl-4′-hexenyl) benzofuran (**47**), 5-hydroxy-2-(1′-hydroxy-5′-methyl-4′-hexenyl) benzofuran (**48**), vermiculin (**49**), 4-oxo-5-phenyl-pentanoic acid (**50**), 5-benzyl-dihydro-furan-2-one (**51**), 3-epi-isopetasol (**52**), 15-hydroxy-3-epi-isopetasol (**53**), two new isomers (**54**, **55**), cordianhydrides A and B (**56**, **57**), altenuene (**58**), chlorogenic acid (**59**), aspergillol B (**60**)

8.4 Nematicidal Compounds Produced by Endophytic Fungi

Among the more than 26,000 known species of nematodes, 4100 are plant parasites, and approximately 100 species from this group are considered phytoparasites of economically important crops (Nicol et al. 2011; Degenkolb and Vilcinskas 2016). These organisms can damage various parts of host plants, especially stems, leaves, flowers, and seeds. Nematode-induced damage results in a range of symptoms, such as stunting, reduced vigor, nutrient deficiency, root lesions, reduced flowering, fruit loss, low yields, and even plant death (Degenkolb and Vilcinskas 2016). Crops of developing countries are affected more severely by nematodes than those of developed countries, and production losses in the tropics and subtropics are generally greater than in temperate zones (Nicol et al. 2011).

To mitigate the damage caused by nematodes in crops, nematicidal chemicals are used. These chemicals must be sufficiently volatile and water-soluble to ensure uniform distribution in the topsoil layer, which requires high-product doses during application (Degenkolb and Vilcinskas 2016). A significant drawback of the widespread use of these nematicides is concern over their toxicity and environmental persistence. For example, organochlorine nematicide dibromochloropropane was banned due to its carcinogenicity (Whorton and Foliart 1983; Seong et al. 2021).

Additionally, other disadvantages include the difficulty of controlling nematodes with these chemicals due to the rapid reproductive cycle of the nematodes and the chemicals' inability to penetrate the protein matrix of nematode cysts (Degenkolb and Vilcinskas 2016). Therefore, alternative approaches are sought to control these organisms effectively and sustainably.

Biological control is a potential alternative for the containment of nematodes. Fungi have become viable biological agents for controlling these organisms, as their large-scale cultivation is possible using a small amount of biomass, making them economically feasible (Barbosa et al. 2019). Endophytic fungi exert significant influence on the host plant by producing antiherbivore products, minimizing the severity of diseases, and inducing plant defense mechanisms (Rudgers et al. 2007; Farhat et al. 2022). Furthermore, fungi possess a vast diversity of secondary metabolic pathways and have yielded various classes of commercially relevant compounds in the food, pharmaceutical, and agricultural sectors (Smedsgaard and Nielsen 2005; Li et al. 2007).

The secondary metabolites produced by these microorganisms have great nematicidal potential (Li et al. 2007). Bogner et al. (2017) tested the ability of *F. oxysporum*, isolated from *Solanum lycopersicum* L., to produce nematicidal compounds. In their study, the compounds 4-hydroxybenzoic acid (**61**), indole-3-acetic acid (IAA) (**62**), and gibepyrone D (**63**) presented LC_{50} values of 104, 117, and 134 μg mL^{-1}, respectively, after 72 h against *Meloidogyne incognita*. IAA is a phytohormone that plays a role in triggering plant resistance to pests, showing its direct action through the death or impairment of nematodes, or indirectly by inducing defense mechanisms against these pathogens in plants (Bogner et al. 2017). Meanwhile, 3-hydroxypropionic acid (**64**), isolated from the fungus *Phomopsis phaseoli* obtained from a leaf of a rainforest tree, showed selective nematicidal activity against *M. incognita* (Schwarz et al. 2004). Li et al. (2007) evaluated the compounds 1-[(2R*,4S*,5S*)-2-chloro-4-methyl-1,3-oxazinan-5-yl]ethanone (**65**), 1-[(2R*,4S*,5R*)-2-chloro-4-methyl-1,3-oxazinan-5-yl]ethanone (**66**), and 1-(2,4-dihydroxyphenyl)ethanone (**67**) isolated from the endophytic fungal lineage *Geotrichum* sp. from *Azadirachta indica* A. Juss against the nematode *Bursaphelenchus xylophilus*. Although they did not surpass the positive control, avermectin, these compounds killed 60% of nematodes at 100 μg mL^{-1} after 48 h. Khan et al. (2019) tested five compounds produced by the endophytic fungus *Chaetomium globosum*, isolated from an unidentified plant, against the root-knot nematode *M. javanica*. The compounds chaetoglobosin A (**68**) and chaetoglobosin B (**69**) exhibited nematicidal activity for the first time, causing 91.6% and 83.8% mortality, respectively, in second-stage juveniles (J2) at a concentration of 200 μg mL^{-1}. The LC_{50} values were 88.4 μg mL^{-1} for chaetoglobosin A and 107.7 μg mL^{-1} for chaetoglobosin B after 72 h.

Endophytic fungi are emerging as agents of bioactive compounds with nematicidal activity. Their use can benefit the current demand for new biomolecules in agriculture and the pharmaceutical industry. Large-scale production of these compounds remains a scientific challenge. However, studying symbiotic relationships between plants and fungi, and analyzing compounds produced by these microorganisms, may offer alternative and sustainable pathways for their use as nematicides (Farhat et al. 2022) (Fig. 8.3).

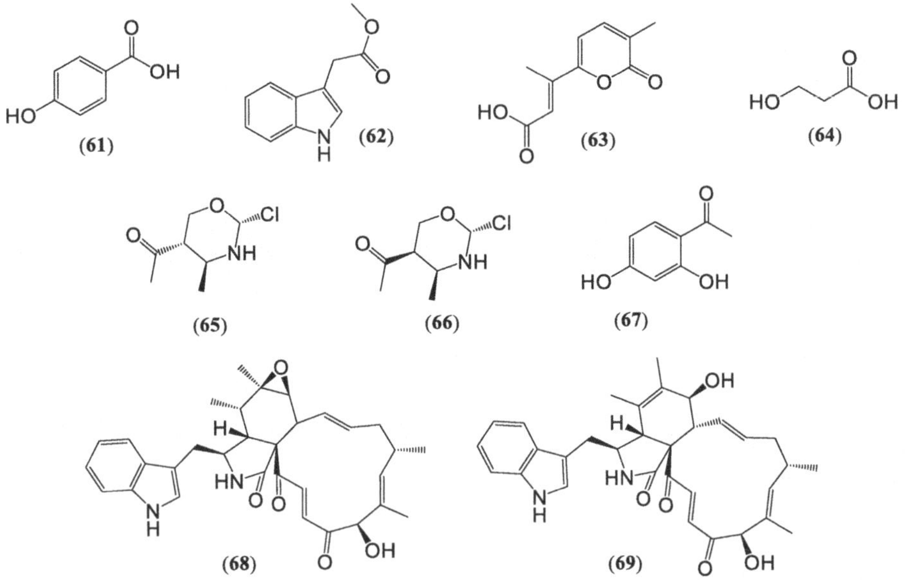

Fig. 8.3 Nematicidal compounds isolated from endophytic fungi. 4-hydroxybenzoic acid (**61**), indole-3-acetic acid (**62**), gibepyrone D (**63**), 3-hydroxypropionic acid (**64**), 1- [(2R*,4S*,5S*)-2-chloro-4-methyl-1,3-oxazinan-5-yl]ethenone (**65**), 1-[(2R*,4S*,5R*)-2-chloro-4-methyl-1,3-oxazinan-5-yl]ethenone (**66**), 1-(2,4-dihydroxyphenyl)- ethenone (**67**), chaetoglobosin A (**68**), chaetoglobosin B (**69**)

8.5 Acaricidal Activity of Bioactive Compounds Isolated from Endophytic Fungi

Endophytic fungi have been attracting increasing interest in the industry, thanks to the diversity of their biotechnological applications and the remarkable adaptations they can offer for the benefit of humans, such as the synthesis of entomotoxic compounds (Preto 2018). One focus is on incorporating these byproducts into acaricide formulations. Additionally, the demonstrated ability of these fungi to significantly reduce attacks by phytophagous mites, which cause damage to crops, reinforces their importance as a reliable source of effective biopesticides. Beyond their role in crop protection, endophytic fungi also exhibit properties that stimulate plant growth, expanding their potential in sustainable agriculture (Alves 2019; Amandio 2022). Within this context, the exploration of these microorganisms presents an ecologically advantageous approach to pest control and the optimization of plant development (Preto 2018; Amandio 2022). Research on acaricidal compounds from endophytic fungi is a particularly niche subject with a limited number of studies. In contrast, most investigations have focused on entomopathogenic fungi, which can be adapted through processing or inoculation into plant tissues to transform them into endophytic agents (Santos 2021). Two notable examples are *Metarhizium* spp. and *Beauveria bassiana*, which are the subjects of analysis in a wide range of scientific articles. Once emulsified or inoculated into plant tissue, these fungi

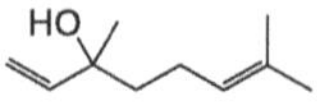

(70)

Fig. 8.4 Chemical structure of Linalool (**70**), an acaricide compound isolated from endophytic fungi

demonstrate the ability to produce byproducts capable of repelling or eliminating mites, emerging as potential acaricidal agents from natural sources (Sousa et al. 2005; Santos 2021).

Sousa et al. (2005) conducted tests to assess the acaricidal potential of compounds present in the tissues of *Melia azedarach* L. and its associated endophytic fungi. The fungi were isolated, cultivated, and used for the extraction and purification of their metabolites; however, they were not identified. It was observed that the acaricidal action developed by the plant is not attributable to an isolated compound or any of the tested compounds alone, suggesting that the combination of the plant and endophytic fungi is essential for this activity. The obtained data supported the findings of a study on which it was based (Borges et al. 2003), demonstrating high larval mortality rates and a reduction in the hatching index of *Boophilus microplus* larvae. Silva (2010) conducted a thorough investigation to determine whether endophytic fungi isolated from *Aniba rosaeodora* Ducke could synthesize linalool (**70**), a substance with acaricidal properties typically produced by plants. The compound is described as a tertiary open-chain alcoholic monoterpene and exists as a racemic mixture in various plant species due to its asymmetric carbon atom. The scientific literature suggests that the fungus utilizes the methyl-erythritol-phosphate metabolic pathway for monoterpene synthesis. Furthermore, in a proto-cooperative relationship with the host plant, there is a possibility of transferring the regulatory enzyme of this metabolic pathway to the fungi, thereby enabling these microorganisms to produce linalool. The research utilized fungi isolated from various parts of *Aniba rosaeodora*. The analysis of the 17 supernatants indicated the presence of linalool. This discovery is of great significance as it indicates that the studied fungi have promising potential as acaricidal agents, filling a gap in the scientific literature dedicated to strict endophytes. Thus, this research opens new perspectives in the field of biotechnology, emphasizing the importance of these fungi as valuable resources for the development of innovative solutions in mite control (Fig. 8.4).

8.6 Phytotoxic Activity of Bioactive Compounds Isolated from Endophytic Fungi

Undesirable plants, commonly known as weeds, have long posed a significant challenge in agriculture. Weeds compete with cultivated crops for essential resources such as nutrients, water, and sunlight, resulting in reduced yields and economic losses for farmers (Krishnan and Rajasenbagam 2021). Moreover, some weed species produce phytotoxins that are toxic to crops (Weston and Duke 2003). Consequently, effective weed management strategies are crucial to ensure the productivity and sustainability of agricultural practices and the preservation of natural environments. Complementary to traditional mechanical and cultural control methods, herbicides have emerged since the 1950s as an essential technology for weed management (Clapp 2021; Matthews 2015). Herbicides are chemical substances formulated to target and control weeds while minimizing adverse effects on desirable plants and the environment. Over the years, significant advancements have been made in herbicide research and development, leading to a diverse range of compounds with varying modes of action. While these compounds have proven to be powerful tools in weed management, their use raises concerns regarding environmental impact and the development of herbicide resistance (Perotti et al. 2020). In this context, the search for new herbicides, primarily derived from natural products, has proven to be a worthwhile alternative. Endophytic fungi are sources of substances with phytotoxic activity, standing out as promising producers to be explored for this purpose (Bendejacq-Seychelles et al. 2023). Few studies on phytoxins from endophytes from 2022 to 2023 were identified, as the majority of phytotoxic compounds mentioned in the literature are associated with phytopathogenic fungi. However, an issue with endophytes producing phytotoxins is that the endophyte can poison its host, suggesting there may not be an evolutionary advantage for endophytes to produce such compounds.

A total of 77 fungi were isolated from the roots, stems, leaves, and tubers of *Pinellia ternata* (Thunb.) Breit. and *P. pedatisecta* Schott (Kong et al. 2023). Biological activity tests indicated that fermentation broths of 22 endophytic fungi showed strong phytotoxic activity against *Echinochloa crus-galli*, with an inhibition rate of 100%. The fermentation broth of *Aspergillus fumigatus* from *P. pedatisecta* exhibited the highest phytotoxic activity among all tested endophytic fungi. *Alternaria alternata* from *P. ternata* also demonstrated strong phytotoxic activity against *E. crus-galli*, with an inhibition rate of 91%. Therefore, both fungi were selected for the identification of active secondary metabolites. Seven compounds were isolated and tested for their ability to inhibit radicle growth in *E. crus-galli* and *Abutilon theophrasti* using a Petri dish bioassay. The results showed that the metabolite citrinin (**71**) exhibited potent phytotoxic activity against *E. crus-galli*, with an inhibition rate of 73% at a concentration of 100 μg mL^{-1}. However, this activity was weaker than that of the positive control, 2,4-D, which had an inhibition rate of 100% at the same concentration.

A new compound, 5,7-dihydroxy-3(R)-methylphthalide (**72**), was isolated from *Xylaria brevipes*, and three known compounds, (−)-altenuene (**73**), alternariol (**74**), and altertoxin I, were also isolated from *Alternaria alternata* (Oktriawan et al. 2023). Both endophytic fungi were isolated from a branch of the Indonesian medicinal plant *Psidium guajava* L. All isolated compounds were tested for phytotoxicity against *Brassica chinensis* var. *parachinensis*, with 2,4-dichlorophenoxyacetic acid used as the positive control. Compounds **72–74** exhibited phytotoxicity with IC_{50}

values (μg mL^{-1}) of 4.6, 23.1, and 77.3, respectively. These results confirm that endophytic fungi can act as symbionts and opportunistic pathogens by producing phytotoxic components. *Aspergillus* sp., an endophytic fungus isolated from *Crassula arborescens* (Mill.) Willd., displayed potent inhibitory activity against the seed germination of *Arabidopsis thaliana* (Ma et al. 2022). From the fermentation extracts of *Aspergillus* sp., nine compounds were isolated and identified. The phytotoxic activities of aspertamarinoic acid (**75**) and astellolide A (**76**) against the seed germination of *A. thaliana* were described for the first time, enriching the structural diversity of phytotoxic natural products.

Two pairs of unusual 3-decalinoyltetramic acid derivatives, decalintetracides A (**77a, b**) and B (**78a, b**), isolated from *Fusarium equiseti*—an endophyte of the plant *Suaeda salsa* (L.) Pall.—inhibited the root and hypocotyl growth of *Amaranthus hybridus* and *A. retroflexus* L., indicating their potential as natural herbicides (Zhao et al. 2022). The mangrove is a unique marine ecosystem that thrives in the intertidal zone of tropical and subtropical coasts, showcasing notable characteristics such as hypoxia tolerance, high salinity, and abundant humidity. In the quest to identify new compounds with potent phytotoxicity, seven pairs of azaphilone E/Z isomers were isolated from the culture broth of the fungus *Penicillium sclerotiorum*, derived from mangrove areas (Wang et al. 2022b). Among these compounds, Sclerotiorin B (**79 a, b**), ochratoxin A (**80 a, b**), and isochromophilone I (**81 a, b**) demonstrated significant phytotoxicity against the growth of radicle and plumule in *Amaranthus retroflexus* L., with EC_{50} values ranging from 235 to 321 μM. This compares to the positive control glufosinate-ammonium, which had EC_{50} values of 555 μM for radicle and 656 μM for plumule growth. Additionally, compounds (**80 a, b**) and (**81 a, b**) exhibited inhibitory effects on the growth of velvetleaf (*Abutilon theophrasti* Medik.), with EC_{50} values ranging from 769 to 1201.5 μM. The results suggest promising compounds for exploration in the research and development of bioherbicides derived from marine endophytes in plants from marine environments. In a study by Wei et al. (2023), equisetin (**82**), derived from the endophytic fungus *Fusarium* sp. isolated from *Mirabilis jalapa*, was noted for its fungicidal activity. Interestingly, the same compound also demonstrated herbicidal properties. Equisetin inhibited root and shoot growth of *E. crus-galli* by 98% and 38%, respectively, at a concentration of 200 μg mL^{-1}. Additionally, it exhibited higher efficacy against *Euphorbia prostrata*, inducing 99% root inhibition and 42% shoot inhibition at a concentration of 100 μg mL^{-1} (Wei et al. 2023).

Research on endophytic fungi that produce phytotoxic metabolites emphasizes their potential for the discovery, development, and synthesis of herbicidal compounds. The diversity of these fungi and the metabolites they produce suggests promising applications in agriculture, promoting sustainable methods for weed control (Fig. 8.5, Table 8.1).

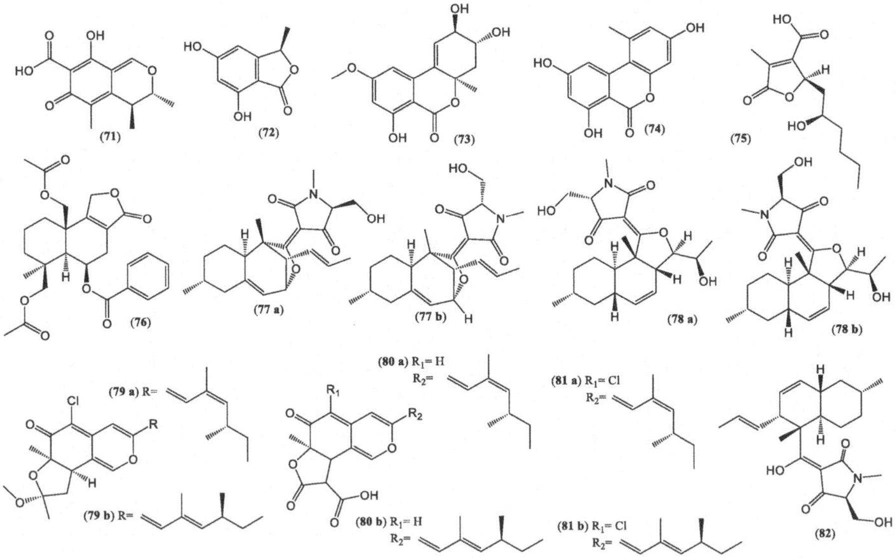

Fig. 8.5 Phytotoxic compounds isolated from endophytic fungi. Citrinin (**71**), 5,7-dihydroxy-3(R)-methylphthalide (**72**), (−)-altenuene (**73**), alternariol (**74**), aspertamarinoic acid (**75**), astellolide A (**76**), decalintetracides A (**77 a, b**) and B (**78 a, b**), sclerotiorin B (**79 a, b**), ochlephilone (**80 a, b**), isochromophilone I (**81 a, b**), equisetin (**82**)

8.7 Conclusion

This chapter discusses compounds with pesticidal activity from endophytic fungi that require further investigation. The molecular targets or modes of action of most of these compounds remain unknown. The urgent need for pesticides with new modes of action should encourage future research to identify the molecular targets of these compounds. Additionally, the studies reported here were conducted under controlled laboratory conditions, which substantially differ from those in the industrial production process. During the cultivation of fungi, several compounds are produced as secondary metabolites, and the compound responsible for the activity is often a minor product of the fungus. Cultivating endophytic fungi on a large scale to isolate the substance can be costly and time-consuming, making chemical synthesis of such compounds a more viable alternative. Therefore, seeking alternatives for the efficient production and isolation of these compounds, as well as conducting tests to ensure safety for humans and the environment, is crucial for making pesticides from fungi viable and marketable products.

Table 8.1 Pesticidal compounds isolated from endophytic fungi

Activity	Host plant	Endophytic fungus	Active compound	Target organism	References
Antifungal	*Mahonia fortune*	*Fusarium decemcellulare*	Enniatin H (**1**) Enniatin I (**2**) Nectriacid B (**3**)	*Colletotrichum musae*	Song et al. (2023)
	Common bean field crop	*Trichoderma erinaceum*	(*Z*)-5-amino-5-(1,1,2-trihydroxybuta-1,3-dienyloxy) pentane-6,7,8,9-tetraol (**4**)	*Pythium ultimum*	Siebatcheu et al. (2023)
	Cinnamomum cassia	*Hypoxylon vinosopulvinatum*	Hypoxylonone B (**5**) Hypoxylonone C (**6**)	*Lasiodiplodia pseudotheobromae*	Lv et al. (2023)
	Uncaria rhynchophylla	*Trichoderma* sp.	Trichodermic acid E (**7**) Trichodermic acid F (**8**) Trichodermic acid G (**9**) Trichodermic acid H (**10**)	*Botrytis cinerea* *Alternaria alternata*	Wu et al. (2023)
			Trichodermic acid (**11**)	*B. cinerea* *A. alternata* *Fusarium moniliforme*	
	Lagopsis supina	*Aspergillus ochraceus*	2-methoxy-6-methyl-1,4-benzoquinone (**12**) Penicillic acid (**13**)	*Fusarium graminearum* *Fusarium oxysporum* *Fusarium moniliforme* *Fusarium stratum* *Botrytis cinerea* *Magnaporthe oryzae* *Verticillium dahliae*	Zhang et al. (2023)
	Sarcandra glabra	*Lasiodiplodia pseudotheobromae*	Lasiodiplodin (**14**)	*Exserohilum turcicum* *Colletotrichum capsici* *Pestalotiopsis theae*	Luo et al. (2023)
	Zingiber officinale	*Penicillium polonicum*	Glaucanic acid (**15**) Dihydrocompactin acid (**16**)	*Colletotrichum gloeosporioides*	Gupta et al. (2023)

	Fritillaria unibracteata var. *wabuensis*	*Cladosporium sphaerospermum*	Cladodionen (**17**)	*Botrytis cinerea*	Pan et al. (2023)
	Stryphnodendron adstringens	*Pseudofusicoccum stromaticum*	Cytochalasin H (**18**) Fomoxanthone A (**19**) Fomoxanthone B (**20**) Paecilin Q (**21**)	*Phyllosticta citricarpa*	Iantas et al. (2023)
	Potato	*Bipolaris eleusines*	Bipolarithizole A (**22**)	*Rhizoctonia solani*	Ai et al. (2022)
	Panax notoginseng	*Emericella* sp.	Emerlactam A (**23**)	*Helminthosporium maydis*	Xian et al. (2022)
			5-alkenylresorcinol (**24**)	*Verticillium dahlia* *Rhizoctonia solani* *Sclerotinia sclerotiorum* *Fusarium oxysporum* *Alternaria fragriae* *Helminthosporium maydis* *Botryosphaeria berengeriana*	
			Emerlactam B (**25**)	*H. maydis* *Sclerotinia sclerotiorum*	
	Camellia sinensis	*Scopulariopsis candelabrum*	Hymeglusin (**26**)	*Exserohilum turcicum*	Tang et al. (2022)
	Euphorbia pekinensis	*Penicillium* sp.	Penicilphilone A (**27**) Penicilphilone C (**28**) 7-epi-WB (**29**)	*Rhizoctonia solani* *Fusarium oxysporum* *Penicillium citrinum* *Phytophthora melonis*	Feng et al. (2022)

(continued)

Table 8.1 (continued)

Activity	Host plant	Endophytic fungus	Active compound	Target organism	References
	Azadirachta indica	*Xylaria* sp.	Xilarioxides A (**30**) Xilarioxides E (**31**) Xilarioxides F (**32**)	*Gibberella saubinetii* *Curvularia lunata* *Colletotrichum gloeosporioides*	Wang et al. (2022a)
	Synsepalum dulcificum	*Aspergillus polyporicola*	11-dehydrocurvularin (**33**)	*Botrytis cinerea* *Fusarium graminearum*	Liu et al. (2022)
			(+)-2,5-dimethyl-3(2H)-benzofuranone (**34**)	*F. graminearum*	
	Kandelia candel	*Aspergillus flavus*	Aflaxanthone A (**35**) Aflaxanthone B (**36**)	*Colletotrichum gloeosporioides* *Fusarium oxysporum* *Collettrichum musae*	Zang et al. (2022)
	Mirabilis jalapa	*Fusarium* sp.	Equisetin (**37**)	*B. cinera* *F. graminearum* *S. sclerotiorum* *R. solani*	Wei et al. (2023)
	Saurauia scaberrinae	*Phoma* sp.	Cercosporamide (**38**)	*Colletotrichum gloeosporioides* *C. scovillei*	Liu et al. (2024)
	Euphorbia geniculate	*Gymnoascus thermotolerans*	Dibutyl phthalate (**39**)	*Fusarium solani*	Kamel et al. (2024)
Insecticidal	Fescue grasses	*Acremonium* sp., *Epichloë* sp., *Neotyphodium* sp.	Loline alkaloids (**40**, **41**, **42**) Peramine (**43**)	Herbivorous insects	Rudgers and Clay (2008), Rohlfs and Churchill (2011), Fe Andres et al. (2017)
	Bontia daphnoides	*Nodulisporium* sp.	Nodulisporic acid (**44**)	Blowfly larvae and other blood-sucking insects	Ondeyka et al. (1997)

	Paullina paullinioides	*Muscodor vitigenus*	Naphthalene (**45**)	*Cephus cinctus*	Daisy et al. (2002)
	Spruce needles	*Phialocephala scopiformis*	Rugulosin (**46**)	*Choristoneura fumiferana*	Fe Andres et al. (2017)
	Gaultheria procumbens L.	Unidentified	5-hydroxy-2-(1′-oxo-5′-methyl-4′-hexenyl) benzofuran (**47**)	*Choristoneura fumiferana*	Findlay et al. (1997)
		Unidentified	5-hydroxy-2-(1′-hydroxy-5′-methyl-4′-hexenyl) benzofuran (**48**)		
	Picea rubens	Unidentified	Vermiculin (**49**)	*Choristoneura fumiferana*	Sumarah et al. (2010)
		Unidentified	4-oxo-5-phenyl-pentanoic acid (**50**)		
		Unidentified	5-benzyl-dihydro-furan-2-one (**51**)		
		Unidentified	3-epi-isopetasol (**52**)		
		Unidentified	15-hydroxy-3-epi-isopetasol (**53**)		
		Unidentified	Two new isomers (**54**, **55**)		
		Unidentified	Cordianhydrides A and B (**56**, **57**)		
	Catharanthus roseus	*Alternaria alternata*	Altenuene (**58**)	*Spodoptera litura*	Bhagat et al. (2016)
	Tinospora cordifolia	*Cladosporium velox*	Chlorogenic acid (**59**)	*Spodoptera litura*	Singh et al. (2016)
	Bertholletia excelsa	*Aspergillus* sp.	Aspergillol B (**60**)	*Aedes aegypti*	Araújo et al. (2022)
Nematicidal	*Solanum lycopersicum*	*Fusarium oxysporum*	4-hydroxybenzoic acid (**61**), Indole-3-acetic acid (**62**), Gibepyrone D (**63**)	*Meloidogyne incognita*	Bogner et al. (2017)
	Leaf of rainforest tree	*Phomopsis phaseoli*	3-hydroxypropionic acid (**64**)	*Meloidogyne incognita*	Schwarz et al. (2004)

(continued)

Table 8.1 (continued)

Activity	Host plant	Endophytic fungus	Active compound	Target organism	References
	Azadirachta indica	*Geotrichum* sp.	1- [(2R*,4S*,5S*)-2-chloro-4-methyl-1,3-oxazinan-5-yl]ethenone (**65**) 1-[(2R*,4S*,5R*)-2-chloro-4-methyl-1,3-oxazinan-5-yl]ethenone (**66**) 1-(2,4-dihydroxyphenyl)- ethenone (**67**)	*Bursaphelenchus xylophilus*	Li et al. (2007)
	Unidentified	*Chaetomium globosum*	Chaetoglobosin A (**68**) Chaetoglobosin B (**69**)	*Meloidogyne javanica*	Khan et al. (2019)
Acaricidal	*Aniba rosaeodora*	Unidentified	Linalool (**70**)	*Unidentified*	Silva (2010)
Phytotoxic	*Pinellia ternata* *P. pedatisecta*	*Alternaria angustiovoidea* *Aspergillus foccosus*	Citrinin (**71**)	*Echinochloa crus-galli*	Kong et al. (2023)
	Psidium guajava	*Xylaria brevipes* *Alternaria alternata*	5,7-dihydroxy-3(R)-methylphthalide (**72**) (−)-altenuene (**73**) Alternariol (**74**)	*Brassica chinensis* var. parachinensis	Oktriawan et al. (2023)
	Crassula arborescens	*Aspergillus* sp.	Aspertamarinoic acid (**75**) Astellolide A (**76**)	*Arabidopsis thaliana*	Ma et al. (2022)
	Suaeda salsa	*Fusarium equiseti*	Decalintetracides A (**77a**, **b**) and B (**78a**, **b**)	*Amaranthus hybrid* *A. retroflexus*	Zhao et al. (2022)
	Unidentified	*Penicillium sclerotiorum*	Sclerotiorin B (**79 a**, **b**) Ochlephilone (**80a**, **b**) Isochromophilone I (**81 a**, **b**)	*Amaranthus retroflexus*	Wang et al. (2022b)
	Mirabilis jalapa	*Fusarium* sp.	Equisetin (**82**)	*E. crus-galli* *Euphorbia protrata*	Wei et al. (2023)

Acknowledgments This work was supported by FAPEMIG, CNPq, CAPES, and Fundação Oswaldo Cruz (PDTIS-FIOCRUZ) and USDA Cooperative Agreement 58-6060-6-015 grant to the University of Mississippi.

References

Abo Nouh FA (2019) Endophytic fungi for sustainable agriculture. Microbial Biosyst 4(1):31–44. https://doi.org/10.21608/MB.2019.38886

Aguiar-Menezes EDL (2005) Inseticidas botânicos: seus princípios ativos, modo de ação e uso agrícola. Embrapa Agrobiologia. Available from: https://www.bdpa.cnptia.embrapa.br/consulta/busca?b=pc&id=626872&biblioteca=vazio&busca=Inseticidas%20bot%C3%A2nicos\:%20seus%20princ%C3%ADpios%20ativos,%20modo%20de%20a%C3%A7%C3%A3o%20e%20uso%20agr%C3%ADcola.%20Embrapa%20Agrobiologia&qFacets=Inseticidas%20bot%C3%A2nicos\:%20seus%20princ%C3%ADpios%20ativos,%20modo%20de%20a%C3%A7%C3%A3o%20e%20uso%20agr%C3%ADcola.%20Embrapa%20Agrobiologia&sort=&paginacao=t&paginaAtual=1

Ai HL, Shi BB, Li W, He J, Li ZH, Feng T, Liu JK (2022) Bipolarithizole A, an antifungal phenylthiazole-sativene merosesquiterpenoid from the potato endophytic fungus Bipolaris eleusines. Org Chem Front 9(7):1814–1819

Alves GS (2019) Fungos promotores do crescimento do feijoeiro, micorriza arbuscular e atributos bioquímicos do solo [Master's thesis, Universidade Federal de Uberlândia]. Repositório Institucional da Universidade Federal de Uberlândia. https://repositorio.ufu.br/handle/123456789/25201

Amandio DTT (2022) Avaliação de isolados de *Metarhizium* spp. (Ascomycota: Hypocreales) nativos de Santa Catarina na promoção de crescimento de gramíneas (Poacea) forrageiras e antagonismo a *Spodoptera frugiperda* (Lepidoptera: Noctuidae) [Master's thesis, Universidade Federal de Santa Catarina]. Repositório Institucional da Universidade Federal de Santa Catarina. https://repositorio.ufsc.br/handle/123456789/242670

Araújo IF, Marinho VHDS, Sena IDS, Curti JM, Ramos RDS, Ferreira RM et al (2022) Larvicidal activity against *Aedes aegypti* and molecular docking studies of compounds extracted from the endophytic fungus *Aspergillus* sp. isolated from *Bertholletia excelsa* Humn. & Bonpl. Biotechnol Lett 44(3):439–459. https://doi.org/10.1007/s10529-022-03220-7

Barbosa ACMS, Silva LPC, Ferraz CM, Tobias FL, de Araújo JV, Loureiro B et al (2019) Nematicidal activity of silver nanoparticles from the fungus *Duddingtonia flagrans*. Int J Nanomedicine:2341–2348. https://doi.org/10.2147/IJN.S193679

Barreto DLC, de Azevedo RN, de Carvalho CR, Ferreira MC, Cantrell CL, Duke SO, Rosa LH (2021) Bioactive compounds produced by neotropical endophytic fungi applied to agriculture. In: Neotropical endophytic fungi: diversity, ecology, and biotechnological applications, pp 257–295. https://doi.org/10.1007/978-3-030-53506-3_12

Bendejacq-Seychelles A, Gibot-Leclerc S, Guillemin JP, Mouille G, Steinberg C (2023) Phytotoxic fungal secondary metabolites as herbicides. Pest Manag Sci. https://doi.org/10.1002/ps.7813

Bhagat J, Kaur A, Kaur R, Yadav AK, Sharma V, Chadha BS (2016) Cholinesterase inhibitor (Altenuene) from an endophytic fungus *Alternaria alternata*: optimization, purification and characterization. J Appl Microbiol 121(4):1015–1025. https://doi.org/10.1111/jam.13192

Bogner CW, Kamdem RS, Sichtermann G, Matthäus C, Hölscher D, Popp J et al (2017) Bioactive secondary metabolites with multiple activities from a fungal endophyte. Microb Biotechnol 10(1):175–188. https://doi.org/10.1111/1751-7915.12467

Borges LMF, Ferri PH, Silva WJ, Silva WC, Silva JG (2003) In vitro efficacy of extracts of *Melia azedarach* against the tick *Boophilus microplus*. Med Vet Entomol 17(2):228–231. https://doi.org/10.1046/j.1365-2915.2003.00426.x

Bush LP, Wilkinson HH, Schardl CL (1997) Bioprotective alkaloids of grass-fungal endophyte symbioses. Plant Physiol 114(1):1. https://doi.org/10.1104/pp.114.1.1

Chatterjee S, Kuang Y, Splivallo R, Chatterjee P, Karlovsky P (2016) Interactions among filamentous fungi *Aspergillus niger*, *Fusarium verticillioides* and *Clonostachys rosea*: fungal biomass, diversity of secreted metabolites and fumonisin production. BMC Microbiol 16:1–13. https://doi.org/10.1186/s12866-016-0698-3

Clapp J (2021) Explaining growing glyphosate use: the political economy of herbicide-dependent agriculture. Glob Environ Chang 67:102239. https://doi.org/10.1016/j.gloenvcha.2021.102239

Coque JJR, Álvarez-Pérez JM, Cobos R, González-García S, Ibáñez AM, Galán AD, Calvo-Peña C (2020) Advances in the control of phytopathogenic fungi that infect crops through their root system. Adv Appl Microbiol 111:123–170. https://doi.org/10.1016/bs.aambs.2020.01.003

Corkley I, Fraaije B, Hawkins N (2022) Fungicide resistance management: maximizing the effective life of plant protection products. Plant Pathol 71(1):150–169. https://doi.org/10.1111/ppa.13467

Daisy BH, Strobel GA, Castillo U, Ezra D, Sears J, Weaver DK, Runyon JB (2002) Naphthalene, an insect repellent, is produced by *Muscodor vitigenus*, a novel endophytic fungus. Microbiology 148(11):3737–3741. https://doi.org/10.1099/00221287-148-11-3737

de León IP, Montesano M (2013) Activation of defense mechanisms against pathogens in mosses and flowering plants. Int J Mol Sci 14(2):3178–3200. https://doi.org/10.3390/ijms14023178

Degenkolb T, Vilcinskas A (2016) Metabolites from nematophagous fungi and nematicidal natural products from fungi as an alternative for biological control. Part I: metabolites from nematophagous ascomycetes. Appl Microbiol Biotechnol 100:3799–3812. https://doi.org/10.1007/s00253-015-7233-6

Dufossé L, Fouillaud M, Caro Y (2021) Fungi and fungal metabolites for the improvement of human and animal nutrition and health. J Fungi 7(4):274. https://doi.org/10.3390/jof7040274

Dutta D, Puzari KC, Gogoi R, Dutta P (2014) Endophytes: exploitation as a tool in plant protection. Braz Arch Biol Technol 57:621–629. https://doi.org/10.1590/S1516-8913201402043

Faeth SH, Shochat E (2010) Inherited microbial symbionts increase herbivore abundances and alter arthropod diversity on a native grass. Ecology 91(5):1329–1343. https://doi.org/10.1890/08-2123.1

FAO (2019) FAO and the African Union Commission launch a new framework to boost farm productivity and agribusiness in Africa. Food and Agriculture Organization of the United Nations. https://www.fao.org/news/story/en/item/1187738/icode/

Farhat H, Urooj F, Sohail N, Ansari M, Ehteshamul-Haque S (2022) Evaluation of nematicidal potential of endophytic fungi associated with healthy plants and GC-MS profiling of metabolites of endophytic *Fusarium solani*. S Afr J Bot 146:146–161. https://doi.org/10.1016/j.sajb.2021.10.011

Fe Andres M, Elisa Diaz C, Gimenez C, Cabrera R, Gonzalez-Coloma A (2017) Endophytic fungi as novel sources of biopesticides: the Macaronesian Laurel forest, a case study. Phytochem Rev 16(5):1009–1022. https://doi.org/10.1007/s11101-017-9514-4

Feng ZY, Huang PZ, Jiang SJ, Shen SJ, Chen JJ (2022) Three new highly oxygenated azaphilones from an endophytic fungus *Penicillium* sp. LZUC-S1. Tetrahedron Lett 113:154260. https://doi.org/10.1016/j.tetlet.2022.154260

Findlay JA, Buthelezi S, Li G, Seveck M, Miller JD (1997) Insect toxins from an endophytic fungus from wintergreen. J Nat Prod 60(11):1214–1215. https://doi.org/10.1021/np970222j

Greenspan MD, Yudkovitz JB, Lo CY, Chen JS, Alberts AW, Hunt VM et al (1987) Inhibition of hydroxymethylglutaryl-coenzyme A synthase by L-659,699. Proc Natl Acad Sci 84(21):7488–7492. https://doi.org/10.1073/pnas.84.21.7488

Gupta S, Choudhary M, Singh B, Kushwaha M, Dhar MK, Kaul S (2023) Green synthesis and biological evaluation of glaucanic acid and dihydrocompactin acid by endophytic fungus *Penicillium polonicum* from Zingiber officinal e. Nat Prod Res 1-5. https://doi.org/10.1080/14786419.2023.2188210

Hamilton CE, Bauerle TL (2012) A new currency for mutualism? Fungal endophytes alter antioxidant activity in hosts responding to drought. Fungal Divers 54:39–49. https://doi.org/10.1007/s13225-012-0156-y

Hassan E, Prijono D (2011) Plants as a source of biopesticides for pest control: a new perspective. In: Survival and sustainability: environmental concerns in the 21st century, pp 1491–1500. https://doi.org/10.1007/978-3-540-95991-5_140

Iantas J, Savi DC, Ponomareva L, Thorson JS, Rohr J, Glienke C, Shaaban KA (2023) Paecilins Q and R: antifungal chromanones produced by the endophytic fungus *Pseudofusicoccum stromaticum* CMRP4328. Planta Med, (AAM). https://doi.org/10.1055/a-2063-5481

Jani AJ, Faeth SH, Gardner D (2010) Asexual endophytes and associated alkaloids alter arthropod community structure and increase herbivore abundances on a native grass. Ecol Lett 13(1):106–117. https://doi.org/10.1111/j.1461-0248.2009.01401.x

Kamel NM, Abdel-Motaal FF, El-Zayat SA, Mohamed AEHH, Ohta S, Hussien TA (2024) A new secondary metabolite with antimicrobial, antioxidant, and cytotoxic activities from an endophytic fungus, Gymnoascus thermotolerans. Egypt J Bot 64(1):107–124. https://doi.org/10.21608/ejbo.2023.212304.2338

Khan B, Yan W, Wei S, Wang Z, Zhao S, Cao L et al (2019) Nematicidal metabolites from endophytic fungus *Chaetomium globosum* YSC5. FEMS Microbiol Lett 366(14):fnz169. https://doi.org/10.1093/femsle/fnz169

Khan RAA, Najeeb S, Hussain S, Xie B, Li Y (2020) Bioactive secondary metabolites from *Trichoderma* spp. against phytopathogenic fungi. Microorganisms 8(6):817. https://doi.org/10.3390/microorganisms8060817

Kim SI, Park C, Ohh MH, Cho HC, Ahn YJ (2003) Contact and fumigant activities of aromatic plant extracts and essential oils against *Lasioderma serricorne* (Coleoptera: Anobiidae). J Stored Prod Res 39(1):11–19. https://doi.org/10.1016/S0022-474X(02)00013-9

Kong K, Huang Z, Shi S, Pan W, Zhang Y (2023) Diversity, antibacterial and phytotoxic activities of culturable endophytic fungi from *Pinellia pedatisecta* and *Pinellia ternata*. BMC Microbiol 23(1):1–13. https://doi.org/10.1186/s12866-022-02741-5

Krishnan GH, Rajasenbagam T (2021) A comprehensive survey for weed classification and detection in agriculture lands. J Inf Technol 3(4):281–289. https://doi.org/10.36548/jitdw.2021.4.004

Li G, Zhang K, Xu J, Dong J, Liu Y (2007) Nematicidal substances from fungi. Recent Pat Biotechnol 1(3):212–233. https://doi.org/10.2174/187220807782330165

Li J, Gu F, Wu R, Yang J, Zhang KQ (2017) Phylogenomic evolutionary surveys of subtilase superfamily genes in fungi. Sci Rep 7(1):45456. https://doi.org/10.1038/srep45456

Lin DJ, Zhang YX, Fang Y, Gao SJ, Wang R (2023) The effect of chlorogenic acid, a potential botanical insecticide, on gene transcription and protein expression of carboxylesterases in the armyworm (*Mythimna separata*). Pestic Biochem Physiol 195:105575. https://doi.org/10.1016/j.pestbp.2023.105575

Liu SZ, Yan X, Tang XX, Lin JG, Qiu YK (2018) New bis-alkenoic acid derivatives from a marine-derived fungus *Fusarium solani* H915. Mar Drugs 16(12):483. https://doi.org/10.3390/md16120483

Liu SS, Huang R, Zhang SP, Xu TC, Hu K, Wu SH (2022) Antimicrobial secondary metabolites from an endophytic fungus *Aspergillus polyporicola*. Fitoterapia 162:105297. https://doi.org/10.1016/j.fitote.2022.105297

Liu R, Paguirigan JA, Hur JS, Kim W (2024) Cercosporamide, a polyketide-derived fungal metabolite, serves as an antifungal agent against phytopathogenic fungi. Mycoscience 65(1):19–27. https://doi.org/10.47371/mycosci.2023.11.002

Luo H, Meng S, Deng Y, Deng Z, Shi H (2023) In vitro antifungal activity of lasiodiplodin, isolated from endophytic fungus *Lasiodiplodia pseudotheobromae* J-10 associated with *Sarcandra glabra* and optimization of culture conditions for lasiodiplodin production. Arch Microbiol 205(4):140. https://doi.org/10.1007/s00203-023-03440-z

Lv J, Zhou H, Dong L, Wang H, Yang L, Yu H et al (2023) Three new furanones from endophytic fungus *Hypoxylon vinosopulvinatum* DYR-1-7 from *Cinnamomum cassia* with their antifungal activity. Nat Prod Res:1–8. https://doi.org/10.1080/14786419.2023.2218530

Ma J, Lu C, Tang Y, Shen Y (2022) Phytotoxic metabolites isolated from *Aspergillus* sp., an endophytic fungus of *Crassula arborescens*. Molecules 27(22):7710. https://doi.org/10.3390/molecules27227710

Matthews G (2015) Pesticides: health, safety and the environment. John Wiley & Sons. Available from: https://books.google.com.br/books?hl=pt-BR&lr=&id=3-vGBwAAQBAJ&oi=fnd&pg=PR7&dq=Matthews+G+(2015)+Pesticides:+health,+safety+and+the+environment.+Wiley&ots=kPz8FTIKsb&sig=eim2bqJdJazdMYOAVXQnISBvxT4#v=onepage&q=Matthews%20G%20(2015)%20Pesticides%3A%20health%2C%20safety%20and%20the%20environment.%20Wiley&f=false

Miller JD, Mackenzie S, Mark FOTO, Adams GW, Findlay JA (2002) Needles of white spruce inoculated with rugulosin-producing endophytes contain rugulosin reducing spruce budworm growth rate. Mycol Res 106(4):471–479. https://doi.org/10.1017/S0953756202005671

Miller JD, Sumarah MW, Adams GW (2008) Effect of a rugulosin-producing endophyte in *Picea glauca* on *Choristoneura fumiferana*. J Chem Ecol 34:362–368. https://doi.org/10.1007/s10886-007-9412-0

Morais LAS, Marinho-Prado JS (2016) Plantas com atividade inseticida. In: Halfeld-Vieira B d A et al (eds) Defensivos Agrícolas Naturais: Uso e Perspectivas. Embrapa, Brasília, pp 542–593

Murmu SB, Mishra HN (2018) Post-harvest shelf-life of banana and guava: mechanisms of common degradation problems and emerging counteracting strategies. Innovative Food Sci Emerg Technol 49:20–30. https://doi.org/10.1016/j.ifset.2018.07.011

Nelson PE (1992) Taxonomy and biology of *Fusarium moniliforme*. Mycopathologia 117:29–36. https://doi.org/10.1007/BF00497276

Nicol JM, Turner SJ, Coyne DL, Nijs Ld, Hockland S, Maafi ZT (2011) Current nematode threats to world agriculture. In: Jones J, Gheysen G, Fenoll C (eds) Genomics and molecular genetics of plant-nematode interactions. Springer, Dordrecht. https://doi.org/10.1007/978-94-007-0434-3_2

Niu S, Tang XX, Fan Z, Xia JM, Xie CL, Yang XW (2019) Fusarisolins A–E, polyketides from the marine-derived fungus *Fusarium solani* H918. Mar Drugs 17(2):125. https://doi.org/10.3390/md17020125

Oktriawan T, Ariefta N, Raharjo TJ, Astuti E, Koseki T, Shiono Y (2023) Phytotoxic and cytotoxic polyketides produced by fungal endophytes isolated from *Psidium guajava*. HAYATI J Biosci 30(3):473–479. https://doi.org/10.4308/hjb.30.3.473-479

Ondeyka JG, Helms GL, Hensens OD, Goetz MA, Zink DL, Tsipouras A et al (1997) Nodulisporic acid A, a novel and potent insecticide from a *Nodulisporium* sp. Isolation, structure determination, and chemical transformations. J Am Chem Soc 119(38):8809–8816. https://doi.org/10.1021/ja971664k

Pan F, Yang N, Zhu X, Yu C, Jiang M, Jiang Y et al (2023) Discovery of a natural hybrid polyketide produced by endophytic *Cladosporium sphaerospermum* for biocontrol of phytopathogenic fungus *Botrytis cinerea*. J Agric Food Chem 71(32):12190–12202. https://doi.org/10.1021/acs.jafc.3c02408

Peng Y, Li SJ, Yan J, Tang Y, Cheng JP, Gao AJ et al (2021) Research progress on phytopathogenic fungi and their role as biocontrol agents. Front Microbiol 12:670135. https://doi.org/10.3389/fmicb.2021.670135

Perotti VE, Larran AS, Palmieri VE, Martinatto AK, Permingeat HR (2020) Herbicide resistant weeds: a call to integrate conventional agricultural practices, molecular biology knowledge and new technologies. Plant Sci 290:110255. https://doi.org/10.1016/j.plantsci.2019.110255

Pipattanapuckdee A, Boonyakait D, Tiyayon C, Seehanam P, Ruangwong OU (2019) *Lasiodiplodia pseudotheobromae* causes postharvest fruit rot of longan in Thailand. Aust Plant Dis Notes 14:1–7. https://doi.org/10.1007/s13314-019-0350-9

Preto ID (2018) Identificação do potencial biotecnológico de microrganismos endofíticos na produção de compostos inseticidas e biorremediação [Doctoral dissertation, Universidade de São Paulo]. Repositório Institucional da Universidade de São Paulo. https://repositorio.usp.br/item/002914776

Rohlfs M, Churchill AC (2011) Fungal secondary metabolites as modulators of interactions with insects and other arthropods. Fungal Genet Biol 48(1):23–34. https://doi.org/10.1016/j.fgb.2010.08.008

Rudgers JA, Clay K (2008) An invasive plant–fungal mutualism reduces arthropod diversity. Ecol Lett 11(8):831–840. https://doi.org/10.1111/j.1461-0248.2008.01201.x

Rudgers JA, Holah J, Orr SP, Clay K (2007) Forest succession suppressed by an introduced plant–fungal symbiosis. Ecology 88(1):18–25. https://doi.org/10.1890/0012-9658(2007)88[18:FSSBAI]2.0.CO;2

Santos LCDS (2021) Potencial de fungos entomopatogênicos endofíticos no controle biológico de insetos-praga [Trabalho de Conclusão de Curso, Universidade Federal de São João Del Rei]. https://www.ufsj.edu.br/portal2-repositorio/File/cobit/TCC/TCCs%20Defendidos/TCC%20Leticia%20Cristine%20Santos%202021.pdf

Schwarz M, Köpcke B, Weber RW, Sterner O, Anke H (2004) 3-Hydroxypropionic acid as a nematicidal principle in endophytic fungi. Phytochemistry 65(15):2239–2245. https://doi.org/10.1016/j.phytochem.2004.06.035

Seong J, Shin J, Kim K, Cho BK (2021) Microbial production of nematicidal agents for controlling plant-parasitic nematodes. Process Biochem 108:69–79. https://doi.org/10.1016/j.procbio.2021.06.006

Siebatcheu EC, Wetadieu D, Youassi Youassi O, Bedine Boat MA, Bedane KG, Tchameni NS, Sameza ML (2023) Secondary metabolites from an endophytic fungus *Trichoderma erinaceum* with antimicrobial activity towards *Pythium ultimum*. Nat Prod Res 37(4):657–662. https://doi.org/10.1080/14786419.2022.2075360

Silva CDJ (2010) Prospecção Química e Biológica de Fungos Endofíticos associados à *Aniba rosaeodora* (Lauraceae) [Master's thesis, Universidade do Estado do Amazonas]. Repositório Institucional da Universidade do Estado do Amazonas. http://repositorioinstitucional.uea.edu.br/handle/riuea/2233

Singh B, Kaur T, Kaur S, Manhas RK, Kaur A (2016) Insecticidal potential of an endophytic *Cladosporium velox* against *Spodoptera litura* mediated through inhibition of alpha glycosidases. Pestic Biochem Physiol 131:46–52. https://doi.org/10.1016/j.pestbp.2016.01.004

Skaff DA, Ramyar KX, McWhorter WJ, Barta ML, Geisbrecht BV, Miziorko HM (2012) Biochemical and structural basis for inhibition of *Enterococcus faecalis* hydroxymethylglutaryl-CoA synthase, mvaS, by hymeglusin. Biochemistry 51(23):4713–4722. https://doi.org/10.1021/bi300037k

Smedsgaard J, Nielsen J (2005) Metabolite profiling of fungi and yeast: from phenotype tometabolome by MS and informatics. J Exp Bot 56(410):273–286. https://doi.org/10.1093/jxb/eri068

Song Z, Sun YJ, Xu S, Li G, Yuan C, Zhou K (2023) Secondary metabolites from the Endophyticfungi *Fusarium decemcellulare* F25 and their antifungal activities. Front Microbiol 14:1127971. https://doi.org/10.3389/fmicb.2023.1127971

Sousa LAD, Soares SF, Pires HB Jr, Ferri PH, Borges LMF (2005) Testes de eficácia sobre fêmeas ingurgitadas de *Boophilus microplus* com compostos e extratos hexânicos de *Melia azedarach*. In: Congresso de Pesquisa, Ensino e Extensão da UFG, Goiânia, Brazil. Recuperado de. https://projetos.extras.ufg.br/conpeex/2005/porta_arquivos/outras/LorenaAlessandraDiasdeSousa_TESTESDEEFIC%C3%81CIASOBREF%C3%8AMEASINGURGITADASDEBOOPHILUSMICROPLU_291.pdf

Soyer JL, Hamiot A, Ollivier B, Balesdent MH, Rouxel T, Fudal I (2015) The APSES transcription factor LmStuA is required for sporulation, pathogenic development and effector gene expression in *Leptosphaeria maculans*. Mol Plant Pathol 16(9):1000–1005. https://doi.org/10.1111/mpp.12249

Sumarah MW, Miller JD (2009) Anti-insect secondary metabolites from fungal endophytes of conifer trees. Nat Prod Commun 4(11):1497–1504. https://doi.org/10.1177/1934578X0900401112

Sumarah MW, Adams GW, Berghout J, Slack GJ, Wilson AM, Miller JD (2008) Spread and persistence of a rugulosin-producing endophyte in *Picea glauca* seedlings. Mycol Res 112(6):731–736. https://doi.org/10.1016/j.mycres.2008.01.007

Sumarah MW, Puniani E, Sørensen D, Blackwell BA, Miller JD (2010) Secondary metabolites from anti-insect extracts of endophytic fungi isolated from *Picea rubens*. Phytochemistry 71(7):760–765. https://doi.org/10.1016/j.phytochem.2010.01.015

Tang XX, Yan X, Fu WH, Yi LQ, Tang BW, Yu LB et al (2019) New β-lactone with tea pathogenic fungus inhibitory effect from marine-derived fungus MCCC3A00957. J Agric Food Chem 67(10):2877–2885. https://doi.org/10.1021/acs.jafc.9b00228

Tang J, Huang X, Cao MH, Wang Z, Yu Z, Yan Y et al (2022) Mono-/bis-alkenoic acid derivatives from an endophytic fungus *Scopulariopsis candelabrum* and their antifungal activity. Front Chem 9:812564. https://doi.org/10.3389/fchem.2021.812564

Tenguria RK, Khan FN, Quereshi S (2011) Endophytes-mines of pharmacological therapeutics. World J Sci Technol 1(5):127–149

Tomoda H, Ohbayashi N, Morikawa Y, Kumagai H, Ōmura S (2004) Binding site for fungal β-lactone hymeglusin on cytosolic 3-hydroxy-3-methylglutaryl coenzyme A synthase. Bioch Biophys Acta 1636(1):22–28. https://doi.org/10.1016/j.bbalip.2003.11.005

Wang JF, Huang R, Song ZQ, Yang QR, Li XP, Liu SS, Wu SH (2022a) Polyhydroxylated sesquiterpenes and ergostane glycosides produced by the endophytic fungus *Xylaria* sp. from *Azadirachta indica*. Phytochemistry 199:113188. https://doi.org/10.1016/j.phytochem.2022.113188

Wang W, Wang M, Wang XB, Li YQ, Ding JL, Lan MX et al (2022b) Phytotoxic azaphilones from the mangrove-derived fungus *Penicillium sclerotiorum* HY5. Front Microbiol 13:880874. https://doi.org/10.3389/fmicb.2022.880874

Wang Z, Wang L, Pan Y, Zheng X, Liang X, Sheng L et al (2023) Research advances on endophytic fungi and their bioactive metabolites. Bioprocess Biosyst Eng 46(2):165–170. https://doi.org/10.1007/s00449-022-02840-7

Wani ZA, Ashraf N, Mohiuddin T, Riyaz-Ul-Hassan S (2015) Plant-endophyte symbiosis, an ecological perspective. Appl Microbiol Biotechnol 99:2955–2965. https://doi.org/10.1007/s00253-015-6487-3

Webber J (1981) A natural biological control of Dutch elm disease. Nature 292(5822):449–451. https://doi.org/10.1038/292449a0

Wei W, Chen P, Khan B, Tian K, Feng Y, Lv B et al (2023) Evaluation of equisetin as an anti-microbial and herbicidal agent from endophytic fungus *Fusarium* sp. JDJR1. Agronomy 14(1):31. https://doi.org/10.3390/agronomy14010031

Weston LA, Duke SO (2003) Weed and crop allelopathy. Crit Rev Plant Sci 22(3-4):367–389. https://doi.org/10.1080/713610861

Westrick NM, Smith DL, Kabbage M (2021) Disarming the host: detoxification of plant defense compounds during fungal necrotrophy. Front Plant Sci 12:651716. https://doi.org/10.3389/fpls.2021.651716

Whorton MD, Foliart DE (1983) Mutagenicity, carcinogenicity and reproductive effects of dibromochloropropane (DBCP). Mutat Res/Rev Genet Toxicol 123(1):13–30. https://doi.org/10.1016/0165-1110(83)90044-1

Wilkinson HH, Siegel MR, Blankenship JD, Mallory AC, Bush LP, Schardl CL (2000) Contribution of fungal loline alkaloids to protection from aphids in a grass-endophyte mutualism. Mol Plant-Microbe Interact 13(10):1027–1033. https://doi.org/10.1094/MPMI.2000.13.10.1027

Wu JL, Yu YH, Yao HZ, Zhao X, Yuan T, Huang YH (2023) Trichodermic acids from an endophytic *Trichoderma* sp. and their antifungal activity against the phytopathogen *Botrytis cinerea*. Phytochem Lett 56:24–29. https://doi.org/10.1016/j.phytol.2023.06.004

Xian PJ, Liu SZ, Wang WJ, Yang SX, Feng Z, Yang XL (2022) Undescribed specialised metabolites from the endophytic fungus *Emericella* sp. XL029 and their antimicrobial activities. Phytochemistry 202:113303. https://doi.org/10.1016/j.phytochem.2022.113303

Yin Z, Ke X, Kang Z, Huang L (2016) Apple resistance responses against *Valsa mali* revealed by transcriptomics analyses. Physiol Mol Plant Pathol 93:85–92. https://doi.org/10.1016/j.pmpp.2016.01.004

Zacharia JT (2011) Identity, physical and chemical properties of pesticides. In: Singh VP, Gupta DK (eds) Pesticides in the modern world - trends in pesticides analysis. CRC Press, pp 1–18. Available from: https://books.google.com.br/books?hl=pt-BR&lr=&id=i-GdDwAAQBAJ&oi=fnd&pg=PA3&dq=Zacharia+JT+(2011)+Identity,+physical+and+chemical+properties+of+pesticides.+In:+Singh+VP,+Gupta+DK+(eds)+Pesticides+in+the+modern+world+-+trends+in+pesticides+analysis.+CRC+Press,+pp+1%E2%80%9318&ots=0byc5_qyZE&sig=EDN_htC9ZaZG4DDSM1TCiTWjLYY#v=onepage&q&f=false

Zang Z, Yang W, Cui H, Cai R, Li C, Zou G et al (2022) Two antimicrobial heterodimeric tetrahydroxanthones with a 7, 7′-linkage from mangrove endophytic fungus *Aspergillus flavus* QQYZ. Molecules 27(9):2691. https://doi.org/10.3390/molecules27092691

Zhang Y, Han T, Ming Q, Wu L, Rahman K, Qin L (2012) Alkaloids produced by endophytic fungi: a review. Nat Prod Commun 7(7):1934578X1200700742. https://doi.org/10.1177/1934578X120070074

Zhang D, Sun W, Xu W, Ji C, Zhou Y, Sun J et al (2023) Antimicrobial and cytotoxic activity of endophytic fungi from *Lagopsis supina*. J Microbiol Biotechnol 33(4):543. https://doi.org/10.4014/jmb.2211.11055

Zhao DL, Liu J, Han XB, Wang M, Peng YL, Ma SQ et al (2022) Decalintetracids A and B, two pairs of unusual 3-decalinoyltetramic acid derivatives with phytotoxicity from *Fusarium equiseti* D39. Phytochemistry 197:113125. https://doi.org/10.1016/j.phytochem.2022.113125

Exploring the Potential Benefits of *Trichoderma* Species in Agro-Industrial Crop Production

9

Heriberto Fortino Ramírez-Cariño,
Paula Cecilia Guadarrama-Mendoza, Teresa Romero-Cortes,
Jaime Alioscha Cuervo-Parra, and Rogelio Valadez-Blanco

Abstract

Trichoderma strains have been widely used as part of strategies directed to attain sustainable agriculture considering that these fungi can reduce or substitute the utilization of agrochemicals. Species of this genus can prevent or eliminate fungal phytopathogens in agricultural crop systems by exerting plant colonization and fungal hyperparasitism. Extensive research has been developed on the biocontrol and biofertilizer activities of species of this genus in agricultural crops at field experimental scale, reporting biocontrol effectiveness from 35 to 100% for different crop-pathogen systems that include the following: *Rhizoctonia* (tomato), *Fusarium* (chickpea, green bean, tomato), *Botrytis* (cucumber), *Sclerotinia* (green bean, brinjal), and *Pythium* (green bean). Fungal inoculants most commonly employed at commercial scale are *Trichoderma harzianum*, *T. viride*, *T. asperellum*, and *T. atroviride*. In this review, a critical assessment was performed regarding the research carried out on the utilization of microbial species of the *Trichoderma* genus as a biological treatment in food crop production directed to plant-disease biocontrol, crop biofertilization, and abiotic stress modulation. In addition, the importance of the use of modern omics techniques on the systematics and taxonomy studies of crop-pathogen systems was addressed in order to build up a more comprehensive background to understand the function-

H. F. Ramírez-Cariño
Universidad de la Sierra Juárez, Ixtlán de Juárez, Oaxaca, Mexico

P. C. Guadarrama-Mendoza · R. Valadez-Blanco (✉)
Universidad Tecnológica de la Mixteca, Huajuapan de León, Oaxaca, Mexico
e-mail: rvaladez@mixteco.utm.mx

T. Romero-Cortes · J. A. Cuervo-Parra (✉)
Escuela Superior de Apan, Universidad Autónoma del Estado de Hidalgo, Apan, Hidalgo, Mexico
e-mail: alioscha@uaeh.edu.mx

© The Author(s), under exclusive license to Springer Nature Singapore Pte Ltd. 2025

K. A. Abd-Elsalam, A. H. Hashem (eds.), *Fungal Endophytes Volume II*,
https://doi.org/10.1007/978-981-97-8804-0_9

ality of the different species and strains in specific agricultural applications. Besides this, safety and production process considerations for inoculant product development are discussed as critical factors to foster the commercialization of *Trichoderma* inoculants and their exploitation in the agronomical sector. In conclusion, the prospects for *Trichoderma*-based inoculants as crop amendments are promising considering the multipurpose use of these fungal strains in agricultural systems and also the wide range of pathogenic biocontrol and their capability of the strains of this genus to establish symbiotic relationships with almost all kinds of plant crops, thus contributing to the development of sustainable agrosystems.

Keywords

Agricultural biological amendment · Commercial inoculants · Fungal phytopathogen biocontrol · Microbial biofertilizers · Plant growth promoter fungi · Sustainable agriculture · *Trichoderma* biological safety

9.1 Introduction

The Food and Agricultural Organization of the United Nations (FAO) has estimated a drastic increase in global human population in the coming years. To meet the food needs of the world population, farmers have implemented intensive agriculture practices. In this regard, the use of agrochemicals to combat pathogenic microorganisms (primarily bacteria and fungi) has increased, thus leading to pathogen resistance (Kubiak et al. 2023). For this reason, the implementation of novel food production models using environmentally friendly techniques has been proposed, in which beneficial microorganisms are employed as biocontrol agents (Adedayo and Babalola 2023). The incorporation of beneficial microorganisms, such as strains of the *Trichoderma* genus, in crop production, is an effective sustainable alternative to agrochemical fungicides. These strains mitigate phytopathogen damage and promote plant growth and development, thus improving crop production and satisfying food global demands without harming the environment.

Trichoderma is a genus of filamentous fungi that includes over 300 species. In general, these species have displayed an outstanding capacity to colonize the surface of plant's roots and also have been reported as endophytes within plant's root tissues, with some species present in the aerial parts of plants. *Trichoderma* species survive in soils across a broad diversity of environments, which is why they have a worldwide geographical distribution (Awad-Allah et al. 2023). In addition, these fungi have been characterized as very efficient spore producers, mainly green conidia, in different substrates, a trait that gives them an important advantage in their competition for space and substrate (Woo et al. 2023). Many *Trichoderma* species have been reported as beneficial microorganisms in the protection and nutrition of plants through various modes of action, which include competition for nutrients and niche, mycoparasitism, and production of lytic enzymes (Tyśkiewicz et al.

2022), as well as rapid root and plant colonization, synthesis of antipathogenic metabolites, and capacity of inducing systemic resistance in plants (Adedayo and Babalola 2023). In addition, it is well documented in the literature the capacity of *Trichoderma* strains as crop biofertilizer or plant-growth-promoting fungi (PGPF). Among the plant-growth-promoting mechanisms reported for these fungi are the following: improvement of nutrient availability in the plant, induction of seed germination, enhancement of tolerance to abiotic stress, and increase in crop yield (Japanis et al. 2022). Among the *Trichoderma* strains mostly reported as efficient biocontrol agents and biofertilizers in agricultural crop systems are *Trichoderma virens*, *T. viride*, *T. harzianum*, *T. koningii*, and *T. asperellum* (Awad-Allah et al. 2023).

Abeed et al. (2022) reported that a strain of *T. harzianum* showed resistance to Cd stress at different concentrations of the heavy metal. In addition, this treatment caused an increase in nutrient uptake (K, Ca, Mg, Fe, P, and nitrate) in sunflower plants (*Helianthus annuus* L.). In addition, Datta et al. (2023) reported that some *Trichoderma* isolates increased the height and tillering in rice plants (*Oryza sativa*) under drought conditions. The inoculant also decreased the time taken to panicle initiation and increased the number of total and filled grains in comparison to the non-treated control. In this review, we report the use of strains of the *Trichoderma* genus in agricultural systems, focusing mainly on their effects on biological control and plant biofertilization. In addition, we present the current state of the art regarding the application of *Trichoderma* strains in agriculture as beneficial microorganisms.

9.2 Direct and Indirect Biocontrol Mechanisms Used by the *Trichoderma* Genus Against Different Phytopathogens

9.2.1 Overview

Fungi of the *Trichoderma* genus are the most widely studied antagonistic microorganisms due to their various mechanisms, both direct and indirect, utilized by these strains to antagonize phytopathogenic microorganisms (Guzmán-Guzmán et al. 2023). The genus includes free-living species, which are common in soils and roots (Kumar et al. 2017) and are often found in habitats with high amounts of organic matter. The species of this genus play an important role as biological control agents by protecting crops against diseases caused by a large number of phytopathogenic fungi (Porteous-Álvarez et al. 2021). The capability of *Trichoderma* strains to act as antagonists of fungal phytopathogens can be related to both direct and indirect mechanisms. The direct biocontrol effects include the inactivation of enzymes of plant pathogens, competition for plant colonization, synthesis of antibiotic compounds, and enzymatic attack known as mycoparasitism. On the other hand, the main indirect mechanism by which *Trichoderma* fungi protect plants from pathogens is through the induction of the plant's defense systems. In addition, *Trichoderma* fungi can act as crop biofertilizers by different mechanisms: (i) stimulating the plant's immune systems and (ii) increasing the availability of nutrients to the plant (Guzmán-Guzmán et al. 2023). The endophyte fungi present in plant roots can protect the host plant against phytopathogens causing foliar diseases (Giodarno et al. 2023). The different antagonistic mechanisms of the *Trichoderma* strains can take place independently or in combination to combat phytopathogenic microorganisms in a particular plant-pathogen system (Mukhopadhyay and Kumar 2020). In practical applications, *Trichoderma*-based products for agricrop production can be formulated from spores and/or mycelia of the antagonistic fungi (Kumar et al. 2017).

Another benefit of biological control agents, compared to synthetic agrochemicals, is that they pose a low risk to induce resistance to the pathogenic strains due to the variety of their antagonistic mechanisms (Guzmán-Guzmán et al. 2023). In addition, biological control agents constitute an alternative for controlling diseases for which there are no chemical products available in the market (Kawasaki et al. 2021). Furthermore, due to the induction of pathogen resistance as a consequence of the irrational application of chemical fungicides, biological formulations based on *Trichoderma* species and their secondary metabolites constitute an alternative strategy to the use of agrochemicals, which is both environmentally friendly and relatively safe for human health (Khan et al. 2020; Mukhopadhyay and Kumar 2020), despite the pathogenicity in humans reported for some *Trichoderma* strains, an issue that will be discussed in more detail in this review. Among the most commercialized *Trichoderma* species are the following: *Trichoderma harzianum*, *T. koningii*, *T. asperellum*, *T. viride*, *T. longibrachiatum*, *T. pseudokoningii*, *T. hamatum*, *T. reesei*, and *T. virens*, which have been employed to combat both soil-borne and foliar plant pathogens (Mukhopadhyay and Kumar 2020; Cuervo-Parra et al. 2022).

9.2.2 Induction of Plant's Resistance Elicitors

Fungi of the *Trichoderma* genus are free-living microorganisms that commonly interact with the roots and leaves of plants (Tyśkiewicz et al. 2022), generating a wide variety of compounds that induce the activation of their defense systems against phytopathogens (Kumar et al. 2017). Derived from this interaction, plants increase their root system for nutrient and water assimilation, as well as their capacity for vegetative development; consequently, crop productivity becomes greatly improved (Fu et al. 2021). *Trichoderma* spp. establish mutualistic endophytic relationships with many plant species mainly through the colonization of the plant's root system, hence inducing systemic resistance in the plant and promoting plant growth (Dautt-Castro et al. 2024). The regulation of the plant-defense genes by *Trichoderma* fungi plays a critical role in the biological control of pathogenic agents. In this regard, Stempien et al. (2020) reported that a strain of *T. atroviride* overexpressed genes associated to pathogenesis-related proteins involved in plant defense through the synthesis of phytoalexins in the leaves of grapevine plants. Similarly, it has been reported that the ABC Taabc2 transporter gene of *T. atroviride* allows the fungus to effectively parasitize the microbes with which it interacts and also make the fungal strain resistant or tolerant to some chemical fungicides and toxins produced by pathogens (Ruocco et al. 2009).

The induction of plant defense systems by *Trichoderma* can be localized or systemic. It has been documented that *Trichoderma* cellulases improve plant-disease resistance; in a study, it was reported that a Thc6 protein regulated two genes, thph1 and thph2, encoding to the synthesis of cellulase enzymes, to stimulate induced systemic resistance (ISR) in the plant host. In addition, the gene Thph2 was associated to the effective colonization of corn (*Zea mays*) roots by a *T. harziaunum* strain. Besides this, the *Trichoderma* strain elicited the overexpression of the plant's defense genes AOS, LOX5, HPL, and OPR1 in the corn plant's leaves as a means of plant protection toward *Cochliobolus heterostrophus*, a fungal phytopathogen that causes the southern leaf blight disease in corn (Lang and Chen 2023). On the other hand, resistance of the antagonist strain to pathogenic attack is usually attained by an increase in the synthesis of defensive enzymes by the plant. In a study, the inoculation of sorghum plants (*Sorghum bicolor*) with a strain of *T. viride* produced a higher accumulation of antioxidant enzymes in comparison with the non-inoculated plants, which in turn contributed to the decrease in the severity of the bacterial stalk rot disease (*Dickeya dadantii*) in the plants (Yadav et al. 2023).

In addition, *Trichoderma* spp. are capable to trigger ISR in the colonized plants. In this regard, it has been reported that the enzyme ethylene-inducing xylanase (EIX), secreted by several *Trichoderma* species, acts as an ISR inducer in colonized plants. Particularly, the EIX enzyme induces the opening of cellular ion channels in host plants. In this regard, studies of the application of *T. harzianum* inoculants in tomato (*Solanum lycopersicum*) and tobacco (*Nicotiana tabacum*) crops in the presence of pathogenic fungal strains of *Botrytis cinerea* and *Sclerotinia sclerotiorum* reported that the response to EIX is controlled by the plant's glycoproteins LeEIX1 and LeEIX2, which are capable of binding to enzyme, thus modulating the defense system of the plant through an ISR mechanism (Leibman-Markus et al. 2021). Therefore, the induction of the plant's immune systems by *Trichoderma* inoculants also improves the capability of the host plants to defend effectively against the attack of pathogenic microorganisms.

9.2.3 Inactivation of Pathogen Attack Systems

Another mode of action used by *Trichoderma* to exert its biocontrol activity in crop-pathogen systems is the inactivation of the pathogen's attack systems, through the excretion of hydrolytic enzymes (proteases, chitinases, and glucanases) that hydrolyze the pathogenic enzymes responsible for degrading the plant's cell walls (Stange et al. 2023). In addition, many *Trichoderma* species are able to produce various defense mechanisms against phytopathogenic microorganisms, which include the release of lytic enzymes and functional peptides, such as xylanases, swollenins, peptaibols, and cerato-platanins (Shoresh et al. 2010).

In this regard, water activity is determinant of the various activities of *Trichoderma* strains toward phytopathogenic microorganisms, including spore germination, germ tube growth, formation of new mycelium, and enzyme production (Rimkus et al. 2023). For example, a higher water activity causes an optimal secretion of the lytic enzymes cellobiohydrolase and β-N-acetylglucosaminidase. In contrast, other enzymes such as glucosidases, xylosidases, and proteases exhibit their maximum activity at water activities lower than that reported for optimal growth (Mukhopadhyay and Kumar 2020).

Therefore, the success of the species of the *Trichoderma* genus as biocontrol agents against a wide variety of phytopathogenic fungi is due to their ability to combine the different direct and indirect antagonistic mechanisms; among which, the secretion of antimicrobial metabolites and hydrolytic enzymes constitutes the main mode of antipathogenic activity by *Trichoderma* strains (Cuervo-Parra et al. 2011; Cubilla-Ríos et al. 2019). On the other hand, Abdelrhim et al. (2024) reported that strains of the species *Trichoderma koningii* and *T. harzianum* demonstrated the ability to grow and sporulate in common bean (*Phaseolus vulgaris*) seedlings under saline conditions (80 mM NaCl). In addition, the strains showed no significant reduction in their capacity to produce hydrolytic enzymes (cellulase and chitinase), thus acting as antagonists against the pathogen *S. sclerotiorum*, a fungal species causing the damping-off and collar-stem rot in bean plants.

9.2.4 Competition for Plant Colonization

An additional mode of action used by *Trichoderma* strains is their competition for nutrients and space with phytopathogenic fungi by using carbon and nitrogen sources in a more efficient and rapid way, thus reducing pathogen population (Mukhopadhyay and Kumar 2020). The appropriate assimilation of iron by plant-associated fungi constitutes a critical factor for effective colonization of fungal phytopathogens in the soil and in the plant. It has been reported that many *Trichoderma* species are able to chelate iron, thus inhibiting the growth of pathogenic fungi. For example, *Trichoderma harzianum* and *T. viride* decreased the severity of the disease caused by *Fusarium oxysporum* in tomato plants through the production of siderophores (hydroxamate-type and carboxylate-type) and other mechanisms of biocontrol such as production of lytic enzymes, phosphate solubilization, and indole-3-acetic acid (IAA) production (Jamil 2021). Similarly, the rapid colonization of areas with necrosis, lesions, or non-infected parts of the plants by *Trichoderma* spp. prevents physical access and subsequent establishment of pathogenic fungi in the plant. In this regard, coating the crop seeds prior to their cultivation constitutes an efficient strategy to protect the seedlings from infection by phytopathogenic microorganisms during their early stages of development (Tyskiewicz et al. 2022).

On the other hand, several *Trichoderma* strains have shown resistance against metabolites released by phytopathogens, as well as to many herbicides and fungicides. In this regard, Limdolthamand et al. (2023) demonstrated the compatibility of *Trichoderma* spp. and certain fungicides such as azoxystrobin (1000 ppm) and difenoconazole + azoxystrobin (500 ppm) when used in combination as part of an integrated pest management treatment in corn plants infected with the northern leaf blight disease caused by the fungal pathogen *Exserohilum turcicum*.

9.2.5 Production of Antibiotic Compounds

In addition, fungal strains of the *Trichoderma* genus exert their biocontrol activity against phytopathogens by secreting low molecular weight antibiotic compounds, a mechanism known as antibiosis (Abdel-lateif 2017). These diffusible secondary metabolites with antimicrobial activity are synthetized and released by *Trichoderma* spp. into the environment when they interact with other foreign and potentially hostile microorganisms, such as the phytopathogen fungi (Kumar et al. 2017). A deep comprehension of the modes of action of different secondary metabolites produced by the fungi will favor the efficient use of antagonistic fungi against various plant pathogens (Khan et al. 2020). Among the secondary metabolites produced by *Trichoderma* species, volatile organic compounds (VOCs) are particularly important in the antibiosis mechanisms of these fungi. These compounds include terpenes (β-acoradiene, β-cubebene, β-cedrene, β-bisabolene), pyrones, diketopiperazines, aromatic compounds, and peptaibols (Jiménez-Bremont et al. 2024). *Trichoderma* species use these metabolites in different combinations depending on the phytopathogen target and the genotype of the crop plant. Schweiger et al. (2021) reported that the interaction of *Trichoderma virens* with corn roots induced the vir4 gene cluster involved in sesquiterpene production, thus improving plant's colonization by the inoculant; in contrast, the absence of the gen synthase vir4 decreased the rate of colonization of *T. virens*. Saadaoui et al. (2023) documented that 111 *Trichoderma* rhizospheric isolates from Florence Aurore wheat plants (*Triticum aestivum*) presented antagonistic activity against *F. culmorum*, a fungus causing the seedling blight disease. The antifungal activity was prompted by the segregation of hydrolytic enzymes as well as antibiotic organic compounds, both diffusible and volatile. Furthermore, it has been reported that *Trichoderma* species are capable of producing several antibiotic compounds simultaneously, thus efficiently exerting biocontrol activity while avoiding the induction of resistant phytopathogenic strains (Stracquadanio et al. 2020).

In this regard, epipolythiodioxopiperazines, characterized by a diketopiperazine ring, owe their toxicity to their disulfide bonds, which bind to thiol groups, thus generating reactive oxygen species that inactivate vital proteins of the pathogens (Gardiner et al. 2005). Similarly, gliotoxin selectively binds to thiol groups of the pathogen plasma membrane (Vargas et al. 2014; Guzmán-Guzmán et al. 2023) and trichorzianines modify the permeability of the pathogen's cellular membrane by forming ion channels (Miguel-Ferrer et al. 2021). These metabolites affect the homeostasis and the control of the transport of nutrients and waste products to and from the cells of the pathogenic fungi. In the same regard, gliotoxins produced by *T. virens* strains displayed antibiotic activity against various plant pathogens, such as *Rhizoctonia bataticola*, *R. solani*, *Macrophomina phaseolina*, *Pythium debaryanum*, *P. aphanidermatum*, and *Sclerotium rolfsii* (Rahman et al. 2023; Shavkiev et al. 2022). In a related study, *T. virens* was effective in the inhibition of the flower blight disease (*Neopestalotiopsis clavispora*) in flamingo flowers (*Anthurium andraeanum*) through the release of volatile organic compounds (VOCs) containing the sesquiterpene antifungal compound germacrene D. The authors reported an inhibition of 51.3% in pathogenic infection as compared to the non-inoculated plants (Athinuwat et al. 2024).

An important secondary metabolite produced by many *Trichoderma* species is pyrone 6-pentyl-2H-pyran-2-one (6-PP), which has a characteristic coconut aroma. This compound, a low molecular weight and volatile metabolite, possesses antifungal and biofertilizer activities (Vinale et al. 2008; Korpi et al. 2009). The 6-PP molecule produced by several strains of *T. harzianum* has shown antifungal effect in treating fungal diseases caused by *B. cinerea* and *Fusarium* spp. in grapevine plants (Mutawila et al. 2016). In other studies, several bioactive analogues of pyrone 6-PP have been reported to be secreted by strains of *T. viride*, *T. harzianum*, and *T. longibrachiatum*. These derivatives were effective against a number of pathogenic fungi such as *Candida albicans*, *Penicillium* spp., *Cryptococcus neoformans*, *Aspergillus fumigatus*, *S. rolfsii* and *Armillaria mellea* (Stracquadanio et al. 2020; Tarus et al. 2003).

In a similar study, two new secondary metabolites of the butenolides group, brasilane A and harzianone, were isolated from a strain of *T. lixii*. The metabolites exhibited good antifungal activity against *Gaeumannomyces graminis* (MIC value of 12.5 μg/mL) and *F. graminearum* (MIC value of 25.0 μg/mL) in wheat plants (Li et al. 2023). Guo et al. (2022) reported a secondary metabolite obtained from a strain of *T. harzianum*, harzianopyridone, which is composed of pyridine rings. The compound presented good antifungal activity against the pathogens *Pythium ultimum*, *G. graminis* (Vinale et al. 2006), as well as towards *R. solani* and *B. cinerea* (Guo et al. 2022). The compound also exhibited antibiotic effects against protists such as *Phytophthora cinnamomi* and pathogenic fungi such as *F. oxysporum*, *S. rolfsii*, and *Leptosphaeria maculans* (Ahluwalia et al. 2015; Vinale et al. 2009).

Azaphilones constitute another group of molecules with antipathogenic activity secreted by *T. harzianum* species, e.g., harziphilone and fleephilone (Khan et al. 2020). Vinale et al. (2006) reported that these secondary metabolites presented antagonist activity against various pathogenic fungi, such as *P. ultimum*, *G. graminis*, and *R. solani*. Additionally, some *Trichoderma* species, including *T. aureoviride*, *T. harzianum*, and *T. koningii*, can synthetize a series of compounds belonging to the group of polyketides, called koninginins (Li et al. 2019). These compounds have been shown to be effective in the control of several pathogenic fungi such as *G. graminis*, *F. oxysporum*, *F. solani*, *Bipolaris sorokiniana*, *P. cinnamomi*, *P middletonii*, and *Alternaria panax* (Khan et al. 2020; Li et al. 2019; Zhang et al. 2021).

Other groups of secondary metabolites that have been shown to be effective for the biocontrol of pathogenic microorganisms include steroids, anthraquinones, and lactones (Khan et al. 2020). Anthraquinones are synthetized by various *T. harzianum* strains, presenting good antifungal activity towards the phytopathogens *R. solani*, *S. rolfsii*, *M. phaseolina*, *F. oxysporum* (Ahluwalia et al. 2015), *G. graminis*, and *P. ultimum* (Vinale et al. 2006). Furthermore, it has been reported that these secondary metabolites can increase the efficiency of *Trichoderma* strains to enhance the resistance of the plant host to pathogenic infection (Liu et al. 2009). Similarly, lactones are compounds that have demonstrated not only good antifungal activity but also the capacity to promote seedling growth (Khan et al. 2020). In this regard, lactones isolated from *Trichoderma* species were effective antagonist of a number of phytopathogenic fungi; *T. cremeum* inoculant: *R. solani*, *B. cinerea*, and *F. oxysporum* (Vinale et al. 2016); *T. arundinaceum* inoculant: *B. cinerea* and *Fusarium sporotrichioides*; and *T. cerinum* inoculant: *Rosellinia necatrix* (Arjona-Girona et al. 2014; Malmierca et al. 2015). Besides this, several genera of fungi including *Trichothecium*, *Stachybotrys*, and *Trichoderma*, produce trichothecenes, compounds that have been reported to inhibit protein synthesis in pathogen fungi (Shentu et al. 2014), thus preventing the formation of peptide bonds in the 60S ribosomal subunit (Degenkolb et al. 2015). These secondary metabolites have been isolated from *T. brevicompactum* and *T. harzianum* cultures showing antibiotic activity against several fungal pathogens (Sha et al. 2013; Shentu et al. 2014).

Peptaibols are linear peptides synthetized by several fungi, including *Trichoderma* strains. These molecules have drawn a lot of attention since they possess great potential as agricultural amendments due to their antibiotic properties and their ability to induce the resistance of plants toward the infection of microbial pathogens (Khan et al. 2020). Khan et al. (2020) identified three peptaibols, tricokonins VI, VII, and VIII, synthetized by a strain of *T. koningii*. The authors reported that these compounds presented important antimicrobial activities against the plant pathogens *R. solani*, *F. oxysporum*, *Verticillium dahliae*, and *B. cinerea* (Khan et al. 2020). In the same regard, the peptaibol trichokonin VI isolated from *T. pseudokoningii* caused apoptosis in cells of the pathogenic fungi *Ascochyta citrullina*, *B. cinerea*, *F. oxysporum*, *Phytophthora parasitica*, and *V. dahliae* (Shi et al. 2012). On the other hand, the peptaibols produced by various *Trichoderma* species, including *T. asperellum* and *T. longibrachiatum*, stimulate the defense mechanisms of plants against several fungal phytopathogens such as *Fusarium* spp., *A. solani*, and *R. solani*. In the study, the concentration of peptaibols increased when the *Trichoderma* strains were co-inoculated with the pathogenic fungus (Tamandegani et al. 2020).

9.2.6 Mycoparasitism

Mycoparasitic activity (hyperparasitism) is a complex strategy used by antagonistic fungi when interacting with phytopathogenic microorganisms. The activation of the process occurs by signal recognition of the phytopathogenic fungus by the *Trichoderma* hyphae (Abdel-lateif 2017). Mycoparasitism consists of the direct attack of the pathogen's cellular structures (somatic and/or reproductive) by the antagonistic fungal strains, with the consequent destruction and utilization of their cellular components as a source of nutrients (Stracquadanio et al. 2020). In this regard, there are two mechanisms of action for fungal mycoparasitism, necrotrophic and biotrophic, which, in both cases, occur by enzymatic hydrolytic processes of the pathogen's cell walls. In this regard, enzymes such as β-glucanases, chitinases, lipases, proteases, cellulases, and antibiotics produced by the *Trichoderma* genus act on the cell walls of the phytopathogenic fungi, thus destroying the pathogens or preventing them from developing new cell walls, thus hampering their development (López-Barrera 2021).

In the case of necrotrophic parasitism, the infected cells are killed before or immediately after the infection of the *Trichoderma* inoculant, so that the latter assimilates the nutrients released by the antagonist's hydrolytic enzymes (Fig. 9.1). On the other hand, in the biotrophic parasitism, the nutrients required by *Trichoderma* are obtained from the living cells of the phytopathogenic fungus. Necrotrophic mycoparasites are very aggressive and can direct their antagonism toward a wide variety of potential hosts; on the contrary, biotrophic mycoparasites present a narrower spectrum and more specialized mechanisms of action, i.e., through the formation of specialized structures for nutrient uptake over the attacked hyphae (Stracquadanio et al. 2020; Mukhopadhyay and Kumar 2020).

It has been reported in the literature that the processes by which *Trichoderma* exerts mycoparasitism on the phytopathogenic fungi begin with the growth of its mycelium towards the pathogenic fungus. Ramírez-Cariño et al. (2020) reported the mycoparasitism of *T. asperelloides* against *Alternaria alternata* (Fig. 9.2) and *F. oxysporum* (Fig. 9.3), in which the inoculant decreased the disease incidence of these phytopathogens in tomato plants. In a similar study, Atanasova et al. (2023) studied the confrontation of the inoculant *T. atroviride* and the phytopathogen *R. solani*. The authors reported that after the two fungal strains are put in contact, cellular recognition of the pathogenic fungus by the *Trichoderma* strain takes place. After pathogen recognition, the fungal antagonist directs its growth toward the host's hyphae leading to its subsequent cell attachment. Following this, *Trichoderma* hyphae coil around the pathogen's hyphae, thus forming mycoparasitic structures called appressoria (Mukhopadhyay and Kumar 2020). Atanasova et al. (2023) reported the overexpression of hydrolytic enzymes in the antagonist cells, allowing the *Trichoderma* strain to destroy the pathogen's cell wall and use its cellular contents.

According to the above, the genes that are activated or inactivated in the different species of *Trichoderma* play a fundamental role during the mycoparasitism process. In this regard, in a study conducted with a strain of *T. atroviride* confronted with

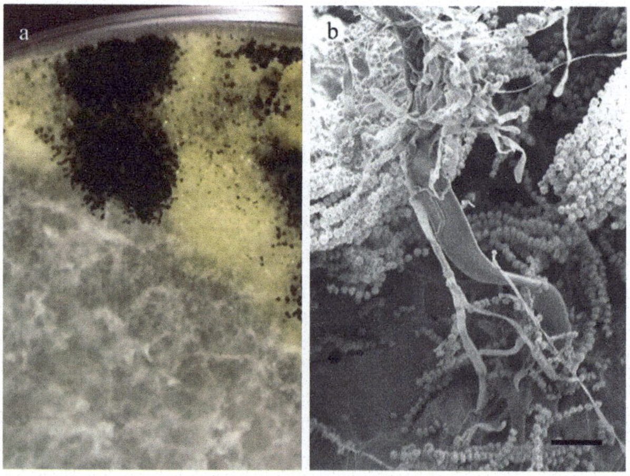

Fig. 9.1 Antagonistic interaction between *Trichoderma asperellum* and *Aspergillus niger*. (**a**) *Trichoderma-Aspergillus* confrontation on PDA medium. (**b**) Scanning electron microscopy micrograph of the mycoparasitic process by the *Trichoderma* hyphae enveloping the pathogen's conidiophore through appressoria (**b**). Bar, 20 μm

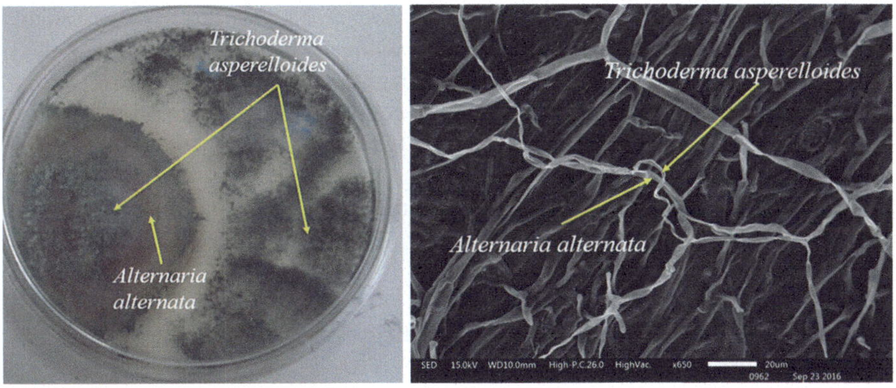

Fig. 9.2 Antagonistic interaction between *Trichoderma asperelloides* and *Alternaria alternata*. Left: *Trichoderma-Alternaria* confrontation on PDA medium showing the direct antagonism of *T. asperelloides*. Right: Scanning electron microscopy micrograph by hyphal penetration into *A. alternata* at 5 days of growth on PDA plates

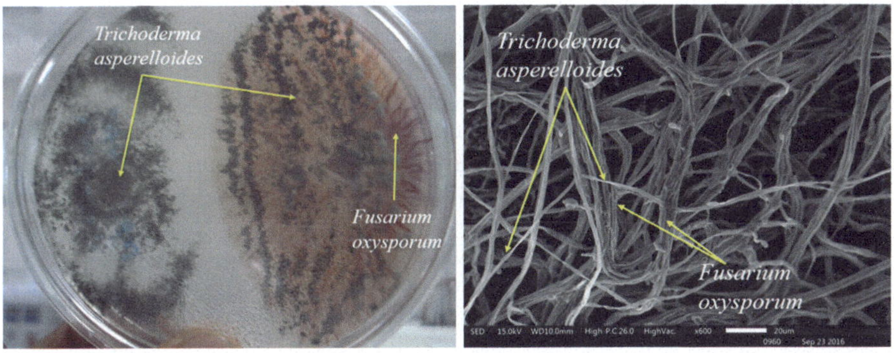

Fig. 9.3 Antagonistic interaction between *Trichoderma asperelloides* and *Fusarium oxysporum*. Left: *Trichoderma-Fusarium* confrontation on PDA medium showing the direct antagonism of *T. asperelloides*. Right: Scanning electron microscopy micrograph showing the direct antagonism of *T. asperelloides* by overgrowth in *F. oxysporum* at 5 days of growth on PDA plates

P. ultimum and *R. solani* in tomato plants (Ruocco et al. 2009), it was observed that the Taabc2 gene, related to resistance mechanisms and cellular detoxification, increased its expression in the presence of metabolites secreted by the phytopathogens; i.e., their mycotoxins and/or some fungicides. Therefore, this gene appears to be a key factor in the antagonism exerted by *Trichoderma* strains against phytopathogens. In this regard, in previous works, it was reported the protective role of a protein attached to the *Trichoderma* cell wall. The synthesis of this protein is mediated by the gene Qid74, which is responsible for protecting *Trichoderma* cells from enzymatic attack and other toxins (Rosado et al. 2007).

Through mycoparasitism, fungi of the *Trichoderma* genus destroy the somatic and reproductive structures of the parasitized pathogens within a period of 4–6 days (Chen et al. 2016; Cuervo-Parra et al. 2022). The *Trichoderma* hyphae grow parallel to the pathogen's hyphae, with the subsequent coiling and penetration of the hyphae onto the pathogen (Ramírez-Cariño et al. 2020). For this reason, it is recommended to apply this group of antagonistic microorganisms as part of a comprehensive chemical-biological plan for the control of pathogenic microorganisms in different crops (Guzmán-Guzmán et al. 2023).

9.2.7 Biocontrol of Fungal Phytopathogens by *Trichoderma* Strains in Field Conditions

Current agricultural practices, based on the production of single crops of economic interest for humans, have caused the loss of biodiversity of the flora in the localities designated for commercial crops. This situation consequently generates an ecological imbalance, which leaves crop agricultural production systems prone to the attack by pathogenic macro- and microorganisms such as bacteria, nematodes, and fungi. These pathogens are responsible for causing considerable losses to crops of economic importance (Mukhopadhyay and Kumar 2020; Hajieghrari et al. 2008). Among these microbes, fungi constitute the main group of phytopathogens that cause diseases in crops, leading to considerable economic losses (Khan et al. 2020). To address this problem, chemical synthetic products have traditionally been used in agriculture, which, despite being effective in the short term, cause negative effects on the agricultural soils and crops, as well as on the environment, and ultimately on human health (Jáquez-Matas et al. 2022; Romero-Cortés et al. 2022). These products are cheap and effective; however, their chemical composition makes them toxic pollutants due to the fact that their prolonged presence in the ecosystem generates adverse environmental effects (Romero-Cortés et al. 2022). In addition, these compounds may induce pathogen's resistance to fungicides, due to the inadequate or excessive doses of these agrochemicals used for the control of plant diseases (Cuervo-Parra et al. 2016).

The problems caused by agrochemicals have generated interest in finding alternative options to control phytopathogens that do not harm the environment and humans. The use of biological control agents from the *Trichoderma* genus, as well as its genes or gene products derived from genetic modification (Saba 2012), offers a safe and eco-friendly alternative for controlling diseases caused by phytopathogenic fungi without using, or by reducing, the utilization of synthetic agrochemicals. In general, treatments based on *Trichoderma* are considered innocuous for the people who apply them and constitute a friendly and safe alternative for the environment since they do not contaminate or generate resistance in the pathogenic microorganisms they control (Kumar et al. 2017). Table 9.1 enlists some *Trichoderma* species that have been tested for the biocontrol of phytopathogenic species and the corresponding crop product of the agricultural system.

Table 9.1 *Trichoderma* species most commonly used to control plant diseases

Trichoderma spp.	Crop	Phytopathogens	References
T. koningii	Bean (*Phaseolus vulgaris*)	*Sclerotinia sclerotiorum*	Abdelrhim et al. (2024)
T. longibrachiatum, T. virens, and T. harzianum	Olive (*Olea europaea*)	*Verticilium dahlie*	Carrasco et al. (2023)
Trichoderma spp.	Soyabean (*Glycine max*)	*Macrophomina phaseolina*	Larran et al. (2023)
T. harzianum, T. koningiopsis,	Chilli (*Capsicum annum*)	*Fusarium solani*	Miguel-Ferrer et al. (2021)
T. agriamazonicum	Chili (*Capsicum chinense*)	*Corynespora cassiicola, Colletotrichum* spp., *Colletotrichum scovillei*	Sousa et al. (2023)
T. asperellum	Corn (*Zea mays*)	*Fusarium verticillioides, Ustilago maydis*	Cuervo-Parra et al. (2022)
T. virens	Eggplant (*Solanum melongena*)	*Verticilium dahliae*	Tomah et al. (2023)
T. afroharzianum	Tomato (*Solanum lycopersicum*)	*Alternaria alternata*	Philip et al. (2024)
T. asperellum	Castor (*Ricinus communis*)	*Fusarium oxysporum f. sp. ricini*	Pradhan et al. (2023)

Since the species of the *Trichoderma* genus are commonly found in all types of climates (temperate and tropical) and soils (forest, agricultural, prairie, marsh, or desert), it is possible to find strains with optimal adaptation traits to the specific climatic conditions of the area where the crop of interest is cultivated (Kumar et al. 2017). *Trichoderma* species regularly exhibit an accelerated growth that allows them to make a more efficient use of resources and space than the pathogens, thus outcompeting the latter in the plant colonizing process. In addition, *Trichoderma* strains can release its repertoire of biocontrol mechanisms when pathogens are present at the colonization sites. Pathogen inactivation is indicated by the vacuolization and denaturation of the newest hyphae of the mycelium, with the consequent disintegration of the pathogen hyphae (Hajieghrari et al. 2008; Jain and Das 2020).

The success of inoculant products based on *Trichoderma* spores as pathogen antagonistic agents relies on the following factors: (i) ability to survive in unfavorable environmental conditions, (ii) efficiency of nutrient utilization, (iii) capacity to adapt to and modify the rhizosphere, (iv) capacity for fast colonization and propagation, and (v) stimulation of plant defense and growth mechanisms (Kumar et al. 2017). A detailed knowledge of the mechanisms of biocontrol and biofertilization used by these strains is essential to increase the effectiveness of the use of this group of antagonistic fungi against a wide spectrum of phytopathogenic microorganisms in different agricultural crop systems (Tyśkiewicz et al. 2022).

9.3 Biofertilizing Properties of *Trichoderma* Species

The use of the *Trichoderma* species has been shown to increase the productivity and quality of vegetables, grains, and cereals. The biofertilizing effect of *Trichoderma* species is related to various mechanisms, among which the most important are the increase in nutrient uptake and the induction of defense hormone synthesis in plants. Awad-Allah et al. (2022) observed that two *Trichoderma* strains (*Trichoderma harzianum* and *T. viridae*, alone or in combination) had positive effects in cherry tomato plants grown in soil infected with the phytopathogen *Fusarium solani* or not. In the study, increases in the weight and root length of the treated plants were reported in comparison with the uninoculated treatments. The use of *Trichoderma* strains increased nutrient concentration (N, P, K, Mg, Ca, Fe, Zn, and Cu), chlorophyll content, and phenol content in leaves of cherry tomato plants cultivated under greenhouse conditions as compared with the control plants without inoculation. Similarly, Sanó et al. (2022) reported that the use of *T. longibrachiatum* increased the pseudostem diameter, number of leaves, and total area as well as the concentrations of Mn, S, and Mg in banana seedlings in comparison with the non-treated control. In another study, Sehim et al. (2023) concluded that *T. asperellum* increased the root length and fresh weight of tomato plant shoots under *Fusarium* infection compared to the non-inoculated treatment. Similarly, Singh et al. (2018) investigated the effects of the inoculation of tobacco plants with *T. asperellum*, which resulted in increased biomass, nitrogen content, nitric oxide production, and cytosolic Ca^{2+} concentration. Furthermore, the lengths of lateral roots and root hairs were greater when the fungal inoculant was supplemented with NO^{3-} and NH^{4+}, as compared to the untreated control.

9.3.1 Solubilization and Nutrient Uptake

It has been well documented in the literature the positive role of *Trichoderma* species toward the improvement of the health and development of agricultural crops by promoting plant growth and enhancing nutrient uptake and its optimal assimilation (Awad-Allah et al. 2023). Bedine et al. (2022) demonstrated that six *Trichoderma* strains solubilized tricalcium phosphate with values ranging from 66.9 to 328.4 µg/mL. The authors reported that *T. gamsii* inoculation significantly improved P uptake and increased plant height by 21.1%; shoot fresh and dry weight by 52.5 and 71.4%, respectively; root length by 21.7%; and root fresh and dry weight by 50 and 66.8%, respectively, in common bean (*Phaseolus vulgaris*) seedlings in comparison to the non-treated control. Similarly, Paul and Rakshit (2021) reported that a *T. viride* strain increased P uptake and root length in three different soybean (*Glycine max*) growth stages with varying fertilizer P doses. In addition, the seed primed with *T. viride* reduced the fertilizer P doses by 20%. Zúñiga-Silgado et al. (2022) reported that isolates of *T. virens* and *Aspergillus tubingensis*, either alone or in combination, increased the solubilization, mineralization, and uptake of P in pepper (*Capsicum annuum*) plants as compared to non-treated plants, thus improving plant growth of the inoculated treatments. The study showed that the combined treatment was more effective than the individual treatments of the inoculants. According to the authors, *T. virens* was very efficient for P mineralization, while *A. tubingensis* was effective for its solubilization.

9.3.2 Phytohormone Production

Phytohormones, such as indole acetic acid (IAA), gibberellic acid (GA), and abscisic acid (ABA), are important for the growth and development of crops. In a study, Zhao and Zhang (2015) observed that a *T. asperellum* strain had the capacity to produce phytohormones and solubilize phosphates. The inoculant increased the concentration of IAA, GA, and ABA in the leaves of cucumber seedlings cultivated under salt stress in comparison to the non-inoculated control. Furthermore, the inoculant enhanced root growth and nutrient uptake in the seedlings due to the increase in phytohormone concentration and phosphate solubilization. Saadaoui et al. (2023) reported that *Trichoderma* strains exhibited biofertilizer traits in wheat (*Triticum aestivum*) cultivation such as ammonia and indole-like compound synthesis. The authors reported that the fungal inoculants increased chlorophyll and nitrogen contents in the wheat seedlings as compared to the non-inoculated control, which in turn caused an increase in plant biomass. In a similar study, Mendes et al. (2020) evaluated the biofertilizer effect of nine strains of *Trichoderma* (six *T. asperelloides* strains, one *T. asperellum* strain, and two *T. harzianum* strains) in combination with a strain of *Bradyrhizobium* sp. on cowpea (*Vigna unguiculata*) plants cultivated under greenhouse conditions. The authors reported that all *Trichoderma* isolates solubilized phosphorus and produced IAA. Particularly, the combined treatment of *T. asperelloides* strains with *Bradyrhizobium* sp. increased plant height, stem diameter, shoot dry weight, total dry weight, and root length in comparison to the other treatments and to the non-inoculated control.

In another study, Machado-Rosa et al. (2023) reported that two *Trichoderma* sp. strains increased the root and shoot mass and number of root tips in tomato plants grown under greenhouse conditions, as compared to the non-inoculated control. The authors inferred that the biofertilizing effect was due to the strain's outstanding capability to synthetize IAA. Similarly, Sehim et al. (2023) found that a *T. asperellum* strain caused an increase in the number of leaves, as well as shoot and root length, and weight of tomato plants in comparison to the non-inoculated control. The biofertilization effect of the fungal strain was attributed to IAA production and phosphate solubilization. In a study by Pedrero-Méndez et al. (2021), the biofertilization effects of four *Trichoderma* strains (two *T. harzianum*, one *T. simmonsii*, and one *T. afroharzianum*) were evaluated in wheat plants. The results showed that *T. simmonsii* had a positive effect in plant height and spike weight under water stress conditions, which was correlated with an increase in the production of IAA and ABA, as well as in the enzymatic activity of 1-aminocyclopropane-1-carboxylic acid deaminase (ACC deaminase), in wheat plants as compared to the non-treated control. The most common biofertilizer mechanisms reported for *Trichoderma* sp. inoculants in crop nutrition and protection are shown in Table 9.2.

Table 9.2 Biofertilizing and biocontrol effects and mechanisms of *Trichoderma* species inoculated in crop plants

Trichoderma species	Crop	Biofertilizing trait	References
T. hamatum	Tomato (*Solanum lycopersicum*)	Increased shoot and root length and chlorophyll content Induced systemic resistance against tobacco mosaic virus	Abdelkhalek et al. (2022)
Trichoderma spp.	Tomato	Increased macro/micronutrients uptake, pigment content, and total phenol. Suppressed *Fusarium* wilt disease	Awad-Allah et al. (2022)
Trichoderma sp.	Bean (*Phaseolus vulgaris*)	Increased plant height, shoot and root weight, phosphorus uptake, pigments, and total protein	Bedine et al. (2022)
T. harzianum	Pepper (*Capsicum annuum*)	Increased root's fresh and dry weight, P solubilization, and proline content	Cañada-Coyote et al. (2021)
T. longibrachiatum and *T. asperellum*	Banana (*Musa* spp.)	Promoted pseudostem diameter, increased the number of leaves and total leaf area, and enhanced fresh and dry mass	Sanó et al. (2022)
T. viride	Soybean (*Glycine max*)	Increased phosphatase activity, dehydrogenase activity, and P content in vegetative parts	Paul and Rakshit (2021)
T. asperellum	Tomato	Stimulated the plant immune system. Enhanced shoot and root length; shoot and root dry weight and number of leaves, and increased P solubilization and IAA[a] production	Sehim et al. (2023)
T. virens and *Aspergillus tubingensis*	Pepper	Improved germination percentage, P content in leaves, root elongation, and plant weight and height	Zúñiga-Silgado et al. (2022)
Trichoderma spp.	Tomato	Increased IAA production, root and shoot weight, and number of root tips	Machado-Rosa et al. (2023)
T. harzianum and *T. simmonsii*	Wheat (*Triticum aestivum*)	Increased plant height and spike weight under water stress, as well as IAA and ABA[b] contents. Increased in ACC deaminase[c] activity	Pedrero-Méndez et al. (2021)

[a]IAA, indole acetic acid
[b]ABA, abscisic acid
[c]ACC deaminase, 1-aminocyclopropane-1-carboxylic acid deaminase

9.3.3 Accumulation of Antioxidant Substances in the Plants and Fruits

The effect of *Trichoderma* strains on the synthesis of antioxidant molecules has been documented in the literature. Abdelrhim et al. (2023) reported that a *Trichoderma viride* strain increased the resistance-related enzymes and compounds when tested in coriander (*Coriandrum sativum* L.) treated with the damping-off and root rot disease caused by *Rhizoctonia solani*. The study attested that *T. viride* stimulated the synthesis of salicylic acid, phenolics, and antioxidant enzymes, which enhanced the resistance of coriander plants to *R. solani*. Similarly, Aseel et al. (2023) demonstrated that inoculating potato plants with a *T. viride* strain increased the levels of peroxidase, polyphenol oxidase, total proteins, and chlorophyll, which in turn enhanced the systemic resistance of potato plants against the *Potato virus* Y (PVY). The authors reported that the potato plants inoculated with the fungal strain showed a high expression of the defense-related genes (PR-1, POD, PAL, CHS, and HQT), that encode for antioxidant enzymes, in comparison with those of the non-inoculated plants.

In related work, the use of the combined inoculants *T. viride* and *B. subtilis* increased the antioxidant enzymatic activity (superoxide dismutase and catalase) in strawberry plants infected with *R. fragariae*, a pathogenic fungal strain causing the black root rot disease (Abied et al. 2023). The authors reported that *T. viride* individually or in combination with *B. subtilis* increased the shoot and root growth of the strawberry plants under greenhouse and field conditions. The inoculants increased the number of fruits by two- and four-fold in plants under greenhouse and field conditions, respectively, in comparison to the non-inoculated control. Similarly, Behiry et al. (2023) reported that the inoculation of *T. pubescens* increased the contents of chlorophyll and total phenolic compounds in tomato plants infected with *R. solani*, which causes the root rot, as compared to the non-inoculated plants. In the study, it was reported that *T. pubescens* increased the expression levels of defense-related genes (PAL, CHS, and HQT) that encoded for antioxidant enzymes in the inoculated plants. In a similar study, *T. asperellum* was shown to increase the chlorophyll and carotenoid contents by 130 and 40%, respectively, in white onion plants in comparison with the control plants (Rodríguez-Hernández et al. 2023). The authors noted that the fungal strain not only enhanced plant growth but also increased the antioxidant activity by 15–70% and the phenolic compound content compared with non-inoculated plants. The authors attributed this effect primarily to the enhanced photosynthetic activity and increased protein and macronutrient content resulting from the use of the inoculant.

Applications of *Trichoderma* strains in combination with another microbial inoculant or with the addition of biological amendments or chemical compounds have been evaluated in the literature. Gulzar et al. (2023) reported that the use of a combination of *T. harzianum* with melatonin in the cultivation of pak choi (*Brassica campestris* sub sp. *chinensis*) crop increased plant growth and the activity of antioxidant enzymes such as ascorbic acid peroxidase (APX), superoxide dismutase (SOD), catalase (CAT), glutathione reductase (GR), and peroxidase (POX) in comparison with the control treatment without inoculation. The fungal treatment also reduced the levels of malondialdehyde (MDA) and hydrogen peroxide on pak choi plants affected by clubroot disease caused by *Plasmodiophora brassicae*. As a result of the study, the disease was controlled by 74.8% in treated plants in comparison with the untreated plants. In a similar study, Shalaby et al. (2022) demonstrated that the use of *T. harzianum* either individually or in combination with paclobutrazol (a fungicide) increased the chlorophyll content, antioxidant enzyme activity, and total antioxidant content in tomato seedlings infected with damping-off disease caused by *Rhizoctonia solani*, in comparison to the non-treated seedlings. The combined treatment of the *Trichoderma* strain and the fungicide resulted in the highest reduction in the disease severity over two seasons, with 12.2 and 10.0%, respectively, with respect to the non-inoculated control, which had severities of 85.4 and 90.23%, respectively. Additionally, the combined treatment improved growth parameters; with the chlorophyll content increased by 38 and 37%, and stem diameter by 26 and 24% for each season respectively, in comparison to the untreated control.

9.4 Enhancing the Plant's Tolerance to Abiotic Stress

Several studies have been conducted to assess the effect of *Trichoderma* species on the alleviation of abiotic stress in crop plants, such as drought, temperature, salinity, and the presence of heavy metals. The main studies concerning these effects, along with the reported mechanisms of action, are listed in Table 9.3.

9.4.1 Drought Stress

Beneficial *Trichoderma* species have been used as a strategy for reducing the negative effects caused by drought stress on plant growth, which is one of the major abiotic stress factors that generates severe losses in crop production in modern agriculture. In this regard, Tripathi et al. (2021) studied the effect of the application of *Trichoderma koningii* on tomato plants grown under thermal stress conditions. The authors evaluated four treatments: T1 (untreated and unstressed seeds), T2 (untreated seeds + heat shock), T3 (*T. koningii* treated and unstressed seeds), and T4 (*T. koningii* treated + heat shock). Eight seeds were sown in each pot, and five pots were used for each treatment. The plants were maintained in greenhouse conditions for 30 days. For the induction of heat stress, plants were grown inside a plant growth chamber and then subjected to a heat shock at 45 °C for 5 h to study the capability of the *Trichoderma* strain to alleviate the thermal effects on the plants. Five tomato plants were randomly uprooted from each treatment for biochemical analysis. The authors reported that thermal stress elevated the activities of superoxide dismutase (SOD), peroxidase (POD), polyphenol oxidase (PPO), phenylalanine ammonia-lyase (PAL), and proline. However, these activities were reduced by the application of *T. koningii* in the stressed seedlings. Additionally, the inoculant increased the

Table 9.3 Beneficial traits and mechanisms of *Trichoderma* species involved plant resistance to abiotic stress

Trichoderma species	Crop	Abiotic stress	Enhancement of abiotic stress tolerance	References
T. longibrachiatum	Mason pine (*Pinus massoniana*)	Drought	Promoted absorption of nitrogen, phosphorus, and potassium. Alleviated damage to cell membranes, increased antioxidant enzymatic activity and photosynthesis, and raised the content of osmotic substances	Yu et al. (2023)
T. asperelloides	Tomato (*Solanum lycopersicum*)	Drought	Increased shoot weight, reduced water stress	Rawal et al. (2022)
T. asperellum	Wheat (*Triticum aestivum*)	Drought	Modulated the expression of ABA[a]-dependent water stress markers, increased plant SOD activity, and promoted tolerance to drought	Illescas et al. (2022)
T. asperellum and *Bacillus amyloliquefaciens*	Soybean (*Glycine max*)	Drought	Decreased the harmful effects of water and salt stress. Maintained midday water potential similar to that of the control, increased photosynthetic rate, and maintained stomatal conductance	da Silva et al. (2023)
T. harzianum and *Fusarium oxysporum*	Corn (*Zea mays*)	Drought	Increased fresh and dry weight, root and shoot ratio, leaf area, relative water content, and membrane stability index. Increased total phenolic, proline, and soluble sugar contents. Improved pigments and gas exchange	Bakhshi et al. (2023)
T. longibrachiatum	Cowpea (*Vigna unguiculata*)	Salt	Reduced ROS accumulation in tissues, improved the K^+/Na^+ ratio, alleviated ion toxicity from salts, and enhanced root colonization	Liu et al. (2023)
T. asperellum	Corn	Salt	Increased the abundance of beneficial microbes, mitigated microbial community degradation, and enhanced crop yield	Fu et al. (2020)

(continued)

Table 9.3 (continued)

Trichoderma species	Crop	Abiotic stress	Enhancement of abiotic stress tolerance	References
T. harzianum	Corn and rice (*Oryza sativa*)	Salt	Increased relative water content, stomatal conductance, photosynthetic activity, and pigment contents. Improved CAT[b] and SOD[c] activity. Increased proline content	Yasmeen and Siddiqui (2017)
T. viride	Tomato	Salt	Increased proline content, total antioxidant capacity, total phenolics, and flavonoids. Decreased malondialdehyde (MDA) and hydrogen peroxide content	Metwally and Soliman (2023)
T. koningii	Tomato	Thermal stress	Decreased SOD[c], POD[d], and PPO[e] activities. Increased starch, protein, and total phenol content. Decreased hydrogen peroxide and lignin content	Tripathi et al. (2021)
T. asperellum	Corn	Heavy metal	Increased height, dry weight, root length, chlorophyll content, and antioxidant enzymatic activity in plants under Cd stress	He et al. (2021)

[a]ABA, abscisic acid
[b]CAT, catalase
[c]SOD, superoxide dismutase
[d]POD, peroxidase
[e]PPO, polyphenol oxidase

contents of starch, protein, and total phenols in the treated plants, in comparison to the non-inoculated control plants. Rawal et al. (2022) reported the application of *T. asperelloides* in the cultivation of two different tomato genotypes susceptible to drought. The authors observed an increase in shoot weight (from 30 to 37%) in the treated tomato plants cultivated under water-deficit conditions with respect to the non-inoculated control. In a similar study, the application of *T. longibrachiatum* in pine (*Pinus massoniana*) cultivation promoted plant growth and increased the absorption of nitrogen, phosphorus, and potassium in the pine seedlings cultivated under drought stress conditions in comparison to the non-inoculated control (Yu et al. 2023). Additionally, in the pine seedlings subjected to drought conditions, the inoculant reduced cell membrane damage, increased the activity of antioxidant enzymes and photosynthesis, and enhanced the contents of osmolytes. According to the authors, *T. longibrachiatum* regulated the soil microbial community by increasing the abundance of dominant taxa.

In another work, Illescas et al. (2022) studied the effect of *T. asperellum* on wheat (*Triticum aestivum*) seedlings cultivated under water stress conditions. The application of the fungal strain decreased the expression of water stress response genes, the antioxidant activity, and the contents of proline, hydrogen peroxide, and malondialdehyde in wheat plants cultivated under severe drought stress, in comparison with the non-inoculated control. Additionally, *T. asperellum* increased the superoxide dismutase activity and elicited the expression of a plant abscisic-acid (ABA)-dependent water-stress modulators in wheat seedlings under moderate drought conditions, with respect to the untreated plants. In a similar study, da Silva et al. (2023) evaluated the effects of *T. asperellum* and *Bacillus amyloliquefaciens* applied individually and in combination on soybean (*Glycine max*) under drought and saline stress. The inoculants attenuated the negative effects of drought and salt conditions, maintaining a midday water potential similar to that of the control plants without stress. Moreover, the treated plants maintained a higher photosynthetic rate without altering stomatal conductance, which prevented tissue dehydration and increased carbohydrate concentration in inoculated soybean plants as compared to the non-inoculated control. In another report, *T. harzianum* increased biomass, leaf area, relative water content, and cell membrane stability in corn plants (Bakhshi et al. 2023). Additionally, the application of the combined inoculants increased total phenolic, proline, and soluble sugar contents by 52.0, 40.3, and 55.6%, respectively, in comparison with the non-inoculated control. The authors also reported that photosynthetic parameters such as chlorophyll and carotenoids as well as gas exchange parameters were enhanced with the application of the inoculants.

9.4.2 Salt Stress

The positive impact of microbes on the mitigation of the harmful effects of saline soils has been widely documented in current literature. Fu et al. (2021) reported that the application of *Trichoderma asperellum* reduced soil pH by 2.2–5.8% and sodium adsorption by 22.7–54.1% in the rhizosphere of corn (*Zea mays*) seedlings cultivated under saline conditions. Consequently, the soil's nutrient content and enzymatic activity of the plants increased, thus improving the growth of the seedlings as compared to the non-inoculated control. Additionally, the inoculant increased the levels of osmo-regulating compounds, oxidized glutathione (43.8–88.2%), and oxidized ascorbate (25.3–222.3%), which, in turn, enhanced the scavenging of reactive oxygen species (ROS) in the inoculated seedlings. Liu et al. (2023) studied the effect of inoculating cowpea (*Vigna unguiculata*) plants with a *T. longibrachiatum* strain under salt-stress conditions. The authors reported that the inoculant increased the plant's salt tolerance by reducing the accumulation of ROS in tissues, thereby mitigating the negative effects of salt stress in comparison to the non-inoculated control. Additionally, the application of the fungal inoculant increased the K^+/Na^+ ratio, thus alleviating the effects of salt toxicity and preventing infection of cowpea plants by *Fusarium* wilt when cultivated under saline conditions in comparison to the control plants without inoculation. In another study, it was investigated the effect of an arbuscular mycorrhizal fungus (AMF) and *T. koningii* on basil plants (*Ocimum basilicum*) grown under saline conditions (Saia et al. 2021). The authors reported that, irrespective of the salinity stress level, the inoculants benefited plant growth by increasing the number and area of leaves, as well as the contents of Ca, Mg, B, and p-coumaric and chicoric acids in the plants, in comparison to the non-inoculated control. Under low salt stress, the inoculants stimulated the plant's photosynthetic activity, increased the uptake of Fe and Mn by the plants, and raised the contents of caffeic and rosmarinic acids with respect to the untreated plants.

Yasmeen and Siddiqui (2017) studied the beneficial effects of *T. harzianum* on corn and wheat plants cultivated under three saline concentrations (50, 100, and 150 mM NaCl). The authors reported that the fungal inoculant improved plant growth and induced physiological adjustments in comparison to the untreated plants. The inoculated treatments exhibited higher relative water content, stomatal conductance, photosynthetic activity, and pigment contents as compared to the non-treated plants. In addition, *T. harzianum* increased the activity of catalase (CAT) and SOD antioxidant enzymes, decreased the content of hydrogen peroxide, and increased the proline content with respect to the control treatment. In another study, the application of a strain of *T. viride* ameliorated the negative effects caused by saline conditions in tomato (*Solanum lycopersicum*) plants (Metwally and Soliman 2023). The authors reported that the inoculant increased proline content and antioxidant capacity in tomato seedlings in comparison to the non-inoculated control. Additionally, the fungal strain increased the activity of the antioxidant enzymes CAT, PPO, and ascorbate peroxidase (APX) (74.6, 58.5, and 61.6%, respectively) in tomato plants under salt-stress conditions (50 nM NaCl) with respect to the non-treated plants. At higher salt concentrations (100 mM NaCl), the application of the *T. viride* strain increased the concentrations of total phenolics (17.8%) and flavonoids (33.2%) in the tomato plants in comparison to the non-inoculated seedlings.

9.4.3 Heavy Metal Stress

Exposure of plants to heavy metals such as cadmium (Cd) can cause them severe stress, reducing their growth and potentially leading to the plant's death. In addition to their biofertilizing capabilities, many microorganisms serve as biological agents to mitigate the toxic effects of heavy metals. For instance, in a study with sunflower plants (*Helianthus annuus* L.), Abeed et al. (2022) reported that a strain of *T. harzianum* exhibited a cadmium biosorption capacity of 81.3% at 150 mg Cd/kg soil. The inoculant decreased the bioavailability and, therefore, the toxic effects of Cd in the plants by its accumulation and/or immobilization in the fungal mycelia. According to the authors, the inoculation with *T. harzianum* increased the plant's nutrient absorption under high Cd concentrations, thus enhancing their tolerance to Cd in comparison with the non-inoculated control.

In a similar study, a *T. asperellum* strain was inoculated on corn seedlings grown under Cd stress. The application of the inoculant improved the agronomic parameters of the seedlings, such as plant height, fresh and dry weight, root length, and chlorophyll content, as well as the activity of antioxidant enzymes in comparison with the non-inoculated plants. In addition, the *T. asperellum* strain enhanced the expression of the enzyme-protective related genes ZmWRKY4 (WRKY-transcription factor 4), ZmSOD4 (superoxide dismutase 4), and ZmCAPX (ascorbate peroxidase), with ZmWRKY4 in the roots and leaves, and ZmSOD4 and ZmAPX in response to Cd stress (He et al. 2021). In another study, the application of strains of *Trichoderma* sp. and *Pseudomonas fluorescens*, individually or in combination, to chickpea (*Cicer arietinum*) plants under different Cd concentrations improved the agronomic parameters of the treated plants with respect to the control with increases in root dry biomass, vigor index, as well as in chlorophyll and carotenoid contents of 71.4, 33, 38, and 41%, respectively (Syed et al. 2023). Furthermore, the authors reported that the inoculants induced a reduction in Cd uptake by the plant's root tissues, in comparison to the non-inoculated plants.

9.5 Effect on the Productivity and Quality of Commercial Crops

Shaheed Siddiqui et al. (2022) documented that the inoculation of *Trichoderma harzianum* in combination with foliar application of iron and zinc positively impacted wheat (*Triticum aestivum*) cultivation. The authors indicated that the combined treatment improved foliar micronutrient content, which in turn enhanced key agronomical parameters related to plant growth and development in comparison to the non-inoculated plants: spike length, biomass, grain mass, harvest index, and photosynthetic pigments. In a similar study, Anwar et al. (2023) reported that the use of *T. harzianum* alone as well as in combination with a foliar application of copper and boron in wheat cultivation positively improved plant-growth traits. These traits include plant height, kernels per spike, spike length, harvest index, and pigment content with respect to the control plants without inoculation. In addition, the combined treatment increased the content of micro- and macronutrients in the wheat grains such as Cu, Fe, Zn, K, and P. In another study, Di Mola et al. (2023) evaluated the effects of *T. afroharzianum* and extracts of the algae *Ascophyllum nodosum*, using both single and combined applications, on the yield and quality of tomato crops. The combined treatment resulted in an increase in fruit yield (24.8%) and phenol content (30%) in comparison to the non-treated control. The authors concluded that the applycation of the fungal strain and algae extract can effectively increase the crop yield and improve the quality traits of tomato fruits.

In another work, Duan et al. (2023) studied the effect of a *Trichoderma* strain combined with different doses of phosphorus fertilizer on pepper (*Capsicum annuum*) cultivation. All fungal treatments improved nutrient absorption, plant growth, chlorophyll content, and fruit yield, in comparison with the non-treated control. Additionally, the treatments positively affected the quality of the pepper fruits by increasing the concentration of soluble sugars, soluble proteins, vitamin C, capsaicin, and capsanthin, with respect to the non-treated plants. In a related study,

Soltaniband et al. (2022) reported that the application of a *T. harzianum* strain in strawberry greenhouse cultivation increased the contents of total soluble solids, total polyphenols, and anthocyanins in strawberry fruits in comparison to the control plants.

Jain et al. (2014) compared the effect of strains of *T. harzianum*, *Bacillus subtilis*, and *Pseudomonas aeruginosa*, either individually or combined, on the nutritional quality of pea seeds (*Pisum sativum*). The authors reported that the application of inoculants as a consortium increased the contents of total phenolics, flavonoids, ascorbic acid, and protein in the plant as compared to the non-inoculated treatment. Additionally, increases in the free radical scavenging activity, hydroxyl radical scavenging activity, reducing power, and iron chelation power were observed in the seeds and pericarp of pods treated with the microbial consortium. The combined treatment caused an increase in the content of phenolic compounds (gallic acid, shikimic acid, tannic acid, p-coumaric acid, syringic acid, quercetin, and kaempferol) in the seeds. Younes et al. (2023) studied the effect of the use of *T. album* and *B. megaterium* both individually and in combination, on onion cultivation. The authors reported that the inoculants increased the growth index, photosynthetic pigment content, and bulb weight in comparison with the control plants. In the study, the individual application of *T. album* raised the level of K^+ (105.8%), Ca^{2+} (37.8%), proline (34.2%), and total free amino acids (144.6%) in the bulb tissues. Additionally, the individual application of *T. album* and *B. megaterium* increased the levels of total soluble carbohydrates by 19.1 and 84.0%, respectively, as well as the content of antioxidant substances in the plant. An increase in the activity of antioxidant enzymes was observed with both inoculants, *T. album* and *B. megaterium*, respectively: SOD by 58.5 and 31.3%, CAT by 164.7 and 232%, ascorbate peroxidase by 175.3 and 212.7%, and glutathione S-transferase by 40 and 9.3%. Additionally, the content of antioxidant compounds in the bulb tissues increased with respect to the control plants: ascorbic acid by 19.1 and 44%, glutathione by 6.2 and 33.8%, and total flavonoid by 171.9 and 56.2%, respectively.

9.6 Critical Appraisal and Outlook

The *Trichoderma* genus comprises a diverse number of species that have been found in a wide range of ecosystems and climate zones (dessert, forest, steppe, wetland, etc.), both in natural and human-modified environments (crops, plant roots, soils, decaying woods, composts, etc.) (Kredics et al. 2021; Martinez et al. 2023), as well as in all types of soil types, both cultivable and non-cultivable (Kredics et al. 2021; Kubiak et al. 2023). Additionally, the species of this genus are ubiquitous and geographically distributed in practically all the continents (Kubicek et al. 2008). Among the characteristics of these fungi, which have made them very successful as a cosmopolitan and geographically diverse genus, are their high growth capacity, their outstanding metabolic characteristics, their high tolerance to biotic and abiotic stress, and their functional traits as very efficient plant colonizers and saprophytic microorganisms (Kredics et al. 2014; Kubiak et al. 2023).

Trichoderma strains have been widely used in modern practices directed to sustainable agriculture considering that these fungi have shown outstanding biological attributes as both biocontrol and biofertilizer agents. A great number of scientific publications have documented basic and applied research regarding the use of the strains of this genus as a biological amendment in the agriculture. The mechanisms employed by *Trichoderma* species for the biocontrol of fungal phytopathogens have been extensively studied, in which these species exert their antagonist effect mainly as a potent mycoparasitic agent by exerting plant colonization and fungal hyperparasitism in agricultural crop systems. It has been widely reported that *Trichoderma* species are opportunistic symbiotic in plants, in which colonization takes place in part thanks to the use of their metabolic capability to synthetize antioxidant compounds and enzymes that benefit the plant (Ghazanfar et al. 2018). In addition, hyperparasitism of fungal strains is exerted mainly by the synthesis of hydrolytic enzymes, such as cellulases, proteases, and particularly chitinases (Ghazanfar et al. 2018). Another important antagonistic mechanism for these fungal species is the synthesis of a wide range of antimicrobial compounds (Kredics et al. 2021; Kubiak et al. 2023).

On the other hand, extensive research has been carried out recently regarding the application of *Trichoderma* strains not only as a biocontrol agent but also as a biofertilizer and a phytoremediation inoculant. As biofertilizers, the species of this genus have shown capacity to reduce the pH of the soil and to solubilize phosphate as well as the micronutrients iron, magnesium, and manganese (Ghazanfar et al. 2018). In addition, there are also a number of reports regarding the beneficial effect of *Trichoderma* species to improve the resistance to abiotic stress of the host plants such as temperature, high salinity, drought, and the presence of heavy metals (Kubiak et al. 2023). In this regard, the resistance of these fungi to heavy metals has made them good candidates as agents for phytoremediation (Schmoll et al. 2014).

A considerable number of in vivo studies of the biocontrol and biofertilizer activities in agricultural crops have been reported in the literature. In this review, we considered mainly the studies made at greenhouse and field experimental scales. Ghazanfar et al. (2018) surveyed the use of *Trichoderma* isolates in the biocontrol of different fungal crop diseases, reporting biocontrol effectiveness from 35 to 100%. Among the main phytopathogens that were reported as biocontrol targets are *Rhizoctonia solani* (tomato), *Fusarium oxysporum* (chickpea), *F. solani* (green bean), *Botrytis cinerea* (cucumber), *Sclerotinia sclerotiorum* (green bean, brinjal), and *Pythium* sp. (green bean). In addition, the *Trichoderma* strains tested in the survey were the following: *Trichoderma harzianium* (10), *T. virens* (3), and *T. viride* (2).

In contrast to the above, only a few studies have been published reporting contrary or null findings of *Trichoderma* applications in agricultural crop production. Altintas and Bal (2008) studied the effect of a commercial inoculant based on *T. harzianum* on onion production finding that the inoculant applied at high concentrations was not effective to improve onion bulb characteristics and crop production. Additionally, in a biocontrol study, the application of two commercial *Trichoderma* formulations were tested on lentil genotypes infected with the pathogenic oomycete

Aphanomyces euteiches (Prashar and Vandenberg 2017). The authors reported that the fungal inoculants were not effective to suppress the disease caused by the pathogen. Similarly, Hjeljord et al. (2000) studied the biocontrol effectiveness of *Trichoderma* commercial and isolated products on the treatment of strawberry plants infected with *B. cinerea* and *Mucor piriformis* pathogens in greenhouse conditions. The authors reported that the fungal inoculants were not effective to suppress the growth of the phytopathogens due to the slow germination rates and low colonization capacity of the *Trichoderma* inoculants in comparison with the fungal pathogens. Considering the above reports, we believe that in a similar fashion to the health and clinical research practices, null finding studies in this area are important to have a realistic appraisal of the application and effectiveness of *Trichoderma*-based inoculants, which will aid to make informed decisions regarding the development and release of new inoculant products to the market.

The study of systematics and taxonomy of this genus is very important for the genetic and functional characterization of *Trichoderma* isolates that are used as agricultural amendments. In previous studies, culture-based approaches were mostly used for species identification. However, recent progress in the taxonomic tools, which make use of novel DNA-fingerprinting methods and data analysis, with the arrival of metagenomics and next-generation sequencing techniques, has prompted the reclassification of many strains and the development of a new taxonomic map for the genus (Kredics et al. 2014; Singh et al. 2020). With this constant progress in the analytical methods for systematics and taxonomy, the strains that were previously classified may belong to different species according to the new taxonomy advances in the field, so the consideration of the taxonomic taxa for a specific strain needs to take into account the type of analysis employed for its identification. According to this, the correct classification of new isolates that are intended to be used as soil beneficial inoculants is vital to be able to accurately associate the biocontrol and biofertilization effects of the inoculant to a specific *Trichoderma* species in a particular agricultural application. With this consideration, the progress in the taxonomy of the genus would allow to perform the correct systematic analysis of the beneficial *Trichoderma* species in the particular crop-production system. In addition, the accuracy of the identification of strains related to this genus is important since currently there is greater awareness of the use of *Trichoderma* strains that are pathogenic to both humans and edible fungi cultivation (Kubicek et al. 2008).

Considering the above, the beneficial traits of *Trichoderma* species in agricultural crops have made these fungi excellent candidates to be used as biocontrol and biofertilizer agents (Kubiak et al. 2023). Additionally, taking into account the current global food security crisis, and the need to reduce the environmental impact caused by the use of agrochemicals, the use of biological amendments in current agriculture, such as *Trichoderma* inoculants, is not only desirable but also fundamental for the sustainability of agricultural practices worldwide. In this regard, the use of *Trichoderma*-based inoculants has become prominent in commercial agricultural applications, which according to Kubiak et al. (2023) accounts for 60% of the biological crop amendments available in the global market.

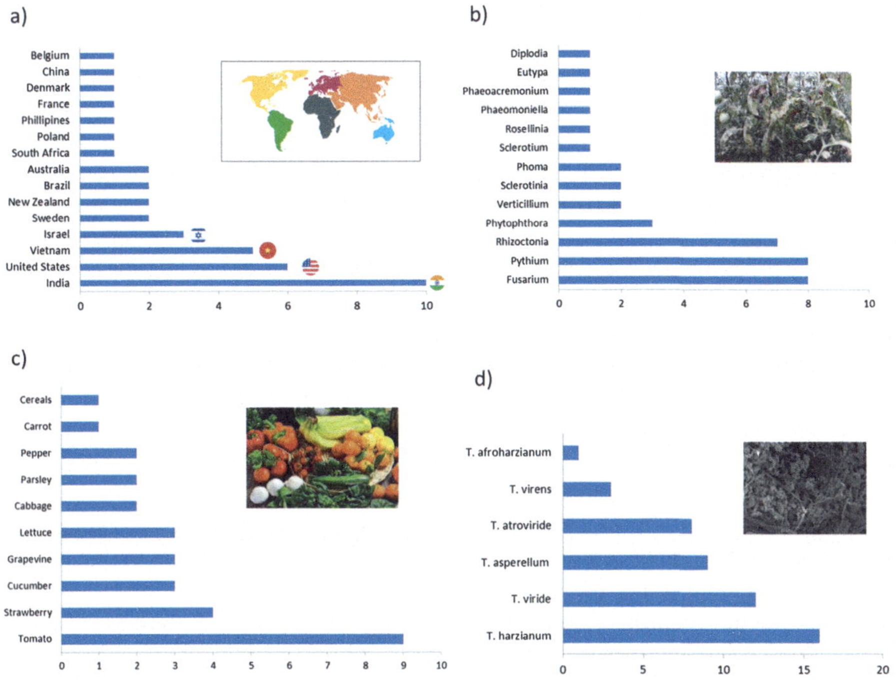

Fig. 9.4 Commercial products formulated with *Trichoderma* inoculants (Ghazanfar et al. 2018; Kubiak et al. 2023; Martinez et al. 2023): (**a**) country of origin; (**b**) targeted phytopathogenic fungi; (**c**) targeted crop products; (**d**) *Trichoderma* species used as inoculants

In this context, several *Trichoderma* products have been commercialized in different countries worldwide. One of the attributes that make *Trichoderma* inoculants particularly attractive to be employed as a crop biological treatment is that the strains of this genus have proved to be effective not only as biocontrol agents but also as biofertilizers. Although a comprehensive list of the marketed products worldwide is very difficult to obtain, Ghazanfar et al. (2018) and Kubiak et al. (2023) surveyed these commercial items and reported the *Trichoderma* species employed as inoculants. Figure 9.4a depicts the number of commercial products based on the country of production, in which it can be seen that according to the survey, the countries with more commercial products are India (10), United States (6), and Vietnam (5). In terms of the target fungal pathogens described by the product manufacturers (Fig. 9.4b), most commercial inoculants are directed to combat the phytopathogens *Fusarium* (8), *Phythium* (8), and *Rhizoctonia* (7) (Kubiak et al. 2023). Fig. 9.4 depicts the crop products on which the inoculants are recommended to be used by the manufacturers (Fig. 9.4c), with most of the applications being directed to the cultivation of tomato (9) and strawberry (4) crops.

On the other hand, the main species used as inoculants in the biological commercial products are the following: *T. harzianum* (16), *T. viride* (12), and *T. asperellum* (9) (Fig. 9.4d) (Kubiak et al. 2023; Ghazanfar et al. 2018; Martinez et al. 2023).

Different spore or cell concentrations have been reported for commercial products, ranging from 10^5 to 10^{10} CFU/g, with shelf lives up to 2 years at both room and refrigeration temperatures (Martinez et al. 2023). Nevertheless, in order to attain these attractive product's shelf lives, the processes of formulation and product design to produce inoculant products based in *Trichoderma* strains are critical for the success and viability of the use of these products in agriculture field applications. In this regard, product development using modern technologies such as microencapsulation is key to extend the shelf life of *Trichoderma* commercial inoculants (Martinez et al. 2023). The use of microencapsulation has a twofold function in product formulation: (i) the protection of the inoculant of the environmental conditions, to extend its shelf life primarily by the reduction in the moisture content and the addition of protective encapsulating materials; and (ii) the conservation of the viability of the inoculant and of its effectiveness in the application to crop cultivation by the slow release of the inoculants into the plant-soil system. Among the most promising technologies for microencapsulation, the most relevant are the ones that use drying techniques such as spray drying and freeze-drying (Martinez et al. 2023).

An important condition for the commercialization at a large scale of a biological product is that it needs to comply with the fundamental biological safety; that is, the use of the inoculant in normal conditions should be not pathogenic for the user or the environment. According to the current research, *Trichoderma* strains are generally safe for crop plants, not presenting virulence against their host; on the contrary, the opportunistic symbiosis of these strains with the host plant is well documented in the literature that makes species of this genus widely recognized as beneficial microorganisms and biofertilizers (Kubiak et al. 2023; Martinez et al. 2023). However, in the scientific literature, there are several reports that document the pathogenic nature of some *Trichoderma* strains toward both humans and edible fungi. In the first case, strains of this genus have been reported to be opportunistic pathogens toward immunocompromised patients, which include people with HIV/AIDS, as well as with cancer and hematological illnesses (Kredics et al. 2021); nevertheless, some studies have reported increasing number of cases of *Trichoderma* infections caused in non-immunocompromised people. Among the most important identified pathogenic strain is *T. longibrachiatum* and its clade, which has been previously reported as an efficient biocontrol and biofertilizer inoculant (Sanó et al. 2022); however, other common species of the genus have been also reported as virulent such as *T. harzianum* sensu stricto, *T. asperellum*, *T. asperellodides*, and *T. atriviride*, so virtually any *Trichoderma* strain can present human pathogenicity (Kredics et al. 2021; Kubicek et al. 2008). In addition, some *Trichoderma* species cause the green-mold epidemics, which affect drastically the production of edible mushrooms, particularly the cultivation of the button mushroom (*Agaricus bisporus*) and oyster mushroom (*Pleurotus ostreatus*). Among the species of the genus involved in the green-mold disease are *T. aggressivum* in the case of the former and *T. pleurotum* and *T. pleuroticola* affecting the oyster mushroom.

Considering the above, for both pathogenic cases of *Trichoderma*, whenever there is a specific isolate with potential as a beneficial crop amendment, a precise identification at the species level should be carry out in order to discard the strain as a possible pathogen toward humans or mushroom cultivation. However, in the case of human pathogenicity, this approach may not be sufficient, since the virulent species can expand to virtually all the species of the genus (Kredics et al. 2021). So in this case, to discard the risk of selecting a pathogenic strain, it is recommended to ensure that the biocontrol candidate cannot grow at temperatures greater than 37 °C, which is the physiological temperature of the human body (Kredics et al. 2021). On the other hand, for the prevention of the green-mold disease during mushroom cultivation, it is important to apply fungicide treatments to the raw materials involved in the production process, which include the use of chemical or biological fungicides, as well as common disinfection and sterilization treatments (Kredics et al. 2021).

The improvement of *Trichoderma* strains has been reported, consisting of the introduction of a gene of interest that increases the capacity of the modified strain for the biocontrol of phytopathogens. Xia et al. (2018) described a *Trichoderma* strain that expressed a transgenic chitinase from *Metarhizium anisopliae*. The authors reported that the overexpression of *M. anisopliae* enhanced the expression of endogenous enzymes (β-1,3-glucanases and polygalacturonase) and increased the antagonistic activity of the *Trichoderma* modified strain against *B. cinerea*. Similarly, Peng et al. (2021) reported that L-amino acid oxidase (ThLAAO) secreted by *T. harzianum* is a flavoenzyme with antimicrobial qualities. The authors studied the transformation of the ThLAAO gene into tobacco (*Nicotiana tabacum*) to observe whether ThLAAO can activate defense mechanisms and increase resistance against phytopathogens. The results showed that the transgenic tobacco stimulated the expression of defense-related genes and those involved in salicylic acid, jasmonic acid, and ethylene biosynthesis, thereby increasing the plant's resistance against *S. sclerotiorum* and *B. cinerea*. On the other hand, Sánchez-Cruz et al. (2021) reported that the overexpression of swollenin (SWO1), a protein with morphogenic activity, in a transgenic strain of *T. atroviride* increased its colonization in chili (*Capsicum annum*) roots. The authors reported that the treatments inoculated with the mutant strain enhanced the plant's growth, its response to pathogen infection, and its resistance to drought conditions.

In conclusion, the application of *Trichoderma* isolates as crop inoculants needs to consider firstly the accurate identification of the specific strain and preferably the results of experimental field studies in the particular system crop-pathogen system. This allows the proper systematic considerations of the candidate strain not only with respect to its biocontrol and biofertilization activities but also regarding its safety traits, moreover considering that the whole genomes of specific *Trichoderma* strains are currently available. In general, the strains of this genus have shown to be non-toxic and non-pathogenic in most of the agricultural systems reported, but problems have arisen regarding the pathogenic nature of these fungi against humans and edible mushrooms. Strategies to cope with these obstacles, thus ensuring the biological safety of *Trichoderma*-based products, have been outlined in this section. The species of *Trichoderma* that have been mostly employed in inoculant formulations at a commercial scale are *T. harzianum*, *T. viride*, and *T. asperellum*. According to the literature, the biocontrol effectiveness of *Trichoderma* strains varies toward the cultivation of different crop products, thus complicating the standardization process of the inoculant products. In addition, a significant obstacle for the commercial launching and consolidation of these biological products is the laborious and expensive process of product development that involves a range of studies from in vitro to field large-scale experiments. In addition, from field experiments, it can be implied that the species of this genus generally have beneficial effects on the biocontrol of a wide range of phytopathogenic fungi and industrial crops, which establishes *Trichoderma*-based inoculants as very promising products for controlling crop diseases in a sustainable-agriculture framework, thus reducing or replacing the use of agrochemicals.

On the other hand, technological process considerations need to be addressed in order to increase the shelf life of the bioproducts and ensure their effectiveness in the particular agricultural systems. In this regard, microbial encapsulation of the fungi using spray drying and freeze-drying have emerged as efficient technologies to ensure the viability and effectiveness of the fungal inoculants in the agricultural applications. The prospects of *Trichoderma*-based inoculants as crop amendments are encouraging considering the dual use of these fungal strains as biocontrol and biofertilizer agents. In addition, the consolidation of these biological products at a commercial scale will require further studies in the systematics of the different species, with the use of omics techniques, metabolomics, proteomics, metagenomics, and transcriptomics, which allows to streamline the process of product development of *Trichoderma* inoculants for particular crop production systems associated to different phytopathogenic fungi.

Fungi of the genus *Trichoderma* have a broad package of mechanisms through which they interact, attack, and mycoparasite the pathogenic microorganisms with which they interact. In addition, many species of this genus are resistant to the biotic and abiotic factors that occur in the agricultural soils where they develop; also presenting certain tolerance to the defense mechanisms used by plants during the host-pathogen interaction. These biological characteristics make *Trichoderma* strains very efficient to colonize spaces and use resources (nutrients) of the plant, thus outcompeting the pathogenic microorganisms present in the plant-soil system. All these qualities have prompted the production of *Trichoderma* inoculants at a commercial scale to be used in agriculture as plant-soil amendment products which foster the biological protection and growth of agricultural crops in a sustainable manner.

Acknowledgements Ramírez-Cariño thanks Consejo Nacional de Humanidades Ciencia y Tecnología (CONAHCYT), Mexico, for the financial support of his postdoctoral fellowship (No. 644716).

References

Abdelkhalek A, Al-Askar AA, Arishi AA, Behiry SI (2022) *Trichoderma hamatum* strain Th23 promotes tomato growth and induces systemic resistance against tobacco mosaic virus. J Fungi 8(3). https://doi.org/10.3390/jof8030228

Abdel-lateif KS (2017) *Trichoderma* as biological control weapon against soil borne plant pathogens. Afr J Biotechnol 16(50):2299–2306. https://doi.org/10.5897/ajb2017.16270

Abdelrhim AS, Abdellatif YMR, Hossain MA, Alamri S, Pessarakli M, Lessy AMN, Dawood MFA (2023) Comparative study of three biological control agents and two conventional fungicides against coriander damping-off and root rot caused by *Rhizoctonia solani*. Plants (Basel) 12(8):1694. https://doi.org/10.3390/plants12081694

Abdelrhim AS, Hemeda NF, Mwaheb MA, Omar MOA, Dawood MFA (2024) The role of *Trichoderma koningii* and *Trichoderma harzianum* in mitigating the combined stresses motivated by *Sclerotinia sclerotiorum* and salinity in common bean (*Phaseolus vulgaris*). Plant Stress 100370. https://doi.org/10.1016/j.stress.2024.100370

Abeed AHA, Mahdy RE, Alshehri D, Hammami I, Eissa MA, Abdel Latef AAH, Mahmoud GAE (2022) Induction of resilience strategies against biochemical deteriorations prompted by severe cadmium stress in sunflower plant when *Trichoderma* and bacterial inoculation were used as biofertilizers. Front Plant Sci 13:1004173. https://doi.org/10.3389/fpls.2022.1004173

Abied MAE, Khalil MI, Awad MF, Youssef SA (2023) Alleviation of black root rot symptoms and alteration of strawberry growth via modulating physiological and biochemical mechanisms using *Trichoderma viride* and *Bacillus subtillis*. Eur J Plant Pathol 167:235–250. https://doi.org/10.1007/s10658-023-02697-w

Adedayo AA, Babalola OO (2023) Fungi that promote plant growth in the rhizosphere boost crop growth. J Fungi (Basel) 9(2):239. https://doi.org/10.3390/jof9020239

Ahluwalia V, Kumar J, Rana VS, Sati OP, Walia S (2015) Comparative evaluation of two *Trichoderma harzianum* strains for major secondary metabolite production and antifungal activity. Nat Prod Res 29(10):914–920. https://doi.org/10.1080/14786419.2014.958739

Altintas S, Bal U (2008) Effects of the commercial product based on *Trichoderma harzianum* on plant, bulb and yield characteristics of onion. Sci Hortic 116(2):219–222. https://doi.org/10.1016/j.scienta.2007.11.012

Anwar S, Ali A, Ullah Z, Binjawhar DN, Sher H, Ali R, Iqbal R, Ali B, Ali I (2023) The impact of *Trichoderma harzianum* together with copper and boron on wheat yield. ACS Agric Sci Technol 3(6):517–527. https://doi.org/10.1021/acsagscitech.3c00076

Arjona-Girona I, Vinale F, Ruano-Rosa D, Lorito M, López-Herrera CJ (2014) Effect of metabolites from different *Trichoderma* strains on the growth of *Rosellinia necatrix*, the causal agent of avocado white root rot. Eur J Plant Pathol 140(2):385–397. https://doi.org/10.1007/s10658-014-0472-z

Aseel DG, Soliman SA, Al-Askar AA, Elkelish A, Elbeaino T, Abdelkhalek A (2023) *Trichoderma viride* isolate tvd44 enhances potato growth and stimulates the defense system against potato virus Y. Horticulturae 9(6). https://doi.org/10.3390/horticulturae9060716

Atanasova L, Marchetti-Deschmann M, Nemes A, Bruckner B, Rehulka P, Stralis-Pavese N, Łabaj PP, Kreil DP, Zeilinger S (2023) Mycoparasitism related targets of Tmk1 indicate stimulating regulatory functions of this MAP kinase in *Trichoderma atroviride*. Sci Rep 13(1). https://doi.org/10.1038/s41598-023-47027-6

Athinuwat D, Ruangwong OU, Harishchandra DL, Pitija K, Sunpapao A (2024) Biological control activities of rhizosphere fungus *Trichoderma virens* T1-02 in suppressing flower blight of flamingo flower (*Anthurium andraeanum* Lind.). J Fungi 10(1). https://doi.org/10.3390/jof10010066

Awad-Allah EFA, Shams AHM, Helaly AA, Ragheb EIM (2022) Effective applications of *Trichoderma* spp. as biofertilizers and biocontrol agents mitigate tomato *Fusarium* wilt disease. Agriculture 12(11). https://doi.org/10.3390/agriculture12111950

Awad-Allah EFA, Mohamed IAA, Awd Allah SFA, Shams AHM, Elsokkary IH (2023) *Trichoderma* Species: An overview of current status and potential applications for sustainable agriculture. Indian J Agric Res 57(3):273–282. https://doi.org/10.18805/IJARe.AF-751

Bakhshi S, Eshghi S, Banihashemi Z (2023) Application of candidate endophytic fungi isolated from extreme desert adapted trees to mitigate the adverse effects of drought stress on maize (*Zea mays* L.). Plant Physiol Biochem 202. https://doi.org/10.1016/j.plaphy.2023.107961

Bedine MAB, Iacomi B, Tchameni SN, Sameza ML, Fekam FB (2022) Harnessing the phosphate-solubilizing ability of *Trichoderma* strains to improve plant growth, phosphorus uptake and photosynthetic pigment contents in common bean (*Phaseolus vulgaris*). Biocatal Agric Biotechnol 45. https://doi.org/10.1016/j.bcab.2022.102510

Behiry S, Soliman SA, Massoud MA, Abdelbary M, Kordy AM, Abdelkhalek A, Heflish A (2023) *Trichoderma pubescens* elicit induced systemic resistance in tomato challenged by *Rhizoctonia solani*. J Fungi 9(2). https://doi.org/10.3390/jof9020167

Cañada-Coyote E, Ramírez-Pimentel JG, Aguirre-Mancilla CL, Raya-Pérez JC, Acosta-García G, Iturriaga G (2021) *Trichoderma harzianum* mutants enhance antagonism against phytopathogenic fungi, phosphorus assimilation and drought tolerance in jalapeño pepper plants. Chil J Agric Res 81(3):270–280. https://doi.org/10.4067/S0718-58392021000300270

Carrasco F, Miranda V, Sede SM, Bustos S, González V, Otero L, Fracchia S (2023) Screening for native *Trichoderma* strains as potential biocontrollers of the olive pathogen *Verticillium dahliae*. Arid Land Res Manag. https://doi.org/10.1080/15324982.2023.2233933

Chen JL, Sun SZ, Miao CP, Wu K, Chen YW, Xu LH, Guan HL, Zhao LX (2016) Endophytic *Trichoderma gamsii* YIM PH30019: A promising biocontrol agent with hyperosmolar, mycoparasitism, and antagonistic activities of induced volatile organic compounds on root-rot pathogenic fungi of *Panax notoginseng*. J Ginseng Res 40(4):315–324. https://doi.org/10.1016/j.jgr.2015.09.006

Cubilla-Ríos AA, Ruíz-Díaz-Mendoza DD, Romero-Rodríguez MC, Flores-Giubi ME, Barúa-Chamorro JE (2019) Antibiosis of proteins and metabolites of three species of *Trichoderma* against paraguayan isolates of *Macrophomina phaseolina*. Agron Mesoam 30(1):63–77. https://doi.org/10.15517/am.v30i1.34423

Cuervo-Parra JA, Sánchez-López V, Ramirez-Suero M, Ramírez-Lepe M (2011) Morphological and molecular characterization of *Moniliophthora roreri* causal agent of frosty pod rot of cocoa tree in Tabasco, Mexico. Plant Pathol J 10:122–127. https://doi.org/10.3923/ppj.2011.122.127

Cuervo-Parra JA, Cortés TR, Ramirez-Lepe M (2016) Mosquito-borne diseases, pesticides used for mosquito control, and development of resistance to insecticides. Insecticides Resistance. https://doi.org/10.5772/61510

Cuervo-Parra JA, Pérez España VH, Zavala-González EA, Peralta-Gil M, Aparicio Burgos JE, Romero-Cortes T (2022) *Trichoderma Asperellum* strains as potential biological control agents against *Fusarium verticillioides* and Ustilago maydis in maize. Biocontrol Sci Tech 32(5):624–647. https://doi.org/10.1080/09583157.2022.2042196

da Silva MCAP, Batista PF, Menezes-Silva PE, Sousa LF, Loram-Lourenço L, Alencar KMB, de Abreu IMPG, Alves RDFB, Souza Freire E, Silva FG, da Silva FB, Farnese FS (2023) With a little help from my friends: inoculation with *Bacillus amyloliquefaciens* and *Trichoderma asperellum* alleviates drought and salt stress in soybean. Theor Exp Plant Physiol 35:169–183. https://doi.org/10.1007/s40626-023-00274-4

Datta D, Senapati AK, Behera L, Zaidi NW, Kumar S, Dey P, Kumar S (2023) Alleviating drought stress in rice plant through intervention of *Trichoderma spp*. J Environ Biol 44(3):373–379. https://doi.org/10.22438/JEB/44/3/MRN-4084

Dautt-Castro M, Rebolledo-Prudencio OG, Estrada-Rivera M, Islas-Osuna MA, Jijón-Moreno S, Casas-Flores S (2024) *Trichoderma virens* Big Ras GTPase-1, a molecular switch for transforming a mutualistic fungus to plants in a deleterious microbe. Microbiol Res 278. https://doi.org/10.1016/j.micres.2023.127508

Degenkolb T, Fognielsen K, Dieckmann R, Branco-Rocha F, Chaverri P, Samuels GJ, Thrane U, Vondçhren H, Vilcinskas A, Brückner H (2015) Peptaibol, secondary-metabolite, and hydrophob in pattern of commercial biocontrol agents formulated with species of the *Trichoderma harzianum* complex. Chem Biodivers 12(4):662–684. https://doi.org/10.1002/cbdv.201400300

Di Mola I, Ottaiano L, Cozzolino E, Marra R, Vitale S, Pironti A, Fiorentino N, Mori M (2023) Yield and quality of processing tomato as improved by biostimulants based on *Trichoderma* sp. and *Ascophyllum nodosum* and biodegradable mulching films. Agronomy 13(3). https://doi.org/10.3390/agronomy13030901

Duan X, Zou C, Jiang Y, Yu X, Ye X (2023) Effects of reduced phosphate fertilizer and increased *Trichoderma* application on the growth, yield, and quality of pepper. Plan Theory 12(16). https://doi.org/10.3390/plants12162998

Fu J, Xiao Y, Liu Z, Zhang Y, Wang Y, Yang K (2020) *Trichoderma asperellum* improves soil microenvironment in different growth stages and yield of maize in saline-alkaline soil of the songnen plain. J Soil Plant Environ 66(12):639–647. https://doi.org/10.17221/456/2020-PSE

Fu J, Xiao Y, Wang Y-f, Liu Z-h, Yang K (2021) Saline–alkaline stress in growing maize seedlings is alleviated by *Trichoderma asperellum* through regulation of the soil environment. Sci Rep 11(1). https://doi.org/10.1038/s41598-021-90675-9

Gardiner DM, Waring P, Howlett BJ (2005) The epipolythiodioxopiperazine (ETP) class of fungal toxins: distribution, mode of action, functions and biosynthesis. Microbiol 151(4):1021–1032. https://doi.org/10.1099/mic.0.27847-0

Ghazanfar MU, Raza M, Raza W, Qamar MI (2018) *Trichoderma* as potential biocontrol agent, its exploitation in agriculture: a review. Plant Prot 2(3):109–135.

Giodarno DF, Pastor NA, Marie Rouws LF, Moura de Freitas K, Erazo JG, Del Canto A, da Silva Coelho I, Oddino CM, Mabel Torre A (2023) *Trichoderma harzianum* ITEM 3636 colonizes peanut roots as an endophyte and protects the plants against late leaf spot. Symbiosis 89:337–352. https://doi.org/10.1007/s13199-023-00913-z

Gulzar S, Manzoor MA, Liaquat F, Shah IH, Rehman A, Hameed MK, Arif S, Zhou X, Zhang Y (2023) Effects of melatonin and *Trichoderma harzianum* on pak choi yield, chlorophyll contents and antioxidant defense system under clubroot disease. S Afr J Bot 158:292–300. https://doi.org/10.1016/j.sajb.2023.05.021

Guo R, Li G, Zhang Z, Peng X (2022) Structures and biological activities of secondary metabolites from *Trichoderma harzianum*. Mar Drugs 20(11):701. https://doi.org/10.3390/md20110701

Guzmán-Guzmán P, Kumar A, de los Santos-Villalobos S, Parra-Cota FI, Orozco-Mosqueda M d C, Fadiji AE, Hyder S, Babalola OO, Santoyo G (2023) *Trichoderma* species: our best fungal allies in the biocontrol of plant diseases—A review. Plan Theory 12(3). https://doi.org/10.3390/plants12030432

Hajieghrari B, Torabi-Giglou M, Mohammadi MR, Davari M (2008) Biological potential of some Iranian *Trichoderma* isolates in the control of soil borne plant pathogenic fungi. Afr J Biotechnol 7(8):967–972. http://www.academicjournals.org/AJB

He M, Tian Z, Liu Q, Guo Y (2021) *Trichoderma asperellum* promotes cadmium accumulation within maize seedlings. Biotechnol Biotechnol Equip 35(1):1546–1559. https://doi.org/10.1080/13102818.2021.1997155

Hjeljord LG, Stensvand A, Tronsmo A (2000) Effect of temperature and nutrient stress on the capacity of commercial *Trichoderma* products to control *Botrytis cinerea* and *Mucor piriformis* in greenhouse strawberries. Biol Control 19(2):149–160. https://doi.org/10.1006/bcon.2000.0859

Illescas M, Morán-Diez ME, de Alba ÁEM, Hermosa R, Monte E (2022) Effect of *Trichoderma asperellum* on wheat plants' biochemical and molecular responses, and yield under different water stress conditions. Int J Mol Sci 23(12). https://doi.org/10.3390/ijms23126782

Jain A, Singh A, Chaudhary A, Singh S, Singh HB (2014) Modulation of nutritional and antioxidant potential of seeds and pericarp of pea pods treated with microbial consortium. Food Res Int 64:275–282. https://doi.org/10.1016/j.foodres.2014.06.033

Jain A, Das S (2020) *Trichoderma* Proteome: Multifunctional Role in Plant Defense. In: Manoharachary C, Singh HB, Varma A (eds) *Trichoderma*: Agricultural Applications and Beyond. Soil Biology, vol 61. Springer, Cham. https://doi.org/10.1007/978-3-030-54758-5_14

Jamil A (2021) Antifungal and plant growth promoting activity of *Trichoderma* spp. against *Fusarium oxysporum f. sp. lycopersici* colonizing tomato. J Plant Prot Res 61(3):243–253. https://doi.org/10.24425/jppr.2021.137950

Japanis FG, Vetaryan S, Raja NKK, Mokhtar MAA, Fishal EMM (2022) The impact of *Trichoderma* spp. on agriculture and their identification. Malays Appl Biol 51(6):1–15. https://doi.org/10.55230/mabjournal.v51i6.2198

Jáquez-Matas SV, Pérez-Santiago G, Márquez-Linares MA, Pérez-Verdín G (2022) Economic and environmental impacts of pesticides in corn, alfalfa, and walnut crops in Durango, Mexico. Int J Environ Contam 38:219–233. https://doi.org/10.20937/RICA.54169

Jiménez-Bremont JF, González-Pérez E, Ortega-Amaro MA, Madrigal-Ortiz S, Duque-Ortiz A, Mendoza-Mendoza A (2024) Volatile organic compounds emitted by *Trichoderma*: small molecules with biotechnological potential. Sci Hortic 325:112656. https://doi.org/10.1016/j.scienta.2023.112656

Kawasaki M, Ohara T, Ishida M, Takahata Y, Hatakeyama K (2021) Development of novel clubroot resistant rapeseed lines (*Brassica napus* L.) effective against Japanese field isolates by marker assisted selection. Breed Sci 71(5):528–537. https://doi.org/10.1270/JSBBS.21014

Khan RAA, Najeeb S, Hussain S, Xie B, Li Y (2020) Bioactive secondary metabolites from *Trichoderma* spp. against phytopathogenic fungi. Microorganisms 8(6). https://doi.org/10.3390/microorganisms8060817

Korpi A, Järnberg J, Pasanen AL (2009) Microbial volatile organic compounds. Crit Rev Toxicol 39(2):139–193. https://doi.org/10.1080/10408440802291497

Kredics L, Hatvani L, Naeimi S, Körmöczi P, Manczinger L, Vágvölgyi C, Druzhinina I (2014) Biodiversity of the genus Hypocrea/*Trichoderma* in different habitats. In: Biotechnology and biology of *Trichoderma*. Elsevier, pp 3–24. https://doi.org/10.1016/B978-0-444-59576-8.00001-1

Kredics L, Naeimi S, Hatvani L, Vágvölgyi C, Cai F, Druzhinina IS, Manczinger L (2021) 'The good, the bad and the ugly' in the shades of green: the genus *Trichoderma* in the spotlight. Indian Phytopathol 74(2):403–411. Springer. https://doi.org/10.1007/s42360-021-00352-0

Kubiak A, Wolna-Maruwka A, Pilarska AA, Niewiadomska A, Piotrowska-Cyplik A (2023) Fungi of the *Trichoderma* genus: future perspectives of benefits in sustainable agriculture. Appl Sci 13(11). https://doi.org/10.3390/app13116434

Kubicek CP, Komon-Zelazowska M, Druzhinina IS (2008) Fungal genus Hypocrea/*Trichoderma*: from barcodes to biodiversity. J Zhejiang Univ Sci B 9(10):753–763. https://doi.org/10.1631/jzus.B0860015

Kumar G, Maharshi A, Singh Patel J, Sarma B (2017) *Trichoderma*: a potential fungal antagonist to control plant diseases. SATSA Mukhapatra – annual technical (21):206–218. https://www.researchgate.net/publication/314946122

Lang B, Chen J (2023) *Trichoderma harzianum* cellulase gene thph2 affects *Trichoderma* root colonization and induces resistance to southern leaf blight in maize. J Fungi 9(12). https://doi.org/10.3390/jof9121168

Larran S, Rosa Simón M, Pilar Santamarina M, Roselló Caselles J, Fabiana Consolo V, Perelló A (2023) Endophytic *Trichoderma* strains increase soya bean growth and promote charcoal rot control. J Saudi Soc Agric Sci 22(7):395–406. https://doi.org/10.1016/j.jssas.2023.03.005

Leibman-Markus M, Gupta R, Pizarro L, Gershony O, Rav-David D, Elad Y (2021) Gene editing of the decoy receptor LeEIX1 increases host receptivity to *Trichoderma* Bio-Control. Front Fungal Biol 2:678840. https://doi.org/10.3389/ffunb.2021.678840

Li MF, Li GH, Zhang KQ (2019) Non-volatile metabolites from *Trichoderma* spp. Meta 9(3). https://doi.org/10.3390/metabo9030058

Li CP, Shi ZZ, Fang ST, Song YP, Ji NY (2023) Lipids and terpenoids from the deep-sea fungus *Trichoderma lixii* R22 and their antagonism against two wheat pathogens. Molecules 28(17). https://doi.org/10.3390/molecules28176220

Limdolthamand S, Songkumarn P, Suwannarat S, Jantasorn A, Dethoup T (2023) Biocontrol efficacy of endophytic *Trichoderma spp.* in fresh and dry powder formulations in controlling northern corn leaf blight in sweet corn. Biol Control:181. https://doi.org/10.1016/j.biocontrol.2023.105217

Liu SY, Lo CT, Shibu MA, Leu YL, Jen BOY, Peng KC (2009) Study on the anthraquinones separated from the cultivation of *Trichoderma harzianum* strain Th-R16 and their biological activity. J Agric Food Chem 57(16):7288–7292. https://doi.org/10.1021/jf901405c

Liu Z, Xu N, Pang Q, Khan RAA, Xu Q, Wu C, Liu T (2023) A salt-tolerant strain of *Trichoderma longibrachiatum* HL167 is effective in alleviating salt stress, promoting plant growth, and managing Fusarium wilt disease in cowpea. J Fungi 9(3). https://doi.org/10.3390/jof9030304

López-Barrera GL (2021) Molecular analysis of antagonist fungi isolated from plantations of cocoa (*Theobroma cacao*) from north of Santander. J Eng Comp 23(2):1–11. https://doi.org/10.25100/iyc.v23i2.11154

Machado-Rosa TA, Barbosa ET, Cortes MVB d C, Lobo M (2023) Microtiter method for quantitative assay of IAA from fungal isolates, demonstrated with *Trichoderma*. Rhizosphere 25. https://doi.org/10.1016/j.rhisph.2023.100666

Malmierca MG, Barua J, McCormick SP, Izquierdo-Bueno I, Cardoza RE, Alexander NJ, Hermosa R, Collado IG, Monte E, Gutiérrez S (2015) Novel aspinolide production by *Trichoderma arundinaceum* with a potential role in *Botrytis cinerea* antagonistic activity and plant defense priming. Environ Microbiol 17(4):1103–1118. https://doi.org/10.1111/1462-2920.12514

Martinez Y, Ribera J, Schwarze FWMR, De France K (2023) Biotechnological development of *Trichoderma*-based formulations for biological control. Appl Microbiol Biotechnol 107:5595–5612. https://doi.org/10.1007/s00253-023-12687-x

Mendes JBS, da Costa Neto VP, de Sousa CDA, de Carvalho Filho MR, Rodrigues AC, Bonifacio A (2020) *Trichoderma* and *Bradyrhizobia* act synergistically and enhance the growth rate, biomass and photosynthetic pigments of cowpea (*Vigna unguiculata*) grown in controlled conditions. Symbiosis 80:133–143. https://doi.org/10.1007/s13199-019-00662-y

Metwally RA, Soliman SA (2023) Alleviation of the adverse effects of NaCl stress on tomato seedlings (*Solanum lycopersicum* L.) by *Trichoderma viride* through the antioxidative defense system. Bot Stud 64(1). https://doi.org/10.1186/s40529-023-00368-x

Miguel-Ferrer L, Romero-Arenas O, Andrade-Hoyos P, Sánchez-Morales P, Rivera-Tapia JA, Fernández-Pavía SP (2021) Antifungal activity of *Trichoderma harzianum* and *T. koningiopsis* against *Fusarium solani* in seed germination and vigor of Miahuateco chili seedlings. Mex J Phytopathol 39(2). https://doi.org/10.18781/r.mex.fit.2101-5

Mukhopadhyay R, Kumar D (2020) *Trichoderma*: a beneficial antifungal agent and insights into its mechanism of biocontrol potential. Egypt J Biol Pest Control 30(1). https://doi.org/10.1186/s41938-020-00333-x

Mutawila C, Vinale F, Halleen F, Lorito M, Mostert L (2016) Isolation, production and in vitro effects of the major secondary metabolite produced by *Trichoderma* species used for the control of grapevine trunk diseases. Plant Pathol 65(1):104–113. https://doi.org/10.1111/ppa.12385

Paul S, Rakshit A (2021) Effect of seed bio-priming with *Trichoderma viride* strain BHU-2953 for enhancing soil phosphorus solubilization and uptake in soybean (*Glycine max*). J Soil Sci Plant Nutr 21(2):1041–1052. https://doi.org/10.1007/s42729-021-00420-4

Pedrero-Méndez A, Insuasti HC, Neagu T, Illescas M, Rubio MB, Monte E, Hermosa R (2021) ¿Why is the correct selection of *Trichoderma* strains important? The case of wheat endophytic strains of *T. harzianum* and *T. simmonsii*. J Fungi 7(12). https://doi.org/10.3390/jof7121087

Peng KC, Lin CC, Liao CF, Yu HC, Lo CT, Yang HH, Lin KC (2021) Expression of L-amino acid oxidase of *Trichoderma harzianum* in tobacco confers resistance to *Sclerotinia sclerotiorum* and *Botrytis cinerea*. Plant Sci 303. https://doi.org/10.1016/j.plantsci.2020.110772

Philip B, Behiry SI, Salem MZM, Amer MA, El-Samra IA, Abdelkhalek A, Heflish A (2024) *Trichoderma afroharzianum* TRI07 metabolites inhibit *Alternaria alternata* growth and induce tomato defense-related enzymes. Sci Rep 14(1). https://doi.org/10.1038/s41598-024-52301-2

Porteous-Álvarez AJ, Mayo-Prieto S, Carro-Huerga G, Rodríguez-González A, Álvarez-García S, del Ser S, Lorenzana A, Campelo MP, Maldonado-González MM, Gutiérrez S, Casquero PA (2021) *Trichoderma* species isolated from hop soils in the Órbigo valley, León, Spain. Acta Hortic 1328:63–66. https://doi.org/10.17660/ActaHortic.2021.1328.8

Pradhan DA, Bagagoni P, Makandar R (2023) Assessing rhizosphere *Trichoderma asperellum* strains for root colonizing and antagonistic competencies against *Fusarium* wilt through molecular and biochemical responses in castor. Biol Control 184. https://doi.org/10.1016/j.biocontrol.2023.105280

Prashar P, Vandenberg A (2017) Genotype-specific responses to the effects of commercial *Trichoderma* formulations in lentil (*Lens culinaris* ssp. *culinaris*) in the presence and absence of the oomycete pathogen *Aphanomyces euteiches*. Biocontrol Sci Technol 27(10):1123–1144. https://doi.org/10.1080/09583157.2017.1376035

Rahman M, Borah SM, Borah PK, Bora P, Sarmah BK, Lal MK, Tiwari RK, Kumar R (2023) Deciphering the antimicrobial activity of multifaceted rhizospheric biocontrol agents of solanaceous crops viz., *Trichoderma harzianum* MC2, and *Trichoderma harzianum* NBG. Front Plant Sci 14. https://doi.org/10.3389/fpls.2023.1141506

Ramírez-Cariño HF, Guadarrama-Mendoza PC, Sánchez-López V, Cuervo-Parra JA, Ramírez-Reyes T, Dunlap CA, Valadez-Blanco R (2020) Biocontrol of *Alternaria alternata* and *Fusarium oxysporum* by *Trichoderma asperelloides* and *Bacillus paralicheniformis* in tomato plants. Antonie Van Leeuwenhoek 113(9):1247–1261. https://doi.org/10.1007/s10482-020-01433-2

Rawal R, Scheerens JC, Fenstemaker SM, Francis DM, Miller SA, Benitez MS (2022) Novel *Trichoderma* isolates alleviate water deficit stress in susceptible tomato genotypes. Front Plant Sci 13. https://doi.org/10.3389/fpls.2022.869090

Rimkus A, Namina A, Dzierkale MT, Grigs O, Senkovs M, Larsson S (2023) Impact of growth conditions on the viability of *Trichoderma asperellum* during storage. Microorganisms 11(4). https://doi.org/10.3390/microorganisms11041084

Rodríguez-Hernández AA, Herrera-Alvarez M, Zapata-Sarmiento DH, Becerra-Martínez E, Rodríguez-Monroy M, Sepúlveda-Jiménez G (2023) *Trichoderma asperellum* promotes the development and antioxidant activity of white onion (*Allium cepa* L.) plants. *Hortic Environ Biotechnol* 64:25–39. https://doi.org/10.1007/s13580-022-00467-x

Romero-Cortés T, Tamayo-Rivera L, Morales-Ovando MA, Burgos JEA, España VHP, Peralta-Gil M, Cuervo-Parra JA (2022) Growth and yield of purple kculli corn plants under different fertilization schemes. J Fungi 8(5). https://doi.org/10.3390/jof8050433

Rosado IV, Rey M, Codón AC, Govantes J, Moreno-Mateos MA, Benítez T (2007) QID74 Cell wall protein of *Trichoderma harzianum* is involved in cell protection and adherence to hydrophobic surfaces. Fungal Genet Biol 44(10):950–964. https://doi.org/10.1016/j.fgb.2007.01.001

Ruocco M, Lanzuise S, Vinale F, Marra R, Turrà D, Woo SL, Lorito M (2009) Identification of a new biocontrol gene in *Trichoderma atroviride*: the role of an ABC transporter membrane pump in the interaction with different plant-pathogenic fungi. Mol Plant-Microbe Interact 22(3):291–301. https://doi.org/10.1094/MPMI-22-3-0291

Saadaoui M, Faize M, Bonhomme L, Benyoussef NO, Kharrat M, Chaar H, Label P, Venisse JS (2023) Assessment of Tunisian *Trichoderma* isolates on wheat seed germination, seedling growth and *Fusarium* seedling blight suppression. Microorganisms 11(6). https://doi.org/10.3390/microorganisms11061512

Saba H (2012) *Trichoderma* – a promising plant growth stimulator and biocontrol agent. Mycosphere 3(4):524–531. https://doi.org/10.5943/mycosphere/3/4/14

Saia S, Corrado G, Vitaglione P, Colla G, Bonini P, Giordano M, Di Stasio E, Raimondi G, Sacchi R, Rouphael Y (2021) An endophytic fungi-based biostimulant modulates volatile and non-volatile secondary metabolites and yield of greenhouse basil (Ocimum basilicum l.) through variable mechanisms dependent on salinity stress level. Pathogens 10(7). https://doi.org/10.3390/pathogens10070797

Sánchez-Cruz R, Mehta R, Atriztán-Hernández K, Martínez-Villamil O, Sánchez-Carbente M d R, Sánchez-Reyes A, Lira-Ruan V, González-Chávez CA, Tabche-Barrera ML, Bárcenas-Rodríguez RC, Batista-García RA, Herrera-Estrella A, Balcázar-López E, Folch-Mallol JL (2021) Effects on *Capsicum annuum* plants colonized with *Trichoderma atroviride* P. Karst strains genetically modified in taswo1, a gene coding for a protein with expansin-like activity. Plan Theory 10(9). https://doi.org/10.3390/plants10091919

Sanó L, Oliveira LLB d, Leão MDM, Santos JE d Á d, Medeiros SC d, Schneider F, Sousa ABO d, Taniguchi CAK, Muniz CR, Grangeiro TB, Silva C (2022) *Trichoderma longibrachiatum* as a biostimulant of micropropagated banana seedlings under acclimatization. Plant Physiol Biochem 190:184–192. https://doi.org/10.1016/j.plaphy.2022.09.008

Schmoll M, Seiboth B, Druzhinina I, Kubicek CP (2014) Genomics analysis of biocontrol species and industrial enzyme producers from the genus *Trichoderma*. Fungal Genom 10:233–264. https://doi.org/10.1007/978-3-642-45218-5_10

Schweiger R, Padilla-Arizmendi F, Nogueira-López G, Rostás M, Lawry R, Brown C, Hampton J, Steyaert JM, Müller C, Mendoza-Mendoza A (2021) Insights into metabolic changes caused by the *Trichoderma virens*–maize root interaction. Mol Plant-Microbe Interact 34(5):524–537. https://doi.org/10.1094/MPMI-04-20-0081-R

Sehim AE, Hewedy OA, Altammar KA, Alhumaidi MS, Abd Elghaffar RY (2023) *Trichoderma asperellum* empowers tomato plants and suppresses *Fusarium oxysporum* through priming responses. Front Microbiol 14. https://doi.org/10.3389/fmicb.2023.1140378

Sha S, Lei L, Shu P, Weimin W (2013) Isolation and purification of antifungal components from *Trichoderma harzianum* ferment broth by high-speed counter-current chromatography. Chinese J Biol Control:83–88. http://www.zgswfz.com.cn/EN/Y2013/V29/I1/83

Shaheed Siddiqui Z, Imran M, Khan A, Ali I, Dai D-Q, D-q D, Khan A, Ali A, Ullah Z, Khan N, Khan A, Rahman Al-Tawaha A, Sher H (2022) Iron and zinc micronutrients and soil inoculation of *Trichoderma harzianum* enhance wheat grain quality and yield. Front Plant Sci 13:1–10. https://doi.org/10.3389/fpls.2022.960948

Shalaby TA, Taha N, El-Beltagi HS, El-Ramady H (2022) Combined application of *Trichoderma harzianum* and paclobutrazol to control root rot disease caused by *Rhizoctonia solani* of Tomato Seedlings. Agronomy 12(12). https://doi.org/10.3390/agronomy12123186

Shavkiev J, Karimov HK, Turaeva BI, Azimova NS, Nazirbekov MK, Khamidova KM (2022) Some volatile metabolites produced by the antifungal-*Trichoderma asperellum* UZ-A4 micromycete. Int J Plant Pathol 11(3):239–251. https://doi.org/10.33687/phytopath.011.03.4263

Shentu X, Zhan X, Ma Z, Yu X, Zhang C (2014) Antifungal activity of metabolites of the endophytic fungus *Trichoderma brevicompactum* from garlic. Braz J Microbiol 45(1):248–254. https://doi.org/10.1590/S1517-83822014005000036

Shi M, Chen L, Wang XW, Zhang T, Zhao PB, Song XY, Sun CY, Chen XL, Zhou BC, Zhang YZ (2012) Antimicrobial peptaibols from *Trichoderma pseudokoningii* induce programmed cell death in plant fungal pathogens. Microbiology 158(1):166–175. https://doi.org/10.1099/mic.0.052670-0

Shoresh M, Harman GE, Mastouri F (2010) Induced systemic resistance and plant responses to fungal biocontrol agents. Annu Rev Phytopathol 48:21–43. https://doi.org/10.1146/annurev-phyto-073009-114450

Singh BN, Dwivedi P, Sarma BK, Singh GS, Singh HB (2018) *Trichoderma asperellum* T42 reprograms tobacco for enhanced nitrogen utilization efficiency and plant growth when fed with N nutrients. Front Plant Sci 9. https://doi.org/10.3389/fpls.2018.00163

Singh SK, Singh PN, Maurya DK, Rana S (2020) Advances in systematics, taxonomy, and conservation of *Trichoderma* species. In: Varma A (ed) *Trichoderma*: agricultural applications and beyond, Soil biology, vol 61. Springer, Cham. https://doi.org/10.1007/978-3-030-54758-5_1

Soltaniband V, Brégard A, Gaudreau L, Dorais M (2022) Biostimulants promote plant development, crop productivity, and fruit quality of protected strawberries. Agronomy 12(7). https://doi.org/10.3390/agronomy12071684

Sousa TF, Vieira Reça BNP, Castro GS, da Silva IJS, Caniato FF, de Araújo Júnior MB, Yamagishi MEB, Koolen HHF, Bataglion GA, Hanada RE, da Silva GF (2023) *Trichoderma agriamazonicum* sp. nov. (Hypocreaceae), a new ally in the control of phytopathogens. Microbiol Res 275. https://doi.org/10.1016/j.micres.2023.127469

Stange P, Seidl S, Karl T, Benz JP (2023) Evaluation of *Trichoderma* isolates as biocontrol measure against *Claviceps purpurea*. Eur J Plant Pathol 167(4):651–675. https://doi.org/10.1007/s10658-023-02716-w

Stempien E, Pierron RJG, Adendorff I, Van Jaarsveld WJ, Halleen F, Mostert L (2020) Host defence activation and root colonization of grapevine rootstocks by the biological control fungus *Trichoderma atroviride*. Phytopathol Mediterr 59(3):615–626. https://doi.org/10.14601/Phyto-11137

Stracquadanio C, Quiles JM, Meca G, Cacciola SO (2020) Antifungal activity of bioactive metabolites produced by *Trichoderma asperellum* and *Trichoderma atroviride* in liquid medium. J Fungi 6(4):1–18. https://doi.org/10.3390/jof6040263

Syed A, Elgorban AM, Bahkali AH, Eswaramoorthy R, Iqbal RK, Danish S (2023) Metal-tolerant and siderophore producing *Pseudomonas fluorescence* and *Trichoderma spp.* improved the growth, biochemical features and yield attributes of chickpea by lowering Cd uptake. Sci Rep 13(1). https://doi.org/10.1038/s41598-023-31330-3

Tamandegani PR, Marik T, Zafari D, Balázs D, Vágvölgyi C, Szekeres A, Kredics L (2020) Changes in peptaibol production of *Trichoderma* species during in vitro antagonistic interactions with fungal plant pathogens. Biomol Ther 10(5). https://doi.org/10.3390/biom10050730
Tarus PK, Lang'at-Thoruwa CC, Wanyonyi AW, Chhabra SC (2003) Bioactive metabolites from *Trichoderma harzianum* and *Trichoderma longibrachiatum*. Bull Chem Soc Ethiop 17(2):185–190. https://doi.org/10.4314/bcse.v17i2.61675
Tomah AA, Alamer ISA, Khattak AA, Ahmed T, Hatamleh AA, Al-Dosary MA, Ali HM, Wang D, Zhang J, Xu L, Li B (2023) Potential of *Trichoderma virens* HZA14 in controlling *Verticillium* wilt disease of eggplant and analysis of its genes responsible for microsclerotial degradation. Plan Theory 12(21). https://doi.org/10.3390/plants12213761
Tripathi R, Keswani C, Tewari R (2021) *Trichoderma Koningii* enhances tolerance against thermal stress by regulating ROS metabolism in tomato (*Solanum lycopersicum* L.) plants. J Plant Interact 16(1):116–126. https://doi.org/10.1080/17429145.2021.1908634
Tyśkiewicz R, Nowak A, Ozimek E, Jaroszuk-ściseł J (2022) *Trichoderma*: the current status of its application in agriculture for the biocontrol of fungal phytopathogens and stimulation of plant growth. Int J Mol Sci 23(4). https://doi.org/10.3390/ijms23042329
Vargas WA, Mukherjee PK, Laughlin D, Wiest A, Moran-Diez ME, Kenerley CM (2014) Role of gliotoxin in the symbiotic and pathogenic interactions of *Trichoderma virens*. Microbiology 160:2319–2330. https://doi.org/10.1099/mic.0.079210-0
Vinale F, Marra R, Scala F, Ghisalberti EL, Lorito M, Sivasithamparam K (2006) Major secondary metabolites produced by two commercial *Trichoderma* strains active against different phytopathogens. Lett Appl Microbiol 43(2):143–148. https://doi.org/10.1111/j.1472-765X.2006.01939.x
Vinale F, Sivasithamparam K, Ghisalberti EL, Marra R, Barbetti MJ, Li H, Woo SL, Lorito M (2008) A novel role for *Trichoderma* secondary metabolites in the interactions with plants. Physiol Mol Plant Pathol 72(1–3):80–86. https://doi.org/10.1016/j.pmpp.2008.05.005
Vinale F, Ghisalberti EL, Sivasithamparam K, Marra R, Ritieni A, Ferracane R, Woo S, Lorito M (2009) Factors affecting the production of *Trichoderma harzianum* secondary metabolites during the interaction with different plant pathogens. Lett Appl Microbiol 48(6):705–711. https://doi.org/10.1111/j.1472-765X.2009.02599.x
Vinale F, Strakowska J, Mazzei P, Piccolo A, Marra R, Lombardi N, Manganiello G, Pascale A, Woo SL, Lorito M (2016) Cremenolide, a new antifungal, 10-member lactone from *Trichoderma cremeum* with plant growth promotion activity. Nat Prod Res 30(22):2575–2581. https://doi.org/10.1080/14786419.2015.1131985
Woo SL, Hermosa R, Lorito M, Monte E (2023) *Trichoderma*: a multipurpose, plant-beneficial microorganism for eco-sustainable agriculture. Nat Rev Microbiol 21(5):312–326. Nature Research. https://doi.org/10.1038/s41579-022-00819-5
Xia H, Li YY, Liu Z, Li YQ, Chen J (2018) Transgenic expression of chit42 gene from *Metarhizium anisopliae* in *Trichoderma harzianum* enhances antagonistic activity against *Botrytis cinerea*. Mol Biol 52:668–675. https://doi.org/10.1134/S002689331805014X
Yadav SS, Arya A, Singh V, Singh Y (2023) Elicitation of native bio protective microbial agents associated systemic defense responses and plant growth promotion against bacterial stalk rot pathogen in sorghum (*Sorghum bicolor*). Phytopathol Res 5(1). https://doi.org/10.1186/s42483-023-00202-z
Yasmeen R, Siddiqui ZS (2017) Physiological responses of crop plants against *Trichoderma harzianum* in saline environment. Acta Bot Croat 76(2):154–162. https://doi.org/10.1515/botcro-2016-0054
Younes NA, Anik TR, Rahman MM, Wardany AA, Dawood MFA, Tran LSP, Abdel Latef AAH, Mostofa MG (2023) Effects of microbial biostimulants (*Trichoderma album* and *Bacillus megaterium*) on growth, quality attributes, and yield of onion under field conditions. Heliyon 9(3). https://doi.org/10.1016/j.heliyon.2023.e14203
Yu C, Jiang X, Xu H, Ding G (2023) *Trichoderma longibrachiatum* inoculation improves drought resistance and growth of *Pinus massoniana* seedlings through regulating physiological responses and soil microbial community. J Fungi 9(7). https://doi.org/10.3390/jof9070694

Zhang JL, Tang WL, Huang QR, Li YZ, Wei ML, Jiang LL, Liu C, Yu X, Zhu HW, Chen GZ, Zhang XX (2021) *Trichoderma*: a treasure house of structurally diverse secondary metabolites with medicinal importance. Front Microbiol 12:723828. https://doi.org/10.3389/fmicb.2021.723828

Zhao L, Zhang YQ (2015) Effects of phosphate solubilization and phytohormone production of *Trichoderma asperellum* Q1 on promoting cucumber growth under salt stress. J Integr Agric 14(8):1588–1597. https://doi.org/10.1016/S2095-3119(14)60966-7

Zúñiga-Silgado D, Sánchez-Reyes A, Ortiz-Hernández ML, Otero M, Balcázar-López E, Valencia-Díaz S, Serrano M, Coleman JJ, Sarmiento-López L, De-Bashan LE, Folch-Mallol JL (2022) Soil type influences novel "milpa" isolates of *Trichoderma virens* and *Aspergillus tubingensis* that promote solubilization, mineralization, and phytoabsorption of phosphorus in *Capsicum annuum* L. J Fungi 8(11). https://doi.org/10.3390/jof8111178

10 Endophytic Fungi as Promising Sources of Antiviral Compounds for Plant Virus Pathogens

Anbazhagan Sathiyaseelan, Kumar Vishven Naveen, Zhang Xin, Kiseok Han, Lu Yuting, Zhang Lina, Patrick Brice Defo Deeh, and Myeong-Hyeon Wang

Abstract

Endophytic fungi, which reside in medicinal plants, are considered potential sources of therapeutic production by mimicking the host plant metabolisms. Additionally, they have the capability to produce novel metabolites through complex gene expression. Endophytes, their crude extracts, or active molecules derived from them are employed for plant protection, particularly against plant pathogenic viruses. The utilization of biological agents such as endophytic fungi in disease management and the development of agricultural systems, as opposed to chemical agents, is deemed environmentally safe. Recognizing the significance of endophytes, numerous studies have been conducted. Therefore, this book chapter explores the discussion of endophytes and their active molecules, specifically exploring their antiviral properties against plant viruses, methods for identifying active molecules, molecular mechanisms underlying the inhibition of pathogens, agricultural applications of endophytes and their derivatives, biotechnological approaches in novel metabolite production from endophytes, environmental safety, challenges in the application of endophytes, and potential future applications.

Anbazhagan Sathiyaseelan, Kumar Vishven Naveen, Zhang Xin, Kiseok Han, Lu Yuting, Zhang Lina, Patrick Brice Defo Deeh and Myeong-Hyeon Wang contributed equally with all other contributors.

A. Sathiyaseelan (✉) · K. V. Naveen · Z. Xin · K. Han · L. Yuting · Z. Lina · P. B. D. Deeh
M.-H. Wang (✉)
Department of Bio-Health Convergence, Kangwon National University, Chuncheon, Republic of Korea
e-mail: mhwang@kangwon.ac.kr

© The Author(s), under exclusive license to Springer Nature Singapore Pte Ltd. 2025

K. A. Abd-Elsalam, A. H. Hashem (eds.), *Fungal Endophytes Volume II*,
https://doi.org/10.1007/978-981-97-8804-0_10

Keywords

Medicinal plants · Endophytic fungi · Plant viruses · Antiviral compounds · Agriculture applications

10.1 Introduction

Endophytic fungi (EF) represent distinctive microorganisms inhabiting various plant tissues, predominantly the bark, stem, leaves, roots, and seeds, where they establish colonization in the intercellular/extracellular spaces. EF exhibits associations with a wide range of plant species, and their symbiotic presence contributes significantly to the host plant's growth while concurrently safeguarding it against diverse external stressors, all without inducing diseases (Alam et al. 2021). His symbiotic relationship holds direct relevance to agricultural applications and economic development. Furthermore, EF demonstrates a close interconnection with host plant metabolisms, influencing the production of metabolites. They possess the capability to mimic host plant metabolism through intricate gene expression mechanisms (Alam et al. 2021; Wen et al. 2022). The study revealed that six phytocompounds (5-methyl-6-prenyl-isobenzofuran-1(3H)-one, 5-hydroxymethyl-6-prenyl-isobenzofuran-1(3H)-one, nigrolineabenzopyran A, muralatin H, dehydrozingerone, and (2R)-12-hydoxy-4-methoxy-tremeton) isolated from *Nicotiana tabacum* displayed strong anti-TMV activities (inhibition rates, 16.9–26.2%) (Shen et al. 2016). Therefore, the EF from the medicinal plants could have the metabolites with medicinal properties (Adeleke and Babalola 2021; Kaul et al. 2012). Generally, the EF has various classes of metabolites such as terpenoids, alkaloids, steroids, polyketides, quinone, lactone, azaphilone, isoindolone, etc. (Lacerda et al. 2022; Wen et al. 2022). Hence, EF is considered the reservoir of novel metabolites and microbial cell factories to produce metabolites with various therapeutic properties (antioxidant, anti-inflammatory, anticancer, antibacterial, antifungal, antiviral, etc.) (Adeleke and Babalola 2021). EF such as *Aspergillus* sp., *Rhizopus* sp., *Phoma* sp., *Phomopsis* sp., *Diaporthe* sp., *Penicillium* sp., etc. have been reported widely (Luo et al. 2018; Xu et al. 2021a).

Antiviral compounds isolated from EF and exerted their antiviral activity against various viruses. For example, 14-nordrimane isolated from *Phoma* sp. exhibited inhibitory activity against Influenza A (H1N1) (Liu et al. 2019) and dankasterone A isolated from *Phomopsis* sp. exhibited antiviral activity against Influenza A (H5N1) (Ma et al. 2014). Besides, altertoxins V and VI isolated from *Alternaria tenuissima* showed antiviral activity against HIV (Bashyal et al. 2014); *pestalotiopsone* F from *Diaporthe* sp. showed inhibitory activity against H1N1 (Luo et al. 2018); periconiasin B, from *Periconia* sp., displayed the anti-HIV activity (Zhang et al. 2016a); and hydroanthraquinone derivative (7S)-7-hydroxy-3,7-dimethyl-isochromene-6,8-dione isolated from *Nigrospora* sp. exhibited anti-H1N1 activity (Zhang et al. 2016b). Also, emerimidines A and B, emeriphenolicins A and D, aspernidines A and B, austin, austinol, dehydroaustin, and acetoxydehydroaustin isolated from

Emericella sp. displayed the anti-H1N1 activity (Zhang et al. 2011). Further, the study reported that isolating the alternariol and alternariol-(9)-methyl ether from the *Pleospora tarda* associated with the medicinal plant *Ephedra aphylla* have antiviral activity against herpes simplex virus type-2 (HSV-2) and vesicular stomatitis virus (VSV) (Selim et al. 2018). Further, the study found that *Chaetomium globosum* associated with *Adiantum capillus-veneris* exhibited potent antioxidant activity (Selim et al. 2018). However, kheiric acid isolated from *Curvularia papendorfii* showed antibacterial activity but not inhibitory activity against human coronavirus HCoV 229E and a norovirus surrogate, the feline coronavirus FCV F9 (Khiralla et al. 2020).

Nevertheless, the primary focus of this book chapter is to address the subject matter of endophytic fungi (EF) holding antiviral compounds, particularly those effective against plant viruses. Consequently, the chapter is structured into distinct subtopics, encompassing the diversity of endophytic fungi exhibiting antiviral potential, elucidation of the mechanisms underlying the antiviral actions facilitated by endophytic fungi, the identification and characterization of antiviral compounds, potential applications in agriculture and crop protection, biotechnological approaches for harnessing the antiviral potential, as well as an exploration of the challenges and opportunities inherent in the development of antiviral agents. Furthermore, each subtopic is thoroughly examined with a detailed presentation of experimental results.

10.2 Diversity of Endophytic Fungi with Antiviral Potential

Antiviral compounds from endophytic fungi associated with medicinal plants could be an alternative therapy for the treatment of various ailments, including viral infections in plants. Many studies reported the effectiveness of some endophytic fungi-derived compounds against common viral infections in humans (Lacerda et al. 2022). However, only a few studies on the beneficial effects of active compounds from endophytic fungi against plant viruses have been found in the literature (Table 10.1). A recent study conducted by Yang et al. (2023) revealed the effectiveness of some new isochromenes against the tobacco mosaic virus (TMV) with inhibition rates of 26.2–30.5% (Yang et al. 2023). Yang et al. (2022) reported the anti-tobacco mosaic virus activity of aspergillines F–J, aspergilline A, aspergilline C, and cyclopiamide E, extracted from the *Nicotiana tabacum*-derived fungus *Aspergillus versicolor*. According to these authors, aspergillines G and J exhibited the strongest antiviral activity with inhibition rates of 41.2 and 56.8%, respectively (Yang et al. 2022). Moreover, butyrolactones aspernolides C and D obtained from an endophytic fungi *Aspergillus versicolor* associated with the Chinese plant *Paris polyphylla* have been reported as potent antiviral compounds for the treatment of TMV, aspernolide D being the most effective (IC50 values: 64.2 μM) (Zhou et al. 2015).

Table 10.1 Compounds from endophytic fungi associated with medicinal plants against plant viruses

Compounds	Endophytic fungus	Host plant	Antiviral activity against	References
(5-Methoxy-7-prenyl-1H-isochromen-3-yl)methanol, 3-(3-(Hydroxymethyl)-5-methoxy-1H-isochromen-7-yl) propan-1-ol, 5-Methoxy-7-methyl-1H-isochromen-3-yl)methanol	*Aspergillus versicolor*	*Nicotiana tabacum*	Anti-tobacco mosaic virus	Yang et al. (2023)
Aspergillines F–J, aspergilline A, aspergilline C, and cyclopiamide E	*Aspergillus versicolor*	*Nicotiana tabacum*	Anti-tobacco mosaic virus	Yang et al. (2022)
Butyrolactones aspernolides C and D	*Aspergillus versicolor*	*Paris polyphylla*	Anti-tobacco mosaic virus	Zhou et al. (2015)

10.3 Mechanisms of Antiviral Action by Endophytic Fungi

Fungal endophytes reside within plant tissues without causing noticeable symptoms in their hosts. When plants are associated with fungal endophytes, their defense mechanisms are triggered more swiftly in response to external stimuli. It is believed that endophytic fungi share a symbiotic relationship with the host plants, namely, acting as antagonists to plant diseases, supporting rhizosphere inhabitants, enhancing tolerance to environmental conditions, and promoting plant growth (Rodriguez et al. 2009). The explanation behind the increased tolerance to pathogens in endophyte-containing plants remains primarily speculative, with the hypothesis that secondary compounds generated by fungal endophytes might contribute partially to this phenomenon. When endophytes inhabit plants, they generate enzymes and secondary metabolites that actively suppress the actions of plant pathogens, which is recognized as a significant factor in plant defense mechanisms against attacks by insects and pathogens (Gao et al. 2010). For instance, the fungal metabolite 3-methyl-2-arylbenzofurans, isolated from endophytic *Phomopsis* sp., are reported to exhibit *Tobacco mosaic virus* (TMV) infection inhibitory activities through some unknown mechanisms (Du et al. 2017). However, it is thought that the plants, endophytes, and viruses may interact in ways that prevent viral growth or multiplication to induce resistance to viral infections.

Alterations in the characteristics of host plants could impact viral diseases in two ways: directly through the influence of host metabolites and indirectly through the effects of plant quality on insect vectors responsible for transmitting the viruses (Muvea et al. 2018). The systemic colonization of endophytic fungi within plants, moving through intercellular spaces and vascular xylem elements, can impede or disrupt the systemic spread of plant viruses from one cell to another, which ultimately leads to delayed duplication and spread of plant viruses (Martelli 1980). In that context, a case study investigating the spread and replication of *Zucchini yellow mosaic virus* (ZYMV) discovered reduced levels of the virus in plants (squash) colonized by endophytic *Beauveria bassiana* compared to plants without these endophytes (Jaber and Salem 2014). Moreover, according to Lehtonen et al., there was a decreased incidence of *barley yellow dwarf virus* (BYDV) infections observed in meadow ryegrass (*Lolium pretense*) plants colonized by endophytic *Neotyphodium uncinatum* compared to those lacking endophytes. This suggests that endophyte colonization may serve as a protective factor against virus infections and subsequent duplication in plants (Lehtonen et al. 2006). Similarly, in a recent study, endophytic *Trichoderma harzianum* and *Trichoderma atroviride* colonization in maize plants resulted in different degrees of protection (1.4-fold and 2.7-fold, respectively) against *Sugarcane mosaic virus* (SCMV). This protection manifested as delayed symptom development, decreased disease severity, and lowered levels of the virus within the plants (Kiarie et al. 2020).

Furthermore, the presence of endophytes within plants is acknowledged to impact the population dynamics of disease-causing insect vectors. For instance, aphids acquire the virus through feeding on contaminated plants and act as medium for transmitting disease to other plants but do not pass the infection to their offspring. An investigation by Muvea et al. observed that the presence of endophytic *Hypocrea lixii* significantly decreased the feeding behavior, egg-laying activity, and overall survival rates of thrips (*Thrips tabaci*) in onion plants (Muvea et al. 2014). Further, replication of *Iris yellow spot virus* (IYSV) was substantially reduced in endophytic *Hypocrea lixii*-colonized onion plants than in endophyte-free plants (Muvea et al. 2018). This indicates that the combination of minimal feeding by viruliferous thrips (on endophyte-colonized plants) and lowered ability of the virus to duplicate in endophyte-colonized plants are involved simultaneously in catering to IYSV infection in onion plants. In this context, Lehtonen et al. also observed decreased transmission and reproduction of the aphid-transmitted *Luteovirus*, BYDV, in plants colonized by endophytes (*Neotyphodium uncinatum*) compared to non-colonized plants and hypothesized that the decreased virus infection frequency in endophyte-colonized meadow ryegrass was likely due to alterations in aphid feeding behaviors on these plants (Lehtonen et al. 2006). This also could be supported by the fact that the activation of gene expression in defense pathways by endophytic fungi leads to an augmented production of defensive compounds in the plant (Pieterse et al. 2014; Shrivastava et al. 2015). Similarly, the simultaneous presence of endophytic *Beauveria bassiana* and mycorrhizae in tomato plants can notably enhance the terpenoid levels in leaves. This increase in terpenoids results in a decrease in herbivores feeding on the foliage, which also corroborates the fact that alkaloids released by endophytic fungi exhibit antiviral properties (Selim et al. 2012).

The antiviral action of endophytic fungi relies on multiple elements, such as plant species, fungal diversity, virus types, and environmental conditions, which

Fig. 10.1 Disease symptom development in maize plants inoculated with fungal isolates, (A) (*Metarhizium anisopliae*-ICIPE 69), (B) (*Trichoderma atroviride*-F5S21), (C) (*Hypocrea lixii*-F3ST1), and (D) (*Trichoderma harzianum*-F2L4) following artificial inoculation with sugarcane mosaic virus (SCMV). Panel (E) represents the positive control, where untreated maize plants were challenge-inoculated with SCMV, while (F) represents the negative control, where plants were neither treated with fungal endophytes nor challenge-inoculated with SCMV. The arrows indicate the onset of symptom expression in various treatments. Challenge inoculation with sugarcane mosaic virus was conducted 21 days after planting (Kiarie et al. 2020)

influence the complex interactions among the host plant, endophytes, viruses, and insect vectors. For example, in the presence of high-intensity light, the endophytic *Piriformospora indica* colonized tomato plant appeared with a lowered concentration of *Pepino mosaic virus*-Sav E397 (PMV) and with increased fruit biomass compared to control plants (Fakhro et al. 2010). Notably, the light conditions (low/high intensity) influence the photosynthetic rates, and so do the carbohydrate contents within the cells and subsequently induce systemic resistance against viruses. On the other hand, endophytic *Trichoderma harzianum* and *Trichoderma atroviride* colonization protected the maize plants against SCMV but did not confer protection against *Maize chlorotic mottle virus* (MCMV) infection (Kiarie et al. 2020). The lack of resistance against MCMV could result from the fungal endophytes' specificity in protecting against particular plant viruses (Fig. 10.1). Further, some studies involving endophytic fungi-mediated plant protection against plant viruses are

Table 10.2 Summary of the reports suggesting endophytic fungi involvement in the management of plant viruses

Endophytic fungi	Isolation source	Targeted plant virus	Targeted diseased plant	References
Beauveria bassiana	*Busseola fusca* (Fuller) and *Pachnoda interrupta* (Olivier)	*Zucchini yellow mosaic virus* (ZYMV)	Squash plants	Jaber and Salem (2014)
Neotyphodium uncinatum	Meadow ryegrass (*Lolium pretense*)	*Barley yellow dwarf virus* (BYDV)	*Lolium pratense*	Lehtonen et al. (2006)
Trichoderma harzianum and *Trichoderma atroviride*	Onion	*Sugarcane mosaic virus* (SCMV)	Maize	Kiarie et al. (2020)
Piriformospora indica	–	*Pepino mosaic virus*-Sav E397	Tomato	Fakhro et al. (2010)
Hypocrea lixii	Maize	*Iris yellow spot virus* (IYSV)	Onion	Muvea et al. (2018)

summarized in Table 10.2. Taken altogether, further research is necessary to delve deeper into the mechanisms by which endophytic fungi contribute to an antiviral response in managing viral diseases affecting plants.

10.4 Identification and Characterization of Antiviral Compounds

The infection of plant viruses can threaten the yield and application of many food and cash crops by causing serious economic losses (Miklos et al. 2018). Affected plants are usually discolored, necrotic, and deformed. To date, chemical agents have been widely used to control plant viruses. This method is highly efficient, but the widespread use of chemical reagents causes environmental pollution (Brewer et al. 2018). The application of endophytic fungi in plant virus control can control both plant diseases and minimize pollution. Specifically, the main anti-plant virus components in endophytic fungi are polysaccharides, peptides, and proteins. The direct use of such substances increases the efficiency of endophytic fungi against plant viruses. However, isolation and extraction, preliminary screening tests for anti-plant virus activity, identification, and characterization are required before use (Lehtonen et al. 2006). Isolation of components from endophytic fungi with antiviral effects in plants for direct application significantly improves the efficiency of application without the existence of endophytic fungi-related toxicity (Kiarie et al. 2020). The general procedure is as follows: (1) culturing the endophytic fungi without contamination at the required time interval, and (2) the crude extract is extracted from fungal mycelium using appropriate solvents such as ethyl acetate, methanol, and n-butanol. Before extraction, the mycelium can be homogenized with homogenizer and liquid nitrogen, and then solvent extraction can be carried out to improve the extraction efficiency. (3) The whole extraction process needs to be repeated 2–3 times. (4) Preliminary screening for antiviral activity test was carried out to ensure the antiviral components in the crude extract. (5) Finally, the characterization was carried out using various analytical techniques such as UV spectroscopy, IR Fourier, nuclear magnetic resonance spectroscopy (NMR), fast atom bombardment ionization-mass spectrometry (FAB-MS), and electrospray (ESI-MS) ionization (Kiarie et al. 2020; Lehtonen et al. 2006; Muvea et al. 2018).

10.4.1 Identification and Characterization Methods

10.4.1.1 Thin-Layer Chromatography (TLC)

TLC is a laboratory test technique for the rapid separation and qualitative analysis of small amounts of substances. It is often used for the identification of pharmaceuticals, the detection of impurities, and the determination of content. The application of thin-layer chromatography can be briefly described as the separation of a mixture of antiviral crude by a glass plate or aluminum coated to form a homogeneous silica thin layer. After spotting, the antiviral components are identified based on the specific antiviral actives and isolated by using the same method. The principle is that each compound/antiviral compound has a different adsorption capacity to the same adsorbent to achieve mutual separation. The process involves adsorption, desorption, re-adsorption, and re-desorption (Lisboa 1969; Skipski et al. 1964).

10.4.1.2 High-Performance Liquid Chromatography (HPLC)

HPLC is often used in the fields of medicine, chemistry, agronomy, industry, and testing. It is often used to separate mixtures, analyze quantitatively, and detect purity. The advantages of this method include high pressure, high speed, high efficiency, high sensitivity, a wide range of applications, reusable columns, low sample content, and easy recovery. Commonly used instruments are high-performance liquid chromatography. It mainly consists of several parts such as a liquid reservoir, pump, injector, column, detector, recorder, and so on. Its working principle is that the component material in the mobile phase and the stationary phase do relative movement and repeated times of adsorption and desorption so that the components in the speed of movement produce a large difference. As a result, the components are separated into individual components and passed through the detector, and the concentration of each component is finally detected (Horváth 2013; Reuhs 2017).

10.4.1.3 Ultraviolet-Visible Spectroscopy (UV-vis)

UV-vis is a method for studying the relative intensity of light absorption by molecules of a substance. It consists of five main parts: light source, monochromator, absorption cell, detector, and signal display system. The ultraviolet absorption spectrum is the absorption spectrum produced by the electronic energy level jump after certain groups in the sample molecule absorb the ultraviolet-visible region. As various substances have their different molecules, atoms, and different space between the atoms, their absorption of light energy will not be the same; therefore, each substance has its own unique, fixed absorption spectrum curve, according to the absorption spectrum of certain characteristic wavelengths of the content that can be identified or measured (Akash et al. 2020; Lai et al. 2012).

10.4.1.4 Infrared Spectra Analysis (IR)

Infrared spectroscopy refers to the formation of an infrared absorption spectrum by irradiating a beam of infrared rays of different wavelengths onto the molecules of a target substance, where some specific infrared rays are absorbed. The method is the structural analysis and identification of the molecules of the target substance (Haris and Chapman 1995). The infrared spectrometer typically comprising a light source, a monochromator, a detector, and a computer processing system. The working principle can be briefly described as follows: when the sample material is irradiated with infrared light of varying frequencies, molecular vibrations within the sample absorb specific frequencies that match their vibrational energies. This absorption, due to vibrational resonance, results in an infrared spectrum unique to the molecular structure of the sample, as each functional group within the molecule absorbs at characteristic frequencies (Hsu 1997).

10.4.1.5 Nuclear Magnetic Resonance Spectroscopy (NMR)

NMR spectroscopy is a technique that applies the phenomenon of nuclear magnetic resonance to determine the structure of molecules. The instrument used was an NMR spectrometer. The principle of operation can be described simply as the splitting of energy levels in the nucleus in a strong magnetic field, and when external electromagnetic radiation is absorbed, a jump in the nuclear energy level occurs. Similar to infrared spectroscopy, NMR spectroscopy not only provides information on the type and number of chemical functional groups in the antiviral components but also allows the analysis of the structure and conformation of proteins, nucleic acids, and sugars. Moreover, the method is not limited to the solids state but also to the liquids state (Bovey et al. 1988; Hatzakis 2019).

10.4.1.6 Electrospray Ionization-Mass Spectrometry (ESI-MS)

ESI is a technique in which an electrospray applies a high voltage to a liquid to generate an aerosol to produce ions for mass spectrometry. It can be broadly divided into three stages: droplet formation, desolation, and gas-phase ion formation. The principle can be briefly described as follows: An antiviral component is subjected to high voltage to form a charged droplet. The charge on the surface of the droplet increases as the solvent evaporates, forming one or more charged ions. Eventually, the antiviral component enters the gas phase as a single or multiple charge and becomes a gas phase ion. ESI offers sensitivity, accuracy, and sophistication for the ionization of highly polar, thermally unstable proteins and the detection of large organic molecules particularly suitable for the characterization of plant antiviral compounds (e.g., proteins, peptides) and in endophytic fungi (Chupina et al. 2022). High-resolution electrospray ionization mass spectrometry (HRESI-MS) was also applied to characterize the antiviral components. The addition of high-resolution mass spectrometry improves the accuracy of ESI-MS for identifying unknowns.

10.4.1.7 Gas Chromatography-Mass Spectrometry (GC-MS)

GC-MS is the category that can be used to identify antiviral compounds. Headspace injection is often used. It is suitable for the analysis of volatile components. The instrument used is a gas chromatograph. The principle is that when a multicomponent antiviral mixture of components enters the chromatographic column, due to the adsorbent's adsorption force on each component is different, the speed of each component in the chromatographic column is different, and the sequence of leaving the chromatographic column is also different. In this way, the different components of the antiviral components can be separated from each other in the chromatographic column and sequentially enter the detector to be detected and recorded (Semreen et al. 2020). From the current research, studies on identifying and characterizing anti-plant virus components isolated from endophytic fungi are fewer and in their infancy. Du et al. (2017) identified and characterized anti-TMV components of endophytic fungus *Phomopsis* sp. Their structural form is shown in Fig. 10.2. The 1, 2, and 3 belong to the new arylbenzofurans and 4–7 are known. The methods used in the identification and characterization of the three new arylbenzofurans were RP-HPLC, UV, IR, NMR, ESI-MS, and HRESI-MS.

Seven compounds were isolated by RP-HPLC and then compared using ^{1}H and ^{13}C NMR analyses to determine the compound species. The molecular mass was determined using HR-ESI-MS. For example, the molecular formula of component 1 was $C_{22}H_{24}O_3$. The UV maximum absorption wavelengths were found at 210, 275, and 315 nm. The IR adsorptions were found at 3416, 1625, 1498, and 1455 cm^{-1}.

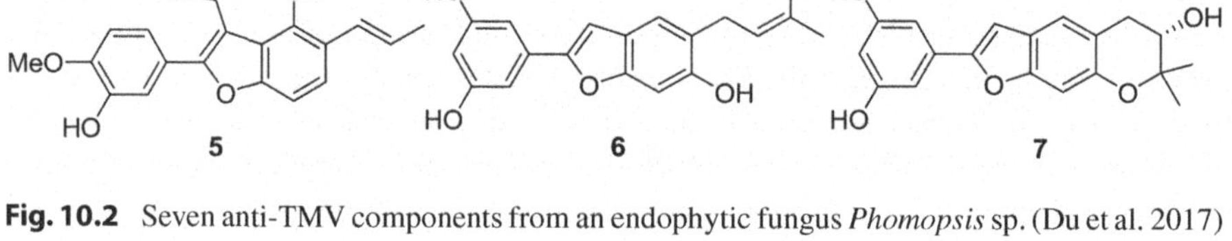

Fig. 10.2 Seven anti-TMV components from an endophytic fungus *Phomopsis* sp. (Du et al. 2017)

Fig. 10.3 The key HMBC correlations of component 1 (Du et al. 2017)

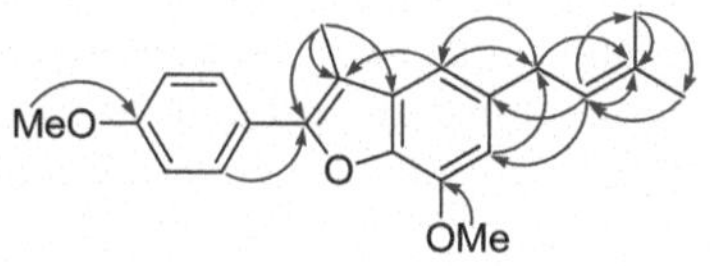

And, from the ^{1}H and ^{13}C NMR and DEPT analysis, 22 carbon and 24 proton signals were displayed. The 3-methyl-2-arylbenzofuran skeleton of component 1 was also assigned by the HMBC correlations (Fig. 10.3).

Overall, proficiency in these identification and characterization techniques is beneficial for furthering our understanding of plant virus-resistant components in endophytic fungi and improving the efficiency of their utilization.

10.5 Potential Applications in Agriculture and Crop Protection

Uncontrolled and indiscriminate use of pesticides and fertilizers is directly linked to environmental pollution and human health problems. Chemical residues from crops can be found in vegetables, fruits, and grains. In particular, excessive use of fertilizers can cause soil acidification, reservoir contamination, and heavy metal accumulation (Savci 2012). In addition, residual pesticides in crops can cause respiratory and endocrine disorders when humans are exposed and have a lasting effect on numerous organisms existing in the soil and aquatic environment (van Lexmond et al. 2015). To solve cost problems related to pollution emissions and misuse of chemical pesticides and to respond to demands for increased agricultural productivity, various studies are being conducted to reduce the use of pesticides and fertilizers or replace them through environmentally friendly methods (Ahmad et al. 2018).

Among various methods for sustainable agriculture, discussions on the connection between EF and the agricultural industry have recently received more attention and are being studied in-depth (Rai et al. 2014; Rodriguez et al. 2009). 'Sustainable agriculture' refers to a variety of ideas and technologies that increase or maintain food production levels while minimizing environmental pollution and threats to human health (Le Cocq et al. 2017). Among various methods for sustainable agriculture, the method of utilizing EF is attracting attention concerning improving plant growth and tolerance to various biological and physical stresses (Fig. 10.4). The method using EF is biological control agents (BCAs) that interact symbiotically within plant tissue without causing significant damage to the host plant. Both the host and EF benefit from the interaction. EF are protected from external environmental stresses and benefit nutritionally from plants. In addition, EF can help plants grow against negative factors such as pests and viruses through compounds such as alkaloids (Peixoto Neto et al. 2002). The direct benefits of interaction with endophytes include the acquisition of nutrients and an increase in the number of phytohormones in the plant, which is directly related to increased biomass production, expanded root system development, plant height, weight reproduction, and yield (Bamisile et al. 2018).

Various studies are being conducted to confirm that crops growing in various regions have tolerance to specific types of stress to which they are exposed through EF so that they can be applied to crop farming (Table 10.3). The types of stress that crops can experience can be broadly divided into two types. The first is biological stress such as pests, animals, nematodes, viruses, etc. Second, there are abiotic

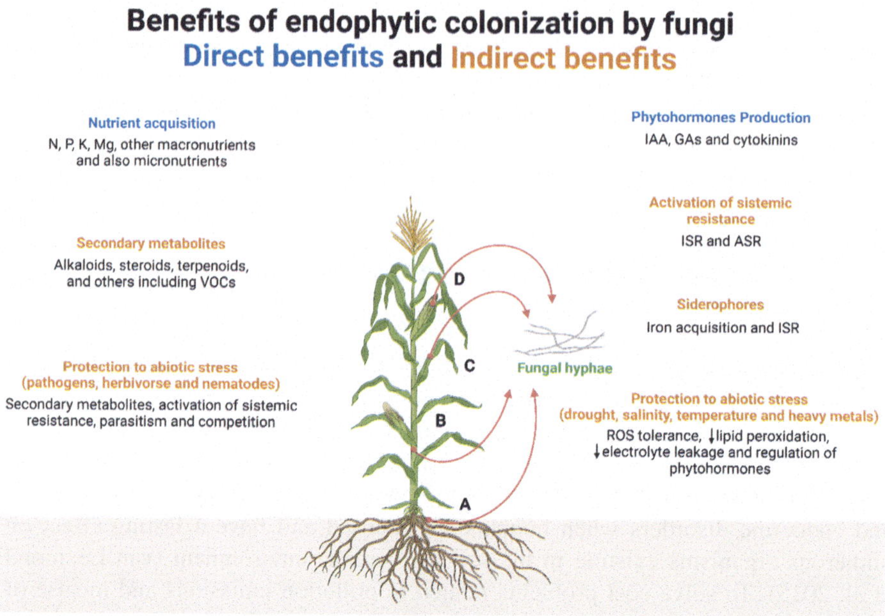

Fig. 10.4 Benefits provided to plants by endophytic colonization with fungi. Endophytic colonization can occur in the tissues of one or more parts of the host plant, including roots (A), stems (B), leaves (C), reproductive systems, and fruits (D) (Baron and Rigobelo 2022)

Table 10.3 Plant's biotic stress management through endophytic fungi

Plant's biotic stress management through endophytic fungi	Fungal strain	References
The nematicidal ability of endophytic fungi for host plant	*Fusarium oxysporum*	Diedhiou et al. (2003), Zum Felde et al. (2009)
Phytohormone production ability of endophytic fungi for host plant	*Fusarium tricinctum* *Alternaria alternate* *Piriformospora indica*	Johnson et al. (2014), Khan et al. (2014)
Plant's salinity stress management through endophytic fungi	*Piriformospora indica* *Fusarium culmorum*	Baltruschat et al. (2008), Rodriguez et al. (2008)
Plant's temperature stress management by endophytic fungi	*Curvularia* spp. *Curvularia protuberata*	Kaur (2020), Redman et al. (2011)

stresses such as drought, temperature fluctuations, and chemical stress (Seleiman et al. 2021). The biocontrol properties exhibited by fungal endophytes could be due to the secretion of chitinolytic enzymes, competition for space and nutrients, mycoparasitism, and the production of inhibitory compounds. *Piriformospora indica*, *Sebacina vermifera*, and *Trichoderma* species help in the biological control of plant pathogens such as *Gaeumannomyces graminis*, which causes wheat diseases (Melo and Faull 2000). More than 20,000 secondary metabolites are produced by fungi, and approximately 80% of them are bioactive compounds. In particular, among secondary metabolites with antiviral activity from endophytic fungi, they are excellent producers of compounds with activity against pathogens and herbivores, including alkaloids, steroids, terpenoids, peptides, polyketones, flavonoids, quinols, phenols, chlorinated compounds, and volatile organic compounds (Card et al. 2016). ZhiNengCong (ZNC), an endophyte extract, was obtained from *Pseudomonas syringae* pv. Tomato (Pst) DC3000 is widely used as it protects crops and makes them resistant to viruses (Peng et al. 2020a). Global climate change such as global warming and the gradual deterioration of the agricultural environment are making it increasingly difficult to increase plant growth and yield. EF are a promising alternative for protecting plants from abiotic stresses including drought, high and low temperatures, and salinity (Aly et al. 2011).

When using EF in agriculture, (1) biological safety, (2) low environmental pollution and human health risk, (3) specificity with target pests, (4) high efficiency even in small quantities, (5) proliferation, (6) as advantages such as no resistance to chemicals, and (7) no promotion of the selection of resistant pests become known, it is being chosen as a more promising method (Berg 2009). In the case of entomopathogens, fungi are used as biological control agents to overcome the side effects of existing chemical control agents and are becoming widespread as a sustainable alternative. This means that EF do not require complex technologies or formulas to be applied in the field (Card et al. 2016). EF are beneficial because they provide several direct and indirect benefits to crops, protecting crops from biotic stresses such as viruses and pests and abiotic stresses such as drought and extreme temperatures. Therefore, the use of endophytic fungi not only has great potential in agriculture but is also an excellent alternative for sustainable agriculture.

10.6 Biotechnological Approaches for Harnessing Antiviral Potential

The viruses multiply exclusively in living cells, inducing severe symptoms and death and causing serious outbreaks on all continents. Elevated resistance of current viruses to existing antiviral drugs and the continued emergence of new serotypes in the viral population with high mutation rates and low fidelity of viral replication increases the difficulty of combating these viruses (Manganyi and Ateba 2020). The number of antiviral drugs available for the virus is minimal and there is an imbalance between the development of vaccines and the mutation of the virus. Therefore, it is essential to search for new alternatives to antiviral drugs.

10.6.1 Genetic Engineering of Endophytic Fungi

There are almost 900 plant viruses that are known to exist. Most currently known plant viruses are acute, transmitted by plant-feeding insects, and continue to cause acute infections until the death or recovery of the host (Roossinck 2013). Plant viruses invade their hosts through a multistep process that includes entry into the plant cell, nucleic acid decapitation, translation of viral proteins, replication of viral nucleic acids, assembly of daughter viroid, intercellular motility, systemic motility, and plant-to-plant motility (Kang et al. 2005). Plant viruses (DNA viruses and RNA viruses) process and replicate viral sequences via mRNA in the host cell. After undergoing complex interactions with proteins and gene sequences in the host cell, they are transported by motor proteins and undergo cell-to-cell transport (Nelson and Citovsky 2005). Recent data suggest that for plant viruses, microfilaments are involved in the intercellular movement of viruses, whereas microtubules and microtubule-associated proteins may be involved in the degradation of viral movement proteins (Kragler et al. 2003). Microfilaments and microtubules therefore play a major role in the intercellular movement of plant viruses. Disruption of microfilaments by drug or virus-induced gene silencing impairs cytotoxic virus transmission between cells (Dunoyer and Voinnet 2005). RNA silencing is a conserved surveillance mechanism that plays a key role in plant defense against invasive nucleic acid. Salicylic acid (SA) induces RNA silencing-associated gene resistance to plant RNA pathogens. Recent studies have found that endophytic fungi have inhibited the cell-to-cell spread of plant viruses by inducing SA and contributing to RNA silencing (Peng et al. 2020b). Accumulation of H_2O_2 levels affects SA accumulation, inducing each other in many biological processes (Choudhury et al. 2017). Virus invasion induces a hypersensitivity reaction in plants and produces large amounts of hydrogen peroxide, which accelerates cells into programmed death. Endophytic fungi promote inducers that induce plant defense responses and protect against pathogen infection while upregulating H_2O_2 levels. These inducers are flagellin, chitin, oligosaccharides, and so on (Benedetti et al. 2015). The inducers interact with plant receptors, thereby activating immunity triggered by pathogen-associated molecular pattern (PAMP) (Dangl et al. 2013). The general concentration of inducers is high compared to endophytic fungal extracts. Thus, endophytic fungi show superior antiviral potential against plant viruses.

In recent years, with the further development of DNA sequencing technology, more and more endophytic fungal genomes have been sequenced. However, most of the gene clusters associated with secondary metabolite production are silent or cryptic (Mei et al. 2019). Because of the low yield of endophytic fungal secondary metabolites, gene knockout and editing for endophytic fungi have become popular for research.

10.6.1.1 Silent Gene Clusters

The biosynthetic gene cluster (BGC) is a group of genes involved in the production of secondary metabolites. The genes and regulatory components needed for the production of physiologically active compounds are found in the BGC genome.

Additionally, it has resistance genes to stop self-destruction and transport genes to export the secondary metabolites generated (Tran et al. 2019). In recent years, more novel BGCs have been discovered in endophytic fungi. However, most BGCs are regarded as silent genes because they do not stimulate endophytic fungi to produce significant concentrations of metabolites under laboratory conditions (Keller 2019). Among them, the deletion of the epigenetic regulator HdaA activated the expression of meleagrin/roquefortine BGC and promoted the production of bioactive compounds in endophytic fungi as well as the generation of a diversity of new natural products from different chemotypes and biosynthetic sources (Ding et al. 2020; Mao et al. 2015). In addition to this, external stimulation of the host in the presence of endophytic fungi can stimulate the BGC genome. Activation of silenced BGC genes by changing culture conditions and medium composition (OSMAC method) completely altered the metabolic profile of various filamentous fungi and changes in the salt content of the medium, and carbon catabolite metabolite repression (CCR) was involved in affecting the metabolites (Almeida et al. 2019). Histones are closely related to BGC expression. Deletion of histone deacetylase (HDAC) leads to the induced expression of two telomere-proximal BGCs in *Aspergillus oryzae* (Shwab et al. 2007). Post-translationally modified (PTM) histones have an impact on the expression or activity of BGCs and the proteins these clusters encode. Histones (H2A, H2B, H3, and H4) control chromatin-dependently the expression of BGCs and non-histone proteins (Xue et al. 2023).

10.6.1.2 Manipulation of Gene Expression by Chemical Epigenetic Modifiers

Numerous structurally unique tiny molecular weight chemicals from various chemical groups can effectively facilitate endophytic fungal epigenetic modification. These modifiers activate groups of cryptic biosynthetic genes to increase the synthesis of secondary metabolites. 5-Azacitidine (5-Aza C) functions as an inhibitor of DNA methyltransferase and lowers DNA methylation to manipulate the endophytic fungus and increase secondary metabolites (Varasteh-kojourian et al. 2023). In addition, histones can be modified by small-molecule chemical modifiers. Gene activation and repression in endophytic fungi are mediated by histone acetyltransferases (HAT) and histone deacetylases (HDAC), which add or remove acetyl groups from histones, respectively. For example, Trichostatin A (TSA), apicidin, and sodium butyrate act as HDAC inhibitors, especially TSA that can promote the production of endophytic fungal metabolites (Shwab et al. 2007; Smith and Edlind Thomas 2002). Moreover, nicotinamide and quercetin act as histone deacetylase inhibitors (Moore et al. 2012).

10.6.1.3 CRISPR-Cas9

An instrument for editing DNA, the CRISPR-Cas9 system, was inspired by the bacterial adaptive immune system. It has been frequently utilized in living things because of its effectiveness and simplicity. The CRISPR/Cas9 system allows for precise and efficient modification of cellular DNA (Verma et al. 2023). This technology can be used in endophytic fungi to have beneficial effects on their hosts.

Short guide RNA (sgRNA) and Cas9 endonuclease are used for gene targeting and cleavage in the CRISPR-Cas9 system. Short guide RNA (sgRNA) directs Cas9 endonuclease to the location of the target gene in the endophytic fungal genome and prevents the gene from being expressed (Xu et al. 2021b). Numerous CRISPR-based methods, such as transcriptional control, genome editing, and epigenetic modifications, have been effectively applied to modify secondary metabolite pathways.

10.6.2 Metabolic Engineering for Enhanced Antiviral Activity

Bacteria, fungi, and eukaryotes present in soil, plant interiors, oceans, etc. are capable of producing secondary metabolites. The most widely known of these are aminoglycoside antibiotics, acid enol antibiotics, beta-lactam antibiotics, erythromycin, and so on (Yadav et al. 2019). Endophytic fungi produce different groups of secondary metabolites in the fight against plant viruses. Endophytic fungi have the inherent property of mimicking host metabolic pathways to similarly produce bioactive molecules. The secondary metabolites they produce promote plant growth and resistance to exotic viruses and bacteria. In addition to this, endophytic fungi provide a new dimension to drug discovery and development. For example, paclitaxel has a strong medicinal value as a typical secondary metabolite. In general, the metabolites of endophytic fungi are divided into several categories. Currently known are terpenoids (sesquiterpenes, diterpenes, triterpenes), steroidal compounds and alkaloids (amines, amides), phenols and phenolic acids, isocoumarin derivatives, flavonoids, lipids, peptides, and others (Mousa and Raizada 2013). Recent studies have found that endophytic alkaloids within meadow ryegrass (*Lolium pratense*) plants may block barley yellow dwarf virus (Lehtonen et al. 2006). Plant virus-infected meadow ryegrass produces alkaloids identified as ryegrass, and alkaloids affect and discourage herbivorous insects, which inhibit virus transmission (Lehtonen et al. 2005). Terpenoids are inhibited by plant virus infestation. Terpenoid production is influenced by terpenoid synthesis genes (Luan et al. 2013). Among them, JA signaling plays an important role in terpenoid synthesis. Related studies have found that terpenoids also mediate the defense response of tobacco plants against viruses (Seo et al. 2003). Monoterpenes are synthesized in the cytoplasm via the methylerythritol phosphate (MEP) pathway, while sesquiterpenes are synthesized in the cytoplasm via the mevalonate (MVA) pathway (Fontana et al. 2009) (Fig. 10.5). These two pathways provide versatile precursors for terpene biosynthesis. Phosphorus compounds are converted to monoterpenes and sesquiterpenes catalyzed by terpene synthase, respectively. The induced production of monoterpenes and sesquiterpenes can occur at two levels: by increasing the level of substrate through an inducible pathway or by increasing the level of terpene synthase (Shrivastava et al. 2015). When the host encounters an external stimulus, the endophytic fungus is stimulated along with it, thus triggering a defense mechanism to protect itself and the host.

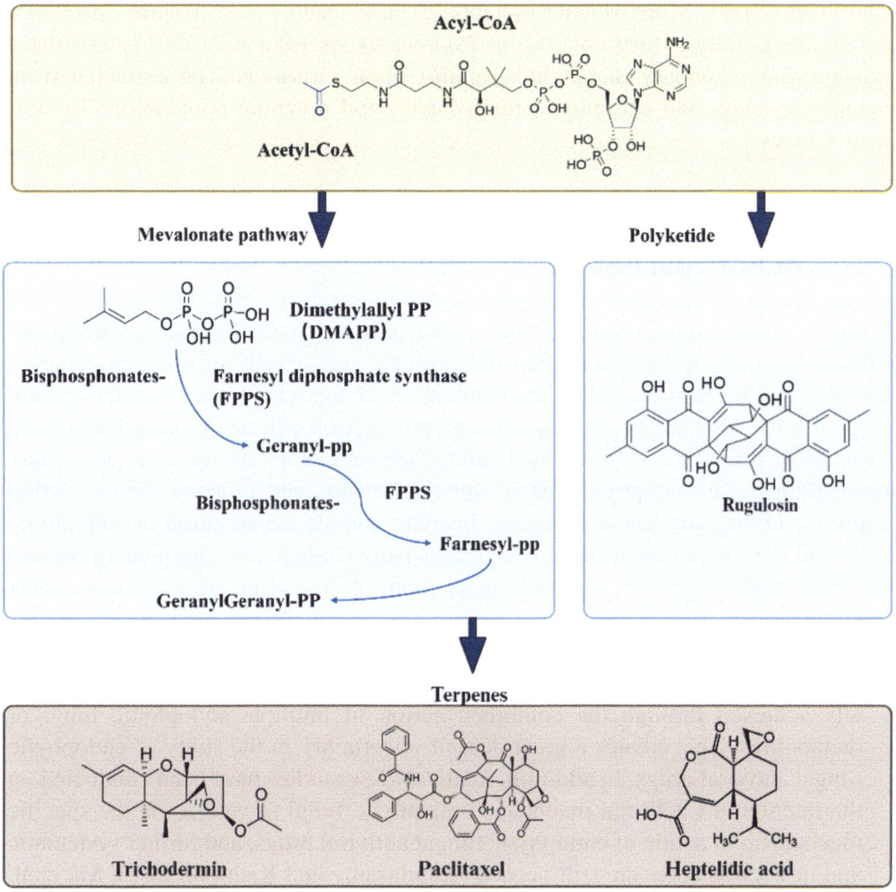

Fig. 10.5 Schematic diagram of terpene production by endophytic fungi through the methylerythritol phosphate (MEP) pathway

10.6.3 Synthetic Biology Applications

In recent years, the natural products produced by organisms following external stimuli or infections have received a great deal of attention. In particular, plants contain a large number of active substances, and endophytic fungi accelerate the synthesis of secondary metabolites (flavonoids, terpenoids, phenols, and so on) after receiving the same stimuli as the host. Endophytic fungi can produce a wide range of extracellular enzymes involved in biodegradation and hydrolysis processes to resist infection, such as pectinases, cellulases, lipases, amylases, lactases, and proteases (Paramanantham et al. 2019). Thus, due to their highly active secondary metabolites, endophytic fungi show excellent resistance to plant viruses. The cyclic pentapeptide Malformin A 1 produced during fermentation by the endophytic fungus *Aspergillus tubularis* FJBJ11 is highly effective against tobacco mosaic virus

(Tan et al. 2015). Moreover, its metabolites have been found to have anti-TMV activity, especially cytosporone U, in *Phomopsis* sp. (Tan et al. 2017). All these indicate that secondary metabolites against plant viruses can be extracted from endophytic fungi and endophytic fungi have good potential applications in anti-plant viruses.

10.7 Challenges and Opportunities in the Development of Antiviral Agents

In recent years, researchers have conducted extensive studies on the antiviral properties of endophytic fungi and their metabolites, and significant research progress has been made. However, the development of endophytic fungus-based antiviral drugs still has challenges in many aspects; this section will be organized into three areas, basic research, environmental safety, and regulatory compliance, describing the challenges of endophytic fungal antiviral drugs, and propose corresponding solutions. During the last decade, the diversity and characterization of endophytic fungi and their applications have been extensively studied, and significant progress has been made. However, research on endophytic fungal antiviral drugs remains underdeveloped.

1. There is a wide variety of endophytic fungi in plants and antiviral action is usually achieved through the combined action of multiple endophytic fungi or metabolites. This creates a great deal of uncertainty in the study of endophytic fungal antiviral drugs. In addition, relatively few studies have been conducted on the interactions between plants and endophytic fungi as well as on the specific mechanism of action of endophytic fungal antiviral drugs, and further systematic and in-depth studies are still needed (Kandasamy and Kathirvel 2023; Ma et al. 2016, 2019). Although gene editing has been used to study the mechanisms of synthesis of endophytic fungi and their metabolites, current studies are limited to filamentous fungi and metabolites have not yet been involved in antiviral drugs (Cao et al. 2018; Chowdhary et al. 2022; El-Sayed et al. 2017; Xu et al. 2021b). Therefore, the application of gene editing technology for the synthesis of antiviral drugs in endophytic fungi remains to be further investigated.
2. Furthermore, the widespread use of endophytic fungi and their metabolites is largely dependent on their ability to be cultured and maintained under laboratory conditions (Mishra et al. 2021). However, there is insufficient research data on culture parameters (such as medium, pH, temperature, and incubation time) for specific endophytic fungi, and there are still major difficulties in the isolation and purification of endophytic fungi, simulation of natural growth conditions, and metabolite extraction and purification, thus limiting the practical application of endophytic fungi to research (Patil et al. 2016).
3. Since endophytes usually survive only within a specific range of conditions, endophytic fungal drugs when applied against plant viruses, the strength of wind, precipitation, temperature, and solar radiation, as well as seasonal changes,

can cause endophytic fungi to become unsuitable for survival. Therefore the application of endophytic fungal antiviral drugs in practical production still needs a lot of theoretical data support (Grabka et al. 2022).

4. Endophytic fungal metabolites suffer from the disadvantages of poor stability, potential toxicity, and poor drug compatibility, which require further rational optimization of their structures, thus posing greater difficulties for their basic research and application (Verma et al. 2022).

10.7.1 Environmental and Safety Concerns

Endophytic fungi and their metabolites have many advantages in the fight against plant viruses. However, factors such as potential toxicity and dose also need to be considered and monitored to assess their impact in terms of environmental safety.

1. For host plants, moderate levels of endophytic fungi and their metabolites can be beneficial for growth and virus defense. When using endophytic fungi and metabolite levels that are too high, it may have a negative effect (Omran and Baek 2022).
2. With the widespread use of endophytic fungi and their metabolites, there will be an increasing loss of endophytic fungal antiviral drugs in the soil and water environment, which may affect the survival of other microorganisms, flora, and fauna and thus damage ecosystems (Verma et al. 2022).
3. Whether endophytic fungi and their metabolites can lead to viral drug resistance and induce viral mutation to produce super-viruses are also issues to be considered (Lefeuvre et al. 2019). All of these factors affect the environmental safety of endophytic fungi and their metabolites and are important challenges that limit the use of endophytic fungal antiviral drugs in practical production and application.

10.7.2 Regulatory Considerations

Currently, there are no standards and regulations for endophytic antiviral drugs worldwide. The development of effective standards and regulations is an important prerequisite for avoiding the risk of endophytic antiviral drugs and ensuring the stable development of the industry. Relevant standards and regulatory considerations relate mainly to the production and preparation of antiviral drugs, as well as their use and distribution.

1. In the area of endophytic fungal derived antiviral drug production, preparation, and safe production management, researchers focus on optimizing methods to ensure both efficacy and safety. Furthermore, regulations and standards need to be considered to protect against risks such as leakage of endophytic fungi and their metabolites, biosecurity and health hazards for production personnel, low

levels of production processes, and insufficient knowledge of process risks (De Silva et al. 2019).
2. Concerning the use and sale of endophytic fungal antiviral drugs, the question of whether the use of endophytic fungal antiviral drugs would jeopardize the lives and health of citizens by affecting foodstuffs and products used by human beings daily is also an issue that needs to be considered. All of the above issues require regulations to ensure the continued and stable development of endophytic fungal antiviral drugs (Verma et al. 2022).

10.7.3 Strategies for Overcoming Obstacles

In response to the challenges facing endophytic fungal antivirals, a range of responses need to be developed to advance the field, including the following:

1. Strengthening research on the antiviral mechanism and application of endophytic fungi. Studies on antiviral drugs targeting endophytic fungi. It can be centered on isolation and purification, identification of active ingredients, analysis of the molecular mechanism of action, in vitro culture of endophytic fungi, and metabolite scale synthesis. The development of new research methods is also essential to achieving these research objectives (Burragoni and Jeon 2021; Gaiero et al. 2013; Kaul et al. 2016; White et al. 2019), for example, accurate identification of endophytic fungi using multilocus typing technology; sequencing of specific genes using microarray multiplex technology; combining genomics, transcriptomics, and gene sequencing to investigate the interaction between endophytic fungi and host plants and isotope-labeled nucleic acid detection to investigate the effect of the host plant on endophytic colonies; and combining genomics and metabolomics to identify the active constituents of endophytic fungi (Anand et al. 2023). The discovery of the potential of endophytes to produce antiviral compounds is still in its infancy, and the number of compounds available as endophyte antiviral agents is limited. However, a few isolated metabolites of endophytic fungi have been reported to have antiviral functions, which provide an effective reference pathway for antiviral drug production (Hashem et al. 2023). In this regard, separation, purification, and identification of endophytic fungal metabolites by chromatography-mass spectrometry are used to provide an effective reference for the precise design and artificial scale preparation of endophytic fungal antiviral drugs, in order to better guide the application of endophytic fungal antiviral drugs and to better avoid possible risks while improving the effectiveness of their application (Saravanakumar et al. 2021; Singh et al. 2020; Tayung et al. 2011; Wang et al. 2014).
2. The rapid development of gene editing technology in recent years has also provided powerful tools for the study of endophytic fungi. CRISPR/Cas9 technology, with the advantages of simple operation, good targeting, and high safety, is the technology of choice for gene editing in endophytic fungi and has very important application prospects for studying the functions of genes related to the

biosynthesis of antiviral drugs in endophytic fungi, applying biotechnological means to genetically select them, and enhancing the ability of endophytic fungi to synthesize bioactive substances. In addition, as the editing efficiency of the CRISPR/Cas9 system improves in future studies, CRISPR/Cas9 achieves more important applications in the research of endophytic fungi and antiviral drugs (Verma et al. 2023).

3. Strengthening experimental studies on pharmacokinetics and toxicology of endophytic fungi and their metabolites. Pharmacokinetic studies can help to understand the distribution, accumulation, and metabolism of drugs in the organism, which is very important for improving drug efficacy and reducing toxic side effects; toxicology tests, on the other hand, are an important basis for assessing the environmental and human safety of drugs (Lacerda et al. 2022). Pharmacokinetics of endophytic fungal antiviral drugs is mainly focused on drug dosage, time of application, application cycle, plant body distribution, and changes in the metabolism of plant tissues and organs, for optimal administration regimen. Toxicology experiments, on the other hand, are needed to assess the rate of drug release and its distribution in the environment. At the same time, the effects of endophytic fungi and their metabolites on other species (especially microorganisms) in the environment were determined to assess the environmental safety of endophytic fungal antiviral drugs (Li and Song 2017). In addition, there is a need for animal or human physiological models for toxicological risk assessment of endophytic fungal antiviral drug exposures to explore their effects on human health. Endophytic fungal antiviral drugs may be pushed into large-scale practical applications only after environmental safety and human safety have been guaranteed (Katoch et al. 2014; Khiralla et al. 2016).
4. Establishment of standardized operating procedures for endophytic fungal antiviral drugs and sound regulatory legislation for related products. The standardized workflow is an important guide for the safe research and production of antiviral drugs, and it is of great significance in promoting the development of the related industry. Government departments, scientific research institutions, and enterprises should cooperate and play their respective roles to jointly promote the construction of standardized workflow for endophytic fungal antiviral drugs. In addition, the government should strengthen the regulation of endophytic fungal-derived antiviral drugs in research and development, production, use, distribution, and other aspects. The formulation of relevant laws and regulations can be used as a reference for the regulation of pesticides to protect the lives and health of consumers (Karimi-Maleh et al. 2023).

10.8 Conclusion

This chapter presents a conclusion based on the findings that endophytic fungi, which have been isolated from medicinal plants, exhibit notable antiviral activity against both human and plant viruses. Moreover, these endophytic fungi play a pivotal role in safeguarding plants from diverse external stimuli by initiating a defense mechanism that involves complex gene expression within the host plant. Additionally, they generate a variety of metabolites by emulating the metabolic system of the host plant. The utilization of endophytic fungi holds significant promise for their potential development into fungal cell factories through genetic modification. However, concerning environmental safety and health considerations, it is imperative to assess the potential toxicity of fungal spores and their metabolites before progressing to industrial-scale applications. It is noteworthy that there is a scarcity of studies examining the antiviral activity of compounds derived from endophytic sources against plant viruses. Consequently, there is an imperative need for research focused on isolating anti-plant viral compounds from endophytic fungi, aiming to mitigate the introduction of hazardous chemicals into an industrial crop, thereby enhancing the economic strategies of the agricultural system.

Acknowledgments This work was supported by the National Research Foundation of Korea (2021R1I1A1A01057742; 2022R1A2C2091029).

References

Adeleke BS, Babalola OO (2021) Pharmacological potential of fungal endophytes associated with medicinal plants: a review. J Fungi 7(2):147. Retrieved from https://www.mdpi.com/2309-608X/7/2/147

Ahmad M, Pataczek L, Hilger TH, Zahir ZA, Hussain A, Rasche F et al (2018) Perspectives of microbial inoculation for sustainable development and environmental management. Front Microbiol 9:2992

Akash MSH, Rehman K, Akash MSH, Rehman K (2020) Ultraviolet-visible (UV-VIS) spectroscopy. In: Essentials of pharmaceutical analysis, pp 29–56. Springer, Singapore.

Alam B, Lǐ J, Gě Q, Khan MA, Gōng J, Mehmood S et al (2021) Endophytic fungi: from symbiosis to secondary metabolite communications or vice versa? Front Plant Sci 12:3060

Almeida EL, Kaur N, Jennings LK, Carrillo Rincón AF, Jackson SA, Thomas OP, Dobson ADW (2019) Genome mining coupled with OSMAC-based cultivation reveal differential production of surugamide A by the marine sponge isolate Streptomyces sp. SM17 when compared to its terrestrial relative S. albidoflavus J1074. Microorganisms 7(10):394. Retrieved from https://www.mdpi.com/2076-2607/7/10/394

Aly AH, Debbab A, Proksch P (2011) Fungal endophytes: unique plant inhabitants with great promises. Appl Microbiol Biotechnol 90:1829–1845

Anand K, Kumar V, Prasher IB, Sethi M, Raj H, Ranjan H et al (2023) Bioactive molecules from fungal endophytes and their applications in pharmaceutical industries: challenges and future scope. J Basic Microbiol 63(7):690–708

Baltruschat H, Fodor J, Harrach BD, Niemczyk E, Barna B, Gullner G et al (2008) Salt tolerance of barley induced by the root endophyte Piriformospora indica is associated with a strong increase in antioxidants. New Phytol 180(2):501–510

Bamisile BS, Dash CK, Akutse KS, Keppanan R, Wang L (2018) Fungal endophytes: beyond herbivore management. Front Microbiol 9:544

Baron NC, Rigobelo EC (2022) Endophytic fungi: a tool for plant growth promotion and sustainable agriculture. Mycology 13(1):39–55. https://doi.org/10.1080/21501203.2021.1945699

Bashyal BP, Wellensiek BP, Ramakrishnan R, Faeth SH, Ahmad N, Gunatilaka AAL (2014) Altertoxins with potent anti-HIV activity from Alternaria tenuissima QUE1Se, a fungal endophyte of Quercus emoryi. Bioorg Med Chem 22(21):6112–6116. https://doi.org/10.1016/j.bmc.2014.08.039

Benedetti M, Pontiggia D, Raggi S, Cheng Z, Scaloni F, Ferrari S et al (2015) Plant immunity triggered by engineered in vivo release of oligogalacturonides, damage-associated molecular patterns. Proc Natl Acad Sci 112(17):5533–5538
Berg G (2009) Plant–microbe interactions promoting plant growth and health: perspectives for controlled use of microorganisms in agriculture. Appl Microbiol Biotechnol 84:11–18
Bovey FA, Mirau PA, Gutowsky HS () Nuclear magnetic resonance spectroscopy. Elsevier, Netherlands
Brewer HC, Hird DL, Bailey AM, Seal SE, Foster GD (2018) A guide to the contained use of plant virus infectious clones. Plant Biotechnol J 16(4):832–843
Burragoni SG, Jeon J (2021) Applications of endophytic microbes in agriculture, biotechnology, medicine, and beyond. Microbiol Res 245:126691
Cao F, Cheng J-T, Chen X-A, Li Y-Q, Mao X-M (2018) Development of an efficient genetic system in a gene cluster-rich endophytic fungus Calcarisporium arbuscula NRRL 3705. J Microbiol Methods 151:1–6
Card S, Johnson L, Teasdale S, Caradus J (2016) Deciphering endophyte behaviour: the link between endophyte biology and efficacious biological control agents. FEMS Microbiol Ecol 92(8):fiw114
Choudhury FK, Rivero RM, Blumwald E, Mittler R (2017) Reactive oxygen species, abiotic stress and stress combination. Plant J 90(5):856–867
Chowdhary K, Arora H, Sharma S (2022) CRISPR/Cas9-based genome editing as a way ahead for inducing production of bioactive metabolites in endophytes. Natl Acad Sci Lett 45(3):275–280
Chupina AV, Yanshole VV, Sulyaeva VS, Kokovkin VV, Abramov PA, Sokolov MN (2022) Self-assembly patterns of non-metalloid silver thiolates: structural, HR-ESI-MS and stability studies. Dalton Trans 51(2):705–714
Dangl JL, Horvath DM, Staskawicz BJ (2013) Pivoting the plant immune system from dissection to deployment. Science 341(6147):746–751
De Silva NI, Brooks S, Lumyong S, Hyde KD (2019) Use of endophytes as biocontrol agents. Fungal Biol Rev 33(2):133–148
Diedhiou P, Hallmann J, Oerke E-C, Dehne H-W (2003) Effects of arbuscular mycorrhizal fungi and a non-pathogenic Fusarium oxysporum on Meloidogyne incognita infestation of tomato. Mycorrhiza 13(4):199–204
Ding Z, Zhou H, Wang X, Huang H, Wang H, Zhang R et al (2020) Deletion of the histone deacetylase HdaA in endophytic fungus Penicillium chrysogenum Fes1701 induces the complex response of multiple bioactive secondary metabolite production and relevant gene cluster expression. Molecules 25(16):3657. Retrieved from https://www.mdpi.com/1420-3049/25/16/3657
Du G, Wang Z-C, Hu W-Y, Yan K-L, Wang X-L, Yang H-M et al (2017) Three new 3-methyl-2-arylbenzofurans from the fermentation products of an endophytic fungus Phomopsis sp. and their anti-TMV activity. Phytochem Lett 21:287–290. https://doi.org/10.1016/j.phytol.2016.04.003
Dunoyer P, Voinnet O (2005) The complex interplay between plant viruses and host RNA-silencing pathways. Curr Opin Plant Biol 8(4):415–423
El-Sayed AS, Abdel-Ghany SE, Ali GS (2017) Genome editing approaches: manipulating of lovastatin and taxol synthesis of filamentous fungi by CRISPR/Cas9 system. Appl Microbiol Biotechnol 101:3953–3976
Fakhro A, Andrade-Linares DR, von Bargen S, Bandte M, Büttner C, Grosch R et al (2010) Impact of Piriformospora indica on tomato growth and on interaction with fungal and viral pathogens. Mycorrhiza 20:191–200
Fontana A, Reichelt M, Hempel S, Gershenzon J, Unsicker SB (2009) The effects of arbuscular mycorrhizal fungi on direct and indirect defense metabolites of Plantago lanceolata L. J Chem Ecol 35:833–843
Gaiero JR, McCall CA, Thompson KA, Day NJ, Best AS, Dunfield KE (2013) Inside the root microbiome: bacterial root endophytes and plant growth promotion. Am J Bot 100(9):1738–1750
Gao F-k, Dai C-c, Liu X-z (2010) Mechanisms of fungal endophytes in plant protection against pathogens. Afr J Microbiol Res 4(13):1346–1351

Grabka R, d'Entremont TW, Adams SJ, Walker AK, Tanney JB, Abbasi PA, Ali S (2022) Fungal endophytes and their role in agricultural plant protection against pests and pathogens. Plan Theory 11(3):384

Haris PI, Chapman D (1995) The conformational analysis of peptides using Fourier transform IR spectroscopy. Biopolymers 37(4):251–263

Hashem AH, Attia MS, Kandil EK, Fawzi MM, Abdelrahman AS, Khader MS et al (2023) Bioactive compounds and biomedical applications of endophytic fungi: a recent review. Microb Cell Factories 22(1):1–23

Hatzakis E (2019) Nuclear magnetic resonance (NMR) spectroscopy in food science: a comprehensive review. Compr Rev Food Sci Food Saf 18(1):189–220

Horváth C (2013) High-performance liquid chromatography: advances and perspectives

Hsu CPS (1997) Infrared spectroscopy. In: Handbook of instrumental techniques for analytical chemistry, Willey-VCH Verlag GmbH, D69469Weinheim

Jaber LR, Salem NM (2014) Endophytic colonisation of squash by the fungal entomopathogen Beauveria bassiana (Ascomycota: Hypocreales) for managing Zucchini yellow mosaic virus in cucurbits. Biocontrol Sci Tech 24(10):1096–1109

Johnson JM, Alex T, Oelmüller R (2014) Piriformospora indica: the versatile and multifunctional root endophytic fungus for enhanced yield and tolerance to biotic and abiotic stress in crop plants. J Trop Agric 52(2):103–122

Kandasamy GD, Kathirvel P (2023) Insights into bacterial endophytic diversity and isolation with a focus on their potential applications–a review. Microbiol Res 266:127256

Kang B-C, Yeam I, Jahn MM (2005) Genetics of plant virus resistance. Annu Rev Phytopathol 43(1):581–621. https://doi.org/10.1146/annurev.phyto.43.011205.141140

Karimi-Maleh H, Ghalkhani M, Dehkordi ZS, Singh J, Wen Y, Baghayeri M et al (2023) MOF-enabled pesticides as developing approach for sustainable agriculture and reducing environmental hazards. J Ind Eng Chem 129:105–123

Katoch M, Singh G, Sharma S, Gupta N, Sangwan PL, Saxena AK (2014) Cytotoxic and antimicrobial activities of endophytic fungi isolated from Bacopa monnieri (L.) Pennell (Scrophulariaceae). BMC Complement Altern Med 14(1):1–8

Kaul S, Gupta S, Ahmed M, Dhar MK (2012) Endophytic fungi from medicinal plants: a treasure hunt for bioactive metabolites. Phytochem Rev 11(4):487–505. https://doi.org/10.1007/s11101-012-9260-6

Kaul S, Sharma T, Dhar K, M. (2016) "Omics" tools for better understanding the plant–endophyte interactions. Front Plant Sci 7:955

Kaur T (2020) Fungal endophyte-host plant interactions: role in sustainable agriculture. In: Sustainable crop production, IntechOpen, United Kingdom

Keller NP (2019) Fungal secondary metabolism: regulation, function and drug discovery. Nat Rev Microbiol 17(3):167–180

Khan N, Khan M, Asgher M, Fatma M, Masood A, Syeed S (2014) Salinity tolerance in plants: revisiting the role of sulfur metabolites. J Plant Biochem Physiol 2(120):2

Khiralla A, Mohamed IE, Tzanova T, Schohn H, Slezack-Deschaumes S, Hehn A et al (2016) Endophytic fungi associated with Sudanese medicinal plants show cytotoxic and antibiotic potential. FEMS Microbiol Lett 363(11):fnw089

Khiralla A, Spina R, Varbanov M, Philippot S, Lemiere P, Slezack-Deschaumes S et al (2020) Evaluation of antiviral, antibacterial and antiproliferative activities of the endophytic fungus Curvularia papendorfii, and isolation of a new polyhydroxyacid. Microorganisms 8(9). https://doi.org/10.3390/microorganisms8091353

Kiarie S, Nyasani JO, Gohole LS, Maniania NK, Subramanian S (2020) Impact of fungal endophyte colonization of maize (Zea mays L.) on induced resistance to thrips-and aphid-transmitted viruses. Plan Theory 9(4):416

Kragler F, Curin M, Trutnyeva K, Gansch A, Waigmann E (2003) MPB2C, a microtubule-associated plant protein binds to and interferes with cell-to-cell transport of tobacco mosaic virus movement protein. Plant Physiol 132(4):1870–1883

Lacerda ÍC d S, Polonio JC, Golias HC (2022) Endophytic fungi as a source of antiviral compounds–a review. Chem Biodivers 19(6):e202100971

Lai Q, Zhu S, Luo X, Zou M, Huang S (2012) Ultraviolet-visible spectroscopy of graphene oxides. AIP Adv 2(3):032146

Le Cocq K, Gurr SJ, Hirsch PR, Mauchline TH (2017) Exploitation of endophytes for sustainable agricultural intensification. Mol Plant Pathol 18(3):469–473. https://doi.org/10.1111/mpp.12483

Lefeuvre P, Martin DP, Elena SF, Shepherd DN, Roumagnac P, Varsani A (2019) Evolution and ecology of plant viruses. Nat Rev Microbiol 17(10):632–644

Lehtonen P, Helander M, Wink M, Sporer F, Saikkonen K (2005) Transfer of endophyte-origin defensive alkaloids from a grass to a hemiparasitic plant. Ecol Lett 8(12):1256–1263

Lehtonen PT, Helander M, Siddiqui SA, Lehto K, Saikkonen K (2006) Endophytic fungus decreases plant virus infections in meadow ryegrass (Lolium pratense). Biol Lett 2(4):620–623

Li X-y, Song B-a (2017) Progress in the development and application of plant-based antiviral agents. J Integr Agric 16(12):2772–2783

Lisboa BP (1969) [1] Thin-layer chromatography of steroids, sterols, and related compounds. In: Methods in enzymology, vol 15. Elsevier, pp 3–158

Liu S-S, Jiang J-X, Huang R, Wang Y-T, Jiang B-G, Zheng K-X, Wu S-H (2019) A new antiviral 14-nordrimane sesquiterpenoid from an endophytic fungus Phoma sp. Phytochem Lett 29:75–78. https://doi.org/10.1016/j.phytol.2018.11.005

Luan JB, Yao DM, Zhang T, Walling LL, Yang M, Wang YJ, Liu SS (2013) Suppression of terpenoid synthesis in plants by a virus promotes its mutualism with vectors. Ecol Lett 16(3):390–398

Luo X, Yang J, Chen F, Lin X, Chen C, Zhou X et al (2018) Structurally diverse polyketides from the mangrove-derived fungus Diaporthe sp. SCSIO 41011 with their anti-influenza A virus activities. Front Chem 6:282. https://doi.org/10.3389/fchem.2018.00282

Ma KX, Shen XT, Huang R, Wang T, Xie XS, Liu SW et al (2014) Bioactive metabolites produced by the endophytic fungus Phomopsis sp. YM355364. Nat Prod Commun 9(5):669–670

Ma Y, Rajkumar M, Zhang C, Freitas H (2016) Beneficial role of bacterial endophytes in heavy metal phytoremediation. J Environ Manag 174:14–25

Ma Y, Vosátka M, Freitas H (2019) Beneficial microbes alleviate climatic stresses in plants. Front Plant Sci 10:595

Manganyi MC, Ateba CN (2020) Untapped potentials of endophytic fungi: a review of novel bioactive compounds with biological applications. Microorganisms 8(12):1934

Mao XM, Xu W, Li D, Yin WB, Chooi YH, Li YQ et al (2015) Epigenetic genome mining of an endophytic fungus leads to the pleiotropic biosynthesis of natural products. Angew Chem 127(26):7702–7706

Martelli GP (1980) Ultrastructural aspects of possible defence reactions in virus-infected plant cells. Microbiologica 25

Mei Y-Z, Zhu Y-L, Huang P-W, Yang Q, Dai C-C (2019) Strategies for gene disruption and expression in filamentous fungi. Appl Microbiol Biotechnol 103:6041–6059

Melo IS d, Faull JL (2000) Parasitism of Rhizoctonia solani by strains of Trichoderma spp. Sci Agric 57:55–59

Miklos DB, Remy C, Jekel M, Linden KG, Drewes JE, Hübner U (2018) Evaluation of advanced oxidation processes for water and wastewater treatment–a critical review. Water Res 139:118–131

Mishra S, Sahu PK, Agarwal V, Singh N (2021) Exploiting endophytic microbes as micro-factories for plant secondary metabolite production. Appl Microbiol Biotechnol 105(18):6579–6596

Moore JM, Bradshaw E, Seipke RF, Hutchings MI, McArthur M (2012) Chapter 18: Use and discovery of chemical elicitors that stimulate biosynthetic gene clusters in streptomyces bacteria. In: Hopwood DA (ed) Methods in enzymology, vol 517. ACADEMC PRESS, INC. San Diego, CA pp 367–385

Mousa WK, Raizada MN (2013) The diversity of anti-microbial secondary metabolites produced by fungal endophytes: an interdisciplinary perspective. Front Microbiol 4:65

Muvea AM, Meyhöfer R, Subramanian S, Poehling H-M, Ekesi S, Maniania NK (2014) Colonization of onions by endophytic fungi and their impacts on the biology of Thrips tabaci. PLoS One 9(9):e108242
Muvea AM, Subramanian S, Maniania NK, Poehling H-M, Ekesi S, Meyhöfer R (2018) Endophytic colonization of onions induces resistance against viruliferous thrips and virus replication. Front Plant Sci 9:1785
Nelson RS, Citovsky V (2005) Plant viruses. Invaders of cells and pirates of cellular pathways. Plant Physiol 138(4):1809–1814. https://doi.org/10.1104/pp.104.900167
Omran BA, Baek K-H (2022) Control of phytopathogens using sustainable biogenic nanomaterials: recent perspectives, ecological safety, and challenging gaps. J Clean Prod 372:133729
Paramanantham P, Pattnaik S, Siddhardha B (2019) Natural products from endophytic fungi: synthesis and applications. In: Advances in endophytic fungal research: present status and future challenges. Springer, Cham, pp 83–103
Patil RH, Patil MP, Maheshwari VL (2016) Bioactive secondary metabolites from endophytic fungi: a review of biotechnological production and their potential applications. Stud Nat Prod Chem 49:189–205
Peixoto Neto P d S, Azevedo JL, Araújo W d (2002) Microrganismos endofíticos. Biotecnologia Ciência & Desenvolvimento 29:62–77
Peng C, Zhang A, Wang Q, Song Y, Zhang M, Ding X et al (2020a) Ultrahigh-activity immune inducer from Endophytic Fungi induces tobacco resistance to virus by SA pathway and RNA silencing. BMC Plant Biol 20(1):169. https://doi.org/10.1186/s12870-020-02386-4
Peng C, Zhang A, Wang Q, Song Y, Zhang M, Ding X et al (2020b) Ultrahigh-activity immune inducer from Endophytic Fungi induces tobacco resistance to virus by SA pathway and RNA silencing. BMC Plant Biol 20(1):1–14
Pieterse CMJ, Zamioudis C, Berendsen RL, Weller DM, Van Wees SCM, Bakker PAHM (2014) Induced systemic resistance by beneficial microbes. Annu Rev Phytopathol 52:347–375
Rai M, Rathod D, Agarkar G, Dar M, Brestic M, Pastore GM, Junior MRM (2014) Fungal growth promotor endophytes: a pragmatic approach towards sustainable food and agriculture. Symbiosis 62:63–79
Redman RS, Kim YO, Woodward CJ, Greer C, Espino L, Doty SL, Rodriguez RJ (2011) Increased fitness of rice plants to abiotic stress via habitat adapted symbiosis: a strategy for mitigating impacts of climate change. PLoS One 6(7):e14823
Reuhs BL (2017) High-performance liquid chromatography. Springer, Germany
Rodriguez RJ, Henson J, Van Volkenburgh E, Hoy M, Wright L, Beckwith F et al (2008) Stress tolerance in plants via habitat-adapted symbiosis. ISME J 2(4):404–416
Rodriguez R, White J Jr, Arnold A, Redman a R a (2009) Fungal endophytes: diversity and functional roles. New Phytol 182(2):314–330
Roossinck MJ (2013) Plant virus ecology. PLoS Pathog 9(5):e1003304
Saravanakumar K, Sriram B, Sathiyaseelan A, Hu X, Mariadoss AVA, MubarakAli D, Wang M-H (2021) Molecular identification, volatile metabolites profiling, and bioactivities of an indigenous endophytic fungus (Diaporthe sp.). Process Biochem 102:72–81
Savci S (2012) Investigation of effect of chemical fertilizers on environment. APCBEE Procedia 1:287–292. https://doi.org/10.1016/j.apcbee.2012.03.047
Seleiman M, Al-Suhaibani N, Ali N, Akmal M, Alotaibi M, Refay Y, Battaglia M (2021) Drought stress impacts on plants and different approaches to alleviate its adverse effects. Plan Theory 10:259
Selim KA, El-Beih AA, AbdEl-Rahman TM, El-Diwany AI (2012) Biology of endophytic fungi. Curr Res Environ Appl Mycol 2(1):31–82
Selim KA, Elkhateeb WA, Tawila AM, El-Beih AA, Abdel-Rahman TM, El-Diwany AI, Ahmed EF (2018) Antiviral and antioxidant potential of fungal endophytes of Egyptian medicinal plants. Fermentation 4(3):49. Retrieved from https://www.mdpi.com/2311-5637/4/3/49
Semreen MH, Alniss H, Cacciatore S, El-Awady R, Mousa M, Almehdi AM et al (2020) GC–MS based comparative metabolomic analysis of MCF-7 and MDA-MB-231 cancer cells treated with Tamoxifen and/or Paclitaxel. J Proteome 225:103875

Seo S, Seto H, Koshino H, Yoshida S, Ohashi Y (2003) A diterpene as an endogenous signal for the activation of defense responses to infection with tobacco mosaic virus and wounding in tobacco. Plant Cell 15(4):863–873

Shen Q, Xu X, Zhang F, Xiang N, He P, Si X et al (2016) Two new benzolactones from the leaves of Nicotiana tabacum and their anti-tobacco mosaic virus activities. Nat Prod Res 30(13):1464–1469. https://doi.org/10.1080/14786419.2015.1104675

Shrivastava G, Ownley BH, Augé RM, Toler H, Dee M, Vu A et al (2015) Colonization by arbuscular mycorrhizal and endophytic fungi enhanced terpene production in tomato plants and their defense against a herbivorous insect. Symbiosis 65:65–74

Shwab EK, Bok JW, Tribus M, Galehr J, Graessle S, Keller NP (2007) Histone deacetylase activity regulates chemical diversity in Aspergillus. Eukaryot Cell 6(9):1656–1664

Singh P, Sharma A, Bordoloi M, Nandi SP (2020) Molecular identification of endophytic fungi isolated from medicinal plant. Biointerface Res Appl Chem 10:6436–6443

Skipski VP, Peterson RF, Barclay M (1964) Quantitative analysis of phospholipids by thin-layer chromatography. Biochem J 90(2):374

Smith WL, Edlind Thomas D (2002) Histone deacetylase inhibitors enhance Candida albicans sensitivity to azoles and related antifungals: correlation with reduction in CDR and ERG upregulation. Antimicrob Agents Chemother 46(11):3532–3539. https://doi.org/10.1128/aac.46.11.3532-3539.2002

Tan Q-W, Gao F-L, Wang F-R, Chen Q-J (2015) Anti-TMV activity of malformin A1, a cyclic penta-peptide produced by an endophytic fungus Aspergillus tubingensis FJBJ11. Int J Mol Sci 16(3):5750–5761. Retrieved from https://www.mdpi.com/1422-0067/16/3/5750

Tan Q-W, Fang P-H, Ni J-C, Gao F, Chen Q-J (2017) Metabolites produced by an endophytic Phomopsis sp. and their anti-TMV activity. Molecules 22(12):2073. Retrieved from https://www.mdpi.com/1420-3049/22/12/2073

Tayung K, Barik B, Jha D, Deka D (2011) Identification and characterization of antimicrobial metabolite from an endophytic fungus, Fusarium solani isolated from bark of Himalayan yew. Mycosphere 2(3):203–213

Tran PN, Yen M-R, Chiang C-Y, Lin H-C, Chen P-Y (2019) Detecting and prioritizing biosynthetic gene clusters for bioactive compounds in bacteria and fungi. Appl Microbiol Biotechnol 103(8):3277–3287. https://doi.org/10.1007/s00253-019-09708-z

van Lexmond MB, Bonmatin J-M, Goulson D, Noome DA (2015) Worldwide integrated assessment on systemic pesticides. Environ Sci Pollut Res 22(1):1–4. https://doi.org/10.1007/s11356-014-3220-1

Varasteh-kojourian M, Ganjeali A, Asili J, Malekzadeh-Shafaroudi S, Taleghani A (2023) 5-azacytidine increases tanshinone production in endophytic fungi of Salvia abrotanoides (Kar.) Sytsma. J Cell Mol Res 14(2):78–84

Verma A, Shameem N, Jatav HS, Sathyanarayana E, Parray JA, Poczai P, Sayyed R (2022) Fungal endophytes to combat biotic and abiotic stresses for climate-smart and sustainable agriculture. Front Plant Sci 13:953836

Verma V, Batta A, Singh HB, Srivastava A, Garg SK, Singh VP, Arora PK (2023) Bioengineering of fungal endophytes through the CRISPR/Cas9 system. Front Microbiol 14:1146650

Wang J, Wang G, Zhang Y, Zheng B, Zhang C, Wang L (2014) Isolation and identification of an endophytic fungus Pezicula sp. in Forsythia viridissima and its secondary metabolites. World J Microbiol Biotechnol 30:2639–2644

Wen J, Okyere SK, Wang S, Wang J, Xie L, Ran Y, Hu Y (2022) Endophytic fungi: an effective alternative source of plant-derived bioactive compounds for pharmacological studies. J Fungi 8(2). https://doi.org/10.3390/jof8020205

White JF, Kingsley KL, Zhang Q, Verma R, Obi N, Dvinskikh S et al (2019) Endophytic microbes and their potential applications in crop management. Pest Manag Sci 75(10):2558–2565

Xu T-C, Lu Y-H, Wang J-F, Song Z-Q, Hou Y-G, Liu S-S et al (2021a) Bioactive secondary metabolites of the genus Diaporthe and anamorph Phomopsis from terrestrial and marine habitats and endophytes: 2010–2019. Microorganisms 9(2). https://doi.org/10.3390/microorganisms9020217

Xu X, Huang R, Yin W-B (2021b) An optimized and efficient CRISPR/Cas9 system for the endophytic fungus Pestalotiopsis fici. J Fungi 7(10):809

Xue M, Hou X, Fu J, Zhang J, Wang J, Zhao Z et al (2023) Recent advances in search of bioactive secondary metabolites from fungi triggered by chemical epigenetic modifiers. J Fungi 9(2):172

Yadav AN, Kour D, Rana KL, Yadav N, Singh B, Chauhan VS et al (2019) Metabolic engineering to synthetic biology of secondary metabolites production. In: New and future developments in microbial biotechnology and bioengineering. Elsevier, Netherlands, pp 279–320

Yang G-Y, Dai J-M, Mi Q-L, Li Z-J, Li X-M, Zhang J-D et al (2022) Cyclopiazonic acid type indole alkaloids from Nicotiana tabacum-derived fungus Aspergillus versicolor and their anti-tobacco mosaic virus activities. Phytochemistry 198:113137. https://doi.org/10.1016/j.phytochem.2022.113137

Yang F-X, Dai J-M, Liu H-Y, Mi Q-L, Wang J, Zhang J-D et al (2023) Isochromenes from the Nicotiana tabacum-derived endophytic fungus Aspergillus versicolor and their anti-tobacco mosaic virus activities. Nat Prod Res 37(10):1608–1616. https://doi.org/10.1080/14786419.2022.2103554

Zhang G, Sun S, Zhu T, Lin Z, Gu J, Li D, Gu Q (2011) Antiviral isoindolone derivatives from an endophytic fungus Emericella sp. associated with Aegiceras corniculatum. Phytochemistry 72(11):1436–1442. https://doi.org/10.1016/j.phytochem.2011.04.014

Zhang D-W, Tao X-Y, Liu J-M, Chen R-D, Zhang M, Fang X-M et al (2016a) A new polyketide synthase–nonribosomal peptide synthetase hybrid metabolite from plant endophytic fungus Periconia sp. Chin Chem Lett 27(5):640–642. https://doi.org/10.1016/j.cclet.2016.02.005

Zhang S-P, Huang R, Li F-F, Wei H-X, Fang X-W, Xie X-S et al (2016b) Antiviral anthraquinones and azaphilones produced by an endophytic fungus Nigrospora sp. from Aconitum carmichaeli. Fitoterapia 112:85–89. https://doi.org/10.1016/j.fitote.2016.05.013

Zhou M, Lou J, Li Y-K, Wang Y-D, Zhou K, Ji B-K et al (2015) Butyrolactones from the endophytic fungus Aspergillus versicolor and their anti-tobacco mosaic virus activity. J Braz Chem Soc 26:545–549

Zum Felde A, Mendoza A, Cabrera J, Kurtz A, Schouten A, Pocasangre L, Sikora R (2009) The burrowing nematode of banana: strategies for controlling the uncontrollable. Acta Hortic 828(1):101–107

Biosynthesis of Copper Nanoparticles from Endophytic Fungi and Their Role in Plant Disease Management

11

Dounia Elfadil, Amr H. Hashem, Mohamed S. Attia, and Gharieb S. El-Sayyad

Abstract

This chapter delves into the emerging realm of biosynthesis of copper nanoparticles (CuNPs) using endophytic fungi and explores their crucial role in the management of plant diseases. The intricate mechanisms of the biosynthesis process are examined, unraveling the impact of fungal enzymes, metabolic pathways, and environmental factors. The chapter discusses the selection and screening of optimal endophytic fungal strains, highlighting criteria for their suitability and identification techniques. Techniques such as UV-visible spectroscopy and microscopic analysis are employed for characterization, offering insights into the size, shape, and properties of the synthesized CuNPs.

Emphasis is placed on the applications of these nanoparticles in plant disease management, specifically focusing on their antimicrobial properties and their

D. Elfadil
Biology and Chemistry Department, Faculty of Sciences and Techniques, Hassan II University of Casablanca, Mohammedia, Morocco

A. H. Hashem (✉) · M. S. Attia
Botany and Microbiology Department, Faculty of Science, Al-Azhar University, Cairo, Egypt
e-mail: amr.hosny86@azhar.edu.eg; drmohamedsalah92@azhar.edu.eg

G. S. El-Sayyad (✉)
Medical Laboratory Technology Department, Faculty of Applied Health Sciences Technology, Badr University in Cairo (BUC), Cairo, Egypt

Department of Microbiology and Immunology, Faculty of Pharmacy, Galala University, New Galala City, Suez, Egypt

Drug Microbiology Lab., Drug Radiation Research Department, National Center for Radiation Research and Technology (NCRRT), Egyptian Atomic Energy Authority (EAEA), Cairo, Egypt
e-mail: Gharieb.Elsayyad@gu.edu.eg

© The Author(s), under exclusive license to Springer Nature Singapore Pte Ltd. 2025
K. A. Abd-Elsalam, A. H. Hashem (eds.), *Fungal Endophytes Volume II*,
https://doi.org/10.1007/978-981-97-8804-0_11

ability to stimulate plant defense mechanisms. Real-world case studies are presented to underscore the successful integration of CuNPs, showcasing their effectiveness in mitigating plant diseases. The chapter addresses challenges, including environmental impact and regulatory considerations, and proposes strategies to overcome these obstacles. It concludes by outlining future directions, stressing the optimization of formulations, and advocating for the integration of CuNPs in sustainable agricultural practices. In summary, this chapter provides a comprehensive overview of the biosynthesis of copper nanoparticles from endophytic fungi and highlights their promising role in advancing strategies for managing plant diseases.

Keywords

Copper nanoparticle · Biosynthesis · Endophytic fungi · Plant disease management

11.1 Introduction

Since the 1940s, antibiotics have been employed to treat bacterial diseases, prevent infections in immunocompromised patients, and enhance the health of livestock (Fadiji et al. 2022). Unfortunately, the overuse of antibiotics has led to the emergence of antibiotic-resistant microbes, posing a risk to global health (Elfadil et al. 2022). Recently, nanotechnologies have revolutionized various scientific disciplines, offering unprecedented opportunities to tackle complex challenges through the manipulation of materials at the nanoscale (Hasanin et al. 2023a; Lashin et al. 2023; Abd Elkodous et al. 2021). From among the myriad of nanoparticles, copper nanoparticles have emerged as versatile entities with remarkable properties, including high reactivity, stability, and powerful antimicrobial characteristics. With the scientific community increasingly recognizing the importance of sustainable practices, nanoparticle biosynthesis has emerged as an environmentally friendly alternative to traditional chemical methods (Mani et al. 2021). In the light of this, the role of endophytic fungi as biogenic agents in the biosynthesis of copper nanoparticles takes center stage, offering a novel and environmentally friendly approach (Fadiji et al. 2022).

The endophytic fungi that reside in the internal tissues of plants have long been recognized for their symbiotic relationships with host plants (Abdelaziz et al. 2023; Yan et al. 2019). Besides their ecological importance, these fungi have proved to be efficient mediators of nanoparticle synthesis. Their ability to harness nutrients from their host plants and convert them into metallic nanoparticles underscores the potential for sustainable, biocompatible nanoparticle production. In this chapter, we delve into the fascinating world of copper nanoparticle biosynthesis by endophytic fungi and explore the multiple applications of these bioengineered nanomaterials, focusing on their role in plant disease control (Fadiji et al. 2022).

The unique properties of copper oxide nanoparticles (CuO-NPs) are attracting attention, including their thermal and chemical stability, low toxicity, affordability, high catalytic activity, and compatibility with a variety of materials such as polymers (Mani et al. 2021). Their applications include antibacterial, antifungal, antifouling, anticancer, antihelminthic, antioxidant, antidiabetic, anti-insect, medical textiles, agriculture, gas sensors, superconductors, magnetism-resistant materials, lubricants, solar cells, lithography, wastewater treatment, and photocatalysis (Hasanin et al. 2022, 2023b; Saied et al. 2023). Economically speaking, CuO-NPs are ten times less expensive than Ag-NPs and are less toxic in biomedical and biotechnological applications. In particular, CuO-NPs display powerful antimicrobial properties against bacteria, fungi, and viruses, making them promising for the development of new treatments against infectious diseases, especially drug-resistant strains (Hashem et al. 2023a). The biosynthesis of copper nanoparticles by endophytic fungi represents a convergence of the disciplines of nanotechnology, mycology, and plant pathology. Its synergy opens up new avenues for innovative and sustainable solutions to plant diseases, a persistent challenge in agriculture. Copper nanoparticles, thanks to their inherent antimicrobial properties, offer considerable potential for plant disease management. Understanding the complex mechanisms governing the biosynthesis process allows us not only to unravel the mysteries of nanobiotechnology but also to effectively exploit the synthesized nanoparticles (Fadiji et al. 2022; Mani et al. 2021).

This chapter intends to explore in detail the biosynthesis of copper nanoparticles facilitated by endophytic fungi, providing an overview of the underlying mechanisms and influencing factors. In addition, it aims to elucidate the specific role of these biogenic copper nanoparticles in plant disease mitigation. By bridging the gap between nanotechnology and agriculture, this chapter strives to provide valuable insights that could pave the way for sustainable, environmentally friendly strategies in plant disease management. As we embark on this exploration, we expect to uncover new insights and applications that highlight the transformative potential of endophytic fungus-mediated biosynthesis in nanobiotechnology and agriculture.

11.2 Endophytic Fungi

11.2.1 Definition and Characteristics

Endophytic fungi are a unique group of microorganisms that reside in the internal tissues of plants without causing any apparent damage to their hosts. Unlike pathogens, these fungi form symbiotic relationships with plants, displaying a harmonious coexistence (Mani et al. 2021; Nassar et al. 2023). They are characterized by their capacity to reside asymptomatically within plant tissues, encompassing roots, stems, and leaves. They play a vital role in improving the overall well-being of the host plant by facilitating nutrient absorption, fostering growth, and aiding in stress tolerance. The interactions between endophytic fungi and plants are often complex, involving intricate biochemical signaling and the exchange of metabolites. This ecological relationship is prevalent in nature, with diverse endophytic fungi identified across a range of plant species. The exploration of endophytic fungi holds substantial potential for applications in agriculture, biotechnology, and environmental management, rendering them subjects of growing interest in scientific research (Mani et al. 2021; Rehana et al. 2017).

11.2.2 Role of Endophytic Fungi in Plants

Endophytic fungi play crucial roles in promoting the health and well-being of plants through a variety of mechanisms. Some key roles include:

- *Enhanced nutrient uptake*: Endophytic fungi enhance plant nutrient absorption, including crucial elements like nitrogen and phosphorus, thereby promoting overall plant growth and development (Rehana et al. 2017).
- *Stress tolerance*: Endophytic fungi boost the host plant's resilience to environmental stresses like drought, salinity, and pathogens by producing bioactive compounds (Nassar et al. 2023; Rehana et al. 2017).
- *Disease resistance*: Endophytic fungi can provide disease resistance by directly antagonizing plant pathogens or inducing systemic resistance in the host plant, thereby reducing reliance on chemical pesticides in agriculture (Rehana et al. 2017; Yan et al. 2019).
- *Promotion of growth*: Certain endophytic fungi produce growth-promoting substances like auxins, cytokinins, and gibberellins, enhancing plant growth, development, and productivity (Bogas et al. 2022; Fadiji et al. 2022).
- *Phytoremediation*: Some endophytic fungi aid in soil pollutant degradation, participating in phytoremediation to detoxify contaminated environments (Fadiji et al. 2022; Rasool and Hemalatha 2017; Yan et al. 2019).
- *Biocontrol*: Endophytic fungi serve as biocontrol agents by producing secondary metabolites with antifungal, antibacterial, or insecticidal properties, safeguarding the host plant from pathogens and herbivores (Hassan et al. 2018; Yan et al. 2019).
- *Symbiotic relationships*: The symbiotic association between endophytic fungi and plants is often mutualistic, benefiting both parties. The fungi receive nutrients and a protected environment, while the plants gain improved fitness, resilience, and adaptability (Akram et al. 2023).

Understanding and harnessing the roles of endophytic fungi in plants have significant implications for sustainable agriculture, environmental conservation, and the development of bio-based solutions for crop protection and growth promotion.

11.3 Methods Used for the Study of Endophytic Fungi

11.3.1 Selection of Plant

Plant selection is the first and crucial step in studies of fungal endophytes because the true selection of plant leads to extraction of novel bioactive compounds from fungal endophytes (dos Reis and Lorenzi 2022). There are several criteria to select the plant including:

1. Plants that reside in a unique ecological niche and grow in atypical habitats.
2. Plants possessing a significant ethnobotanical history.
3. Plants that exhibit a wide distribution, possess outstanding longevity, or lived on a specific prehistoric landmass.
4. Plants that live in regions characterized by high levels of biodiversity often possess a variety of endophytes.
5. Plants that are surrounded by pathogen-infected areas however do not exhibit any symptoms and serve as a habitat for endophytes.

In line with existing literature, it is worth noting that not only are leaves, flowers, and fruits effective but that stems and roots also show considerable efficacy in the fields of ethnobotany and pharmacognosy. The collected plant samples can be stored at a temperature of 4 °C. It is necessary to quickly begin isolation procedures subsequent to sample collection in order to avoid contamination (Abubakar and Haque 2020). Following the process of plant selection, authentication by consulting experts and depositing voucher specimens in a herbarium are carried out (Rodriguez et al. 2009).

11.3.2 Surface Sterilization of Plant Samples

A key obstacle to the study of endophytic fungi is the elimination of epiphytic microorganisms. Sterilization of the plant surface can be chemical or physical, following several washes with tap water. The chemical method is the most preferred and comprises three steps: Firstly, the plant sample is immersed in ethanol; secondly, the tissue is immersed in the primary sterilizing agent (sodium hypochlorite, mercury chloride, hydrogen peroxide, or peroxyacetic acid); and thirdly, several washes with sterilized distilled water are performed (dos Reis and Lorenzi 2022; Khalil et al. 2021). Ultraviolet (UV) light or sonication can be used to disinfect physical plant surfaces. In the context of sonication, the cleaned plant tissue is put into a sonicator or sonication bath supplemented with a buffer solution. This will allow sound waves to spread throughout the area, removing and lysing any microbial cells on the plant sample surface (Saldierna Guzmán et al. 2020). This technique of surface disinfecting is effective at removing of epiphytes on the sample surface but needs more time and affected the plant tissue. The plant sample must be repeatedly rinsed in tap water beforehand to eliminate as much dirt from the plant tissue as possible before being exposed to UV rays for 20 min in order to disinfect surfaces using UV light (Sarsaiya et al. 2020).

11.3.3 Culture Medium for Growth of Fungal Endophytes

Potato dextrose agar (PDA) is the most common medium used for isolation of fungal endophytes (dos Reis and Lorenzi 2022; Jayatilake and Munasinghe 2020). Moreover, other media are used for isolation such as malt extract agar, Hagem minimal medium, Czapek medium, tryptone soybean agar, tryptone bovine extract agar, and Luria–Bertani medium (Chand and Shah 2020; Li et al. 2016a).

11.3.4 Growth Conditions and Isolation

The cultivation parameters of endophytic fungi in artificial culture media vary depending on the study's goals. Petri dishes are typically incubated at a temperature between 25 and 28 °C for a period of 3 to 20 days, which may be increased to up to 6 weeks if required (Chand and Shah 2020; Du et al. 2020; Li et al. 2016a; Yao et al. 2017).

The process of fragmenting plant tissue can be effectively accomplished through the utilization of a scalpel, eyelet pliers, or scissors. Subsequently, these fragments can be transferred onto a suitable culture medium, where they will undergo incubation. Fragments measuring 5 mm × 5 mm are frequently employed in experimental settings. The recommended approach involves the isolation of hyphae from the outer edges of fungal colonies, which are subsequently inoculated into a new culture medium, either with or without the presence of antibiotics. This step is repeated many times to get a pure fungi (Ibrahim et al. 2017; Li et al. 2016b).

11.3.5 Identification and Preservation

Initially, classical identification involves both macroscopic and microscopic examinations. Macroscopic classification groups isolates into morphotypes based on observable characteristics like surface colony color, reverse color, texture, and pigmentation. Meanwhile, microscopic classification categorizes isolates based on morphological characteristics, including the shape and dimensions of fungal structures (Du et al. 2020; Li et al. 2016a; Yao et al. 2017).

Secondly, molecular identification is carried out to confirm classical identification. Molecular identification is performed either by 18S rRNA gene or ITS regions, where ITS includes ITS1 (located between 18S and 5.8S rDNA) and ITS2 (between 5.8S and 28S rDNA).

In order to mitigate the risk of contamination-induced losses in isolated colonies, it is advisable to perform re-isolation on inclined cryotubes prepared with potato dextrose agar (PDA), followed by preserving them in a refrigerator at 4 ° C and a freezer at −80 ° C (Wang et al. 2019). In order to preserve fungi for long time, it is recommended to transfer mycelia and spores into a solution consisting of 20% glycerol in ultrapure distilled water. Subsequently, these samples should be stored at a temperature of −80 °C (Wang et al. 2019).

11.4 Bioactive Compounds from Fungal Endophytes

Fungal endophytes have been documented to synthesize a diverse group of bioactive compounds that includes alkaloids, flavonoids, terpenes, phenols, steroids, quinones, and other compounds as illustrated in Fig. 11.1. These compounds often have antimicrobial, antiviral, antifungal, anticancer, antioxidant, and antidiabetic activities (Hashem et al. 2023b).

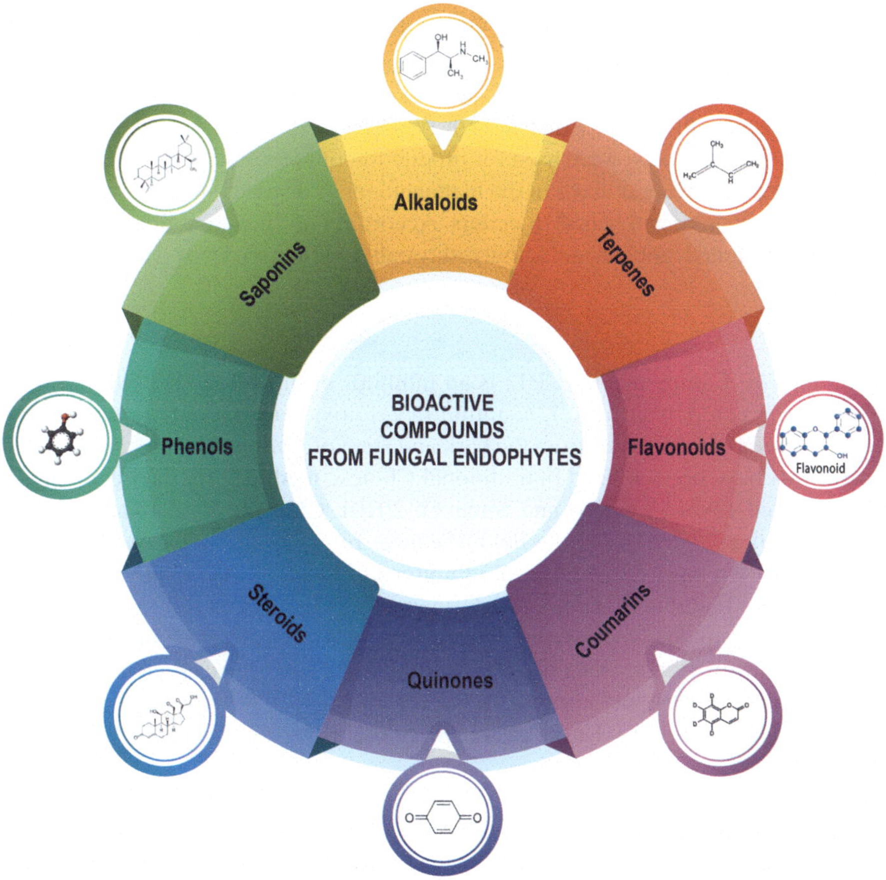

Fig. 11.1 A diagram shows some common bioactive compounds from fungal endophytes

11.4.1 Alkaloids

Alkaloids are a category of naturally existing organic compounds characterized by the presence of a nitrogen atom, typically within a heterocyclic ring structure. They are found in various plants, fungi, and some animals and often exhibit potent physiological effects on humans and other organisms (Rajput et al. 2022). Alkaloids include many compounds such as aconitine, berberine, camptothecin, huperzine A, and others.

Aconitine Aconitine is a diterpenoid alkaloid found in *Aconitum* spp. It is a highly toxic alkaloid, where it is considered one of the most potent naturally occurring neurotoxins and cardiotoxins (Anadón et al. 2016). Moreover, aconitine is produced from endophytic fungus *Cladosporium cladosporioides*, which was isolated from *Aconitum leucostomum* plant (Yang et al. 2013a).

Berberine Berberine is isoquinoline alkaloid compound that is found in several plants, including the roots, rhizomes, and stems of some plants (such as *Berberis vulgaris*) as well as endophytic fungi (Salehi et al. 2019). The endophytic *Alternaria* sp. and *F. solani*, which were isolated from *Phellodendron amurense* and *Coscinium fenestratum*, respectively, were used for production of berberine (Duan et al. 2009). Berberine is extensively utilized in the management of diseases such as hyperglycemia, hyperlipidemia, and neural disorders. According to Singh and Sharma (2018), berberine exhibited anticancer activity through binding to DNA and inhibiting replication.

Camptothecin Camptothecin (CPT) is an inhibitor of quinoline alkaloid that specifically targets topoisomerase I, resulting in cytotoxic effects. CPT is a naturally occurring compound derived from the bark and stem of *Camptotheca acuminata*. Its constituents have been utilized in traditional Chinese medicine for the treatment of cancer-related cases (Venditto and Simanek 2010). CPT and its derivatives have gained significant attention in the field of cancer research and chemotherapy. This substance is extremely effective as an inhibitor of topoisomerase I, an enzyme that plays a role in both the replication and repair of DNA. CPT causes DNA damage and ultimately cell death because it binds to the DNA-topoisomerase I complex that leads to the destruction of DNA (Pommier 2006). CPT has been extracted from various fungal endophytes, including *F. solani*, *Entrophospora infrequens*, *Neurospora* sp. and *Nodulisporium* sp. (Kusari et al. 2009; Puri et al. 2005; Rehman et al. 2008).

Huperzine A Huperzine A is a natural compound derived from the Chinese club moss plant (*Huperzia serrata*). Also, many endophytic fungi such as *Aspergilli*, *Colletotrichum*, *Phyllosticta*, *Hypoxylon*, *Xylaria*, and *Nigrospora* have been shown to synthesize novel strong acetylcholinesterase inhibitors (AChEIs) in their metabolite extracts (Wang et al. 2016). Huperzine A works primarily as an acetylcholinesterase inhibitor. Acetylcholinesterase is an enzyme that breaks down the neurotransmitter acetylcholine, which is involved in memory, learning, and cognitive skills. By inhibiting this enzyme, huperzine A increases the levels of acetylcholine in the brain, which may help improve cognitive skills and memory (Walczak-Nowicka Ł and Herbet 2021).

11.4.2 Flavonoids

Flavonoids are a class of pigments found in edible plants, characterized by the presence of two benzene rings adjacent to a three-carbon ring. Endophytic fungi produced many flavonoid compounds such as apigenin, cajanol, chrysin, curcumin, kaempferol, quercetin, rutin, and silymarin. Gao et al. (2012) isolated *Chaetomium globosum* from *Cajanus cajan* plant where it had ability to produce apigenin that exhibited antioxidant activity. Moreover, endophytic *Hypocrea lixii* has ability to produce cajanol with anticancer activity (Zhao et al. 2013).

Moreover, chrysin was derived from three fungal species, namely, *A. alternata*, *Colletotrichum capsici*, and *Colletotrichum taiwanense*, where results proved that chrysin has anticancer activity (Seetharaman et al. 2017). Additionally, endophytic *Chaetomium globosum* has been used for curcumin production (Wang et al. 2012). Curcumin has gained significant attention for its potential health benefits and has been studied for its anti-inflammatory, antioxidant, and anticancer properties (Hewlings and Kalman 2017).

Kaempferol is a natural flavonoid compound found in plants (fruits, vegetables, herbs) as well as endophytic fungi. Keampferol was produced from endophytic *Fusarium chlamydosporum* that was isolated from *Tylophora indica* plant (Chaturvedi et al. 2014). Moreover, quercetin, a class of flavonoids, has received much attention in scientific research due to its properties as antioxidant, anti-inflammatory agent, and immune-modulator (Qi et al. 2022). Ebada et al. (2016) successfully isolated three quercetin compounds from an endophytic *Nigrospora oryzae*, which isolated from the leaves of *Loranthus micranthus*.

11.4.3 Terpenes

Terpenes are a broad and heterogeneous group of organic compounds that are synthesized by a diverse range of plant species and endophytic fungi. According to Masyita et al. (2022), the chemical structures and biological activity of fungal terpenes are extremely varied. Terpenes are classified to sesquiterpenes, diterpenes, and meroterpenes.

Sesquiterpenes are a category of terpenes characterized by their composition of 15 carbon atoms, which are obtained from the combination of 3 isoprene units. Xylarenones A and B, along with xylarenic acid, were extracted from *Xylaria* sp. NCY2 (Hu et al. 2008). Silva et al. (2006) reported the isolation of several cadinane sesquiterpene derivatives from *Phomopsis cassiae*, which was isolated from *Cassia spectabilis*.

Diterpenes are a subclass of terpenes characterized by their composition of 20 carbon atoms, which are formed by the combination of four isoprene units. Paclitaxel (Taxol) is the most prevalent diterpene. Taxol, a diterpenoid with significant functional groups, is present in all species of the *Taxus* genus, although its first isolation occurred from *Taxus brevifolia*. There are many endophytic fungi that have ability to produce Taxol such as *Fusarium solani*, *Alternaria* sp., and *Aspergillus* sp., which isolated *Taxus chinensis*, *Ginkgo biloba*, and *Podocarpus* sp., respectively (Liu et al. 2009). Furthermore, Zhao et al. (2010) reported that various fungal species, including *Phyllosticta spinarum* and *Bartalinia robillardoides*, have been identified as producers of Taxol.

Meroterpenes are a class of natural compounds that are hybrid molecules containing both terpenoid and non-terpenoid moieties. They are formed through the combination of terpenoid precursor units with other biosynthetic building blocks, such as polyketides, shikimate pathway intermediates, or amino acids (Shen et al. 2020). Fill et al. (2007) reported that *Penicillium* sp. has ability to produce meroterpenoids.

11.4.4 Phenolics

Phenolics have frequently been extracted from endophytic fungi derived from diverse host plants. In a previous study, endophytic *Pestalotiopsis adusta* was used for production of pestalachlorides A and B (Li et al. 2008). Also, Hoffman et al. (2008) successfully extracted a collection of phenolic acids from the culture broth of a *Phoma* sp., which was identified as an endophytic fungus residing within a Guinea plant species known as *Saurauia scaberrinae*. Numerous novel phenols and phenolic acids, which have been extracted from endophytic cultures in recent research, have frequently exhibited significant antimicrobial properties (Caruso and Palombo 2022). Through conducting additional research on phenol and phenolic acids, there exists significant potential for the discovery of novel and highly efficacious antibiotics.

11.4.5 Steroids

Steroids are a class of organic compounds that have a specific molecular structure consisting of four interconnected rings of carbon atoms. They are found in plants, animals, as well as microbes. Steroids are characterized by their diverse range of physiological effects, including regulatory roles in metabolism, growth, reproduction, and immunity (Tong 2013). Zhang et al. (2009) reported that *Colletotrichum* sp. was used for production of ring B aromatic steroids. Two new steroid compounds were extracted from an endophytic *Acrocalymma* sp. (Yang et al. 2022). The steroid compound aspergilolide was extracted from *Aspergillus* sp. MBL1612

(Zhang et al. 2019). Furthermore, two new steroid compounds were isolated from endophytic *Phomopsis* sp. (Wu et al. 2013).

11.4.6 Quinones

Quinones represent a category of organic compounds that exhibit a distinctive bicyclic structure, comprising two adjacent carbon atoms with carbonyl (C=O) groups. Quinones within the natural environment can be observed across various taxa, including plants, animals, fungi, and bacteria. Quinones play important roles in various biological processes and have diverse applications in industry and medicine (Christiansen et al. 2021). Many endophytic fungi were used for quinone compound production. *A. fumigatus* that was isolated from *Rumex nepalensis* has ability to produce anthraquinone with anticancer and antioxidant activities (Yang et al. 2013b). Moreover, emodin (6-methyl-1,3,8-trihydroxyanthraquinone) was produced by *Talaromyces* sp. *and Aspergillus versicolor* where it had antibacterial, anticancer, and anti-inflammatory activities (Hawas et al. 2012; Liu et al. 2010). Furthermore, lawsone (2-hydroxy-1,4-naphthoquinone) with cytotoxic activity was extracted from endophytic *Gibberella moniliformis* (Sarang et al. 2017).

11.4.7 Other Compounds

There are other compounds that can be produced from endophytic fungi, such as trichodermin, saponins (diosgenin, ginsenoside, and gymnemagenin) and coumarins (bergapten, meranzin, isofraxidin, marmesin, mellein, scopoletin, and umbelliferone) (Cheng et al. 2012; Huang et al. 2012; Lie-Zhong et al. 2007; Ligang et al. 2004; Pongcharoen et al. 2007).

11.4.8 Potential for Nanoparticle Biosynthesis

The potential of endophytic fungi in nanoparticle biosynthesis is an exciting and burgeoning field of research with diverse applications. Endophytic fungi possess unique biochemical capabilities that make them excellent candidates for the green synthesis of nanoparticles, offering an environmentally friendly and sustainable alternative to conventional methods. Here are some key aspects of their potential in this context:

- *Biocompatibility*: Endophytic fungi can produce nanoparticles with inherent biocompatibility, making them suitable for various biomedical applications, including drug delivery systems, imaging agents, and tissue engineering (Fadiji et al. 2022).
- *Versatility in nanoparticle types*: These fungi have demonstrated the ability to synthesize a variety of nanoparticles, including metallic nanoparticles (such as

silver, gold, and copper), metal oxide nanoparticles, and carbon-based nanoparticles. This versatility opens up a wide range of potential applications in different industries (Fadiji et al. 2022).

- *Eco-friendly synthesis*: The green synthesis of nanoparticles using endophytic fungi typically involves using plant extracts or fungal biomass as reducing and stabilizing agents. This eco-friendly approach eliminates the need for harsh chemicals and high-energy processes, reducing the environmental impact of nanoparticle production (Hassan et al. 2018; Nassar et al. 2023; Rasool and Hemalatha 2017).
- *Controlled synthesis parameters*: Researchers can manipulate various parameters in the synthesis process, such as culture conditions and substrate composition, to control the size, shape, and composition of the synthesized nanoparticles. This control is crucial for tailoring nanoparticles to specific applications (Hassan et al. 2018).
- *Biogenic synthesis mechanisms*: Endophytic fungi employ complex biochemical pathways and enzymes during nanoparticle biosynthesis. Understanding these mechanisms can provide insights into the green synthesis process and help optimize the production of nanoparticles with desired properties (Fadiji et al. 2022).
- *Antimicrobial properties*: Nanoparticles synthesized by endophytic fungi often exhibit inherent antimicrobial properties, making them potential candidates for applications in antimicrobial coatings, wound healing, and combating drug-resistant microorganisms (Hassan et al. 2018).
- *Agricultural applications*: Endophytic fungi-derived nanoparticles have shown promise in agricultural settings, with potential applications in crop protection, nutrient delivery, and soil remediation (Fadiji et al. 2022).

Harnessing the potential of endophytic fungi in nanoparticle biosynthesis aligns with the principles of green nanotechnology, emphasizing sustainability, reduced environmental impact, and biomedical and industrial applications.

11.5 Endophytic Fungi: Nature's Nanoparticle Factories—Biosynthesis of Copper Nanoparticles.

11.5.1 General Process of Nanoparticle Biosynthesis

Nanoparticle biosynthesis represents a fascinating convergence between biology and nanotechnology, offering a sustainable and environmentally friendly alternative to traditional chemical methods (Mani et al. 2021). The typical process involves using living organisms such as bacteria, fungi, plants, or other biological entities to facilitate the reduction of metal ions, leading to the formation of nanoparticles. In the context of microbial biosynthesis (Fig. 11.2), microorganisms often act as efficient nano-factories, secreting biomolecules with reducing capabilities that convert

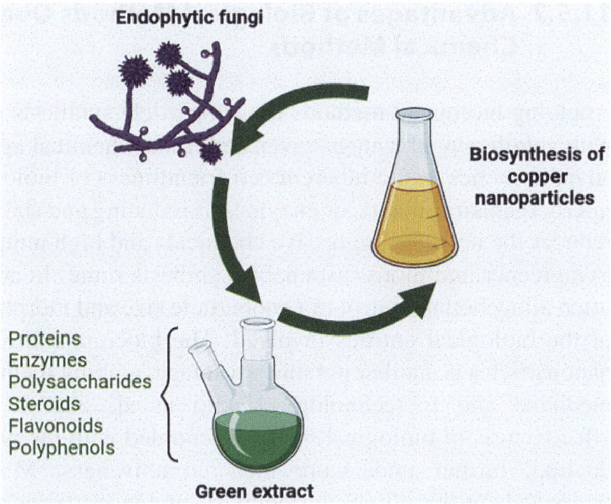

Fig. 11.2 Biosynthesis of copper nanoparticles from endophytic fungi

metal precursors into nanoparticles. Biosynthesis based on plants exploits the reducing potential of plant extracts, while fungus-based processes take advantage of the enzymatic machinery of fungi. The specific biomolecules involved, such as enzymes, proteins, and secondary metabolites, play a crucial role in determining the size, shape, and stability of the resulting nanoparticles. This biological approach makes it possible not only to precisely control the characteristics of nanoparticles but also to tailor their properties to a variety of applications, from medicine and catalysis to agriculture and environmental cleanup. The general process of nanoparticle biosynthesis illustrates nature's ingenuity in orchestrating the formation of nanomaterials, paving the way for sustainable, environmentally friendly advances in nanotechnology.

As an illustrative example of a copper nanoparticle synthesis protocol reported by Nassar et al. (2023), the endophytic fungal strain BR.1 was cultivated in 100 mL of PD broth media at 25 ± 2 °C for 5 days. Following centrifugation, the fungal biomass was washed and mixed with distilled water for 24 h in the dark, and the resulting fungal biomass filtrate served as a biocatalyst for CuO-NP biosynthesis. Gas chromatography–mass spectrometry (GC–MS) was employed to analyze the components of the fungal biomass filtrate, offering insights into the biomolecules present. For CuO-NP biosynthesis, 100 μg of Cu(CH3COO)2·H2O was dissolved in 5 mL of distilled water, mixed with 95 mL of fungal biomass filtrate, and stirred for 1 h at 100 rpm, and the pH was adjusted to 8 using 1 N NaOH. The discernible color change from colorless to greenish indicated CuO-NP formation. The as-formed NPs were collected, washed, and subjected to calcination at 200 °C for 2 h before preservation. This streamlined description outlines the key steps in the biosynthesis of CuO-NPs using the cell-free filtrate of the BR.1 fungal strain.

11.5.2 Advantages of Biological Methods Over Chemical Methods

Applying biological methods to nanoparticle synthesis represents a paradigm shift with significant advantages over traditional chemical approaches. One of the main advantages lies in the inherent eco-friendliness of biological processes. The use of microorganisms, plants, or enzymes as reducing and stabilizing agents considerably reduces the need for aggressive chemicals and high temperatures, thus contributing to a greener and more sustainable synthesis route. In addition, biological methods often allow better control of nanoparticle size and morphology due to the specificity of the biological entities involved. The biocompatibility of these biosynthesized nanoparticles is another notable advantage, making them suitable for applications in medicine and biotechnology (Fadiji et al. 2022). The simplicity and cost-effectiveness of biological methods, coupled with the potential for large-scale production, further underscore their attractiveness. Moreover, the versatility of biological entities allows the synthesis of nanoparticles under mild conditions, widening the scope of materials that can be employed for nanoparticle fabrication. Collectively, these advantages position biological methods as promising alternatives, aligning with the growing emphasis on sustainable and green nanotechnology practices (Bogas et al. 2022).

11.5.3 Mechanisms Involved in Copper Nanoparticle Biosynthesis

Copper nanoparticle biosynthesis by endophytic fungi involves a series of complex mechanisms governed by the unique biochemical composition and metabolic activities of these fungi. The unique biochemical composition and metabolic activities inherent to these fungi govern the complex processes involved. Enzymatic reduction appears to be a key mechanism, in which enzymes such as reductases and nitrate reductases serve as bioreduction tools, orchestrating the controlled synthesis of nanoparticles from copper ions (Mani et al. 2021). Metabolite-mediated reduction further contributes to the process, with secondary metabolites acting as pivotal capping and stabilizing agents, preventing the agglomeration of nanoparticles. The interplay of intracellular and extracellular processes, involving the use of enzymes and metabolites within fungal cells and the secretion of bioactive molecules into the external environment, adds another layer of complexity. Nucleation and growth of copper nanoparticles are facilitated by the fungi, guiding the assembly of reduced copper atoms to determine the size and shape of the resulting nanoparticles. Fungal strain specificity introduces variability in mechanisms and efficiencies, impacting the ultimate properties of the nanoparticles (Mani et al. 2021; Rehana et al. 2017). The influence of pH and temperature conditions, coupled with the redox potential maintained by the fungi, significantly shapes the biosynthesis process by affecting enzymatic activity and reducing agent availability and nanoparticle stability. This multifaceted understanding of mechanisms is paramount for optimizing the biosynthesis process, tailoring the properties of copper nanoparticles, and unlocking the

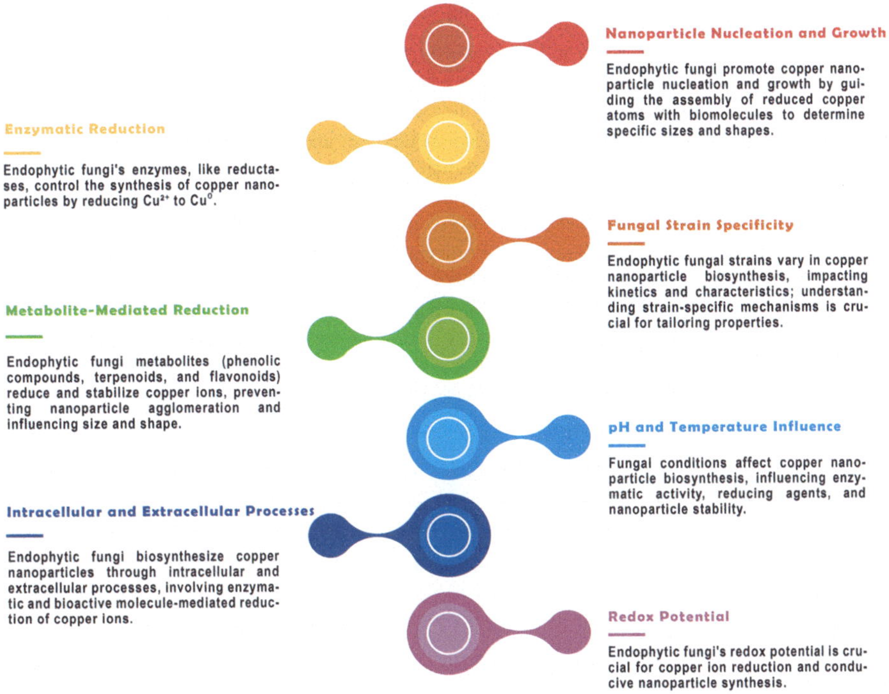

Fig. 11.3 Graphical representation of key mechanisms in copper nanoparticle biosynthesis by endophytic fungi

eco-friendly applications of these bioengineered nanomaterials facilitated by endophytic fungi (Akram et al. 2023; Bogas et al. 2022; Nassar et al. 2023; Rasool and Hemalatha 2017; Shankar Naik 2020).

Figure 11.3 depicts a graphical representation illustrating the essential mechanisms linked to the biosynthesis of copper nanoparticles through the utilization of endophytic fungi.

Elucidating these mechanisms is essential for optimizing the biosynthesis process, tailoring the properties of copper nanoparticles, and expanding the applications of these bioengineered nanomaterials. Investigating the interplay of these factors provides valuable insights into harnessing the potential of endophytic fungi for sustainable and eco-friendly nanoparticle synthesis.

11.5.4 Examples of Endophytic Fungi with Nanoparticle Biosynthesis Potential

A more economical alternative to produce nanoparticles, specifically CuNPs, emerges as a viable option compared to the costlier counterparts, namely, AuNPs and AgNPs. Research indicates that the fungus *Aspergillus terreus*, isolated from

Table 11.1 Endophytic fungi and their biosynthesis potential for nanoparticles

Endophytic fungi	Characteristics and biosynthesis potential
Fusarium spp.	Versatile in various biological processes, including the reduction of metal ions for nanoparticle biosynthesis
Penicillium spp.	Known for unique secondary metabolites and enzymatic activities facilitating the reduction of metal ions and nanoparticle formation
Phoma spp.	Exhibits nanoparticle biosynthesis capabilities with metabolic diversity, making them promising for green synthesis approaches
Alternaria spp.	Explored for nanoparticle biosynthesis due to the presence of bioactive compounds aiding in nanoparticle reduction and stabilization
Trichoderma spp.	Recognized for diverse biological processes, including nanoparticle biosynthesis, attributed to extracellular enzymes and secondary metabolites
Colletotrichum spp.	Shows promise in nanoparticle biosynthesis with unique characteristics, including adaptability and enzymatic activities, suitable for green synthesis

Origanum majorana and *Aegle marmelos*, exhibits the capability to generate spheroidal CuO NPs from CuSO4 (Mousa et al. 2021). The utilization of both biomass and cell extract of this fungus results in the formation of CuO NPs with diverse sizes. Other sources of CuNPs, measuring less than 100 nm, include marine actinomycetes extracted from seaweeds (Rasool and Hemalatha 2017) and *Streptomyces capillispiralis*, isolated from the leaves of *Calendula arvensis* (Hassan et al. 2018). *Aspergillus terreus*, isolated from *Origanum majorana* and *Aegle marmelos* (Mousa et al. 2021), has demonstrated the ability to produce spheroidal CuNPs using CuSO4 as the reagent. Previous studies have highlighted the feasibility of altering the shape of CuNPs by switching the source of Cu2+ ions to Cu(NO3)2 and employing *Phaeoacremonium* sp. to yield nanorods with an approximate diameter of 97 nm (Bogas et al. 2022).

Many endophytic fungi have demonstrated significant potential for nanoparticle biosynthesis, including copper nanoparticles. Here are examples of endophytic fungi known for their nanoparticle biosynthesis capabilities as shown in Table 11.1:

These examples underscore the diversity of endophytic fungi that can be explored for nanoparticle biosynthesis. It's important to note that the nanoparticle synthesis potential can vary even among strains within the same fungal species. Understanding the specific capabilities of each fungal strain is crucial for harnessing their biosynthetic potential in a targeted and efficient manner (Fadiji et al. 2022; Nassar et al. 2023; Shankar Naik 2020).

11.6 Factors Affecting Biosynthesis

11.6.1 Environmental Factors

The biosynthesis of copper nanoparticles by endophytic fungi is influenced by various environmental factors that play a crucial role in shaping the characteristics of the nanoparticles, impacting overall efficiency, yield, and properties (Bogas et al.

2022). These factors are detailed in Table 11.1, providing a comprehensive overview of the diverse elements significantly affecting the biosynthesis process. Encompassing the pH of the growth medium, temperature, nutrient availability, copper ion concentration, incubation period, oxygen levels, light conditions, and agitation with aeration, these factors play crucial roles in determining the characteristics of the synthesized nanoparticles. Temperature affects growth rate and morphology, and nutrient availability and copper ion concentration have an impact on metabolic activities and precursor availability. The incubation period balances nanoparticle production and degradation problems, while oxygen levels influence redox potential. Lighting conditions and agitation with aeration contribute to the complex interplay of factors affecting overall efficiency (Fadiji et al. 2022; Nassar et al. 2023). This comprehensive understanding is essential for optimizing copper nanoparticle biosynthesis using endophytic fungi and ensuring tailored characteristics for specific applications through experimental fine-tuning (Table 11.2).

Understanding and optimizing these environmental factors are essential for fine-tuning the biosynthesis of copper nanoparticles using endophytic fungi. The interplay of these factors can be intricate, and experimental optimization is often necessary to achieve the desired nanoparticle characteristics for specific applications.

Table 11.2 Environmental factors influencing copper nanoparticle biosynthesis by endophytic fungi

Environmental factors	Impact on copper nanoparticle biosynthesis
pH of the growth medium	Influences enzymatic activity and metabolic processes, with optimal pH conditions determined for each fungal strain to maximize nanoparticle production
Temperature	Affects growth rate, metabolism, and enzymatic activities, influencing the kinetics, size, and morphology of synthesized copper nanoparticles
Nutrient availability	Impacts metabolic activities, with nutrient sources like carbon and nitrogen influencing the production of biomolecules involved in nanoparticle reduction and stabilization
Copper ion concentration	Critical for nanoparticle formation, optimizing copper ion concentration ensures sufficient material for synthesis without reaching toxic levels
Incubation period	Crucial duration of fungal interaction with copper ions, balancing longer periods for increased nanoparticle production with potential degradation or agglomeration concerns
Oxygen levels	Influences redox potential, essential for reduction processes in nanoparticle biosynthesis, requiring adequate oxygen for enzymatic activities and efficient nanoparticle production
Light conditions	Can influence fungal growth and metabolic activities, with variations impacting the synthesis of biomolecules involved in nanoparticle biosynthesis
Agitation and aeration	Contribute to nutrient distribution, enhancing interactions between fungal biomass and copper ions for improved overall efficiency in nanoparticle biosynthesis

11.6.2 Fungal Strain-Specific Factors

Fungal strain-specific factors play a pivotal role in the biosynthesis of copper nanoparticles from endophytic fungi, influencing the kinetics, yield, and properties of the resulting nanoparticles (Bogas et al. 2022; Fadiji et al. 2022; Nassar et al. 2023). Different fungal strains exhibit unique enzymatic profiles, impacting the rate and efficiency of copper ion reduction to nanoparticles. Additionally, variations in secondary metabolite production, such as phenolic compounds, terpenoids, and flavonoids, contribute to differences in nanoparticle size, shape, and stability. Fungal strains' inherent reduction capacity, as well as their tolerance to high copper ion concentrations, also influences the efficiency of copper ion conversion to nanoparticles. Growth characteristics, including growth rate and biomass production, vary from strain to strain, affecting nanoparticle production potential. Strain preferences for extracellular or intracellular biosynthetic pathways and distinct pH and temperature preferences contribute to the diversity of nanoparticle characteristics (Nassar et al. 2023). Genetic variability among strains, reflected in unique genes encoding enzymes or regulatory proteins, adds another layer of complexity. Understanding these strain-specific factors is crucial for tailoring the biosynthesis process and selecting an appropriate fungal strain to optimize copper nanoparticle production for specific applications.

11.6.3 Substrate and Culture Conditions

The substrate and culture conditions utilized in the biosynthesis of copper nanoparticles by endophytic fungi play a pivotal role in influencing the overall process, impacting aspects such as nanoparticle yield, size, stability, and characteristics (Fadiji et al. 2022). A number of key effects are associated with the substrate and culture conditions used in this biosynthesis process. These include the choice and composition of the growth medium, which profoundly affects fungal metabolism and nanoparticle synthesis by influencing the availability of nutrients, carbon sources, and metal precursors such as copper ions. Adequate nutrient availability is considered essential for the growth and metabolic activities of endophytic fungi, with nitrogen, carbon, and other essential nutrients in the growth medium influencing the synthesis of enzymes and biomolecules crucial for copper ion reduction and nanoparticle stabilization. The initial concentration of copper ions in the growth medium is identified as a critical factor, where higher concentrations can increase nanoparticle production, but careful optimization is required to balance yield and stability. The pH of the growth medium has a significant impact on enzymatic activity and metabolic processes, affecting the stability and size distribution of the nanoparticles produced (Bogas et al. 2022). The temperature at which endophytic fungi are cultured influences growth rate and metabolic activities, thereby influencing the kinetics, size, and morphology of copper nanoparticles. The duration of incubation, or exposure time to copper ions, influences the rate of nanoparticle synthesis, with longer incubation periods potentially increasing production, though optimal durations must be determined to avoid degradation or agglomeration. Adequate aeration and agitation of the growth medium contribute to the uniform distribution of nutrients and facilitate interactions between fungal biomass and copper ions, with proper mixing enhancing overall efficiency. The type of substrate or host plant used for cultivating endophytic fungi can impact the availability of precursor compounds and secondary metabolites, with different host plants providing diverse bioactive compounds influencing copper ion reduction and nanoparticle stabilization. Light exposure can influence the growth and metabolic activities of endophytic fungi, potentially impacting the synthesis of biomolecules involved in nanoparticle biosynthesis. Oxygen availability in the growth environment influences redox potential, crucial for the reduction processes involved in nanoparticle biosynthesis, with adequate levels being crucial for maintaining enzymatic activities required for efficient nanoparticle production (Mousa et al. 2021; Rehana et al. 2017). Optimizing these substrate and culture conditions is deemed essential for achieving efficient and controlled biosynthesis of copper nanoparticles from endophytic fungi, with careful consideration required to tailor the nanoparticle characteristics for specific applications (Fadiji et al. 2022; Mani et al. 2021; Mousa et al. 2021).

11.7 Characterization of Copper Nanoparticles

The characterization of copper nanoparticles synthesized from endophytic fungi is essential for understanding their properties and potential applications. Various analytical techniques are employed to explore the size, shape, structure, and surface properties of these nanoparticles. The Table 11.3 below outlines key methods employed in the characterization process (Bogas et al. 2022; Fadiji et al. 2022; Nassar et al. 2023; Rasool and Hemalatha 2017):

Comprehensive characterization using these techniques allows researchers to gain a holistic understanding of copper nanoparticles synthesized by endophytic fungi, guiding their potential applications in fields such as medicine, catalysis, and materials science.

11.7.1 UV-Vis Spectroscopy

The greenish color intensity was measured using a spectrophotometer in the 200–700 nm range to detect the maximum surface plasmon resonance (SPR) of CuO-NPs. The UV-vis chart revealed a peak at 280 nm, indicating the SPR of CuO (Fig. 11.4a). The SPR absorption peaks of CuO-NPs synthesized by different methods varied, showing sensitivity to factors such as crystallite size, shape, and metal precursors (Nassar et al. 2023).

Table 11.3 Characterization techniques for copper nanoparticles

Technique	Principle	Insights
Transmission electron microscopy (TEM)	Uses electron beams to visualize nanoparticles at high resolution	Provides detailed information on the size, shape, and distribution of copper nanoparticles. Essential for understanding their morphology
Scanning electron microscopy (SEM)	Scans the surface of nanoparticles using electron beams, offering 3D images	Complements TEM by providing information on surface morphology, aggregation patterns, and structural details
X-ray diffraction (XRD)	Analyzes the diffraction pattern of X-rays to determine the crystalline structure	Identifies the crystallographic phases of copper nanoparticles, aiding in understanding their structural properties
UV-visible spectroscopy	Utilizes the absorbance of light by nanoparticles	Reveals the surface plasmon resonance of copper nanoparticles, offering initial insights into their optical properties and concentration
Fourier-transform infrared spectroscopy (FT-IR)	Examines molecular vibrations and identifies functional groups on the nanoparticle surface	Unveils the biomolecules involved in capping and stabilizing the copper nanoparticles, contributing to their unique characteristics
Dynamic light scattering (DLS)	Measures the size distribution and surface charge of nanoparticles in solution	DLS provides information on hydrodynamic size, while zeta potential reveals nanoparticle stability and potential interactions
Energy-dispersive X-ray spectroscopy (EDS)	Detects the elemental composition of nanoparticles	Confirms the presence of copper in the synthesized nanoparticles and provides information about other elemental constituents l
Thermogravimetric analysis (TGA)	Measures the weight change of nanoparticles with temperature	Helps determine the thermal stability and composition of the nanoparticles, including the presence of capping agents
Biological characterization	Biocompatibility studies: Evaluate the impact of copper nanoparticles on living organisms	Antimicrobial assays: Assess the efficacy of copper nanoparticles against pathogenic microorganisms
Surface area and porosity analysis (BET method)	Measures the specific surface area and porosity of nanoparticles	Provides information on the physical properties of the nanoparticles relevant for applications like catalysis
Magnetic characterization (if applicable)	Examines the magnetic properties of nanoparticles	Relevant if the nanoparticles exhibit magnetic behavior, useful in applications like drug delivery
Cytotoxicity and cellular uptake studies (if applicable)	Cytotoxicity and cellular uptake studies (if applicable)	Evaluate the impact of copper nanoparticles on cell viability and study their cellular internalization for biomedical applications

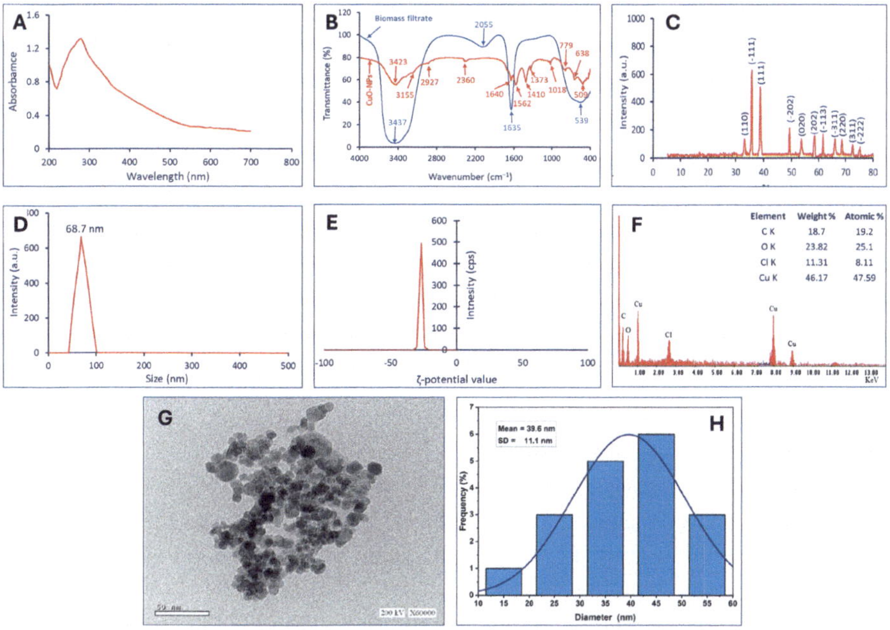

Fig. 11.4 (**a**) UV-vis spectroscopy of CuO-NPs fabricated by the endophytic fungal strain *A. terreus* BR.1, revealing maximum surface plasmon resonance (SPR) at a wavelength of 280 nm. (**b**) FT-IR of fungal biomass filtrate compared to biosynthesized CuO-NPs, highlighting various functional groups. Dynamic light scattering (**d**) and zeta potential analysis (**e**) of synthesized CuO-NPs using the endophytic fungal strain *A. terreus* BR.1. Morphological and crystallographic investigation of CuO-NPs fabricated by the same fungal strain: (**g**) Transmission electron microscopy (TEM) displaying a spherical shape, (**h**) illustrating size distribution, (**f**) presenting EDX analysis, and (**c**) showcasing X-ray diffraction, confirming the crystalline nature. (Adapted from Nassar et al. (2023))

11.7.2 FT-IR

Functional groups in fungal biomass filtrate and synthesized CuO-NPs were investigated using FT-IR (Fig. 11.4b). Peaks at 3437, 2055, 1635, and 539 cm^{-1} in the biomass filtrate shifted after CuO-NP synthesis. These shifts indicated changes in hydroxyl groups, aromatic compounds, polysaccharides, and C-S, C = S groups. The presence of various functional groups confirmed the reduction and capping roles of biomass filtrate components (Nassar et al. 2023).

11.7.3 Morphological Characteristics

Transmission electron microscopy (TEM) displayed spherical CuO-NPs with well-arranged structures (Fig. 11.4g). The sizes ranged from 15 to 55 nm with an average size of 39.6 ± 11.1 nm (Fig. 11.4b). EDX analysis confirmed the presence of Cu and

O ions, with additional peaks for C and Cl (Fig. 11.4f). The elemental mapping indicated successful Cu–O formation and the scattering of capping agents on the CuO-NPs surface (Nassar et al. 2023).

11.7.4 Crystallographic Investigation

X-ray diffraction (XRD) analysis revealed a face-centered cubic (FCC) phase for the fungal-mediated CuO-NPs, with 11 Bragg's diffraction peaks corresponding to specific planes (Fig. 11.4c). The sharp and well-defined peaks confirmed the pure crystalline nature of CuO-NPs, and the average crystallite size was calculated as 31 nm based on the sharpest peak (Nassar et al. 2023).

11.7.5 Zeta Potential Analysis

Zeta potential analysis showed a ζ value of −26.6 mV, indicating high stability in the colloidal solution (Fig. 11.4e). The negative charge on CuO-NPs prevented agglomeration or aggregation, enhancing stability. The presence of diverse capping agents on the NPs' surface contributed to this stability, aligning with established guidelines for assessing stability based on ζ values (Nassar et al. 2023).

11.8 Role of Copper Nanoparticles in Plant Disease Management

11.8.1 Antimicrobial Properties

Recently, the attention of plant pathologists has turned to thinking of effective modern methods that help in combating disease without harming the environment. The scientists also turned to many modern means of resistance that try to avoid environmental pollution, such as the use of genetically engineered plants that carry the characteristics of resistance, as well as the use of biological control methods (Attia et al. 2022; El-Batal et al. 2023; Gerhardson 2002). Biosynthesis of nanomaterials using endophytic fungal extracts as a rich source of natural compounds such as carbohydrates, phenols, flavonoids, tannins, and alkaloids (Albalawi et al. 2022) that could act as safe reducing and stabilizing agents in addition to their ability to maintaining aseptic environments throughout the process (Elkhodary et al. 2023; Mohammadi et al. 2016). Nanotechnology applications are the latest trend in revolutionizing the field of therapeutic plant nutrition. Nanoparticles are a unique phenomenon of chemical reactions to control the properties of physical, chemical, and biological materials (El-Batal et al. 2020; Mu et al. 2014). CuO-NPs used to meet the challenges facing the world in the field of agricultural production and food safety (Bakshi and Kumar 2022; He et al. 2019). One of the most important applications of CuO-NPs in the agricultural sectors is finding integrated solutions for plant nutrition, combating agricultural diseases and pests, and treating water and soil (Maity et al. 2022; Rajput et al. 2021). One of the most important antibacterial and antifungal agents is nanocopper oxide. In several studies, biofilms produced by the most common pathogenic bacteria in potatoes, such as *Clavibacter michiganensis*, *Dickeya solani*, and *Ralstonia solanacearum*, were estimated in the absence and presence of CuO NP nanoparticles (El-Batal et al. 2020; Omran and Baek 2022; Rabea et al. 2023). On the other hand, some investigations can be gathered from literature reviews. The results performed by Mishra et al. (2022) indicated that the synthesized CuO NPs exogenously interacted with bacterial pathogens through electrostatic attraction and defeated the bacteria cells via membrane leakage. CuO nanoparticles (NPs) have been shown in bacterial cells to boost oxidative pressure (Applerot et al. 2012; Meghana et al. 2015). They also replace bacterial cell membranes abruptly due to strong ROS generation (Ali et al. 2019). Therefore, it has been demonstrated that CuO NPs can efficiently cause oxidative stress in human blood cells in addition to harming their mitochondria and lysosomes (Ansarifard et al. 2021; Naz et al. 2020). The majority of the impacts of CuO NPs can be ascribed to the oxidative stress that the NPs generate, which can be brought on by the NPs' size, shape, and dissolution (Du et al. 2021). Free radicals were produced as a result of the rise in oxidative stress, which the antioxidant system was unable to fully combat (Abd El 2012). The results conducted by Gaber et al. (2023) showed high activity of CuO NPs against *Fusarium* fungus, which causes wilt disease. This indicates that the antifungal activity of CuO NPs is very high compared to CuO (Al-Dhabaan et al. 2017; Doolotkeldieva et al. 2022; Nadeem et al. 2019). CuO NPs' mechanism of action may be related to electrostatic interaction, which damages cell membranes, disrupts proteins and enzymes, produces reactive oxygen species (ROS) and oxidative stress, binds proteins to disturb homeostasis (disturbs the electron transport chain), inhibits signal transduction, and causes genotoxicity (Farshori et al. 2022; Sajjad et al. 2023). Additionally, the effectiveness could be ascribed to the roughness of the nanoparticles' outer surface, which damages the cell wall and allows the particles to pass across to the plasma membrane (Mubeen et al. 2021).

11.8.2 Induction of Plant Defense Mechanisms

Plants become resistant to some harmful bacteria, viruses, and fungi due to a phenomenon known as systemic resistance, which is caused by certain microbes interacting with their roots (Gaber et al. 2023). ISR triggers the plant's defense mechanisms and protects those parts of the plant that are not exposed from upcoming pathogenic invasions by insects and microbes (Elbasuney et al. 2023). To create novel formulations of pesticides, insecticides, and insect repellents, CuO NPs can be employed (Attia et al. 2021). CuO NPs plays an essential role in respiratory processes and stimulates the activity of enzymes (ascorbate oxidase, cytochrome oxidase, and polyphenol oxidase), which improves the synthesis of photosynthetic pigments, especially anthocyanins (Jampílek and Kráľová 2019). The use of CuO

NPs strongly stimulates the plant to produce a substance called phytoalexins, which are low molecular weight substances that are anti-growth of pathogenic microorganisms (Dimkpa et al. 2012). CuO NPs are created and accumulated in plant tissues and thus improve plant immunity and increase systemic resistance against pathogens (Kamel et al. 2022). Resorting to the use of CuO NPs compounds as environmentally friendly stimuli that enhance plant physiological immunity is among the most integrated control strategies currently available in protecting crops and increasing production (Agrawal et al. 2022; Vaishnavi et al. 2023). CuO NPs also function well in conjunction with specific proteins and enzymes to support photosynthesis, mitochondrial respiration, cell wall metabolism, lignification, pollen formation, and oxidative stress responses (Cakmak et al. 2023). However, several bacterial and fungal diseases, such as *Pythium ultimum*, *Fusarium oxysporum*, and *Rhizoctonia solani*, attack sugar beet plants, reducing their yields and resulting in financial losses at various plant stages (Abou-Salem et al. 2022; Kalia et al. 2020).

11.8.3 Impact on Pathogenic Microorganisms

Copper nano-fertilizers are one of the most important solutions resulting from the applications of nanotechnology (Abbasifar et al. 2020). CuO-NPs are characterized by being environmentally friendly, economically viable, and highly efficient, which makes it an effective alternative to traditional copper fertilizers (Hafeez et al. 2015). CuO-NPs achieve many advantages, allowing it to be used in smaller quantities (with high efficiency) and its high stability under different conditions (Badawy et al. 2021). Consequently, the possibility of using CuO-NPs as antibacterial agents has been brought up in a number of studies. CuO-NPs have demonstrated outstanding efficacy against plant pathogenic fungi such as *Phoma destructiva*, *Curvularia lunata*, and *Alternaria alternata* by a well diffusion assay (Kanhed et al. 2014). Additionally, Shende et al. (2015) showed strong inhibitory activity against *F. oxysporum*, *F. culmorum*, *Pseudomonas aeruginosa*, and *Klebsiella pneumonia*. The results carried out by El-Batal et al. (2020) proved that *Clavibacter michiganensis*-treated tubes treated with CuO NPs (10.0 lg/ml) showed a negative and antibacterial effect. Although a high inhibition of *Clavibacter michiganensis* bacteria was recorded after treatment with copper nanoparticles (Rivas-Cáceres et al. 2018), the mechanism of inhibition requires further studies. The antibacterial role of copper oxide nanoparticles can be explained by the presence of positive CuO NPs that can rapidly interact electrostatically with the negative cell wall. *Clavibacter michiganensis* is easily spread according to Bibi et al. (2021) and Noshad et al. (2019).

11.8.4 Potential for Disease Resistance Enhancement

The primary factor influencing systemic resistance in plants treated with CuO NPs is disease severity. As the present results show, *Fusarium* infection was the cause of 87.5% of infection incidence. These results concur with those of Beyene (2014) and

Guo et al. (2020). The most significant indication that resistance has developed in the plant is the reduction in infection severity. According to some, nano-fertilizers work like magic to boost antioxidant levels, improve plant growth, and help plants withstand biological stresses—especially fungus-related plant diseases—all of which lead to higher productivity (El-Ghamry et al. 2018; Iqbal and Umar 2019; Jiang et al. 2022; Tripathi et al. 2022).

In order to fight *Fusarium* in the lab, researchers employed copper oxide nanoparticles (CuO NPs), and their application was found to limit fungal growth (Ashraf et al. 2021; Shah et al. 2022). According to numerous scientific studies, applying zinc nanoparticles to infected plants reduced the severity of the pathogenic infection (Abdelaziz et al. 2021; Abdelaziz et al. 2022; Abdelkhalek and Al-Askar 2020; Awan et al. 2019; Cai et al. 2019; Imran et al. 2022). These findings make sense because zinc plays a crucial role in encouraging the plant to produce growth hormones and correct the hormonal imbalance brought on by infection. CuO nanoparticles were also applied to a large number of afflicted plants, and the outcomes showed a decrease in the disease's severity of infection (Ma et al. 2019). Copper micronutrient is essential for the synthesis of chlorophyll, the synthesis of carbohydrates, cell respiration, and nitrogen absorption (Tripathi et al. 2015). It promotes systemic resistance against infections and symptoms of dietary deficiencies (Djoko et al. 2015). CuO NPs supplementation may result in higher levels of chlorophyll in diseased plants. This is because CuO is essential for cell divisions in plants (Murali et al. 2021; Patel et al. 2016) and in the formation of chlorophyll (Badawy et al. 2021; Tiwari et al. 2019) and because studies have demonstrated that chlorophyll, which is primarily connected to chloroplasts, contains up to 70% of the copper contained in plants (Arnon 1949; Yruela 2009). Copper is part of the plastocyanin, which forms a link in the electron transport chain and is involved in photosynthesis (Ubbink et al. 1998). *Fusarium* creates a significant disruption in growth hormones, which results in a blatant flaw in cell biology, which explains these results (Basco et al. 2017; Rivera-Jiménez et al. 2018). Possible causes of the negative effects include disruptions in the growth hormone pathways (Attia et al. 2017; Attia et al.; Faheed et al. 2005; Orcutt 2000).

11.9 Conclusion Challenges and Future Perspectives

11.9.1 Conclusion

In conclusion, the biosynthesis of copper nanoparticles (CuNPs) from endophytic fungi is at the forefront of sustainable approaches to plant disease management. This chapter has highlighted the potential for exploiting the unique capabilities of endophytic fungi for CuNP synthesis, presenting an environmentally friendly alternative to conventional methods. The multifaceted antimicrobial properties of CuNPs, their compatibility with biological systems, and the potential for targeted applications make this method of synthesis a promising avenue for revolutionizing agriculture.

11.9.2 Challenges

Nevertheless, this emerging field encounters several challenges that demand careful consideration. One of these challenges involves the necessity for a deeper understanding of the intricate interaction between CuNPs and diverse plant pathogens. It is essential to unravel the molecular mechanisms governing antifungal and antibacterial activities to refine formulations and ensure optimal efficacy. Additionally, addressing concerns regarding the potential cytotoxicity of CuNPs to nontarget organisms and the environment is imperative for the safety and sustainability of this approach.

The scalability and reproducibility of CuNP biosynthesis also present challenges. Overcoming critical hurdles, such as establishing standardized protocols for large-scale production and maintaining CuNP stability under various environmental conditions, is essential for practical implementation in agriculture. Economic considerations, including the cost-effectiveness of the biosynthesis process and scalability to meet agricultural demands, pose additional challenges that necessitate innovative solutions.

Future Prospects

Moving forward, research should center on perfecting the biosynthesis process to enhance CuNP yield, stability, and uniformity. This entails optimizing fungal culture conditions, exploring different endophytic strains, and refining reduction and stabilization processes. Collaborative efforts among biologists, chemists, and agronomists are essential to establish standardized protocols easily adoptable by the farming community.

Thorough ecological studies of CuNPs are necessary, especially concerning soil health, microbial communities, and potential bioaccumulation in plants. Developing comprehensive mitigation strategies and guidelines for responsible application will be crucial for ensuring the long-term sustainability of CuNP-based disease management.

Advancements in nanotechnology could lead to innovative formulations and delivery systems, enabling targeted and controlled CuNP release. Encapsulation in biodegradable materials or using nanocarriers could address environmental concerns while improving the overall efficacy of CuNP-based treatments.

In conclusion, despite challenges, CuNP biosynthesis from endophytic fungi holds tremendous promise for sustainable plant disease management. Addressing current limitations and pursuing future research directions can unlock the full potential of this technology, offering farmers a powerful, environmentally friendly tool to preserve crop health globally amid evolving agricultural challenges.

References

Abbasifar A, Shahrabadi F, ValizadehKaji B (2020) Effects of green synthesized zinc and copper nano-fertilizers on the morphological and biochemical attributes of basil plant. J Plant Nutr 43:1104–1118

Abd El HAHM (2012) Lipid peroxidation end-products as a key of oxidative stress: effect of antioxidant on their production and transfer of free radicals. In: Lipid peroxidation. IntechOpen

Abd Elkodous M et al (2021) Recent advances in waste-recycled nanomaterials for biomedical applications: Waste-to-wealth. Nanotechnol Rev 10:1662–1739

Abdelaziz AM et al (2021) Protective role of zinc oxide nanoparticles based hydrogel against wilt disease of pepper plant. Biocatal Agric Biotechnol 35:102083

Abdelaziz AM, Salem SS, Khalil A, El-Wakil DA, Fouda HM, Hashem AH (2022) Potential of biosynthesized zinc oxide nanoparticles to control Fusarium wilt disease in eggplant (Solanum melongena) and promote plant growth. BioMetals:1–16

Abdelaziz AM, El-Wakil DA, Hashem AH, Al-Askar AA, AbdElgawad H, Attia MS (2023) Efficient role of endophytic Aspergillus terreus in biocontrol of Rhizoctonia solani causing damping-off. Disease of Phaseolus vulgaris and Vicia faba. Microorganisms 11:1487

Abdelkhalek A, Al-Askar AA (2020) Green synthesized ZnO nanoparticles mediated by Mentha spicata extract induce plant systemic resistance against Tobacco mosaic virus. Appl Sci 10:5054

Abou-Salem E, Ahmed AR, Elbagory M, Omara AE-D (2022) Efficacy of biological copper oxide nanoparticles on controlling damping-off disease and growth dynamics of sugar beet (Beta vulgaris L.). Plants Sustain 14:12871

Abubakar AR, Haque M (2020) Preparation of medicinal plants: Basic extraction and fractionation procedures for experimental purposes. J Pharm Bioallied Sci 12:1–10. https://doi.org/10.4103/jpbs.JPBS_175_19

Agrawal S, Kumar V, Kumar S, Shahi SK (2022) Plant development and crop protection using phytonanotechnology: a new window for sustainable agriculture. Chemosphere 299:134465

Akram S et al (2023) Uniting the Role of Endophytic Fungi against Plant Pathogens and Their Interaction. Journal of Fungi 9:72

Albalawi MA, Abdelaziz AM, Attia MS, Saied E, Elganzory HH, Hashem AH (2022) Mycosynthesis of silica nanoparticles using aspergillus niger: control of Alternaria solani causing early blight disease, induction of innate immunity and reducing of oxidative stress in eggplant. Antioxidants 11:2323

Al-Dhabaan FA, Shoala T, Ali AA, Alaa M, Abd-Elsalam K, Abd-Elsalam K (2017) Chemically-produced copper, zinc nanoparticles and chitosan-bimetallic nanocomposites and their antifungal activity against three phytopathogenic fungi. Int J Agric Technol 13:753–769

Ali K et al (2019) Comparative in situ ROS mediated killing of bacteria with bulk analogue, Eucalyptus leaf extract (ELE)-capped and bare surface copper oxide nanoparticles. Mater Sci Eng 100:747–758

Anadón A, Martínez-Larrañaga MR, Ares I, Martínez MA (2016) Interactions between nutraceuticals/nutrients and therapeutic drugs. In: Nutraceuticals. Elsevier, pp. 855–874

Ansarifard E, Zareshahrabadi Z, Sarafraz N, Zomorodian K (2021) Evaluation of antimicrobial and antibiofilm activities of copper oxide nanoparticles within soft denture liners against oral pathogens. Bioinorg Chem Appl 2021:1–7

Applerot G, Lellouche J, Lipovsky A, Nitzan Y, Lubart R, Gedanken A, Banin E (2012) Understanding the antibacterial mechanism of CuO nanoparticles: revealing the route of induced oxidative stress. Small 8:3326–3337

Arnon DI (1949) Copper enzymes in isolated chloroplasts. Polyphenoloxidase in Beta vulgaris. Plant Physiol 24:1

Ashraf H et al (2021) Inhibition mechanism of green-synthesized copper oxide nanoparticles from Cassia fistula towards Fusarium oxysporum by boosting growth and defense response in tomatoes. Environ Sci 8:1729–1748

Attia MS, Younis AM, Ahmed AF, Elaziz AMA (2016) Comprehensive management for wilt disease caused by Fusarium oxysporum in tomato plant. Int J Innov Sci Eng Technol 4:2348–7968

Attia MS, Sharaf AE-MM, Zayed AS (2017) Protective action of some bio-pesticides against early blight disease caused by Alternaria solani in tomato plant. JISET Int J Innov Sci Eng Tech 4:67–94

Attia MS et al (2021) Protective role of copper oxide-streptomycin nano-drug against potato brown rot disease caused by ralstonia solanacearum. J Clust Sci:1–14

Attia MS, Salem MS, Abdelaziz AM (2022) Endophytic fungi Aspergillus spp. reduce fusarial wilt disease severity, enhance growth, metabolism and stimulate the plant defense system in pepper plants. Biomass Convers Bioref:1–11

Awan ZA, Shoaib A, Khan KA (2019) Crosstalk of Zn in combination with other fertilizers underpins interactive effects and induces resistance in tomato plant against early blight disease. Plant Pathol J 35:330

Badawy AA, Abdelfattah NA, Salem SS, Awad MF, Fouda A (2021) Efficacy assessment of biosynthesized copper oxide nanoparticles (cuo-nps) on stored grain insects and their impacts on morphological and physiological traits of wheat (triticum aestivum l.). Plant Biol 10:233

Bakshi M, Kumar A (2022) Applications of copper nanoparticles in plant protection and pollution sensing: Toward promoting sustainable agriculture. In: Copper nanostructures: next-generation of Aagrochemicals for sustainable agroecosystems. Elsevier, pp. 393–413

Basco M, Bisen K, Keswani C, Singh H (2017) Biological management of Fusarium wilt of tomato using biofortified vermicompost. Mycosphere 8:467–483

Beyene AT (2014) Genetic analysis and characterization of faba bean (vicia faba) for resistance to chocolate spot (Botrytis fabae) disease and yield in the Ethiopian highlands

Bibi H et al (2021) Green synthesis of multifunctional carbon coated copper oxide nanosheets and their photocatalytic and antibacterial activities. Sci Rep 11:10781

Bogas AC, Henrique Rodrigues S, Gonçalves MO, De Assis M, Longo E, Paiva De Sousa C (2022) Endophytic microorganisms from the tropics as biofactories for the synthesis of metal-based nanoparticles: Healthcare applications. Front Nanotechnol 4. https://doi.org/10.3389/fnano.2022.823236

Cai L, Liu C, Fan G, Liu C, Sun X (2019) Preventing viral disease by ZnONPs through directly deactivating TMV and activating plant immunity in Nicotiana benthamiana. Environ Sci 6:3653–3669

Cakmak I et al (2023) Micronutrients. In: Marschner's mineral nutrition of plants. Elsevier, pp. 283–385

Caruso DJ, Palombo EA (2022) Exploring the promise of endophytic Fungi: a review of novel antimicrobial compounds. Microorganisms 10. https://doi.org/10.3390/microorganisms10101990

Chand K, Shah S (2020) Isolation, characterization, and plant growth-promoting activities of endophytic fungi from a wild orchid Vanda cristata. Plant Signal Behav 15:1744294. https://doi.org/10.1080/15592324.2020.1744294

Chaturvedi P, Gajbhiye S, Roy S, Dudhale R, Chowdhary A (2014) Determination of Kaempferol in extracts of Fusarium chlamydosporum, an endophytic fungi of Tylophora indica (Asclepeadaceae) and its anti-microbial activity. J Pharm Biol Sci 9:1–51

Cheng M-J, Wu M-D, Yuan G-F, Chen Y-L, Su Y-S, Hsieh M-T, Chen I-S (2012) Secondary metabolites and cytotoxic activities from the endophytic fungus Annulohypoxylon squamulosum. Phytochem Lett 5:219–223. https://doi.org/10.1016/j.phytol.2011.12.012

Christiansen JV et al (2021) Fungal quinones: diversity, producers, and applications of quinones from Aspergillus, Penicillium, Talaromyces, Fusarium, and Arthrinium. Appl Microbiol Biotechnol 105:8157–8193. https://doi.org/10.1007/s00253-021-11597-0

Dimkpa CO, McLean JE, Britt DW, Anderson AJ (2012) Bioactivity and biomodification of Ag, ZnO, and CuO nanoparticles with relevance to plant performance in agriculture. Indus Biotechnol 8:344–357

Djoko KY, Cheryl-lynn YO, Walker MJ, McEwan AG (2015) The role of copper and zinc toxicity in innate immune defense against bacterial pathogens. J Biol Chem 290(18954–18):961

Doolotkeldieva T, Bobusheva S, Zhasnakunov Z, Satybaldiev A (2022) Biological activity of Ag and Cu monometallic nanoparticles and Ag-Cu bimetallic nanocomposites against plant pathogens and seeds. J Nanomater 2022:1190280. https://doi.org/10.1155/2022/1190280

dos Reis JBA, Lorenzi AS, do Vale HMM (2022) Methods used for the study of endophytic fungi: a review on methodologies and challenges, and associated tips. Arch Microbiol 204:675 doi:https://doi.org/10.1007/s00203-022-03283-0.

Du W et al (2020) Diversity and antimicrobial activity of endophytic fungi isolated from Securinega suffruticosa in the Yellow River Delta. PloS one 15:e0229589

Du J, Fu L, Li H, Xu S, Zhou Q, Tang J (2021) The potential hazards and ecotoxicity of CuO nanoparticles: an overview. Toxin Rev 40:460–472

Duan L, Liwei G, Hong Y (2009) Isolation and identification of producing endophytic fungi of berberine from the plant Phellodendron amurense. J Anhui Agric Sci

Ebada SS, Eze P, Okoye FB, Esimone CO, Proksch P (2016) The fungal endophyte Nigrospora oryzae produces quercetin monoglycosides previously known only from plants. Chem Sel 1:2767–2771

Elbasuney S, El-Sayyad GS, Abdelaziz AM, Rizk SH, Tolba MM, Attia MS (2023) Stable colloidal iron oxide nanoparticles: a new green nanofertilizer and therapeutic nutrient for eggplant immune response against Fusarium Wilt disease. J Clust Sci:1–15

El-Batal AI, Balabel NM, Attia MS, El-Sayyad GS (2020) Antibacterial and antibiofilm potential of mono-dispersed stable copper oxide nanoparticles-streptomycin nano-drug: implications for some potato plant bacterial pathogen treatment. J Clust Sci 31:1021–1040

El-Batal AI, Al-shammari BM, El-Sayyad GS, Rizk SH, Abdelaziz AM, Nofel MM, Attia MS (2023) Gum Arabic-assisted biomass synthesis of bimetallic ZnO-CuO nanoparticles using gamma rays for controlling potato post-harvest tuber rots-causing Alternaria solani: towards improving food safety. Biomass Convers Bioref:1–18

Elfadil D, Elkhatib WF, El-Sayyad GS (2022) Promising advances in nanobiotic-based formulations for drug specific targeting against multidrug resistant microbes and biofilm-associated infections. Microbial Pathogen:105721

El-Ghamry A, Mosa AA, Alshaal T, El-Ramady H (2018) Nanofertilizers vs. biofertilizers: new insights. Environment, Biodiversity and Soil Security 2:51–72

Elkhodary BH, Attia MS, El-Sayyad GS, Salem MS (2023) Effectiveness of bimetallic ZnO-B2O3 nanoparticles produced by Streptomyces gancidicus as prospective antifungal agents and therapeutic nutrients to enhance pea plant immunity against damping off-causing Pythium irregulare: in vivo and in vitro investigations. Biomass Convers Bioref:1–24

Fadiji AE, Mortimer PE, Xu J, Ebenso EE, Babalola OO (2022) Biosynthesis of nanoparticles using endophytes: a novel approach for enhancing plant growth and sustainable agriculture. Sustainability 14. https://doi.org/10.3390/su141710839

Faheed FA, Abd-Elaah GA, Mazen A (2005) Alleviation of disease effect on tomato plants by heat shock and salicylic acid infected with Alternaria solani. Int J Agric Biol 7:783–789

Farshori NN et al (2022) Copper oxide nanoparticles exhibit cell death through oxidative stress responses in human airway epithelial cells: a mechanistic study. Biol Trace Elem Res 200:5042–5051

Fill TP, Pereira GK, Santos RMG, Rodrigues-Fo E (2007) Four additional meroterpenes produced by Penicillium sp. found in association with Melia azedarach. Possible biosynthetic intermediates to austin. Zeitschrift für Naturforschung B 62:1035–1044

Gaber SE, Hashem AH, El-Sayyad GS, Attia MS (2023) Antifungal activity of myco-synthesized bimetallic ZnO-CuO nanoparticles against fungal plant pathogen Fusarium oxysporum. Biomass Convers Bioref:1–15

Gao Y et al (2012) Antioxidant properties, superoxide dismutase and glutathione reductase activities in HepG2 cells with a fungal endophyte producing apigenin from pigeon pea [Cajanus cajan (L.) Millsp.]. Food Res Int 49:147–152. https://doi.org/10.1016/j.foodres.2012.08.001

Gerhardson B (2002) Biological substitutes for pesticides. Trends Biotechnol 20:338–343

Guo Y, Lv J, Zhao Q, Dong Y, Dong K (2020) Cinnamic acid increased the incidence of Fusarium wilt by increasing the pathogenicity of Fusarium oxysporum and reducing the physiological and biochemical resistance of faba bean, which was alleviated by intercropping with wheat. Front Plant Sci 11:608389

Hafeez A, Razzaq A, Mahmood T, Jhanzab HM (2015) Potential of copper nanoparticles to increase growth and yield of wheat. J Nanosci Adv Technol 1:6–11

Hasanin M, Al Abboud MA, Alawlaqi MM, Abdelghany TM, Hashem AH (2022) Ecofriendly synthesis of biosynthesized copper nanoparticles with starch-based nanocomposite: antimicrobial, antioxidant, and anticancer activities. Biol Trace Elem Res 200:2099–2112. https://doi.org/10.1007/s12011-021-02812-0

Hasanin M, Hashem AH, Lashin I, Hassan SA (2023a) In vitro improvement and rooting of banana plantlets using antifungal nanocomposite based on myco-synthesized copper oxide nanoparticles and starch. Biomass Convers Bioref 13:8865–8875

Hasanin MS, Hashem AH, Al-Askar AA, Haponiuk J, Saied E (2023b) A novel nanocomposite based on mycosynthesized bimetallic zinc-copperoxide nanoparticles, nanocellulose and chitosan: characterization, antimicrobial and photocatalytic activities. Electron J Biotechnol 65:45–55. https://doi.org/10.1016/j.ejbt.2023.05.001

Hashem AH, Al-Askar AA, Haponiuk J, Abd-Elsalam KA, Hasanin MS (2023a) Biosynthesis, characterization, and antifungal activity of novel trimetallic copper oxide-selenium-zinc oxide nanoparticles against some mucorales fungi. Microorganisms 11:1380

Hashem AH et al (2023b) Bioactive compounds and biomedical applications of endophytic fungi: a recent review. Microb Cell Fact 22:107. https://doi.org/10.1186/s12934-023-02118-x

Hassan SELD, Salem SS, Fouda A, Awad MA, El-Gamal MS, Abdo AM (2018) New approach for antimicrobial activity and bio-control of various pathogens by biosynthesized copper nanoparticles using endophytic actinomycetes. J Radiat Res Appl Sci 11:262–270. https://doi.org/10.1016/j.jrras.2018.05.003

Hawas UW, El-Beih AA, El-Halawany AM (2012) Bioactive anthraquinones from endophytic fungus Aspergillus versicolor isolated from red sea algae. Arch Pharma Res 35:1749–1756. https://doi.org/10.1007/s12272-012-1006-x

He X, Deng H, Hwang H-m (2019) The current application of nanotechnology in food and agriculture. J Food Drug Anal 27:1–21

Hewlings SJ, Kalman DS (2017) Curcumin: a review of its effects on human health. Foods (Basel, Switzerland) 6. https://doi.org/10.3390/foods6100092

Hoffman AM et al (2008) Purification, identification and activity of phomodione, a furandione from an endophytic Phoma species. Phytochemistry 69:1049–1056. https://doi.org/10.1016/j.phytochem.2007.10.031

Hu ZY, Li YY, Huang YJ, Su WJ, Shen YM (2008) Three new sesquiterpenoids from Xylaria sp. NCY2. Helvet Chim Acta 91:46–52

Huang Z, Yang J, Cai X, She Z, Lin Y (2012) A new furanocoumarin from the mangrove endophytic fungus Penicillium sp. (ZH16). Nat Prod Res 26:1291–1295. https://doi.org/10.1080/14786419.2011.569502

Ibrahim M, Kaushik N, Sowemimo A, Chhipa H, Koekemoer T, Van de Venter M, Odukoya OA (2017) Antifungal and antiproliferative activities of endophytic fungi isolated from the leaves of Markhamia tomentosa. Pharm Biol 55:590–595

Imran M, Abo-Elyousr KA, El-Sharnouby ME, Ali EF, Sallam N, Bagy HM, Abdel-Rahim IR (2022) Biocontrol potential of Trichoderma harzianum and zinc nanoparticles to mitigate gray mold disease of tomato. Gesunde Pflanzen:1–13.

Iqbal M, Umar S (2019) Nano-fertilization to enhance nutrient use efficiency and productivity of crop plants. In: Nanomaterials and plant potential. Springer, pp. 473–505.

Jampílek J, Kráľová K (2019) Impact of nanoparticles on photosynthesizing organisms and their use in hybrid structures with some components of photosynthetic apparatus. In: Plant Nanobionics: Volume 1, Advances in the understanding of nanomaterials research and applications, pp. 255–332.

Jayatilake PL, Munasinghe H (2020) Antimicrobial activity of cultivable endophytic and rhizosphere fungi associated with "mile-a-Minute". BioMed Res Int 2020:5292571. https://doi.org/10.1155/2020/5292571

Jiang Y et al (2022) Green synthesis of metal-based nanoparticles for sustainable agriculture. Environ Pollut:119755

Kalia A, Abd-Elsalam KA, Kuca K (2020) Zinc-based nanomaterials for diagnosis and management of plant diseases: Ecological safety and future prospects. J Fungi 6:222

Kamel SM et al (2022) Antifungal activity of copper oxide nanoparticles against root rot disease in cucumber. J Fungi 8:911

Kanhed P et al (2014) In vitro antifungal efficacy of copper nanoparticles against selected crop pathogenic fungi. Mater Lett 115:13–17

Khalil A, Abdelaziz A, Khaleil M, Hashem A (2021) Fungal endophytes from leaves of Avicennia marina growing in semi-arid environment as a promising source for bioactive compounds. Lett Appl Microbiol 72:263–274

Kusari S, Zühlke S, Spiteller M (2009) An endophytic fungus from Camptotheca acuminata that produces camptothecin and analogues. J Nat Prod 72:2–7. https://doi.org/10.1021/np800455b

Lashin I, Hasanin M, Hassan SA, Hashem AH (2023) Green biosynthesis of zinc and selenium oxide nanoparticles using callus extract of Ziziphus spina-christi: Characterization, antimicrobial, and antioxidant activity. Biomass Convers Bioref 13:10133–10146

Li E, Jiang L, Guo L, Zhang H, Che Y (2008) Pestalachlorides A-C, antifungal metabolites from the plant endophytic fungus Pestalotiopsis adusta. Bioorg Med Chem 16:7894–7899. https://doi.org/10.1016/j.bmc.2008.07.075

Li JL, Sun X, Chen L, Guo LD (2016a) Community structure of endophytic fungi of four mangrove species in Southern China. Mycology 7:180–190. https://doi.org/10.1080/21501203.2016.1258439

Li P, Wu Z, Liu T, Wang Y (2016b) Biodiversity, phylogeny, and antifungal functions of endophytic fungi associated with Zanthoxylum bungeanum. Int J Mol Sci 17:1541

Lie-Zhong C, Jian-Ming C, Xu-Song Z, Jue-Feng Z, Xiao-Ping Y (2007) Identification and antifungal activity of the metabolite of endophytic fungi isolated from Llex cornuta. 农药学学报 9:143–148

Ligang Z, Xiaodong C, Chengzong Y, Xuehong W, Liqun Z (2004) Endophytic fungi of {\sl Paris polyphylla} var.{\sl Yunnanensis} and steroid analysis in the fungi Nat Prod Res Dev 16:198–200.

Liu K, Ding X, Deng B, Chen W (2009) Isolation and characterization of endophytic taxol-producing fungi from Taxus chinensis. J Indus Microbiol Biotechnol 36:1171–1177. https://doi.org/10.1007/s10295-009-0598-8

Liu F et al (2010) The bioactive metabolites of the mangrove endophytic fungus Talaromyces sp. ZH-154 isolated from Kandelia candel (L.) Druce. Planta Med 76:185–189. https://doi.org/10.1055/s-0029-1186047

Ma C, Borgatta J, De La Torre-Roche R, Zuverza-Mena N, White JC, Hamers RJ, Elmer WH (2019) Time-dependent transcriptional response of tomato (Solanum lycopersicum L.) to Cu nanoparticle exposure upon infection with Fusarium oxysporum f. sp. lycopersici. ACS Sustain Chem Eng 7:10064–10074

Maity D, Gupta U, Saha S (2022) Biosynthesized metal oxide nanoparticles for sustainable agriculture: next-generation nanotechnology for crop production, protection and management. Nanoscale 14:13950–13989

Mani VM et al (2021) Copper oxide nanoparticles synthesized from an endophytic fungus Aspergillus terreus: Bioactivity and anti-cancer evaluations. Environ Res 201:111502. https://doi.org/10.1016/j.envres.2021.111502

Masyita A et al (2022) Terpenes and terpenoids as main bioactive compounds of essential oils, their roles in human health and potential application as natural food preservatives. Food Chem X 13:100217. https://doi.org/10.1016/j.fochx.2022.100217

Meghana S, Kabra P, Chakraborty S, Padmavathy N (2015) Understanding the pathway of antibacterial activity of copper oxide nanoparticles. RSC Adv 5:12293–12299

Mishra A et al (2022) Metal nanoparticles against multi-drug-resistance bacteria. J Inorg Biochem:111938

Mohammadi S, Poursedi S, Amini A (2016) Green synthesis of silver nanoparticles with a long lasting stability using colloidal solution of cowpea seeds (Vigna sp. L). J Environ Chem Eng 4:2023–2032

Mousa SA, El-Sayed E-SR, Mohamed SS, Abo El-Seoud MA, Elmehlawy AA, Abdou DA (2021) Novel mycosynthesis of Co 3 O 4, CuO, Fe 3 O 4, NiO, and ZnO nanoparticles by the endophytic Aspergillus terreus and evaluation of their antioxidant and antimicrobial activities. Appl Microbiol Biotechnol 105:741–753

Mu Q, Jiang G, Chen L, Zhou H, Fourches D, Tropsha A, Yan B (2014) Chemical basis of interactions between engineered nanoparticles and biological systems. Chem Rev 114:7740–7781

Mubeen B et al (2021) Nanotechnology as a novel approach in combating microbes providing an alternative to antibiotics. Antibiotics 10:1473

Murali M et al (2021) Genotoxic and cytotoxic properties of zinc oxide nanoparticles phyto-fabricated from the obscure morning glory plant Ipomoea obscura (L.) Ker Gawl. Molecules 26:891

Nadeem A, Naz S, Ali JS, Mannan A, Zia M (2019) Synthesis, characterization and biological activities of monometallic and bimetallic nanoparticles using Mirabilis jalapa leaf extract. Biotechnol Rep 22:e00338

Nassar A-RA, Atta HM, Abdel-Rahman MA, El Naghy WS, Fouda A (2023) Myco-synthesized copper oxide nanoparticles using harnessing metabolites of endophytic fungal strain Aspergillus terreus: an insight into antibacterial, anti-Candida, biocompatibility, anticancer, and antioxidant activities. BMC Complement Med Ther 23:261. https://doi.org/10.1186/s12906-023-04056-y

Naz S, Gul A, Zia M (2020) Toxicity of copper oxide nanoparticles: a review study. IET nanobiotechnol 14:1–13

Noshad A, Hetherington C, Iqbal M (2019) Impact of AgNPs on seed germination and seedling growth: A focus study on its antibacterial potential against Clavibacter michiganensis subsp. michiganensis infection in Solanum lycopersicum. J Nanomater 2019:1–12

Omran BA, Baek K-H (2022) Control of phytopathogens using sustainable biogenic nanomaterials: recent perspectives, ecological safety, and challenging gaps. J Clean Prod:133729

Orcutt DM (2000) The physiology of plants under stress: soil and biotic factors, vol 2. Wiley

Patel P, Kansara K, Senapati VA, Shanker R, Dhawan A, Kumar A (2016) Cell cycle dependent cellular uptake of zinc oxide nanoparticles in human epidermal cells. Mutagenesis 31:481–490

Pommier Y (2006) Topoisomerase I inhibitors: camptothecins and beyond Nature reviews. Cancer 6:789–802. https://doi.org/10.1038/nrc1977

Pongcharoen W, Rukachaisirikul V, Phongpaichit S, Sakayaroj J (2007) A new dihydrobenzofuran derivative from the endophytic fungus Botryosphaeria mamane PSU-M76. Chem Pharma Bull 55:1404–1405

Puri SC, Verma V, Amna T, Qazi GN, Spiteller M (2005) An endophytic fungus from Nothapodytes foetida that produces camptothecin. J Nat Prod 68:1717–1719. https://doi.org/10.1021/np0502802

Qi W, Qi W, Xiong D, Long M (2022) Quercetin: its antioxidant mechanism, antibacterial properties and potential application in prevention and control of toxipathy. Molecules 27. https://doi.org/10.3390/molecules27196545

Rabea A, Naeem E, Balabel NM, Daigham GE (2023) Management of potato brown rot disease using chemically synthesized CuO-NPs and MgO-NPs. Bot Stud 64:20

Rajput VD et al (2021) Nano-enabled products: challenges and opportunities for sustainable agriculture. Plants 10:2727

Rajput A, Sharma R, Bharti R (2022) Pharmacological activities and toxicities of alkaloids on human health. Mater Today 48:1407–1415. https://doi.org/10.1016/j.matpr.2021.09.189

Rasool U, Hemalatha S (2017) Marine endophytic actinomycetes assisted synthesis of copper nanoparticles (CuNPs): Characterization and antibacterial efficacy against human pathogens. Mater Lett 194:176–180. https://doi.org/10.1016/j.matlet.2017.02.055

Rehana D, Mahendiran D, Kumar RS, Rahiman AK (2017) Evaluation of antioxidant and anticancer activity of copper oxide nanoparticles synthesized using medicinally important plant extracts. Biomed Pharmacother 89:1067–1077

Rehman S et al (2008) An endophytic Neurospora sp. from Nothapodytes foetida producing camptothecin. Prikladnaia biokhimiia i mikrobiologiia 44:225–231

Rivas-Cáceres RR et al (2018) Bactericidal effect of silver nanoparticles against propagation of Clavibacter michiganensis infection in Lycopersicon esculentum. Mill Microbial Pathogenesis 115:358–362

Rivera-Jiménez MN, Zavaleta-Mancera HA, Rebollar-Alviter A, Aguilar-Rincón VH, García-de-los-Santos G, Vaquera-Huerta H, Silva-Rojas HV (2018) Phylogenetics and histology provide insight into damping-off infections of 'Poblano'pepper seedlings caused by Fusarium wilt in greenhouses. Mycol Progr 17:1237–1249

Rodriguez RJ, White JF Jr, Arnold AE, Redman RS (2009) Fungal endophytes: diversity and functional roles. New Phytol 182:314–330. https://doi.org/10.1111/j.1469-8137.2009.02773.x

Saied M, Hasanin M, Abdelghany TM, Amin BH, Hashem AH (2023) Anticandidal activity of nanocomposite based on nanochitosan, nanostarch and mycosynthesized copper oxide nanoparticles against multidrug-resistant. Candida Int J Biol Macromol 242:124709. https://doi.org/10.1016/j.ijbiomac.2023.124709

Sajjad H, Sajjad A, Haya RT, Khan MM, Zia M (2023) Copper oxide nanoparticles: In vitro and in vivo toxicity, mechanisms of action and factors influencing their toxicology. Comp Biochem Physiol C Toxicol Pharmacol:109682

Saldierna Guzmán JP, Nguyen K, Hart SC (2020) Simple methods to remove microbes from leaf surfaces. J Basic Microbiol 60:730–734

Salehi B et al (2019) Berberis plants—Drifting from farm to food applications, phytotherapy, and phytopharmacology. Foods (Basel, Switzerland) 8:522

Sarang H, Rajani P, Vasanthakumari MM, Kumara PM, Siva R, Ravikanth G, Uma Shaanker R (2017) An endophytic fungus, Gibberella moniliformis from Lawsonia inermis L. produces lawsone, an orange-red pigment. Antonie Van Leeuwenhoek 110:853–862. https://doi.org/10.1007/s10482-017-0858-y

Sarsaiya S et al (2020) Molecular identification of endophytic fungi and their pathogenicity evaluation against Dendrobium nobile and Dendrobium officinale. Int J Mol Sci 21:316

Seetharaman P, Gnanasekar S, Chandrasekaran R, Chandrakasan G, Kadarkarai M, Sivaperumal S (2017) Isolation and characterization of anticancer flavone chrysin (5,7-dihydroxy flavone)-producing endophytic fungi from Passiflora incarnata L. leaves. Ann Microbiol 67:321–331. https://doi.org/10.1007/s13213-017-1263-5

Shah IH et al (2022) Green synthesis and characterization of Copper oxide nanoparticles using Calotropis procera leaf extract and their different biological potentials. J Mol Struct 1259:132696

Shankar Naik B (2020) Nanoparticles from endophytic fungi and their efficacy in biological control. In: Thangadurai D, Sangeetha J, Prasad R (eds) Nanotechnology for food, agriculture, and environment. Springer International Publishing, Cham, pp 161–179. https://doi.org/10.1007/978-3-030-31938-0_8

Shen X, Wang X, Huang T, Deng Z, Lin S (2020) Naphthoquinone-based meroterpenoids from marine-derived streptomyces sp. B9173. Biomolecules 10 https://doi.org/10.3390/biom10081187

Shende S, Ingle AP, Gade A, Rai M (2015) Green synthesis of copper nanoparticles by Citrus medica Linn.(Idilimbu) juice and its antimicrobial activity. World J Microbiol Biotechnol 31:865–873

Silva GH et al (2006) Cadinane sesquiterpenoids of Phomopsis cassiae, an endophytic fungus associated with Cassia spectabilis (Leguminosae). Phytochemistry 67:1964–1969

Singh N, Sharma B (2018) Toxicological effects of Berberine and Sanguinarine. Front Mol Biosci 5. https://doi.org/10.3389/fmolb.2018.00021

Tiwari PK et al (2019) Liquid assisted pulsed laser ablation synthesized copper oxide nanoparticles (CuO-NPs) and their differential impact on rice seedlings. Ecotoxicol Environ Saf 176:321–329

Tong W (2013) Biotransformation of terpenoids and steroids. In: Natural Products. Springer, p. 2733–2759.

Tripathi DK, Singh S, Singh S, Mishra S, Chauhan D, Dubey N (2015) Micronutrients and their diverse role in agricultural crops: advances and future prospective. Acta Physiol Plant 37:1–14

Tripathi D, Singh M, Pandey-Rai S (2022) Crosstalk of nanoparticles and phytohormones regulate plant growth and metabolism under abiotic and biotic stress. Plant Stress:100107

Ubbink M, Ejdebäck M, Karlsson BG, Bendall DS (1998) The structure of the complex of plastocyanin and cytochrome f, determined by paramagnetic NMR and restrained rigid-body molecular dynamics. Structure 6:323–335

Vaishnavi PA, Prakash NU, Jayanthi M, Suganthi M, Thenmozhi M (2023) Impact of metal oxide nanoparticles against biotic stress in plants. In: Nanometal oxides in horticulture and agronomy. Elsevier, pp 101–140

Venditto VJ, Simanek EE (2010) Cancer therapies utilizing the camptothecins: a review of the in vivo literature. Mol Pharma 7:307–349. https://doi.org/10.1021/mp900243b

Walczak-Nowicka ŁJ, Herbet M (2021) Acetylcholinesterase inhibitors in the treatment of neurodegenerative diseases and the role of acetylcholinesterase in their pathogenesis, Int J Mol Sci:22. https://doi.org/10.3390/ijms22179290

Wang Y, Xu L, Ren W, Zhao D, Zhu Y, Wu X (2012) Bioactive metabolites from Chaetomium globosum L18, an endophytic fungus in the medicinal plant Curcuma wenyujin. Phytomedicine 19:364–368. https://doi.org/10.1016/j.phymed.2011.10.011

Wang Y, Lai Z, Li XX, Yan RM, Zhang ZB, Yang HL, Zhu D (2016) Isolation, diversity and acetylcholinesterase inhibitory activity of the culturable endophytic fungi harboured in Huperzia serrata from Jinggang Mountain. China World J Microbiol Biotechnol 32:20. https://doi.org/10.1007/s11274-015-1966-3

Wang L, Ren L, Li C, Gao C, Liu X, Wang M, Luo Y (2019) Effects of endophytic fungi diversity in different coniferous species on the colonization of Sirex noctilio (Hymenoptera: Siricidae). Sci Rep 9:5077. https://doi.org/10.1038/s41598-019-41419-3

Wu SH, Huang R, Miao CP, Chen YW (2013) Two new steroids from an endophytic fungus Phomopsis sp. Chem Biodiv 10:1276–1283

Yan L, Zhu J, Zhao X, Shi J, Jiang C, Shao D (2019) Beneficial effects of endophytic fungi colonization on plants. Appl Microbiol Biotechnol 103:3327–3340

Yang K, Liang J, Li Q, Kong X, Chen R, Jin Y (2013a) Cladosporium cladosporioides XJ-AC03, an aconitine-producing endophytic fungus isolated from Aconitum leucostomum. World J Microbiol Biotechnol 29:933–938. https://doi.org/10.1007/s11274-012-1246-4

Yang Y et al (2013b) Anthraquinone derivatives from Rumex plants and endophytic Aspergillus fumigatus and their effects on diabetic nephropathy. Bioorg Med Chem Lett 23:3905–3909. https://doi.org/10.1016/j.bmcl.2013.04.059

Yang T et al (2022) Acrocalysterols A and B, two new steroids from endophytic fungus Acrocalymma sp. Phytochem Lett 48:77–80. https://doi.org/10.1016/j.phytol.2022.02.003

Yao YQ, Lan F, Qiao YM, Wei JG, Huang RS, Li LB (2017) Endophytic fungi harbored in the root of Sophora tonkinensis Gapnep: Diversity and biocontrol potential against phytopathogens. Microbiol Open 6:e00437

Yruela I (2009) Copper in plants: acquisition, transport and interactions. Funct Plant Biol 36:409–430

Zhang W, Draeger S, Schulz B, Krohn K (2009) Ring B aromatic steroids from an endophytic fungus, Colletotrichum sp. Nat Prod Commun 4:1449–1454

Zhang P-L et al (2019) Aspergilolide, a steroid lactone produced by an endophytic fungus Aspergillus sp. MBL1612 isolated from Paeonia ostii. Nat Prod Res 33:2133–2138. https://doi.org/10.1080/14786419.2018.1488706

Zhao J, Zhou L, Wang J, Shan T, Zhong L, Liu X, Gao X (2010) Endophytic fungi for producing bioactive compounds originally from their host plants. Curr Res Technol Educ Trop Appl Microbiol Microbial Biotechnol 1:567–576

Zhao J et al (2013) Hypocrea lixii, novel endophytic fungi producing anticancer agent cajanol, isolated from pigeon pea (Cajanus cajan [L.] Millsp.). J Appl Microbiol 115:102–113. https://doi.org/10.1111/jam.12195

Biosynthesis of Silver Nanoparticles from Endophytic Fungi and Their Role in Plant Disease Management

12

Mohamed T. Selim, Mohamed K. Y. Soliman, Amr H. Hashem, and Salem S. Salem

Abstract

In recent years, authors from all around the world have become interested in the connections between nanomaterials and plant growth. Consequently, a novel area of nanotechnology known as "green nanotechnology" emerged, fusing biological concepts with physical and chemical processes to create nanoparticles with targeted applications that are environmentally benign. A renewable and environmentally acceptable way to produce nanoparticles (NPs) is currently being investigated and improved through the use of microbes and plants through biochemical synthesis techniques. The popularity of biological synthesis has increased as a potential remedy for the shortcomings associated with both chemical and physical synthesis techniques. Because they produce huge quantities of proteins, have high productivity, are easy to handle, and produce residues with minimal toxic effects, fungi are interesting as reduction and stabilization agents in the naturally occurring production of silver nanoparticles (AgNPs). The usage of AgNPs as bactericides and fungicides as well as nanofertilizers to improve plant health has been incorporated into disease management techniques. Only Ag has received significant attention thus far, despite reports of over 18 different single-element and carbon nanomaterials impacting disease and/or plant infections. Certain AgNPs work primarily as nutritional modifiers that change the host's nutritional state and trigger defensive mechanisms, while others act as direct antibacterial agents. Therefore, the methods by which diseases are controlled by AgNPs are intricate and involve a range of direct and indirect effects on the pathogen, the plant, and their relationship to one another.

M. T. Selim · M. K. Y. Soliman · A. H. Hashem · S. S. Salem (✉)
Botany and Microbiology Department, Faculty of Science, Al-Azhar University, Cairo, Egypt
e-mail: salemsalahsalem@azhar.edu.eg

© The Author(s), under exclusive license to Springer Nature Singapore Pte Ltd. 2025

K. A. Abd-Elsalam, A. H. Hashem (eds.), *Fungal Endophytes Volume II*,
https://doi.org/10.1007/978-981-97-8804-0_12

Keywords

Silver nanoparticles · Endophytic fungi · Plants · Agriculture

12.1 Introduction

Nanotechnology is the production, synthesis, and manipulation of particles, mass matter, and molecules in the field of material science. Since nanoscale materials enhanced science and mankind, the uses of nanotechnology have become more and more significant (Salem and Fouda 2021; Salem et al. 2023). Due to their massive surface space-to-volume ratio as well as quantum impact, nanomaterials exhibit special mechanical, thermal, physiochemical, and biological properties (Salem 2023b; Dezfuli et al. 2023). Science has been very interested in metallic nanoparticles because of their diversity in regard to assembly, dimensions, form, structure, and even optical properties (Thakur et al. 2022). Scientists are interested in AgNPs because of their remarkable properties, as their conductivity, chemical stability, and catalytic potential (Abu-Elghait et al. 2023; Said et al. 2023; Soliman et al. 2023; Salem 2022a). The numerous distinct types of nanoparticles that may be produced using different metals are selenium, platinum, copper, gold, iron, titanium, zinc, and magnesium (Salem 2022b; Nilesh Rai et al. 2023; Yaqoob et al. 2020; Soliman et al. 2022; El-Khawaga et al. 2023; Hammad et al. 2022; Shehabeldine et al. 2023; Abdelfattah et al. 2023). Although there are many ways to make nanoparticles from several materials, they may be broadly considered into different ways: biological, chemical, as well as physical approach (Alabssawy et al. 2024; Soliman et al. 2024a, b; Attia et al. 2024; Almuhayawi et al. 2024; Salem et al. 2024a). The ones that show the most potential for creating nanoparticles are biological ones (Sandhu et al. 2017; Al-Rajhi et al. 2022; Salem et al. 2020; Salem et al. 2022). High pressure, radiation, and temperature can be used in physical processes and have the ability to seriously injure the environment and living organisms. Because chemical techniques can quickly produce a large number of homogenous nanoparticles with control over their size and distribution, they are commonly used in several fields (Sandhu et al. 2017) (He et al. 2015). However, chemical processes use hazardous ingredients, use a lot of energy, and result in harmful pollutants that pose a serious danger to the environment. Furthermore, physical-chemical approaches for creating nanoparticles have low yields because of the frequent employment of potentially hazardous substances and radioactivity-damaging radiation in physical and chemical processes. The processes of manufacturing were slow, costly, and resulted in the discharge of hazardous chemicals into the environment (Naik 2020). Because of this, a completely new field known as "nanobiotechnology" has developed, fusing biotechnology with nanotechnology to offer techniques for creating biosynthetic NPs that are inexpensive, simple to use, and environmentally friendly (Salem 2023a; Salem and Mekky 2024). Additionally, because they don't generate any harmful waste, they don't need a lot of energy (Ovais et al. 2018b; Shah et al. 2018). NPs are produced through numerous biological agents, as plants, algae, bacteria, and fungus

(Aref and Salem 2020; Al-Zahrani et al. 2022; El-Refaey and Salem 2023; Alsharif et al. 2020; Fadiji et al. 2022). The study of nanoparticles made by fungi is referred to as myconanotechnology, a topic that combines nanotechnology with mycology. Because of their ability to tolerate metals and to bioaccumulate them, fungi is thought to be so perfect way in the biosynthesis of NPs (Salem et al. 2024b; Hashem et al. 2021). This is the basis for the notion of employing fungi. In addition to creating biogenic AgNPs, fungi are chosen over bacteria and other microorganisms due to their easier-to-handle biomass (Ranjani et al. 2020a). It has been demonstrated that nitrate reductase-mediated synthesis is the fundamental mechanism by which fungus bioreduces silver ions. The extracellular production of nanoparticles via NADH dependent reductase was supported by several investigations (Anil Kumar et al. 2007; Ingle et al. 2008). AgNPs mediated by extracellular fungal extract were created among *Lasiodiplodia pseudotheobromae* that is endophytic fungus from *Eupatorium triplinerve*. These AgNPs show strong antibacterial and toxicological properties (Ranjani et al. 2020b). In an antimicrobial susceptibility test, the following fungi *Penicillium funiculosum* GS2 and *Alternaria solani* GS1 from *Gloriosa superba* L produced AgNPs with average sizes of 5–10 nm and 5–20 nm. These AgNPs demonstrated substantial antibacterial sensitivity (Nameirakpam Nirjanta Devi and Singh 2013). Numerous endophytic fungal species have been found to be disclosed in scientific literature that exhibit a benefit biologically, including antimicrobial, anti-candida, antioxidant, anticancer, mosquitocidal, carcinogenic, and photocatalytic properties. These fungi include *Botryodiplodia theobromae*, *Penicillium crustosum*, *Alternaria alternata*, *Aspergillus flavus*, *Phoma* sp., *T. atroviride*, *A. niger*, *T. viride*, and *T. reesei* (Melappa et al. 2022; Sonbol et al. 2022; Abd-Elkhalek et al. 2024; Janakiraman et al. 2019; Abdel-Azeem et al. 2020; Gemishev et al. 2022). AgNPs' antimicrobial effects against plant pathogens are characterized by their ability to prevent or stop the development of plant infections without causing harm to nearby plant tissues (Panja et al. 2021). The most interesting metal nanoparticles with sufficient biological activity seem to be AgNPs (Xi-Feng Zhang et al. 2016). When present in particular amounts, they offer a variety of advantageous benefits for plants (Haroon et al. 2019), as well as stop the harmful fungi that are connected from growing (Parveen and Rao 2015). Vegetative growth (Saha and Dutta Gupta 2018), AgNPs have demonstrated increased germination (Latif et al. 2017).

12.2 Endophytic Fungi

Endophytic fungal organisms are tiny fungi that inhabit plants without harming or infecting their hosts (Azevedo et al. 2000). They are present in practically every kind of vascular plant and grass that has been investigated too far (Hua Wei Zhang et al. 2006). Apart from fungus, numerous other classes of organisms, including bacteria, actinomycetes, and mycoplasma, have also been designated as plant endophytes (Bandara et al. 2006). It has been more than a century since their presence was discovered. The Greek terms "endon," which means "within," as well as "phyton," which means "plant," are the sources of the term "endophyte" (Sandhu et al. 2017). Many investigations are carried out to determine if endophytic fungi may have an impact on the morphological and physiological of plants since they are of significance to applied microbiology (Toju et al. 2016; Hiruma et al. 2018). These studies demonstrate that some root fungi stimulate plant development by strengthening the rhizosphere's nutritional quality or lowering infections (Rodriguez et al. 2009; Newsham 2011). Many elements are important, such as the kind of soil, fungal strains, and compatibility, along with the host (plant) (Hiruma et al. 2018) (Svenningsen et al. 2018). Plant-produced bioactive chemicals may be replaced by those produced by endophytic fungi. These chemicals exhibit a multitude of antibacterial, anticancer, antioxidant, and immunomodulatory activities (Ding et al. 2009; Netala et al. 2016). Although it was also substantially documented that endophytes can produce nanoparticles, they have attracted the most attention (Musarrat et al. 2010; Fathima and Balakrishnan 2014). The creation of a trustworthy, eco-friendly method for biosynthesis of novel nanoparticles is a crucial step for using nanotechnology. A promising technique is applied on endophytes, which are microscopic fungus that live in plant tissues (Abdel-Azeem et al. 2020).

12.3 Biologically Active Substances Derived from Endophytic Fungi

Numerous significant chemicals with possible bioactivities have been derived from endophytic fungus. The yew *Taxus brevifolia* was the source for isolation of *Taxomyces andreanae* and had been used a new anticancer chemical taxol in 1993 (Huang et al. 2008). *Pestalotiopsis microspora* of *Terminalia morobensis* was used to separate the inflammatory-fighting chemicals pestacin and isopestacin (Strobel et al. 2002). From the isolation of *A. clavatonanicus* that was fungus from *Torreya mairei*, an antibacterial chemical called clavatol was produced (Verma et al. 2009). It has been possible to distinguish two new chemicals, cytonic acids A and B, within *Cytonaema* sp. The substances in question suppress the human cytomegalovirus (hCMV) and protease, demonstrating their antiviral properties (Guo et al. 2000). Multiple investigators subsequently isolated, refined, and described a great number of these chemicals through endophytic fungus.

12.4 Biogenic Synthesis of Metallic Silver Nanoparticles by Endophytic Fungi

The construction of a dependable, eco-friendly method for biosynthesis of NPs is novel way for nanotechnology. A promising approach was depended on endophytes, which are microscopic fungus that live in plant tissues (Manoharachary and Nagaraju 2016). Nonetheless, silver-based nanofabrication has been enhanced and improves the stability and effectiveness of the bioactive substances of fungal endophytes. Because they are stores of extracellular enzymes, fungi endophytes operate as biofactories in the production of green nanoparticles. Fungus secretes a protein that functions as a stabilizing as well as reducing agent. It interacts with metal ions to promote reduction, which forms nanoparticles. This helps to synthesize AgNPs quickly and sustainably (Alghuthaymi et al. 2015). Mycelial mass along with fungal aqueous filtrate also contains the bioactive chemicals. Metal reductions and the process of nanomanufacturing are mediated by both extracellularly and intracellularly generated enzymes. This makes it possible for fungus to synthesize AgNPs through two different mechanisms: extracellular and intracellular; the precursors internalize into the fungal biomass and induce the reduction of (Ag^+) to (Ag^0) elemental silver, making NPs. This is one aspect of the intracellular process for producing silver nanoparticles. Fungal endophyte cell-free filtrate is a way to the extracellular mechanism to synthesize AgNPs (Guilger-Casagrande and Lima 2019; Awual et al. 2019). The divisions of Ascomycota and Basidiomycota include the majority of the fungal endophytes that can be implemented in the environmentally friendly manufacture of AgNPs (Khan et al. 2018). Fungal synthesis of AgNPs with possible biological activity has been reported in several publications. Research describes the production of AgNPs by using *Chiliadenus montanus* as source of isolation of *Trichoderma atroviride*. The produced AgNPs show antibacterial feature against a numerous pathogenic microorganisms as bacteria and fungi (Abdel-Azeem et al. 2020). A variety of kinds of bacteria had been susceptible against antibacterial impacts of mycogenic AgNPs made from *Aspergillus aureoles* found in *Curculigo orchioides* (Bhalla et al. 2021).

12.5 The Technique that Uses Endophytic Fungi to Produce AgNPs

Despite the fact that the precise process underlying the synthesis of NPs remains unclear, a numerous of recommendations had been put up by study specialists. Microorganisms primarily use both external and internal enzymes to produce nanoparticles (Ovais et al. 2018a; Rahman et al. 2019). Exogenous stimuli, such as exposure to metal-containing ions that are poisonous, temperature changes, predators, etc., can cause certain fungus to create metabolites extracellularly as enzymes (Mehra and Winge 1991). The fungal organism has a tendency to transform silver ions into nanoparticles despite the fact that it wouldn't exist absent these metabolites plus enzymes. Enzymes produced during the above procedure can reduce Ag^+ to Ag^0 during synthesis intracellularly (Zhao et al. 2018). Different enzymes along with metabolites, besides compounds engaged in NP production (Barabadi et al. 2017). It has also been reported that fungi exhibit remarkable redox characteristics and function as an electron shuttle during the decomposition of metal. As stated by Durán and Seabra (2012) and Zhao et al. (2018), electrostatic interaction among residual cysteine or unbound amine groups and carboxylate groups is how the stabilization is carried out. Through the production of AgNPs by cell wall-based enzymes, this interaction additionally reduces Ag^+ to silver nuclei on mycelia

(Salunke et al. 2016). PH, concentration, incubation period with substrate, and temperature are some of the variables that can affect how quickly and in what size nanoparticles are synthesized (Singh et al. 2014; Alghuthaymi et al. 2015; Sandhu et al. 2017).

12.6 Morphological Characterizations of Nanomaterials

12.6.1 UV-Visible Spectrometry

The primary factors influencing the activity, biodistribution, safety, and efficacy of nanoparticles are their physicochemical properties. Different methods were used to manufacture different nanoparticles, and UV-visible spectrometry had been used to investigate the produced NPs. A consistent peak climb observed along elevation of salt ion concentration and reaction time is a clear indicator that nanoparticles are being created. The absorption spectra of nanoscale particles in the UV-visible range exhibit peaks that align with the surface plasmon resonance (Shaheen et al. 2021). Using a method that measures the absorption at wavelengths not less 200–800 nm for the solution that was reduced, the existence of AgNPs is confirmed. Silver nanoparticles, which are brown solutions with reduced $AgNO_3$ in aqueous extracts of fungus, have been demonstrated to exhibit magnetic dipole oscillation as well as to create a high peak in absorption between 390 and 420 nm when given ultraviolet light. The group used as a control, on the other hand, shows an absence of absorption peak throughout this frequency range; it is a pale white solution that lacks $AgNO_3$ in deionized Milli-Q water. This indicates that the amount of $AgNO_3$ in endophytic fungal water extracts is lower (Kleemann 1993; Abdullah et al. 2020). Using *phomosis*, *A. conicus*, and *P. janthinellum*, the biosynthesis of AgNPs was seen in the UV-vis spectrophotometer. Following 24, 48, 72, and 96 h of incubation, the UV-vis showed a prominent peak at 420 nm, suggesting AgNPs must exist (Bharathidasan and Panneerselvam 2012). A UV-vis spectrophotometer was used to evaluate the resulting combinations, which included the cell-free filtrate of *Pestalotia* sp., after treatment with aqueous silver ions. The results showed a sharp peak at 415 nm, which is within the specific range for AgNPs (390–420 nm) (Raza et al. 2015). The synthesis of AgNPs through *A. clavatus* came about *Azaidrecta indica* stem. The following was verified by comparing the mixture of colloidal AgNPs solution's prominent peak at 415 nm to that of the $AgNO_3$ solution using UV spectroscopy (Verma et al. 2007). *Aspergillus tamarii* PFL2, *Penicillium ochrochloron* PFR8, and *Aspergillus niger* PFR6 mycelia free filtrates were also subjected to UV-visible spectroscopy analysis. Reduced AgNPs had been present in these samples because they produced sharp absorption peaks at 419 and 430 nm for the last two, respectively (Lamabam Sophiya Devi and Joshi 2015).

12.6.2 Scanning Electron Microscopy (SEM)

There are multiple methods that scanning electron microscopy provides a clear picture of size measurement and morphology. They do, however, only offer a scant amount of information on the size distribution and population mean. To prepare the nanoparticle solution for SEM examination, it must first be ground into a powder. This is done by depositing the solution onto a sample holder and then using a sputter coater to coat it with gold used as a conductive metal. Subsequently, a concentrated, precise particle electron scanned resonance is applied to the material (Jores et al. 2004). The surface qualities of the object being studied are ascertained by analyzing the secondary electrons emitted by its surface. A top-mounted electron cannon, an electron column with coils and lenses that are electromagnetically linked, a chamber for the specimen, detectors, and vacuum pumps are all necessary parts of a SEM apparatus. The size determined through SEM is equal to the dynamic light scattering data. Moreover, this approach is costly and time-consuming and sometimes requires further data on the size distribution (Molpeceres et al. 2000). The following organisms *A. cinerus*, *P. janthinellum*, and *Phomosis* produced silver nanoparticles, which were measured for size using a SEM. The investigation revealed that the NPs' sizes were 80 and 120 μm (Bharathidasan and Panneerselvam 2012). Similarly, the use of SEM verified the oval form of the silver nanoparticles that are produced through the endophytic fungus *Penicillium* sp. (Nameirakpam Nirjanta Devi et al. 2012).

12.6.3 Transmission Electron Microscopy (TEM)

TEM and SEM are similar in most aspects, as suggested by the titles they bear. The main difference is that in TEM, the material is passed through by the electron beam rather than SEM. This method is frequently used to determine the sample's internal composition; as opposed to the SEM, it only obtains a picture of the material's uppermost layer for analysis (Ebnesajjad and Landrock 2014). Electron diffraction or dispersion along the planes of the lattice produces phase recognition when an intensively concentrated electron beam having an electric field as high as 80–200 KeV goes through an incredibly thin solid sample with an average thickness of 100–200 nm in a TEM (Akhtar 2012). Based on the concept of electron transmission, TEM may offer information about the majority of substance at magnification levels that vary from very low to high. Moreover, TEM analysis is available for the quadrupolar hollow-shell arrangement of NPs as well as additional substances that include more than two sheets (Mohamed et al. 2019). *Amylomyces rouxii* strain KSU-09 is endophytic fungus that produced the silver nanoparticles, that appeared primarily spherical but varied in appearance according to the transmission electron microscopy picture (Musarrat et al. 2010). Comparable to this, *Pestalotia* sp. was found to be present in the reaction mixture where sphere and polydispersive AgNPs were eventually validated by TEM investigation. The tiny particles varied in size from 10 to 40 nm, with a median dimension of 12.40 nm (Raheman et al. 2011).

A transmission electron microscope was used to examine silver nanoparticles that the *A. tenuissima* PGL#71 had produced. The resulting particles have been shown to have the characteristics of polydispersed, spherical, as sized between 2 and 20 nm (Shukla and Sandhu 2017).

12.6.4 FTIR Spectroscopy (Fourier Transform Infrared)

FTIR may be used to get accuracy, consistency, and high signal-to-noise ratio performance. FTIR spectroscopy has ability to detect a scale of 10–3 and measure minute variations in absorbance, which facilitates the use of difference spectroscopy to discriminate between the large ambient reflections reflecting all of the proteins and the minute of absorption Al bands of involved residues The efficiency of attaching functional compounds to substances such as graphene, silver, gold, and carbon nanotube nanoparticles, as well as chemical reactions that take place during the process of catalysis involving an enzymatic agent and the substrate, have all been verified by FTIR analysis of nanoscaled materials (Baudot et al. 2010; Barth and Zscherp 2002). Ultimately, dispersive spectrometers are inferior than FTIR spectrometers in terms of rapid gathering of information, strong signal, high signal-to-noise ratio, and reduced specimen warming (Kumar and Barth 2010). In comparison to conventional FTIR, ATR-FTIR makes sample preparation simpler and enables us to analyze the chemical characteristics of the outermost layer of polymers (Johal and Johnson 2018; Kazarian and Chan 2006). According to FTIR spectroscopy, proteins' peptides and amino acid residues have a greater repulsion for metal, which means that proteins may form a protective film around the metal nanoparticles, regulating the silver nanoparticles to keep them from clumping together and restoring stability in the medium in which they exist (Basavaraja et al. 2008). FTIR offers an easy, inexpensive, noninvasive, practical, and helpful technique for determining the role that biological substances play in the conversion of $AgNO_3$ into AgNPs.

12.6.5 X-Ray Diffraction

A method for examining a sample's crystalline and molecular arrangement without damaging it is called X-ray diffraction, or XRD. XRD is the most effective technique for analyzing materials that are amorphous (non-crystalline), crystalline, and polycrystalline. The light wavelengths of the X-rays used in XRD are on the level of the nanometer. XRD is the term for the elastic scattering of X-rays by material's particles (no energy is wasted during a collision). The basis for the operation of XRD depends on the interfering of dispersed waves. The trajectory difference, also known as a change in the distance of the waves' pathways, or a path difference (phase or angle), determines the final amplitude of a scattered wave (J Aravind Kumar et al. 2021). After a protracted incubation period (more than 70 h), AgNPs combine to form sizable aggregates. The crystalline character of AgNPs is attributed to these aggregates, which have distinct morphologies and are further supported by XRD. By using XRD, it was possible to see that *A. clavatus* generated silver nanoparticles that were crystalline. Peaks at 38.06° (111) and 44.1392° (200) were found, which correlate to the silver diffraction facets and show that the precipitate is made entirely of pure crystallized silver (Verma et al. 2010). The *Penicillium* sp. generated AgNPs that had not been well-crystallized, according to XRD examination. The crystalline maximum is connected with the cubic morphologies of the particles, which had an average dimension of grain approximately 32 nm (Nameirakpam Nirjanta Devi et al. 2012).

12.6.6 Dark-Field Microscopy (DFM)

A kind of microscopy called DFM is used in both light and electron microscopy, in which the specimen being examined remains the only thing illuminated by the UV ray or electron beam, and surrounding specimen field is kept completely dark. Compared to bright-field microscopy, DFM imaging has several advantages such as reduced background, accurate single-particle assessment, great spatial and temporal resolution, as well as excellent sensitivity. Within the year 2000, when single-particle dispersion spectra as well as high-quality color photographs were widely available, the most notable developments and applications of modern DFM took place (Schultz et al. 2000). The use of dark-field spectroscopy in conjunction with a spectrometer to identify the scattering spectra of individual nanoparticles has become an essential technique in plasmon spectroscopy. Conventional DFM may be used in regular studies to image PNPs larger than 30 nm; however, direct investigation of gold nanoparticles bound to functioning surfaces as tiny as 10 nm is done using innovative optical structures (Weigel et al. 2014).

12.7 Silver Nanoparticles in Plant Disease Management

12.7.1 AgNPs in Bacterial Disease Management

AgNPs have important antibacterial capabilities against a variety of microorganisms, such as fungi, viruses, and bacteria (Sre et al. 2015). AgNPs with a size range of 10–100 nm exhibit potent antibacterial activity against a variety of Gram-positive and Gram-negative bacteria (Braithwaite et al. 2014). AgNPs are well recognized for their effectiveness as antimicrobials against *Streptococcus pyogenes*, *S. aureus*, and *E. coli* bacteria (Bose and Chatterjee 2016).

12.7.1.1 Mechanism of Action of AgNPs Against Plant Pathogenic Bacteria

By combining and denaturing the DNA, silver ions stop the replication of bacterial DNA. This process involves the interaction of Ag ions with thiol proteins and results in DNA condensation, which kills the cell (Maharashtra K Rai et al. 2012). In

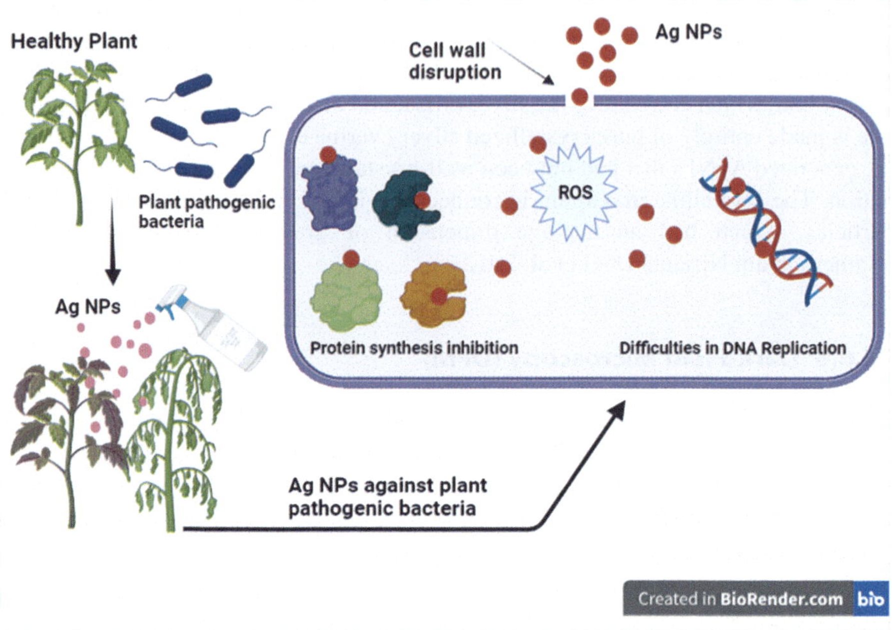

Fig. 12.1 AgNPs' mode of action against plant pathogenic bacteria

Fig. 12.1 we explain the comparative mode of action of AgNPs on plant pathogen bacteria. The primary mechanism underlying AgNPs' antibacterial action is the dissolution of silver cations, which may attract proteins of bacteria specifically to thiol (– SH) and impair their physiological cell functions, culminating in necrobiosis (Bilal Ahmed et al. 2020). AgNPs use a Trojan horse technique to kill bacteria: They first bind to the cell's surface, which changes and breaks down the membrane, and then they release metal silver ions into the cell, which are toxic (Burduşel et al. 2018).

AgNPs' wide surface layers, which promote increased interaction between NPs and bacteria cells and inhibit cell growth at extremely low concentrations in the environment, are primarily responsible for their antibacterial properties (Palanisamy et al. 2017). Gram-negative bacteria, such as *E. coli*, have an exterior lipopolysaccharide (LPS) layer on their surface that limits the entry of chemicals, which may explain their lower susceptibility to NPs. However, to have the same effect on Gram-positive bacteria, higher doses of antibacterial drugs are needed (da Rosa et al. 2020). It is also believed that AgNPs mediated production of reactive oxygen species has a bactericidal effect (ROS) (Wang et al. 2016). There have been two possible AgNPs toxicity pathways investigated (Wong and Liu 2010). One is that AgNPs have significant surface contact with bacteria, which can help the membranes of the cells attach to the particles and allow the particles to enter the bacterium. The disruption of the respiratory chain of the cell membrane by AgNPs or dissolved ions is another mechanism (Fig. 12.2). AgNPs' antibacterial properties

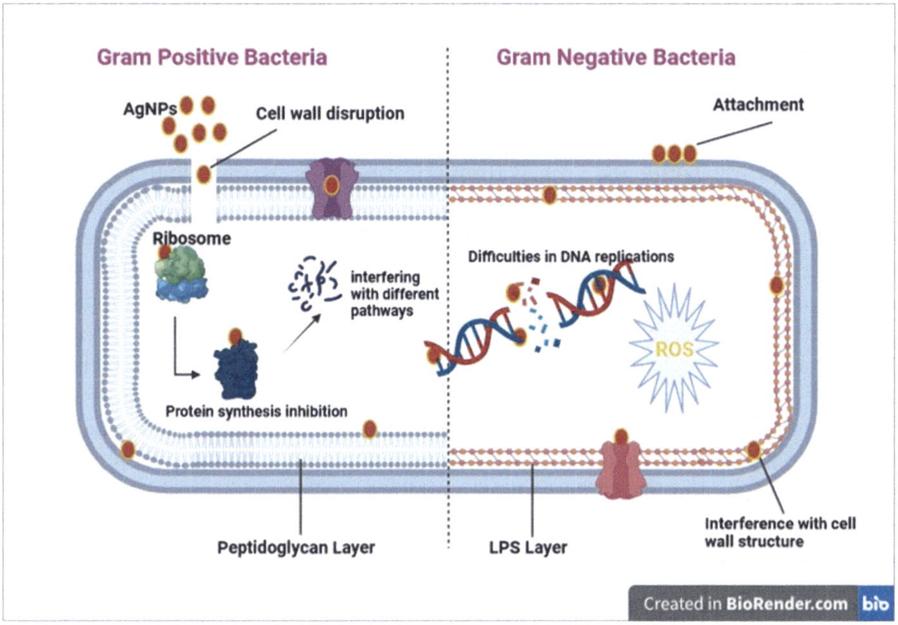

Fig. 12.2 Comparative mechanism illustration of AgNPs on bacteria according to different cell wall structure

include silver ions, as numerous studies have shown (Wu et al. 2015; Yuan et al. 2013).

12.7.2 AgNPs in Viral Disease Management

AgNPs can be applied as antivirals in agricultural applications. It is believed that viruses have a role in over 50% of all plant illnesses that evolve; these diseases can be challenging to treat in certain circumstances. Appropriate and efficient management is required in cases of viral infections. The viral replication cycle consists of viral replication within cells, viral component assembly, viral biosynthesis, and viral particle release from the host cell (Mazzon and Marsh 2019). El-Shazly et al. (2017) discovered that potato plants sprayed with AgNPs 24 h after being infected with tomato bushy stunt virus (TBSV) showed a reduction in viral concentration and disease percentage. *Cyamopsis tetragonoloba* treated with AgNPs and sun hemp rosette virus (SHRV) showed a notable reduction in illness by reducing viral multiplication (Jain and Kothari 2014). Tomato mosaic virus and potato virus Y therapy reduced disease severity and virus concentration when AgNPs were sprayed on diseased tomatoes for 7 days prior to treatment. This was because the NPs bound to the virus and prevented viral reproduction (Noha et al. 2018). Mahfouze et al.

(2020) discovered that BBTV-infected banana plants treated with 50 ppm AgNPs did not exhibit any external symptoms, despite the infection rate being 36%.

12.7.2.1 Antiviral Mechanism of AgNPs

AgNPs can be used to stop multiple viral infections by either directly inactivating the viruses or preventing the viruses from infecting cells (Baram-Pinto et al. 2009). AgNPs' large surface area makes it easier for them to come into touch with virus particles, which contributes to their strong antiviral action. Antiviral drugs often attach to proteins coat viral and obstruct interactions of structural or activities to directly affect viral particles. Here, interactions between viruses and nanoparticles are greatly impacted by the size of AgNPs (Khandelwal et al. 2014). AgNPs may have an antiviral mechanism, as seen in Fig. 12.3. The genome of the virus is eventually destroyed or prevented from getting through the barrier of the cell as a result of AgNPs' first interactions with the viral surface. AgNPs also attach themselves to the virus in order to prevent the virus from interacting with the cell membrane. AgNPs also function inside the cell as inhibitors of the nucleocapsids of the viral organism. Further, AgNPs interact with the viral genomic material to stop the host cell's genome from copying itself (Ratan et al. 2021). Ultimately, they obstruct biological functions like protein synthesis in order to stop the virus from multiplying. AgNPs are the most effective metals, even if other metals have antiviral qualities as well; they have demonstrated strong antiviral activity against a variety of viruses, independent of the virus family.

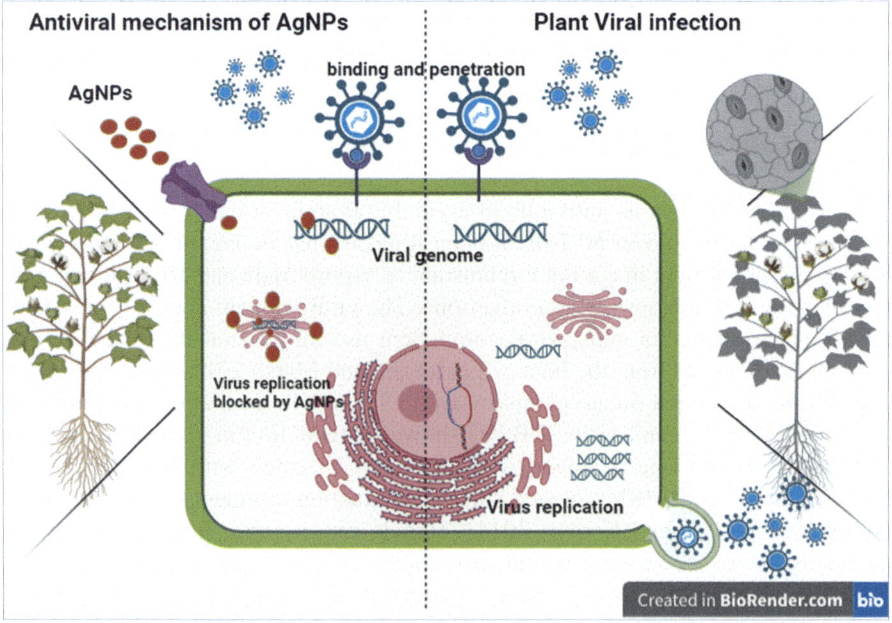

Fig. 12.3 A diagram showing how AgNPs work as antivirals

AgNPs also may have antiviral properties through contact with gp120 antagonistic effects on virus adhering to cells, pre-entry particle destruction, and interactions with double-stranded DNA (Galdiero et al. 2011). Moreover, AgNPs may combine with the polypeptide envelopes of the tomato mosaic virus (ToMV) and potato virus Y (PVY) (Noha et al. 2018). AgNPs directly inactivate TMV by interacting with envelope glycoproteins, which results in mediated impairment of TMV envelope proteins, aggregation, and even breaking down.

12.7.3 AgNPs in Fungal Disease Management

Using AgNPs as a more effective substitute for synthetic fungicides is a relatively new strategy. When AgNPs are implemented, they can be utilized to control plant diseases. The application of AgNPs against phytopathogenic fungi has only been the subject of a small number of research to date (Jo et al. 2015). The development of resistance in diseased fungi often drives demand for the newest antifungal medications (Nyilasi et al. 2010). AgNPs produced with *A. flavus* prevented the growth of *Trichoderma* spp. (Fatima et al. 2016). AgNPs also exhibited antifungal characteristics against *F. oxysporum* (Gopinath and Velusamy 2013), *C. albicans*, and *C. tropicalis* (Mallmann et al. 2015). It has been showed that AgNPs have a broad spectrum of antifungal characteristics (Mahendra Rai et al. 2014), including the capacity to inhibit spread of severe fungal diseases, including *C. albicans*, *C. tropicalis*, *C. parapsilosis*, *C. glabrata* (Rahisuddin et al. 2015), *Trichophyton rubrum* (Xia et al. 2016), *Trichosporon asahii*, *A. niger*, *Rhizoctonia solani*, *Curvularia lunata*, *Colletotrichum* sp., and *Fusarium* sp. (Bera et al. 2014; Balakumaran et al. 2015).

12.7.3.1 Antifungal Mechanism of AgNPs

The antifungal activity of biogenic AgNPs is greatly influenced by their dimensions, form, and capping agents (Mussin and Giusiano 2022). Determining a single mechanism of action is consequently challenging due to the wide diversity of biogenic AgNPs. So, AgNPs exhibit multiple mods of antifungal action, containing binding to DNA-phosphate groups (Matsumura et al. 2003) and membrane interactions. These mechanisms are involved in proton dispersion and cell death, as well as in the degradation of the transport chain and groups of enzyme, disruption of membrane proton motive force, and groups of phosphate (Feng et al. 2000). AgNPs adhere to the fungus' surface by electrostatic attraction (Fig. 12.4). AgNPs accumulate extracellularly, which causes a kinetic release of Ag ions. Ag^{+} then actively passes through the cell, increasing the dose inside it and facilitating the production of AgNPs inside (Vazquez-Muñoz et al. 2014; Lara et al. 2015; Mussin et al. 2019; Mussin and Giusiano 2022). As of right now, no known cell receptors or membrane channels are involved in the absorption of silver. However, it has been found that Ag+ is imported by the highly specific copper transporter (Ruta et al. 2018; Horstmann et al. 2019).

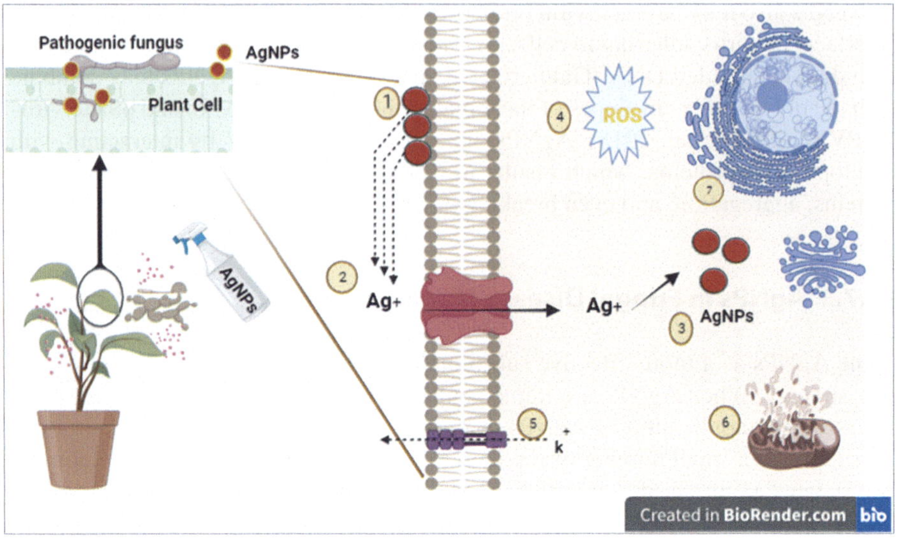

Fig. 12.4 An illustration of antifungal mechanisms of AgNPs. (1): AgNPs adhere to the fungus' surface by electrostatic attraction, (2) dynamic release of Ag^+ and import into the cell interior, (3) intracellular biosynthesis of AgNPs, (4) accumulation of intracellular ROS, (5) alteration of efflux system and induction of the release of potassium ions, (6) decrease activity of copper -depend cytochrome C- oxidase, and (7) modulation of the transcriptome, epigenome, and metabolome

12.8 General Mechanisms of AgNPs Against Plant Pathogenic Microorganisms

All plant protectants that use nanoparticles as active ingredients to improve their ability to prevent disease and their biocidal qualities are considered nano-pesticides (Iavicoli et al. 2017; Jovanović et al. 2016). Because of their amazing microbicidal action, nanoparticles have shown outstanding performance as antibacterial treatments against plant diseases (Temoor Ahmed et al. 2023). Recent studies have shown that various NP treatments have an impact on metrics linked to growth and production, including average branches, stem diameter, leaf number, pod number, plant height, and total yield in both positive and negative ways. AgNPs are one type of NP that has attracted more attention lately. NPs can be used to manage phytopathogenic infections and plant diseases. Because of their strong suppression of bacteria, viruses, and fungi as well as their broad range of antimicrobial capabilities, AgNPs are the most studied antibacterial nano-agent (Zhang et al. 2022). The methods by which disease is controlled by silver nanoparticles are intricate and involve a range of in-direct and direct effects on the plant, the pathogen, and their interactions. To maximize the effectiveness and safety of NP-based strategies for the plant disease management, more study is required to completely comprehend the underlying mechanisms.

12.8.1 Direct Interaction with Pathogens

AgNPs have demonstrated promising outcomes when interacting directly with plant pathogens (Fig. 12.5), causing the cell membrane to rupture and pierce, resulting in inhibitory activity that lowers infection (Ahmad et al. 2020). Furthermore, after being exposed to NPs, the generation of reactive oxygen species (ROS) as superoxide anions, hydrogen peroxide, and hydroxyl radical damages DNA, preventing the synthesis of mRNA and proteins, and finally results in pathogen death (Ogunsona et al. 2020). Numerous studies have revealed that biogenic NPs' large surface area, nanosize, ease of cell penetration, and other unique qualities can greatly enhance their antimicrobial effects (Ali et al. 2020). Because of their larger surface area and increased capacity for cell surface attachment, small-sized AgNPs are more effective (Bindhu et al. 2020). Additionally compared to other NP solutions, the composition of AgNPs solution is more stable. Moreover, antimicrobial methods are postulated for the regulated release of protein-capped silver ions from nanocrystals (Ahmed et al. 2021b; Al-Khattaf 2021). As a result, the combination between AgNPs and cells damages the structure of cellular membranes, depletes antioxidants, and prevents microorganisms from absorbing nutrients (Ahmed et al. 2021a). Ibrahim and colleagues discovered, for instance, that green AgNPs derived from *B. siamensis* showed a strong bactericidal effect on the rice infection *X. oryzae* pv.

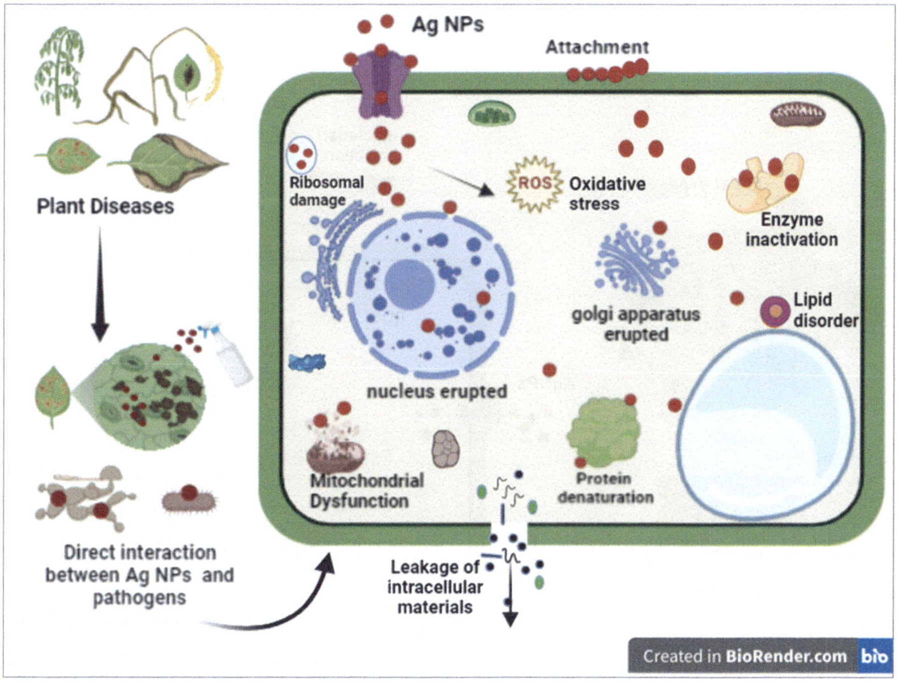

Fig. 12.5 Schematic diagram of the *direct interaction* of AgNPs with plant pathogens

oryzae, which has the potential to cause the crop illness known as bacterial leaf blight (Ibrahim et al. 2019). In another study, Kumari et al. (2017) reported the in vitro fungicidal activity of produced AgNPs against *A. solani* in a different investigation. The NPs inhibited germination and reduced biomass cell by 100% after 7 days. The utilization of NP for the disease management of virus has also been established by numerous recent research. For example, AgNPs produced from *P. fluorescens* were reported to be used in a study to manage tobacco mosaic virus in tobacco plants (Ahsan 2020).

12.8.2 Indirect Interaction with Pathogens

AgNPs' impacts on growth of plant and defense mods have demonstrated potential for indirect plant disease control (Fig. 12.6). For example, AgNPs have been demonstrated to boost plant uptake and consume of nutrients, resulting in increased yield and growth (Ejaz et al. 2018). As a result, plants may produce enzymes, phytohormones, and other defense materials that slow the pathogen growth and increase their ability to fend off infections. AgNPs pass via plasmodesmata to infiltrate the cell membrane and pass plant tissues through the pathway of apoplastic (Ma et al. 2010). In rice plants subjected to biotic stress, AgNPs and $AgNO_3$ enhanced the number, area, and physiological parameters (Ejaz et al. 2018). Biologic AgNPs greatly accelerated the germination of *Solanum lycopersicum* seeds and seedling

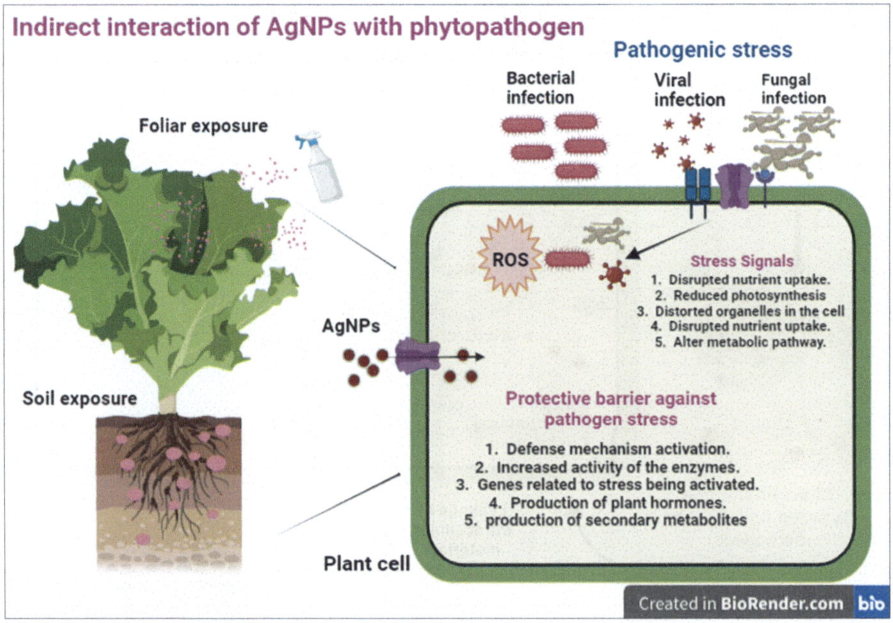

Fig. 12.6 Schematic diagram of the *indirect interaction* of AgNPs with plant pathogens

development (Noshad et al. 2019). Additionally, it has been shown that AgNPs improved the biomass, net photosynthesis, and seedling height of two bean cultivars, "Bali" and "Delfina." Additionally, it has been shown that AgNPs improved the biomass cell, net photosynthesis, and height of the seedlings of two bean varieties, "Delfina" and "Bali" (Prażak et al. 2020). AgNPs expanded the number and length of roots and enhanced the amount of nitrogen and phosphorus in the leaves of tomato seedlings (Guzmán-Báez et al. 2021; Vu et al. 2021). Through their influence on plant defense mechanisms, biogenic AgNPs also indirectly reduce plant disease (Jiang et al. 2022). It has been shown that some NPs cause plants to produce ROS, which can activate defense mechanisms and lessen the severity of illness (Yan et al. 2022). Furthermore, NP may promote the synthesis of plant hormones that are essential to plant defense against infections, such as salicylic acid and jasmonate (El-Shetehy et al. 2021; Luo et al. 2023). Earlier research has shown that biogenic NPs can be effective in disease management plants (Nandini et al. 2023; Omran and Baek 2022). Overall, by promoting plant development and defense systems, AgNPs present a viable strategy for the indirect prevention of plant disease. To completely comprehend their safety and environmental effects, as well as to maximize their application in agriculture, more research is necessary.

12.9 Conclusion

Utilizing endophytic fungal cells has become a popular method for producing metal nanoparticles through biosynthesis. The advantages of this approach include excellent synthesis reproducibility, a simple procedure for scaling up, economic feasibility, tolerance for high doses of metal NPs in the medium being used, ease of administration for large-scale nanoparticle fabrication, good dispersion of nanoparticles, and significantly greater numbers. Therefore, fungi are chosen over other technologies to perform the large-scale synthesis of NPs. In order to increase agricultural sustainability, the current chapter explores the possibility of employing AgNPs produced by green method to improve plant health and disease resistance.

References

Abdel-Azeem A, Nada AA, O'Donovan A, Thakur VK, Elkelish A (2020) Mycogenic silver nanoparticles from endophytic Trichoderma atroviride with antimicrobial activity. J Renew Mater 8(2):171–185

Abdelfattah NAH, Yousef MA, Badawy AA, Salem SS (2023) Influence of biosynthesized magnesium oxide nanoparticles on growth and physiological aspects of cowpea (Vigna unguiculata L.) plant, cowpea beetle, and cytotoxicity. Biotechnol J 18(12):2300301. https://doi.org/10.1002/biot.202300301

Abdullah JAA, Eddine LS, Abderrhmane B, Alonso-González M, Guerrero A, Romero A (2020) Green synthesis and characterization of iron oxide nanoparticles by pheonix dactylifera leaf extract and evaluation of their antioxidant activity. Sustain Chem Pharm 17:100280

Abu-Elghait M, Soliman MKY, Azab MS, Salem SS (2023) Response surface methodology: optimization of myco-synthesized gold and silver nanoparticles by Trichoderma saturnisporum. Biomass Convers Biorefinery. https://doi.org/10.1007/s13399-023-05188-4

Ahmad SA, Das SS, Khatoon A, Ansari MT, Afzal M, Hasnain MS et al (2020) Bactericidal activity of silver nanoparticles: a mechanistic review. Mater Sci Energy Technol 3:756–769

Ahmed B, Ameen F, Rizvi A, Ali K, Sonbol H, Zaidi A et al (2020) Destruction of cell topography, morphology, membrane, inhibition of respiration, biofilm formation, and bioactive molecule production by nanoparticles of Ag, ZnO, CuO, TiO2, and Al2O3 toward beneficial soil bacteria. ACS Omega 5(14):7861–7876

Ahmed T, Luo J, Noman M, Ijaz M, Wang X, Masood HA et al (2023) Microbe-mediated nanoparticle intervention for the management of plant diseases. 3 Biotech 1(1):3

Ahmed T, Noman M, Luo J, Muhammad S, Shahid M, Ali MA et al (2021a) Bioengineered chitosan-magnesium nanocomposite: a novel agricultural antimicrobial agent against Acidovorax oryzae and Rhizoctonia solani for sustainable rice production. Int J Biol Macromol 168:834–845

Ahmed T, Wu Z, Jiang H, Luo J, Noman M, Shahid M et al (2021b) Bioinspired green synthesis of zinc oxide nanoparticles from a native Bacillus cereus strain RNT6: characterization and antibacterial activity against rice panicle blight pathogens Burkholderia glumae and B. gladioli. Nanomaterials (Basel) 11(4):884

Ahsan T (2020) Biofabrication of silver nanoparticles from Pseudomonas fluorescens to control tobacco mosaic virus. Egypt J Biol Pest Control 30(1):66

Akhtar S (2012) Transmission electron microscopy of graphene and hydrated biomaterial nanostructures: novel techniques and analysis. Acta Universitatis Upsaliensis

Al-Khattaf FS (2021) Gold and silver nanoparticles: green synthesis, microbes, mechanism, factors, plant disease management and environmental risks. Saudi J Biol Sci 28(6):3624–3631

Al-Rajhi AMH, Salem SS, Alharbi AA, Abdelghany TM (2022) Ecofriendly synthesis of silver nanoparticles using Kei-apple (Dovyalis caffra) fruit and their efficacy against cancer cells and clinical pathogenic microorganisms. Arab J Chem 15(7):103927. https://doi.org/10.1016/j.arabjc.2022.103927

Al-Zahrani FAM, Salem SS, Al-Ghamdi HA, Nhari LM, Lin L, El-Shishtawy RM (2022) Green synthesis and antibacterial activity of Ag/Fe2O3 nanocomposite using Buddleja lindleyana extract. Bioengineering 9(9):452. https://www.mdpi.com/2306-5354/9/9/452

Alabssawy AN, Abu-Elghait M, Azab AM, Khalaf-Allah HM, Ashry AS, Ali AO, Sabra AB, Salem SS (2024) Hindering the biofilm of microbial pathogens and cancer cell lines development using silver nanoparticles synthesized by epidermal mucus proteins from Clarias gariepinus. BMC Biotechnol 24(1):28

Alghuthaymi MA, Almoammar H, Rai M, Said-Galiev E, Abd-Elsalam KA (2015) Myconanoparticles: synthesis and their role in phytopathogens management. Biotechnol Biotechnol Equip 29(2):221–236

Ali MA, Ahmed T, Wu W, Hossain A, Hafeez R, Islam Masum MM et al (2020) Advancements in plant and microbe-based synthesis of metallic nanoparticles and their antimicrobial activity against plant pathogens. Nanomaterials (Basel) 10(6):1146

Almuhayawi MS, Alruhaili MH, Soliman MK, Tarabulsi MK, Ashy RA, Saddiq AA, Selim S, Alruwaili Y, Salem SS (2024) Investigating the in vitro antibacterial, antibiofilm, antioxidant, anticancer and antiviral activities of zinc oxide nanoparticles biofabricated from Cassia javanica. Plos one 19(10):e0310927

Alsharif SM, Salem SS, Abdel-Rahman MA, Fouda A, Eid AM, El-Din Hassan S et al (2020) Multifunctional properties of spherical silver nanoparticles fabricated by different microbial taxa (Article). Heliyon 6(5). https://doi.org/10.1016/j.heliyon.2020.e03943

Anil Kumar S, Abyaneh MK, Gosavi S, Kulkarni SK, Pasricha R, Ahmad A et al (2007) Nitrate reductase-mediated synthesis of silver nanoparticles from AgNO 3. Biotechnol Lett 29:439–445

Aref MS, Salem SS (2020) Bio-callus synthesis of silver nanoparticles, characterization, and antibacterial activities via Cinnamomum camphora callus culture. Biocatal Agric Biotechnol 27:101689. https://doi.org/10.1016/j.bcab.2020.101689

Attia MS, Salem SS, Elakraa AA, Abdel-Maksoud MA, Malik A, Kiani BH, Malash MN, El-Sayyad GS (2024) Promising antagonistic effect of bimetallic silver-selenium nanoparticles against Ralstonia solanacearum-causing wilt disease in eggplant (Solanum melongena L.). Physiol Mol Plant Pathol 133:102369

Awual MR, Hasan MM, Islam A, Rahman MM, Asiri AM, Khaleque MA et al (2019) Offering an innovative composited material for effective lead (II) monitoring and removal from polluted water. J Clean Prod 231:214–223

Azevedo JL, Maccheroni W Jr, Pereira JO, De Araújo WL (2000) Endophytic microorganisms: a review on insect control and recent advances on tropical plants. Electron J Biotechnol 3(1):15–16

Balakumaran M, Ramachandran R, Kalaichelvan PJ (2015) Exploitation of endophytic fungus, Guignardia mangiferae for extracellular synthesis of silver nanoparticles and their in vitro biological activities. Microbiol Res 178:9–17

Bandara WMS, Seneviratne G, Kulasooriya SA (2006) Interactions among endophytic bacteria and fungi: effects and potentials. J Biosci 31:645–650

Barabadi H, Ovais M, Shinwari ZK, Saravanan M (2017) Anti-cancer green bionanomaterials: present status and future prospects. Green Chem Lett Rev 10(4):285–314

Baram-Pinto D, Shukla S, Perkas N, Gedanken A, Sarid R, c. (2009) Inhibition of herpes simplex virus type 1 infection by silver nanoparticles capped with mercaptoethane sulfonate. Bioconjug Chem 20(8):1497–1502

Barth A, Zscherp C (2002) What vibrations tell about proteins. Q Rev Biophys 35(4):369–430

Basavaraja S, Balaji S, Lagashetty A, Rajasab A, Venkataraman A (2008) Extracellular biosynthesis of silver nanoparticles using the fungus Fusarium semitectum. Mater Res Bull 43(5):1164–1170

Baudot C, Tan CM, Kong JC (2010) FTIR spectroscopy as a tool for nano-material characterization. Infrared Phys Technol 53(6):434–438

Bera R, Mandal S, Raj CR (2014) Antimicrobial activity of fluorescent Ag nanoparticles. Lett Appl Microbiol 58(6):520–526

Bhalla K, Tiwari G, Varghese M, O'Neill B, Bangdiwala SI (2021) Protecting the vulnerable by design–a tribute to Dinesh Mohan (1945–2021). Taylor & Francis, pp 277–289

Bharathidasan R, Panneerselvam A (2012) Biosynthesis and characterization of silver nanoparticles using endophytic fungi Aspergillus concius, Penicillium janthinellum and Phomosis sp. Int J Pharm Sci Res 3(9):3163

Bindhu M, Umadevi M, Esmail GA, Al-Dhabi NA, Arasu MV (2020) Green synthesis and characterization of silver nanoparticles from Moringa oleifera flower and assessment of antimicrobial and sensing properties. J Photochem Photobiol B 205:111836

Bose D, Chatterjee S (2016) Biogenic synthesis of silver nanoparticles using guava (Psidium guajava) leaf extract and its antibacterial activity against Pseudomonas aeruginosa. j 6(6):895–901

Braithwaite MC, Tyagi C, Tomar LK, Kumar P, Choonara YE, Pillay V (2014) Nutraceutical-based therapeutics and formulation strategies augmenting their efficiency to complement modern medicine: an overview. J Funct Foods 6:82–99

Burdușel A-C, Gherasim O, Grumezescu AM, Mogoantă L, Ficai A, Andronescu E (2018) Biomedical applications of silver nanoparticles: an up-to-date overview. Nanomaterials (Basel) 8(9):681

da Rosa CG, de Melo APZ, Sganzerla WG, Machado MH, Nunes MR, Maciel MV et al (2020) Application in situ of zein nanocapsules loaded with Origanum vulgare Linneus and Thymus vulgaris as a preservative in bread. Food Hydrocoll 99:105339

Devi LS, Joshi S (2015) Ultrastructures of silver nanoparticles biosynthesized using endophytic fungi. J Microsc Ultrastruct 3(1):29–37

Devi NN, Shankar D, Sutha S (2012) Biomimetic synthesis of silver nanoparticles from an endophytic fungus and their antimicrobial efficacy. Int J Biomed Adv Res 5:637–642

Devi NN, Singh MS (2013) GC-MS Analysis of metabolites from endophytic fungus Colletotrichum gloeosporioides isolated from Phlogacanthus thyrsiflorus Nees. Int J Pharm Sci 23(2):392–395

Dezfuli AAZ, Abu-Elghait M, Salem SS (2023) Recent insights into nanotechnology in colorectal cancer. Appl Biochem Biotechnol. https://doi.org/10.1007/s12010-023-04696-3

Ding G, Li Y, Fu S, Liu S, Wei J, Che Y (2009) Ambuic acid and torreyanic acid derivatives from the endolichenic fungus Pestalotiopsis sp. J Nat Prod 72(1):182–186

Melappa G, Manasa DJ, Vinaykiya V, Bhoomika V, Dutta S, Pawar R, Raghavendra VB (2022) Screening of antibacterial and antioxidant activity of biogenically synthesized silver nanoparticles from alternaria alternata, endophytic fungus of dendrophthoe falcata-a parasitic plant. BioNanoScience 12(1):128–141

Durán N, Seabra AB (2012) Metallic oxide nanoparticles: state of the art in biogenic syntheses and their mechanisms. Appl Microbiol Biotechnol 95:275–288

Ebnesajjad S, Landrock AH (2014) Adhesives technology handbook. William Andrew

Ejaz M, Raja NI, Mashwani ZUR, Ahmad MS, Hussain M, Iqbal M (2018) Effect of silver nanoparticles and silver nitrate on growth of rice under biotic stress. IET Nanobiotechnol 12(7):927–932

El-Khawaga AM, Elsayed MA, Gobara M, Suliman AA, Hashem AH, Zaher AA et al (2023) Correction to: green synthesized ZnO nanoparticles by Saccharomyces cerevisiae and their antibacterial activity and photocatalytic degradation. Biomass Convers Biorefinery. https://doi.org/10.1007/s13399-023-05131-7

El-Refaey AA, Salem SS (2023) Algae materials for bionanopesticides: nanoparticles and composites. Algae Materials (pp. 219–230). Academic Press. https://doi.org/10.1016/B978-0-443-18816-9.00004-6

El-Shazly MA, Attia YA, Kabil FF, Anis E, Hazman M (2017) Inhibitory effects of salicylic acid and silver nanoparticles on potato virus Y-infected potato plants in Egypt. Middle East J Agric Res 6(3):835–848

El-Shetehy M, Moradi A, Maceroni M, Reinhardt D, Petri-Fink A, Rothen-Rutishauser B et al (2021) Silica nanoparticles enhance disease resistance in Arabidopsis plants. Nat Nanotechnol 16(3):344–353

Fadiji AE, Mortimer PE, Xu J, Ebenso EE, Babalola OO (2022) Biosynthesis of nanoparticles using endophytes: a novel approach for enhancing plant growth and sustainable agriculture. Sustain For 14(17):10839

Fathima BS, Balakrishnan RM (2014) Biosynthesis and optimization of silver nanoparticles by endophytic fungus Fusarium solani. Mater Lett 132:428–431

Fatima F, Verma SR, Pathak N, Bajpai P (2016) Extracellular mycosynthesis of silver nanoparticles and their microbicidal activity. J Glob Antimicrob Resist 7:88–92

Feng QL, Wu J, Chen GQ, Cui F, Kim T, Kim JO (2000) A mechanistic study of the antibacterial effect of silver ions on Escherichia coli and Staphylococcus aureus. J Biomed Mater Res 52(4):662–668

Abd-Elkhalek HF, Badawy AA, Al-Askar AA, Abd Elgawad H, Hashem AH, Salem SS (2024) Biosynthesis and characterization of selenium and silver nanoparticles using Trichoderma viride filtrate and their impact on Culex pipiens. Green Process Synth 13(1):20240025

Galdiero S, Falanga A, Vitiello M, Cantisani M, Marra V, Galdiero M (2011) Silver nanoparticles as potential antiviral agents. Molecules 16(10):8894–8918

Gemishev O, Panayotova M, Gicheva G, Mintcheva N (2022) Green synthesis of stable spherical monodisperse silver nanoparticles using a cell-free extract of Trichoderma reesei. Materials 15(2):481

Gopinath V, Velusamy P (2013) Extracellular biosynthesis of silver nanoparticles using Bacillus sp. GP-23 and evaluation of their antifungal activity towards Fusarium oxysporum. Spectrochim Acta A Mol Biomol Spectrosc 106:170–174

Guilger-Casagrande M, Lima RD (2019) Synthesis of silver nanoparticles mediated by fungi: a review. Front Bioeng Biotechnol 7:287

Guo B, Dai J-R, Ng S, Huang Y, Leong C, Ong W et al (2000) Cytonic acids A and B: novel tridepside inhibitors of hCMV protease from the endophytic fungus Cytonaema species. J Nat Prod 63(5):602–604

Guzmán-Báez GA, Trejo-Téllez LI, Ramírez-Olvera SM, Salinas-Ruíz J, Bello-Bello JJ, Alcántar-González G et al (2021) Silver nanoparticles increase nitrogen, phosphorus, and potassium concentrations in leaves and stimulate root length and number of roots in tomato seedlings in a hormetic manner. Dose Response 19(4):15593258211044576

Hammad EN, Salem SS, Mohamed AA, El-Dougdoug W (2022) Environmental impacts of eco-friendly iron oxide nanoparticles on dyes removal and antibacterial activity. Appl Biochem Biotechnol 194(12):6053–6067. https://doi.org/10.1007/s12010-022-04105-1

Haroon M, Zaidi A, Ahmed B, Rizvi A, Khan MS, Musarrat J (2019) Effective inhibition of phytopathogenic microbes by eco-friendly leaf extract mediated silver nanoparticles (AgNPs). Indian J Microbiol 59:273–287

Hashem AH, Khalil AMA, Reyad AM, Salem SS (2021) Biomedical Applications of Mycosynthesized Selenium Nanoparticles Using Penicillium expansum ATTC 36200. Biol Trace Elem Res 199, 3998–4008. https://doi.org/10.1007/s12011-020-02506-z

He M, Protesescu L, Caputo R, Krumeich F, Kovalenko MV (2015) A general synthesis strategy for monodisperse metallic and metalloid nanoparticles (In, Ga, Bi, Sb, Zn, Cu, Sn, and their alloys) via in situ formed metal long-chain amides. Chem Mater 27(2):635–647

Hiruma K, Kobae Y, Toju H (2018) Beneficial associations between Brassicaceae plants and fungal endophytes under nutrient-limiting conditions: evolutionary origins and host–symbiont molecular mechanisms. Curr Opin Plant Biol 44:145–154

Horstmann C, Campbell C, Kim DS, Kim K (2019) Transcriptome profile with 20 nm silver nanoparticles in yeast. FEMS Yeast Res 19(2):foz003

Huang Y-F, Chang H-T, Tan W (2008) Cancer cell targeting using multiple aptamers conjugated on nanorods. Anal Chem 80(3):567–572

Iavicoli I, Leso V, Beezhold DH, Shvedova AA (2017) Nanotechnology in agriculture: opportunities, toxicological implications, and occupational risks. Toxicol Appl Pharmacol 329:96–111

Ibrahim E, Fouad H, Zhang M, Zhang Y, Qiu W, Yan C et al (2019) Biosynthesis of silver nanoparticles using endophytic bacteria and their role in inhibition of rice pathogenic bacteria and plant growth promotion. RSC Adv 9(50):29293–29299

Ingle A, Gade A, Pierrat S, Sonnichsen C, Rai M (2008) Mycosynthesis of silver nanoparticles using the fungus Fusarium acuminatum and its activity against some human pathogenic bacteria. Curr Nanosci 4(2):141–144

Jain D, Kothari S (2014) Green synthesis of silver nanoparticles and their application in plant virus inhibition. J Mycol Plant Pathol 44(1):21–24

Janakiraman V, Govindarajan K, CR, M. (2019) Biosynthesis of silver nanoparticles from endophytic fungi, and its cytotoxic activity. BioNanoScience 9:573–579

Jiang L, Xiang S, Lv X, Wang X, Li F, Liu W et al (2022) Biosynthesized silver nanoparticles inhibit Pseudomonas syringae pv tabaci by directly destroying bacteria and inducing plant resistance in Nicotiana benthamiana. Phytopathol Res 4(1):43

Jo Y-K, Cromwell W, Jeong H-K, Thorkelson J, Roh J-H, Shin D-B (2015) Use of silver nanoparticles for managing Gibberella fujikuroi on rice seedlings. Crop Prot 74:65–69

Johal MS, Johnson LE (2018) Understanding nanomaterials. CRC Press

Jores K, Mehnert W, Drechsler M, Bunjes H, Johann C, Mäder K (2004) Investigations on the structure of solid lipid nanoparticles (SLN) and oil-loaded solid lipid nanoparticles by photon correlation spectroscopy, field-flow fractionation and transmission electron microscopy. J Control Release 95(2):217–227

Jovanović B, Bezirci G, Çağan AS, Coppens J, Levi EE, Oluz Z et al (2016) Food web effects of titanium dioxide nanoparticles in an outdoor freshwater mesocosm experiment. Nanotoxicology 10(7):902–912

Kazarian S, Chan K (2006) Applications of ATR-FTIR spectroscopic imaging to biomedical samples. Biochim Biophys Acta - Biomembr 1758(7):858–867

Khan N, Bano A, Zandi P (2018) Effects of exogenously applied plant growth regulators in combination with PGPR on the physiology and root growth of chickpea (Cicer arietinum) and their role in drought tolerance. J Plant Interact 13(1):239–247

Khandelwal, N., Kaur, G., Kumar, N., Tiwari, AN. (2014). Application of silver nanoparticles in viral inhibition: a new hope for antivirals. Digest J Nanomater Biostruct 9(1)

Kleemann W (1993) Random-field induced antiferromagnetic, ferroelectric and structural domain states. Int J Mod Phys B 7(13):2469–2507

Kumar JA, Krithiga T, Manigandan S, Sathish S, Renita AA, Prakash P et al (2021) A focus to green synthesis of metal/metal based oxide nanoparticles: various mechanisms and applications towards ecological approach. J Clean Prod 324:129198

Kumar S, Barth A (2010) Following enzyme activity with infrared spectroscopy. Sensors 10(4):2626–2637

Kumari M, Pandey S, Bhattacharya A, Mishra A, Nautiyal CS (2017) Protective role of biosynthesized silver nanoparticles against early blight disease in Solanum lycopersicum. Plant Physiol Biochem 121:216–225

Lara HH, Romero-Urbina DG, Pierce C, Lopez-Ribot JL, Arellano-Jiménez MJ, Jose-Yacaman M (2015) Effect of silver nanoparticles on Candida albicans biofilms: an ultrastructural study. J Nanobiotechnol 13(1):1–12

Latif H, Ghareib M, Tahon M (2017) Phytosynthesis of silver nanoparticles using leaf extracts from Ocimum basilicum and Mangifira indica and their effect on some biochemical attributes of Triticum aestivum. Gesunde Pflanzen 69(1)

Luo X, Wang Z, Wang C, Yue L, Tao M, Elmer WH et al (2023) Nanomaterial size and surface modification mediate disease resistance activation in cucumber (Cucumis sativus). ACS Nano 17(5):4871–4885

Ma X, Geiser-Lee J, Deng Y, Kolmakov A (2010) Interactions between engineered nanoparticles (ENPs) and plants: phytotoxicity, uptake and accumulation. Sci Total Environ 408(16):3053–3061

Mahfouze HA, El-Dougdoug NK, Mahfouze SA (2020) Virucidal activity of silver nanoparticles against Banana bunchy top virus (BBTV) in banana plants. Bull Natl Res Cent 44(1):1–11

Mallmann EJJ, Cunha FA, Castro BN, Maciel AM, Menezes EA, Fechine PBA (2015) Antifungal activity of silver nanoparticles obtained by green synthesis. Rev Inst Med Trop Sao Paulo 57:165–167

Manoharachary C, Nagaraju D (2016) Medicinal plants for human health and welfare. Ann Phytomed 5(1):24–34

Matsumura Y, Yoshikata K, Kunisaki S-I, Tsuchido T (2003) Mode of bactericidal action of silver zeolite and its comparison with that of silver nitrate. Appl Environ Microbiol 69(7):4278–4281

Mazzon M, Marsh M (2019) Targeting viral entry as a strategy for broad-spectrum antivirals. F1000Res:8

Mehra RK, Winge DR (1991) Metal ion resistance in fungi: molecular mechanisms and their regulated expression. J Cell Biochem 45(1):30–40

Mohamed AA, Fouda A, Abdel-Rahman MA, Hassan SE-D, El-Gamal MS, Salem SS et al (2019) Fungal strain impacts the shape, bioactivity and multifunctional properties of green synthesized zinc oxide nanoparticles. Biocatal Agric Biotechnol 19:101103

Molpeceres J, Aberturas M, Guzman M (2000) Biodegradable nanoparticles as a delivery system for cyclosporine: preparation and characterization. J Microencapsul 17(5):599–614

Musarrat J, Dwivedi S, Singh BR, Al-Khedhairy AA, Azam A, Naqvi A (2010) Production of antimicrobial silver nanoparticles in water extracts of the fungus Amylomyces rouxii strain KSU-09. Bioresour Technol 101(22):8772–8776

Mussin J, Giusiano G (2022) Biogenic silver nanoparticles as antifungal agents. Front Chem 10:1023542

Mussin JE, Roldán MV, Rojas F, de Los Ángeles Sosa M, Pellegri N, Giusiano G (2019) Antifungal activity of silver nanoparticles in combination with ketoconazole against Malassezia furfur. AMB Express 9(1):1–9

Naik BS (2020) Biosynthesis of silver nanoparticles from endophytic fungi and their role in plant disease management. In: Microbial endophytes, pp 307–321

Nandini B, Krishna L, Jogigowda SC, Nagaraja G, Hadimani S, Ali D et al (2023) Significance of Bryophyllum pinnatum (Lam.) for green synthesis of anti-bacterial copper and selenium nanoparticles and their influence on soil microflora. Appl Nanosci 13(5):3609–3623

Netala VR, Bethu MS, Pushpalatha B, Baki VB, Aishwarya S, Rao JV et al (2016) Biogenesis of silver nanoparticles using endophytic fungus Pestalotiopsis microspora and evaluation of their antioxidant and anticancer activities. Int J Nanomedicine:5683–5696

Newsham KK (2011) A meta-analysis of plant responses to dark septate root endophytes. New Phytol 190(3):783–793

Noha K, Bondok A, El-Dougdoug K (2018) Evaluation of silver nanoparticles as antiviral agent against ToMV and PVY in tomato plants. Sciences 8(01):100–111

Noshad A, Hetherington C, Iqbal MJ (2019) Impact of AgNPs on seed germination and seedling growth: a focus study on its antibacterial potential against Clavibacter michiganensis subsp. michiganensis infection in Solanum lycopersicum. J Nanomat 2019:1–12

Nyilasi I, Kocsubé S, Krizsán K, Galgóczy L, Pesti M, Papp T et al (2010) In vitro synergistic interactions of the effects of various statins and azoles against some clinically important fungi. FEMS Microbiol Lett 307(2):175–184

Ogunsona EO, Muthuraj R, Ojogbo E, Valerio O, Mekonnen TH (2020) Engineered nanomaterials for antimicrobial applications: a review. Appl Mater Today 18:100473

Omran BA, Baek K (2022) Control of phytopathogens using sustainable biogenic nanomaterials: recent perspectives, ecological safety, and challenging gaps. J Clean Prod:133729

Ovais M, Khalil AT, Ayaz M, Ahmad I, Nethi SK, Mukherjee S (2018a) Biosynthesis of metal nanoparticles via microbial enzymes: a mechanistic approach. Int J Mol Sci 19(12):4100

Ovais M, Zia N, Ahmad I, Khalil AT, Raza A, Ayaz M et al (2018b) Phyto-therapeutic and nanomedicinal approaches to cure Alzheimer's disease: present status and future opportunities. Front Aging Neurosci 10:284

Palanisamy S, Rajasekar P, Vijayaprasath G, Ravi G, Manikandan R, Prabhu NM (2017) A green route to synthesis silver nanoparticles using Sargassum polycystum and its antioxidant and cytotoxic effects: an in vitro analysis. Mater Lett 189:196–200

Panja A, Mishra AK, Dash M, Pandey NK, Singh SK, Kumar B (2021) Silver nanoparticles–a review. Eurasian J Med Oncol 5(2):95

Parveen A, Rao S (2015) Effect of nanosilver on seed germination and seedling growth in Pennisetum glaucum. J Clust Sci 26:693–701

Prażak R, Święciło A, Krzepiłko A, Michałek S, Arczewska M (2020) Impact of Ag nanoparticles on seed germination and seedling growth of green beans in normal and chill temperatures. Agriculture 10(8):312

Raheman F, Deshmukh S, Ingle A, Gade A, Rai M (2011) Silver nanoparticles: novel antimicrobial agent synthesized from an endophytic fungus Pestalotia sp. isolated from leaves of Syzygium cumini (L). Nano Biomed Eng 3(3):174–178

Rahisuddin, AL-Thabaiti SA, Khan Z, Manzoor N (2015) Biosynthesis of silver nanoparticles and its antibacterial and antifungal activities towards Gram-positive, Gram-negative bacterial strains and different species of Candida fungus. Bioprocess Biosyst Eng 38:1773–1781

Rahman S, Rahman L, Khalil AT, Ali N, Zia D, Ali M et al (2019) Endophyte-mediated synthesis of silver nanoparticles and their biological applications. Appl Microbiol Biotechnol 103:2551–2569

Rai M, Kon K, Ingle A, Duran N, Galdiero S, Galdiero M (2014) Broad-spectrum bioactivities of silver nanoparticles: the emerging trends and future prospects. Appl Microbiol Biotechnol 98:1951–1961

Rai MK, Deshmukh S, Ingle A, Gade AK (2012) Silver nanoparticles: the powerful nanoweapon against multidrug-resistant bacteria. J Appl Microbiol 112(5):841–852

Rai N, Gupta P, Verma A, Singh S, Tiwari H, Kumar R et al (2023) Fungal endophyte-mediated green synthesis of silver nanoparticles as potential anticancer agent: current perspective and challenges. In: Handbook of oncobiology: from basic to clinical sciences. Springer, pp 1–24

Ranjani S, Shariq Ahmed M, Adnan M, Senthil Kumar N, Ruckmani K, Hemalatha S (2020a) Synthesis, characterization and applications of endophytic fungal nanoparticles. Inorg Nano-Met Chem 51:280–287

Ranjani S, Shariq Ahmed M, MubarakAli D, Ramachandran C, Senthil Kumar N, Hemalatha S (2020b) Toxicity assessment of silver nanoparticles synthesized using endophytic fungi against nosacomial infection. Inorg Nano-Met Chem 51(8):1080–1085

Ratan ZA, Mashrur FR, Chhoan AP, Shahriar SM, Haidere MF, Runa NJ et al (2021) Silver nanoparticles as potential antiviral agents. Pharmaceutics 13(12):2034. https://www.mdpi.com/1999-4923/13/12/2034

Raza ZA, Rehman A, Mohsin M, Bajwa SZ, Anwar F, Naeem A et al (2015) Development of antibacterial cellulosic fabric via clean impregnation of silver nanoparticles. J Clean Prod 101:377–386

Rodriguez R, White J Jr, Arnold A, Redman RS (2009) Fungal endophytes: diversity and functional roles. New Phytol 182(2):314–330

Ruta LL, Banu MA, Neagoe AD, Kissen R, Bones AM, Farcasanu IC (2018) Accumulation of Ag (I) by Saccharomyces cerevisiae cells expressing plant metallothioneins. Cells 7(12):266

Saha N, Dutta Gupta S (2018) Promotion of shoot regeneration of Swertia chirata by biosynthesized silver nanoparticles and their involvement in ethylene interceptions and activation of antioxidant activity. Plant Cell Tissue Organ Cult 134:289–300

Said A, Abu-Elghait M, Atta HM, Salem SS (2023) Antibacterial activity of green synthesized silver nanoparticles using *Lawsonia inermis* against common pathogens from urinary tract infection. Appl Biochem Biotechnol. https://doi.org/10.1007/s12010-023-04482-1

Salem SS (2022a) Baker's yeast-mediated silver nanoparticles: characterisation and antimicrobial biogenic tool for suppressing pathogenic microbes. BioNanoScience 12(4):1220–1229. https://doi.org/10.1007/s12668-022-01026-5

Salem SS (2022b) Bio-fabrication of selenium nanoparticles using Baker's yeast extract and its antimicrobial efficacy on food borne pathogens. Appl Biochem Biotechnol 194(5):1898–1910. https://doi.org/10.1007/s12010-022-03809-8

Salem SS (2023a) Application of nano-materials. In: Raja R, Hemaiswarya S, Narayanan M, Kandasamy S, Jayappriyan KR (eds) Haematococcus: biochemistry, biotechnology and biomedical applications. Springer, Singapore, pp 149–163. https://doi.org/10.1007/978-981-99-2901-6_10

Salem SS (2023b) A mini review on green nanotechnology and its development in biological effects. Arch Microbiol 205(4):128. https://doi.org/10.1007/s00203-023-03467-2

Salem SS, Ali OM, Reyad AM, Abd-Elsalam KA, Hashem AH (2022) Pseudomonas indica-mediated silver nanoparticles: antifungal and antioxidant biogenic tool for suppressing mucormycosis fungi. J Fungi 8(2):126. https://www.mdpi.com/2309-608X/8/2/126

Salem SS, EL-Belely EF, Niedbała G, Alnoman MM, Hassan SE-D, Eid AM et al (2020) Bactericidal and in-vitro cytotoxic efficacy of silver nanoparticles (Ag-NPs) fabricated by Endophytic actinomycetes and their use as coating for the textile fabrics. Nano 10(10):2082. https://www.mdpi.com/2079-4991/10/10/2082

Salem SS, Fouda A (2021) Green synthesis of metallic nanoparticles and their prospective biotechnological applications: an overview. Biol Trace Elem Res 199(1):344–370. https://doi.org/10.1007/s12011-020-02138-3

Salem SS, Hammad EN, Mohamed AA, El-Dougdoug W (2023) A comprehensive review of nanomaterials: types, synthesis, characterization, and applications (article). Biointerface Res Appl Chem 13(1). https://doi.org/10.33263/BRIAC131.041

Salem SS, Mekky AE (2024) Biogenic nanomaterials: synthesis, characterization, and applications. In: Shah MP, Bharadvaja N, Kumar L (eds) Biogenic nanomaterials for environmental sustainability: principles, practices, and opportunities. Springer, Cham, pp 13–43. https://doi.org/10.1007/978-3-031-45956-6_2

Salem SS, Saied E, Shah MP (2024a) Advanced (nano) materials. InDevelopment in wastewater treatment research and processes 2024 Jan 1 (pp. 93-115). Amsterdam, Elsevier. https://doi.org/10.1016/B978-0-323-99278-7.00011-0

Salem SS, Soliman MKY, Azab MS, Abu-Elghait M (2024b) Optimization Growth Conditions of Fusarium pseudonygamai for Myco-Synthesized Gold and Silver Nanoparticles Using Response Surface Methodology. Bio Nano Sci. https://doi.org/10.1007/s12668-024-01349-5

Salunke BK, Sawant SS, Lee S-I, Kim BS (2016) Microorganisms as efficient biosystem for the synthesis of metal nanoparticles: current scenario and future possibilities. World J Microbiol Biotechnol 32:1–16

Sandhu SS, Shukla H, Shukla S (2017) Biosynthesis of silver nanoparticles by endophytic fungi: its mechanism, characterization techniques and antimicrobial potential. Afr J Biotechnol 16(14):683–698

Schultz S, Smith DR, Mock JJ, Schultz DA (2000) Single-target molecule detection with non-bleaching multicolor optical immunolabels. Proc Natl Acad Sci 97(3):996–1001

Shah A, Lutfullah G, Ahmad K, Khalil AT, Maaza M (2018) Daphne mucronata-mediated phytosynthesis of silver nanoparticles and their novel biological applications, compatibility and toxicity studies. Green Chem Lett Rev 11(3):318–333

Shaheen TI, Salem S, Fouda A (2021) Current advances in fungal nanobiotechnology: mycofabrication and applications. In: Microbial nanobiotechnology: principles and applications, pp 113–143

Shehabeldine AM, Amin BH, Hagras FA, Ramadan AA, Kamel MR, Ahmed MA et al (2023) Potential antimicrobial and antibiofilm properties of copper oxide nanoparticles: time-kill kinetic essay and ultrastructure of pathogenic bacterial cells. Appl Biochem Biotechnol 195(1):467–485. https://doi.org/10.1007/s12010-022-04120-2

Shukla H, Sandhu SS (2017) Mycofabrication and characterization of silver nanoparticles by using some endophytic fungi with special reference to their antimicrobial potential. Int J Nanotechnol Appl 7:7–22

Singh D, Rathod V, Ninganagouda S, Hiremath J, Singh AK, Mathew J (2014) Optimization and characterization of silver nanoparticle by endophytic fungi Penicillium sp. isolated from Curcuma longa (turmeric) and application studies against MDR E. coli and S. aureus. Bioinorg Chem Appl 2014

Soliman MKY, Abu-Elghait M, Salem SS, Azab MS (2022) Multifunctional properties of silver and gold nanoparticles synthesis by Fusarium pseudonygamai. Biomass Convers Biorefinery. https://doi.org/10.1007/s13399-022-03507-9

Soliman MKY, Salem SS, Abu-Elghait M, Azab MS (2023) Biosynthesis of silver and gold nanoparticles and their efficacy towards antibacterial, antibiofilm, cytotoxicity, and antioxidant activities. Appl Biochem Biotechnol 195(2):1158–1183. https://doi.org/10.1007/s12010-022-04199-7

Soliman MK, Hashem AH, Al-Askar AA, AbdElgayed G, Salem SS (2024a) Green synthesis of silver nanoparticles from Bauhinia variegata and their biological applications. Green Process Synth 13(1):20240099

Soliman MK, Amin MAA, Nowwar, AI, Hendy MH, Salem SS (2024b) Green synthesis of selenium nanoparticles from Cassia javanica flowers extract and their medical and agricultural applications. Sci Rep 14:26775. https://doi.org/10.1038/s41598-024-77353-2

Sonbol H, Mohammed AE, Korany SM (2022) Soil fungi as biomediator in silver nanoparticles formation and antimicrobial efficacy. Int J Nanomedicine 17:2843–2863

Sre PR, Reka M, Poovazhagi R, Kumar MA, Murugesan K (2015) Antibacterial and cytotoxic effect of biologically synthesized silver nanoparticles using aqueous root extract of Erythrina indica lam. Spectrochim Acta A Mol Biomol Spectrosc 135:1137–1144

Strobel G, Ford E, Worapong J, Harper JK, Arif AM, Grant DM et al (2002) Isopestacin, an isobenzofuranone from Pestalotiopsis microspora, possessing antifungal and antioxidant activities. Phytochemistry 60(2):179–183

Svenningsen NB, Watts-Williams SJ, Joner EJ, Battini F, Efthymiou A, Cruz-Paredes C et al (2018) Suppression of the activity of arbuscular mycorrhizal fungi by the soil microbiota. ISME J 12(5):1296–1307

Thakur M, Sharma A, Chandel M, Pathania D (2022) Modern applications and current status of green nanotechnology in environmental industry. In: Green functionalized nanomaterials for environmental applications. Elsevier, pp 259–281

Toju H, Yamamoto S, Tanabe AS, Hayakawa T, Ishii HS (2016) Network modules and hubs in plant-root fungal biomes. J R Soc Interface 13(116):20151097

Vazquez-Muñoz R, Avalos-Borja M, Castro-Longoria E (2014) Ultrastructural analysis of Candida albicans when exposed to silver nanoparticles. PLoS One 9(10):e108876

Verma VC, Gond SK, Kumar A, Kharwar RN, Strobel G (2007) The endophytic mycoflora of bark, leaf, and stem tissues of Azadirachta indica A. Juss (Neem) from Varanasi (India). Microb Ecol 54:119–125

Verma VC, Kharwar RN, Gange AC (2010) Biosynthesis of antimicrobial silver nanoparticles by the endophytic fungus Aspergillus clavatus. Nanomedicine 5(1):33–40

Verma VC, Kharwar RN, Strobel GA (2009) Chemical and functional diversity of natural products from plant associated endophytic fungi. Nat Prod Commun 4(11) 1934578X0900401114

Vu NQH, Hong HTK, Quang HT (2021) Effects of different treatments of silver nanoparticles (AgNPs) on the growth & physiological characteristics of lotus (Nelumbo nucifera). IOP Conf Ser Earth Environ Sci 947:012038

Wang Y-W, Tang H, Wu D, Liu D, Liu Y, Cao A et al (2016) Enhanced bactericidal toxicity of silver nanoparticles by the antibiotic gentamicin. Environ Sci Nano 3(4):788–798

Weigel A, Sebesta A, Kukura P (2014) Dark field microspectroscopy with single molecule fluorescence sensitivity. ACS Photonics 1(9):848–856

Wong KK, Liu X (2010) Silver nanoparticles—the real "silver bullet" in clinical medicine? MedChemComm 1(2):125–131

Wu J, Yu C, Tan Y, Hou Z, Li M, Shao F et al (2015) Effects of prenatal exposure to silver nanoparticles on spatial cognition and hippocampal neurodevelopment in rats. Environ Res 138:67–73

Xia Z-K, Ma Q-H, Li S-Y, Zhang D-Q, Cong L, Tian Y-L et al (2016) The antifungal effect of silver nanoparticles on Trichosporon asahii. J Microbiol Immunol Infect 49(2):182–188

Yan X, Chen S, Pan Z, Zhao W, Rui Y, Zhao L (2022) AgNPs-triggered seed metabolic and transcriptional reprogramming enhanced rice salt tolerance and blast resistance. ACS Nano 17(1):492–504

Yaqoob AA, Ahmad H, Parveen T, Ahmad A, Oves M, Ismail IM et al (2020) Recent advances in metal decorated nanomaterials and their various biological applications: a review. Front Chem 8:341

Yuan Z, Li J, Cui L, Xu B, Zhang H, Yu C-P (2013) Interaction of silver nanoparticles with pure nitrifying bacteria. Chemosphere 90(4):1404–1411

Zhang HW, Song YC, Tan RX (2006) Biology and chemistry of endophytes. Nat Prod Rep 23(5):753–771

Zhang J, Wang F, Yalamarty SSK, Filipczak N, Jin Y, Li X (2022) Nano silver-induced toxicity and associated mechanisms. Int J Nanomedicine 17:1851–1864

Zhang X-F, Liu Z-G, Shen W, Gurunathan S (2016) Silver nanoparticles: synthesis, characterization, properties, applications, and therapeutic approaches. Int J Mol Sci 17(9):1534

Zhao X, Zhou L, Riaz Rajoka MS, Yan L, Jiang C, Shao D et al (2018) Fungal silver nanoparticles: synthesis, application and challenges. Crit Rev Biotechnol 38(6):817–835

Biocontrol of Blight Diseases Using Endophytic Fungi

13

Mohamed S. Attia, Amr H. Hashem,
and Amer M. Abdelaziz

Abstract

Agriculture is greatly affected by blight diseases, which cause devastating losses in crop productivity and quality. The destructive nature of blight diseases can have considerable negative economic effects on farmers and agricultural production systems. To reduce these losses and preserve food security, research must establish efficient blight diseases management approaches. Biocontrol products are useful for controlling the blight diseases. Endophytic fungi have an amazing capacity to colonize plants and form symbiotic relationships with them, making them a desirable alternative to conventional pesticides in agricultural settings. In this chapter, we'll concentrate on the role of endophyte fungi and discuss their effectiveness as biocontrol agents against blight diseases. The difficulties posed by strain selection, host specificity, formulation, treatment, and environmental considerations will also be covered. High-throughput screening tools, such as molecular methods, can help us understand the mechanisms driving endophytic fungi. Thus the use of endophytic fungi as biocontrol agents for blight diseases is a sustainable and eco-friendly approach to crop protection.

Keywords

Endophytic fungi · Blight diseases · Biocontrol · Colonization

M. S. Attia (✉) · A. H. Hashem · A. M. Abdelaziz
Botany and Microbiology Department, Faculty of Science, Al-Azhar University, Cairo, Egypt
e-mail: drmohamedsalah92@azhar.edu.eg

© The Author(s), under exclusive license to Springer Nature Singapore Pte Ltd. 2025

K. A. Abd-Elsalam, A. H. Hashem (eds.), *Fungal Endophytes Volume II*,
https://doi.org/10.1007/978-981-97-8804-0_13

13.1 Introduction

Fungal diseases are a major concern in agriculture, causing losses in crop yield and quality worldwide and incurring enormous economic costs (Hashem et al. 2021). Fungal plant pathogens can cause rapid and widespread issues such as blight, wilting, browning, rot, scab, rust, canker, stunting, gall, and damping off. These issues lead to degradation of plant tissues, resulting in decreased yield, poor quality, and even plant death (Hariharan and Prasannath 2021). These diseases are caused by pathogenic fungi that permeate plant tissues and establish colonies, impeding normal physiological processes (Pal and Kumar 2023). Fungal infections significantly impact on agriculture as they can affect essential crops, vegetables plants, and trees. For example, in the mid-nineteenth century, *Phytophthora infestans*, the pathogen responsible for potato blight, caused the Irish Potato Famine, leading to widespread starvation and social unrest (Stukenbrock and McDonald 2008). The significant impact of blight diseases is not only from their direct financial losses but also from the environmental effects of their control. Chemical fungicides, which can be harmful to the environment, nontarget living organisms, and human health, are widely used in traditional treatments for blight diseases (Attia et al. 2023). Fungicides have become less effective over time due to the emergence of blight disease strains resistant to common treatments. Consequently, there is a growing demand for environmentally friendly and sustainable methods of controlling blight diseases that maintain crop yield and reduce potential risks to the environment (Attia et al. 2022a).

Biocontrol is an approach to controlling plant diseases and pests that involves the use of living organisms or their products (Abdelaziz et al. 2023b). It involves using beneficial organisms to control the populations of pathogens of pests that cause damage to plants (Abdelaziz et al. 2023c). Biocontrol has grown in relevance in the investigation of plant disease due to its eco-friendliness and potential for long-term sustainable pest management (Upadhyay and Rai 2023). Parasitism, competition, and antibiosis are various biocontrol mechanisms. Pest-specific biocontrol agents cause little harm to the environment, beneficial insects, or humans. Consequently, biocontrol is a long-term and sustainable tactic. Biocontrol is an integral part of Integrated Pest Management, a comprehensive approach that combines a variety of pest management techniques (Grégoire and Gould 2023). Choosing biocontrol agents based on their sensitivity to particular infections offers another advantage of biocontrol. With this modified approach, the target pathogen can be managed accurately while avoiding unfavorable effects on beneficial organisms or the environment. Additionally, because they are live organisms or their products, biocontrol agents do not leave harmful residues, ensuring safer food production and consumption (Abdelaziz et al. 2022b). Furthermore, biocontrol is an important component of organic agricultural systems, which aim to reduce synthetic inputs while promoting natural processes. The use of biocontrol agents adheres to organic agriculture standards by offering effective pest management without harming environmental sustainability (Kaari et al. 2023). The initial cost of biocontrol agents may be higher than that of conventional pesticides, but the long-term cost-effectiveness can be significant. In the field, biocontrol agents can develop self-sustaining populations, minimizing the need for repeated applications and limiting overall input costs. Biocontrol promotes the preservation of natural enemies and a robust ecosystem, thus fostering biodiversity in agricultural areas. This promotes the preservation of biodiversity in agricultural areas. The discovery of new biocontrol agents, the improvement of their application techniques, and the integration of these agents into advanced pest management programs are primary goals of ongoing research and development (Neal 2023). A recent study demonstrated the ability of endophytic microorganisms isolated from sugar beet plants to control plant pathogens (*Fusarium* spp.) through different tactics: (1) direct influence on the pathogenic fungi in the laboratory (Dual culture), (2) the production of some substances that act to support the host plant against biotic and abiotic stresses (Petrović et al. 2024), (3) increasing the concentration of plant phytohormones (Vanama et al. 2024), and (4) the production of enzymes such as lipases, cellulases, laccases, and amylases (Toppo et al. 2024).

Endophytic fungi have been described as an extensive category of fungi with the potential to function as biocontrol agents against various phytopathogens including fugal pathogens (Attia et al. 2022b). Endophytic fungi are an attractive alternative to chemical pesticides in agricultural settings due to their exceptional ability to colonize and develop symbiotic relationships with host plants (Attia et al. 2022b). Secondary metabolites, such as antibiotics, enzymes, toxins, and volatile organic compounds, are produced by them (Sharaf et al. 2022; Hashem et al. 2023a). Pesticides, herbicides, and insecticides are a few of the agrochemicals that prevent the growth and development of phytopathogens; thus endophytic fungi have the great potential to be used in biocontrol of blight diseases (Kumar et al. 2024).

A recent study demonstrated the ability of endophytic fungi to combat the *Phytophthora* fungus in vitro, where 17 fungal isolates were isolated from the roots of the kauri plant, 5 isolates of which demonstrated high efficiency against the pathogen in vitro by dual culture method (Lawrence et al. 2024). Also, the *Penicillium* fungus was used as a biocontrol tool against some plant-pathogenic fungi (*Sclerotinia sclerotiorum*, *Rhizoctonia solani*, *Exserohilum turcicum*, and *Fusarium graminearum*) in vitro, and it proved highly efficient and effective (Sun et al. 2024). The main goal of this chapter is to focus on maximizing the use of endophytic fungi as promising environmentally friendly biological agents to combat plant blight diseases.

13.2 Endophytic Fungi Selection

The sampling technique used depends mainly on the plant species, tissue (leaves, stems, seed, fruits, and roots), and study objectives. The use of techniques and sampling procedures has enhanced the study of the distribution of endophytic fungi and their potential as biocontrol agents against blight diseases (Hashem et al. 2023b). There are many techniques including (1) surface sterilization method: This method involves eliminating epiphytic organisms before sampling endophytic fungi. This method, often used for leaves and stems, involves washing the tissues with sterilizing substances such as ethanol (70%) and/or 4% sodium hypochlorite. Then, the tissues are sliced into tiny fragments and plated onto sterilized solid media (Khalil et al. 2021). (2) The direct plating method involving cutting fragments of plant and directly depositing them onto growth media: This method is suitable for plant tissues that are naturally sterilized, such as seeds and fruits (Ananda and Sridhar 2002). (3) Tissue maceration method: This method involves crushing plant tissues in a blender or mixer to release endophytic fungi. The macerated tissue is then plated on the culture media. This technique is commonly used for roots because surface sterilization is difficult (Hallmann et al. 2006). (4) The soil dilution method: This involves diluting soil samples and inoculating culture media. Endophytic fungi, which reside in the soil and colonize plant roots through the rhizosphere, are often isolated from roots using this approach (Unterseher and Schnittler 2009). (5) Molecular performances: Molecular techniques including PCR and DNA sequencing can be used to identify endophytes directly from plant tissues. The aforementioned techniques are helpful for exploring the variety, distribution, and interactions between endophytic fungi and their host. Recognizing the diversity and distribution of endophytic fungi as well as their potential as biocontrol agents has improved by the use of molecular techniques (Sohrabi et al. 2023).

13.3 Evaluation of Endophytic Fungi as a Biocontrol Agent

When evaluating endophytic fungi potential as biocontrol agents, cultivation and identification are key initial steps. The following steps are encompassed in these processes (see Fig. 13.1): (1) isolation of endophytic fungi: The first step in culturing endophytic fungi is to isolate them from plant tissue using one of the isolation procedures. Pure cultures ensure that only one fungal species grows uncontaminated by other species. These pure fungal cultures can be stored for long-term preservation by freezing or lyophilization. (2) Morphological investigation involves macroscopic examination, which observes the visual characteristics of a fungal colony, including color, texture, and shape. Microscopic examination is carried out by observing samples under a microscope. Microscopic inspections are carried out by observing under microscopes. Microscopic characteristics include mycelium, hyphae, spores, (vegetative and reproductive components). (3) Molecular identification: as DNA sequencing that can be obtained from fungal cultures or directly from plant tissues that have endophytic fungi. The polymerase chain reaction (PCR) is used to amplify certain DNA sequences, which can then be sequenced and compared to databases of known fungi in order to identify the species of fungus. Molecular techniques provide accurate and precise identification (Hashem et al. 2022).

Depending on the phytopathogens, several criteria might be used to evaluate endophytic fungi as biocontrol agents. However, here are some common parameters that are often considered during the screening process: (1) Development characteristics include growth rate, mycelial density, and spore production ability. The growth rate provides an indication of the aptitude of endophytes' ability to inhabit and

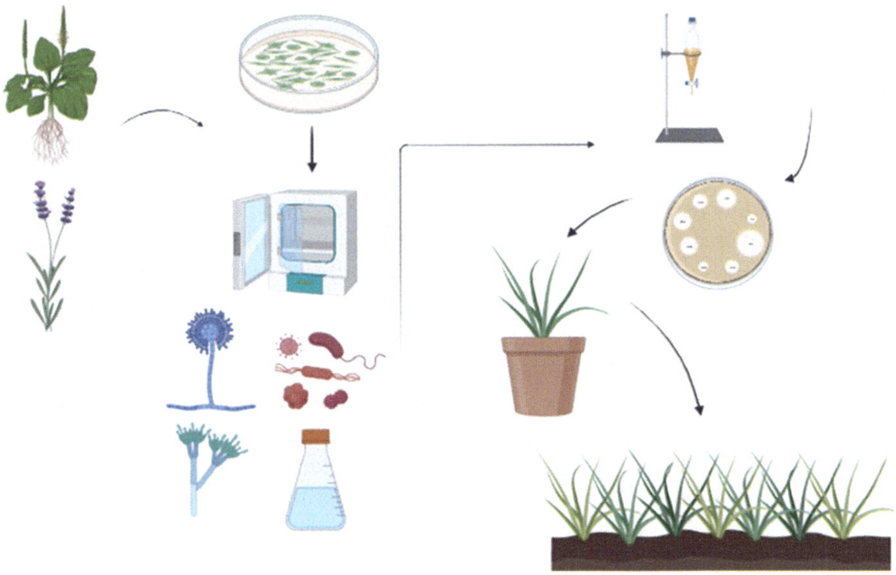

Fig. 13.1 Steps for evaluating endophytic fungi as biocontrol agents

spread within the host (Anyasi and Atagana 2019). (2) In vitro bioactivity assays: The bioactivity of endophytes against target pathogens can involve conducting in vitro assays to evaluate their antagonistic effects and the production of active metabolites that can suppress phytopathogens (Card et al. 2021). Endophytic fungi have the capacity to produce substances that promote plant growth, such as phytohormones and indole acetic acid, solubilize phosphate, generate hydrogen cyanide, and trigger plant defense systems through acquired systemic resistance (Attia et al. 2022b). (3) In vivo disease suppression (protection percent): The efficacy of the endophytic fungi to suppress blight disease is evaluated in greenhouse or field trials (Attia et al. 2022a). (4) Host specificity: The screening procedure involves assessing the host range of endophytic fungi and ensuring they do not harm nontarget organisms. This also involves testing the activity of the endophytes on distribution of beneficial organisms, such as other plants, insects, or animals. (5) Ecological limitations and safety: The safety of using the endophytic fungi as biocontrol agents is a chief factor. Potential risks, such as toxicity to people, animals, or the environment, should be assessed. (6) Production viability: The ease of cultivation, production costs, and stability of the fungal endophytes are assessed (Kusari et al. 2013; Priyashantha et al. 2023).

13.4 Modes of Endophytic Application

The effect of biotic elements and nonliving factors such as temperature and relative humidity on endophytic fungi colonization of plants have been revealed (Custodio et al. 2022). Endophytic fungi can be applied in various ways, including seed treatments, seedling treatments, foliar applications, and soil treatments, all of which have been widely used and shown to be beneficial. Seed bio-priming is a pre-sowing procedure that creates a physiological state allowing endophytic fungi to effectively come into contact with the seed. The prepared seedlings grew more quickly when compared to seedlings developed from unprimed seeds, and young seedlings from bio-primed seeds are frequently stronger and resistant to stresses (Singh et al. 2020). The effectiveness of various artificial inoculation techniques, such as foliar spray, seedling dipping, soil soaking, seed inoculation, and direct injection, has been previously studied. In separate investigations, researchers have proposed various strategies that are most suited for specific fungi. Dipping plant roots in conidial solutions has proven efficient in banana tissue culture, offering a successful alternative method for endophytic fungi colonization (Akello et al. 2007). The way of application of endophytic fungi also affected colonization. Foliar treatment has been shown to facilitate endophytic fungi colonization in various economically important crop plants, including cauliflower (Bamisile et al. 2018). Conidial suspensions are applied around the seedlings' root zones in an artificial endophytic fungi colonization technique that has been successful for a range of agricultural plants that uses soil soaking as the inoculation method (Greenfield et al. 2016). The foliar spray of biotic agents (*Desmonostoc muscorum*) increases the induction of systemic resistance in pepper plants against *Fusarium* wilt more effectively than soil treatment. It offers 90% protection and reduces the *Fusarium* wilt index to 8.3%, compared to 54.99% protection and a reduction of the *Fusarium* wilt index to 37.5% with soil treatment (Abdelaziz et al. 2022a). In another our study, encouragement of immune reactions in pepper plant by endophytes *Aspergillus alabamensis*, *Aspergillus oryzae*, and *Aspergillus tubingensis* against *Fusarium* wilt explained that the efficacy of endophytic fungi depends on mode of supplying. *Aspergillus tubingensis* and *Aspergillus alabamensis* protected the pepper in soil application (83.33%, 61.11%) more than foliar (72.22%, 58.33%) respectively, while *Aspergillus oryzae* protect pepper plant in foliar (66.67%) more than soil (63.88%) (Attia et al. 2022b). The formulation of endophytic fungi requires the creation of suitable carriers or to improve their stability, viability, and efficacy. Several criteria must be considered during formulation, including the carrier's compatibility with the endophytic fungi, simplicity of application, and long-term survival of the fungi. Liquid suspensions, granules, powders, and seed coatings are all common formulations. Each formulation has advantages and disadvantages that must be carefully considered in regard to the disease triangle (Limdolthamand et al. 2023).

13.5 Mechanisms of Endophytic Fungal Plant Colonization

The biocontrol of fungal diseases critically depends on interactions between plants and endophytic fungi. For efficient biocontrol measures against fungal diseases, it is essential to understand the colonization patterns and interactions between endophytic fungi and their hosts. Research on the processes by which endophytic fungi colonize plant tissues and create symbiotic partnerships has been considerable. Endophytic fungi employ a multitude of tactics to enter plant tissues and establish their presence within the host. The initial identification of and adsorption to plant surfaces entail the specific of particular plant-derived signals, such as elements of cell walls and chemical cues. Fungal endophytes produce a variety of enzymes, including cellulase, pectinases, and lipases, once they have attached to break down and enter the plant cell wall. Toxins that break down plant cell walls and tiny compounds that promote tissue invasion are produced in addition to this enzymatic activity (Shankar Naik et al. 2019). The process of colonization is sped up by the creation of clingy molecules that facilitate fungal attachment, such as extracellular polysaccharides. Additionally, endophytic fungi demonstrate chemotaxis toward plant root exudates, allowing them to locate niches within the host that are nutrient-rich (Timmusk 2003). Endophytic fungi improve plant stress tolerance through the creation of bioactive components and promotion of nutrient uptake, which contribute to their fitness. Signaling chemicals, hormones, and nutrients are exchanged between the fungus and the plant to facilitate these interactions (Rai and Golinska 2023). Endophytic fungal colonization methods in plants are complex and diverse. Enzymatic activities, physical interactions, chemotaxis, and molecular signaling are all involved. Understanding the endophytic lifestyle and exploiting it involves studying how endophytes benefit plants by enhancing stress resilience, nutrient acquisition, and pathogen resistance, which can be applied to sustainable agriculture and biotechnology (Das et al. 2022).

In order to facilitate communication and coordination between endophytic fungi and their host plant, signaling molecules (also known as signaling compounds or secondary metabolites) are crucial. By prompting a variety of physicochemical pathways in both parties, they help to establish and maintain a beneficial relationship (Pang et al. 2021). Mycorrhizal symbiosis describes the relationships that arise between plants and arbuscular mycorrhizal fungus (AMF). Plant secretions of strigolactones send signals to the AMF, which, upon respect, releases mycorrhizal lipo-chitooligosaccharides, and can result in the formation of the symbiotic relationships between host plants and arbuscular mycorrhizal fungus (Jamil et al. 2022). Generally, there are some fundamental points about signaling molecule exchange that can be summarized as the following:

1. *Mycotoxins and defense responses*: Endophytic fungi can produce mycotoxins, which are secondary metabolites with significant anti-pathogenic activity. These signaling molecules enhance the expression of genes involved in defense, activate defense pathways, and increase resistance to phytopathogens in plants (Rauf et al. 2022).

2. *Phytohormones*: Endophytic fungi can produce or alter phytohormones such as auxins, cytokines, gibberellins, and ethylene. These phytohormones can influence numerous aspects of plant growth and development. Endophytic fungi may regulate plant hormone signaling pathways, modify plant development, boost nutrient acquisition, and promote stress tolerance by producing or modifying phytohormones (Mukherjee et al. 2022a).
3. *Organic volatile compounds (OVCs)*: In their host plant, endophytic fungi can release OVCs. These OVCs may impact both the fungus and the host plant by acting as signaling molecules. The ability of endophytic fungi to manufacture OVCs to control plant development and produce systemic resistance has been demonstrated (Yang et al. 2021). Gliovirin compound is widely produced from *Trichoderma virens* that increase their efficiency as biocontrol agent (Howell 1998). Because gliotoxin is a broad-spectrum antibiotic, that correlated with the biocontrol efficacy of control *Pythium* cotton damping-off (Howell 1998). Previous research demonstrated that fugal endophytes' biocontrol activity increased when enzymes and antibiotics were combined (Monte 2001; Howell 1998). Howell (2003) reported the synergetic effect between endochitinase from *T. harzianum* and gliotoxin in inhibition conidial germination of *Botrytis cinerea*.
4. *Nutrient exchange*: Plants and endophytic fungi can form a symbiotic relationship to enhance nutrient transfer and absorption. Signaling molecules are also a part of this food exchange. Some endophytic fungi, for instance, can produce organic acids including citric acid or malic acid, which can chelate and solubilize nutrient sources in the soil, enhancing their availability to the host plant. In return, the plant offers the fungus nutrients including sugars and amino acids (Hoysted et al. 2023). Endophytic fungi release enzymes that decompose organic debris, which increases plant productivity and soil fertility by delivering nutrients to plants. Lignocellulolytic fungi, particularly *Trichoderma harzianum*, break down complex molecules and release nutrients into the soil, which are absorbed by plants (Ahmed et al. 2018).

13.6 Mechanisms of Biocontrol of Blight Diseases by Endophytic Fungi

Endophytic fungi employ various strategies to combat blight diseases, including direct antagonism through the production of antifungal substances such alkaloids, terpenoids, phenolics, and peptides. By interfering with crucial cellular functions like the development of cell walls, the integrity of membranes, and enzyme activity, these secondary metabolites eliminate infections. Understanding these pathways is critical for realizing the potential of endophytic fungi as long-term solutions to blight disease management. Generally there are two main different tactics (Fig. 13.2):

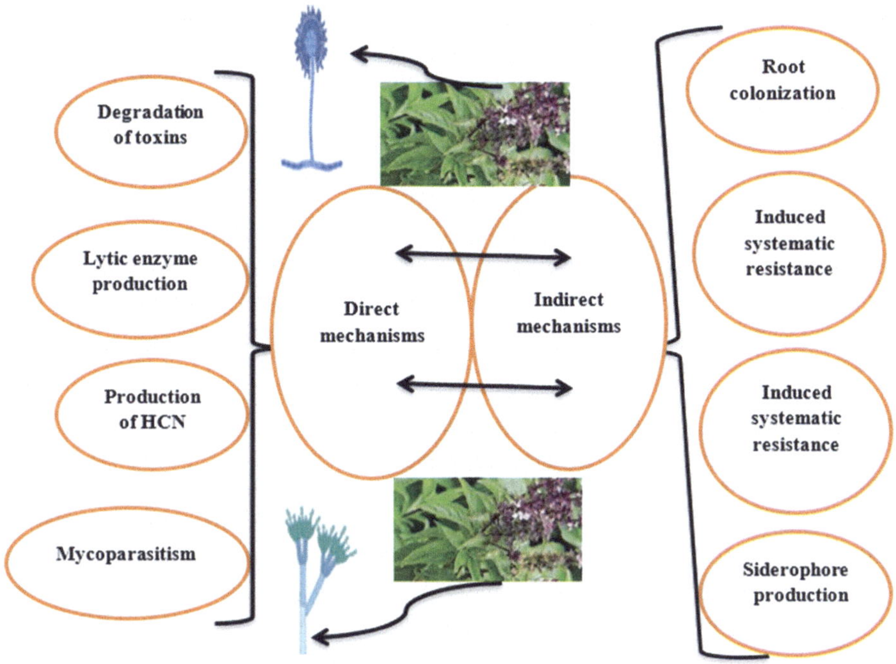

Fig. 13.2 Mechanisms of endophytic fungi against plant diseases

13.6.1 Direct Mechanisms

These mechanisms directly target fungal pathogens, prohibiting their growth, development, and establishment in the host plant. This includes antifungal chemical synthesis, mycoparasitic activity, and resource competition. Understanding and leveraging these direct mechanisms is critical for creating efficient biocontrol measures that can reduce dependency on chemical fungicides and encourage environmentally friendly ways to fungal disease management. Space competition, infection sites, and other growth factors can all be exploited by endophytic fungi to control blight diseases.

13.6.1.1 Mycoparasitism and Fungicidal Compound Production

Mycoparasitism is a process where one fungus attacks another pathogen directly. It involves identifying the host and attacking, penetrating, and ultimately killing it. Mycoparasitism involves the development of several morphological modifications, such as coiling and the production of an appressorium-like structure that aids in penetrating the host (Mcintyre et al. 2004). One of the most successful methods for controlling blight diseases is through the use of biocontrol agents that generate fungicides. Di(2-ethylhexyl) phthalate, stearic acid, TMS derivative, and 1-docosine were produced from the *Aspergillus terreus* fungus, which showed high antifungal activity against *Rhizoctonia solani* in vitro. This effect was demonstrated in vitro study by the appearance of an inhibition zone of 22 mm at a concentration of 5 mg/ml (Abdelaziz et al. 2023a). Endophytic fungi *Colletotrichum* spp. are found effective against fungal phytopathogens such as *Rhizoctonia* sp., and *Phytophthora* sp. (Li et al. 2000). Endophytic fungi *A. nidulans*, *A. fumigatus*, and *A. flavus* isolated from *Ocimum basilicum* produce biologically active natural substances (fatty acids, fatty acid esters, tetrahydrofurans, and sterols) that have been demonstrated to be an excellent source as antifungal components (Sharaf et al. 2022). In the study of Attia et al. (2022a), an endophytic fungi were isolated from the *Ocimum basilicum* plant and identified it morphologically and molecularly as *Aspergillus terreus* ON380424 for combating *Alternaria solani*, the causal agent of eggplant early blight. The results showed that *A. terreus* ethyl acetate extract has fungicidal activity toward *A. solani* with MIC 1.25 mg/ml.

13.6.1.2 Production of Hydrogen Cyanide

One of the most significant substances that can be employed to control phytopathogens is HCN. HCN's ability to block electron transport and inhibit cytochrome oxidase is believed to be the mechanism behind its toxic effects, disrupting the energy acquisition of phytopathogenic cells. Endophyte *A. terreus* isolated from tomato plant can produce HCN (Javed et al. 2020). Recent study revealed the fungicidal activity of *Aspergillus chevalieri* OP593083 and *Aspergillus egyptiacus* OP593080 against *Alternaria solani* by synthesis of HCN (Daigham et al. 2023). Endophytic fungi (*Aspergillus alabamensis*, *Aspergillus oryzae*, and *Aspergillus tubingensis*) were recorded antifungal potential by production of hydrocyanic acid. TEM analysis of fungal pathogen ultrastructure revealed that management with these endophytic fungi caused disrupt alterations in the sexual elements (conidia) and vegetative components (mycelium and cell wall) (Attia et al. 2022b).

13.6.1.3 Lytic Enzyme Production

Once attached to pathogenic fungal cell wall, it secretes lytic enzymes that damage cell wall of the host fungus and releasing components of the cell wall (Howell 2003). The fungal cell wall is composed of chitin, glucan, and proteins (Garcia-Rubio et al. 2020); thus chitinase, glucanase, and protease can degrade fungal cell wall biopolymers during biocontrol process.

Chitinase: Endophytic fungi as *Trichoderma* spp. have high numbers of chitinolytic genes that are responsible for degrading chitin and chitosan during mycoparasitism process (Kubicek et al. 2019).

Glucanases: Fungal glucanases are classified to two types α 1,3-glucanases and β 1,3-glucanases. The exo-α-1,3-glucanases AGN13.1 and AGN13.2 are produced by *T. harzianum* CECT 2413 and *T. asperellum* CECT20539, respectively, to inhibit cell wall of *Botrytis cinerea*, where AGN13.1 has lytic tool to degrade the cell wall (Mukherjee et al. 2022b). Anasontzis et al. (2019) reported the importance of β-glucanases in mycoparasitism process in of oomycetes, whose cell walls of cellulose and β-1,3- and β-1,6-glucans. Marco et al. (2003) confirmed that *Trichoderma* spp. have glucanolytic enzymes that degrade cell walls.

Cellulase: One of the most economically significant enzymes is cellulase, which is employed in numerous industrial processes, most notably the creation of biofuels. *Trichoderma* sp. secretes enzymes called cellulase that break down plant pathogens' cell walls and trigger plant defense mechanisms (Gruber and Seidl-Seiboth 2012; Ghasemi et al. 2020). The cellulase gene is upregulated during the interaction with the pathogen, according to research on the expression of the gene in *T. reesei*, *T. atroviride*, and *T. virens* during mycoparasitism against *Rhizoctonia solani* (Atanasova et al. 2013). This emphasizes how *Trichoderma* sp. and pathogen components might facilitate the process and activate the cellulase regulator gene (Bischof et al. 2013).

Proteases: During antagonistic processes, the fungus that has antagonistic activity secretes proteases to degrade proteins of the fungal pathogen cell wall (Garcia-Rubio et al. 2020). Howell (2003) reported that secretion of proteases by *T. harzianum* is a critical step in degrading the fungal cell wall in biocontrol of *B. cinerea*. Protease genes that are differentially regulated occur in the fungal cell walls during mycoparasitism (Seidl et al. 2009; Suárez et al. 2007; Troian et al. 2014).

Endophytes' production of lytic enzymes is an example of a direct biocontrol strategy because it directly targets and destroys phytopathogen pathogens (Morales-Cedeño et al. 2021). Endophytic fungal capacity to create lytic enzymes, which are vital in dissolving and dismantling pathogens' cell walls, is an intriguing feature of their biocontrol capacities. Chitinases, glucanases, proteases, cellulase, and lipases are only a few examples of the wide variety of enzymes known as lytic enzymes produced by fungal endophytes (El-Gendi et al. 2021). Endophytic fungi can damage the integrity of the pathogen's cell wall by destroying these structural elements, which results in the pathogen's lysis and eventual death (Hashem et al. 2022). Endophytic fungi's production of lytic enzymes is influenced by the genetic makeup of the fungal strain, the host plant's environment, and the existence of pathogenic challenges. Studies reveal that some endophytic fungi produce lytic enzymes on a constitutive basis, whereas others only do so in response to pathogen attack or other external stimuli (Priyashantha et al. 2023). This dynamic regulation enables endophytic fungi to quickly adjust to changing environmental conditions and maximize the production of lytic enzymes for effective blight biocontrol. Endophytic fungi's ability to produce lytic enzymes has important effects on plant health and fungal disease prevention. These enzymes can stop pathogen growth, colonization, and disease development by breaking down their cell walls. Additionally, lytic enzymes have the ability to activate host plants' systemic resistance responses, enhancing their defense mechanisms and tolerance to pathogen attacks. It is possible for endophytic fungi to biocontrol a number of plant diseases, including blights, thanks to the dual role of lytic enzymes. Along with their potential for disease biocontrol, lysozymes made by endophytic fungus have attracted interest for their biotechnological applications. These enzymes have been used for various purposes, including as industrial procedures, waste disposal, and the creation of products with added value (Ferreira and Musumeci 2021; Benedetti et al. 2019).

13.6.1.4 Degradation of Toxins

Plant pathogenic fungi not only cause direct destruction to crop but can also produce toxins that further aggravate the impact on plant health. These toxins can harm plant physiological processes and result in stunted development, yield losses, and potential risks to human health. Numerous toxins generated by fungi that cause blight have been shown to be degraded by endophytic fungi. These toxins may be breakdown or transformed to less harmful or entirely detoxified by endophytic fungal enzymes including esterase, oxidases, and hydrolases (Venkatesh and Keller 2019).

The biodegradation of fungal toxins by endophytic fungi offers numerous benefits as the following: Firstly, it lessens the toxic load on the host plant, lessens the destructive physiological properties, and restores normal plant vegetative growth. Additionally, the degradation of toxins can interrupt the virulence of fungal pathogens. Biodegrading these toxins mitigated environmental dangers, reducing the potential for contamination and associated health concerns. The ability of endophytes to degrade toxins may be influenced by environmental factors, such as nutrient availability, pH, and temperature. Therefore, identifying and characterizing endophytic fungi with high efficient toxin biodegradation capabilities is an imperative goal of research (Dai et al. 2010). Abdelaziz et al. (2022b) isolated two strains of fungi, *A. fumigatus* AM1 MW444550 and *A. terreus*, from *Moringa oleifera* leaves and then identified. Our findings are explained by the ability of *A. fumigatus* to convert aflatoxin groups into less or harmless components. Our results showed that *A. fumigatus* and *A. terreus* both inhibited the growth of *A. flavus*, while *A. fumigatus* was more successful than *A. terreus*. Additionally, *A. fumigatus* inhibited *A. flavus* with a 77.60% efficiency compared to *A. terreus* with a 39.70% efficiency.

13.6.2 Indirect Mechanisms

According to updated literatures, some endophytic fungi have the ability to produce antimicrobial substances and develop systemic resistance in their host plants. Through these techniques, endophytes can reduce the spread of blight diseases, lessen the severity of the disease, and improve plant health (Abdelaziz et al. 2023b). Endophytes provide a long-lasting and sustainable method of managing fungal disease by developing symbiotic associations with their host plants through different mechanisms.

13.6.2.1 Siderophore Production

Siderophores are low-molecular-weight metabolites that have a high affinity for Fe^{3+}. When endophytic fungus recognizes on Fe^{3+} through a particular siderophore receptor protein, siderophore can chelate Fe^{3+} from surroundings and deliver the iron into microbial cells. By chelating the available iron and rendering it unavailable to infections, siderophore-producing organisms have been shown to protect plant roots from pathogenic organisms. Through competition for iron with diseases that create fewer siderophores with lower affinity for iron, endophytic fungi can be selected for biological control in a mechanism known as siderophore-mediated suppression of plant infections. Inhibition of growth, decreased DNA and RNA synthesis, sporulation reduction, morphological changes, and changes in the energy were needed for the tricarboxylic acid cycle (Prabhakar 2020). *A. flavus* and *Mucor* sp. performed best for siderophore production in the recent study (Abd Alhakim et al. 2022), followed by *A. niger* and *Pencillium* sp. for scavenging iron from the extracellular environment. As a result of membrane receptors, siderophore-iron complexes are transported into the cell.

13.6.2.2 Induced Systematic Resistance (ISR)

Induced systemic resistance (ISR) is stimulated by nonpathogenic fungal endophytes. Salicylic acid (SA) and lipopolysaccharides also play a role in ISR. Some endophytic fungi use SA-dependent SAR mechanisms to activate resistance, while other processes require the perception of jasmonic acid and ethylene. Pathogenesis-related proteins (PR) are stimulated together with SAR. ISR led to a variety of changes, including the following: (1) promoting the formation of new barriers, such as lignin and phenolics, outside of infection sites, as well as epidermal and cell walls; (2) increasing the activity of enzymes such chitinase, polyphenol oxidase, and peroxidase; (3) encouraging the creation of phytoalexins; and (4) enhancing the expression of genes related to stress. Biocontrol ability to prevent disease and its reliability in the field are typically insufficient to compete with traditional disease control strategies. Additionally, the combination of ISR and SAR resulted in improved defense against blight pathogens (Abdelaziz et al. 2023b).

Elicitors, secondary metabolites, and cell wall-degrading enzymes are just a few of the bioactive chemicals that endophytic fungi can release that the immune system of the host plant can identify. This recognition activates several signal transduction pathways, which in turn trigger defense genes, resulting in the production of antimicrobial compounds, disease-related proteins, and phytohormones (Lu et al. 2021). Endophytic fungi can prime plants by changing the expression of defense-related genes, boosting the production of protective chemicals, and altering hormone signaling pathways. As a result, the plant is better equipped to generate a powerful defense reaction when it comes into contact with a disease (Poveda et al. 2020). The activation of SAR by endophytic fungus affects biocontrol in a number of aspects. Firstly, it provides a comprehensive, long-lasting defense response throughout the plant. As endophytic fungi colonies establish tissues, a network of induced resistance is built up, extending protection beyond the site of initial infection. This systemic defense has the potential to significantly reduce the growth and spread of diseases throughout the plant. Additionally, SAR mediated by endophytic fungi can provide all-around resistance to a number of diseases. Endophytic fungi can enhance resistance not only to the specific pathogen they are associated with but also to other diseases that utilize similar infection strategies or signaling pathways by activating defense mechanisms in the host plant. This impact of cross-protection can be especially helpful in agricultural settings where a variety of diseases threaten the crop health (Poveda et al. 2020). The study of Pieterse et al. (2014) reported that endophytic fungi caused host plants to express defense-related genes and create defense-related compounds.

13.6.2.3 Root Colonization

The first stage in the pathogenesis of blight fungal diseases, as well as a requirement for the beneficial use of endophytic fungi, is rhizosphere colonization. Endophytic fungi often promote plant growth by colonizing the root system and reducing or blocking the establishment of damaging rhizosphere pathogens. The endophytic fungi must successfully colonize the plants' rhizosphere and be able to outcompete the pathogens. Colonization is a well-known prerequisite for biocontrol and a biocontrol agent should colonize the root surface (El-Saadony et al. 2022; Saeed et al. 2021).

Suppression of plant disease has a strong correlation with colonization of the biocontrol agent. Pathogen adhesion to roots is influenced by cell surface features and may be required for effective colonization. Certain mutations that influence secondary metabolite accumulation have an effect on root colonization in the field. Due to their capacity to colonize the surfaces of plant roots, endophytic fungi affect the metabolism of plants (Harman et al. 2004). This may strengthen resistance to blight diseases, promote plant growth, and increase nutrient availability (Lugtenberg et al. 2001). *Trichoderma* secretes elicitors that stimulate plant defense, plant growth enhancement, and nutrient availability; this impact expands the colonization zone (Hanson and Howell 2004; Harman et al. 2004). In the study of Attia et al. (2022a), an endophytic fungi were isolated from the *Ocimum basilicum* plant offering 66.67% of the defense against *Alternaria solani*, and *Aspergillus terreus* considerably aided the recovery of early blight disease in vivo. In this study the treatment of *A. terreus* increased plant shoot length, root length, and total leaf number by 66.61%, 20.12%, and 65.49%, respectively, in comparison to infected control plants. Also, the treatment of eggplant infected seedlings with *A. terreus* demonstrated an obvious enhancement in photosynthetic pigments. For details, findings showed that the *A. terreus* treatment caused improvement of chlorophylls a and b (147.95% and 83.279%, respectively). However related to infected plants, the quantity of carotenoids has decreased by 44%. For more, use of *A. terreus* reduced the adverse impacts of carbohydrates in the pathogenic *A. solani* on the affected eggplants by 27.13% and protein levels by about 6.36% of the diseased eggplants. When *A. terreus* was applied, phenolics significantly decreased by 24.64% in comparison to infected control plants. Lastly, proline drastically declined by 11.88% in the infected plants as a result of the endophytic fungi application. The use of *A. terreus* caused a noticeable drop in MDA, which was reduced by 16.6%. For more, the treatment with *A. terreus* raised the activity of the antioxidant enzymes SOD, CAT, POD, and PPO by 55.15%, 3.8%, 23.27%, and 54.88%, respectively.

13.7 Effectiveness of Endophytic Fungi in Biocontrolling Blight Diseases

The previous studies proved the highly activity of endophytic fungi against blight diseases (Table 13.1).

The choice of appropriate biocontrol agents, understanding of pathogen variability, consideration of environmental conditions, appropriate timing and application methods, compatibility with other management practices, dosage and persistence, and regular monitoring are all necessary for biocontrol to be successful in managing blight diseases. By taking care of these issues, biocontrol systems can be improved to offer blight diseases control that is sustainable and successful by the following criteria: (1) *selection of endophytic fungi*: The agent's compatibility with the target pathogen, its capacity to colonize the plant or soil, and its endurance under a variety of environmental circumstances are all things to take into perspective (Ribes et al. 2018). (2) *Pathogen variability and diversity*: To achieve long-term success, it is crucial to choose endophytic fungi that exhibit broad-spectrum activity against many strains or species of blight disease pathogen (Hashemi et al. 2022). (3) *Environmental conditions*: The effectiveness of endophytic fungi can be considerably impacted by environmental conditions. Depending on the soil, climate, humidity, light, and other conditions, both the pathogen and the endophytic fungi can develop and interact in different ways. Thus, selecting agents that can withstand environmental changes and are suited for the situation is crucial (Nega 2014). (4) *Application methods and time*: To guarantee successful colonization and

Table 13.1 Activity of endophytic fungi against blight diseases

Host plant	Endophytic fungi	Fungal pathogens	References
Ocimum basilicum	*Aspergillus terreus* ON380424	*Alternaria solani*	Attia et al. (2022a)
Common bean	*Aspergillus terreus* OQ338187	*Rhizoctonia solani*	Abdelaziz et al. (2023a)
Wheat	*Aspergillus chevalieri* AGH-1 OP593083 and *Aspergillus egyptiacus* AGH-4 OP593080	*Alternaria solani*	Daigham et al. (2023)
Oilseed rape	*Alternaria alternata Aspergillus flavipes CanS-34A, Chaetomium globosum CanS-73, Clonostachys rosea CanS-43,* and *Leptosphaeria biglobosa CanS-51 a*	*Sclerotinia sclerotiorum and Botrytis cinerea*	Zhang et al. (2014)
Rice	*Trichoderma virens, Trichoderma harzianum, Curvularia lunata, Aspergillus fumigatus,* and *Aspergillus awamori*	*Rhizoctonia solani J.G. Kühn*	Safari Motlagh et al. (2022)
Siam weed (*Chromolaena odorata*)	*Trichoderma phayaoense*	*S. cucurbitacearum*	Nuangmek et al. (2021)

establishment of the biocontrol agent, appropriate application techniques, such as foliar sprays, root drenches, or seed treatments, must be taken into attention (Lahlali et al. 2022). (5) *Relationship to other management techniques*: Biocontrol techniques are frequently used in conjunction with other disease control techniques, like cultural customs or chemicals. To prevent interference or harmful interactions that can reduce the effectiveness of disease management, it is essential to ensure compatibility between endophytic fungi and these procedures (Alabouvette et al. 2006). (6) *Dosage and persistence*: The quantity and concentration of biocontrol agents can impact their efficacy. A concentration that is either ineffective or detrimental to the plant or the species around it depends on how low or high it is. It is crucial to assess the duration of activity for the biocontrol agent in the environment to ensure long-term suppression of blight diseases (Savazzini et al. 2009; Dorner et al. 1998). (7) *Monitoring and adaptation*: The biocontrol procedure must be regularly monitored to accurately assess its effectiveness and make any necessary corrections. This involves evaluating the health of the plant and the prevalence of the blight diseases. Biocontrol tactics can be adjusted based on monitoring results to enhance their effectiveness (Jenkins et al. 1998).

13.8 Conclusion

In conclusion, the use of endophytic fungi for biocontrol in blight disease management shows promise as a sustainable and eco-friendly approach in agriculture. Ongoing research and implementation efforts aimed at minimizing reliance on chemical pesticides can foster a more balanced and resilient agricultural system. These efforts contribute to the development of realistic and successful management strategies for blight diseases by formulating and studying the cytotoxic effects of endophytic fungi on human health.

References

Abd Alhakim A, Hashem A, Abdelaziz AM, Attia MS (2022) Impact of plant growth promoting fungi on biochemical defense performance of tomato under fusarial infection. Egypt J Chem 65(13):291–301

Abdelaziz AM, Attia MS, Salem MS, Refaay DA, Alhoqail WA, Senousy HH (2022a) Cyanobacteria-mediated immune responses in pepper plants against fusarium wilt. Plan Theory 11(15):2049

Abdelaziz AM, El-Wakil DA, Attia MS, Ali OM, AbdElgawad H, Hashem AH (2022b) Inhibition of Aspergillus flavus growth and aflatoxin production in Zea mays L. using endophytic Aspergillus fumigatus. J Fungi 8(5):482

Abdelaziz AM, El-Wakil DA, Hashem AH, Al-Askar AA, AbdElgawad H, Attia MS (2023a) Efficient role of Endophytic aspergillus terreus in biocontrol of Rhizoctonia solani causing damping-off disease of Phaseolus vulgaris and Vicia faba. Microorganisms 11(6):1487

Abdelaziz AM, Hashem AH, El-Sayyad GS, El-Wakil DA, Selim S, Alkhalifah DH, Attia MS (2023b) Biocontrol of soil borne diseases by plant growth promoting rhizobacteria. Trop Plant Pathol 48(2):105–127

Abdelaziz AM, Sharaf MH, Hashem AH, Al-Askar AA, Marey SA, Mohamed FA, Abdelstar MN, Abdelgawad H, Attia MS (2023c) Biocontrol of Fusarium wilt disease in pepper plant by plant growth promoting Penicillium expansum and Trichoderma harzianum. Notulae Botanicae Horti Agrobotanici Cluj-Napoca 51(3):13302–13302

Ahmed S, Rahman S, Hasan M, Paul N, Sajib AA (2018) Microbial degradation of lignocellulosic biomass: discovery of novel natural lignocellulolytic bacteria. BioTechnologia J Biotechnol Comput Biol Bionanotechnol 99(2)

Akello J, Dubois T, Gold CS, Coyne D, Nakavuma J, Paparu P (2007) Beauveria bassiana (Balsamo) Vuillemin as an endophyte in tissue culture banana (Musa spp.). J Invertebr Pathol 96(1):34–42

Alabouvette C, Olivain C, Steinberg C (2006) Biological control of plant diseases: the European situation. Eur J Plant Pathol 114:329–341

Ananda K, Sridhar K (2002) Diversity of endophytic fungi in the roots of mangrove species on the west coast of India. Can J Microbiol 48(10):871–878

Anasontzis GE, Lebrun MH, Haon M, Champion C, Kohler A, Lenfant N, Martin F, O'Connell RJ, Riley R, Grigoriev IV (2019) Broad-specificity GH131 β-glucanases are a hallmark of fungi and oomycetes that colonize plants. Environ Microbiol 21(8):2724–2739

Anyasi RO, Atagana HI (2019) Endophyte: understanding the microbes and its applications. Pak J Biol Sci 22:154–167

Atanasova L, Druzhinina IS, Jaklitsch WM, Mukherjee P, Horwitz B, Singh U (2013) Two hundred Trichoderma species recognized on the basis of molecular phylogeny. In: Trichoderma: biology and applications. CABI International, London, pp 10–42

Attia MS, Hashem AH, Badawy AA, Abdelaziz AM (2022a) Biocontrol of early blight disease of eggplant using endophytic Aspergillus terreus: improving plant immunological, physiological and antifungal activities. Bot Stud 63(1):26

Attia MS, Salem MS, Abdelaziz AM (2022b) Endophytic fungi Aspergillus spp. reduce fusarial wilt disease severity, enhance growth, metabolism and stimulate the plant defense system in pepper plants. Biomass Conv Bioref 4:16603–16613

Attia MS, Abdelaziz AM, Hassanin MM, Al-askar AA, Marey SA, Abdelgawad H, Hashem AH (2023) Eco-friendly preparation of thyme essential oil nano emulsion: characterization, antifungal activity and resistance of Fusarium wilt disease of Foeniculum vulgare. Notulae Botanicae Horti Agrobotanici Cluj-Napoca 51(3):13312

Bamisile BS, Dash CK, Akutse KS, Keppanan R, Afolabi OG, Hussain M, Qasim M, Wang L (2018) Prospects of endophytic fungal entomopathogens as biocontrol and plant growth promoting agents: an insight on how artificial inoculation methods affect endophytic colonization of host plants. Microbiol Res 217:34–50

Benedetti M, Locci F, Gramegna G, Sestili F, Savatin DV (2019) Green production and biotechnological applications of cell wall lytic enzymes. Appl Sci 9(23):5012

Bischof R, Fourtis L, Limbeck A, Gamauf C, Seiboth B, Kubicek CP (2013) Comparative analysis of the Trichoderma reeseitranscriptome during growth on the cellulase inducing substrates wheat straw and lactose. Biotechnol Biofuels 6(1):1–14

Card SD, Bastías DA, Caradus JR (2021, 1997) Antagonism to plant pathogens by Epichloë fungal endophytes—a review. Plan Theory 10(10)

Custodio V, Gonin M, Stabl G, Bakhoum N, Oliveira MM, Gutjahr C, Castrillo G (2022) Sculpting the soil microbiota. Plant J 109(3):508–522

Dai C-c, Tian L-s, Zhao Y-t, Chen Y, Xie H (2010) Degradation of phenanthrene by the endophytic fungus Ceratobasidum stevensii found in Bischofia polycarpa. Biodegradation 21:245–255

Daigham GE, Mahfouz AY, Abdelaziz AM, Nofel MM, Attia MS (2023) Protective role of plant growth-promoting fungi Aspergillus chevalieri OP593083 and Aspergillus egyptiacus OP593080 as biocontrol approach against Alternaria leaf spot disease of Vicia faba plant. Biomass Conv Bioref:1–17

Das PP, Singh KR, Nagpure G, Mansoori A, Singh RP, Ghazi IA, Kumar A, Singh J (2022) Plant-soil-microbes: a tripartite interaction for nutrient acquisition and better plant growth for sustainable agricultural practices. Environ Res 214:113821

Dorner JW, Cole RJ, Blankenship PD (1998) Effect of inoculum rate of biological control agents on preharvest aflatoxin contamination of peanuts. Biol Control 12(3):171–176

El-Gendi H, Saleh AK, Badierah R, Redwan EM, El-Maradny YA, El-Fakharany EM (2021) A comprehensive insight into fungal enzymes: structure, classification, and their role in mankind's challenges. J Fungi 8(1):23

El-Saadony MT, Saad AM, Soliman SM, Salem HM, Ahmed AI, Mahmood M, El-Tahan AM, Ebrahim AA, El-Mageed A, Taia A (2022) Plant growth-promoting microorganisms as biocontrol agents of plant diseases: mechanisms, challenges and future perspectives. Front Plant Sci 13:923880

Ferreira FV, Musumeci MA (2021) Trichoderma as biological control agent: scope and prospects to improve efficacy. World J Microbiol Biotechnol 37(5):90

Garcia-Rubio R, de Oliveira HC, Rivera J, Trevijano-Contador N (2020) The fungal cell wall: Candida, Cryptococcus, and Aspergillus species. Front Microbiol 10. https://doi.org/10.3389/fmicb.2019.02993

Ghasemi S, Safaie N, Shahbazi S, Shams-Bakhsh M, Askari H (2020) The role of cell wall degrading enzymes in antagonistic traits of Trichoderma virens against Rhizoctonia solani. Iran J Biotechnol 18(4):e2333. https://doi.org/10.30498/IJB.2020.2333

Greenfield M, Gómez-Jiménez MI, Ortiz V, Vega FE, Kramer M, Parsa S (2016) Beauveria bassiana and Metarhizium anisopliae endophytically colonize cassava roots following soil drench inoculation. Biol Control 95:40–48

Grégoire J-C, Gould JR (2023) Forest insect–natural enemy interactions. In: Forest entomology and pathology: volume 1: entomology. Springer, Cham, pp 141–167

Gruber S, Seidl-Seiboth V (2012) Self versus non-self: fungal cell wall degradation in Trichoderma. Microbiology 158(1):26–34

Hallmann J, Berg G, Schulz B (2006) Isolation procedures for endophytic microorganisms. In: Microbial root endophytes. Springer, pp 299–319

Hanson L, Howell C (2004) Elicitors of plant defense responses from biocontrol strains of Trichoderma viren. Phytopathology 94(2):171–176

Hariharan G, Prasannath K (2021) Recent advances in molecular diagnostics of fungal plant pathogens: a mini review. Front Cell Infect Microbiol 10:600234

Harman GE, Howell CR, Viterbo A, Chet I, Lorito M (2004) Trichoderma species—opportunistic, avirulent plant symbionts. Nat Rev Microbiol 2(1):43–56

Hashem AH, Abdelaziz AM, Askar AA, Fouda HM, Khalil AM, Abd-Elsalam KA, Khaleil MM (2021) Bacillus megaterium-mediated synthesis of selenium nanoparticles and their antifungal activity against Rhizoctonia solani in faba bean plants. J Fungi 7(3):195

Hashem AH, Shehabeldine AM, Abdelaziz AM, Amin BH, Sharaf MH (2022) Antifungal activity of endophytic Aspergillus terreus extract against some fungi causing mucormycosis: ultrastructural study. Appl Biochem Biotechnol 194(8):3468–3482

Hashem AH, Abdelaziz AM, Hassanin MM, Al-Askar AA, AbdElgawad H, Attia MS (2023a) Potential impacts of clove essential oil nanoemulsion as bio fungicides against Neoscytalidium blight disease of Carum carvi L. Agronomy 13(4):1114

Hashem AH, Attia MS, Kandil EK, Fawzi MM, Abdelrahman AS, Khader MS, Khodaira MA, Emam AE, Goma MA, Abdelaziz AM (2023b) Bioactive compounds and biomedical applications of endophytic fungi: a recent review. Microb Cell Factories 22(1):1–23

Hashemi M, Tabet D, Sandroni M, Benavent-Celma C, Seematti J, Andersen CB, Grenville-Briggs LJ (2022) The hunt for sustainable biocontrol of oomycete plant pathogens, a case study of Phytophthora infestans. Fungal Biol Rev 40:53–69

Howell C (1998) The role of antibiosis in biocontrol. Trichoderma Gliocladium 2:173–184

Howell C (2003) Mechanisms employed by Trichoderma species in the biological control of plant diseases: the history and evolution of current concepts. Plant Dis 87(1):4–10

Hoysted GA, Field KJ, Sinanaj B, Bell CA, Bidartondo MI, Pressel S (2023) Direct nitrogen, phosphorus and carbon exchanges between Mucoromycotina 'fine root endophyte'fungi and a flowering plant in novel monoxenic cultures. New Phytol 238(1):70–79

Jamil F, Mukhtar H, Fouillaud M, Dufossé L (2022) Rhizosphere signaling: insights into plant–rhizomicrobiome interactions for sustainable agronomy. Microorganisms 10(5):899

Javed A, Shah AH, Hussain A, Shinwari ZK, Khan SA, Khan W, Jan SA (2020) Potential of endophytic fungus Aspergillus terreus as potent plant growth promoter. Pak J Bot 52(3):1083–1086

Jenkins NE, Heviefo G, Langewald J, Cherry AJ, Lomer CJ (1998) Development of mass production technology for aerial conidia for use as mycopesticides. Biocontrol News Inf 19:21N–32N

Kaari M, Manikkam R, Annamalai KK, Joseph J (2023) Actinobacteria as a source of biofertilizer/biocontrol agents for bio-organic agriculture. J Appl Microbiol 134(2):lxac047

Khalil A, Abdelaziz A, Khaleil M, Hashem A (2021) Fungal endophytes from leaves of Avicennia marina growing in semi-arid environment as a promising source for bioactive compounds. Lett Appl Microbiol 72(3):263–274

Kubicek CP, Steindorff AS, Chenthamara K, Manganiello G, Henrissat B, Zhang J, Cai F, Kopchinskiy AG, Kubicek EM, Kuo A (2019) Evolution and comparative genomics of the most common Trichoderma species. BMC Genomics 20(1):1–24

Kumar S, Sindhu SS, Kumar R (2024) Microbial endophytes: prospects in biological control of plant pathogens and plant growth stimulation for sustainable agriculture. In: Plant endophytes and secondary metabolites. Elsevier, pp 375–422

Kusari P, Kusari S, Spiteller M, Kayser O (2013) Endophytic fungi harbored in Cannabis sativa L.: diversity and potential as biocontrol agents against host plant-specific phytopathogens. Fungal Divers 60:137–151

Lahlali R, Ezrari S, Radouane N, Kenfaoui J, Esmaeel Q, El Hamss H, Belabess Z, Barka EA (2022) Biological control of plant pathogens: a global perspective. Microorganisms 10(3):596

Lawrence P, Jackson L, Padamsee M, Lee K, Lacap-Bugler D (2024) Screening of Agathis australis endophytes as biological control agents against kauri dieback pathogen Phytophthora agathidicida. J Plant Pathol:1–13

Li J, Strobel G, Harper J, Lobkovsky E, Clardy J (2000) Cryptocin, a potent tetramic acid antimycotic from the endophytic fungus Cryptosporiopsis cf. q uercina. Org Lett 2(6):767–770

Limdolthamand S, Songkumarn P, Suwannarat S, Jantasorn A, Dethoup T (2023) Biocontrol efficacy of endophytic Trichoderma spp. in fresh and dry powder formulations in controlling northern corn leaf blight in sweet corn. Biol Control 181:105217

Lu H, Wei T, Lou H, Shu X, Chen Q (2021) A critical review on communication mechanism within plant-endophytic fungi interactions to cope with biotic and abiotic stresses. J of Fungi 7(9):719

Lugtenberg BJ, Dekkers L, Bloemberg GV (2001) Molecular determinants of rhizosphere colonization by Pseudomonas. Annu Rev Phytopathol 39(1):461–490

Marco JLD, Valadares-Inglis MC, Felix CR (2003) Production of hydrolytic enzymes by Trichoderma isolates with antagonistic activity against Crinipellis perniciosa, the causal agent of witches' broom of cocoa. Braz J Microbiol 34(1):33–38

Mcintyre M, Nielsen J, Arnau J, Brink H, Hansen K, Madrid S (2004) Proceedings of the 7th European conference on fungal genetics. Copenhagen Denmark

Monte E (2001) Understanding Trichoderma: between biotechnology and microbial ecology. Int Microbiol 4(1):1–4

Morales-Cedeño LR, del Carmen O-MM, Loeza-Lara PD, Parra-Cota FI, de Los S-VS, Santoyo G (2021) Plant growth-promoting bacterial endophytes as biocontrol agents of pre-and post-harvest diseases: fundamentals, methods of application and future perspectives. Microbiol Res 242:126612

Mukherjee A, Gaurav AK, Singh S, Yadav S, Bhowmick S, Abeysinghe S, Verma JP (2022a) The bioactive potential of phytohormones: a review. Biotechnol Rep 35:e00748

Mukherjee PK, Mendoza-Mendoza A, Zeilinger S, Horwitz BA (2022b) Mycoparasitism as a mechanism of Trichoderma-mediated suppression of plant diseases. Fungal Biol Rev 39:15–33. https://doi.org/10.1016/j.fbr.2021.11.004

Neal JC (2023) Biological control of weeds in turfgrass: opportunities and misconceptions. Pest Manag Sci

Nega A (2014) Review on concepts in biological control of plant pathogens. J Biol Agric Healthcare 4(27):33–54

Nuangmek W, Aiduang W, Kumla J, Lumyong S, Suwannarach N (2021) Evaluation of a newly identified endophytic fungus, Trichoderma phayaoense for plant growth promotion and biological control of gummy stem blight and wilt of muskmelon. Front Microbiol 12:634772
Pal A, Kumar V (2023) AgriDet: plant leaf disease severity classification using agriculture detection framework. Eng Appl Artif Intell 119:105754
Pang Z, Chen J, Wang T, Gao C, Li Z, Guo L, Xu J, Cheng Y (2021) Linking plant secondary metabolites and plant microbiomes: a review. Front Plant Sci 12:621276
Petrović M, Janakiev T, Grbić ML, Unković N, Stević T, Vukićević S, Dimkić I (2024) Insights into endophytic and rhizospheric bacteria of five sugar beet hybrids in terms of their diversity, plant-growth promoting, and biocontrol properties. Microb Ecol 87(1):1–25
Pieterse CM, Zamioudis C, Berendsen RL, Weller DM, Van Wees SC, Bakker PA (2014) Induced systemic resistance by beneficial microbes. Annu Rev Phytopathol 52:347–375
Poveda J, Abril-Urias P, Escobar C (2020) Biological control of plant-parasitic nematodes by filamentous fungi inducers of resistance: trichoderma, mycorrhizal and endophytic fungi. Front Microbiol 11:992
Prabhakar PK (2020) Bacterial siderophores and their potential applications: a review. Curr Mol Pharmacol 13(4):295–305
Priyashantha AKH, Karunarathna SC, Lu L, Tibpromma S (2023) Fungal endophytes: an alternative biocontrol agent against phytopathogenic fungi. Encyclopedia 3(2):759–780
Rai M, Golinska P (2023) Myconanotechnology: emerging trends and applications
Rauf M, Ur-Rahman A, Arif M, Gul H, Ud-Din A, Hamayun M, Lee I-J (2022) Immunomodulatory molecular mechanisms of Luffa cylindrica for downy mildews resistance induced by growth-promoting endophytic fungi. J Fungi 8(7):689
Ribes S, Fuentes A, Talens P, Barat JM (2018) Prevention of fungal spoilage in food products using natural compounds: a review. Crit Rev Food Sci Nutr 58(12):2002–2016
Saeed Q, Xiukang W, Haider FU, Kučerik J, Mumtaz MZ, Holatko J, Naseem M, Kintl A, Ejaz M, Naveed M (2021) Rhizosphere bacteria in plant growth promotion, biocontrol, and bioremediation of contaminated sites: a comprehensive review of effects and mechanisms. Int J Mol Sci 22(19):10529
Safari Motlagh MR, Jahangiri B, Kulus D, Tymoszuk A, Kaviani B (2022) Endophytic fungi as potential biocontrol agents against Rhizoctonia solani JG Kühn, the causal agent of rice sheath blight disease. Biology 11(9):1282
Savazzini F, Longa CMO, Pertot I (2009) Impact of the biocontrol agent Trichoderma atroviride SC1 on soil microbial communities of a vineyard in northern Italy. Soil Biol Biochem 41(7):1457–1465
Seidl V, Song L, Lindquist E, Gruber S, Koptchinskiy A, Zeilinger S, Schmoll M, Martínez P, Sun J, Grigoriev I (2009) Transcriptomic response of the mycoparasitic fungus Trichoderma atroviride to the presence of a fungal prey. BMC Genomics 10(1):1–13
Shankar Naik B, Abrar S, Krishnappa M (2019) Industrially important enzymes from fungal endophytes. In: Recent advancement in white biotechnology through fungi: volume 1: diversity and enzymes perspectives, pp 263–280
Sharaf MH, Abdelaziz AM, Kalaba MH, Radwan AA, Hashem AH (2022) Antimicrobial, antioxidant, cytotoxic activities and phytochemical analysis of fungal endophytes isolated from ocimum basilicum. Appl Biochem Biotechnol 194:1271–1289
Singh S, Singh UB, Malviya D, Paul S, Sahu PK, Trivedi M, Paul D, Saxena AK (2020) Seed biopriming with microbial inoculant triggers local and systemic defense responses against Rhizoctonia solani causing banded leaf and sheath blight in maize (Zea mays L.). Int J Environ Resarch Public Health 17(4):1396
Sohrabi M, Samsampour D, Bagheri A (2023) Molecular identification of fungal endophytes of medicinal plant Citrullus colocynthis (L.) Schrad as a medicinal plant: role of tissue type and sampling location on the diversity. Mol Biotechnol 66:424–431
Stukenbrock EH, McDonald BA (2008) The origins of plant pathogens in agro-ecosystems. Annu Rev Phytopathol 46:75–100

Suárez MB, Vizcaíno JA, Llobell A, Monte E (2007) Characterization of genes encoding novel peptidases in the biocontrol fungus Trichoderma harzianum CECT 2413 using the TrichoEST functional genomics approach. Curr Genet 51(5):331–342. https://doi.org/10.1007/s00294-007-0130-5

Sun X, Ma X, Gong R, Zhang D, Wang X (2024) Screening and identification of antagonistic fungi against tomato early blight and research on its pot-plant control effects. Pak J Bot 56(1):359–365

Timmusk S (2003) Mechanism of action of the plant growth promoting bacterium Paenibacillus polymyxa. Acta Universitatis Upsaliensis

Toppo P, Jangir P, Mehra N, Kapoor R, Mathur P (2024) Bioprospecting of endophytic fungi from medicinal plant Anisomeles indica L. for their diverse role in agricultural and industrial sectors. Sci Rep 14(1):588

Troian RF, Steindorff AS, Ramada MH, Arruda W, Ulhoa CJ (2014) Mycoparasitism studies of Trichoderma harzianum against Sclerotinia sclerotiorum: evaluation of antagonism and expression of cell wall-degrading enzymes genes. Biotechnol Lett 36(10):2095–2101. https://doi.org/10.1007/s10529-014-1583-5

Unterseher M, Schnittler M (2009) Dilution-to-extinction cultivation of leaf-inhabiting endophytic fungi in beech (Fagus sylvatica L.)–different cultivation techniques influence fungal biodiversity assessment. Mycol Res 113(5):645

Upadhyay R, Rai B (2023) Biocontrol agents of plant pathogens: their use and practical constraints. In: Biocontrol of plant diseases. CRC Press, pp 15–36

Vanama S, Gopalan NR, Pesari M, Baskar M, Gali UD, Lakshmi DL, Koteshwar P, Jesudasu G, Rathod S, Prasad MS (2024) Native bio-control agents from the rice fields of Telangana, India: characterization and unveiling the potential against stem rot and false smut diseases of rice. World J Microbiol Biotechnol 40(1):2

Venkatesh N, Keller NP (2019) Mycotoxins in conversation with bacteria and fungi. Front Microbiol 10:403

Yang Y, Chen Y, Cai J, Liu X, Huang G (2021) Antifungal activity of volatile compounds generated by endophytic fungi Sarocladium brachiariae HND5 against Fusarium oxysporum f. sp cubense. PloS One 16(12):e0260747

Zhang Q, Zhang J, Yang L, Zhang L, Jiang D, Chen W, Li G (2014) Diversity and biocontrol potential of endophytic fungi in Brassica napus. Biol Control 72:98–108

Endophytic Fungi: Biocontrol Agents for Postharvest Disease-Causing Pathogenic Fungi

14

Julekha Bagum and Debdulal Banerjee

Abstract

Today's modern agriculture system is facing a major problem regarding the storage of foods after harvesting. These postharvest fungal infections negatively affect food security as more than 30% of food is getting wasted during postharvest operations. In this case, fungal pathogens are playing a major role behind these losses. So, they are needed to be controlled at the earliest. Synthetic fungicides can prevent these pathogens, but their toxicity toward humans, animals, and the environment is raising serious questions. The solitary solution could be green remedies like metabolites extracted from natural entities. Other than phytometabolites, microbial metabolites are also effective in controlling phytopathogens. Biocontrol agents are the good option against this problem as they are known to have fungicidal ability, mainly for protecting postharvest products like fruits, vegetables, and grains. Nowadays, secondary metabolites from untapped sources like endophytic fungi are acting as the rich hub for antimicrobial metabolites with multidomain utility in the field of agriculture. These metabolites are safe and are applied in the form of foliar/root spray. Endophytic fungi can also be a promising biocontrol agent as it protects the host plant from pest infection for its entire life cycle. So, inevitably endophytic fungi and their metabolites can prevent postharvest pathogens and become a permanent alternative to synthetic

J. Bagum
Centre for Life Sciences, Vidyasagar University, Midnapore, India

D. Banerjee (✉)
Centre for Life Sciences, Vidyasagar University, Midnapore, India

Microbiology and Microbial Biotechnology Laboratory, Department of Botany and Forestry, Vidyasagar University, Midnapore, West Bengal, India
e-mail: db@mail.vidyasagar.ac.in

© The Author(s), under exclusive license to Springer Nature Singapore Pte Ltd. 2025

K. A. Abd-Elsalam, A. H. Hashem (eds.), *Fungal Endophytes Volume II*,
https://doi.org/10.1007/978-981-97-8804-0_14

fungicides. This chapter extensively summarizes the role of endophytic fungi in preventing postharvest fungal diseases in cash crops.

Keywords

Biocontrol · Endophytic fungi · Postharvest pathogens · Antifungal compounds · Sustainable agriculture

14.1 Introduction

Postharvest loss (PHL) is the major problem in modern agriculture, which becomes a great threat to food security. The Ministry of Food Processing Industries conducted various primary surveys to find out postharvest losses in various agriculture commodities. Recent study was held in 2022 by NABARD Consultancy Service Pvt. Ltd. (NABCONS) entitled "Study to Determine Post-Harvest Losses of Agri Produce in India"—as per this study, the losses are 3.89–5.92% in cereals, 5.65–6.74% in pulses, 2.87–7.51% in oil seeds, 6.02–15.05% in fruits, 4.87–11.61% in vegetables, and 1.29–7.33% in cultivated crops and spices (MoFPI 2022). There are so many causes behind postharvest loss. The stage at which postharvest loss occurs depends on location, the supply chain, and a number of other factors (Fig. 14.1). Among all other factors, postharvest pathogens have been identified as

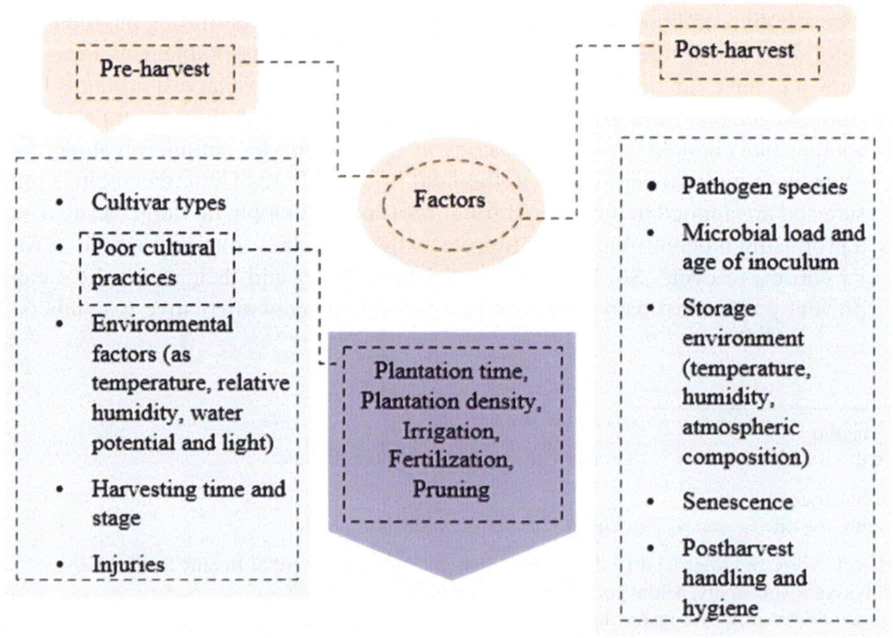

Fig. 14.1 Factors affecting postharvest disease

the major one responsible for PHL. Yearly 40–60% of postharvest losses introduced due to the disease are caused by pathogens (Tripathi et al. 2022). Fungal pathogens are the main culprits behind the agricultural crop loss (Table 14.1). Fungal pathogens like *Aspergillus*, *Alternaria*, *Geotrichum*, *Monilinia*, *Penicillium*, *Fusarium*, *Gloeosporium*, *Mucor*, *Botrytis*, and *Rhizopus* are the most frequent contributor in developing postharvest disease in the storage place of grains, fruits, vegetables, cereals, and pulses (Oufensou et al. 2023). Though fungi are the main culprit behind most of the postharvest diseases, it also has been proven that they have the immense potential to control pest and pathogens. In many studies, endophytic fungi are documented as promising biocontrol agent of several postharvest diseases. Considering the adverse effects of chemical fungicides on multidrug resistance, climate change, and human health, they can be a potential substitute of chemical fungicides during preharvest and postharvest storage conditions (Kumari et al. 2021; Tripathi et al. 2022). Endophytic fungi are predominant in nature that reside asymptomatically within the tissues of plants. They protect the plants from several biotic and abiotic stresses (Santra and Banerjee 2023). Directly and indirectly, they can build up interspecies interactions with pathogens. Direct mechanisms involve competition, antimicrobial effects, and parasitic relationships, which are facilitated by the production of various types of bioactive metabolites, volatile compounds, or enzymes, as well as by indirect mechanisms through inducing resistance in plants that help them to escape pathogen invasion (Akram et al. 2023). Endophytes exhibit a symbiotic relationship with plants and evolved together over time. That help them to share some valuable genes which leads to the production of metabolites either similar or superior to plant's metabolites. These may be processes due to the vertical or horizontal gene transfer across two different but closely related species (Santra and Banerjee 2023). These fungal endophytes are well-suited for agricultural applications due to their explicit method of delivery, improved formulation, overexpression of endogenous toxins, pre-characterized pathogenic strains, and simple engineering techniques (Senthil et al. 2011; Berkodia et al. 2018).

This chapter deals with the role of endophytic fungi in preventing postharvest diseases and several strategies by which endophytic fungi eliminate their targeted pathogens.

14.2 Eco-Friendly Approaches to Control Postharvest Fungal Diseases

The proven toxicity of chemical fungicides to human health increases the need to develop and explore alternative routes of postharvest disease control. There are some eco-friendly strategies to prevent postharvest disease, which are mentioned in Fig. 14.2 and discussed below. Among them, biocontrol is the most safe and efficacious.

Table 14.1 List of fungal pathogens of different postharvest produce

Pathogenic fungi	Host	Symptoms	Reference
Alternaria tenuis, *Botrytis cinerea*	Peach (*Prunus persica* L.)	Rot and decay	Zhang et al. (2019)
Penicillium digitatum	Tangerine (*Citrus reticulata* Blanco)	Rot	Yang et al. (2019)
Aspergillus flavus, *A. parasiticus*	Rice grains	Discoloration, destruction, off-flavors, and off-odors of grains	Douksouna et al. (2019)
Talaromyces minioluteus	Quince, tomato, orange fruit, onion bulbs	Brown color watery spots, blue/green sporulation on host's surface	Stošić et al. (2020)
Talaromycesminioluteus, *T. rugulosus*, *Penicillium italicum*	Pear fruit (*Pyrus communis*)	Brown color round spots, sometimes sunken spots, watery tissue, and discoloration on host surface	Stošić et al. (2021)
Neoscytalidium dimidiatum	Guava (*Psidium guajava* L.)	Small, circular brown spots (5–20 mm in diameter) on host surface. Gradually softening and rotting occur	Ismail et al. (2021)
Lasiodiplodia theobromae	Passion fruit (*Passiflora edulis* Sims)	Irregular, brown to black color watery lesions appear	Zhang et al. (2021a)
Diaporthe hongkongensis	Peach (*Prunus persica* (L.) Batsch)	Brown necrotic spots appear. Fruit rot takes place	Zhang et al. (2021b)
Diaportheeres	Hardy kiwifruit (*Actinidia arguta*)	Small, soft, chlorotic spots to light brown lesions and then watery brown lesions	Liu et al. (2021)
Lasiodiplodia theobromae	Guava (*Psidium guajava* L.)	Fiest irregular, brown color, watery lesions appear upper side then spread to the peduncle. Later the whole fruit surface covered with numerous black pycnidia and grayish mycelium	Zee et al. (2021)
Penicillium expansum, *P. polonicum*	"Golden delicious," apple (*Malus domestica*)	Blue mold lesion and decay	Bartholomew et al. (2022)
Colletotrichum gloeosporioides	Avocado	Pepper spots and blackish spots with raised margin appear first as infection proceeds; they coalesced	Geremu and Kidane (2022)
Gilbertellapersicaria	Pitahaya (*Hylocereus* spp.)	Light brown color watery lesions appear and later spread over the fruit	Carrillo-Fasio et al. (2022)

(continued)

Table 14.1 (continued)

Pathogenic fungi	Host	Symptoms	Reference
Rhizopus arrhizus	Garlic bulb	Soft water-soaked lesions on the surface and brown and soft in interior	Zhang et al. (2022a)
Fusarium citri	Chayote (*Sechium edule*) var. nigrum spinosum	Watery brownish lesions covered by white mycelium	Montecinos-Pedro et al. (2022)
Alternaria alternata	Pepino (*Solanum muricatum*)	Small brown spots gradually changed to irregular black or deep brown lesions; later gray-white mold noticed in the center of the lesions	Chen et al. (2022)
Fusarium concentricum	Nanfeng tangerine (*Citrus reticulata* Blanco)	Dark-brown lesions noticed in fruit center and pulp causing total fruit rot	Xiao et al. (2022a)
Diaporthefusicola	Peach (*Prunus persica* (L.) Batsch)	Circular, rotten, pale brown to brown, necrotic lesions, surrounded with small spore masses and white hyphae	Xiao et al. (2022b)
Fusarium tricinctum	Navel orange (*Citrus sinensis* (L.) Osbeck)	The pedicles became brown color; softening and yellowish-brown lesions appear on surface	Yang et al. (2022)
Colletotrichum fructicola	Pineapple (*Ananas comosus* (L.) Merr.)	Causing rot and leaf-tip dieback	Armand et al. (2023)
Rhizopus stolonifer	Jackfruit (*Artocarpus heterophyllus*)	Premature soft rot	Haque et al. (2023)
Phytophthora nicotianae	Navel orange (*Citrus sinensis* Osbeck cv. Newhall)	Circular, water-soaked, pale brown to brown, slightly sunken lesions noticed first, later lesions turned into dark-brown color and cover the fruit surface	Xiao et al. (2023)
Alternaria alternata	Persimmon (*Diospyros kaki* L. cv. Yangfeng)	Small circular dark-brown spots appear and rotting takes place	Xiang et al. (2023)
Fusarium graminearum	Ponkan (*Citrus reticulata* Blanco cv. Ponkan)	Brown necrotic spots spread to the entire fruit and the fruit rotted	Wang et al. (2023)
Diaporthe unshiuensis	Gannan navel orange (*Citrus sinensis* Osbeck cv. Newhall)	Small circular, light brown, watery spots with slightly indented edges appear and rotting takes place	Yuan et al. (2023)

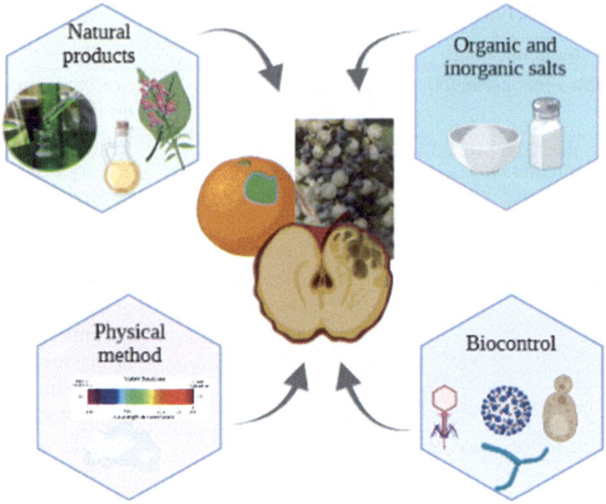

Fig. 14.2 Eco-friendly strategies for postharvest disease prevention

14.2.1 Biocontrol

Biological control, often known as biocontrol, is defined as the use of biological agents like bacteria, fungi, actinobacteria, and viruses to combat harmful organisms (including pathogens, pests, and weeds) for a variety of purposes that benefit humans (Stenberg et al. 2021). Biocontrol of postharvest diseases of agricultural products implementing antagonistic microorganisms is regarded as one of the most effective alternatives to conventional synthetic fungicide application. The application of microorganisms to control pre- and postharvest pathogens is often evaluated in two ways: spraying the crop with an antagonist suspension and submerging the crop in an antagonist suspension. Some pathogens attack at an early stage and remain quiescent until the environment is advantageous for their growth; in such cases, preharvest application of the antagonist is very effective in disease control; obviously, the environment must be favorable for the antagonist to establish and grow. As a result, the antagonist may be able to regulate the infection in its early phases (Dukare et al. 2019; Ji et al. 2020). However, in some studies, the postharvest treatment of antagonists is referred to as more effective in controlling disease because of its stable environments (Guigón-López et al. 2021; Sukmawati et al. 2021). Most often utilized microbial antagonists against postharvest pathogens are *Streptomyces griseoviridis* against bacterial soft rot, gray mold, *Phytophthora*; *Bacillus subtilis* against *Botrytis* spp., *Penicillium* sp., and *Rhizopus stolonifera*; *Coniothyrium minitans* against *Sclerotinia* sp.; *Trichoderma* sp. against *Fusarium*, *Rhizoctonia*, *Pythium*, and *Phytophthora*; *Trichoderma harzianum* against *Botryodiplodia theobromae* (stem end rot), *Colletotrichum gloeosporioides* (anthracnose), and *Gliocephalotrichum microchlamydosporum* (brown spot); and *Trichoderma atroviride* against *Botrytis cinerea* (Sivakumar et al. 2000; Punja et al. 2016; Tripathi et al. 2022). In a comparison with epiphytes and soil microbes, endophytes especially endophytic fungi have received less attention in postharvest fungal disease prevention. However, in recent years, its significance has been emphasized. In many studies, the role of endophytic fungi in increasing plant growth and decreasing various abiotic and biotic stress has been already established. As a result, the number of studies on postharvest disease resistance by endophytic fungi is expanding both at home and abroad. This chapter will go more into the role of endophytic fungi in postharvest disease prevention (Huang et al. 2021).

14.2.1.1 Endophytic Fungi as Biocontrol Agent of Postharvest Fungal Disease

Endophytes are endosymbionts that colonize the endosphere of plants during their life cycle. Endophytes are more beneficial as compared to epiphytes in the field of biocontrol. The plant provides shelter and nourishment to endophytic microorganisms, and they grow in less competitive environments within plant tissues and safeguarding the plant interior against phytopathogens (Hardoim et al. 2015). Endophytes are grouped into fungi, bacteria, and actinomycetes. The majority of endophytic fungi are Ascomycota, but some belong to Basidiomycota, Zygomycota, and Oomycota. All of these fungi silently and symbiotically live in the fruits, flowers, leaves, and stems of the plant without any external indications (Sinclair and Cerkauskas 1996). The diversity of endophytic fungi is different between plant species and even between varieties (Munir et al. 2020). Endophytes' behavior in plants can vary depending on a variety of circumstances, including genotype, surrounding environment, and complicated interactions with plants (Hardoim et al. 2015). Endophytes produce numerous kinds of volatile organic compounds and diffusible substances. During the primary and secondary metabolism, hydrocarbons (low molecular weight), alcohol, lactones, aldehydes, peptides, and inorganic volatile compounds such as hydrogen cyanide (HCN) are secreted by endophytes. Some of these compound act as signaling molecules that stimulate the activity of other microorganisms within the same ecological niche and proved advantageous as it enhances productivity and provides protection to plants (Shweta et al. 2013). Table 14.2 provides the list of bioactive compounds produced from endophytic fungi having potential to control the postharvest pathogens (Table 14.2).

14.2.1.1.1 Screening of Endophytic Fungi Having Biocontrol Potential

Endophytes with high antagonistic activity are better agents for developing microbial endophyte-based biocontrol that work both in the lab and in the field. That's why screening procedure is vital to identify the most effective one. Here we discuss some methods of screening and detecting endophytic fungi with maximum biocontrol potential (Fig. 14.3). Primarily antagonistic fungi are screened by dual culture assay. In the dual culture method, a mycelial disk of endophytic fungi and a mycelial disk of pathogenic fungi are placed in a petri plate at different locations and cocultured. The distance between two mycelial discs is 3 cm. Petri dishes are kept at 26 °C and measurements are recorded after 7–10 days. After incubation period, the growth diameter of the pathogen is measured between control and treatment. The decrease in diameter growth compared to the control is calculated. A higher percentage of inhibition means more antagonistic activity by endophytic fungi (Elad

Table 14.2 List of endophytic fungi produced active compounds and their use in different postharvest disease management

Endophytic antagonistic fungi	Mechanism of action	Postharvest pathogen	Postharvest commodity	Name of diseases/ symptoms	References
Hypoxylonanthochroum	Volatile organic compound (VOC) eucalyptol	*Fusarium oxysporum*	Cherry tomatoes	Fruit rots	Macías-Rubalcava et al. (2018)
Trichoderma spirale	Cell wall-degrading enzymes (glucanase, chitinase), VOC (6-pentyl-2H-pyran-2- one)	*Corynespora cassiicola*	Lettuce	Leaf spot	Baiyee et al. (2019)
Yarrowia lipolytica	Defensive enzymes like polyphenol oxidase (PPO), peroxidase (POD), catalase (CAT), phenylalanine ammonia-lyase (PAL), flavonoid compounds, and phenols	*Penicillium digitatum* and *P. Italicum*	*Citrus reticulata* (mandarin oranges)	Green mold, blue mold	Zhu et al. (2019)
Candida intermedia	VOCs (1,3,5,7-cyclooctatetraene and 3-methyl-1-butanol)	*Botrytis cinerea* Pers.	Strawberry	Fruit rot	Huang et al. (2011)
Trichoderma atroviride	VOCs (3-methyl-1-butanol, 2-methyl-1-propanol, and acetoin)	*Phytophthora infestans*	Potato tubers	Late blight of potato	Elsherbiny et al. (2020)
Pichia kudriavzevii KKP *Pichia occidentalis* KKP *Meyerozyma guilliermondii*	VOCs (ethyl esters, phenylethyl alcohol)	*Mucor* spp. *Penicillium chrysogenum*, *P. expansum*, *Aspergillus flavus*, *Fusarium cereals*, *F. poae*, *Botrytis cinerea*	Rye grains, grape berries	Molds	Choińska et al. (2020)
Trichoderma piluliferum	Lytic enzymes (chitinase, protease glucosaminidase, glucanase)	*Colletotrichum musae*	Banana	Anthracnose	da Costa et al. (2021)
Wickerhamomyces anomalus	Defensive enzymes (PPO, POD, CAT, PAL)	*Botrytis cinerea*	Cherry tomatoes (*Lycopersicon esculentum* L.)	Gray mold	Raynaldo et al. (2021)

Rhodotorula glutinis	Degrades and/or metabolizes aflatoxin B_1	*Aspergillus* sp.	Wheat grains	Seed quality↓, mycotoxins↑	Solanki et al. (2021)
Trichoderma afroharzianum	Phenyl ethyl alcohol	*Fusarium oxysporum*, *F. proliferatum*	Chili fruits	Postharvest decay	Khruengsai et al. (2021a)
Daldinia eschscholtzii	Elemicin, benzaldehyde dimethyl acetal, ethyl sorbate, methyl geranate, trans-sabinene hydrate, 3,5-dimethyl-4-heptanone	*Colletotrichum acutatum*	Strawberry fruits	Anthracnose disease	Khruengsai et al. (2021b)
Meyerozyma guilliermondii	Defense-related enzyme, antioxidant increased in host. Cell wall-degrading enzyme activity decreased in pathogens	*Botryosphaeria dothidea*, *Diaporthe actinidiae*	Kiwifruit	Soft rot	Pan et al. (2022)
Saturnisporadiversa	POD and PPO and chitinase	*Colletotrichum gloeosporioides*	Loquat fruit	Anthracnose	Yan et al. (2022)
Aspergillus aculeatus GC-09	Reactive oxygen species (ROS) increased in host. CAT and POD decreased in pathogen	*Penicillium italicum*	Citrus fruits	Blue mold	Zhang et al. (2022b)
Aureobasidium pullulans	Z-3-hexen-1-ol, benzyl alcohol and nonanal	*Colletotrichum acutatum*	Olive	Anthracnose	Sdiri et al. (2022)
Wickerhamomyces anomalus	2-Phenethyl acetate and 1-butanol, 3-methyl-	*Penicillium expansum*, *Botrytis cinerea*	Kiwifruit	Blue mold, gray mold	Zhao et al. (2023)

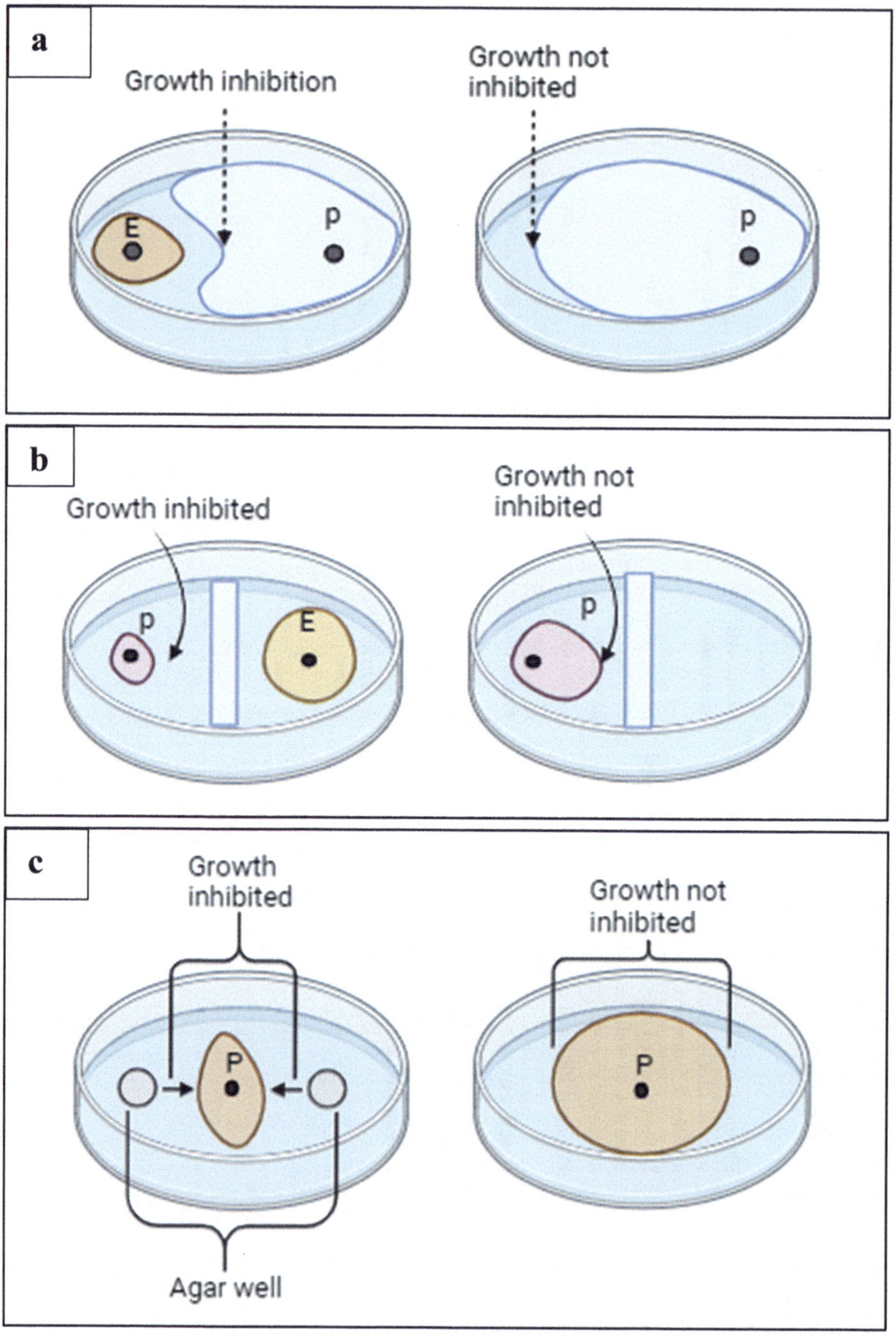

Fig. 14.3 Screening procedure of antagonistic fungi; (**a**) Dual culture assay; (**b**) Split plate method for VOC test; (**c**) Anti-fungal test by ager well diffusion method; P- Pathogen, E- Endophyte

et al. 1980; Sivakumar et al. 2000). Next, we can detect antifungal activity of culture filtrate by agar well diffusion method. Initially, the pathogen is grown on PDA plates at 24 °C for 2–4 days. Then two holes in the PDA plates are made by cork-borers at a distance from pathogen inoculation site. The holes are filled with culture extracts from endophytic fungi, and the plates are then placed in an incubator at 30 °C for a period of 3–5 days. Following the incubation, any observable clear zone around the wells is noted as an indication of inhibition. Any standard antifungal compound can be used for control plates. More inhibition zone means more activity (Santra and Banerjee 2023). To assess the production of volatile components by endophytic fungi, split plate method is used. In case of split plate method, agar is cut through the center of a normal petri plate of PDA, resulting in two halves of agar. One-half is inoculated with the endophytic fungus and incubated at 24 °C for 4–8 days. On the fourth day, the other half is inoculated with the test pathogenic fungus. The petri plates are then kept in incubation for 3–6 days at 24 °C after carefully wrapping with parafilm. A control plate without the fungal endophyte is also prepared. After 6 days of incubation, the growth of fungal pathogens is observed, and the level of inhibition (%) is calculated by comparing it to the control plate. Pathogen growth inhibition means positive in volatile compounds assay (Mitchell et al. 2009; Santra and Banerjee 2023). Mycoparasitism testing is another way for determining antagonistic activity of fungi. In this test, a sterilized glass slide is placed within a petri dish on two L-shaped glass bars. The PDA media is then cut into squares and placed on a sterile glass slide, after which pathogens and endophytes are placed on two opposite sides of the square and a cover slip is placed on top. Petri dishes must be kept at 28 °C. Take over the cover slip and add lactophenol as soon as the mycelia of the fungi reach each other. Then, under a light and polarized microscope, fungal interactions in the contact zones of the two colonies is seen (Fenice and Gooday 2006). Further screening of antioxidant activity, various hydrolytic enzyme activities like protease, chitinase, and glucanase in endophytic fungi could be done.

14.2.1.1.2 Action Mechanisms of Endophytic Fungi in Preventing Fungal Pathogen on Postharvest Products

Endophytes have been shown to have a variety of defense mechanisms against phytopathogens, including direct competition and alterations in the molecular structure of the host plant (Fig. 14.4). Endophytes as biocontrol agents for postharvest pathogens is a recent area of study that requires an in-depth literature analysis to unravel the hidden molecular mechanisms used against these pathogens. Endophytes may use the following techniques to resist pathogenic attacks on harvested commodities.

14.2.1.1.2.1 Direct Competition for Space and Nutrients

In an interaction between host-pathogen-endophyte, the host is limited of nutrient and space. In this interaction, the survival and the growth of both endophytes and pathogens is dependent on various types of crucial macro- and microelements (Kumari et al. 2020). Endophytes rapidly occupy exposed fruit surfaces due to rapid

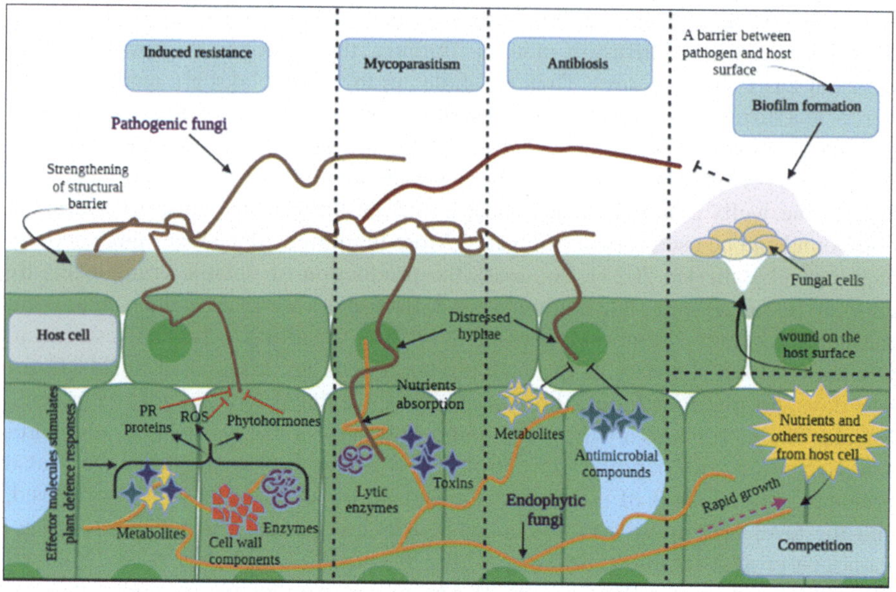

Fig. 14.4 Action mechanisms of endophytic fungi to control pathogenic fungi in postharvest products; PR proteins, pathogenesis-related proteins

growth and colonization and outcompete pathogens for space and utilization of nutrient resources (Spadaro and Droby 2016). Like when yeast cells come into contact with wound area of the fruit surface, they divided multiple times and grow rapidly by utilizing nutrients, leading to a depletion of resources that would otherwise support the growth of pathogens (Liu et al. 2013). Interactions between *Pichia guilliermondii* and *Botrytis cinerea* in apple and *Colletotrichum* spp. on pepper were found to be involved in sugar and nitrate competition (Chanchaichaovivat et al. 2008; Spadaro and Droby 2016).

14.2.1.1.2.2 Competition for Iron and Production of Siderophore

Iron is essential for growth, survival, and pathogenicity of a pathogen. Endophytes produce siderophores, a secondary metabolite that can combine with available iron to create a rigid and secure$[Fe\ (H_2O)_6]^{3+}$ complex (Miethke and Marahiel 2007). Numerous studies have shown that siderophore-producing microorganisms are essential for disease resistance. Endophytes have fastest colonizing ability than pathogens in their fight for survival and excrete a range of siderophores that chelate available iron consequently depriving the opposing pathogen of its iron source (Chowdappa et al. 2020; Sood et al. 2020). The yeasts *Metschnikowia pulcherrima* and *Moniliniafructicola*, which produce the siderophore pulcherrimin, showed effective results in biocontrol of pathogens of apple (*Botrytis cinerea*, *Alternaria alternata*, and *Penicillium expansum*) after harvesting (Saravanakumar et al. 2008). *Trichoderma* spp. is recognized for its biocontrol ability against the pathogens of citrus and apple fruits after harvesting by the production of hydroxamate siderophore (Sood et al. 2020). Although endophytic *Trichoderma* spp. is still in its early stages for postharvest fruit disease control, it may provide a new and sustainable way to control postharvest fruit diseases. By the proper optimization of different factors responsible for siderophore production, we can enhance the efficacy of an endophytic fungi against postharvest pathogens (Kumari et al. 2022).

14.2.1.1.2.3 Antibiosis

Pathogens can be directly inhibited by substances produced by biocontrol agents, which is referred to as antibiosis (Di Pietro 1995). Various bioactive metabolites produced by endophytic fungi can suppress competing microorganisms (Thines et al. 2004). The diversity of endophytic fungi within a plant can trigger the production of numerous secondary metabolites by both which inhibits the growth of pathogens on the host (Kusari et al. 2012). In most of the cases, the endophytes and host involve in same pathways and collaborate to produce some metabolites, promote one other's metabolism, or metabolize one other's products (Kusari et al. 2013; Ludwig-Müller 2015). A notable instance is the identification of the anticancer drug Taxol, derived from endophytic fungi *Taxomyces andreanae*, which was isolated from the inner bark of the Pacific yew tree *Taxus brevifolia* (Stierle et al. 1993). Meanwhile, Heinig et al. (2013) came to the conclusion that a lot of endophyte strains weren't truly independent producers. Different bioactive compounds isolated from endophytic fungi having antimicrobial activity are of different chemical nature such as phenols, steroids, alkaloids, quinones, flavonoids, and terpenoids. Some antimicrobial compounds are volatile, and most of them are terpenoids and polyketides (Latz et al. 2018). A variety of substances have also been tested for their inhibitory effect against phytopathogens. Our focus should be on cultivating fungi on a large scale to extract their antimicrobial compounds, with the goal of commercializing these products as biocontrol agents (Schulz et al. 2002; Thines et al. 2004; Daguerre et al. 2017).

14.2.1.1.2.3.1 Alkaloids

Natural products like alkaloids are the key source of antimicrobial compounds. Numerous alkaloid substances, such as acylamine, pyridine, pyrrole, organic amine, quinoline and iso-quinoline, and quinazoline, can be produced by endophytes (Ma and Feng 2009; Bernardi et al. 2019). Some alkaloids fumigaclavine C and pseurtotin A, produced by the endophytic fungus *Aspergillus* sp. EJC08, exhibit strong antimicrobial activity (Pinheiro et al. 2013). Endophytes from *Melia azedarach* bark produce alkaloid compounds with antimicrobial properties that effectively combat postharvest pathogens. These endophytic fungi not only suppress *Fusarium* pathogens but also demonstrate inhibitory effects on other harmful fungi like *Botrytis cinerea*, *Alternaria solani*, *Alternaria alternata*, *Colletotrichum gloeosporioides*, and more (Li et al. 2012).

14.2.1.1.2.3.2 Volatile Organic Compounds (VOCs)

Volatile organic compounds (VOCs) are minute compounds with a molecular weight below 300 Da, capable of evaporating under normal atmospheric pressure and temperature. The majority of VOC mixtures primarily include ester, aromatic and aliphatic hydrocarbons, alcohols, aldehydes, nitrides, and sulfides, all of these possess potent antimicrobial properties (Strobel 2011; Lemfack et al. 2018). VOCs produced by endophytic fungi exhibit significant antimicrobial activity, and that's way those can be used as biocontrol agent against fruit decay. *Daldinia eschscholtzii* MFLUCC 19–0493, an endophytic fungus derived from *Barleria prionitis* plants grown in Thailand, displayed the antifungal activity against the pathogen *Colletotrichum acutatum* on strawberry fruits. The prominent compounds identified in the VOCs were ethyl sorbate, elemicin, benzaldehyde dimethyl acetal, trans-sabinene hydrate, methyl geranate, and 3,5-dimethyl-4-heptanone (Khruengsai et al. 2021a, b). The VOCs produced by *Muscodor suthepensis* strain CMU-Cib462 exhibit complete inhibition of *Penicillium digitatum* mycelial growth in laboratory conditions (Suwannarach et al. 2016). Similarly, VOCs from *M. albus* are effective biocontrol agent of decay in apples and peaches (Mercier and Jiménez 2004). Furthermore, some endophytic fungi such as *Trichoderma* and *Oxyporus latemarginatus* also produce VOCs that exhibit inhibitory effects on postharvest pathogens (Lee et al. 2009; Ajith and Lakshmidevi 2010).

14.2.1.1.2.3.3 Antifungal Metabolites

Fusarium solani-F4–1007, an endophyte, was isolated from the medicinal plant *Solenostemma argel*. This endophyte was tested for antifungal activity against *Cochliobolus spicifer* causal organism of rot disease in okra (*Abelmoschus esculentus*). Mainly terpenoid and phenolic compounds were identified from the metabolites extracted from endophytic fungus which inhibit the sporulation and conidia formation of pathogenic fungus (Tayung et al. 2010). *Paecilomyces* sp., an endophyte, which exhibited remarkable antagonistic activity against *Rhizoctonia*, a pathogen responsible for causing a 30% loss in potato yield, was isolated from *Moringa oleifera* leaves. Using gas chromatography-mass spectrometry (GC-MS), many bioactive components like methyl ester; 9, cis-13-octadecenoic acid; 1-nonadecene; cyclotetracosane; 1-heptacosanol; behenic alcohol; 1,2-benzenedicarboxylic acid; di-sec-butyl phthalate; 1-hexadecanol; butyl 2-methylpropyl ester; 1-nonadecene; *n*-nonadecanol-1, oleic acid, *n*-heptadecanol-1; dodecanoic acid; *n*-pentadecanol; 2,3-dihydroxypropyl ester; and 9-octadecenal were identified from the extract of this endophyte (Hawar et al. 2023).

14.2.1.1.2.4 Mycoparasitism and Production of Lytic Enzymes

Endophytic fungi utilize mycoparasitism as a crucial mechanism to combat pathogenic fungi, by the production of lytic enzymes. These lytic enzymes play crucial role to demise fungal propagules completely or in the breakdown of their structures. The mycoparasitism encompasses several steps. First interaction between the antagonist and pathogen. Then mutual recognition occurs, therefore antagonist release their lytic enzymes that helps them to penetrate into the host and grow actively within the host tissue. Ultimately the parasite exits the host after completing the parasitic cycle (Spadaro et al. 2004; Talibi et al. 2014). Some chemicals act as signaling compounds such as lectins that help the antagonist to recognize its host—further, antagonistic fungi release lytic enzymes such as proteinases, chitinases, lipases, and β-1,3-glucanases to degrade cell wall for successful penetration (Vos et al. 2015). Due to enzymatic activity, mycelial deformation, and lysis, cellular deformities, cytological damage takes place. As a result membrane permeability of cells increase and cytoplasmic matter ooze out. *Talaromyces acidophilus* AUN-1 emerge as potential antagonist against *B. cinerea*. It secretes lytic enzymes like chitinase, protease, and lipase (Abdel-Rahim and Abo-Elyousr 2018). *Choiromyces aboriginum*, an endophyte, effectively restrained the postharvest pathogen *Pythium* sp. by generating b-1,3-glucanases. These enzymes played a crucial role in breaking down the pathogenic cytoplasm that was entwined around the hyphae of the fungus (Cao et al. 2009). *Trichoderma* is a very effective fugal genus in inhibiting several phytopathogens through parasitism. In a study, *Trichoderma* sp. inhibits the fungal pathogen *Fusarium oxysporum* by producing lytic enzymes resulting in hyphal coiling of the fungal pathogen (Rajani et al. 2021).

14.2.1.1.2.5 Production of Killer Toxins and Some Bioactive Compounds Derived from Fatty Acids

Some yeast produces killer toxins which are often glycosylated proteins, which can disrupt some crucial fungal cell wall constituents such as β-1,3-D-glucans, β-1,6-D-glucans, glycoproteins, and chitin; as a consequence, lysis of fungal cells occurs (Mannazzu et al. 2019; Villalba et al. 2020). The working process of killer toxins is to first bind with the cell membrane—next, they established the interaction with secondary receptors present on cell membrane leading to the alterations of the permeability of cell membrane, and then they inhibit the synthesis of DNA, disrupt the cell cycle, and fragment the RNA molecules (Belda et al. 2017; Crabtree et al. 2019). In a scientific study, it was revealed that the genes behind the production of killer toxins in *Saccharomyces cerevisiae* are inherited from the cytoplasm. These genes are present in double-stranded linear RNA (dsRNA) plasmids, and these plasmids are encapsulated to form noninfectious viruslike particles (VLP) within the cell cytoplasm (Nally et al. 2015). In acidic environment, production of killer toxin increases with increasing pH. The production of killer toxin severely affect the yeast cells (Abu-Mejdad et al. 2020).

14.2.1.1.2.6 Modulating the Redox Homeostasis of Harvested Commodities

Most of all postharvest pathogens modulate redox homeostasis in host body, and they develop ROS scavenging system to overcome host defense responses, and additively they produce fungal ROS to damage host cell more. For example, *Penicillium digitatum*, a very well-known pathogen of harvest produce, secrete catalase enzyme that scavenges hydrogen peroxide that helps them to escape from host defense and to establish successful infection (Macarisin et al. 2007). Endophytes help their host to combat various types of biotic and abiotic stresses (White et al. 2019). However, the role of endophytic fungi in modulating stress is not much explored yet. In certain studies, fungi such as *Paraburkholderia phytofirmans* PsJN were found to upregulate the activity of genes associated with the removal of reactive oxygen species (ROS). This led to the detoxification of ROS produced by fungal pathogens and modulates the signaling pathways. Endophytes play a role in assisting plants to mitigate biotic stress by reducing lipid peroxidation and limiting the accumulation of proline (Spadaro and Droby 2016; Pacifico et al. 2019). There is considerable literature on the interactions among plants, pathogens, and endophytes—additional investigation is needed to understand the involvement of endophytes in managing the redox homeostasis in stored fruits. This particular domain requires focused attention to comprehend the potential advantages of endophytes in preserving and sustaining the quality of fruits.

14.2.1.1.2.7 Biofilm Adhesion

Formation of biofilm is the very advantageous quality of fungal pathogens to evade immune response and for further infection (Zhao et al. 2020; Eladawy et al. 2021). In such types of pathogenicity, yeast is a very good option to control pathogens. Yeast forms biofilm on wounded area—this is how they block the connection between pathogen and host, and because of low nutrient availability, pathogen can't grow properly (Spadaro and Droby 2016; Lei et al. 2022). A study analyzed the biofilm-forming ability of endophytic yeast, *Vishniacozyma victoriae*, isolated from kiwifruit, against the phytopathogen *Botrytis cinerea*. And the result was positive—yeasts form biofilm against *Botrytis cinerea* (Nian et al. 2023). Similarly, Lei et al. (2022) confirmed that the yeast *Debaryomyces nepalensis* having antagonistic potential against the pathogen causing black spot in jujube fruits after harvesting can form biofilm to hinder the growth of that pathogen. This finding is a proof of the potential role of biofilm in controlling postharvest pathogens in fruits.

14.2.1.1.2.8 Modulation and Synthesis of Phytohormones

Endophytic microorganisms can produce diverse types of phytohormones. These phytohormones play a significant role in enhancing the natural defense mechanisms of postharvest plants against biotic stresses, thereby aiding them to prevent wilting caused by stress (Ali et al. 2017). Notably, endophytes not only synthesize plant hormones but also have the capability to regulate different metabolic pathways in plants related to stress. *Piriformospora indica*, an endophyte, induces the production of phytohormone such as cytokinin, ethylene, auxins, gibberellins, jasmonate, salicylic acid (SA), and brassinosteroids. These hormones contribute to increase the stress tolerance ability of higher plants (Xu et al. 2018).

14.2.1.1.2.9 Induced Resistance

Induced resistance is defined as the phenomenon where plants develop resistance against disease-causing pathogens as a result of being triggered or stimulated by external factors. After harvest, fruit and vegetables' ability to resist pathogens is reduced as a result of the impact of both their internal and exterior environments. So, they can be treated with antagonistic endophytes in postharvest condition to increase their resistance (Ali et al. 2018). Endophyte treatment can promote the expression of proteins related to pathogenesis (PR proteins) and boost the enzymes related to defense of the host. Yeasts can function indirectly to resist fungal pathogens in the host by inducing resistance. When a fungus first infiltrates into a host tissue, the host begins to exhibit a hypersensitive reaction (HR) in that area, and gradually that area becomes dead, which helps to restrict the pathogens from being spread to the remaining healthy parts of the host. HR can be induced by numerous substances known as inducers, including synthetic products, plant pathogens, non-harmful microorganisms, ultraviolet radiation, insects, and many more. When exposed to an inducer, the plant exhibits an upregulation of genes and enzymes associated with its defense mechanisms, which produces various substances, including protein which is related to pathogenesis; phytoalexins like coumarins, flavonoids, and diterpenes; reactive oxygen species (ROS); and cell wall-degrading enzymes like proteases, glucanases, and chitinases. These responses occur simultaneously and participates in plant's defense mechanism (Petrov et al. 2015; Ali et al. 2018; Nandakumar et al. 2021; Das et al. 2021). These substances demonstrate inhibitory effects and can cause necrosis in the cells of pathogenic fungus. PR proteins are the proteins that are generally absent in the host tissue, but if they are present, that is also very low concentration. When a pathogen attacks the host tissue, at that time PR proteins appear and their amount also increases considerably (van Loon et al. 2006). *Pichia membranifaciens* triggered the expression of PR-9 and PR-10 in peach fruit (Xu et al. 2008). Nevertheless, the responses of PR proteins can be highly variable depending on the specific host tissue and microbial stimuli. In apple, *Candida oleophila* increased the expression of PR-8 against *Botrytis cinerea* (Zhang et al. 2020). In a research conducted by Nantawanit et al., in 2010, they illustrated the stimulation of phytoalexin and lytic enzyme production in response to a phytopathogen in a host organism. The study disclosed that the application of *Pichia guilliermondii* to chili plants substantially increased the levels of lytic enzymes such as chitinase and glucanase and increased the activity of phenylalanine ammonia-lyase (PAL), along with the accumulation of capsidiol and phytoalexin in the tissues of chili plants (Nantawanit et al. 2010). PAL is the precursor of numerous substances belonging to phenylpropanoid pathway which are related to plant defense and also precursor of phenols, lignin, and phytoalexins (MacDonald and D'Cunha 2007). Furthermore, antioxidant enzyme activity can be increased by biocontrol yeast agents through generating ROS to reduce the damages caused by oxidation in response to pathogen invasion. Peroxidase (POD) activity increased following *P. membranaefaciens* treatment of cherries, whereas catalase (CAT) and superoxide dismutase (SOD) activity decreased (Chan and Tian 2006). Yeast antagonists simultaneously boost a number of resistance-inducing pathways. For instance, the number and actions of antioxidant enzymes and defense-related enzymes were increased by the presence of *Rhodosporidium* sp., *Pichia* sp., and *Cryptococcus laurentii* (Zhao and Yin 2018; Lai et al. 2018).

14.2.1.1.2.10 Modulation of Native Microbiota

The presence of endophytic microbes influences the inherent microbial diversity in leaves, fruits, roots, and soil, thereby playing a role in the establishment of a sustainable agricultural production system. This has considerable economic importance (Sturz et al. 2000; Baghel et al. 2020). Nevertheless, this part needed more scientific researches to understand the effects of endophytes on the native microbiota of harvested fruits. Many studies reported that the treatments done on the vegetables and fruits while being on and off the field have an impact on the surrounding microbiome (Wisniewski and Droby 2019). In a study, the impact of *Wickerhamomyces anomalus* was examined on the native microbiota of postharvest kiwifruit. The use of *W. anomalus* had a much bigger effect on the fungal community than the bacterial one (Shi et al. 2022; Zhao et al. 2023). It's possible that the used antagonistic strain was a fungus and that pathogenic fungi compete with the antagonistic yeast for nutrients and available space, which generates the antagonistic yeast's mechanism. In comparison with control and treated kiwifruits with yeast, it was noticed that the alpha diversity of the existing microbiota was much greater in control kiwifruit than the treated one. Analysis of this finding reveals that introduction of any biocontrol agents or pathogens frequently reduces microbial diversity and influences fruit health (Wu et al. 2019; Zhao et al. 2023).

14.2.1.1.2.11 Control of Mycotoxin Production

Mycotoxins are a key factor behind the major economic losses in case of most of the stored fruits. Antagonistic fungi produce mycotoxins like deoxynivalenol, alternariol, aflatoxin, and patulin, which are harmful to fruits as well as human (Bartholomew et al. 2021). Many studies have shown that the endophytes and their metabolites are capable to control mycotoxins in in vitro as well as in many crop plants (Abdallah et al. 2018). However, there are few researches available on endophytes' effects on postharvested fruits and vegetables. Endophytes should be applied prior to harvest in order to prevent mycotoxins from causing postharvest damage (Sarrocco and Vannacci 2018). To efficiently manage mycotoxins in stored fruits, endophytic fungi's VOCs can be added to edible biofilms or used as an ingredient in packaging elements (Mari et al. 2016).

14.3 Commercial Upscale Production and Hurdles Ahead

Endophytic fungi having antagonistic potential to control postharvest pathogens have emerged as a modern form of pesticides. For endophytes and their products to be successfully commercialized, however, a number of obstacles must be overcome. Although the microbial community has a lot of potential for preventing and controlling postharvest illnesses in fruits and vegetables, an appropriate design and effective application strategy are also necessary for improved impact.

The Derwent Innovation database listed 163 patents worldwide from 2009 to 2021 that included yeasts as biocontrol agent. 73.68% of the records among 163 patents have specified the genus and species name of the yeast, whereas the rest have only used the word "yeast" as one of its constituents. Furthermore, the majority of patents (84.21%) are held by companies, with Bayer® being the most dominant player. Academic institutions own a smaller percentage of patents (15.78%). This observation indicates limited involvement of academics from higher education institutions, potentially due to their primary focus on teaching and the lack of adequate resources for large-scale production and focused marketing efforts for the new products (Hernandez-Montiel et al. 2021).

To effectively use antagonistic endophytes against postharvest diseases and achieve higher production, it is crucial for these endophytes to possess stress tolerance, ensuring their extended shelf life and sustained antipathogenic activity. Therefore, a comprehensive screening process should be conducted to identify stress-tolerant endophytes and assess their stress amelioration potential both in vitro and in vivo. This step is essential to facilitate the transition of endophytes from the laboratory to field applications.

Certain endophytes establish a close association with their host plants, which assist them to tolerate stress and to produce various bioactive compounds. Consequently, it is important to evaluate their ability to withstand stress conditions independently, without the presence of host plants and in the postharvest fruit niche, before considering their commercial use (Khare et al. 2018).

Very few biocontrol agents are successful in field application against postnatal pathogens due to their volatility and unpredictability. But bioformulation stands out as a highly effective measure to address this issue. Bioformulation ensures that biocontrol agents retain their effectiveness over an extended duration and remain stable even in challenging environmental conditions (Carmona-Hernandez et al. 2019). Typically, inert carriers such as talc, coffee grounds, alginate, oil, and nanoparticles are used in the formulation to provide nourishment and stability to the antagonist. It is always possible that the antagonist's biocontrol abilities could be improved or hampered through formulation (White et al. 2021). The compositions are often applied as liquids or powders or pastes. However, the dried form offers an extended shelf life. Silica-based compounds are excellent formulation additives due to their hydrophilicity, anticaking properties, and wettability enhancement (Jones and Burges 1998). Boniprotect, Yield plus, Nexy, Avogreen, Aspire, Biosave, Bligh Ban, Candifruit, Shemer, Pantovital, Bioact, and Trichodex are examples the of biocontrol agents available in the market (Spadaro and Droby 2016; White et al. 2021).

Another successful technique is nanoencapsulation, which extends the shelf life of stored crops, vegetables, and fruits. It is also important to investigate nanoencapsulation of the products and nano emulsions for the continuous release of bioactive secondary metabolites from the endophytes (de Carvalho and Conte-Junior 2022).

The application of endophytes before harvest and after harvest is also an important aspect of disease management. Optimizing the application method is a crucial factor in achieving large-scale endophyte production for the management of postharvest disease. The range and impact of a single endophyte are limiting, but a mixture of various microbial agents is more effective and more able to withstand environmental pressure than a single species. Incorporation of specific organic factors like growth regulators, fungicides in low concentration, organic matter like chitosan, and inorganic salts can enhance the effectiveness of endophytes as the biocontrol agents (Huang et al. 2021; Kumari et al. 2022).

Endophyte-derived secondary metabolites and VOCs encounter challenges in economically scalable manufacturing and sustainable release. Several factors contribute to the extensive and continuous production of secondary metabolites from endophytes, including the types of strains, the nature of culturable media, and the application of metabolic engineering (Sah et al. 2020; Taritla et al. 2021).

Endophytes as new biocontrol agents of postharvest disease require close examination to ensure both consumer and environmental safety. Unfortunately, certain rhizobacteria that were previously investigated for their bioactive secondary metabolites have displayed pathogenic behavior to human or are coming out as environmental contaminants. To prevent encountering similar issues with endophytes, it is crucial to assess the safety of these substances in animal models and evaluate their environmental impact when subjected to elevated dosages. By conducting these assessments, we can ensure the responsible and secure use of endophytes in various applications while minimizing potential risks to human health and the environment (Keswani et al. 2019).

The most important factor in implementing endophytes as an alternative to traditional pesticides in controlling postharvest pathogens is to generate awareness among the users and the individuals in the distribution chain. To achieve this, continuous efforts should be made to organize outreach programs and workshops focusing on these novel concepts. Such initiatives will help disseminate knowledge and promote the adoption of endophyte-based biopesticides in practice Any attempt to deploy new technologies in the agricultural sector without the support of government, policymakers, and funding organizations is unsuccessful. The National Biocontrol Network Programme (NBNP) was previously introduced by the Department of Biotechnology (DBT), India, with the goal of popularizing and commercializing more than 30 biopesticides (Kumari et al. 2020; Kumari et al. 2022). Similar initiatives and funding should be made to promote the most affordable, efficient, and environmentally friendly products to manage postharvest disease.

14.4 Conclusion

Currently eco-friendly technologies are seeking more attention because of fungicide-resistant pathogens as well as the prevalence of hazardous chemicals in fruits and vegetables. The practice of integrative and biological methods for controlling postharvest disease has improved dramatically during the past few decades. Endophyte is an essential asset for developing solutions related to the safety and quality of postharvest commodities and control of foodborne pathogens. Compared to epiphytes, the impact of endophytes on plants such as growth promoting activity and plant defense has been extensively researched. Endophytes have remarkable ability to effectively initiate and stimulate the stress response of the host plant against various types of phytopathogens in a very efficient way. In order to effectively control postharvest fruit and vegetables, endophytes should be a key component of the control system. Genes associated with disease management can also be identified using next-generation omics. Modern molecular tools like proteomic, transcriptomic, genomic, and other OMIC methods help in the characterization of infections. The optimal antagonist against a certain phytopathogen can be chosen with the help of information on an antagonist's effectiveness and persistence.

Detailed study of microbial antagonist genomes helps to improve gene transcription investigations, discover gene families with unknown activities, compare genomes with those of other biological control agents, and provide insight into possible genes involved in biocontrol action. Still now this topic requires thorough investigation to achieve a comprehensive understanding. Subsequent research endeavors ought to explicitly establish the connections among endophytes, harvested fruits and vegetables, pathogens, and microbial communities. Additionally, it is important to explore the impact of antagonistic endophytes on postharvest quality and the commercial viability of products.

References

Abdallah FM, De Boevre M, Landschoot S, De Saeger S, Haesaert G, Audenaert K (2018) Fungal endophytes control *Fusarium graminearum* and reduce trichothecenes and zearalenone in maize. Toxins 10(12):493

Abdel-Rahim IR, Abo-Elyousr KA (2018) *Talaromyces pinophilus* strain AUN-1 as a novel mycoparasite of *Botrytis cinerea*, the pathogen of onion scape and umbel blights. Microbiol Res 212:1–9

Abu-Mejdad NMJA, Al-Saadoon AH, Al-Badran AI, Minati MH (2020) Optimum conditions of killer toxins produced by *Torulasporadelbrueckii* and *Wickerhamomycesanomalus* and their action as antifungal agents. Bull Natl Res Cent 44(1):1–9

Ajith PS, Lakshmidevi N (2010) Effect of volatile and non-volatile compounds from Trichoderma spp. against *Colletotrichum capsici* incitant of anthracnose on bell peppers. Nat Sci 8(9):265–269

Akram S, Ahmed A, He P, He P, Liu Y, Wu Y, Munir S, He Y (2023) Uniting the role of Endophytic fungi against plant pathogens and their interaction. J Fungi 9(1):72

Ali S, Charles TC, Glick BR (2017) Endophytic phytohormones and their role in plant growth promotion. In: Functional importance of the plant microbiome: implications for agriculture, forestry and bioenergy, pp 89–105

Ali S, Ganai BA, Kamili AN, Bhat AA, Mir ZA, Bhat JA, Tyagi A, Islam ST, Mushtaq M, Yadav P, Rawat S, Grover A (2018) Pathogenesis-related proteins and peptides as promising tools for engineering plants with multiple stress tolerance. Microbiol Res 212:29–37

Armand A, Hyde KD, Jayawardena RS (2023) First report of *Colletotrichum fructicola* causing Fruit Rot and Leaf-Tip Dieback on Pineapple in Northern Thailand. Plan Theory 12(4):971

Baghel V, Thakur JK, Yadav SS, Manna MC, Mandal A, Shirale AO, Sharma P, Sinha NK, Mohanty M, Singh AB, Patra AK (2020) Phosphorus and potassium solubilization from rock minerals by endophytic *Burkholderia* sp. strain FDN2-1 in soil and shift in diversity of bacterial endophytes of corn root tissue with crop growth stage. Geomicrobiol J 37(6):550–563

Baiyee B, Ito SI, Sunpapao A (2019) *Trichoderma asperellum* T1 mediated antifungal activity and induced defense response against leaf spot fungi in lettuce (*Lactuca sativa* L.). Physiol Mol Plant Pathol 106:96–101

Bartholomew HP, Bradshaw M, Jurick WM, Fonseca JM (2021) The good, the bad, and the ugly: mycotoxin production during postharvest decay and their influence on tritrophic host–Pathogen–Microbe interactions. Front Microbiol 12:611881

Bartholomew HP, Lichtner FJ, Bradshaw M, Gaskins VL, Fonseca JM, Bennett JW, Jurick WM (2022) Comparative *Penicillium* spp. transcriptomics: conserved pathways and processes revealed in ungerminated conidia and during postharvest Apple fruit decay. Microorganisms 10(12):2414

Belda I, Ruiz J, Alonso A, Marquina D, Santos A (2017) The biology of *Pichia membranifaciens* killer toxins. Toxins 9(4):112

Berkodia M, Joshi U, Rami NV, Wati L (2018) Endophytes: A hidden treasure inside plant. IJCS 6(5):1660–1665

Bernardi DI, das Chagas FO, Monteiro AF, Dos Santos GF, de Souza Berlinck RG (2019) Secondary metabolites of endophytic Actinomycetes: isolation, synthesis, biosynthesis, and biological activities. Prog Chem Org Nat Prod 108:207–296

Cao R, Liu X, Gao K, Mendgen K, Kang Z, Gao J, Wang X (2009) Mycoparasitism of endophytic fungi isolated from reed on soilborne phytopathogenic fungi and production of cell wall-degrading enzymes in vitro. Curr Microbiol 59:584–592

Carmona-Hernandez S, Reyes-Pérez JJ, Chiquito-Contreras RG, Rincon-Enriquez G, Cerdan-Cabrera CR, Hernandez-Montiel LG (2019) Biocontrol of postharvest fruit fungal diseases by bacterial antagonists: A review. Agronomy 9(3):121

Carrillo-Fasio JA, Nuñez-García PR, Salazar-Mesta RJ, Mora-Romero GA, Leyva-Madrigal KY, Cruz-Lachica I, Tovar-Pedraza JM (2022) First report of soft rot of pitahaya fruit caused by *Gilbertellapersicaria* in Mexico. Plant Dis 106(12):3218

Chan Z, Tian S (2006) Induction of H2O2-metabolizing enzymes and total protein synthesis by antagonistic yeast and salicylic acid in harvested sweet cherry fruit. Postharvest Biol Technol 39(3):314–320

Chanchaichaovivat A, Panijpan B, Ruenwongsa P (2008) Putative modes of action of *Pichia guilliermondii* strain R13 in controlling chilli anthracnose after harvest. Biol Control 47(2):207–215

Chen M, Jia MS, Li SC, Xiao LH, Wang YB, Peng WW, Chen J, Xiang ML (2022) First report of Postharvest Fruit Rot in *Solanum muricatum* Caused by *Alternaria alternata* in Southwest China. Plant Dis 106(9):2520

Choińska R, Piasecka-Jóźwiak K, Chabłowska B, Dumka J, Łukaszewicz A (2020) Biocontrol ability and volatile organic compounds production as a putative mode of action of yeast strains isolated from organic grapes and rye grains. Antonie Van Leeuwenhoek 113:1135–1146

Chowdappa S, Jagannath S, Konappa N, Udayashankar AC, Jogaiah S (2020) Detection and characterization of antibacterial siderophores secreted by endophytic fungi from *Cymbidium aloifolium*. Biomol Ther 10(10):1412

Crabtree AM, Kizer EA, Hunter SS, Van Leuven JT, New DD, Fagnan MW, Rowley PA (2019) A rapid method for sequencing double-stranded RNAs purified from yeasts and the identification of a potent K1 killer toxin isolated from *Saccharomyces cerevisiae*. Viruses 11(1):70

da Costa AC, de Miranda RF, Costa FA, Ulhoa CJ (2021) Potential of *Trichoderma piluliferum* as a biocontrol agent of *Colletotrichum musae* in banana fruits. Biocatal Agric Biotechnol 34:102028

Daguerre Y, Edel-Hermann V, Steinberg C (2017) Fungal genes and metabolites associated with the biocontrol of soil-borne plant pathogenic fungi. Fungal Metabol:33–104

Das J, Yadav SK, Ghosh S, Tyagi K, Magotra A, Krishnan A, Jha G (2021) Enzymatic and non-enzymatic functional attributes of plant microbiome. Curr Opin Biotechnol 69:162–171

de Carvalho APA, Conte-Junior CA (2022) Nanoencapsulation application to prolong postharvest shelf life. Curr Opin Biotechnol 78:102825

Di Pietro A (1995) Fungal antibiosis in biocontrol of plant disease

Douksouna Y, Masanga J, Nyerere A, Runo S, Ambang Z (2019) Towards managing and controlling aflatoxin producers within *Aspergillus* species in infested rice grains collected from local markets in Kenya. Toxins 11(9):544

Dukare AS, Paul S, Nambi VE, Gupta RK, Singh R, Sharma K, Vishwakarma RK (2019) Exploitation of microbial antagonists for the control of postharvest diseases of fruits: a review. Crit Rev Food Sci Nutr 59(9):1498–1513

Elad Y, Chet I, Katan J (1980) Trichoderma harzianum: A biocontrol agent effective against *Sclerotium rolfsii* and *Rhizoctonia solani*. Phytopathology 70(2):119–121

Eladawy M, El-Mowafy M, El-Sokkary MM, Barwa R (2021) Antimicrobial resistance and virulence characteristics in ERIC-PCR typed biofilm forming isolates of P. aeruginosa. Microb Pathog 158:105042

Elsherbiny EA, Amin BH, Aleem B, Kingsley KL, Bennett JW (2020) Trichoderma volatile organic compounds as a biofumigation tool against late blight pathogen *Phytophthora infestans* in postharvest potato tubers. J Agric Food Chem 68(31):8163–8171

Fenice M, Gooday GW (2006) Mycoparasitic actions against fungi and oomycetes by a strain (CCFEE 5003) of the fungus *Lecanicilliummuscarium* isolated in Continental Antarctica. Ann Microbiol 56:1–6

Geremu MA, Kidane B (2022) Associated fungal pathogens and causes of postharvest losses in the avocado value chain in Hossana town market, Ethiopia. Afr J Food Agric Nutr Dev 22(6)

Guigón-López C, Holguín-Ibarra PD, Torres-Zapien JH, García-Cruz I, Villapando I, Salas-Salazar NA (2021) *Metarhizium anisopliae* reduces conidial germination and mycelium growth of the apple graymold*Botrytiscinerea*. Biol Control 160:104660

Haque MMU, Rahman MA, Ador MAH, Ahmed R (2023) First report of *Rhizopus stolonifer* causing Premature Soft Rot of Jackfruit in Bangladesh. Plant Dis

Hardoim PR, Van Overbeek LS, Berg G, Pirttilä AM, Compant S, Campisano A, Doring M, Sessitsch A (2015) The hidden world within plants: ecological and evolutionary considerations for defining functioning of microbial endophytes. Microbiol Mol Biol Rev 79(3):293–320

Hawar SN, Taha ZK, Hamied AS, Al-Shmgani HS, Sulaiman GM, Elsilk SE (2023, 2023) Antifungal activity of bioactive compounds produced by the Endophytic fungus *Paecilomyces* sp. (JN227071. 1) against *Rhizoctonia solani*. Int J Biomater

Heinig U, Scholz S, Jennewein S (2013) Getting to the bottom of Taxol biosynthesis by fungi. Fungal Divers 60:161–170

Hernandez-Montiel LG, Droby S, Preciado-Rangel P, Rivas-García T, González-Estrada RR, Gutiérrez-Martínez P, Ávila-Quezada GD (2021) A sustainable alternative for postharvest disease management and phytopathogens biocontrol in fruit: antagonistic yeasts. Plan Theory 10(12):2641

Huang R, Li GQ, Zhang J, Yang L, Che HJ, Jiang DH, Huang HC (2011) Control of postharvest Botrytis fruit rot of strawberry by volatile organic compounds of *Candida intermedia*. Phytopathology 101(7):859–869

Huang X, Ren J, Li P, Feng S, Dong P, Ren M (2021) Potential of microbial endophytes to enhance the resistance to postharvest diseases of fruit and vegetables. J Sci Food Agric 101(5):1744–1757

Ismail SI, Ahmad Dahlan K, Abdullah S, Zulperi D (2021) First report of *Neoscytalidiumdimidiatum* causing fruit rot on guava (*Psidium guajava*) in Malaysia. Plant Dis 105(1):220

Ji ZL, Peng S, Zhu W, Dong JP, Zhu F (2020) Induced resistance in nectarine fruit by Bacillus licheniformis W10 for the control of brown rot caused by *Monilinia fructicola*. Food Microbiol 92:103558

Jones KA, Burges HD (1998) Technology of formulation and application. In: Formulation of microbial biopesticides: beneficial microorganisms, nematodes and seed treatments. Springer Netherlands, Dordrecht, pp 7–30

Keswani C, Prakash O, Bharti N, Vílchez JI, Sansinenea E, Lally RD, Borriss R, Singh SP, Gupta VK, Fraceto FL, Lima DR, Singh HB (2019) Re-addressing the biosafety issues of plant growth promoting rhizobacteria. Sci Total Environ 690:841–852

Khare E, Mishra J, Arora NK (2018) Multifaceted interactions between endophytes and plant: developments and prospects. Front Microbiol 9:2732

Khruengsai S, Pripdeevech P, D'Souza PE, Panuwet P (2021a) Biofumigation activities of volatile compounds from two *Trichoderma afroharzianum* strains against *Fusarium* infections in fresh chilies. J Sci Food Agric 101(14):5861–5871

Khruengsai S, Pripdeevech P, Tanapichatsakul C, Srisuwannapa C, D'Souza PE, Panuwet P (2021b) Antifungal properties of volatile organic compounds produced by *Daldiniaeschscholtzii* MFLUCC 19-0493 isolated from *Barleriaprionitis* leaves against *Colletotrichum acutatum* and its post-harvest infections on strawberry fruits. PeerJ 9:e11242

Kumari M, Kamat S, Dixit R, Pandey S, Giri VP, Mishra A (2020) Microbial formulation approaches in postharvest disease management. In: Food security and plant disease management, pp 279–305

Kumari M, Kamat S, Dixit R, Pandey S, Giri VP, Mishra A (2021) Microbial formulation approaches in postharvest disease management. In: Food security and plant disease management. Woodhead Publishing, pp 279–305

Kumari M, Qureshi KA, Jaremko M, White J, Singh SK, Sharma VK, Singh KK, Santoyo G, Puopolo G, Kumar A (2022) Deciphering the role of endophytic microbiome in postharvest diseases management of fruits: opportunity areas in commercial up-scale production. Front Plant Sci 13:4390

Kusari S, Hertweck C, Spiteller M (2012) Chemical ecology of endophytic fungi: origins of secondary metabolites. Chem Biol 19(7):792–798

Kusari S, Pandey SP, Spiteller M (2013) Untapped mutualistic paradigms linking host plant and endophytic fungal production of similar bioactive secondary metabolites. Phytochemistry 91:81–87

Lai J, Cao X, Yu T, Wang Q, Zhang Y, Zheng X, Lu H (2018) Effect of *Cryptococcus laurentii* on inducing disease resistance in cherry tomato fruit with focus on the expression of defense-related genes. Food Chem 254:208–216

Latz MA, Jensen B, Collinge DB, Jørgensen HJ (2018) Endophytic fungi as biocontrol agents: elucidating mechanisms in disease suppression. Plant Ecol Divers 11(5–6):555–567

Lee SO, Kim HY, Choi GJ, Lee HB, Jang KS, Choi YH, Kim JC (2009) Mycofumigation with *Oxyporuslatemarginatus* EF069 for control of postharvest apple decay and *Rhizoctonia* root rot on moth orchid. J Appl Microbiol 106(4):1213–1219

Lei X, Deng B, Ruan C, Deng L, Zeng K (2022) Phenylethanol as a quorum sensing molecule to promote biofilm formation of the antagonistic yeast *Debaryomycesnepalensis* for the control of black spot rot on jujube. Postharvest Biol Technol 185:111788

Lemfack MC, Gohlke BO, Toguem SMT, Preissner S, Piechulla B, Preissner R (2018) mVOC 2.0: a database of microbial volatiles. Nucleic Acids Res 46(D1):D1261–D1265

Li B, Wang W, Zong Y, Qin G, Tian S (2012) Exploring pathogenic mechanisms of *Botrytis cinerea*secretome under different ambient pH based on comparative proteomic analysis. J Proteome Res 11(8):4249–4260

Liu J, Sui Y, Wisniewski M, Droby S, Liu Y (2013) Utilization of antagonistic yeasts to manage postharvest fungal diseases of fruit. Int J Food Microbiol 167(2):153–160

Liu J, Guo X, Zhang H, Cao Y, Sun Q (2021) First report of postharvest fruit rot disease of hardy kiwifruit caused by *Diaportheeres* in China. Plant Dis 105(11):3745

Ludwig-Müller J (2015) Plants and endophytes: equal partners in secondary metabolite production? Biotechnol Lett 37:1325–1334

Ma Y, Feng C (2009) Progress in the studies on alkaloids from endophytic fungi. Chin J Organ Chem 29(08):1182

Macarisin D, Cohen L, Eick A, Rafael G, Belausov E, Wisniewski M et al (2007) *Penicillium digitatum* suppresses production of hydrogen peroxide in host tissue during infection of citrus fruit. Phytopathology 97:1491–1500

MacDonald MJ, D'Cunha GB (2007) A modern view of phenylalanine ammonia lyase. Biochem Cell Biol 85(3):273–282

Macías-Rubalcava ML, Sánchez-Fernández RE, Roque-Flores G, Lappe-Oliveras P, Medina-Romero YM (2018) Volatile organic compounds from *Hypoxylon anthochroum* endophytic strains as postharvest mycofumigation alternative for cherry tomatoes. Food Microbiol 76:363–373

Mannazzu I, Domizio P, Carboni G, Zara S, Zara G, Comitini F, Budroni M, Ciani M (2019) Yeast killer toxins: from ecological significance to application. Crit Rev Biotechnol 39(5):603–617

Mari M, Bautista-Banos S, Sivakumar D (2016) Decay control in the postharvest system: role of microbial and plant volatile organic compounds. Postharvest Biol Technol 122:70–81

Mercier J, Jiménez JI (2004) Control of fungal decay of apples and peaches by the biofumigant fungus *Muscodor albus*. Postharvest Biol Technol 31(1):1–8

Miethke M, Marahiel MA (2007) Siderophore-based iron acquisition and pathogen control. Microbiol Mol Biol Rev 71(3):413–451

Mitchell AM, Strobel GA, Moore E, Robison R, Sears J (2009) Volatile antimicrobials from *Muscodorcrispans*, a novel endophytic fungus. Microbiology 156(1):270–277

MoFPI (2022) Study to determine post-harvest losses of agri produce in India. In NABARD

Montecinos-Pedro LA, Ayala-Escobar V, Arévalo-Galarza MDL, Cadena-Iñiguez J, Leyva-Madrigal KY, Mora-Romero GA, Tovar-Pedraza JM (2022) First report of *Fusarium citri* causing Postharvest Fruit Rot of Chayote in Mexico. Plant Dis 107:2226

Munir S, Li Y, He P, Huang M, He P, He P, He Y (2020) Core endophyte communities of different citrus varieties from citrus growing regions in China. Sci Rep 10(1):1–12

Nally MC, Pesce VM, Maturano YP, Assaf LR, Toro ME, De Figueroa LC, Vazquez F (2015) Antifungal modes of action of *Saccharomyces* and other biocontrol yeasts against fungi isolated from sour and grey rots. Int J Food Microbiol 204:91–100

Nandakumar M, Malathi P, Sundar AR, Viswanathan R (2021) Host-pathogen interaction in sugarcane and red rot pathogen: exploring expression of phytoalexin biosynthesis pathway genes. Indian Phytopathol 74:529–535

Nantawanit N, Chanchaichaovivat A, Panijpan B, Ruenwongsa P (2010) Induction of defense response against *Colletotrichum capsici* in chili fruit by the yeast *Pichia guilliermondii* strain R13. Biol Control 52(2):145–152

Nian L, Xie Y, Zhang H, Wang M, Yuan B, Cheng S, Cao C (2023) *Vishniacozymavictoriae*: an endophytic antagonist yeast of kiwifruit with biocontrol effect to *Botrytis cinerea*. Food Chem 411:135442

Oufensou S, Ul Hassan Z, Balmas V, Jaoua S, Migheli Q (2023) Perfume guns: potential of yeast volatile organic compounds in the biological control of mycotoxin-producing fungi. Toxins 15(1):45

Pacifico D, Squartini A, Crucitti D, Barizza E, Lo Schiavo F, Muresu R, Carimi F, Zottini M (2019) The role of the endophytic microbiome in the grapevine response to environmental triggers. Front Plant Sci 10:1256

Pan H, Zhong C, Wang Z, Deng L, Li W, Zhao J et al (2022) Biocontrol ability and action mechanism of *Meyerozyma guilliermondii* 37 on Soft Rot Control of Postharvest Kiwifruit. Microorganisms 10(11):2143

Petrov V, Hille J, Mueller-Roeber B, Gechev TS (2015) ROS-mediated abiotic stress-induced programmed cell death in plants. Front Plant Sci 18;6:69

Pinheiro EAA, Carvalho JM, dos Santos DCP, Feitosa ADO, Marinho PSB, Guilhon GMSP et al (2013) Antibacterial activity of alkaloids produced by endophytic fungus *Aspergillus* sp. EJC08 isolated from medical plant *Bauhinia guianensis*. Nat Prod Res 27(18):1633–1638

Punja ZK, Rodriguez G, Tirajoh A (2016) Effects of Bacillus subtilis strain QST 713 and storage temperatures on post-harvest disease development on greenhouse tomatoes. Crop Prot 84:98–104

Rajani P, Rajasekaran C, Vasanthakumari MM, Olsson SB, Ravikanth G, Shaanker RU (2021) Inhibition of plant pathogenic fungi by endophytic *Trichoderma* spp. through mycoparasitism and volatile organic compounds. Microbiol Res 242:126595

Raynaldo FA, Dhanasekaran S, Ngea GLN, Yang Q, Zhang X, Zhang H (2021) Investigating the biocontrol potentiality of *Wickerhamomyces anomalus* against postharvest gray mold decay in cherry tomatoes. Sci Hortic 285:110137

Sah B, Kumari M, Subban K, Chelliah J (2020) Evaluation of the anticancer activity of enzymatically synthesized Baccatin III: an intermediate precursor of Taxol®. 3 Biotech 10:1–10

Santra HK, Banerjee D (2023) Antifungal activity of volatile and non-volatile metabolites of endophytes of Chloranthus elatior Sw. Front Plant Sci 14:1156323

Saravanakumar D, Ciavorella A, Spadaro D, Garibaldi A, Gullino ML (2008) *Metschnikowia pulcherrima* strain MACH1 outcompetes *Botrytis cinerea*, *Alternaria alternata* and *Penicillium expansum* in apples through iron depletion. Postharvest Biol Technol 49(1):121–128

Sarrocco S, Vannacci G (2018) Preharvest application of beneficial fungi as a strategy to prevent postharvest mycotoxin contamination: a review. Crop Prot 110:160–170

Schulz B, Boyle C, Draeger S, Römmert AK, Krohn K (2002) Endophytic fungi: a source of novel biologically active secondary metabolites. Mycol Res 106(9):996–1004

Sdiri Y, Lopes T, Rodrigues N, Silva K, Rodrigues I, Pereira JA, Baptista P (2022) Biocontrol ability and production of volatile organic compounds as a potential mechanism of action of olive endophytes against *Colletotrichum acutatum*. Microorganisms 10(3):571

Senthil KR, Joo KH, Byung-Ki H (2011) Taxol-producing fungal endophyte, *Pestalotiopsis* species isolated from *Taxus cuspidata*. J Biosci Bioeng 111(6):731–731

Shi Y, Yang Q, Zhao Q, Dhanasekaran S, Ahima J, Zhang X, Zhou X, Droby S, Zhang H (2022) *Aureobasidium pullulans* S-2 reduced the disease incidence of tomato by influencing the postharvest microbiome during storage. Postharvest Biol Technol 185:111809

Shweta S, Gurumurthy BR, Ravikanth G, Ramanan US, Shivanna MB (2013) Endophytic fungi from *Miquelia dentata*Bedd., produce the anti-cancer alkaloid, camptothecine. Phytomedicine 20(3–4):337–342

Sinclair JB, Cerkauskas RF (1996) Latent infection vs. endophytic colonization by fungi. In: Endophytic fungi in grasses and woody plants, pp. 3–29

Sivakumar D, Wijeratnam RW, Wijesundera RLC, Marikar FMT, Abeyesekere M (2000) Antagonistic effect of *Trichoderma harzianum* on postharvest pathogens of rambutan (*Nephelium lappaceum*). Phytoparasitica 28:240–247

Solanki MK, Abdelfattah A, Sadhasivam S, Zakin V, Wisniewski M, Droby S, Sionov E (2021) Analysis of stored wheat grain-associated microbiota reveals biocontrol activity among microorganisms against mycotoxigenic fungi. J Fungi 7(9):781

Sood M, Kapoor D, Kumar V, Sheteiwy MS, Ramakrishnan M, Landi M, Araniti F, Sharma A (2020) *Trichoderma*: the "secrets" of a multitalented biocontrol agent. Plan Theory 9(6):762

Spadaro D, Droby S (2016) Trends in Food Science & Technology Development of biocontrol products for postharvest diseases of fruit: the importance of elucidating the mechanisms of action of yeast antagonists. Trends Food Sci Technol [Internet] 47:39–49

Spadaro D, Garibaldi A, Gullino ML (2004) Control of *Penicillium expansum* and *Botrytis cinerea* on apple combining a biocontrol agent with hot water dipping and acibenzolar-S-methyl, baking soda, or ethanol application. Postharvest Biol Technol 33(2):141–151

Stenberg JA, Sundh I, Becher PG, Björkman C, Dubey M, Egan PA, Friberg H, Gil FJ, Jensen DF, Jonsson M, Karlsson M, Khalil S, Ninkovic V, Rehermann G, Vetukuri RR, Viketoft M (2021) When is it biological control? A framework of definitions, mechanisms, and classifications. J Pest Sci 94(3):665–676

Stierle A, Strobel G, Stierle D (1993) Taxol and taxane production by *Taxomyces andreanae*, an endophytic fungus of Pacific yew. Science 260(5105):214–216

Stošić S, Ristić D, Gašić K, Starović M, Ljaljević Grbić M, Vukojević J, Živković S (2020) *Talaromyces minioluteus*: New postharvest fungal pathogen in Serbia. Plant Dis 104(3):656–667

Stošić S, Ristić D, Savković Ž, Grbić ML, Vukojević J, Živković S (2021) *Penicillium* and *Talaromyces species* as postharvest pathogens of pear fruit (*Pyrus communis*) in Serbia. Plant Dis 105(11):3510–3521

Strobel G (2011) *Muscodor* species-endophytes with biological promise. Phytochem Rev 10:165–172

Sturz AV, Christie BR, Nowak J (2000) Bacterial endophytes: potential role in developing sustainable systems of crop production. Crit Rev Plant Sci 19(1):1–30

Sukmawati D, Family N, Hidayat I, Sayyed RZ, Elsayed EA, Dailin DJ, Hanapi SZ, Wadaan MA, Enshasy HE (2021) Biocontrol activity of *Aureubasidium pullulans* and *Candida orthopsilosis* isolated from *Tectona grandis* L. phylloplane against *Aspergillus* sp. in post-harvested citrus fruit. Sustain For 13(13):7479

Suwannarach N, Bussaban B, Nuangmek W, Pithakpol W, Jirawattanakul B, Matsui K, Lumyong S (2016) Evaluation of *Muscodorsuthepensis* strain CMU-Cib462 as a postharvest biofumigant for tangerine fruit rot caused by *Penicillium digitatum*. J Sci Food Agric 96(1):339–345

Talibi I, Boubaker H, Boudyach EH, Aoumar AB, A. (2014) Alternative methods for the control of postharvest citrus diseases. J Appl Microbiol 117(1):1–17

Taritla S, Kumari M, Kamat S, Bhat SG, Jayabaskaran C (2021) Optimization of physicochemical parameters for production of cytotoxic secondary metabolites and apoptosis induction activities in the culture extract of a Marine Algal–derived Endophytic Fungus *Aspergillus*sp. Front Pharmacol 12:542891

Tayung K, Barik BP, Jha DK (2010) Antifungal activity and biocontrol potential of metabolite produced by an endophytic *Fusarium* (MTCC-9622) against some postharvest pathogens. J Agric Technol 6(3):409–419

Thines E, Heidrun ANKE, Weber RW (2004) Fungal secondary metabolites as inhibitors of infection-related morphogenesis in phytopathogenic fungi. Mycol Res 108(1):14–25

Tripathi AN, Tiwari SK, Behera TK (2022) Postharvest diseases of vegetable crops and their management. In: Postharvest technology-recent advances, new perspectives and applications. Intech Open London

van Loon LC, Rep M, Pieterse CM (2006) Significance of inducible defense-related proteins in infected plants. Annu Rev Phytopathol 44:135–162

Villalba ML, Mazzucco MB, Lopes CA, Ganga MA, Sangorrín MP (2020) Purification and characterization of *Saccharomyces eubayanus* killer toxin: biocontrol effectiveness against wine spoilage yeasts. Int J Food Microbiol 331:108714

Vos CM, De Cremer K, Cammue BP, De Coninck B (2015) The toolbox of *Trichoderma* spp. in the biocontrol of *Botrytis cinerea* disease. Mol Plant Pathol 16(4):400–412

Wang Y, Kang N, Yang X, Ouyang D, Wu H, Chen J, Xiang ML, Chen M (2023) First report of Fruit Rot caused by *Fusarium graminearum* on Ponkan (*Citrus reticulata Blanco* cv. Ponkan) in China. Plant Dis 107:2532

White JF, Kingsley KL, Zhang Q, Verma R, Obi N, Dvinskikh S, Elmore MT, Verma SK, Gond SK, Kowalski KP (2019) Endophytic microbes and their potential applications in crop management. Pest Manag Sci 75(10):2558–2565

White JF, Chang X, Kingsley KL, Zhang Q, Chiaranunt P, Micci A, Kowalski KP (2021) Endophytic bacteria in grass crop growth promotion and biostimulation. Grass Res 1(1):1–9

Wisniewski M, Droby S (2019) The postharvest microbiome: the other half of sustainability. Biol Control 137:104025

Wu W, Lei J, Hussain M, Cao S, Du B, Wang R (2019) Structure and function of the fruit microbiome in healthy and diseased kiwifruit. Pak J Agric Sci 56(3)

Xiang M, Wang Y, Wang D, Fu Y, Zeng J, Ouyang D, Chen J, Chen M (2023) First report of Black Spot Disease Caused by *Alternaria alternata* on Persimmon Fruit in China. Plant Dis 107:2525

Xiao L, Luo Z, Fu Y, Zeng J, Xiang M, Chen J, Chen M (2022a) First report of Postharvest Fruit Rot on Citrus reticulata Blanco Caused by *Fusarium concentricum* in China. Plant Dis

Xiao Y, Lai H, Zhou W, Yang C, Cui C (2022b) First report of *Diaporthefusicola* causing Postharvest Fruit Rot of Peach (*Prunus persica*) in China. Plant Dis 107:2231

Xiao Y, Luo S, Guo M, Ai S, Huo G, Cui C (2023) First report of Brown Rot caused by *Phytophthora nicotianae* on Navel Orange Fruit in China. Plant Dis 107(4):1245

Xu X, Qin G, Tian S (2008) Effect of microbial biocontrol agents on alleviating oxidative damage of peach fruit subjected to fungal pathogen. Int J Food Microbiol 126(1–2):153–158

Xu D, Deng Y, Han T, Jiang L, Xi P, Wang Q, Jiang Z, Gao L (2018) In vitro and in vivo effectiveness of phenolic compounds for the control of postharvest graymold of table grapes. Postharvest Biol Technol 139:106–114

Yan F, Zhang D, Ye X, Wu Y, Fang T (2022) Potential of *Saturnispora diversa* MA as a postharvest biocontrol agent against anthracnose in loquat fruit. Biol Control 173:105006

Yang Q, Qian X, Dhanasekaran S, Boateng NAS, Yan X, Zhu H, He F, Zhang H (2019) Study on the infection mechanism of *Penicillium digitatum* on postharvest citrus (*Citrus reticulata* Blanco) based on transcriptomics. Microorganisms 7(12):672

Yang X, Wang Y, Li S, Wu F, Luo Z, Chen J, Xiang M, Chen M (2022) First report of *Fusarium tricinctum* causing fruit rot on navel orange (*Citrus sinensis* (L.) Osbeck) in China. Plant Dis 107:567

Yuan X, Li ZM, Wen L, Huo G, Cui C (2023) First report of Postharvest Fruit Rot on Gannan Navel Orange Caused by *Diaportheunshiuensis* in Jiangxi Province, China. Plant Dis

Zee KY, Asib N, Ismail SI (2021) First report of *Lasiodiplodia theobromae* causing postharvest fruit rot on guava (*Psidium guajava*) in Malaysia. Plant Dis 105(9):2716

Zhang S, Zheng Q, Xu B, Liu J (2019) Identification of the fungal pathogens of postharvest disease on peach fruits and the control mechanisms of *Bacillus subtilis* JK-14. Toxins 11(6):322

Zhang Z, Chen Y, Li B, Chen T, Tian S (2020) Reactive oxygen species: a generalist in regulating development and pathogenicity of phytopathogenic fungi. Comput Struct Biotechnol J 18:3344–3349

Zhang W, Niu XL, Yang JY (2021a) First report of postharvest fruit rot on passion fruit (*Passiflora edulis*) caused by *Lasiodiplodia theobromae* in Mainland China. Plant Dis 105(4):1198

Zhang Z, Zhang ZB, Huang YT, Wang FX, Hu WH, Dai LY, Zhong J, Liu Y, Zhu JZ (2021b) First report of *Diaporthehongkongensis* causing fruit rot on peach (*Prunus persica*) in China. Plant Dis 105(7):2017

Zhang YN, Wang ZJ, Swingle B, Niu BY, Xu J, Ma X, Wei HL, Gao M (2022a) First report of *Rhizopus arrhizus* (syn. *R. oryzae*) causing Garlic Bulb Soft Rot in Hebei Province, China. Plant Dis 107:949

Zhang J, He L, Guo C, Liu Z, Kaliaperumal K, Zhong B, Jiang Y (2022b) Evaluation of *Aspergillus aculeatus* GC-09 for the biological control of citrus blue mold caused by *Penicillium italicum*. Fungal Biol 126(3):201–212

Zhao Y, Yin J (2018) Effects of *Pichia guilliermondii* and hot air treatment on the postharvest preservation of red fuji apple quality attributes. J Food Prot 81(2):186–194

Zhao F, Yang HX, Bi DZ, Khaledi A, Qiao MQ (2020) A systematic review and meta-analysis of antibiotic resistance patterns, and the correlation between biofilm formation with virulence factors in uropathogenic *E. coli* isolated from urinary tract infections. Microb Pathog 144

Zhao Q, Shi Y, Ngea GLN, Zhang X, Yang Q, Zhang Q, Xu X, Zhang H (2023) Changes of the microbial community in kiwifruit during storage after postharvest application of *Wickerhamomycesanomalus*. Food Chem 404:134593

Zhu H, Zhao L, Zhang X, Foku JM, Li J, Hu W, Zhang H (2019) Efficacy of *Yarrowia lipolytica* in the biocontrol of green mold and blue mold in *Citrus reticulata* and the mechanisms involved. Biol Control 139:104096

The Multifaceted Biocontrol Mechanisms of Endophytic Fungi

15

Mirza Abid Mehmood, Areeba Rauf, Kamel A. Abd-Elsalam, Muhammad Ashfaq, Samah Bashir Kayani, and Sania Javeed

Abstract

To enhance the global food supply and address human demand, it is crucial to improve crop production and sustainability. The growth of the human population and various human activities pose threats to food security. While traditional agricultural systems have long relied on agrochemical inputs to increase crop yields, these practices harm the environment. Therefore, there is a current emphasis on exploring soil-derived biological resources and plant microbes to develop environmentally friendly agricultural methods. The microbiota of the rhizosphere and endosphere are influenced by biotic and abiotic factors, respectively, to some extent. Plant microbes and their hosts have a mutually dependent relationship. Endophytic fungi (EF) are fascinating fungal species associated with hosts, as they can penetrate both intercellular and intracellular domains of host tissues and provide benefits to their hosts. Endophytic fungi employ various strategies to regulate phytopathogens, including inducing and enhancing plant resistance. This improves plants' ability to withstand stress through gene regulation, synthesis of fungal metabolites, and production of hormone signaling molecules. Exploring the potential of beneficial endophytes as bioinoculants for biological fertilizers and their integration into plant breeding and disease management programs appears to be a promising alternative to agricultural chemicals. Beneficial

M. Abid Mehmood (✉) · A. Rauf · M. Ashfaq · S. Javeed
Plant Pathology, Institute of Plant Protection, Muhammad Nawaz Shareef University of Agriculture, Multan, Punjab, Pakistan
e-mail: abid.mehmood@mnsuam.edu.pk

K. A. Abd-Elsalam
Plant Pathology Research Institute, Agricultural Research Center (ARC), Giza, Egypt

S. B. Kayani
Govt. Graduate College (W) Mumtazabad, Multan, Punjab, Pakistan

© The Author(s), under exclusive license to Springer Nature Singapore Pte Ltd. 2025

K. A. Abd-Elsalam, A. H. Hashem (eds.), *Fungal Endophytes Volume II*,
https://doi.org/10.1007/978-981-97-8804-0_15

endophytic fungal organisms have the ability to enhance plant health by limiting the growth of pathogenic fungi. This disease-prevention method depends on endophytic fungus and host plants coexisting. Understanding beneficial endophytic fungal species along with their interactions with pathogens and host plants, as well as making judgements about them, is a significant challenge associated with endophyte biocontrol techniques. The investigation of the antibiosis activity of fungal endophytes can be conducted through the development of biocontrol agents, contributing to the integrated management of plant diseases. Research on endophytic fungi as bioinoculants is growing. The biological defense mechanism of fungal endophytes against plant diseases is the primary focus of this chapter. It also discusses the problems and major obstacles that still require investigation.

Keywords

Endophytic fungi · Secondary metabolites · Host-associated fungal communities · Anthropogenic activities · Antibiosis

15.1 Introduction

Agricultural productivity and the safety of the global food supply face significant challenges. According to estimates, the global population is expected to surpass 9 billion individuals by the year 2050. This demographic trend would need a substantial increase of approximately 70% in the production of food in order to meet the growing demand (Data 2020). As a result, protecting the viability of farmed crops is essential for many different economic sectors. In addition to food, plants contribute in essential resources including wood, fiber, medicines, and bioenergy. Plant diseases cause significant financial losses by reducing both the quantity and quality of production (Savary et al. 2019; Yoshida et al. 2013; Goss et al. 2014). In cases where genetic resistance is available, the disease can be controlled through the practice of chemical pesticides and genetic resistance. Genetic resistance is commonly employed and generally proves to be effective in most situations (Pirrello et al. 2023; Thakur et al. 2023). But there have been disadvantages to this tactic.

Pathogen populations are becoming resistant to the active components as a result of this strategy's widespread application throughout time (Lucas et al. 2015; Delp 1988). Because of this, studies on supplemental and alternative management techniques have increased recently and are now an important aspect of integrated management programs. Biological modulation and induced resistance are two strategies that have garnered interest (Köhl et al. 2019; Llorens et al. 2017; O'Brien 2017). Numerous scientific studies have documented instances of effective pathogen control that arise from the examination, application, and analysis of these two methodologies within industrial environments. The relationship between both plants and microbes in nature has a major impact on the growth, development, and defense system of plants against biotic as well as abiotic threats (Nataraja et al. 2019).

Endophytic fungi are a fascinating group of microorganisms that have great potential for utilization in induced resistance and biological control. These microorganisms coexist with plants and can be found inside the plant tissues, such as the seed, stem, leaf, and root (Yadav and Saxena 2018). Unlike pathogens, endophytes do not harm the host plant. In fact, they can promote plant growth by releasing phytohormones. In addition, they enhance the transportation of nutrients in both directions, thereby enhancing nutrition. Furthermore, they contribute to the overall well-being of plants by providing protection against diseases (Shen et al. 2019; Andreozzi et al. 2019). Plants and their endophytes interact to defend themselves against environmental factors like heavy metals and drought (Khan et al. 2014; Kushwaha et al. 2020). Endophytic microorganisms thrive throughout numerous plant hosts and have recently garnered significance owing to their capacity to synthesize bioactive chemicals and enzymes that hold relevance in the field of biotechnology (Maheshwari et al. 2017; Khan et al. 2004).

Endophytes usually lead to a substantial increase in plant biomass when introduced (Shen et al. 2019; Santoyo et al. 2016). Due to the fact that endophytes are capable of producing secondary metabolites with a variety of beneficial properties, they have become more important in both biotechnology and trade. These consist of immunosuppressive, antitumor, antibacterial, and biological control properties. Additionally, endophytes have the ability to release antibiotics, insecticidal substances, organic antioxidants, antiviral compounds, and molecules that prevent diabetes. Over the past 20 years, these microorganisms derived from plants have been recognized as a valuable source of naturally occurring chemical compounds with diverse applications in commercial and agricultural sectors (Das et al. 2016; Yadav and Saxena 2018). Enzymes can be used as substitutes for toxic substances. They thrive in stable temperature conditions and require a pH of 7. Over the years, scientists have recognized the potential for microbial enzymes in various sectors such as agriculture and healthcare.

Recent studies have demonstrated that microbes from harsh environments have significant biotechnological applications (Sahay et al. 2017; Padaria 2015; Singh et al. 2016; Archna et al. 2015). According to Reshma et al. (2019), endophytic fungal organisms and the plants they colonize play a role in promoting plant growth and managing diseases. They achieve this through invasion and the production of biological control agents. Studies conducted by Zhang et al. (2019) have shown that the majority of endophytic microorganisms in the plant's endosphere contribute to disease management and influence how plants adapt to environmental stresses. Recent research (Zhang et al. 2020) suggests that agricultural biotechnology can be used to produce endophytic microorganisms, aiming to enhance plant persistence and integrate them into crop development. Previously, farmers have used agrochemicals to increase food production, but because of their negative impacts on the biosphere, they are not sustainable (Marinova and Bogueva 2022; Maldani et al. 2022). The concept states that investigating endophytic microorganisms and reducing ecothreats can assist in creating a stable ecosystem and encourage pathogen-free plants that produce more crops (Akanmu et al. 2021).

Research on endophytic microorganisms has focused on their potential as bioinoculants, offering alternatives to synthetic pesticides/chemicals in modern agriculture (Orozco-Mosqueda et al. 2021). The chapter is mostly concerned with managing plant diseases. In order to develop biocontrol agents in this field, it is also worthwhile to investigate the understanding regarding endophytic fungal organisms and their antibiosis activity. The endosphere indicates distinct interior tissue regions inside plants, whereas the microorganisms inhabiting these regions are commonly referred to as endophytic organisms (Dubey et al. 2020). It's interesting to note that microbial endophytes have antagonistic or mutualistic relationships with host plants depending on how similar or different their genomes are. Plant growth-promoting endophytes are beneficial strains that possess specific characteristics that promote plant growth. These include the production of phytohormones, nutrient acquisition, the release of biocontrol agents, and the ability to induce stress tolerance. Importantly, these endophytes do not exhibit any pathogenic effects (Adeleke and Babalola 2021).

Because of their near proximity to the root endosphere, certain underground endophytic fungi can rapidly change their shape and become endophytes. The plant organ's position influences this adaptation. It has been noted that fungal endophytes penetrate the endosphere, or interior plant tissues, and colonize the rhizosphere, or external root environment. The overall number of endophytic microbes rises as a result (Yan et al. 2019). These endophytic microorganisms contribute to the growth and health of plants by directly or indirectly aiding in metabolic pathways, as determined by specific genes in their genomes (Baghel et al. 2020). The utilization of endophytic fungal species as biocontrol agents is facilitated by their capacity to infiltrate plant tissues, discharge exopolysaccharides as well as hydrogen cyanide, and trigger unique genes associated with secretory processes and the production of secondary metabolites (Singh et al. 2021b).

The objective of this study is to investigate the possible application of endophytic fungi in the management of plant diseases through the utilization of their bioactive secondary metabolites as biopesticides. By incorporating these substances into plant disease management, the reliance on synthetic pesticides can be reduced, leading to improved crop production. The study focuses on updating our understanding of the unique qualities of endophytic fungi and the mechanisms through which they contribute to plant defense against phytopathogens. The ability of endophytes to function as biological pest regulators and to stimulate plant defense mechanisms against abiotic and biotic stressors has drawn attention to them. Because of these two functions, they could be considered for both additional approaches to traditional control and integrated disease management systems. Both methods typically yield positive outcomes. In 2018, Brazil's biological pesticide market saw a remarkable growth of over 70%, generating revenue exceeding US$127.2 million.

The purpose of this chapter is to highlight the several defense mechanisms that endophytes use to shield their host plants, with an emphasis on endophytic fungi. It also explores intriguing biological activities associated with these strategies. Given their potential to enhance global food safety and combat malnutrition, the analysis of fungal-plant relationships in modern agriculture is gaining increasing importance, necessitating the use of advanced technological approaches (Sharma et al. 2021). Given estimates that the world population index will rise by 2050, the use of biological approaches to alleviate food demand stress, famine, and future food shortages is being closely studied (Sahu and Mishra 2021). Although this figure was higher, the international market predicted that the sector will bring in US$5 billion in the year 2020 and US$11 billion in 2025 (França 2022). In this chapter, we investigate the potential role of endophytic fungi in assisting and enhancing plant defense mechanisms.

15.2 Endophytic Fungi

Endophytic microbes colonize plants, particularly perennials, without causing any apparent damage or morphological changes to their hosts. They reside within plant tissues throughout their life cycle or for a specific period of time. Research indicates that endophytes can be found in both the phyllosphere and the rhizosphere. These microbes, which include fungus and bacteria, cohabit with pathogens. Based on their colonization behavior, endophytic microorganisms can be divided into facultative as well as obligatory forms. At various phases of their life cycles, facultative endophytes may live outside of the host plant with the goal of establishing an association with the soil within the rhizosphere of the plant (Meena et al. 2021). During certain stages of plant life cycles, plants can be colonized by certain species. In contrast, obligatory species remain inside plants throughout their entire lives. These species often spread through vertical transmission and may exploit or alter plant metabolic processes to ensure their own survival (Alam et al. 2021; Gouda et al. 2016).

Researchers have shown significant interest in endophytic microbes, including fungal endophytic species, owing to their unique reservoirs of cytotoxic compounds. These substances include anticarcinogenic and antibacterial compounds (Singh and Dubey 2015; Tsivileva et al. 2022). Additionally, these microbes have the ability to synthesize biological stimulants that contribute to essential oil biosynthesis. They can enhance plant growth, accelerate nutrient dissolution in the plant's root zone, act as biological control agents, and stimulate systemic plant defense against both abiotic and biotic factors (Mehta et al. 2019; Cui et al. 2021; Ayyash et al. 2023).

We provided a thorough summary of the biological properties of EFs in this chapter. Understanding these characteristics is crucial to comprehending their adaptability, spread, and complex interactions with different ecosystem components. We devote particular attention to their interactions with pathogens, epiphytes, and plant hosts. We additionally explore the function of secondary substances in these intricate relationships and how they develop by controlling gene expression and modifying the production of phytochemicals in host plants. Readers will have a thorough understanding of EFs and their secondary metabolites from this chapter (Brader et al. 2017b).

Endophytes, due to their significant role in plant development and survival in harsh conditions, have gained attention from researchers worldwide (Shen et al.

2019). These microbes provide bioactive substances that are used in a range of fields, such as business, medicine, and agriculture. Andreozzi et al. (2019) highlight the influence of plants on the composition of microorganisms found in their root exudates. Thus, the capacity of fungal endophyte to utilize the exudates released by plant roots as an energy source plays a fundamental role in their relationship with plants (Kandel et al. 2017).

Endophytes can stimulate plant development by employing various processes and enhancing resistance to unfavorable environmental conditions (Yadav and Saxena 2018). Microbial biotechnology has expanded its focus beyond the production of ethanol and butanol to mitigate environmental contamination. Various methods such as biological remediation, waste disposal, and composting are now used to transform many chemicals. In recent years, there has been a momentous emphasis on developing innovative medications using microorganisms, animals, and plants (Latz et al. 2018; Gouda et al. 2016). These natural products are cost-effective and pose fewer risks.

The potential for endophytic fungi to release various bioactive substances is significant. Endophytes are valuable sources of fungal diversity, and research by Oono et al. (2015) and Ek-Ramos et al. (2013) suggests that they could provide promising strains for biological control purposes. A brief overview of "endophytes" is provided here to help with comprehension of the subject matter that follows in this chapter. Numerous studies have reported that endophytes are microorganisms that live within plants for a part of their life cycles without inducing any symptoms (Clay et al. 2016; Kusari et al. 2012). Endophytes can spread within the plant cells or outside them, both locally and throughout the host plant, without showing any signs of infection or disease (Su et al. 2013). Endophytes, as defined by Busby et al. (2016), refer to microbes that invade plants internally. Ek-Ramos et al. (2013) state that all plants harbor endophytes. The diversity of these organisms is believed to be influenced by various factors, such as host plant species, plant density, nutrient availability, geographical location, and interactions with soil microorganisms. Many aboveground as well as belowground plant tissues, such as liverworts and others including ferns, spermatophytes, hornworts, mosses, as well as lycophytes, possess endophytes (Jaiswal et al. 2022; Kusari et al. 2012).

According to Fig. 15.1, there are two major categories of fungal endophytes: the clavicipitaceous (C) and non-clavicipitaceous (NC) endophytes. These groups are generally classified based on taxonomy, plant-host range, and ecological relevance (O'Hanlon et al. 2012). Among the family *Clavicipitaceae* endophytes that are commonly found in relation to grasses in the Poaceae family are *Balansia*, *Atkinsonella*, *Neotyphodium*, *Myriogenospora*, *Parepichloe*, *Epichloe*, *Echinodothis*, *Balansiopsis*, as well as *Echinodothis* species. According to O'Hanlon et al. (2012), mutualist species rely on their host throughout their entire life cycle. Many terrestrial plants have non-clavicipitaceous endophytes, such as *Phomopsis* spp., *Xylaria* spp., *Fusarium* spp., and *Colletotrichum* spp. However, these endophytes may not remain inside the host plant for its entire lifespan (Sharma et al. 2020; Roy and Banerjee 2018; Uzma et al. 2019).

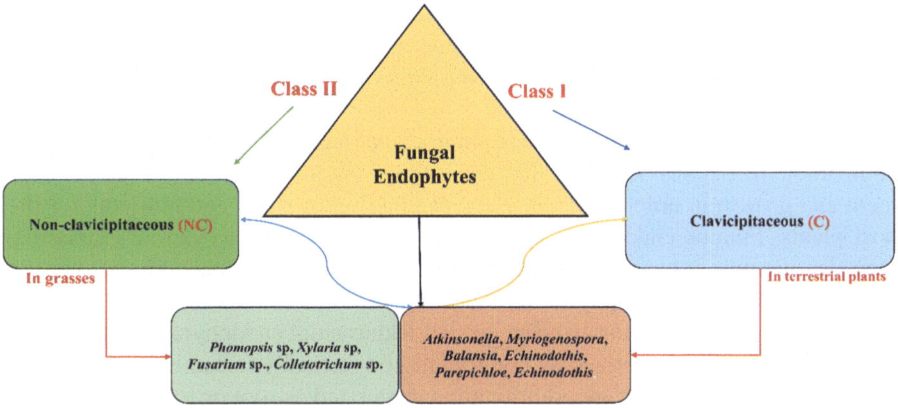

Fig. 15.1 Fungal endophytes are classed based on numerous variables, including their host plants, lifestyle choices, and taxonomic categorization

15.3 Classification and Attributes of Fungal Endophytes in the Endosphere

Due to the fact that endophytes are excellent at controlling a variety of plant diseases, Abaya et al. (2021) have lately expanded the identification of endophytic fungus as a biocontrol agent. The study conducted by Wang et al. (2022) demonstrates the positive effects of endophytic microorganisms on plant development, disease resistance, and carbon sequestration. Adeleke et al. (2022) have also discovered that endophytic fungi inhabiting various plant components have a role in plant growth and development through both direct and indirect mechanisms. Endophytes have a regulatory role in the direct process by modulating the levels of ethylene, auxin, and several other plant hormones. They improve nitrogen fixation, siderophore production, iron solubility, and the accessibility of significant nutrients through the Earth's soil, including phosphorus.

By producing volatile compounds, hydrogen cyanide, enzymes, and antibacterial compounds, endophytes safeguard their host plants in the indirect method. These compounds increase systemic resistance while obstructing pathogen action. Fungal endophytes, such as *Macrophomina phaseolina*, *T. koningii*, *T. erinaceum*, *Daldinia eschscholtzii*, as well as *Diaporthe phaseolorum*, are reported by Suebrasri et al. (2020) to produce enzymes (cellulase, protease, amylase, and xylanase) from sunchoke and other medicinal plants.

Research conducted by Suebrasri et al. (2020) reveals an intriguing finding that *M. phaseolina*, a well-known plant pathogen, can actually have beneficial effects on plants. This discovery opens up the possibility of further investigation to enhance plant protection strategies. Furthermore, Suebrasri et al. (2020) identified a distinct strain of *T. koningii* (ST-KKU1), different from the previously studied strain SMF2, which has shown effectiveness in combating phytopathogens according to Xiao-Yan et al. (2006).

Endophytic fungus organisms from the family Hypocreales have been shown to protect plants from insect invasion by producing peramine. Peramine prohibits flies, worms, and various other pests from destroying the plants. Researchers have found that endophytic fungi release substances that are toxic to dangerous microbes that can affect the host plants, such as sesquiterpene along with diacetamide, which are comparable to indoles. Additionally, these fungi assist in enhancing root development and growth in order to enhance the availability of water and nutrients for the host plants. Fungus endophytes can be classified according to various standards. One classification is based on their ecology, dividing them into clavicipitaceous and non-clavicipitaceous endophytes. Another classification is based on their mode of reproduction, distinguishing between sexual and asexual endophytes. Additionally, endophytes can be categorized as horizontally or vertically transmitted, depending on how they spread. They are additionally categorized as root or foliar endophytes based on the body component they target. Ultimately, endophytes can be classified as either necrotrophic or biotrophic according to the way they manifest their infection and the source of their nourishment (Bamisile et al. 2021).

15.4 Role of Fungal Endophyte in Biocontrol

The concept of biocontrol encompasses the utilization of advantageous microorganisms, their genetic material, and by-products to mitigate the detrimental impacts caused by plant diseases and stimulate favorable reactions in plants (Hashem et al. 2019; Vinale et al. 2008). Biocontrol, as defined by the International Biocontrol Manufacturers Association (IBMA), encompasses the utilization of naturally occurring substances or products to regulate the population growth of pests and pathogens in agricultural settings. Illustrative instances of these items or products encompass macro- and microbes, chemical mediators, as well as natural substances (Nicot et al. 2016).

Despite the perception of biocontrol as an environmentally friendly approach, there has been limited popularity and adoption of biocontrol products. Consequently, the progress in this technology has been relatively stagnant (Rajput et al. 2022). It has been confirmed that *Aspergillus* species, *Gliocladium* species, *Fusarium* species, and nonpathogenic *Trichoderma* species serve as significant biocontrol agents (Boro et al. 2022; Lahlali et al. 2022). Recently, non-clavicipitaceous endophytes have gained attention in various scientific fields due to their ability to transition between endophytic, infectious, and saprobic habitats (Jones 2016).

Clavicipitaceous endophytes, on the other hand, are transmitted vertically from parent plants to offspring through seeds, as demonstrated by De Silva et al. (2019). Systemic growth of clavicipitaceous endophytic fungus develops within the interior tissues of plants when seeds germinate and form seedlings. Hardoim et al. (2015) hypothesize that clavicipitaceous endophytes serve as intercellular symbionts, exhibiting the ability to migrate from the parent plant to the progeny and disperse evenly throughout the stems and leaves. According to Arnold et al. (2003), endophytes in woody angiosperm leaves are dispersed by both sexual and asexual spores.

These endophytes have the potential to serve as biological control agents by altering interactions with pathogens and pests. Ek-Ramos et al. also discuss this potential impact of endophytes. The discovery made in 2013 revealed that endophytic fungi such as *Acremonium alternatum* have the ability to augment the resistance of tomato plants against the powdery mildew fungal disease caused by *Leveillula taurica*. Additionally, it confers defense against *Plutella xylostella* disease in beans.

In a study conducted by Ek-Ramos et al. (2013), it was demonstrated that the presence of the fungal organism *Phomopsis* sp., which was isolated from cotton plants, resulted in a decrease in the number of caterpillar herbivores in cotton plantations. Some endophytic species have the ability to act as biocontrol agents and also promote the growth of the host plant, providing additional benefits. *Neotyphodium* species, for example, contribute to increased growth, health, and stress tolerance in the host plant, thereby protecting it from pathogens and pests (Clay et al. 2016; Oono et al. 2015). The population of the pathogenic fungus *Sclerotium rolfsii* was diminished through the presence of endophytic fungi, specifically *Penicillium citrinum* as well as *Aspergillus terreus*. The decrease in the abundance of pathogens resulted in a corresponding augmentation in the overall biomass of sunflower plants (Tang et al. 2022; De Silva et al. 2019).

The aforementioned examples highlight the importance of fungal endophytes as biological control agents in the context of environmental security policies and their impact on plant productivity. The inquiry on the mechanisms by which endophytes mitigate the occurrence of diseases and pests holds significant importance. We will explore the close relationship between endophytes and their host before discussing various biocontrol strategies. The use of biological methods for plant pest control has become increasingly common. As previously mentioned, these agents have demonstrated the ability to coexist with plants or establish strong relationships with them.

A significant number of effective biological control endophytes have been found among different crops, including some species of *Trichoderma*. *Trichoderma* species have been extensively studied and can colonize various crops such as cotton, wheat, soybean, and corn. This fungus has the potential to induce plant resistance to diseases and has demonstrated multiple disease control mechanisms while also parasitizing plant pathogens (Fontana et al. 2021). It has been found that *Trichoderma* spp. parasitize the hyphae of *S. sclerotiorum* and *Rhizoctonia solani* (Borah et al. 2023). Zhang et al. (2017) report that some strains have shown the capacity to infiltrate and parasitize *Heterodera avenae* cysts that are extremely resistant, *M. javanica* second stage juveniles, and eggs. *Trichoderma* species exhibit a wide range of hosts, including plants and various phytopathogens in both parasitic and symbiotic relationships. Many fungal endophytes not only directly parasitize plant pathogens but also produce and release various effector chemicals to counteract them, including poisons, antibiotics, and fungi toxic metabolites.

Endophytes produce volatile chemical compounds such as chitinase, cellulose, glucanase, and lactones, which aid in the suppression of infections (Fontana et al. 2021; Dimkić et al. 2022). It has been discovered that many endophytes use this method. For instance, *P. lilacinum*, a kind of biocontrol agent that has historically been employed to combat worms and insects that parasitize plants, has demonstrated the capacity to inhibit a number of plant diseases by secreting effectors. Leucinostatins, which are antibiotics, have been shown to inhibit *P. infestans* and *P. capsici* (Wang et al. 2016). The culture filtrate of *P. lilacinum* containing effectors has been found to reduce *S. sclerotiorum* and activate defense mechanisms in common beans (Elsherbiny et al. 2019). This was demonstrated by Daneshkhah et al. (2013). The infection of *Heterodera schachtii* in *Arabidopsis* was found to be decreased by the culture filtrate as well as the cell wall extract of *Piriformospora indica.*

It is crucial to understand that the nematode is not parasitized by this fungus. These illustrations show how cell-free filtrate produced from fungal endophytes can be used as a form of biocontrol to lessen diseases of plants in contemporary farming practices. Plant pathogens that are not infectious, like *F. oxysporum*, may interact with plants to induce wilting or root rot. Certain noninfectious strains of *F. oxysporum*, when present as endophytes in plants, have the ability to provide protection against soilborne diseases such as *Pythium ultimum* along with *Verticillium dahlia* (De Lamo and Takken 2020).

F. oxysporum sensu lato not only competes for nutrition and root niches but can also inhibit fungal diseases of plants (De Lamo and Takken 2020; Fadiji and Babalola 2020). Other mechanisms may also play a role. As an illustration, it has been discovered that *Fusarium* endophytes in tomato plant suppress *F. oxysporum* by inducing resilience through the activity of JA, ET along with SA. The next section will provide more detailed information on endophyte-mediated induced resistance (Constantin et al. 2019). Recent studies have shown that nontoxic strains of *Aspergillus flavus* can effectively prevent the growth of toxin-producing *Aspergillus* in cereals. These noninfectious *A. flavus* strains were specifically released to combat toxic *Aspergillus* and prevent contamination with aflatoxin before harvest (Sarrocco and Vannacci 2018).

Numerous biological control agents have been used to manage harmful *Fusarium* strains in maize crops (Sarrocco et al. 2019). However, the challenge of isolating and selecting fungal endophytes for plant disease control remains. The plant's biome may hold the key to solving this problem, as the microorganisms associated with plants can effectively control most diseases related to this host plant. In a study conducted by Halecker et al. (2020), the primary objective was the development of a biocontrol agent utilizing an endophyte fungus as a means to address the issue of ash dieback, which is attributed to the pathogen *Hymenoscyphus fraxineus*. A comprehensive collection of 340 endophytic fungal species was obtained from the host plant tree, *Fraxinus excelsior*. The fungi underwent thorough examination and were cocultivated in order to identify an appropriate biocontrol agent. The study conducted by Rubini et al. (2005) focused on the examination of various fungal species present on *Theobroma cacao* plants with the aim of effectively controlling *Moniliophthora perniciosa*, the pathogen responsible for the development of witches' broom disease. The efficacy of *Gliocladium catenulatum* treatment was demonstrated by its ability to significantly decrease the prevalence of witches' broom disease in roughly 70% of the afflicted plants. This underscores the prospective application of harnessing the phytobiome for the purpose of mitigating plant diseases.

Numerous endophyte species from the genera *Aspergillus*, *Phyllosticta*, *Colletotrichum*, *Gliocladium*, as well as *Diaporthe* have been studied for the antifungal properties of their secondary metabolites. *R. solani* infection of seed was decreased by 64%, 60%, and 55%, respectively, when *T. longibrachiatum*, *T. asperellum*, as well as *T. atroviride* isolates from soybean (*Glycine max*) were treated as a solution to infested soils (Grabka et al. 2022). These *Trichoderma* species also produced hydrolyzing enzymes such as pectinase and chitinase, which can dissolve cell wall constituents (Khairah et al. 2023). Additionally, they produced IAA, a plant growth regulator, and siderophores that limit the availability of iron to infectious fungal organisms (Kour et al. 2023). *Chaetomium globosum* is an exceptional fungal endophyte known for its ability to synthesize a diverse range of antibacterial chemicals with broad-spectrum activity. The substance has efficacy against a range of harmful microorganisms and moreover possesses qualities that inhibit the formation of biofilms (Oliveira et al. 2016).

Endophytic fungal organisms are being used as a therapy for seeds on agricultural plants, as well as through root injection, to protect against worms (Nagendran et al. 2014). The full extent of their effectiveness as nematode antagonists is still unknown, but endophytic fungi have the potential to contribute to the development of effectual biocontrol approaches. We must comprehend the bioactive compounds produced by endophytes or encouraged within the host plant in order to gain a better understanding how endophytes limit nematode development. The usefulness of previously discovered species of nematode-parasitic fungi as biocontrol agents for managing various nematode species requires further investigation. Biological activities of fungal endophytes against various pathogens has been demonstrated in (Table 15.1).

15.5 Systematic Defensive Networking of Endophytic Fungi for Plant Protection

Current research endeavors have been primarily directed toward the examination of the intricate dynamics occurring between microorganisms and plants, with a specific emphasis on the root endosphere. Both biotic and abiotic factors can influence microbial interactions in this area (Adeleke and Babalola 2021). According to Ananda and Sridhar (2002), environmental factors frequently lead to a greater number of microbes belowground than in the phyllosphere. Adeleke et al. (2022) reported that the plant's underground sections displayed a greater amount of microbial diversity than the stem. This can be attributed to the substantial rhizodeposition of organic compounds, which stimulated microbial activities in that area.

Previous studies have focused on isolating fungal pathogens from plant environments to support plant development and health, but there has been limited attention given to phytopathogen management. This lack of attention may stem from a limited understanding of their sporadic interaction at the plant-root interfaces and the

Table 15.1 Biocontrol of various phytopathogens by endophytic fungi

Sr. no.	Endophytic fungi/ biocontrol agent	Pathogen targets	Reference
01	*Chaetomium globosum*	*Pyrenophora* sp.	Istifadah and McGee (2006)
02	*Beauveria bassiana*	*F.avenaceum*, *F. moniliforme*, *F. graminearum*, *Alternaria tennus*, *Aspergillus paraziticus*, *F.oxysporum*,	Sahab (2012)
03	*T. virens*	*F.oxysporum* f. sp. *lycopersici*	Jogaiah et al. (2018)
04	*T. harzianum*	*Alternaria solani*, *Botrytis cinerea Phytophthora infestans*	Chowdappa et al. (2013), Martínez-Medina et al. (2013)
05	*Arisaema erubescens*	*Fusarium oxysporium*, *R. solani*, *Magnaporthe oryzae*, and *Colletotrichum gloeosporioides*	Wang et al. (2012), Akram et al. (2023)
06	*Trichoderma citrinoviride*	*B. cinerea* and *Cylindrocarpon destructans*	Park et al. (2019)
07	*Trichoderma phayaoense*	*F. equiseti*	Nuangmek et al. (2021)
08	*Trichoderma longibrachiatum*, *T. atroviride*, and *T. asperellum*	*Rhizoctonia solani*	Sallam et al. (2021)
09	*Phialemonium inflatum*	*Meloidogyne incognita*	Zhou et al. (2018)
10	*Pochonia chlamydosporia*	*Meloidogyne javanica*	Escudero et al. (2017)
11	*Trichoderma atroviride*	*M. javanica*	Medeiros et al. (2017)
12	*Hypocrea lixii*	*Iris yellow spot virus*	Muvea et al. (2018)
13	*T. asperellum*	*Cucumer mosaic virus*	Elsharkawy et al. (2013)
14	*T. hamatum*	*Xanthomonas euvesicatoria*	Fontana et al. (2021)

metabolites they produce. Fungi that reside on plant roots are considered to be a natural part of the microflora in the endo-rhizosphere. Sylvia and Chellemi (2001) demonstrated that these fungi primarily reside in the soil-root environment. Researchers can now determine their functional characteristics through in vitro assays, thanks to studies on the relationships between endophytes that inhabit plant roots (Vélez et al. 2017). Rhizosphere, a distinct group of root endophytes, can easily invade plant roots from the surrounding soil and establish colonization (Ghaffari et al. 2019).

Further investigation should be conducted to comprehensively understand the potential use of endophytic fungi in mitigating plant diseases, as these fungi possess the capacity to enhance a plant's resilience against environmental stressors and phytopathogens. Furthermore, learning about the diverse functions these microbes do in the environment may assist in our comprehension of the advantageous plant-microbe relationship that boosts biomass yield.

Recent studies have documented insights into the community structure and mode of existence of endophytic fungi in different plants, aiming to evaluate their functional characteristics (Manzotti et al. 2020). Nadeem et al. (2014) found that environmental conditions, such as salinity levels, dryness, and high temperatures, can affect the interaction of fungus communities at plant-root interfaces. Root exudate secretion and substrate metabolism also play a significant role in controlling the metabolic processes and diversification of fungal communities below the surface (Woźniak et al. 2019). The soil-root interface is recognized as a crucial area for the growth of microbial communities and the colonization of root-associated endophytic microorganisms. This is primarily due to the abundance of nutrients available in this region (Liu et al. 2019).

In order to comprehend the intricate dynamics and extensive array of interactions that occur between endophytic microorganisms and host plants, it is imperative to conduct a comprehensive examination of metabolite production (Adeleke and Babalola 2022). According to Qu et al. (2020), there is a proposition that plants emit a substantial quantity of exudates through their root systems. The incorporation of diverse methodologies in the examination of endosphere biology has the potential to make valuable contributions toward the formulation of a sustainable framework for the management of fungal biodiversity in plants, as well as the comprehension of plant-fungal interactions. Agricultural biological prospecting has led to the identification of numerous advantageous endophytes in various plant species, spanning a wide range of climatic and geographic situations (Dubey et al. 2020).

Using methods unique to a certain culture or by direct inspection, fungi are capable of being isolated and identified. Fungus inside plant tissues can be directly observed and visualized through the use of optical and electron microscopes. The method described in this study is particularly advantageous in the identification of endophytic fungi in addition to those that exhibit limited growth on regular media (Nazir and Rahman 2018). However, direct observation has limitations as it does not provide taxonomic categorization and can only identify endophytic fungi based on their hyphal structure. Therefore, a cultivation-dependent approach is necessary to complement direct observation. The fungal endophyte can be extracted from plant tissues and examined using either conventional or molecular methods, such as the culture-dependent approach. In place of traditional morphological characterization, the molecular exploration of ribosomal DNA internal transcribed spacer sequencing has been utilized (Nazir and Rahman 2018).

Endophytic microorganisms that are plant-specific exhibit dual functionality by providing protection against plant diseases and facilitating plant development through diverse mechanisms. The provision of this protection serves to augment agricultural productivity and guarantee the sustenance of food supplies. The utilization of plant endophytic fungi, known for their efficacy in combating plant insects, for the purpose of manufacturing plant insecticides along with the associated metabolites for commercial applications, has the potential to significantly augment food safety (Bamisile et al. 2021).

15.6 Mechanisms Employed by Fungal Endophytes in Phytopathogen Control

Endophytic fungi retain a tendency to inhabit the inner tissues of both monocotyledonous and dicotyledonous plants, hence augmenting their adaptability and facilitating their growth across diverse ecological conditions (Waqas et al. 2012). These fungal endophytes employ different mechanisms to control pathogens, as illustrated in Fig. 15.3. While most endophytes primarily colonize the roots, they establish a protective system that extends to other parts of the plant (Adeleke and Babalola 2021). The host plant's chemical and physical defenses serve as a defense mechanism, actively promoting the development of induced resistance to plant diseases (Wani et al. 2016). These defense mechanisms elicit an intrinsic response to possible pathogenic hazards specifically fungus, by initiating interchangeable signals within the host organism. The initiation of an induced reaction is frequently triggered by specific chemical agents that promote the selective activation of genes, metabolic changes, and the synthesis of proteins. The incidence of the disease has been observed to decrease due to modifications in the plant's metabolic rate and enhanced integration with the host organism (Latz et al. 2018).

Abiotic as well as biotic factors can induce local or systemic host reactions. Priming refers to the activation of a plant's natural defenses against infections (Martinez-Medina et al. 2016). Plant-induced resistance is often associated with the ability to mobilize cellular defensive responses against foreign agents. Molitor et al. (2011) have reported that *S. indica* elicits plant defense responses against barley powdery mildew utilizing an array of strategies. The researchers postulated that the application of *S. indica* promotes the augmentation of localized cellular apoptosis and the development of papilla structures in barley. Additionally, it was observed

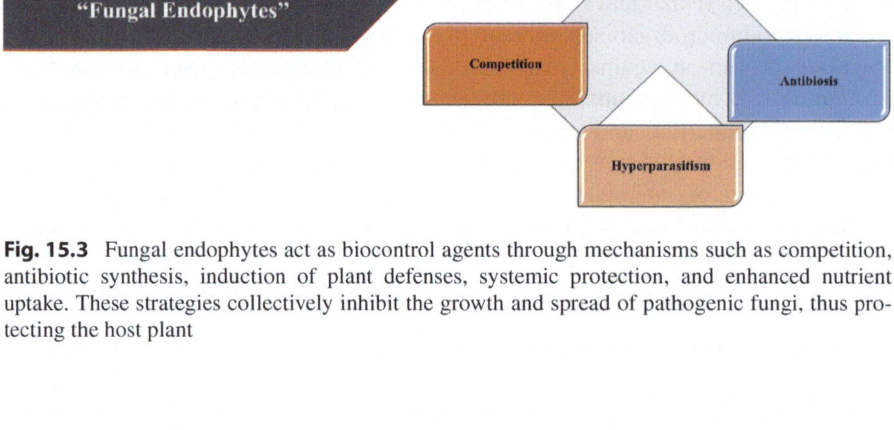

Fig. 15.3 Fungal endophytes act as biocontrol agents through mechanisms such as competition, antibiotic synthesis, induction of plant defenses, systemic protection, and enhanced nutrient uptake. These strategies collectively inhibit the growth and spread of pathogenic fungi, thus protecting the host plant

that the expression of HvPR17b, a pathogenesis-related gene, was upregulated in the leaves. These physiological responses are believed to reduce pathogen penetration and contribute to the induced resistance against powdery mildew. Additionally, variations in the expression of PR genes were observed in the roots. Genes for PR2, BCI-7, PR1, and Hsp70 form a group that promotes protein synthesis, activating defense mechanisms in barley crops treated with *Blumeria graminis* f. sp. *hordei* (Molitor et al. 2011). This PR complex achieves both direct and indirect plant growth advancement as well as antifungal effects. The inhibitory effect of *Harpophora oryzae* on *Pyricularia oryzae*'s influence on rice was observed in a study conducted by Ashrafi et al. (2023).

Polonio et al. (2015) reported that the virulence of *Guaco* (*Mikania glomerata* Spreng.)-induced pathogens, specifically *Fusarium solani* as well as *Didymella bryoniae*, is modulated by the presence of the endophytic fungus *Diaporthe citri*. The endophytic fungus in question has been observed to promote plant development and demonstrate antibacterial capabilities against multiple diseases. Furthermore, in a study conducted by Latz et al. (2018), it was observed that the upregulation of OsWRKY4, a transcriptional gene that is dependent on salicylic acid and is correlated with resistance to rice blast, occurred when the infection of *Pyricularia oryzae* in rice was inhibited through the use of the fungal endophyte, *Harpophora oryzae*. Efforts have been undertaken to ascertain the fungal species accountable for the production of compounds that are linked to plant defense mechanisms in industrial contexts, specifically in the context of combatting phytopathogens (Davies et al. 2021).

According to Latz et al. (2018), a variety of bioproducts derived by endophytes, including steroid hormones flavonoids, peptides, terpenoids, polyketides, as well as quinones, have been found to possess antibacterial properties against plant diseases. The research carried out by Rojas et al. (2020) revealed that the interaction between various microorganisms present in the root endosphere of plants might lead to synergistic effects and the production of biocontrol agents. These compounds are utilized for the purpose of hindering the spread of phytopathogens. The aforementioned process bears resemblance to the commencement of metabolism observed in metabolites generated by endophytes and is contingent upon specificity. One instance illustrating this phenomenon is the recent finding that a majority of endophytic fungal species found in *Taxus brevifolia* are capable of synthesizing anticancer chemicals (El-Bialy and El-Bastawisy 2020). Nevertheless, numerous studies have shown evidence that a substantial number of these endophytes had the capacity to engage in mutualistic interactions with various species, hence contributing to the development of resistance mechanisms that might be harnessed in industrial contexts to combat phytopathogens (Heinig et al. 2013).

The confirmation of the efficacy of an antibiotic chemical synthesized by an endophytic microbe against diseases relies primarily on the establishment of direct contact with the pathogen. However, it can be challenging to determine the rate at which metabolites are formed due to the presence of endophytic fungal organisms in the plant's endosphere. Nonetheless, plant endophytes can reach the pathogens' source through metabolites produced by microbes, and organic substances released can efficiently reach the site of infection (Ahmed et al. 2023). The adequacy of substance release at the infection site for impeding plant pathogen invasion, as well as the potential use of alternative strategies to mitigate plant diseases, remains uncertain at now (Lahlali et al. 2022).

15.6.1 Fungal-Induced Plant Resistance

The term “plant resistance” pertains to the capacity of a host organism to impede or restrict the incursion and progression of diseases. The phenomenon encompasses a range of defense mechanisms, comprising both biological and structural aspects. This resistance is a process that occurs over time. The plant’s immune system includes both intrinsic and induced defensive mechanisms, which can be augmented by both abiotic and biotic stressors. Induced resistance pertains to the heightened capacity for protection against diseases and pests. The application of this biological phenomenon can be utilized for the aim of investigating the resistance and susceptibility of plants to plant pathogens, as well as its potential implementation as a strategy for the control and prevention of plant diseases (Oliveira et al. 2016; Burketova et al. 2015; Llorens et al. 2017).

Induced resistance exhibits potential efficacy against a diverse range of pathogens such as fungus, viruses, phytonematodes, bacteria, as well as abiotic stresses, hence offering several beneficial results. Nonspecificity, systematicity, tenacity, and grafting transfer are a set of mechanisms for resistance that collectively contribute to the establishment and maintenance of induced resistance, hence enhancing stability in the system (Pascholati and Dalio 2018). The “priming” state is the optimal outcome of induced resistance. It triggers a heightened state of alert in stimulated plants, leading to stronger expression of their defense systems when exposed to stressors (Pascholati and Dalio 2018; Mauch-Mani et al. 2017). “Induced resistance” typically refers to a protective response that is localized to the tissues exposed to the inducing medication. However, it can also encompass a systemic reaction that affects other areas (El-Shetehy et al. 2021).

Through the means of antibiosis, formation of lytic enzymes, parasitism, paralysis, as well as competition for space, endophytic fungal organisms mitigate the incursion of plant-parasitic nematodes. The other mode of action mechanism includes the stimulation of plant defenses, specifically the initiation of induced systemic resistance as well as systemic required resistance by *Trichoderma*. These effects also seem to be heritable. Additionally, arbuscular mycorrhizal fungi have the capacity to alter the exudates that are released from the plant roots, along with the formation of strigolactones, numerous enzymes, and plant secondary metabolites (Poveda et al. 2020). Furthermore, the initiation of induced systemic resistance and systemic acquired resistance within the plant root system along with the formation of strigolactones by fungal endophytes and the translocation of chemical defense compounds, has been clearly demonstrated in (Fig. 15.2).

Plant extracts, cultured yeast, rhizobacteria, fungi, avirulent pathogens, and endophytic fungi are examples of substances that can activate a plant’s defense

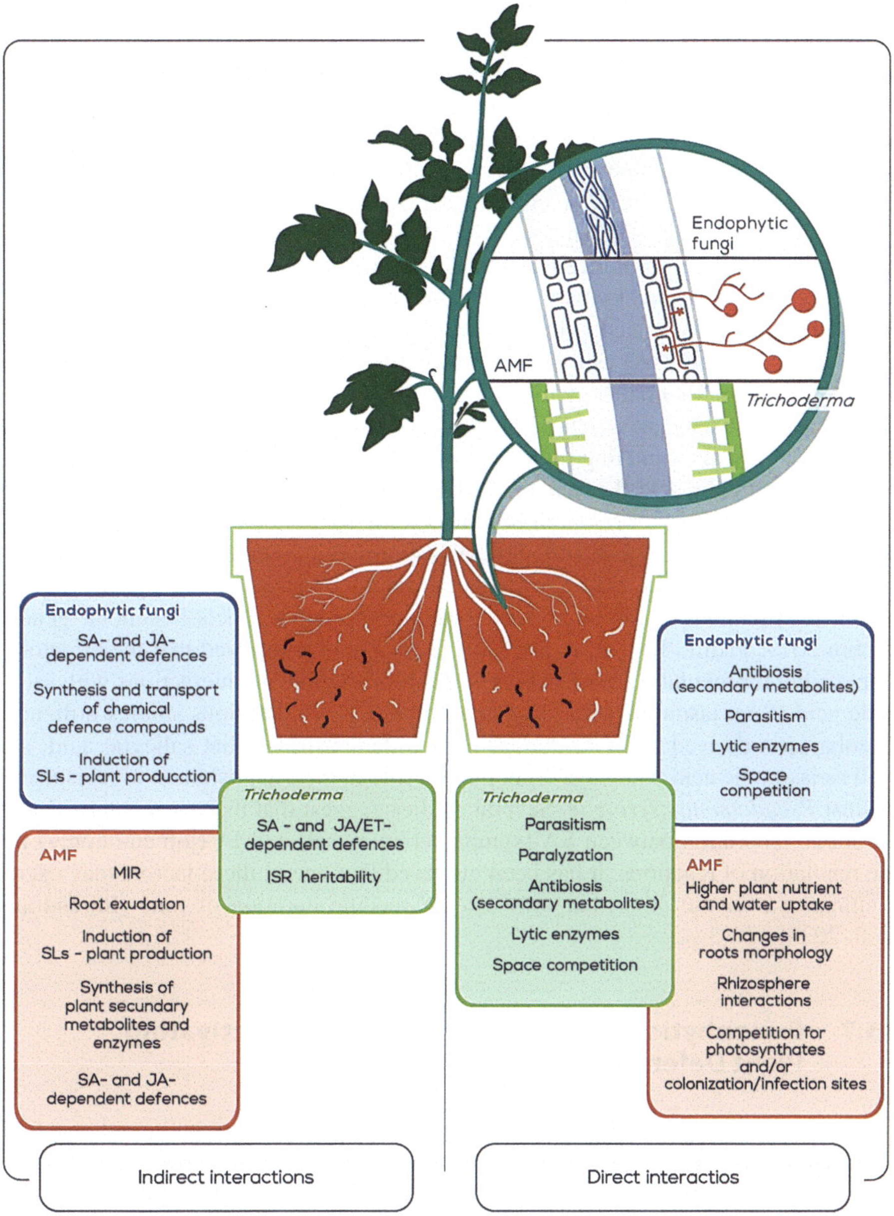

Fig. 15.2 The above figure shows the survival mechanism of filamentous fungi residing in a split-root system against the plant-parasitic nematodes. Different fungi practice various tactics for colonizing specific tissues, which alter the development of nematodes, help in reducing the rate of infection, and elevate the mortality of J2. The resistance provided by fungal species is characterized by different terminologies, i.e., systemic acquired resistance (SAR), jasmonic acid (JA), salicylic acid (SA), arbuscular mycorrhizal fungi (AMF), and mycorrhizal induced resistance (MIR). (Cited from Poveda et al. 2020)

mechanism (Schulz and Boyle 2005; Vardharajula et al. 2017; Haas and Défago 2005; Ab Rahman et al. 2018; Khoshru et al. 2023). Induced resistance is a technique that involves using external substances to trigger resistance without altering the plant's genome (Sundin et al. 2016). This phenomenon is achieved through the activation of genes associated with diverse defensive reactions, including oxidative bursts, hypersensitive reactions, the accumulation of pathogenesis-related (PR) proteins, enzymes in the phenylpropanoid pathway, and lipid peroxidation-inducing enzymes (Nawrot 2017; Ghosh et al. 2020; Silva et al. 2022; Stroud 2023).

Additional types of induced resistance include systemic acquired resistance (SAR), which is initiated by biotrophic pathogens and largely involves the production of pathogenesis-related (PR) proteins, as well as nonpathogenic bacteria that utilize salicylic acid (SA) as their main signaling molecule. This categorization is based on the signaling pathway that triggers the activation of defense mechanisms. In the case of resistance provided by bacteria or diseases promoting rhizospheric growth, the primary signaling molecules are ethylene and jasmonic acid, which do not rely on salicylic acid (Lucena et al. 2023; Salwan et al. 2023).

This is an overview of research findings on the bacteria *Phakopsora pachyrhizi*, which causes Asian soybean rust. Studies suggest that when these bacteria invade host tissue, it interferes with SA-mediated defense systems and triggers the jasmonic acid pathway. This alteration in gene expression affects beneficial genes (Cabre 2019; Padilla-Roji et al. 2023). It is widely hypothesized that the effectors responsible for regulating gene expression exhibit significant interactions with salicylic acid (SA), jasmonic acid (JA), and ethylene (ET). Previous studies utilizing rhizobacteria derived from *Arabidopsis* have demonstrated that salicylic acid, as well as jasmonic acid pathways both, plays a role in the establishment of resistance against *Pseudomonas syringae*. Several studies suggest that there is a potential for partial collaboration between SA (sympathetic arousal) and JA (job autonomy) in the regulation of reactions. It has been observed that one of these factors may exert dominance over the other, but only after a specific duration of time (Rajendran et al. 2023).

15.7 Endophytic Fungi-Derived Compounds Activating Plant Defenses

Microbes that are both useful and pathogenic can enter their host without damaging the plant's plasma membrane in order to obtain nutrients. Plants possess an inherent defense mechanism that reacts to molecules linked to microorganisms or other harmful agents (known as MAMPs and PAMPs) in addition to their effectors. Plant pathogens secrete a diverse array of potent molecules known as effectors, which facilitate the establishment and progression of pathogenic infections. Nevertheless, certain effectors have the potential to inadvertently trigger the activation of plant immune receptors. Pattern-triggered immunity (PTI) serves as the initial defense mechanism that can be induced upon the identification of extracellular damage-associated molecular patterns (DAMPs) and microbe-associated molecular patterns (MAMPs/PAMPs) (Loo et al. 2022; Hou et al. 2019).

The implementation of pathogen transmission interventions (PTIs) is of utmost importance in mitigating the dissemination of evolved pathogens to susceptible hosts and safeguarding the host population against illness caused by non-adapted microorganisms. Plants exploit receptor-like kinases (RLKs) as well as receptor-like proteins (RLPs) as receptors for pattern recognition (PRRs) for the purpose of detecting chemical patterns originating from both host organisms and microbes (Ali et al. 2023; Oelmüller et al. 2023). Plants, akin to human responses, possess the ability to discern nonself/microbial constituents as discrete external entities. Through the initiation of defense responses, host plants are able to promptly anticipate and make necessary preparations in response to potentially detrimental microorganisms. While the host plant identifies both endophytes and pathogens in a similar manner, its reactions to those two foreign organisms are significantly different (Wani et al. 2016).

The utilization of fungal-induced resistance by endophytic microorganisms has been observed to successfully reduce diseases, as evidenced by the effective coexistence of endophytes as well as host plants. Host plants recognize specific components of endophytic fungi, such as protein compounds, cell walls, lipids, and volatile compounds like biocontrol agents. These compounds and elements from endophytes activate defense mechanisms in plants. MAMPS (microbe-associated molecular patterns) and PAMPS (pathogen-associated molecular patterns) are terms used to describe microbial mechanisms (Nürnberger and Kemmerling 2009). They are in charge of inducing immunological reactions referred to as MAMP-/PAMP-triggered reactions (Land 2023). MAMPs, including chitin as well as β-glucans, are found in fungal cell walls and function by recognizing receptors to cause immunological responses in plants. Researchers generally agree that endophytic secretions, such as proteins and peptides, act as triggers for plant reactions (Rojas et al. 2020). In addition, disease-induced production of enzymes like chitinases, cellulase, and xylanases can also induce plant defense mechanisms. The degraded by-products of these enzymes are easily detected by the host (Sood et al. 2020; Bulgari et al. 2023).

Endophytic and pathogenic interactions result in the production of cysteine-rich proteins and fungal effectors, which promote compatibility with the host plant by triggering both physiological and defensive responses (Puranik et al. 2023; Gurung et al. 2023). However, research has shown that substances developed to inhibit the growth of competing microbes can also contribute to resilience (Jayaraman et al. 2021). The endophytic fungi-derived compounds mentioned earlier possess complex structures and have the ability to stimulate plant defense responses, which are necessary for reducing the risks associated with plant diseases and other soil-related abnormalities. The early phase of the plant cell immune system response can differentiate between pathogens and endophytes through a MAMP-mediated evaluation.

Terrestrial plant health and survival in a microbe-rich habitat are dependent on a critical genetic network that regulates the number and variety of endophytic organisms (MIN7 and CAD1 proteins, Igiehon and Babalola 2018). Additionally, this framework incorporates immune signaling triggered by plant patterns. Lipid compartments comprise plant extracellular vesicles (EVs), which serve to facilitate the intercellular transport of signaling lipids, metabolites, and RNA. In addition, they assume a pivotal function in the regulation of interplant-microbe interactions and demonstrate nonpathogenic properties when subjected to stress. EVs are produced by plant cells in response to immune stimuli. They have the ability to inhibit fungal gene expression and stress responses (Rutter and Innes 2018). Furthermore, plants possess the ability to release plant hormones as well as secondary compounds, including triterpenes, flavonoids, alkaloid compounds, organic substances, and amino acids, via root exudates. These compounds have the ability to function as stimulants, attractants, repellents, inhibitors, and signaling agents. Their functions include the prevention of pathogen-caused diseases and the inhibition of herbivore activity (Lu et al. 2021; Al-Khayri et al. 2023). Bioactive compounds produced by various endophytic fungi has been demonstrated in (Table 15.2).

Table 15.2 Endophytic fungus and their diverse array of bioactive compounds/metabolites

Sr. no	Fungal endophyte	Bioactive compound/ metabolites	Reference
01	*Phyllostica* sp.	Heptelidic acid, hydroheptelidi, and chlorinated metabolites	Anamika et al. (2018)
02	*Phomopsis* sp.	Cycloepoxylactone, phomopsichalasin, and cycloepoxytriol B	Anamika et al. (2018), Mousa and Raizada (2013)
03	*Seimatosporium* sp.	Seimatorone	Zhao et al. (2020)
04	*Colletotrichum* sp.	Colletrotic acid	Anamika et al. (2018)
05	*Talaromyces purpureogenus*	Talaromyolide K	Cao et al. (2020)
06	*T. asperellum*, *Trichoderma longibrachiatum*, and *T. atrovirid*	Chitinase, siderophore, Pectinase, and IAA	Sallam et al. (2021)
07	*Nodulisporium* sp.	Sporothriolide	Cao et al. (2016)
08	*Trichoderma harzianum*, *T. lentiforme*, *Epicoccum purpurascens*	Secondary metabolites, Microbial antagonist	González et al. (2020)
09	Ampelomyces	Altersolanol A	Patil et al. (2016)
10	A. *alternata* (*KT380662*)	Flavone chrysin (5, 7-dihydroxy flavone)	Chatterjee et al. (2019)
11	*A. awamori*	Peptide lectin (N acetylgalactosamine	Adeleke and Babalola (2021)
12	*A. niger* (PN2)	Monacolin K/lovastatin	Patil et al. (2016)
13	*Chaetomium globosum*	Chaetoglobosin and Chaetoglobosin	Zhao et al. (2017)
14	*Colletotrichum gloeosporioides*	Colletotric acid	Zou et al. (2000)

15.8 Assessing Induced Resistance Using a Plant's Defensive Mechanisms

Over the past two decades, several studies have investigated how plants respond to parasite and pathogen invasions using various subcategories (Mehmood et al. 2023). Among these, induced systemic resistance (ISR) and systemic acquired resistance have been the most extensively studied. ISR is regulated by ethylene and nonpathogenic rhizobacteria that produce jasmonic acid, which is not associated with the accumulation of pathogenesis-related (PR) proteins. Pathogen-induced diseases lead to systemic acquired resistance (SAR), which is regulated by salicylic acid (SA) and is associated with the production of PR proteins (Tripathi et al. 2020).

A multitude of enzymes, including chitinases and 1, 3-glucanases, which are present within these PR proteins, efficiently degrade invading cells. Additionally, they promote cell mortality, fortify cell wall borders, and bolster resistance to invasion (Rani 2008). ISR is stimulated by *Fusarium solani*, an important endophyte found in tomato roots, in response to *Septoria lycopersici*, the organism accountable for foliar disease in tomatoes. Moreover, it enhances the operational capacity of the PR genes PR7 and PR5 located in the root system (Singh et al. 2020b; Anand et al. 2023).

Greater amounts of peroxidase along with deposition of lignin were observed when *Citrullus lanatus* and *Cucumis sativus* were treated with a nonpathogenic mutant variant of *C. magna* (Redman et al. 1999). These changes help protect the plants from diseases caused by *F. oxysporum* and *C. orbiculare*. *Neotyphodium lolii* has been found to combat four different diseases, resulting in reduced lesions on leaves. This effect is likely due to increased peroxidase and superoxide dismutase activities in the host plant (Zhang et al. 2021; Prakash 2023). After introducing the pathogen to the plant, it is important to observe the expression of defense responses to confirm its presence. The utilization of the exclusion concept is a legitimate methodology for evaluating the efficacy of induced resistance in safeguarding plants. Nevertheless, neglecting a direct examination of resistance induced in vitro presupposes that the influence of the pathogen on the observed response is unintended. An additional conserved activity that potentially plays a role is the plant's capacity to fortify cell wall appositions and boost barriers that protect against pathogen invasion (Hu 2023). The occurrence of this result has been attributed to the inciting factors, according to a multitude of studies.

An investigation conducted on *Trichoderma harzianum* (T-22) unveiled that maize displayed a notably elevated concentration of phenylalanine ammonia-lyase (PAL), an enzyme responsible for lignin production. Induced resistance occurs when a plant's defense mechanisms are stimulated in order to augment its resistance against diseases. These mechanisms work together to strengthen the physical barriers of plants and deter phytopathogens (Meena et al. 2020; Shanmugaiah et al. 2023). Pathogen-repellent proteins and enzymes are secreted as pathogen repellents. In order to classify and evaluate plant reactions and ascertain whether a fungal endophyte induces a defense mechanism against a particular disease, two criteria are employed (Fig. 15.3). Firstly, the response should effectively control the targeted plant pathogen(s). Demonstrating the efficacy of fungi in pathogen eradication is imperative, with any ensuing implications strongly associated with pathogen infection prevention. Further, a correlation must exist between Koch's postulate as well as the phytopathogen eradication mechanism. It was discovered that the existence of endophytic fungus inside the cells of a distinct strain of *T. harzianum* (T-203) induced a mechanism that fortified the epidermal as well as cortical wall of cells of cucumber fruit (Shafi et al. 2023).

Antimicrobial proteins and metabolites that are crucial for the effective management of plant diseases have been stimulated and activated via systematic administration of inducing agents. An example of such beneficial metabolites is phytoalexin-like substances (Oliveira et al. 2016). Endophytic fungal colonization results in the production of pathogen-related proteins, which serve additional functions such as stress response and antibacterial abilities. These proteins include lipid-protein transporters, thionins, and thaumatin-like proteins (Sodhi and Saxena 2023).

Increased activity of the plant-related protein (PR2) 1, 3-glucanase was found in oilseed rape plants infected by *Plasmodiophora brassicae* via the endophytic fungal microbe, *Heteroconium chaetospira*, according to studies conducted by Lahlali et al. (2022). Dudeja et al. (2021) discovered systemic resistance in *Cephalotaxus harringtonia* infected with *Fusarium oxysporum*, caused by the endophytic fungus *Paraconiothyrium variabile*. Beauvericin as well as 13-oxo-9, 11-octadecadienoic acid, two metabolic products that prevent growth and harmful effects, were produced as a result of this resistance. In tomato plants, *F. solani* strain "Fs-K" invasion triggered an induced resistance that increased the production of endo-proteinase (PR7) and thaumatin-like (PR5) enzymes (Latz et al. 2018).

In their study, Huang et al. (2016) put forth a hypothesis suggesting that *Blumeria graminis*, a fungal organism, will exhibit resistance toward the antifungal properties associated with the upregulation of the protein HvPR17b in barley that is colonized by the endophytic *Serendipita indica*. In a study conducted by Deshmukh and Kogel (2007), it was demonstrated that the activity and significance of PR genes may vary across different systems. They investigated the synergistic influence of *F. graminearum* and *S. indica* on barley, which resulted in a decrease in the expression of genes related to pathogens (PR1b and PR5).

15.9 Antibiotics and Secondary Metabolite Production

A considerable proportion of endophytes possess the ability to synthesize secondary substances, a subset of which have been scientifically validated to exhibit antibacterial and antifungal characteristics, hence impeding the proliferation of phytopathogenic microorganisms (Hashem et al. 2023; Singh and Kumar 2023). Endophytes have been shown to produce various secondary compounds with potential biological effects against pathogens or pests, as demonstrated by Kusari et al. (2012) and Perumal et al. (2023). Examples of bioactive metabolites include camptothecin, paclitaxel, deoxypodophyllotoxin, and podophyllotoxin (Rutkowska et al. 2023).

The endophyte *Aspergillus fumigatus* derived from *Juniperus communis* synthesizes deoxypodophyllotoxin, a compound with antibacterial properties effective against various harmful microorganisms (Choudhury et al. 2023). The fungal endophyte *Eupenicillium parvum*, discovered in *Azadirachta indica*, produces the natural insecticide azadirachtin. This compound possesses antifeedant and growth regulating properties against insects (Kharwar et al. 2020). Furthermore, the endophytic *Phialocephala fortinii* extracted lignan podophyllotoxin from the plant *Podophyllum peltatum*. This compound is a clinically useful precursor for anticancer medicines (Biswas et al. 2020; Eyberger et al. 2006; Silva et al. 2022). Endophytic *Fusarium solani* has been identified as the origin of important precursors for anticancer drugs, such as camptothecin and 10-hydroxycamptothecin (Aly et al. 2013).

The induction of host defense-associated genes in *Theobroma cacao* by the fungal organism *Colletotrichum tropicale* has been documented by Winters (2022) as well as Samanta et al. (2023). Mejía et al. (2014) found that *C. tropicale* also alters expression of gene related to the host's physiology, metabolism, growth, and resistance to pathogens and herbivores. Specific genes involved in pathogenesis include PR4 protein and proline-rich proteins responsible for cell wall formation, modification, and degradation. Additionally, signaling proteins such as peroxidases, receptor kinases, and members of the jasmonate defense pathway play a role in disease response.

Endophytes play a crucial role in combating infections through diverse mechanisms. According to Dutilloy et al. (2022), endophytes exhibit various behaviors that enhance their defense against pathogens like *Puccinia triticina*. These behaviors include increasing levels of harmful phenolic substances, producing other defense-related compounds, showing direct chemical and physical antagonism, and reducing the pathogen's carbon utilization. Furthermore, *Harpophora oryza*e, an endophytic fungus that establishes symbiotic relationships inside the roots of rice plants, employs a variety of mechanisms to effectively inhibit the growth and development of the biotrophic rice blast disease that is caused by the pathogenic fungus *Magnaporthe oryzae*.

Following infection, *Harpophora oryzae* induces an elevation in the accumulation of hydrogen peroxide (H_2O_2) inside the plant tissues. The induction of programmed cell death is initiated by the passage of this particular reactive oxygen species across biological membranes, encompassing certain epidermal as well as outer cortical cells. In addition, *Harpophora oryzae* effectively impedes the accessibility of *Magnaporthe oryzae* to essential resources like water and nutrients, hence thwarting its invasion (Su et al. 2013). Antioxidant enzymes play a crucial role in safeguarding plant cells from reactive oxygen species and maintaining their optimal functionality (Su et al. 2013). Inducing ROS-detoxifying enzyme activity in rice would trigger a systemic defense response against *Magnaporthe oryzae* (Su et al. 2013).

Numerous research fields are still actively pursuing the discovery of endophyte metabolites for potential commercial use. Studies conducted by Suryanarayanan (2013) and Daguerre et al. (2016) have explored various bioactive substances that show promise in inhibiting phytopathogens. In addition, the existence of antimicrobial endophyte metabolites, including terpenoids, flavonoids, peptides, polyketides, alkaloids, phenols, quinones, as well as steroids, has been emphasized in recent studies conducted by Lugtenberg et al. (2016) and Mousa and Raizada (2013). Endophytic fungi represent a significant reservoir of bioactive substances that confer advantageous effects on their host plants, hence facilitating the synthesis of defensive molecules.

Plants synthesize a variety of compounds, such as alkaloids, volatile organic compounds, peptides, flavonoids, phenols, steroids, as well as polyketones which exhibit characteristics that might effectively combat pathogens and herbivores (Lugtenberg et al. 2016). Furthermore, Latz et al. (2018) examine the process of synthesizing compounds that exhibit antifungal, antiviral, and antibacterial properties against insects. The phenomenon of fungal endophytes releasing secondary chemicals, exemplified by *Epichloë* species synthesizing alkaloids in diverse grass species, is widely recognized in scientific communities. Alkaloids, which accumulate in plants, have harmful effects on several insect pests, including vertebrates (Riaz et al. 2023). *Nodulisporium* sp. produces nodulisporic acid, which plays a crucial role in deterring insect herbivory. This acid stimulates glutamate within chloride channels found in insect nerve and muscle cells. The activation of glutamate leads to the influx of chlorine ions through specific pathways, resulting in flaccid paralysis (Bills et al. 2012).

Endophytic fungi are known to produce a wide range of secondary metabolites, including volatile organic molecules. These compounds have been found to possess bioactivity against various diseases. However, despite extensive research, there are still metabolic pathways and compounds produced by endophytic fungi that remain poorly understood. Recent studies by Baron and Rigobelo (2022) and Tabbasum et al. (2022) have identified over 300 different secondary chemicals produced by these fungi. They function as special agents in the interaction between fungi and various organisms, including plants. The biocontrol approach known as mycofumigation, which was developed by Strobel et al. (2001), is employed for the purpose of controlling diseases in harvested fruits as well as tubers. Kaddes et al. (2019) conducted a comprehensive study on the volatile organic compound (VOC) production of *Muscodor*, revealing a diverse range of metabolites produced by this organism. Furthermore, Giorgio et al. (2015) have reported that this fungus has been employed as a soil inoculant with the purpose of suppressing the proliferation of deleterious fungi. The *Nodulisporium* genus, like *Muscodor*, has also been utilized for similar purposes and is known to produce VOCs with antifungal properties (Strobel 2018).

Secondary metabolites are synthesized by organisms to fulfill various functions, such as signaling, defense, or other objectives, with the aim of establishing a link with the host plant. Furthermore, they possess the capacity to exert an influence on the secondary metabolic by-products produced by host, hence potentially impacting the pathogenicity of pathogens. These goods are unlikely to be commercialized in the near future due to the harmful effects of synthetic compounds used in agriculture on humans, livestock, and the environment. Additionally, many of these compounds are already prohibited (Lugtenberg et al. 2016). Consequently, research and businesses are focusing on utilizing and modifying biological products as a more sustainable alternative for agriculture.

15.10 Role of Hormones in Inducing Resistance

Plant hormones play a crucial role in providing comprehensive and structural inherent resistance in plants. There are two distinct types of systematically induced resistance: systemically acquired resistance (SAR) and induced systemic resistance (ISR). The induction of systemic resistance is principally facilitated by the secretions of ethylene and hormonal jasmonic acid, whereas systemically acquired resistance is predominantly mediated by SA (salicylic acid) (Latz et al. 2018). The simultaneous occurrence of ethylene, JA, and SA in plants establishes a consistent link between hormonal reactions and promotes defense mechanisms (Tiwari et al. 2021; Kaniganti et al. 2022). For example, *Penicillium* sp., *Serendipita indica*, and *Trichoderma asperellum* have all induced defensive responses leading to the production of ethylene and JA-dependent systemic resistance. These responses are crucial for limiting the survival of host pathogens (Llorens et al. 2022). In addition to ethylene and jasmonic acid (JA), *S. indica* has been found to induce defense mechanisms for several pathophysiological processes unrelated to ethylene/JA. An alternative scenario involves the activation of a salicylic acid (SA)-dependent mechanism in a plant that has been invaded by *T. asperellum* (Konappa et al. 2020).

This study demonstrated the complexity and diversity of hormonal responses and relationships with the host, as well as the usual cross-communication involved in plant-induced reactions. Sometimes microbial colonization affects not just one but several hormonal responses in a plant, changing both the typical response and profile. The coexistence of microbial strains with the host organism affects variations in hormonal responses. Even while research on induced resistance has advanced significantly, it is still difficult to comprehend the precise roles that each hormone plays in signal transmission, especially in complex systems. To detect induced resistance, a plant's defense mechanisms must be strengthened and used as biomarkers.

15.11 Antibiosis Activities of Fungal Endophytes Against Phytopathogen

Endophytic fungi induce antibiosis, a type of antagonistic interaction wherein they exert an effect on potential diseases of plants through the consumption of their metabolic products. The apoplastic fluid of red fescue inoculated with endophytes contains a purified version of Efe-AfpA, which has shown antiparasitic activity against *Sclerotinia homoeocarpa* (Adeleke et al. 2022). The recombinant plant expressing the *Pichia pastoris* Efe-AfpA gene also exhibited similar results. The presence of the protein Efe-AfpA, which is produced as a result of the endophytic association between *Epichloë festucae* and *Festuca rubra* sp., was seen in 6% of the transcriptome analysis conducted on the fungal specimens. The research conducted by Tian and Chun (2017) revealed that the product shares similar characteristics with the substance produced when *Aspergillus* spp. and *Penicillium* sp. interact.

Endophytes produce various secondary compounds, some of which have antifungal and antibacterial properties that help prevent the growth of disease-causing microorganisms in plants (Mousa and Raizada 2013; Khan et al. 2020). Antibacterial metabolites found in endophytes include peptides, flavonoids, polyketides, steroids, phenols, alkaloids, quinones, and terpenes (Munir et al. 2020). It is necessary to investigate the secondary compounds produced by fungal endophytes for industrial applications (Bolívar-Anillo et al. 2020; Egamberdieva and Jabborova 2020). Interactions between hosts and endophytes enhance chemical production and release, which have various applications, such as pathogen suppression (Yan et al. 2018; Adetunji et al. 2019). However, certain endophytes possess gene clusters that utilize the same precursors to produce active metabolites (Khare et al. 2018).

Several studies have shown that many endophytic strains cannot independently produce substances essential for host plant resistance against pathogen attacks (Alvin et al. 2014).

Trichoderma phayaoense was used to suppress wilt and other diseases of muskmelon which are caused by *Fusarium equiseti* and *S. cucurbitacearum*. The biocontrol potential of endophytic fungal isolate (UP-L113) was assessed against *F. equiseti* and *S. cucurbitacearum* (Fig. 15.4). The biocontrol potential of *T. phayaoense* was assessed in both in vivo and in vitro conditions. Moreover, the ability of this fungus to grow in stress posed by certain herbicides, fungicides, and insecticides was tested in artificial growth media. This study provided sufficient evidence for biological control potential of *T. phayaoense*, which would reduce the reliance on chemical-based fungicides (Nuankaew et al. 2022).

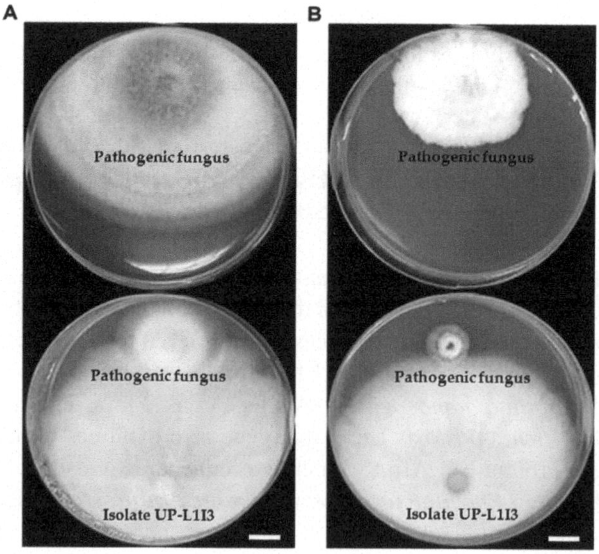

Fig. 15.4 The endophytic fungus isolate (UP-L1I3) was exposed to a dual culture experiment using potato dextrose agar (PDA). At 25 °C, under direct light, the development of *Fusarium equiseti* (**a**) and *Stagonosporopsis cucurbitacearum* (**b**) mycelia was observed after 5 days of incubation. On the above side, a control plate was placed, while an experimental plate was positioned below. The scale bars have a length of 10 millimeters

Furthermore, this particular antagonist has the ability to eradicate *V. dahliae* in both soil and laboratory settings by releasing a substance. This substance, known as glucose oxidase, was identified due to its interaction with glucose and the acetone precipitable component. The mechanism of action involves the production of hydrogen peroxide, which effectively destroys the pathogen's sclerotia. It is worth mentioning that the inclusion of either glucose or glucose oxidase results in a decrease in the quantity of viable microsclerotia in soil. However, the use of either substance individually does not entirely eradicate them. Both lactobacillin and hydrogen peroxide have been associated (Huang et al. 2023).

Fadiji and Babalola (2020) investigated the beneficial relationship between *Paraconiothyrium* strain (SSM001), an endophytic fungus, and a yew tree that produces Taxol to combat fungal organisms involved in wood decay. However, the yew tree is prone to developing bark cracks, which provide entry points for diseases. The investigation documented the dispersion of the fungal endophyte toward the aforementioned gaps, which may have the capacity to impede the buildup of Taxol. Additionally, it has been determined that two essential genes in the *Taxus* species, namely, DXP reductoisomerase and taxadiene synthase, which play a significant role in the production of Taxol, exhibited a downregulation in their transcriptional activity. The results of in vitro experiments done on the endophytic fungal strain SSM001 showed that the application of Taxol did not have a discernible effect on the growth of prominent wood-decaying fungi, which includes *Heterobasidion annosum*, *Phaeolus schweinitzii*, as well as *Perenniporia subacida* (Soliman et al. 2013).

15.12 Mycoparasitism

A fungal organism that depends on another fungal organism for sustenance is classified as parasitic. The phenomenon of mycoparasitism can manifest in both necrotrophic and biotrophic interactions. Necrotrophs feed on dead cells of their host, while biotrophs feed on living host cells (Kim and Vujanovic 2016). Detecting nutrient transfer between microbes makes it challenging to confirm mycoparasitism in plants. The majority of research that provides evidence for mycoparasitism primarily relies on circumstantial observations. The intimate association between the two fungal organisms is commonly known as a fungicolous symbiosis, as opposed to being classified as mycoparasitism. Mycoparasitism has the potential to manifest through both direct and indirect mechanisms. The concept of indirect fungal parasitism refers to the process by which a parasitic fungus obtains essential nutrients from a host organism through the utilization of a metabolite produced by the fungus. In contrast, direct mycoparasitism involves engaging in interactions with the prey (Latz et al. 2018).

In both scenarios, the parasite excretes several substances, including poisons, antibiotics, and enzymes, which facilitate the degradation of cell walls and the liberation of nutrients from the host (Kim and Vujanovic 2016). A study conducted by Chamoun et al. (2015) showed that the interaction between *Manatephorus cucumeris* and *S. elegans* resulted in the production of specific substances, with *S. elegans* preying on *T. cucumeris*. Many studies often confuse mycoparasitism and antibiosis due to their interconnectedness, making it challenging to differentiate between these interactions. Nevertheless, this study provides evidence that microorganisms involved in parasitic relationships can employ a diverse range of predation strategies.

Conventional methodologies, such as microscopy techniques and cultivation in Petri dishes, facilitate the investigation of microbial interactions. This simplifies the study of inhibitory relationships between microbes, making it unnecessary to conduct in planta screening. Mycoparasites have been found to have a close relationship with their hosts, which has been observed using simple microscopic techniques. They either directly interact with the host or wrap around the prey's hyphae to enhance nutrient absorption. This interaction was first identified in a study conducted by Donayre and Dalisay (2016). In our own research, we have discovered a direct mycoparasitic interaction between *Thanatephorus cucumeris*, a pathogen found in soil, and *Geotrichum* sp., an endophytic fungus isolated from *Echinochloa glabrescens*.

Three fungal endophytic strains originating from *Phragmites australis* have been found to enclose and penetrate into the hyphae of infectious soil microbes, resulting in the disruption of their cytoplasm. The procedure entailed the application of extracellular enzymes responsible for the degradation of cellular barriers, including −1, 3-glucanase, as well as other degrading enzymes. It is believed that a wide range of endophytic microorganisms inhabit the vast array of approximately three million plant species on Earth, particularly woody plants. The research on these various species offers significant possibilities for additional exploration (Nisa et al. 2015). In order to enhance the quality of novel products, it is imperative to elucidate and ascertain the structures of the metabolites exhibiting the highest levels of activity. It is noteworthy that chemicals inside a crude extract frequently have limited microbial activity when employed in isolation, as they synergistically interact with supplementary molecules generated by the microorganism. The incorporation of fungal endophytes into practices particularly integrated management is a beneficial tool for agribusiness, owing to the clarification of their action mechanisms and their significant contribution to plant safety.

Endophytic microorganisms have evolved a complex mechanism of communication with their host plants, enabling them to establish colonization and confer various benefits that might be harnessed in agricultural environment. Effector molecules, typically proteins, play a crucial role in signalization. Upon reaching the host plant, they initiate positive effects such as enhanced growth and activation of resistance. Numerous studies have focused on the omics strategy (genomics, metabolomics, proteomics, transcriptomics, etc.) to unravel this intricate network (Xu et al. 2023; Dikobe et al. 2023).

To select effective strains for biological pesticides and plant growth promoters, it is beneficial to employ the omics approach to identify and study proteins and genes. Moreover, metagenomics analysis has been used to examine the diverse range of microbes associated with plants in different environmental conditions

(Anguita-Maeso et al. 2023). This analysis also enables investigation into the role of the microbiome in maintaining the health of agricultural and natural systems. While the primary emphasis of this chapter does not lie on the omics approach, it is crucial to comprehend the correlation between endophytes and plants, along with their prospective agricultural implications. The use of endophytic microorganisms is driven by society's demand for sustainable food production through biotechnological methods.

15.13 Competition

Endophytes are thought to compete with diseases to prevent them from invading plant tissues (Hastuti et al. 2012; Mahmoud et al. 2023). These endophytic fungi can colonize plant tissues both locally and systemically, allowing for the observation of rapid inhabitation, nutrient consumption, and potential pathogen encroachment. A study conducted in 2009 by Mohandoss and Suryanarayanan examined the impact of fumigation on mango leaves. The findings revealed that the fumigation process caused the mortality of specific endophytes, thereby facilitating the dissemination of diseases. The incorporation of additional mechanisms in phytopathological management measures includes the utilization of competitive exclusion as well as endophytic colonization within the internal structures of plants. One instance of reduced impact of *Phytophthora* sp. on plant leaves was observed when a sterilized seed was subjected to treatment with a fungal endophyte derived from a cocoa tree (Arnold et al. 2003).

Lahlali et al. (2022) reported a negative correlation between the presence of symptoms related to clubroot disease and the invasion of soybean roots by the endophytic fungus *Heteroconium chaetospira*. Moreover, it was observed that a higher inoculum of the virus had a weaker control effect, highlighting the limitations of competition. For the purpose of studying competitive exclusion in an effective manner, it is advisable to utilize in-plant microscopic assessment and quantify the biomass of endophytes in relation to the management of phytopathogens. When evaluating strains of pathogens and fungal biological control agents (BCAs), visualizing the interaction between endophytes and pathogens using microscopy is advisable (Latz et al. 2018). According to Floros et al. (2017), molecular 3D mapping mass spectrometry can be used to analyze the distribution of metabolites, identify microbes present, and assess the genetic impact of mining compounds in situ.

15.14 How Might Endophytic Fungus Assist in Biofertilization and Stress Tolerance?

In certain regions, there are issues concerning unsustainable soil usage, forest loss, inadequate irrigation practices, reduced water accessibility, increased soil salinization, and desertification (Ferreira et al. 2023). To tackle these problems, breeding wild plants can be a solution to produce new plant varieties. Nevertheless, the mechanisms underlying stress tolerance in plants at the genetic level are still not fully understood, and the potential role of symbiotic interactions between plants and microorganisms in this context has not been adequately explored (Harris et al. 2020). Plants suffer from loss when their beneficial association with fungi is disrupted, leading to a decline in their ability to cope with environmental challenges. Similarly, fungi are negatively impacted when they lose their protected habitat. Consequently, the population of fungal endophytes thriving in wild plants may undergo significant changes, and some species may disappear during cultivation (Lugtenberg et al. 2016).

A comprehensive analysis of the factors contributing to the potential loss of endophytes during plant breeding was conducted by Lugtenberg et al. (2016). Endophytes reside in the innermost layer of plants, providing them with a protected environment that minimizes competition with other microorganisms. Additionally, this niche supplies the necessary nutrients for fungal growth and survival. Consequently, fungi enhance the overall health and viability of plants (Chitnis et al. 2020).

Differentiating between direct and indirect benefits of fungal endophyte colonization in plants can be challenging (Afzal et al. 2019). Direct benefits include the production of phytohormones and improved nutrient uptake, which promote plant growth. Indirect benefits involve enhanced resistance to abiotic and biotic stressors, including defense against pathogenic microbes (Nadarajah and Abdul Rahman 2023; Anand et al. 2023).

Researchers hypothesize that fungal endophytes can enhance plant development and survival by improving nutrient uptake through various pathways. This hypothesis is based on a comprehensive review of numerous scientific studies. Fungal endophytes have the ability to modify nutrient absorption, distribution, and utilization based on the needs of their hosts. Multiple studies have shown that these endophytic fungi employ different mechanisms to enhance nutrient uptake. These mechanisms include altering root morphology, increasing the production of nutrient transporter proteins, and producing unique secondary metabolites that aid in chelating and dissociating nutrients (García-Latorre et al. 2021).

Fungal endophytes play a dual role in promoting plant growth and enhancing their ability to withstand different stresses. This makes them highly valuable as biofertilizers in various agricultural sectors (Verma et al. 2022). The utilization of microbial agents, such as fungal endophytes, holds great potential for the advancement of sustainable agriculture, which heavily relies on biofertilizers.

Stress tolerance Abiotic stressors greatly impede plant growth, so plants need a resilient system or mechanism to withstand them. Numerous microorganisms regularly derive nutrition from plants, and their interactions with the host can be either advantageous or detrimental. According to Ripa et al. (2019), endophytes are microorganisms that are capable of residing inside plant tissues without inflicting substantial harm or obtaining any form of compensation beyond their residency. All plants possess a number of endophytes, which contribute to the provision of nutrients, enhancement of environmental adaptation, and facilitation of the host plant's resilience against both living organisms and nonliving factors. During its life cycle, a fungal endophyte resides in the petiole, leaves, and stem of a plant. Research conducted by Chouhan et al. (2022) and Vandenkoornhuyse et al. (2015) has shown that these highly resilient endophytic fungi can persist throughout the plant's entire lifespan, potentially aiding the plant in overcoming environmental limitations. The potential of endophytes to mimic disease responses in plant cells has made them a valuable alternative source of biotic stimulants.

Endophytes are unique in their ability to produce secondary substances in the tissues of their host plants (Fontana et al. 2021). These metabolites can influence the activity of antioxidant enzymes, initiate defensive signaling pathways, and regulate gene expression for key enzymes involved in secondary substance production (Khanna et al. 2023). Endophyte connection has been shown to promote tolerance to abiotic stressors in various studies (Sadeghi et al. 2020; Biswas et al. 2016; Malik et al. 2020; Gul Jan et al. 2019). This is particularly important in the face of increasing global climate change, which calls for innovative agricultural solutions to enhance crop productivity in hot regions. Examples of plants, such as *Helianthus annuus*, *Adiantum capillus-veneris*, and *Glycine max*, have demonstrated the ability of endophytes to adapt to heat tolerance (Ismail et al. 2020). The endophytic fungus *Thermomyces* (CpE), known for its ability to thrive in high temperatures, was utilized to treat cucumber plants. This treatment aimed to alleviate the negative impacts of heat stress on the plants while maintaining their photosynthesis rate, photosystem II quantum capacity, and water use efficiency. Additionally, the application of CpE to cucumber plants under heat stress resulted in increased accumulation of sugars, flavonoids, protein, saponins, and antioxidant enzyme activity (Ali et al. 2018). Furthermore, cucumber plants treated with *Thermomyces* sp. exhibited significantly longer roots compared to untreated plants. This phenological reaction is a crucial adaptation characteristic in desert ecosystems. It enables the plant to efficiently absorb and acquire nutrients and moisture from the soil, especially in water-scarce conditions.

Plants increase their stomatal conductance quickly when faced with temperature stress, leading to increased transpiration. However, when treated with endophytes, these plants exhibit a slow activation of stomatal pores and a lower rate of transpiration (Pirasteh-Anosheh et al. 2016). One such endophyte, *Thermomyces* sp., enhances water consumption efficiency by maintaining leaf moisture content in challenging conditions. Thermophilic fungi play a crucial role in minimizing moisture loss from plant tissue. They achieve this by naturally closing stomata, which helps to retain water and prevent further damage caused by high temperatures (Nahar et al. 2015). These fungi enhance the production of primary and secondary metabolites. They also contribute to the increased presence of sugars and flavonoids in plant tissues, which serve as scavengers for reactive oxygen species (ROS) and signaling molecules in different plant-microbe interactions. This promotes plant growth and helps plants withstand both biotic and abiotic stresses (Abdelrahman et al. 2015). *Triticum aestivum*, *Cucumis sativus*, *Zea mays*, *Oryza sativa*, *Solanum lycopersicum*, *Saccharum officernaum*, as well as *Citrus reticulata* have collectively demonstrated the significant involvement of endophytes in the regulation of stress-inducible genes (Guler et al. 2016; Hosseini et al. 2019; Waqas et al. 2012; Ripa et al. 2019; Pandey et al. 2016; Vargas et al. 2014; Prema Sundara Valli and Muthukumar 2018).

An increase in bioactive substances, carotenoid content, enzymes that produce antioxidants, and chlorophyll fluorescence may impact the ability of endophytes to alleviate water stress. The endophytes *Aureobasidium pullulans*, *Dothideomycetes* sp., and *Penicillium citrinum* modified these parameters in *C. reticulata* plants and also promoted plant growth (Ripa et al. 2019). Endophytes are believed to aid plants in adapting to drought by producing phytohormones, reactive oxygen species, exopolysaccharides, and volatile compounds, modifying root system structures, synthesizing anti-stress substances, and positively regulating stress-related genes in host plants (Eid et al. 2019). Furthermore, soluble substances in endophytic plants have been observed to accumulate in grasses under water stress.

The utilization of carbon dioxide produced by endophytes to enhance photosynthesis is a significant factor contributing to the ability of host plants to withstand water stresses. The closure of stomata leads to a reduction in carbon dioxide shortage in plants. Research conducted on *Lycopersicon esculentum*, *Glycine max*, as well as various other plant species has demonstrated the significant role played by endophytes in augmenting resistance to heavy metal stressors (Bilal et al. 2020; Zhu et al. 2018). In one study, it was found that the host plant absorbed 2.7% of the CO_2 produced by the endophyte *Populus deltoides*' roots during photosynthesis. For instance, the fungal endophyte *P. roqueforti* produced indole acetic acid to support the growth of *T. aestivum* plants in heavy metal-contaminated soil. The resistance exhibited led to the prevention of the transport of toxic metals from the soil toward the plants. Following the application of residual water as well as the endophytic fungus, notable improvements were observed in the growth of wheat plants, as well as an increase in nutrient uptake and a decrease in the presence of heavy metals in both the roots and shoots. On the other hand, wheat plants that were not infected exhibited inhibited growth and exhibited symptoms of chlorosis in the presence of heavy metal-induced stress. The introduction of *P. roqueforti* to the host plant results in the establishment of a mutually advantageous symbiotic association, facilitating the stabilization of heavy metals and enhancing the ability of the host plant to flourish in soil with high levels of contamination (Malik et al. 2020).

In addition to aiding in cell detoxification through enzyme activity, endophytic fungi also enhance the host's ability to accumulate heavy metals. They can directly assist the host plant by increasing the absorption of heavy metals and reducing their harmful effects on plants. Additionally, they can indirectly promote plant growth and enhance stress tolerance.

According to Bashan and De-Bashan (2005), endophytes have the potential to enhance the host plant's nutrient absorption from contaminated soil. Additionally, these fungi can degrade pollutants in the soil, transforming them into nontoxic forms. Endophytes also have the ability to supply exogenous phytohormones, which induce beneficial physiological changes in the host plant, enabling it to better cope with stressful conditions. Khan et al. (2010) explained that endophytic fungi have the ability to solubilize nutrients in heavy metal-contaminated soil, which can enhance the biofertilization capacity of the fungus. This, in turn, increases the accessibility of nutrients to the host plant and is aided by phytohormones.

The importance of these bacterial pathogens in promoting salt stress tolerance has been observed in various plant species such as *T. aestivum*, *S. lycopersicum*, and *Cucumis sativus*. Under salt stress conditions, the fungal endophyte *P. indica* enhanced the growth and productivity of *S. lycopersicum*. This resulted in various morphological and biochemical changes that collectively mitigated the negative impacts of salt stress (Fontana et al. 2021). The presence of the fungal endophyte resulted in a significant enhancement of fruit output, as well as an increase in both dry and fresh plant mass, and branches of roots by 65% when subjected to salt stress conditions. Furthermore, it was observed that the treatment resulted in an augmentation of enzyme activity, specifically in superoxide dismutase and catalase. Additionally, there was an elevation in the concentrations of indole acetic acid as well as chlorophyll. The tomato plants that harbored endophytes exhibited reduced concentrations of proline along with abscisic acid (ABA) in comparison with those that were not colonized (Abdelaziz et al. 2019). The role of ROS-sequestering enzymes in improving salt stress tolerance has been highlighted by previous studies (Hosseini et al. 2017). The *Trichoderma*-11 isolate exhibited the highest efficacy as a biological control agent toward both *Curvularia* leaf spot and *maydis* leaf blight in laboratory conditions, with inhibition level of 64.76% and 72.50%, respectively. Additionally, it had the most significant chitinase activity efficacy, reaching up to 90.6%. On the other hand, the *Fusarium*-2 isolate significantly improved the development of maize seedlings, exhibiting a remarkable phosphate dissolution efficiency of 91.5%. This is further exemplified in Fig. 15.5, which illustrates the proportion of root development to plant height for each isolate (Fauriah et al. 2023).

Endophytic fungal species have the potential to mitigate the proline content in plants. Nevertheless, it has been observed that plants exhibit an increased synthesis of proline in response to salt-induced stress. For example, the reduction of abscisic acid (ABA) levels by around 30% during conditions of salinity-induced stress results in a decline in proline levels. Abscisic acid (ABA) is involved in the regulation of proline synthesis as a mechanism to alleviate cytoplasmic osmotic stress induced by elevated salt levels in the root zone. Reactive oxygen species, which are produced by environmental factors such as oxidative damage, extreme temperatures, salt, drought, and flooding, can cause cell damage and eventual cell death when sustained for a prolonged period of time (Waqas et al. 2012). In plant cells, elevated levels of reactive oxygen species (ROS) lead to oxidative stress, lipid peroxidation, and degradation of RNA and DNA. The ROS signal triggers a response in plants that mitigates the detrimental effects of stress and regulates the levels of metabolic ROS. This response is achieved by directly modulating the redox state of regulatory protein molecules, as well as influencing transcription and translation processes (Choudhury et al. 2017).

Another hypothesis proposes that endophytes activate the host's antioxidant enzymes by releasing small quantities of hydrogen peroxide or other reactive oxygen species. Small levels of reactive oxygen species (ROS) are continuously

Fig. 15.5 The above figure illustrates the evaluation of corn seed development on PDA medium inoculated with fungal endophyte at 7 days postinoculation (bottom) and the root length along with plant height of maize for each analyzed fungal endophyte (top). (Cited from Fauriah et al. 2023)

released, which helps reduce cellular hypersensitivity to ROS, enhance other interactions between the host and endosymbiotic organisms, and facilitate the uptake of nutrients such as calcium, potassium, magnesium, and phosphorus by plants (Overmyer et al. 2003).

Plant tissues predominantly accumulate methionine, proline, flavonoids, and additional phenolic compounds in order to augment their resistance to reactive oxygen species (ROS) produced by endophytes. Tanaka et al. (2006) suggest that the primary factors contributing to the manifestation of such phenomena and hypersensitive reactions include endophyte-host plant interactions, the formation of antioxidants and reactive oxygen species (ROS), and the systemic development of host resistance. Some fungal endophytes increase ROS levels to protect their hosts from environmental factors, while others decrease ROS concentrations to obtain nutrients from host cells and maintain mutualistic interactions with plants (Rodriguez and Redman 2008).

Fungal endophytes play a crucial role in ensuring the sustainability and resilience of agriculture. They possess the ability to efficiently mobilize phosphate and provide nitrogen to their plant hosts, thereby promoting rapid vegetative growth and enhancing the strength of associated plants to withstand challenges. Nevertheless, it has been observed that the production of plant growth-promoting compounds by fungal endophytes is diminished in the presence of stressful circumstances. An instance of this is the *Penicillium* sp. fungus, which thrives on the *Suaeda japonica* plant and produces gibberellins, a type of plant hormone (Patel et al. 2023). In a similar vein, the triterpenoid saponins and ginsenosides produced by different species of fungi such as *Fusarium*, *Aspergillus*, *Verticillium*, *Cladosporium*, *Ascomycete*, as well as *Penicillium* exhibit the capacity to augment root development alongside promoting resilience to environmental stressors (Sahoo et al. 2017).

In their recent study, Keswani et al. (2022) undertook a comprehensive review on the developmental aspects of gibberellins derived from diverse fungal sources, with a particular emphasis on elucidating their beneficial effects on enhancing crop yield. Similar to the fungal species *Phoma* sp., which exhibits robust growth in *T. polysporum*, *T. harzianum*, *Tinospora cordifolia*, as well as *T. atroviride* prevalent in *Phaseolus vulgaris*, it secretes agents that solubilize phosphate, proteolytic enzymes, active volatile compounds, as well as nonvolatile compounds. Similarly, the fungal species *Mucor* sp., residing in *B. campestris*, enhances the secretion of Indole-3 (Sahoo et al. 2017).

Fungal endophytes play a role in promoting plant growth in challenging conditions by producing siderophores, antibiotics, and phytohormones and aiding in nutrient mineralization. Additionally, abiotic stress stimuli lead to the production of jasmonic acid, abscisic acid, and salicylic acid, which serve as defense signaling chemicals. Phosphorus deficiency is common in most soil types and poses a challenge for plant growth. Even when added as an inorganic fertilizer, phosphorus remains insoluble and inaccessible to plants. Consequently, the result is the immobilization of iron and aluminum hydroxides within acidic soil. The insolubility of ferric hydroxide in soil poses a challenge for redox-active iron, despite its crucial role as a necessary component and cofactor in numerous enzymes. To enhance phosphate solubilization and increase iron intake by plants, endophytic fungi produce siderophores with low molecular weight and high affinity for iron. This promotes plant growth by improving nutrient uptake. Additionally, endophytes accelerate biomass production and nitrogen uptake during waste degradation (Idbella et al. 2019).

Natural alternatives possess significant potential in facilitating the advancement of sustainable agriculture, particularly in light of the obstacles encountered in plant breeding for diverse crops and the limitations imposed on genetic engineering in numerous nations owing to apprehensions regarding its repercussions on human, animal, and natural ecosystems.

15.15 Limitations Associated with Biocontrol Strategies of Endophyte

To achieve a comprehensive comprehension, application, and modification of fungal endophytes, it has become imperative to assess the biological foundations that underlie the relationship between the pathogenic agent, the host organism, and the fungal endophyte, alongside the physiological mechanisms implicated. In the field of endophytic fungi as well as biocontrol agents, several well-acknowledged principles exist. The aforementioned factors encompass the endophyte's capacity to stimulate plant defense systems, engage in mycoparasitism, exhibit antibiosis, and compete for nutrients and spatial resources (Latz et al. 2018). In many instances, many systems have the capacity to function concurrently. Enhancing plant health and disease prevention can be accomplished through the acquisition of nutrients to facilitate plant development or by modifying the levels of plant growth hormones (Qin et al. 2023). Analyzing the complex interactions between pathogens, biocontrol agents, host plants, and the mechanisms of pathogen inhibition is challenging.

To better understand the link between these factors, the following questions are posed: (i) Are the mechanisms involved in biocontrol agents accessible within plant tissues? By removing one component of the tripartite relationship, our understanding of the importance of endophytes in disease prevention in plants becomes incomplete. Therefore, analyzing endophyte-pathogen interactions solely in vitro may lead to misleading conclusions. A study conducted in vitro investigated the effects of the endophytic yeast known as *Pseudozyma flocculosa* on the growth of *Blumeria graminis* in barley, indicating the potential of antibiosis as a preventive measure. Upon conducting more research utilizing cellular microscopy as well as transcriptomics techniques, it was shown that the parasitic associations were attributed to mycoparasitism rather than antibiosis (Brader et al. 2017a).

How do biological control agents colonize, and what patterns do they utilize? Acquiring knowledge about the mechanisms underlying the invasion and pattern of BCA (biological control agents) will facilitate the correlation between intracellular and intercellular endophytic morphology in the context of disease suppression (Poveda et al. 2020). Which pathosystem does the biological control agent (BCA) utilize as a component of its mechanism of action?

A thorough understanding of the mechanisms by which biological control agents operate is crucial for the efficient management of phytopathogens. In summary, the development of a biocontrol agent occurs intracellularly when it is derived from an exterior surface and then demonstrates endophytic characteristics upon injection into plants. Understanding the behavior of a biocontrol agent (BCA) within plant tissue is crucial for effectively assessing its ability to suppress phytopathogens (Busby et al. 2016).

15.16 Conclusion and Future Perspective

The goal of this study is to comprehensively investigate the defense mechanisms utilized by endophytes in order to safeguard plants against diseases, thereby facilitating the advancement of sustainable agricultural practices. Endophytic microbes aid plant growth through direct and indirect mechanisms. Using these beneficial microbes in agriculture has the main advantage of reducing the need for agrochemicals like synthetic chemicals, pesticides, and fertilizers, making agriculture more sustainable and productive. Endophytic fungi can activate various plant defense mechanisms against both abiotic and biotic diseases. These fungi have the potential to effectively combat plant diseases through genetic alteration and interaction with the host-fungus. Moreover, they can be utilized as a tool for biological prospecting of new compounds and genetic modification of plants. By associating endophytes with desired crops, it is possible to enhance their tolerance to abiotic challenges.

This approach has demonstrated positive outcomes, enabling plant growth in otherwise unfavorable environments. Endophytes produce secondary metabolites with notable biological activity, making them potentially valuable products. To identify previously undiscovered species that exhibit metabolites with the potential to serve as elicitors for induced resistance, biological disease management, and the initiation of tolerance to abiotic stimuli, it is feasible to isolate and investigate endophytic microorganisms from underexplored plant species. Microorganisms utilize a variety of both direct and indirect techniques to protect plants from diseases and promote their growth. The examination of endophytic bacteria as bioinoculants has the potential to augment plant resilience against both biotic and abiotic challenges through the modification of their physiological and phenotypic traits subsequent to inoculation.

Further research is needed to fully understand the biotechnological importance of beneficial compounds produced by endophytic fungi in plant disease management. This includes investigating the antibiosis effects of these compounds on plant pathogens for effective plant protection. To facilitate this investigation, it is helpful to examine vital genes associated with phytohormone production, secretion systems, biological regulation, cellular elements, metabolic pathways, and secondary metabolic products from endophytic microorganisms. This can be achieved by combining both culture-independent and culture-dependent approaches. Several endophytic fungal species have been identified to possess biological control genes that facilitate the regulation of plant diseases. Numerous studies have effectively proved the capacity of endophytic fungi to regulate diseases in plants. These organisms exhibit significant potential for application in the development of novel biocontrol agents aimed at mitigating the difficulties associated with the management of phytopathogens in plant systems.

Researchers are still searching for answers on how endophytes enter the plant endosphere. The potential of endophytic fungi's diverse biocontrol action in SM biosynthesis can be explored as valuable bioproducts. To determine if certain species of endophytic fungi or uncultured ones can produce essential metabolites for use as biocontrol agents against plant diseases, updated information on these species is necessary. To effectively utilize endophytic fungi in maintaining plant health, it is crucial to understand the variables that mediate their biological activity in reducing diseases. This includes considering the source and type of biological control agents (BCAs), as well as their development process and the dosage required to inhibit pathogens. The physiological processes of plants and the structural composition of agricultural ecosystems are significantly influenced by endophytic fungus.

Fungal endophytes, like MBCA, play a crucial role in managing plant diseases in a sustainable and efficient manner. The allocation of endophytes has long-lasting impacts. Many fungal endophytes provide substantial benefits to their host plants by producing secondary compounds and chemicals that can combat diseases and enhance plant growth. Biological fungal pathogens possess the capacity to outperform plant pathogens, bolster a plant's immune system, augment stress tolerance, and confer protection against viral diseases. Nevertheless, the establishment and proliferation of these viruses in the field are hindered by challenging environmental circumstances. In order to proficiently mitigate plant diseases, many strains of fungi have been developed. To address challenges in using fungal endophytes for disease reduction, further research is suggested on obtaining a specific quantity of beneficial microorganisms from them. Endophytic fungi possess the capacity to stimulate many plant mechanisms, hence providing defense against both abiotic and biotic ailments.

Plant diseases can be effectively controlled through the implementation of biological control methods. Additionally, the utilization of genetic alteration and the ability to interact with host organisms render them valuable tools for the investigation of new substances and genetic modification in plants. Although this technology has been available for a long time, fungi are increasingly being used as biocontrol agents. Scientific research is progressing toward industrializing microbes to address environmental concerns and improve food quality by reducing the use of harmful chemicals in agriculture. One potential approach is the colonization of crops by various fungal species, including those traditionally used for pest control, on an endophytic basis.

Endophytic fungi offer direct as well as indirect benefits to agricultural plants, as discussed in this chapter, making them valuable for sustainable agriculture. No man-made chemical can match the diverse range of beneficial interactions provided by these microbes. Hence, the utilization of endophytic fungi demonstrates their efficacy as valuable tools for scholarly investigation and commercial enterprises in the realms of biological control, biological stimulation, and biofertilization. Fungal endophytes are anticipated to assume a pivotal role in the future agricultural sector due to their inherent involvement in the plant microbiome. Consequently, their integration into disease and pest control techniques is expected to witness a significant rise.

References

Ab Rahman SFS, Singh E, Pieterse CM, Schenk PM (2018) Emerging microbial biocontrol strategies for plant pathogens. Plant Sci 267:102–111

Abaya A, Xue A, Hsiang T (2021) Selection and screening of fungal endophytes against wheat pathogens. Biol Control 154:104511

Abdelaziz ME, Abdelsattar M, Abdeldaym EA, Atia MA, Mahmoud AWM, Saad MM, Hirt H (2019) *Piriformospora indica* alters Na+/K+ homeostasis, antioxidant enzymes and LeNHX1 expression of greenhouse tomato grown under salt stress. Sci Hortic 256:108532

Abdelrahman M, Sawada Y, Nakabayashi R, Sato S, Hirakawa H, El-Sayed M, Hirai MY, Saito K, Yamauchi N, Shigyo M (2015) Integrating transcriptome and target metabolome variability in doubled haploids of *Allium cepa* for abiotic stress protection. Mol Breed 35:1–11

Adeleke BS, Babalola OO (2021) Pharmacological potential of fungal endophytes associated with medicinal plants: a review. J Fungi 7(2):147

Adeleke BS, Babalola OO (2022) Roles of plant endosphere microbes in agriculture-a review. J Plant Growth Regul 41(4):1411–1428

Adeleke BS, Ayilara MS, Akinola SA, Babalola OO (2022) Biocontrol mechanisms of endophytic fungi. Egypt J Biol Pest Control 32(1):1–17

Adetunji CO, Kumar D, Raina M, Arogundade O, Sarin NB (2019) Endophytic microorganisms as biological control agents for plant pathogens: *A panacea* for sustainable agriculture. In: Plant biotic interactions: State of the art, pp 1–20

Afzal I, Shinwari ZK, Sikandar S, Shahzad S (2019) Plant beneficial endophytic bacteria: mechanisms, diversity, host range and genetic determinants. Microbiol Res 22:36–49

Ahmed A, He P, He Y, Singh BK, Wu Y, Munir S, He P (2023) Biocontrol of plant pathogens in omics era—with special focus on endophytic *bacilli*. Crit Rev Biotechnol:1–19

Akanmu AO, Babalola OO, Venturi V, Ayilara MS, Adeleke BS, Amoo AE, Sobowale AA, Fadiji AE, Glick BR (2021) Plant disease management: leveraging on the plant-microbe-soil interface in the biorational use of organic amendments. Front Plant Sci:1590

Akram S, Ahmed A, He P, He P, Liu Y, Wu Y, Munir S, He Y (2023) Uniting the role of endophytic fungi against plant pathogens and their interaction. J Fungi 9(1):72

Alam B, Lǐ J, Gě Q, Khan MA, Gōng J, Mehmood S, Yuán Y, Gǒng W (2021) Endophytic fungi: from symbiosis to secondary metabolite communications or vice versa? Front Plant Sci 12:30–60

Ali AH, Abdelrahman M, Radwan U, El-Zayat S, El-Sayed MA (2018) Effect of Thermomyces fungal endophyte isolated from extreme hot desert-adapted plant on heat stress tolerance of cucumber. Appl Soil Ecol 124:155–162

Ali S, Tyagi A, Bae H (2023) Plant microbiome: an ocean of possibilities for improving disease resistance in plants. Microorganisms 11(2):392

Al-Khayri JM, Rashmi R, Toppo V, Chole PB, Banadka A, Sudheer WN, Nagella P, Shehata WF, Al-Mssallem MQ, Alessa FM (2023) Plant secondary metabolites: the weapons for biotic stress management. Meta 13(6):716

Alvin A, Miller KI, Neilan BA (2014) Exploring the potential of endophytes from medicinal plants as sources of antimycobacterial compounds. Microbiol Res 169(7–8):483–495

Aly A, Debbab A, Proksch P (2013) Fungal endophytes–secret producers of bioactive plant metabolites. Die Pharmazie 68(7):499–505

Anamika, Joshi S, Sahgal M, Sahu S, Prakash A (2018) Fungal endophytes and their secondary metabolites: role in sustainable agriculture. In: Fungi and their role in sustainable development: current perspectives, pp 121–146

Anand U, Pal T, Yadav N, Singh VK, Tripathi V, Choudhary KK, Shukla AK, Sunita K, Kumar A, Bontempi E (2023) Current scenario and future prospects of endophytic microbes: promising candidates for abiotic and biotic stress management for agricultural and environmental sustainability. Microb Ecol:1–32

Ananda K, Sridhar K (2002) Diversity of endophytic fungi in the roots of mangrove species on the west coast of India. Can J Microbiol 48(10):871–878

Andreozzi A, Prieto P, Mercado-Blanco J, Monaco S, Zampieri E, Romano S, Valè G, Defez R, Bianco C (2019) Efficient colonization of the endophytes *Herbaspirillum huttiense* RCA24 and *Enterobacter cloacae* RCA25 influences the physiological parameters of *Oryza sativa* L. cv. Baldo rice. Environ Microbiol 21(9):3489–3504

Anguita-Maeso M, Navas-Cortés JA, Landa BB (2023) Insights into the methodological, biotic and abiotic factors influencing the characterization of xylem-inhabiting microbial communities of olive trees. Plan Theory 12(4):912

Archna S, Priyank V, Nath YA, Kumar SA (2015) Bioprospecting for extracellular hydrolytic enzymes from culturable thermotolerant bacteria isolated from Manikaran thermal springs. Res J Biotechnol 10:4

Arnold AE, Mejía LC, Kyllo D, Rojas EI, Maynard Z, Robbins N, Herre EA (2003) Fungal endophytes limit pathogen damage in a tropical tree. Proc Natl Acad Sci 100(26):15649–15654

Ashrafi S, Wennrich J-P, Becker Y, Maciá-Vicente JG, Brißke-Rode A, Daub M, Thünen T, Dababat AA, Finckh MR, Stadler M (2023) *Polydomus karssenii* gen. nov. sp. nov. is a dark septate endophyte with a bifunctional lifestyle parasitising eggs of plant parasitic cyst nematodes (*Heterodera* spp.). IMA Fungus 14(1):6

Ayyash LM, Theer RM, Saleh HM (2023) Effect of some Nematicides on hatching eggs and viability of second stage juveniles of root-knot nematode, *Meloidogyne Incognita* under laboratory conditions. In: IOP Conference Series: Earth and Environmental Science. IOP Publishing, 072003

Baghel V, Thakur JK, Yadav SS, Manna MC, Mandal A, Shirale AO, Sharma P, Sinha NK, Mohanty M, Singh AB (2020) Phosphorus and potassium solubilization from rock minerals by endophytic *Burkholderia* sp. strain FDN2-1 in soil and shift in diversity of bacterial endophytes of corn root tissue with crop growth stage. Geomicrobiol J 37(6):550–563

Bamisile BS, Akutse KS, Siddiqui JA, Xu Y (2021) Model application of entomopathogenic fungi as alternatives to chemical pesticides: prospects, challenges, and insights for next-generation sustainable agriculture. Front Plant Sci 12:741804

Baron NC, Rigobelo EC (2022) Endophytic fungi: a tool for plant growth promotion and sustainable agriculture. Mycology 13(1):39–55

Bashan Y, De-Bashan L (2005) Plant growth-promoting. Encyclopedia of soils in the environment 1:103–115

Bilal S, Shahzad R, Imran M, Jan R, Kim KM, Lee I-J (2020) Synergistic association of endophytic fungi enhances *Glycine max* L. resilience to combined abiotic stresses: heavy metals, high temperature and drought stress. Ind Crop Prod 143:111931

Bills GF, González-Menéndez V, Martín J, Platas G, Fournier J, Peršoh D, Stadler M (2012) *Hypoxylon pulicicidum* sp. nov.(Ascomycota, Xylariales), a pantropical insecticide-producing endophyte

Biswas T, Kalra A, Mathur A, Lal R, Singh M, Mathur A (2016) Elicitors' influenced differential ginsenoside production and exudation into medium with concurrent Rg3/Rh2 panaxadiol induction in *Panax quinquefolius* cell suspensions. Appl Microbiol Biotechnol 100:4909–4922

Biswas D, Biswas P, Nandy S, Mukherjee A, Pandey D, Dey A (2020) Endophytes producing podophyllotoxin from *Podophyllum* sp. and other plants: a review on isolation, extraction and bottlenecks. S Afr J Bot 134:303–313

Bolívar-Anillo HJ, Garrido C, Collado IG (2020) Endophytic microorganisms for biocontrol of the phytopathogenic fungus Botrytis cinerea. Phytochem Rev 19:721–740

Borah M, Routhu GK, Saikia B, Saikia A, Bhamra GK, Nath PD (2023) Fungal biocontrol agents: a sustainable management option for soybean diseases. In: Fungal resources for sustainable economy: current status and future perspectives. Springer

Boro M, Sannyasi S, Chettri D, Verma AK (2022) Microorganisms in biological control strategies to manage microbial plant pathogens: a review. Arch Microbiol 204(11):666

Brader G, Compant S, Vescio K, Mitter B, Trognitz F, Ma L-J, Sessitsch A (2017a) Ecology and genomic insights into plant-pathogenic and plant-nonpathogenic endophytes. Annu Rev Phytopathol 55:61–83

Brader G, Corretto E, Sessitsch A (2017b) Metagenomics of plant microbiomes. In: Functional metagenomics: tools and applications, pp 179–200

Bulgari D, Alias C, Peron G, Ribaudo G, Gianoncelli A, Savino S, Boureghda H, Bouznad Z, Monti E, Gobbi E (2023) Solid-state fermentation of *Trichoderma* spp.: a new way to valorize the agricultural digestate and produce value-added bioproducts. J Agric Food Chem 71(9):3994–4004

Burketova L, Trda L, Ott PG, Valentova O (2015) Bio-based resistance inducers for sustainable plant protection against pathogens. Biotechnol Adv 33(6):994–1004

Busby PE, Ridout M, Newcombe G (2016) Fungal endophytes: modifiers of plant disease. Plant Mol Biol 90:645–655
Cabre L (2019) Identification and study of promoters induced by Asian soybean rust: application in an artificial cell death system. Université de Lyon
Cao L-L, Zhang Y-Y, Liu Y-J, Yang T-T, Zhang J-L, Zhang Z-G, Shen L, Liu J-Y, Ye Y-H (2016) Anti-phytopathogenic activity of sporothriolide, a metabolite from endophyte *Nodulisporium* sp. A21 in *Ginkgo biloba*. Pestic Biochem Physiol 129:7–13
Cao X, Shi Y, Wu S, Wu X, Wang K, Sun H, He S, Dickschat JS, Wu B (2020) Polycyclic meroterpenoids, talaromyolides E– K for antiviral activity against pseudorabies virus from the endophytic fungus *Talaromyces purpureogenus*. Tetrahedron 76(30):131349
Chamoun R, Aliferis KA, Jabaji S (2015) Identification of signatory secondary metabolites during mycoparasitism of *Rhizoctonia solani* by *Stachybotrys elegans*. Front Microbiol 6:353
Chatterjee S, Ghosh R, Mandal NC (2019) Production of bioactive compounds with bactericidal and antioxidant potential by endophytic fungus *Alternaria alternata* AE1 isolated from *Azadirachta indica A.* Juss. PLoS One 14(4):e0214744
Chitnis VR, Suryanarayanan TS, Nataraja KN, Prasad SR, Oelmüller R, Shaanker RU (2020) Fungal endophyte-mediated crop improvement: the way ahead. Front Plant Sci 11:561007
Choudhury FK, Rivero RM, Blumwald E, Mittler R (2017) Reactive oxygen species, abiotic stress and stress combination. Plant J 90(5):856–867
Choudhury D, Tarafdar S, Parvin N, Rit R, Roy S, Dutta S (2023) Endophytic microbes and their diverse beneficial aspects in various sectors: a critical insight. Plant Sci Today 10(1):96–107
Chouhan S, Agrawal L, Prakash A (2022) Amelioration in traditional farming system by exploring the different plant growth-promoting attributes of endophytes for sustainable agriculture. Arch Microbiol 204(2):151
Chowdappa P, Kumar SM, Lakshmi MJ, Upreti K (2013) Growth stimulation and induction of systemic resistance in tomato against early and late blight by Bacillus subtilis OTPB1 or Trichoderma harzianum OTPB3. Biol Control 65(1):109–117
Clay K, Shearin ZR, Bourke KA, Bickford WA, Kowalski KP (2016) Diversity of fungal endophytes in non-native *Phragmites australis* in the Great Lakes. Biol Invasions 18:2703–2716
Constantin ME, De Lamo FJ, Vlieger BV, Rep M, Takken FL (2019) Endophyte-mediated resistance in tomato to *Fusarium oxysporum* is independent of ET, JA, and SA. Front Plant Sci 10:979
Cui R, Lu X, Chen X, Malik WA, Wang D, Wang J, Wang S, Guo L, Chen C, Wang X (2021) A novel raffinose biological pathway is observed by symbionts of cotton≡ *Verticillium dahliae* to improve salt tolerance genetically on cotton. J Agron Crop Sci 207(6):956–969
Daguerre Y, Edel-Hermann V Steinberg C (2016) Fungal genes and metabolites associated with the biocontrol of soil-borne plant pathogenic fungi. Fungal Metabolites, 1001 p
Daneshkhah R, Cabello S, Rozanska E, Sobczak M, Grundler F, Wieczorek K, Hofmann J (2013) *Piriformospora indica* antagonizes cyst nematode infection and development in Arabidopsis roots. J Exp Bot 64(12):3763–3774
Das G, Sen S, Shin H, Patra J (2016) 699 endophytes: a treasure house of bioactive compounds of medicinal 700 importance. Front Microbiol 7:1538
Data FFA (2020) Available online: http://www.fao.org/faostat/en/#data. QC (accessed on 26 Mar 2021)
Davies CR, Wohlgemuth F, Young T, Violet J, Dickinson M, Sanders J-W, Vallieres C, Avery SV (2021) Evolving challenges and strategies for fungal control in the food supply chain. Fungal Biol Rev 36:15–26
De Lamo FJ, Takken FL (2020) Biocontrol by *Fusarium oxysporum* using endophyte-mediated resistance. Front Plant Sci 11:37
De Silva NI, Brooks S, Lumyong S, Hyde KD (2019) Use of endophytes as biocontrol agents. Fungal Biol Rev 33(2):133–148
Delp CJ (1988) Fungicide resistance in North America. ARS Press, American Phytopathological Society

Deshmukh S, Kogel K-H (2007) *Piriformospora indica* protects barley from root rot caused by *Fusarium graminearum*. J Plant Dis Protect 114(6):263–268

Dikobe T, Masenya K, Manganyi MC (2023) Molecular technologies ending with 'omics': the driving force toward sustainable plant production and protection. F1000Research 12:480

Dimkić I, Janakiev T, Petrović M, Degrassi G, Fira D (2022) Plant-associated bacillus and pseudomonas antimicrobial activities in plant disease suppression via biological control mechanisms-a review. Physiol Mol Plant Pathol 117:101754

Donayre DKM, Dalisay TU (2016) Identities, characteristics, and assemblages of dematiaceous-endophytic fungi isolated from tissues of barnyard grass weed. Philippine J Sci 145(2):37–42

Dubey A, Malla MA, Kumar A, Dayanandan S, Khan ML (2020) Plants endophytes: unveiling hidden agenda for bioprospecting toward sustainable agriculture. Crit Rev Biotechnol 40(8):1210–1231

Dudeja SS, Suneja-Madan P, Paul M, Maheswari R, Kothe E (2021) Bacterial endophytes: molecular interactions with their hosts. J Basic Microbiol 61(6):475–505

Dutilloy E, Oni FE, Esmaeel Q, Clément C, Barka EA (2022) Plant beneficial bacteria as bioprotectants against wheat and barley diseases. J Fungi 8(6):632

Egamberdieva D, Jabborova D (2020) Plant microbiome: source for biologically active compounds. In: Biodiversity and biomedicine, pp 1–9

Eid AM, Salim SS, Hassan SE-D, Ismail MA, Fouda A (2019) Role of endophytes in plant health and abiotic stress management. In: Microbiome in plant health and disease: challenges and opportunities, pp 119–144

Ek-Ramos MJ, Zhou W, Valencia CU, Antwi JB, Kalns LL, Morgan GD, Kerns DL, Sword GA (2013) Spatial and temporal variation in fungal endophyte communities isolated from cultivated cotton (*Gossypium hirsutum*). PLoS One 8(6):e66049

El-Bialy HA, El-Bastawisy HS (2020) Elicitors stimulate paclitaxel production by endophytic fungi isolated from ecologically altered Taxus baccata. J Radiat Res Appl Sci 13(1):79–87

Elsharkawy MM, Shimizu M, Takahashi H, Ozaki K, Hyakumachi M (2013) Induction of systemic resistance against Cucumber mosaic virus in *Arabidopsis thaliana* by *Trichoderma asperellum* SKT-1. Plant Pathol J 29(2):193

Elsherbiny EA, Taher MA, Elsebai MF (2019) Activity of *Purpureocillium lilacinum* filtrates on biochemical characteristics of *Sclerotinia sclerotiorum* and induction of defense responses in common bean. Eur J Plant Pathol 155:39–52

El-Shetehy M, Moradi A, Maceroni M, Reinhardt D, Petri-Fink A, Rothen-Rutishauser B, Mauch F, Schwab F (2021) Silica nanoparticles enhance disease resistance in *Arabidopsis* plants. Nat Nanotechnol 16(3):344–353

Escudero N, Lopez-Moya F, Ghahremani Z, Zavala-Gonzalez EA, Alaguero-Cordovilla A, Ros-Ibañez C, Lacasa A, Sorribas FJ, Lopez-Llorca LV (2017) Chitosan increases tomato root colonization by *Pochonia chlamydosporia* and their combination reduces root-knot nematode damage. Front Plant Sci 8:1415

Eyberger AL, Dondapati R, Porter JR (2006) Endophyte fungal isolates from Podophyllum peltatum produce podophyllotoxin. J Nat Prod 69(8):1121–1124

Fadiji AE, Babalola OO (2020) Elucidating mechanisms of endophytes used in plant protection and other bioactivities with multifunctional prospects. Front Bioeng Biotechnol 8:467

Fauriah R, Djaya E, Djaenuddin N, Muis A, Nonci N (2023) Potential of endophytic fungi as a pathogenic biocontrol agent and growth promoters in corn seedlings. Egypt J Biol Pest Control 33(1):83

Ferreira C, Doursout M-FJ, Balingit JS (2023) Climate change and the risk of future pandemics. In: 2000 Years of pandemics: past, present, and future. Springer

Floros DJ, Petras D, Kapono CA, Melnik AV, Ling T-J, Knight R, Dorrestein PC (2017) Mass spectrometry based molecular 3D-cartography of plant metabolites. Front Plant Sci 8:429

Fontana DC, de Paula S, Torres AG, de Souza VHM, Pascholati SF, Schmidt D, Dourado Neto D (2021) Endophytic fungi: biological control and induced resistance to phytopathogens and abiotic stresses. Pathogens 10(5):570

França KC (2022) Avaliação do efeito do estrato, fração proteica e lectina da casca de Genipa americana L. sobre a broca dos frutos das anonáceas Cerconota anonella (Lepidoptera: Depressaridae)
García-Latorre C, Rodrigo S, Santamaría O (2021) Endophytes as plant nutrient uptake-promoter in plants. Endophytes 3:247–265
Ghaffari MR, Mirzaei M, Ghabooli M, Khatabi B, Wu Y, Zabet-Moghaddam M, Mohammadi-Nejad G, Haynes PA, Hajirezaei MR, Sepehri M (2019) Root endophytic fungus Piriformospora indica improves drought stress adaptation in barley by metabolic and proteomic reprogramming. Environ Exp Bot 157:197–210
Ghosh T, Pradhan C, Das AB (2020) Control of stem-rot disease of rice caused by Sclerotium oryzae catt and its cellular defense mechanism–a review. Physiol Mol Plant Pathol 112:101536
Giorgio A, De Stradis A, Lo Cantore P, Iacobellis NS (2015) Biocide effects of volatile organic compounds produced by potential biocontrol rhizobacteria on *Sclerotinia sclerotiorum*. Front Microbiol 6:1056
González V, Armijos E, Garcés-Claver A (2020) Fungal endophytes as biocontrol agents against the main soil-borne diseases of melon and watermelon in Spain. Agronomy 10(6):820
Goss EM, Tabima JF, Cooke DE, Restrepo S, Fry WE, Forbes GA, Fieland VJ, Cardenas M, Grünwald NJ (2014) The Irish potato famine pathogen *Phytophthora infestans* originated in Central Mexico rather than the Andes. Proc Natl Acad Sci 111(24):8791–8796
Gouda S, Das G, Sen SK, Shin H-S, Patra JK (2016) Endophytes: a treasure house of bioactive compounds of medicinal importance. Front Microbiol 7:1538
Grabka R, d'Entremont TW, Adams SJ, Walker AK, Tanney JB, Abbasi PA, Ali S (2022) Fungal endophytes and their role in agricultural plant protection against pests and pathogens. Plan Theory 11(3):384
Gul Jan F, Hamayun M, Hussain A, Jan G, Iqbal A, Khan A, Lee I-J (2019) An endophytic isolate of the fungus Yarrowia lipolytica produces metabolites that ameliorate the negative impact of salt stress on the physiology of maize. BMC Microbiol 19(1):1–10
Guler NS, Pehlivan N, Karaoglu SA, Guzel S, Bozdeveci A (2016) Trichoderma atroviride ID20G inoculation ameliorates drought stress-induced damages by improving antioxidant defence in maize seedlings. Acta Physiol Plant 38:1–9
Gurung SA, Rai AK, Sunar K, Das K (2023) Plant–endophyte interactions: a driving phenomenon for boosting plant health under climate change conditions. In: Microbial symbionts and plant health: trends and applications for changing climate. Springer
Haas D, Défago G (2005) Biological control of soil-borne pathogens by fluorescent pseudomonads. Nat Rev Microbiol 3(4):307–319
Halecker S, Wennrich J-P, Rodrigo S, Andrée N, Rabsch L, Baschien C, Steinert M, Stadler M, Surup F, Schulz B (2020) Fungal endophytes for biocontrol of ash dieback: the antagonistic potential of *Hypoxylon rubiginosum*. Fungal Ecol 45:100918
Hardoim PR, Van Overbeek LS, Berg G, Pirttilä AM, Compant S, Campisano A, Döring M, Sessitsch A (2015) The hidden world within plants: ecological and evolutionary considerations for defining functioning of microbial endophytes. Microbiol Mol Biol Rev 79(3):293–320
Harris JM, Balint-Kurti P, Bede JC, Day B, Gold S, Goss EM, Grenville-Briggs LJ, Jones KM, Wang A, Wang Y (2020) What are the top 10 unanswered questions in molecular plant-microbe interactions? Mol Plant-Microbe Interact 33(12):1354–1365
Hashem A, Tabassum B, Abd-Allah EF (2019) Bacillus subtilis: a plant-growth promoting rhizobacterium that also impacts biotic stress. Saudi J Biol Sci 26(6):1291–1297
Hashem AH, Attia MS, Kandil EK, Fawzi MM, Abdelrahman AS, Khader MS, Khodaira MA, Emam AE, Goma MA, Abdelaziz AM (2023) Bioactive compounds and biomedical applications of endophytic fungi: a recent review. Microb Cell Factories 22(1):1–23
Hastuti RD, Lestari Y, Suwanto A, Saraswati R (2012) Endophytic Streptomyces spp. as biocontrol agents of rice bacterial leaf blight pathogen (*Xanthomonas oryzae* pv. *oryzae*). *HAYATI*. J Biosci 19(4):155–162
Heinig U, Scholz S, Jennewein S (2013) Getting to the bottom of Taxol biosynthesis by fungi. Fungal Divers 60:161–170

Hosseini F, Mosaddeghi MR, Dexter AR (2017) Effect of the fungus *Piriformospora indica* on physiological characteristics and root morphology of wheat under combined drought and mechanical stresses. Plant Physiol Biochem 118:107–120

Hosseini F, Mosaddeghi MR, Dexter AR, Sepehri M (2019) Effect of endophytic fungus *Piriformospora indica* and PEG-induced water stress on maximum root growth pressure and elongation rate of maize. Plant Soil 435:423–436

Hou S, Liu Z, Shen H, Wu D (2019) Damage-associated molecular pattern-triggered immunity in plants. Front Plant Sci 10:646

Hu L (2023) Molecular mechanisms of the age-related resistance in Arabidopsis thaliana. University of Georgia

Huang J, Yang M, Zhang X (2016) The function of small RNAs in plant biotic stress response. J Integr Plant Biol 58(4):312–327

Huang Y-P, Shi J-Y, Luo S-C, Xu S-Y, Zhang J-D, Molnár I, Yang Q-Q, Zhang B-B (2023) Antimicrobial substances and mechanisms of *lactobacillus rhamnosus* against Gardnerella vaginalis. Probiot Antimicrob Prot 15(2):400–410

Idbella M, Zotti M, Cesarano G, Fechtali T, Mazzoleni S, Bonanomi G (2019) Fungal endophytes affect plant response to leaf litter with contrasting chemical traits. Community Ecol 20:205–213

Igiehon NO, Babalola OO (2018) Rhizosphere microbiome modulators: contributions of nitrogen fixing bacteria towards sustainable agriculture. Int J Environ Res Public Health 15(4):574

Ismail AH, Mehmood A, Qadir M, Husna AI, Hamayun M, Khan N (2020) Thermal stress alleviating potential of endophytic fungus rhizopus oryzae inoculated to sunflower (Helianthus annuus L.) and soybean (Glycine max L.). Pak J Bot 52(5):1857–1865

Istifadah N, McGee P (2006) Endophytic Chaetomium globosum reduces development of tan spot in wheat caused by Pyrenophora tritici-repentis. Australas Plant Pathol 35:411–418

Jaiswal P, Kar S, Misra S, Dixit V, Mishra SK, Chauhan PS (2022) Novel trends in endophytic applications for plant disease management. In: Biocontrol mechanisms of endophytic microorganisms. Elsevier

Jayaraman S, Naorem A, Lal R, Dalal RC, Sinha N, Patra A, Chaudhari S (2021) Disease-suppressive soils—beyond food production: a critical review. J Soil Sci Plant Nutr 21:1437–1465

Jogaiah S, Abdelrahman M, Tran LSP, Ito SI (2018) Different mechanisms of Trichoderma virens-mediated resistance in tomato against *Fusarium* wilt involve the jasmonic and salicylic acid pathways. Mol Plant Pathol 19(4):870–882

Jones A (2016) Variations in endophyte diversity of native and non-native pines in Scottish plantations

Kaddes A, Fauconnier M-L, Sassi K, Nasraoui B, Jijakli M-H (2019) Endophytic fungal volatile compounds as solution for sustainable agriculture. Molecules 24(6):1065

Kandel SL, Joubert PM, Doty SL (2017) Bacterial endophyte colonization and distribution within plants. Microorganisms 5(4):77

Kaniganti S, Bhattacharya J, Petla BP, Reddy PS (2022) Strigolactone, a neglected plant hormone, with a great potential for crop improvement: crosstalk with other plant hormones. Environ Exp Bot 204:105072

Keswani C, Singh SP, García-Estrada C, Mezaache-Aichour S, Glare TR, Borriss R, Rajput VD, Minkina TM, Ortiz, A, Sansinenea E (2022) Biosynthesis and beneficial effects of microbial gibberellins on crops for sustainable agriculture. J Appl Microbiol 132(3):1597–1615.

Khairah M, Mubarik NR, Manaf LA (2023) Bacterial selection and characterization of chitinase enzyme from bacteria controlling Fusarium proliferatum. Biodivers J Biol Divers 24(3)

Khan N, Schisler D, Boehm M, Lipps P, Slininger P (2004) Field testing of antagonists of Fusarium head blight incited by Gibberella zeae. Biol Control 29(2):245–255

Khan MS, Zaidi A, Ahemad M, Oves M, Wani PA (2010) Plant growth promotion by phosphate solubilizing fungi–current perspective. Arch Agron Soil Sci 56(1):73–98

Khan AL, Waqas M, Kang S-M, Al-Harrasi A, Hussain J, Al-Rawahi A, Al-Khiziri S, Ullah I, Ali L, Jung H-Y (2014) Bacterial endophyte *Sphingomonas* sp. LK11 produces gibberellins and IAA and promotes tomato plant growth. J Microbiol 52:689–695

Khan RAA, Najeeb S, Mao Z, Ling J, Yang Y, Li Y, Xie B (2020) Bioactive secondary metabolites from *Trichoderma* spp. against phytopathogenic bacteria and root-knot nematode. Microorganisms 8(3):401

Khanna K, Bhardwaj R, Alam P, Reiter RJ, Ahmad P (2023) Phytomelatonin: A master regulator for plant oxidative stress management. Plant Physiol Biochem

Khare E, Mishra J, Arora NK (2018) Multifaceted interactions between endophytes and plant: developments and prospects. Front Microbiol 9:2732

Kharwar RN, Sharma VK, Mishra A, Kumar J, Singh DK, Verma SK, Gond SK, Kumar A, Kaushik N, Revuru B (2020) Harnessing the phytotherapeutic treasure troves of the ancient medicinal plant Azadirachta indica (Neem) and associated endophytic microorganisms. Planta Med 86(13/14):906–940

Khoshru B, Mitra D, Joshi K, Adhikari P, Rion MSI, Fadiji AE, Alizadeh M, Priyadarshini A, Senapati A, Sarikhani MR (2023) Decrypting the multi-functional biological activators and inducers of defense responses against biotic stresses in plants. Heliyon

Kim SH, Vujanovic V (2016) Relationship between mycoparasites lifestyles and biocontrol behaviors against *Fusarium* spp. and mycotoxins production. Appl Microbiol Biotechnol 100(12):5257–5272

Köhl J, Kolnaar R, Ravensberg WJ (2019) Mode of action of microbial biological control agents against plant diseases: relevance beyond efficacy. Front Plant Sci 845

Konappa N, Krishnamurthy S, Dhamodaran N, Arakere UC, Chowdappa S, Ramachandrappa NS (2020) Opportunistic avirulent plant symbionts Trichoderma: exploring its potential against soilborne phytopathogens. In: Trichoderma: agricultural applications and beyond, pp 219–255

Kour D, Kour H, Khan SS, Khan RT, Bhardwaj M, Kailoo S, Kumari C, Rasool S, Yadav AN, Sharma YP (2023) Biodiversity and functional attributes of Rhizospheric microbiomes: potential tools for sustainable agriculture. Curr Microbiol 80(6):192

Kusari S, Hertweck C, Spiteller M (2012) Chemical ecology of endophytic fungi: origins of secondary metabolites. Chem Biol 19(7):792–798

Kushwaha P, Kashyap PL, Kuppusamy P, Srivastava AK, Tiwari RK (2020) Functional characterization of endophytic *bacilli* from pearl millet (*Pennisetum glaucum*) and their possible role in multiple stress tolerance. Plant Biosyst 154(4):503–514

Lahlali R, Ezrari S, Radouane N, Kenfaoui J, Esmaeel Q, El Hamss H, Belabess Z, Barka EA (2022) Biological control of plant pathogens: a global perspective. Microorganisms 10(3):596

Land WG (2023) The DAMP-driven host immune defense program against pathogens. In: Damage-associated molecular patterns in human diseases: volume 3: antigen-related disorders. Springer

Latz MA, Jensen B, Collinge DB, Jørgensen HJ (2018) Endophytic fungi as biocontrol agents: elucidating mechanisms in disease suppression. Plant Ecol Divers 11(5–6):555–567

Liu T, Li J, Zhang J (2019) Rootzone mixture affects the population of root-invading fungi in zoysiagrass. Urban For Urban Green 37:168–172

Llorens E, García-Agustín P, Lapeña L (2017) Advances in induced resistance by natural compounds: towards new options for woody crop protection. Sci Agric 74:90–100

Llorens E, Scalschi L, Sharon O, Vicedo B, Sharon A, García-Agustín P (2022) Jasmonic acid pathway is required in the resistance induced by *Acremonium sclerotigenum* in tomato against pseudomonas syringae. Plant Sci 318:111–210

Loo EP-I, Tajima Y, Yamada K, Kido S, Hirase T, Ariga H, Fujiwara T, Tanaka K, Taji T, Somssich IE (2022) Recognition of microbe-and damage-associated molecular patterns by leucine-rich repeat pattern recognition receptor kinases confers salt tolerance in plants. Mol Plant-Microbe Interact 35(7):554–566

Lu H, Wei T, Lou H, Shu X, Chen Q (2021) A critical review on communication mechanism within plant-endophytic fungi interactions to cope with biotic and abiotic stresses. J Fungi 7(9):719

Lucas JA, Hawkins NJ, Fraaije BA (2015) The evolution of fungicide resistance. Adv Appl Microbiol 90:29–92

Lucena C, Aroca R, Wang J, Zimmermann SD (2023) Beneficial microbes and the interconnection between crop mineral nutrition and induced systemic resistance, volume II. Front Plant Sci 14:1157296

Lugtenberg BJ, Caradus JR, Johnson LJ (2016) Fungal endophytes for sustainable crop production. FEMS Microbiol Ecol 92(12)

Maheshwari R, Varahalarao V, Ramars A, Suryanarayana Murty U (2017) Endophytic fungi as novel resources of natural therapeutics. Braz Arch Biol Technol 60:e17160542

Mahmoud M, BenRejeb I, Punja ZK, Buirs L, Jabaji S (2023) Understanding bud rot development, caused by *Botrytis cinerea*, on cannabis (*Cannabis sativa* L.) plants grown under greenhouse conditions. Botany

Maldani M, Nassiri L, Ibijbijen J (2022) Biodegradation and remediation of pesticides in contaminated agroecosystems: special reference to glyphosate and paraquat. In: Microbial BioTechnology for sustainable agriculture volume 1. Springer

Malik A, Butt TA, Naqvi STA, Yousaf S, Qureshi MK, Zafar MI, Farooq G, Nawaz I, Iqbal M (2020) Lead tolerant endophyte Trametes hirsuta improved the growth and lead accumulation in the vegetative parts of *Triticum aestivum* L. Heliyon 6(7):e04188

Manzotti A, Bergna A, Burow M, Jørgensen HJ, Cernava T, Berg G, Collinge DB, Jensen B (2020) Insights into the community structure and lifestyle of the fungal root endophytes of tomato by combining amplicon sequencing and isolation approaches with phytohormone profiling. FEMS Microbiol Ecol 96(5):fiaa052

Marinova D, Bogueva D (2022) Sustainability transitions in food production. In: Food in a planetary emergency. Springer

Martínez-Medina A, Fernández I, Sánchez-Guzmán MJ, Jung SC, Pascual JA, Pozo MJ (2013) Deciphering the hormonal signalling network behind the systemic resistance induced by *Trichoderma harzianum* in tomato. Front Plant Sci 4:206

Martinez-Medina A, Flors V, Heil M, Mauch-Mani B, Pieterse CM, Pozo MJ, Ton J, van Dam NM, Conrath U (2016) Recognizing plant defense priming. Trends Plant Sci 21(10):818–822

Mauch-Mani B, Baccelli I, Luna E, Flors V (2017) Defense priming: an adaptive part of induced resistance. Annu Rev Plant Biol 68:485–512

Medeiros HAD, Araújo Filho JVD, Freitas LGD, Castillo P, Rubio MB, Hermosa R, Monte E (2017) Tomato progeny inherit resistance to the nematode Meloidogyne javanica linked to plant growth induced by the biocontrol fungus *Trichoderma atroviride*. Sci Rep 7(1):1–13

Meena M, Swapnil P, Divyanshu K, Kumar S, Harish, Tripathi YN, Zehra A, Marwal A, Upadhyay RS (2020) PGPR-mediated induction of systemic resistance and physiochemical alterations in plants against the pathogens: current perspectives. J Basic Microbiol 60(10):828–861

Meena KK, Bitla U, Sorty AM (2021) Root-endophytes and their contribution to plant abiotic stress tolerance. In: Microbial management of plant stresses. Elsevier

Mehmood N, Saeed M, Zafarullah S, Hyder S, Rizvi ZF, Gondal AS, Jamil N, Iqbal R, Ali B, Ercisli S (2023) Multifaceted impacts of plant-beneficial pseudomonas spp. in managing various plant diseases and crop yield improvement. ACS Omega

Mehta P, Sharma R, Putatunda C, Walia A (2019) Endophytic fungi: role in phosphate solubilization. In: Advances in endophytic fungal research: present status and future challenges, pp 183–209

Mejía LC, Herre EA, Sparks JP, Winter K, García MN, Van Bael SA, Stitt J, Shi Z, Zhang Y, Guiltinan MJ (2014) Pervasive effects of a dominant foliar endophytic fungus on host genetic and phenotypic expression in a tropical tree. Front Microbiol 5:479

Molitor A, Zajic D, Voll LM, Pons-Kühnemann J, Samans B, Kogel K-H, Waller F (2011) Barley leaf transcriptome and metabolite analysis reveals new aspects of compatibility and *Piriformospora indica*–mediated systemic induced resistance to powdery mildew. Mol Plant-Microbe Interact 24(12):1427–1439

Mousa W, Raizada M (2013) The diversity of anti-microbial secondary metabolites produced by fungal endophytes: an interdisciplinary perspective. Front Microbiol 4:65

Munir S, Li Y, He P, He P, Ahmed A, Wu Y, He Y (2020) Unraveling the metabolite signature of citrus showing defense response towards Candidatus *Liberibacter asiaticus* after application of endophyte Bacillus subtilis L1-21. Microbiol Res 234:126–425

Muvea AM, Subramanian S, Maniania NK, Poehling H-M, Ekesi S, Meyhöfer R (2018) Endophytic colonization of onions induces resistance against viruliferous thrips and virus replication. Front Plant Sci 9:1785

Nadarajah K, Abdul Rahman NSN (2023) The microbial connection to sustainable agriculture. Plan Theory 12(12):2307
Nadeem SM, Ahmad M, Zahir ZA, Javaid A, Ashraf M (2014) The role of mycorrhizae and plant growth promoting *rhizobacteria* (PGPR) in improving crop productivity under stressful environments. Biotechnol Adv 32(2):429–448
Nagendran K, Karthikeyan G, Mohammed Faisal P, Kalaiselvi P, Raveendran M, Prabakar K, Raguchander T (2014) Exploiting endophytic bacteria for the management of sheath blight disease in rice. Biol Agric Horticult 30(1):8–23
Nahar K, Hasanuzzaman M, Ahamed KU, Hakeem KR, Ozturk M, Fujita M (2015) Plant responses and tolerance to high temperature stress: role of exogenous phytoprotectants. In: Crop production and global environmental issues, pp 385–435
Nataraja KN, Suryanarayanan T, Shaanker RU, Senthil-Kumar M, Oelmüller R (2019) Plant–microbe interaction: prospects for crop improvement and management. Plant Physiol Rep 24:461–462
Nawrot R (2017) Defense-related proteins from *Chelidonium majus* L. as important components of its latex. Curr Protein Peptide Sci 18(8):864–880
Nazir A, Rahman HA (2018) Secrets of plants: endophytes. Int J Plant Biol 9(1):7810
Nicot PC, Stewart A, Bardin M, Elad Y (2016) Biological control and biopesticide suppression of *Botrytis*-incited diseases. In: Botrytis–the fungus, the pathogen and its management in agricultural systems, pp. 165–187
Nisa H, Kamili AN, Nawchoo IA, Shafi S, Shameem N, Bandh SA (2015) Fungal endophytes as prolific source of phytochemicals and other bioactive natural products: a review. Microb Pathog 82:50–59
Nuangmek W, Aiduang W, Kumla J, Lumyong S, Suwannarach N (2021) Evaluation of a newly identified endophytic fungus, *Trichoderma phayaoense* for plant growth promotion and biological control of gummy stem blight and wilt of muskmelon. Front Microbiol 12:634772
Nuankaew S, Chuaseeharonnachai C, Preedanon S, Somrithipol S, Saengkaewsuk S, Kwantong P, Phookongchai S, Srikitikulchai P, Kobmoo N, Wang X-C (2022) Two novel species of *Talaromyces* discovered in a Karst Cave in the Satun UNESCO Global Geopark of Southern Thailand. J Fungi 8(8):825
Nürnberger T, Kemmerling B (2009) Pathogen-associated molecular patterns (PAMP) and PAMP-triggered immunity. In: Annuak Plant Reviews, Molecular Aspects of Plant Disease Resistance, Oxford: Blackwell Publishing Ltd. pp. 16-47
O'Brien PA (2017) Biological control of plant diseases. Australas Plant Pathol 46:293–304
O'Hanlon KA, Knorr K, Jørgensen LN, Nicolaisen M, Boelt B (2012) Exploring the potential of symbiotic fungal endophytes in cereal disease suppression. Biol Control 63(2):69–78
Oelmüller R, Tseng Y-H, Gandhi A (2023) Signals and their perception for Remodelling, adjustment and repair of the plant cell wall. Int J Mol Sci 24(8):7417
Oliveira M, Varanda C, Félix M (2016) Induced resistance during the interaction pathogen x plant and the use of resistance inducers. Phytochem Lett 15:152–158
Oono R, Lefèvre E, Simha A, Lutzoni F (2015) A comparison of the community diversity of foliar fungal endophytes between seedling and adult loblolly pines (*Pinus taeda*). Fungal Biol 119(10):917–928
Orozco-Mosqueda MDC, Flores A, Rojas-Sánchez B, Urtis-Flores CA, Morales-Cedeño LR, Valencia-Marin MF, Chávez-Avila S, Rojas-Solis D, Santoyo G (2021) Plant growth-promoting bacteria as bioinoculants: attributes and challenges for sustainable crop improvement. Agronomy 11(6):1167
Overmyer K, Brosché M, Kangasjärvi J (2003) Reactive oxygen species and hormonal control of cell death. Trends Plant Sci 8(7):335–342
Padaria GTG (2015) Ajar Nath Yadav, Priyanka Verma, Murugan Kumar, Kamal K. Pal, Rinku Dey, Alka Gupta, Jasdeep Chatrath. Ann Microbiol 65:611–629
Padilla-Roji I, Ruiz-Jiménez L, Bakhat N, Vielba-Fernández A, Pérez-García A, Fernández-Ortuño D (2023) RNAi technology: a new path for the research and management of Obligate Biotrophic Phytopathogenic Fungi. Int J Mol Sci 24(10):9082

Pandey V, Ansari MW, Tula S, Yadav S, Sahoo RK, Shukla N, Bains G, Badal S, Chandra S, Gaur A (2016) Dose-dependent response of *Trichoderma harzianum* in improving drought tolerance in rice genotypes. Planta 243:1251–1264

Park Y-H, Mishra RC, Yoon S, Kim H, Park C, Seo S-T, Bae H (2019) Endophytic Trichoderma citrinoviride isolated from mountain-cultivated ginseng (Panax ginseng) has great potential as a biocontrol agent against ginseng pathogens. J Ginseng Res 43(3):408–420

Pascholati SF, Dalio RJD (2018) Fisiologia do parasitismo: como as plantas se defendem dos patógenos. Manual de fitopatologia: princípios e conceitos 1

Patel HK, Makampara RA, Kalaria RK, Joshi MP (2023) Endophytes: A novel tool for sustainable agriculture. In: Endophytic association: What, why and how. Elsevier

Patil RH, Patil MP, Maheshwari VL (2016) Bioactive secondary metabolites from endophytic fungi: a review of biotechnological production and their potential applications. Stud Nat Prod Chem 49:189–205

Perumal S, Radhakrishnan R, Sathasivam R, Arun M, Song J, Park SU (2023) Endophytes: A vital source of medicine–a review. Trop J Pharm Res 22(5):1133–1142

Pirasteh-Anosheh H, Saed-Moucheshi A, Pakniyat H, Pessarakli M (2016) Stomatal responses to drought stress. In: Water stress and crop plants: A sustainable approach, vol 1, pp 24–40

Pirrello C, Magon G, Palumbo F, Farinati S, Lucchin M, Barcaccia G, Vannozzi A (2023) Past, present, and future of genetic strategies to control tolerance to the main fungal and oomycete pathogens of grapevine. J Exp Bot 74(5):1309–1330

Polonio J, Almeida T, Garcia A, Mariucci G, Azevedo J, Rhoden S, Pamphile J (2015) Biotechnological prospecting of foliar endophytic fungi of guaco (*Mikania glomerata Spreng*.) with antibacterial and antagonistic activity against phytopathogens. Genet Mol Res 14(3):7297–7309

Poveda J, Abril-Urias P, Escobar C (2020) Biological control of plant-parasitic nematodes by filamentous fungi inducers of resistance: *Trichoderma*, mycorrhizal and endophytic fungi. Front Microbiol 11:992

Prakash J (2023) Mechanism of biological control of plant diseases by endophytes. In: Endophytic association: What, why and how. Elsevier

Prema Sundara Valli P, Muthukumar T (2018) Dark septate root endophytic fungus Nectria haematococca improves tomato growth under water limiting conditions. Indian J Microbiol 58:489–495

Puranik S, Bundela V, Shylla A, Elakkya M, Shukla L, Singh SK (2023) Peeking into plant-microbe interactions during plant defense. In: Plant-microbe interaction-recent advances in molecular and biochemical approaches. Elsevier

Qin X, Xu J, An X, Yang J, Wang Y, Dou M, Wang M, Huang J, Fu Y (2023) Insight of endophytic fungi promoting the growth and development of woody plants. Crit Rev Biotechnol:1–22

Qu Q, Zhang Z, Peijnenburg W, Liu W, Lu T, Hu B, Chen J, Chen J, Lin Z, Qian H (2020) Rhizosphere microbiome assembly and its impact on plant growth. J Agric Food Chem 68(18):5024–5038

Rajendran L, Durgadevi D, Kavitha R, Archana T, Harish S, Sendhilvel V, Raguchander T, Karthikeyan G (2023) Deciphering the role of growth-promoting bacterial endophytes in harmonizing plant health. In: Microbial symbionts and plant health: trends and applications for changing climate. Springer

Rajput N, Sharma U, Kaur B, Rani P, Tongkachok K, Dornadula VHR (2022) Current global green marketing standard: changing market and company branding. Int J Syst Assur Eng Manag 13(Suppl 1):727–735

Rani GD (2008) Advances in soil borne plant diseases. New India Publishing

Redman RS, Ranson JC, Rodriguez RJ (1999) Conversion of the pathogenic fungus *Colletotrichum magna* to a nonpathogenic, endophytic mutualist by gene disruption. Mol Plant-Microbe Interact 12(11):969–975

Reshma J, Vinaya C, Linu M (2019) Agricultural applications of endophytic microflora. In: Seed endophytes: biology and biotechnology, pp 385–403

Riaz U, Hassan A, Fatima M, Aziz H, Rasool M, Murtaza G (2023) Plant secondary metabolites and environmental stress: An overview. In: Biology and biotechnology of environmental stress tolerance in plants, pp 3–25

Ripa FA, Cao W-D, Tong S, Sun J-G (2019) Assessment of plant growth promoting and abiotic stress tolerance properties of wheat endophytic fungi. Biomed Res Int 2019:1–12

Rodriguez R, Redman R (2008) More than 400 million years of evolution and some plants still can't make it on their own: plant stress tolerance via fungal symbiosis. J Exp Bot 59(5):1109–1114

Rojas EC, Jensen B, Jørgensen HJ, Latz MA, Esteban P, Ding Y, Collinge DB (2020) Selection of fungal endophytes with biocontrol potential against *Fusarium* head blight in wheat. Biol Control 144:104–222

Roy S, Banerjee D (2018) Diversity of endophytes in tropical forests. In: Endophytes of forest trees: biology and applications, pp 43–62

Rubini MR, Silva-Ribeiro RT, Pomella AW, Maki CS, Araújo WL, Dos Santos DR, Azevedo JL (2005) Diversity of endophytic fungal community of cacao (*Theobroma cacao* L.) and biological control of Crinipellis perniciosa, causal agent of Witches' Broom Disease. Int J Biol Sci 1(1):24

Rutkowska N, Drożdżyński P, Ryngajłło M, Marchut-Mikołajczyk O (2023) Plants as the extended phenotype of endophytes—the actual source of bioactive compounds. Int J Mol Sci 24(12):10096

Rutter BD, Innes RW (2018) Extracellular vesicles as key mediators of plant–microbe interactions. Curr Opin Plant Biol 44:16–22

Sadeghi F, Samsampour D, Seyahooei MA, Bagheri A, Soltani J (2020) Fungal endophytes alleviate drought-induced oxidative stress in mandarin (Citrus reticulata L.): toward regulating the ascorbate–glutathione cycle. Sci Hortic 261:108991

Sahab AF (2012) Antimicrobial efficacy of secondary metabolites of *Beauveria bassiana* against selected bacteria and phytopathogenic fungi. J Appl Sci Res 8(3):1441–1444

Sahay H, Yadav AN, Singh AK, Singh S, Kaushik R, Saxena AK (2017) Hot springs of Indian Himalayas: potential sources of microbial diversity and thermostable hydrolytic enzymes. 3 Biotech 7:1–11

Sahoo S, Sarangi S, Kerry RG (2017) Bioprospecting of endophytes for agricultural and environmental sustainability. In: Microbial biotechnology: Volume 1. Applications in agriculture and environment, pp 429–458

Sahu PK, Mishra S (2021) Effect of hybridization on endophytes: the endo-microbiome dynamics. Symbiosis 84(3):369–377

Sallam N, Ali EF, Seleim MA, Khalil Bagy HM (2021) Endophytic fungi associated with soybean plants and their antagonistic activity against *Rhizoctonia solani*. Egypt J Biol Pest Control 31(1):1–9

Salwan R, Sharma M, Sharma A, Sharma V (2023) Insights into plant beneficial microorganism-triggered induced systemic resistance. Plant Stress:100140

Samanta A, Banerjee S, Datta S (2023) The role of endophytes to boost the plant immunity. In: Endophytic association: What, why and how. Elsevier

Santoyo G, Moreno-Hagelsieb G, del Carmen Orozco-Mosqueda M, Glick BR (2016) Plant growth-promoting bacterial endophytes. Microbiol Res 183:92–99

Sarrocco S, Vannacci G (2018) Preharvest application of beneficial fungi as a strategy to prevent postharvest mycotoxin contamination: A review. Crop Prot 110:160–170

Sarrocco S, Valenti F, Manfredini S, Esteban P, Bernardi R, Puntoni G, Baroncelli R, Haidukowski M, Moretti A, Vannacci G (2019) Is exploitation competition involved in a multitrophic strategy for the biocontrol of fusarium head blight? Phytopathology 109(4):560–570

Savary S, Willocquet L, Pethybridge SJ, Esker P, McRoberts N, Nelson A (2019) The global burden of pathogens and pests on major food crops. Nat Ecol Evol 3(3):430–439

Schulz B, Boyle C (2005) The endophytic continuum. Mycol Res 109(6):661–686

Shafi Z, Ilyas T, Shahid M, Vishwakarma SK, Malviya D, Yadav B, Sahu PK, Singh UB, Rai JP, Singh HB (2023) Microbial Management of Fusarium Wilt in Banana: A comprehensive overview. In: Detection, diagnosis and management of soil-borne phytopathogens, pp 413–435

Shanmugaiah V, Gauba A, Hari SK, Prasad R, Ramamoorthy V, Sharma MP (2023) Effect of silicon micronutrient on plant's cellular signaling cascades in stimulating plant growth by mitigating the environmental stressors. Plant Growth Regul 100(2):391–408

Sharma A, Malhotra B, Kharkwal H, Kulkarni GT, Kaushik N (2020) Therapeutic agents from endophytes harbored in Asian medicinal plants. Phytochem Rev 19:691–720

Sharma P, Kumar T, Yadav M, Gill SS, Chauhan NS (2021) Plant-microbe interactions for the sustainable agriculture and food security. Plant Gene 28:00325

Shen F-T, Yen J-H, Liao C-S, Chen W-C, Chao Y-T (2019) Screening of rice endophytic biofertilizers with fungicide tolerance and plant growth-promoting characteristics. Sustain For 11(4):1133

Silva MDC, Guerra-Guimarães L, Diniz I, Loureiro A, Azinheira H, Pereira AP, Tavares S, Batista D, Várzea V (2022) An overview of the mechanisms involved in Coffee-*Hemileia vastatrix* interactions: plant and pathogen perspectives. Agronomy 12(2):326

Singh R, Dubey A (2015) Endophytic actinomycetes as emerging source for therapeutic compounds. Indo Global J Pharm Sci 5(2):106–116

Singh VK, Kumar A (2023) Secondary metabolites from endophytic fungi: production, methods of analysis, and diverse pharmaceutical potential. Symbiosis:1–15

Singh RN, Gaba S, Yadav AN, Gaur P, Gulati S, Kaushik R, Saxena AK (2016) First high quality draft genome sequence of a plant growth promoting and cold active enzyme producing psychrotrophic *Arthrobacter agilis* strain L77. Stand Genomic Sci 11:1–9

Singh SP, Bhattacharya A, Gupta R, Mishra A, Zaidi F, Srivastava S (2020b) Endophytic Phytobiomes as defense elicitors: current insights and future prospects. In: Phytobiomes: current insights and future vistas, pp 299–334

Singh P, Singh RK, Guo D-J, Sharma A, Singh RN, Li D-P, Malviya MK, Song X-P, Lakshmanan P, Yang L-T (2021b) Whole genome analysis of sugarcane root-associated endophyte Pseudomonas aeruginosa B18—a plant growth-promoting bacterium with antagonistic potential against *Sporisorium scitamineum*. Front Microbiol 12:628376

Sodhi GK, Saxena S (2023) Role of endophytic fungi in promoting plant health. In: Fungal resources for sustainable economy: current status and future perspectives. Springer

Soliman SS, Trobacher CP, Tsao R, Greenwood JS, Raizada MN (2013) A fungal endophyte induces transcription of genes encoding a redundant fungicide pathway in its host plant. BMC Plant Biol 1:1–10

Sood M, Kapoor D, Kumar V, Sheteiwy MS, Ramakrishnan M, Landi M, Araniti F, Sharma A (2020) Trichoderma: the "secrets" of a multitalented biocontrol agent. Plan Theory 9(6):762

Strobel G (2018) The emergence of endophytic microbes and their biological promise. J Fungi 4(2):57

Strobel GA, Dirkse E, Sears J, Markworth C (2001) Volatile antimicrobials from *Muscodor albus*, a novel endophytic fungus. Microbiology 147(11):2943–2950

Stroud EA (2023) Understanding the delicate balance between microbial pathogens and insect pests and optimising protection in Actinidia chinensis. ResearchSpace@ Auckland

Su Z-Z, Mao L-J, Li N, Feng X-X, Yuan Z-L, Wang L-W, Lin F-C, Zhang C-L (2013) Evidence for biotrophic lifestyle and biocontrol potential of dark septate endophyte Harpophora oryzae to rice blast disease. PLoS One 8(4):e61332

Suebrasri T, Harada H, Jogloy S, Ekprasert J, Boonlue S (2020) Auxin-producing fungal endophytes promote growth of sunchoke. Rhizosphere 16:100271

Sundin GW, Castiblanco LF, Yuan X, Zeng Q, Yang CH (2016) Bacterial disease management: challenges, experience, innovation and future prospects: challenges in bacterial molecular plant pathology. Mol Plant Pathol 17(9):1506–1518

Suryanarayanan T (2013) Endophyte research: going beyond isolation and metabolite documentation. Fungal Ecol 6:561–568

Sylvia DM, Chellemi DO (2001) Interactions among root-inhabiting fungi and their implications for biological control of root pathogens

Tabbasum I, Ali A, Shafique U, Bibi I, Khalid B, Ijaz N, Naseer MU (2022) Role of endophytic fungal species in plant growth promotion and in sustainable agricultural productions: a review. J Microbiol Mol Genet 3(1):1–21

Tanaka A, Christensen MJ, Takemoto D, Park P, Scott B (2006) Reactive oxygen species play a role in regulating a fungus–perennial ryegrass mutualistic interaction. Plant Cell 18(4):1052–1066

Tang T, Zhang Y, Wang F, Mao T, Guo J, Guo X, Duan Y, You J (2022) *Paenibacillus terrae* PY8 controls *Botrytis cinerea* and modifies the endophytic microbial community of the medicinal plant, Paris polyphylla. Biol Control 169:104888

Thakur N, Nigam M, Mann NA, Gupta S, Hussain CM, Shukla SK, Shah AA, Casini R, Elansary HO, Khan SA (2023) Host-mediated gene engineering and microbiome-based technology optimization for sustainable agriculture and environment. Funct Integr Genomics 23(1):57

Tian F, Chun HS (2017) Natural products for preventing and controlling aflatoxin contamination of food. In: Aflatoxin-control, analysis, detection and health risks; Abdulra'uf, L., Ed, pp. 13–44

Tiwari RK, Lal MK, Kumar R, Mangal V, Altaf MA, Sharma S, Singh B, Kumar M (2021) Insight into melatonin-mediated response and signaling in the regulation of plant defense under biotic stress. Plant Mol Biol:1–15

Tripathi YN, Divyanshu K, Kumar S, Jaiswal LK, Khan A, Birla H, Gupta A, Singh SP, Upadhyay RS (2020) Biopesticides: current status and future prospects in India. In: Bioeconomy for sustainable development, pp 79–109

Tsivileva OM, Koftin OV, Evseeva NV (2022) Coumarins as fungal metabolites with potential medicinal properties. Antibiotics 11(9):1156

Uzma F, Mohan CD, Siddaiah CN, Chowdappa S (2019) Endophytic fungi: promising source of novel bioactive compounds. In: Advances in endophytic fungal research: present status and future challenges, pp 243–265

Vandenkoornhuyse P, Quaiser A, Duhamel M, Le Van A, Dufresne A (2015) The importance of the microbiome of the plant holobiont. New Phytol 206(4):1196–1206

Vardharajula S, SkZ A, Shiva Krishna Prasad Vurukonda S, Shrivastava M (2017) Plant growth promoting endophytes and their interaction with plants to alleviate abiotic stress. Curr Biotechnol 6(3):252–263

Vargas L, Santa Brigida AB, Mota Filho JP, de Carvalho TG, Rojas CA, Vaneechoutte D, Van Bel M, Farrinelli L, Ferreira PC, Vandepoele K (2014) Drought tolerance conferred to sugarcane by association with Gluconacetobacter diazotrophicus: a transcriptomic view of hormone pathways. PLoS One 9(12):e114744

Vélez JM, Tschaplinski TJ, Vilgalys R, Schadt CW, Bonito G, Hameed K, Engle N, Hamilton CE (2017) Characterization of a novel, ubiquitous fungal endophyte from the rhizosphere and root endosphere of Populus trees. Fungal Ecol 27:78–86

Verma A, Shameem N, Jatav HS, Sathyanarayana E, Parray JA, Poczai P, Sayyed R (2022) Fungal endophytes to combat biotic and abiotic stresses for climate-smart and sustainable agriculture. Front Plant Sci 13:953836

Vinale F, Sivasithamparam K, Ghisalberti EL, Marra R, Woo SL, Lorito M (2008) Trichoderma–plant–pathogen interactions. Soil Biol Biochem 40(1):1–10

Wang L-W, Xu B-G, Wang J-Y, Su Z-Z, Lin F-C, Zhang C-L, Kubicek CP (2012) Bioactive metabolites from Phoma species, an endophytic fungus from the Chinese medicinal plant *Arisaema erubescens*. Appl Microbiol Biotechnol 93:1231–1239

Wang G, Liu Z, Lin R, Li E, Mao Z, Ling J, Yang Y, Yin W-B, Xie B (2016) Biosynthesis of antibiotic leucinostatins in bio-control fungus *Purpureocillium lilacinum* and their inhibition on *Phytophthora* revealed by genome mining. PLoS Pathog 12(7):e1005685

Wang S, Wang J, Zhou Y, Huang Y, Tang X (2022) Correction: prospecting the plant growth–promoting activities of endophytic bacteria Franconibacter sp. YSD YN2 isolated from *Cyperus esculentus* L. var. *sativus* leaves. Ann Microbiol 72(1):1–2

Wani S, Kumar V, Shriram V, Sah S (2016) Phytohormones and their metabolic engineering for abiotic stress tolerance in crop plants. Crop J 4:162–176

Waqas M, Khan AL, Kamran M, Hamayun M, Kang S-M, Kim Y-H, Lee I-J (2012) Endophytic fungi produce gibberellins and indoleacetic acid and promotes host-plant growth during stress. Molecules 17(9):10754–10773

Winters NP (2022) Evolutionary and functional Genetics of disease resistance in Theobroma Cacao and its wild relatives. The Pennsylvania State University

Woźniak M, Grządziel J, Gałązka A, Frąc M (2019) Metagenomic analysis of bacterial and fungal community composition associated with *Paulownia elongata*× *Paulownia fortunei*. Bioresources 14(4):8511–8529

Xiao-Yan S, Qing-Tao S, Shu-Tao X, Xiu-Lan C, Cai-Yun S, Yu-Zhong Z (2006) Broad-spectrum antimicrobial activity and high stability of Trichokonins from *Trichoderma koningii* SMF2 against plant pathogens. FEMS Microbiol Lett 260(1):119–125

Xu M, Xu C, Chen M, Xiao Z, Wang Y, Xu Y, Xu D (2023) Comparative analysis of commonly used bioinformatics software based on omics. Gene Reports:101800

Yadav AN, Saxena AK (2018) Biodiversity and biotechnological applications of halophilic microbes for sustainable agriculture. J Appl Biol Biotechnol 6(1):48–55

Yan L, Zhao H, Zhao X, Xu X, Di Y, Jiang C, Shi J, Shao D, Huang Q, Yang H (2018) Production of bioproducts by endophytic fungi: chemical ecology, biotechnological applications, bottlenecks, and solutions. Appl Microbiol Biotechnol 102:6279–6298

Yan L, Zhu J, Zhao X, Shi J, Jiang C, Shao D (2019) Beneficial effects of endophytic fungi colonization on plants. Appl Microbiol Biotechnol 103:3327–3340

Yoshida K, Schuenemann VJ, Cano LM, Pais M, Mishra B, Sharma R, Lanz C, Martin FN, Kamoun S, Krause J (2013) The rise and fall of the Phytophthora infestans lineage that triggered the Irish potato famine. elife 2:e00731

Zhang S, Gan Y, Ji W, Xu B, Hou B, Liu J (2017) Mechanisms and characterization of Trichoderma longibrachiatum T6 in suppressing nematodes (*Heterodera avenae*) in wheat. Front Plant Sci 8:1491

Zhang Y, Yu X, Zhang W, Lang D, Zhang X, Cui G, Zhang X (2019) Interactions between endophytes and plants: beneficial effect of endophytes to ameliorate biotic and abiotic stresses in plants. J Plant Biol 62:1–13

Zhang L, Zhang W, Li Q, Cui R, Wang Z, Wang Y, Zhang Y-Z, Ding W, Shen X (2020) Deciphering the root endosphere microbiome of the desert plant Alhagi sparsifolia for drought resistance-promoting bacteria. Appl Environ Microbiol 86(11):e02863–e02819

Zhang H, Li X, White JF, Wei X, He Y, Li C (2021) Epichloë endophyte improves ergot disease resistance of host (*Achnatherum inebrians*) by regulating leaf senescence and photosynthetic capacity. J Plant Growth Regul:1–10

Zhao S-S, Zhang Y-Y, Yan W, Cao L-L, Xiao Y, Ye Y-H (2017) *Chaetomium globosum* CDW7, a potential biological control strain and its antifungal metabolites. Microbiol Lett 364(3):287

Zhao Y, Ji X-L, Shen T, Tang W-T, Li S-S, Zhu Z-Q, Kumar J, Li H-Y (2020) Fungal endophytic communities of two wild Rosa varieties and the role of an endophytic *Seimatosporium* sp. in enhancing host plant powdery mildew resistance. Plant Soil 447:553–564

Zhou W, Wheeler TA, Starr JL, Valencia CU, Sword GA (2018) A fungal endophyte defensive symbiosis affects plant-nematode interactions in cotton. Plant Soil 422:251–266

Zhu L, Li T, Wang C, Zhang X, Xu L, Xu R, Zhao Z (2018) The effects of dark septate endophyte (DSE) inoculation on tomato seedlings under Zn and Cd stress. Environ Sci Pollut Res 25:35232–35241

Zou W, Meng J, Lu H, Chen G, Shi G, Zhang T, Tan R (2000) Metabolites of *Colletotrichum gloeosporioides*, an endophytic fungus in *Artemisia mongolica*. J Nat Prod 63(11):1529–1530

Fungal Endophytes and Soilborne Fungi 16

Akhtar Hameed, Muzzamil Qazi, Muhammad Umair Rafiq, and Hira Akhtar

Abstract

This chapter explores the complex relationship between plants and endophytes, focusing on fungal endophytes and soilborne fungi. It explores the multifaceted roles of endophytes in promoting plant growth, enhancing resistance to abiotic stress, and aiding in defense against pests and diseases. The significance of bacterial endophytes, such as their resistance to metal toxicity, is highlighted. The research emphasizes the escalating threat of soilborne phytopathogens, particularly fungal species, to global crop production and food security. Additionally, the biocontrol potential of fungal endophytes, their metabolites, and mechanisms such as competition, induced resistance, mycoparasitism, and antibiotics are discussed, offering promising avenues for sustainable agriculture. This comprehensive chapter focuses on the isolation, identification, and application of fungal endophytes for biocontrol in agriculture. The process involves surface sterilization of plant parts, various isolation methods, and morphological identification. The interaction dynamics between endophytes and host plants are explored emphasizing their potential as biocontrol agents (BCAs). Challenges, such as host specificity and environmental adaptation, are discussed. Successful applications, particularly direct antagonism to pathogens and systemic resistance induction, are highlighted. The chapter underscores the environmental and safety considerations of using fungal endophytes emphasizing reduced chemical use and the need for ongoing research and monitoring. Future prospects include sustainable agriculture, efficacy improvement, and multi-omic approaches.

A. Hameed (✉) · M. Qazi · M. U. Rafiq · H. Akhtar
Institute of Plant Protection, MNS- University of Agriculture, Multan, Multan, Pakistan

© The Author(s), under exclusive license to Springer Nature Singapore Pte Ltd. 2025

K. A. Abd-Elsalam, A. H. Hashem (eds.), *Fungal Endophytes Volume II*,
https://doi.org/10.1007/978-981-97-8804-0_16

Keywords

Fungal endophytes · Biocontrol agents · Plant growth · Soilborne fungi · Sustainable agriculture

16.1 Introduction

The father of modern plant pathology, Anton de Bary first used the term endophyte (endo = within, phyto = plant) in 1866 to refer to "any organism growing within plant tissues" (de Bary 1866). The refined definition that is currently in use only includes organisms that do not cause disease and occasionally even benefit their hosts (Collinge et al. 2022). All kinds of plants contain endophytes, and in fact, microorganisms have taken up endophytic lifestyles in all spheres of life. These endophytes can promote plant growth, improve a plant's capacity to withstand abiotic stress, and help plants defend against pests and diseases (Hardoim et al. 2015). A diverse group of organisms known as endophytes include bacteria, fungi, and archaea (Moissl-Eichinger et al. 2018). Endophytes can infect both aerial and subterranean plant organs, but this infection is benign and asymptomatic (Zimmerman and Vitousek 2012). In fact, most plant-endophyte relationships that have been documented are mutualistic—both parties gain from the relationship (Deshmukh et al. 2006). In symbiotic relationships, endophytes and plants can be facultative, opportunistic, or obligate. Because they lack a free-living stage, obligatory endophytes reside in their hosts and are solely dependent on the metabolism of those hosts to survive. Mycorrhizal fungi are one type of obligate endophyte (Hardoim et al. 2015). Endophytes were found for a variety of reasons from ecological research, which includes adaptation to extreme environments to the search for pertinent activities (such as biocontrol, bioremediation, metabolite production, plant growth, and resistance). For example, endophytes have been shown to influence or take part in the biosynthesis of bioactive compounds in medicinal plants (Maggini et al. 2020).

Bacterial endophytes have developed resistance to the harmful effects of metals in heavy metal-accumulating plants (Mengoni et al. 2010). While the endophytic microbiome of plants with antimicrobial metabolites comprises in many aromatic plants. For example, it exhibits specific adaptation toward the essential oil of various plant chemotypes (Checcucci et al. 2017). It has recently been demonstrated that an *Enterobacter* sp. strain isolated from finger millet (*Eleusine coracana*) can both produce multiple antifungal compounds that kill the fungus and suppress the grass pathogen *Fusarium graminearum* in the root system of its host plants (Mousa et al. 2016).

Global crop production and food security are closely linked to soil health, but several soilborne phytopathogens have put soil health at risk (Chakraborty and Newton 2011). Numerous commercial and agricultural crops are susceptible to infection by soilborne phytopathogens (Möller and Stukenbrock 2017), which has resulted in a 78% reduction in fruit, vegetable, and cereal crop productivity (Zhang 2018). Many of the most aggressive phytopathogens are soilborne fungi, which are expected to multiply up to three times by the year 2050 worldwide

(Delgado-Baquerizo et al. 2020). Controlling plant diseases transmitted by soil is the most difficult (Alabouvette et al. 2009). Soilborne plant pathogens, *Fusarium* spp., infect stems, leaves, fruits, and roots of a wide range of host plants. Even they were spreading diseases to people and animals (Glenn 2007; Zhang et al. 2006). *Fusarium oxysporum* particularly has been found to infect cultures of banana trees leading to the ailment known as "Panama disease" (*Fusarium* wilt of banana).

Fungal species were the first organisms to be characterized as endophytes after that bacteria were discovered in abundant form in the plant tissues having the characteristics of endophytes (Freeman 1904). Fungi perform several vital functions in ecosystems including recycling, decomposition, and the movement of nutrients from one environment to another. Over a million fungal species may exist on Earth according to various assessments, of which only about 3% have been formally identified (Smith et al. 2008). They have a saprophytic or parasitic relationship with their hosts because they are heterotrophic. Nonetheless, the fungi and endophytes exhibited a variety of relationships over time and showed a huge influence on overall plant health. Among these connections are endophytes (Pandi et al. 2010).

16.1.1 Fungal Endophytes

Fungal endophytes have occurred in all sections of plants and are asymptomatic inhabitants of plant tissue (Sarsaiya et al. 2019). Many endophytic species can occur in plants, which can stay localized and provide tissue-specific disease protection (Wu et al. 2013) or can systematically spread throughout herbaceous plants (Chutulo and Chalannavar 2018). There are many different types of symbiotic and possibly mutualistic interactions between endophytes and plants including both cultivated and wild plant species (Arnold et al. 2000). Fungal endophytes of dicots, endophytic fungi, other systemic fungal endophytes, lichen-associated endophytes, endophytic fungi of bryophytes and ferns, endophytic fungi of tree bark, xylem-associated endophytes, roots, galls, and cysts are examples of endophytes. *Trichoderma* species were identified as viable biocontrol candidates against *Fusarium* spp. and other phytopathogenic fungi—because of their strong antifungal activities in both laboratory and greenhouse settings (Bidellaoui et al. 2019; Ihrmark et al. 2010; Mambaeva et al. 2019). A variety of pathogenic bacteria as well as a minimum of 19 genera and 30 species of harmful fungi are antagonistically affected by *Trichoderma* spp. Growth competition and mycoparasitism are the main biocontrol mechanisms of *Trichoderma* spp.—which are followed by the promotion of plant immunity and resistance (Harman 2006). *Trichoderma asperellum* was recently found to be the fungal endophyte that prevented *F. oxysporum* from growing when it was separated from the healthy banana plant's surface-sterilized roots. In comparison with other microbes, efforts to remove fungal endophytes from their natural environments are not as successful (Rodrigues et al. 2008). Though they are unable to explain their origin or pathways, researchers are attempting to isolate novel bioactive compounds for use in industrial, medicinal, and agricultural applications from emerging species of fungal endophytes (Table 16.1).

Table 16.1 Common plant endophytes

Endophyte	Plant species	References
Acremonium sp.	*Huperzia serrata* *Taxus chinensis*	Liu et al. (2009) Glienke-Blanco et al. (2002)
Stemphylium globuliferum	*Avicennia marina*	Moussa et al. (2016)
Aspergillus sp.	*Prosopis chilensis* *Datura stramonium* *Moringa olifera*	Mahdi et al. (2014)
Curvularia sp.	*Datura stramonium* *Moringa olifera*	Mahdi et al. (2014)
Colletotrichum sp. *C. gloeosporiodes*	*Tectona grandis and Samanea saman* *Huperzia serrata* *Triticum aestivum* *Cinnamomum camphora* *Citrus* plants *Cinnamomum camphora* *Pasania edulis* *Ginkgo biloba* L. *Lycopersicum esculentum Mill.*	Wang et al. (2011) Thongsandee et al. (2012) He et al. (2012)
Cladosporium sp. *C. herbarum*	*Cinnamomum Camphora* *Opuntia ficus indica* *Triticumaestivum* *Lycopersicum esculentum Mill.*	Bezerra et al. (2012) He et al. (2012) Larran et al. (2001)
Penicillium sp.	*Huperzia serrata* *Lycopersicum esculentum Mill.*	Wang et al. (2011) Larran et al. (2001)
Bacillus megatherium	*Medicago satavia*	Moussa et al. (2016)

16.2 The Role of Soilborne Fungi in Plant Pathology

There are many records of soilborne fungi particularly mycorrhizae and *Trichoderma* protecting roots against soilborne plant pathogens, though fungal or viral diseases of the green parts are usually more severe in mycorrhizal plants. There are different mechanisms by which mycorrhizae protect themselves against soilborne diseases. Mycorrhizae have received considerable attention in recent years because of several benefits they have for the plant. Disease protection against root pathogens is one benefit that has been intensively studied in various parts of the world. The protective influence of mycorrhizae or mycorrhizal fungi has often been demonstrated.

Research has concentrated on diseases caused by soilborne pathogens. The very few reports related to virus disease indicate that mycorrhizal host plants always exhibit increased disease intensity (Dehne 1982). In a field where Panama disease was present healthy, banana plants were grown, and from the rhizosphere soil (suppressive soil), *Trichoderma* sp. was isolated. *Trichoderma asperellum* was identified as a major strain having antifungal activity against the four plant pathogenic species of fungi *Fusarium oxysporum, fujikuroi, tricinctum* and *cantenulatum*. The antagonistic mechanism of *T. asperellum* against the four fungal plant pathogens was also investigated in vitro through direct and indirect means, including antibiosis and mycoparasitism, and by observing competition in overall growth.

16.3 Exploring the World of Fungal Endophytes

Nevertheless, the range of chemical compounds found in fungal endophytes is enormous encompassing a wide range of structures and bioactivities. Moreover, a million or more distinct species of endophytic fungi may exist—according to some estimates, the vast majority of which have not yet been described (Deshmukh et al. 2015). For this reason, endophytic fungi are the source of the discovery of new natural substances.

16.3.1 Natural Products

Natural products on the other hand are the outcome of evolutionary processes and have a wide variety of structural features. Because of this, they have a strong affinity for their particular target receptors and are incredibly ideal for biologically based roles involving biochemical pathways (Molinari 2009). The same award was given in 1945 to those who discovered penicillin and in 1952 to those who discovered streptomycin (an aminoglycoside). The significant impact that naturally occurring microbiological products have had on contemporary medicine is noteworthy (Caruso et al. 2022).

16.3.2 Fungal Endophytes and Metabolites

Plants provide several essential medications including vinblastine, vincristine, Taxol, quinine, and camptothecin (Ramawat et al. 2009). In contrast, with a fungal origin, bioactive metabolites that are known to exist are about 8500. *Taxomyces andreanae*, a fungal endophyte found in *Taxus baccata*, aids in the search for endophytic fungi linked to beneficial bioactive compounds and their derivatives (Arora and Ramawat 2017). Antibiotics are not the only products that microbial metabolites have an impact on in the pharmaceutical industry. Among the most effective examples are the immunosuppressive drugs cyclosporines and rapamycin, which were first extracted from the fungus *Tolypocladium inflatum* (Caruso et al. 2022).

16.3.3 Antimicrobial Potential of Endophytic Fungi

The amount of research on endophytes and their bioactive byproducts has been steadily increasing in recent years, and numerous reviews have been published, describing a wide variety of bioactive compounds (Wen et al. 2019). For instance, a review described the anticancer substances that were extracted from different endophytic fungi between the years 2010 and 2013. It gives extensive information on the medicinally significant bioactive substances that were separated from endophytes between 2000 and 2019—many of which had antibacterial, antifungal, and anticancer properties (Raimi and Adeleke 2021).

16.3.4 Endophyte Culture Extract with Antimicrobial Properties

To achieve this, endophytes are cultivated in liquid or solid cultures from which an aqueous or crude organic extract can be extracted. A study by Arivudainambi et al. (2011) isolated an endophytic fungus from a medical herb (*Vitex negundo L.*) leaves in India. A report was made concerning a medicinal herb that was gathered in India (Arivudainambi et al. 2011).

16.3.5 Anti-biofilm Compounds Produced by Endophytic Fungi

In natural settings, the majority of bacterial species favor the biofilm growth mode. The reported antimicrobial compounds exhibit efficacy against both free-living and planktonic bacterial cells. Treating bacteria living in biofilms can be challenging or even impossible, due to their greater resistance than that of planktonic bacterial cells to responses of the immune system and antibiotic therapy. The *Aspergillus nidulans* chloroform and ethyl acetate extracts inhibited the formation of biofilm in an oral pathogenic strain of *Candida albicans*. A study reported a medical plant (*Acacia nilotica*) contains a fungal endophyte with a 74.86% and 72.53% association. Respectively, as reported by Meenambiga and Rajagopal (2018).

16.4 Mechanism of Fungal Endophyte-Mediated Biocontrol

Endophytic microorganisms use various mechanisms of action to increase the adaptability of plants. Endophytic fungi frequently use pathogen-inhibition mechanisms, such as competition, mycoparasitism, and antibiosis, to directly inhibit pathogens while indirectly inducing resistance to activate the plant's defenses against the disease.

16.4.1 Competition

Competition is a potential mechanism adopted by endophytes to prevent pathogens from colonizing in host tissues (Nunna and Balachandar 2022). Endophytes use mechanisms to compete, and these mechanisms typically work in concert with one another rather than on their own (Dirihan et al. 2015). Since the majority of the mechanisms have direct effects for enhanced pathogen antagonistic effects, the plant needs to be systemically colonized by an endophyte (Rosenblueth and Martínez-Romero 2006). For example, *Phytophthora* infection was low when an endophytic mixture made from cacao leaves was applied topically a significant amount through competition. Nevertheless, it was also discovered that some strains produced active metabolites indicating that competition might not be the only method for controlling diseases (Arnold and Herre 2003). According to the research, in endophytes that rely only on settings with high pathogen loads, competition may not be an effective biocontrol mechanism (Ownley et al. 2010).

16.4.2 Induced Resistance

It is well known that endophytic fungi help plants defend themselves from pest or pathogen attacks. This activation event, also referred to as "defense priming" or "induced systemic resistance," enhances the plant's ability to fend off pathogen attacks in the future and increases its performance (Poveda et al. 2020). Ethylene or jasmonate acid is frequently used to control induced systemic resistance (ISR)—which does not entail the overexpression of proteins linked to pathogenicity (PR) (Ramamoorthy et al. 2001). On the other hand, systemic acquired resistance (SAR) is mediated by salicylic acid—which is usually associated with pathogen infection and leads to the accumulation of PR protein (Tripathi et al. 2019). Among these PR proteins are different enzymes like beta 1,3-glucanase and chitinase, which immediately dissolve strengthening the cell wall barrier to prevent pathogen cells from entering and establishing the capacity to fend off infection and cellular death (Gao et al. 2010). On the other hand, systemic resistance induced by increased expression of genes linked to pathogenesis has also been linked to endophytes (Mishra et al. 2018). It was reported that *Fusarium solani*, which was isolated from tomato plants, promoted ISR against foliar by turning on the root's PR, PR5, and PR7 genes, the pathogen *Septoria lycopersici* (Kavroulakis et al. 2007).

16.4.3 Mycoparasitism

By directly attacking the pathogen that has been identified or its propagules, endophytes employ this crucial mechanism to safeguard the host ecology (Huang et al. 2020). Endophytic fungi enter pathogenic fungus hyphae and break down the pathogen's cell walls by cell lysing. Four steps can be used to categorize this mechanism. The first two phases involve detecting and classifying fungal pathogens and chemotropic proliferation of the endophytic fungal mycelium towards pathogenic fungi. The third and fourth phases involve endophytic fungi penetrating the target fungal cells and coming into direct physical contact with pathogenic fungi. It breaks down the fungal cell wall (Ram et al. 2018). It was discovered that several *Trichoderma* species involved in biocontrol could pass through the hyphae. *Rhizoctonia solani*, a pathogenic fungus, destroys them (Grosch et al. 2006). In a similar way, isolated endophytic fungi from common reed coiled around to prevent soilborne fungal pathogens from growing their hyphae, which penetrates the cells and degrades the hyphal cytoplasm. The several enzymes that break down cell walls are secreted during the breakdown of fungal hyphae by fungal endophytes (Cao et al. 2009).

16.4.4 Antibiosis

Numerous secondary metabolites are produced by endophytes. Some of these are antifungal and antibacterial and aid in preventing the growth of microorganisms that are harmful to plants (Khan et al. 2020). There have been numerous reports of metabolites with antimicrobial activity including phytoplankton, alkaloids, peptides, polyketides, quinones, phenols, steroids, and terpenes (Munir et al. 2020). Nonetheless, examining the secondary metabolites produced by the endophytic fungi for beneficial purposes is necessary (Bolívar-Anillo et al. 2020). Interactions between the host and endophytes facilitate the production and release of metabolites through a range of processes, including pathogen inhibition (Adetunji et al. 2019) (Fig. 16.1).

The enigmatic endophytic fungi are beginning to reveal their secrets. This includes understanding the underlying mechanisms of endophyte asymptomatic colonization, evading plant immune response, modulating gene expression, and establishing a balanced mutualistic relationship. Endophytic fungi play a crucial role in enhancing plant-resistant against biotic stresses like fungi, bacteria, viruses, nematodes and other phytopathogens. Fungus endophytes depend on plant hosts for nutrients, shelter, and transmission. This figure also shows that endophytic fungi enhance plant health by decreasing and eliminating the detrimental effects of phytopathogens by mycoparasitism, competition for resources or space, and direct or indirect defense mechanisms by generating secondary metabolites and induced systemic resistance (ISR). It shows a beneficial step toward a sustainable approach for effective crop improvement. The importance of different bioactive substances with antibacterial potential against a range of agricultural pathogens has also been covered in this figure. Additionally, endophyte-mediated systemic acquired resistance (SAR) is also highlighted. The relationship between plants and endophytes as well as the mechanisms used by fungal endophytes to evade host defense reactions colonize their hosts and establish a mutualistic relationship that is balanced or outlined in this figure or chapter.

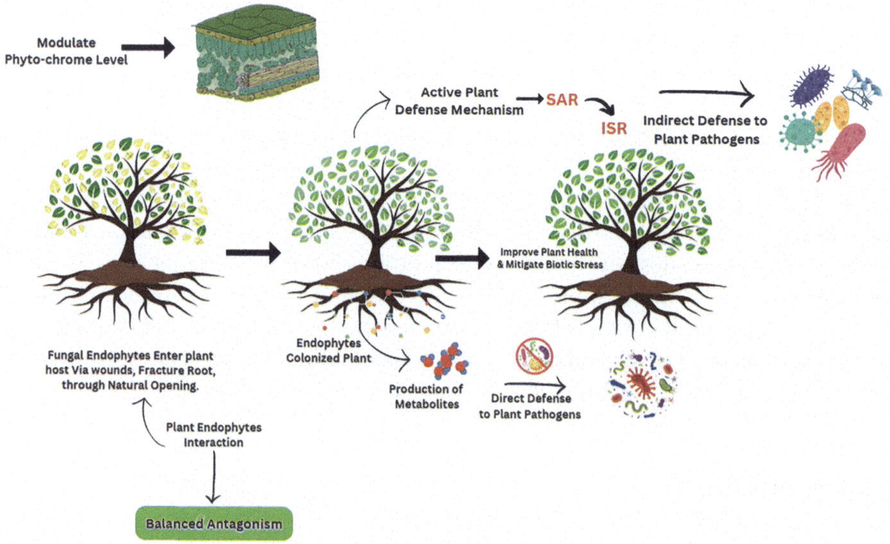

Fig. 16.1 Mechanisms of fungal endophytes interaction with fungal plant pathogens interact with fungal endophytes. Endophytic fungi grow inside plant tissue and typically establish mutualistic relationships with the plants they inhabit

16.5 Isolation and Identification of Fungal Endophytes

The crucial step in terms of plant selection and impediment techniques is the isolation of fungal endophytes. It is important to comprehend the methods and criteria utilized in the selection of plants.

16.5.1 Surface Sterilization of Selected Plant Parts

Plant materials are surface sterilized in laboratories using surface sterilization. Samples are dried and used for the separating process after being cleaned. Different researchers use various host plant species, types of tissue to study, and different ways to clean the surface. Some researchers have closely looked at the effects of different ways to clean surfaces (Rao et al. 2011) and some works on surface sterilization depend on the sample size and distance for isolation (Bok et al. 2006), which are critical factors that significantly influence the effectiveness of the sterilization process and the subsequent isolation of microbes.

16.5.2 Methods for Isolation of Fungal Endophytes

(a) *Isolation by Cutting Plant Parts*

Samples that have been cleaned on the outside are cut into very small pieces (0.2 cm × 0.21 cm). Then they're placed on media at 28 °C (Mane et al. 2017).

(b) *Isolation by Blender Shaft*

Homogenize 10 grams of the sterilized plant part in 90 ml of sterile distilled water using a blender shaft and inoculate the resulting mixture onto suitable ager media to promote fungal growth (Vieira et al. 2012).

(c) *Isolation by Mortar and Pestle*

Put 1 g of sterilized and 1 ml of pure distilled water into a bowl with the sterilized plant sample, and then use a pestle to grind it up. The unused suspension was spread on fake surfaces to help make endophytic fungi grow more (Vieira et al. 2012).

16.5.3 Identification of Fungal Endophytes

Morphological identification of endophytic fungi is an extremely important step (Li et al. 2008). It entails the growth of fungus on common media, analysis of sexual reproduction, and identification of growth traits. Taxonomical groups report that a few new tests, including subculturing, a fast purity test, an analysis of sexual reproduction, conidiogenesis, measurements of media temperature, and radial growth, have been added recently (Li et al. 2008). This identified culture can be used to create more reference sequences (Samarakoon et al. 2020). In a nutshell, DNA sequencing, sequence analysis, ITS region PCR amplification, and genome DNA isolation can all be used to molecularly characterize isolated fungal endophytes (Tiwari 2015).

16.6 Host Plant Fungal Endophyte Interaction

Due to their close interactions with their host, endophytic fungi can multiply within plant tissue, which reduces competition for nutrients and increases their protection against negative alterations to the phyllosphere and rhizosphere environments. Plant roots can have a very diverse microbiota inside the rhizosphere from there (Deng and Cao 2017). Endophytic fungi can coexist with plants as commensals, mutualists, or pathogens that neither hurt nor benefit the host plant (Hardoim et al. 2015). The health of the plant that hosts them is greatly changed by how microbes and plants relate to each other (Cardinale 2014). So, it's very important to study these complex links. This helps us get the good things from fungal endophytes inside plants and solve what we still don't know about them. Using optical and electron microscopes, you can watch the spores, fruiting bodies, and hyphae that are found in fungi (Alves et al. 2013). The connection of fungal threads to plant surfaces is the start in how endophytic fungi and plants interact. This connection causes physical feature expressions like sporulation, invasion of plants' tissues, and growth to more people (Pawlowski and Hartman 2016). Many ways have been used to look at the complex connections that exist between endophytes and plants. By using genetics and measuring enzyme activity, it is possible to discover the genes tied to these connections effectively (Ali et al. 2014) (Fig. 16.2).

Plant endophytic fungal associations entail a complex chemical interaction that involves the regulation of fungal virulence factors, the prevention or suppression of the host plant's defense mechanisms, and the facilitation of coexistence with other fungi and bacteria inside the plant. To deal with competitors from the rest of the microbiome, the host fungus downregulates certain mechanisms and produces chemical mediators or antagonistic secondary metabolites. This promotes the host plant's health by creating several equilibria of balanced antagonists. In the early stages of contact, the interactions between plants and pathogenic and endophytic fungi are identical. This is because the plant responds by producing defense mechanisms in response to the pathogenic fungus and endophyte, which both produce virulence factors that will aid in plant colonization. Endophytic fungi will modulate the production of virulence factors and colonize the plant without causing disease symptoms, in contrast to pathogenic fungi, which will continue to produce virulence factors until they colonize the plant and cause disease.

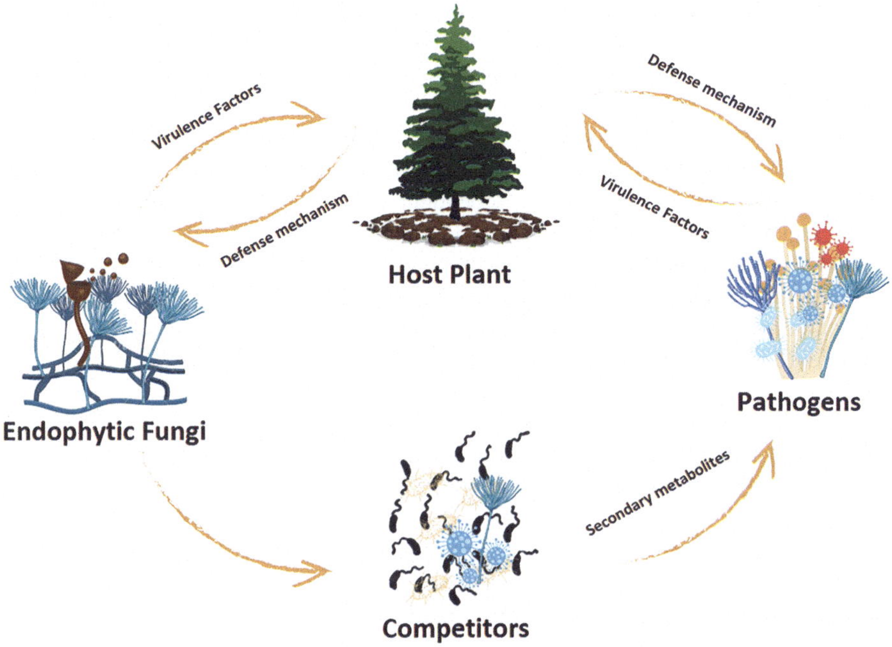

Fig. 16.2 Multiple balanced antagonism in endophytic fungi and plant interaction. Endophytic fungi and plants interact through a complex chemical process, regulating fungal virulence factors, preventing defense mechanisms, and facilitating coexistence. Host fungi downregulate mechanisms and produce antagonistic secondary metabolites to promote plant health. In early contact, plants produce defense mechanisms against pathogenic and endophytic fungi, with endophytic fungi modulating virulence factors and colonizing without disease symptoms, unlike pathogenic fungi

16.7 Field Application of Fungal Endophytes

BCAs are considered an environmentally benign way to increase the host's immunity and reduce vulnerability to infection by triggering the production of various bioactive compounds, including antibiotics, siderophores, and signaling molecules. However, there are additional benefits to using endophytic strains as BCAs because they colonize more successfully than other strains. Many factors have been taken into account before microbial antagonistic therapy is used, such as the stability and durability of BCAs against the main factors as well as inside the host. These factors also significantly affect marketability and practical applications. In the laboratory trial, a large number of endophytic and epiphytic strains including yeast, bacteria, and fungi showed antagonistic potential either in vivo or in vitro. Very few microbial BCAs, nevertheless, have been applied in the business world. The primary reasons for these errors are BCAs' lack of stability and survivability. However, only a very small number of microbial strains especially those from the *Bacillus* and

Trichoderma species have been made available for purchase. These two microbial groups have become popular choices because of their ability to form endospores, which gives them superior stability and survivability over other microbial strains that do not form endospores.

In the last 20 years, BCAs have become widely available and are now used globally to reduce yield losses. Application of BCAs against both the plant and the pathogen after screening is heavily dependent on the pathogen's growth stage, the host immune system, metabolic processes, and signaling molecules. Cultivars and plant physiology are important factors in the effective application of BCAs, in addition to genetic composition. The host-pathogen-BCA relationship is influenced by environmental factors as well, but for biocontrol techniques to work as well as to be possible, they must be standardized.

16.8 Challenges and Limitations in Using Fungal Endophytes for Biocontrol

Despite being the fungal endophytes that are gaining attention to control many plant diseases as biocontrol agents, there are some challenges and limitations given below that need to be covered up when we use them.

16.8.1 Isolation and Identification

Isolating and identifying useful and valuable fungal endophytes is one of the major obstacles. As we know, there are many types and varieties of fungal endophytes present in plants, but we have to find the one which have our desired requirements. The efficacy of endophytes depends upon the accurate and authentic identification of these fungal endophytes.

16.8.2 Host Specification

Fungal endophytes are mostly host-specific which means that they can only work against a specific plant pathogen or some particular plant diseases. Their limited approaches make them less beneficial. To overcome this challenge, we have to find and isolate fungal endophytes with a broad spectrum of host compatibility. It is essential to make them more selective against a wide range of target pathogens and diseases.

16.8.3 Competitive Interactions

The biocontrol ability of fungal endophytes is directly proportional to their competitive ability with pathogens of plants or other endophytes. The selective endophytes must be able to compete with them for space and food to survive and establish themselves. Successful biocontrol implementation needs an understanding of ecological interactions between various endophytes and their associated plants (Zabalgogeazcoa 2008).

16.8.4 Adaptation of Environment

To establish themselves in diverse plants and for effective pathogen compatibility, there must be adaptations of fungal endophytes to environmental conditions. They need to be resistant to abiotic stresses like changes in temperature swings, pH, moisture content, and availability of nutritional components. Furthermore, the long-term effectiveness and viability of biocontrol depends upon the adaptation ability of fungal endophytes to environmental stress (Gao and Tao 2012).

16.8.5 Formulation and Delivery

Creating suitable formulation and the delivery methods of fungal endophytes are challenges that need to be addressed properly. The activity, stability, and long-term viability are guaranteed by the selected formulation. Either they are made as systemic biocontrol agents or contact. They are designed to be tailored to their mode of action, similar to their delivery method. For optimum efficacy, effective techniques for getting them into plant tissues, including foliar sprays or seed treatments, must also be refined.

16.8.6 Regulatory Considerations

Before their wide applications as a biocontrol agent, fungal endophytes must satisfy safety and efficacy standards to remove or reduce their possible side effects on non-target creatures in plants, biodiversity, and human health; this needs comprehensive testing.

16.8.7 Limited Knowledge

The knowledge about the molecular characterization of fungal endophytes, their mode of action, and the interaction between plants, endophytes, and pathogens is very limited. The researchers still have to find these mechanisms for their more beneficial usage in developing a sustainable agriculture, and these studies will create major impact factors for the improvement in agriculture.

By addressing these challenges and limitations, fungal endophytes have considerable promise as environmentally acceptable and sustainable biocontrol solutions. Multidisciplinary cooperation between researchers, agricultural specialists, and regulatory agencies is necessary to overcome these obstacles (Fig. 16.3).

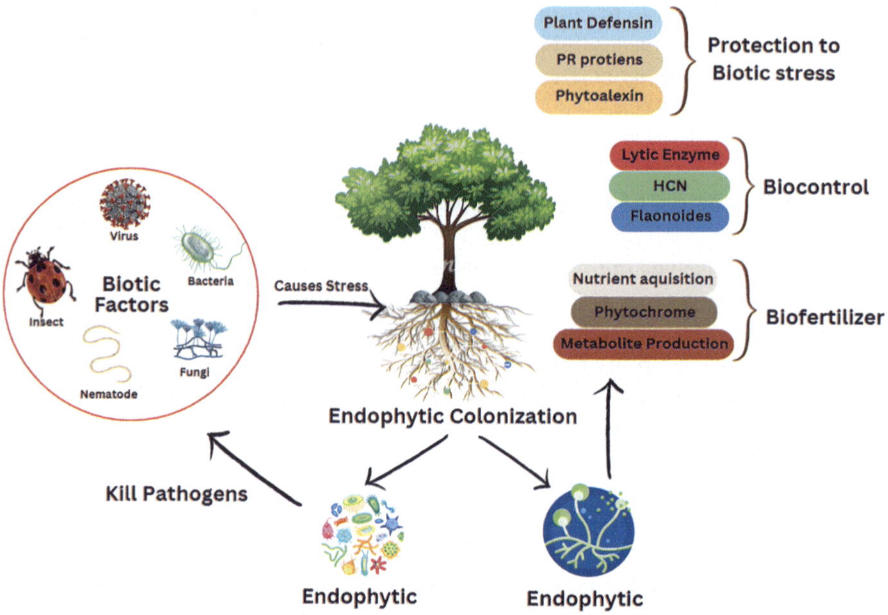

Fig. 16.3 Endophytes play a crucial role in agriculture by acting as biofertilizers, biocontrol, and biotic stress management. They can directly or indirectly promote plant growth by controlling essential nutrients and hormone levels. They can also indirectly aid in phytopathogen biocontrol by decreasing iron access, controlling defense mechanisms, and inhibiting pathogens with volatile compounds. These endophytes can also indirectly affect plant development through direct mechanisms

16.9 Future Prospects and Research Direction

Because of the numerous applications for which the compounds released by EF-host symbiotic interactions have enormous potential in biomedicine and sustainable agriculture, endogenous biology is currently gaining more and more attention. The fundamental processes of endophytism and its biological and ecological functions are of interest to scientists. Numerous studies in the field of endophytic biological research demonstrate its significance. Despite the increased interest in EF research, there are still a lot of problems that will need to be solved in the coming years, for example, selecting suitable host plants and their healthy organs or tissues to identify and isolate new EFs and examine the mutualism with which they are associated. Biocontrol agents are famous as environmentally friendly and are the most reliable sources for plant disease management, and this technique has a very bright future for the development of sustainable agriculture. As we discussed earlier, having environmentally friendly conditions will keep attracting pathologist researchers to reduce the side effects of chemical pesticides. The following are key research areas:

16.9.1 Sustainable Agriculture

It is a ecosystem-based agricultural services in which we systemically do agriculture practices to meet society's needs. A wide use of agrochemicals leads to loss of sustainability and also disturbs biodiversity. We can manage plant diseases smartly when endophytes play a major role having antagonistic properties against plant pathogens by releasing different metabolites, antioxidants, antibiotics, or other degrading enzymes. In this way, we can manage plant diseases with biological control agents without creating side effects on other creatures. In the future, there will be an era of sustainable agriculture, and through biological control, we can achieve it. Researchers are becoming increasingly interested in using environmentally friendly alternatives to meet the growing demand for agricultural production. The negligent use of chemical pesticides has given rise to grave concerns about poisonous residues, resistance development in plant pathogens, environmental pollution, and unintended effects on beneficial microbes (Yadav 2020). To maximize food production while reducing harmful health consequences, sustainable agriculture needs to employ various strategies. Because they enhance endophytes to restore plant health and shield it from disease threats, they are crucial to sustainable agriculture.

16.9.2 Efficacy Improvement

It is the main research area that is under consideration as a prospect. To bring efficacy, we will have to do proper isolation and identification to understand the proper mechanism of endophytes against plant pathogens. This research study will help the world to control plant disease through biocontrol agents without disturbing the environment.

16.9.3 Metabolite and Enzyme Study

Endophytes produce many multifunctioning metabolites that not only directly harm pathogens by inhibiting their growth but also help host plants promote their growth and development. Fungal endophytes produce some lytic enzymes that cause death to microbial cells by degradation of the cell wall. They regulate the growth and development of plants like chitinases, cellulases, etc.

16.9.4 Compatibility and Symbiosis

Fungal endophytes establish compatibility with the plant as they are present in the tissues of the plant and exchange their nutrients with the plants. The study area is "what types of fungal endophytes make compatibility symbiosis relationship with what types of plants." By researching this interaction, researchers can induce resistance and boost the defense mechanism of plants by applying specific endophytes to specific plants (Strobel 2018).

16.9.5 Multi-omic Approaches

Multi-omic approaches like genomics, proteomics, transcriptomics, and metabolomics are concerned with molecular characterization through sequencing of genes, proteins, transcription, and metabolism, and this study is becoming important to understand the functions and structures of body cells. In endophytes, researchers are dealing with omic approaches regarding metabolites to find their effects of interaction with plant pathogens and their host. Different genes of fungal endophytes are getting identification and characterization, which have positive interactions with plants. Through these techniques, metabolite-based pesticides can be tailored for improved performance and reduced environmental impact (Crandall et al. 2020).

16.9.6 Commercialization and Scale-Up

Plant disease management through fungal endophytes is limited to only in vitro conditions. Endophyte-based biopesticides have not reached to market, and farmers are not approaching it. Researchers must conduct field trials and make easy availability to farmers for sustainable agriculture. Cost-effective production and application methods must be developed.

16.9.7 Regulatory Frameworks

Establishment of regulatory consideration for their safe use as biocontrol agents, human health, and ecological concerns. As we know nothing is more important than human health, researchers have to study their effects on health while using these fungal endophytes. Similar to the case with their effects on the ecosystem as these are biological controls, what types of impacts they create on the environment will be the topic where researchers have to work.

Research in these study areas will contribute to developing environmentally friendly biocontrol solutions using fungal endophytes ultimately which will benefit agriculture and food security. Finding possible endophytes is the primary objective of agricultural researchers. These possible endophytes function well across a variety of climatic environments where the crop is grown in addition to being symbionts in that particular crop. In comparison with crops missing potential endophytes, crops with an appropriate endophyte fraction exhibit higher resistance and faster growth. Additionally, the ability of the host plants to withstand biotic and abiotic stresses such as drought, poor soil, and herbivory is increased by the fungal endophytes. The capacity of endophytes to enhance secondary metabolite production or improve plant nutrient availability plays a significant role in promoting plant health and resistance against biotic and abiotic stresses.

16.10 Research Insights and Applications

16.10.1 Successful Application of Fungal Endophytes as a Biocontrol Agent

In recent years, the control of plant diseases caused by different plant pathogens has been boosted. In the above discussion, the ability of biocontrol agents for the colonization of plants has been detailed. Fungal endophytes help the host plants in plant disease management in two ways:

(a) Direct antagonism to pathogen
(b) By induction of resistance in host

16.10.1.1 Direct Antagonism to Pathogen

In direct antagonism, they affect the pathogen directly. Fungal endophytes, such as *Trichoderma* spp., can directly parasitize disease causing pathogens line *Rhizoctonia solani* and *S. sclerotiorum* eggs of plant-parasitic nematodes like *M. javanica* and *Heterodera avenae*. Additionally, they form symbiotic relationships with their host plants, enhancing plant health and resilience. Many BCE (biocontrol endophytes) synthesize and release a variety of effecter molecules such as antibiotics, fungi toxins, poisons, and metabolites against plant pathogens in addition to their direct parasitism. By producing lactones, glucanase, chitinase, and cellulase enzymes (volatile for pathogens), BCEs suppress infection. For instance, *P. lilacinum* is a well-known biocontrol agent for plant parasitic nematodes and insects, but it has also shown promise in suppressing several plant diseases by utilizing effector. Effectors containing culture filtrates of *P. lilacinum* inhibit *S. sclerotiorum* and trigger defense responses in common beans. The nonpathogenic strains of plant pathogens are another notable class of endophytes. For example, root rot or wilting is caused by *Fusarium oxysporum* sensu lato, which interacts with plants as a pathogen, while certain avirulent strains of *Fusarium oxysporum* sensu lato live in plants as endophytes and shield them from soilborne pathogen like *Verticillium dahlia* and *Pythium ultimum*.

16.10.1.2 Systemic Resistance Inducer

To deal with biotic and abiotic stressors, plants have a variety of biochemical and physiochemical defenses. In times of stress, the plant emits a variety of signal molecules to control its enzymatic reactions and metabolic behavior, strengthening its resistance to pressures. However, the plant system defends itself against biotic stressors like pathogen invasion by strengthening its immune system. There are two types of resistance induction: ISR and SAR (Jacob et al. 2020). Necrotizing infections damage tissues where they are located, causing disease growth which then moves to plants using signals. These common words make it easier for a broader audience to understand the information provided. On the other hand, ISR makes plants stronger against many germ attacks. ISR is typically brought on by BCAs. On the other hand, signaling molecules mediate the induction of the plant defense system, which is recognized by pathogenicity-related proteins and receptors such as PRRs. Numerous studies have documented the method of endophytes trigger ISR through various mechanisms, including the initiation of pathogenic protein synthesis and the control of signaling pathways in relation to biotic stress (Shi et al. 2011).

16.10.1.3 Field Application and Commercialization

BCAs are thought to be an environmentally friendly method of lowering the vulnerability to infection by boosting the host's immunity by inducing an immunological response through the production of several kinds of bioactive compounds, such as siderophores, signaling molecules, and antibiotics. However, because endophytic strains colonize more effectively than other strains, using them as BCAs has certain further advantages. Before the actual use of microbial antagonistic, several variables were considered, encompassing BCA's resilience and stability against the main pathogens as well as their ability to work both inside the host and outside of it. These elements also have a big impact on marketability and real-world applications. Several endophytic and epiphytic strains, such as bacteria, fungi, and plants, were tested in vivo or in vitro, and yeast demonstrated the ability to be antagonistic; however, very few microbial BCAs have been applied in industry. The main causes of these failures are BCAs' lack of stability and survivability. However, only a very small number of microbial strains especially those from the *Bacillus* and *Trichoderma* species have been made available for purchase (Gimenez et al. 2007). These two microbial groups have become popular choices because of their ability to form endospores, which gives them superior stability and survivability over other microbial strains that do not form endospores. The commercialization of BCAs has advanced significantly over the past 20 years, and various regions of the world have employed these microbial antagonistic agents to effectively regulate fresh produce loss both before and after harvest. Application of BCAs against the pathogen and the plant after screening is heavily dependent on the host immune system, signaling molecules, metabolic processes, and most importantly the pathogen's growth stage. Cultivars and plant physiology, in addition to genetic makeup, are important factors in the effective use of BCAs. While environmental factors influence the host-pathogen-BCA relationship as well, they must be standardized for biocontrol techniques to function as effectively as possible.

16.11 Environmental and Safety Considerations

Using fungal endophytes as biocontrol agents offers several environmental and safety benefits, but it also comes with certain considerations:

- *Reduced Chemical Use*: As we know, it acts as biological control so it can help in reducing the use of chemical pesticides to control plant diseases that pollute the environment. The reduced use of chemicals leads to sustainable agriculture.

- *Biodiversity*: Chemicals have hazardous effects on biodiversity, so the promotion of biodiversity in agricultural ecosystems can be achieved by the proper management of fungal endophytes which can target specific pests without causing harm to nontarget organisms (Gimenez et al. 2007).
- *Sustainable Agriculture*: By reducing the ecological impact of conventional pest management techniques, they can support sustainable farming practices (Le Cocq et al. 2017).
- Safety Considerations
- *Host Specificity*: It is important to make sure that fungal endophytes only harm the right pest and not other things in nature.
- *Risk Assessment*: Check carefully for any dangers from using fungal endophytes. This involves examining how they communicate with other living beings and their surroundings.
- *Human Health*: See if fungi living inside plants are safe for people, because some might create harmful things when eaten or inhaled.
- *Regulations*: It's crucial to stick with instructions and guidelines for using microbes that fight against pests. This is needed so the environment and people stay safe from harm.
- *Monitoring and Research*: Need to keep watching and learning all the time, so we can understand how fungal endophyte use affects us in the long run. We also have to change our plans if unexpected problems happen because of it.

16.12 Conclusion and Implications

Endophytic fungi are essential to the physiology of plants and the operations of agricultural ecosystems. As MBCA, fungal endophytes offer a visually appealing, practical, and ecologically sustainable way to manage plant diseases. Endophytes are widely distributed and have been shown to have long-term effects. The majority of fungal endophytes have clear benefits over their host plants, such as the ability to produce a large number of secondary metabolites and chemicals that can be used to promote plant growth and fight diseases. To fortify plants, they can combat plant diseases and defenses against disease and pathogens and raise their capacity to withstand stress. There are several environmental and safety benefits to using fungal endophytes as biocontrol agents; nevertheless, their effective application necessitates careful consideration of host specificity, risk assessment, human health, regulatory compliance, and continuous study and monitoring.

The use of fungal endophytes as BCAs has significant implications, highlighting the need to support the researchers and institutions in this field. By promoting this can reduce the reliance on the chemical treatments, benefiting both crop health and the environment. The work on the field applications of fungal endophytes as BCA must be done as soon as possible so that farmers can get benefits from their use. Plant disease management and induction of systemic resistance in plants must be focused on while using them.

References

Adetunji CO, Kumar D, Raina M, Arogundade O, Sarin NB (2019) Endophytic microorganisms as biological control agents for plant pathogens: a panacea for sustainable agriculture. In: Plant biotic interactions: state of the art, pp 1–20

Alabouvette C, Olivain C, Migheli Q, Steinberg C (2009) Microbiological control of soil-borne phytopathogenic fungi with special emphasis on wilt-inducing Fusarium oxysporum. New Phytol 184(3):529–544

Ali S, Duan J, Charles TC, Glick BR (2014) A bioinformatics approach to the determination of genes involved in endophytic behavior in Burkholderia spp. J Theor Biol 343:193–198

Alves E, Lucas GC, Pozza EA, de Carvalho Alves M (2013) Scanning electron microscopy for fungal sample examination. In: Laboratory protocols in fungal biology: current methods in fungal biology, pp 133–150

Arivudainambi UE, Anand TD, Shanmugaiah V, Karunakaran C, Rajendran A (2011) Novel bioactive metabolites producing endophytic fungus Colletotrichum gloeosporioides against multidrug-resistant Staphylococcus aureus. FEMS Immunol Med Microbiol 61(3):340–345

Arnold AE, Herre EA (2003) Canopy cover and leaf age affect colonization by tropical fungal endophytes: ecological pattern and process in Theobroma cacao (Malvaceae). Mycologia 95(3):388–398

Arnold AE, Maynard Z, Gilbert GS, Coley PD, Kursar TA (2000) Are tropical fungal endophytes hyperdiverse? Ecol Lett 3(4):267–274

Arora J, Ramawat KG (2017) An introduction to endophytes. In: Endophytes: biology and biotechnology: volume 1, pp 1–23

Bezerra J, Santos M, Svedese V, Lima D, Fernandes M, Paiva L, Souza-Motta C (2012) Richness of endophytic fungi isolated from Opuntia ficus-indica Mill. (Cactaceae) and preliminary screening for enzyme production. World J Microbiol Biotechnol 28:1989–1995

Bidellaoui B, Segarra G, Hakkou A, Isabel Trillas M (2019) Beneficial effects of Rhizophagus irregularis and Trichoderma asperellum strain T34 on growth and fusarium wilt in tomato plants. J Plant Pathol 101:121–127

Bok JW, Noordermeer D, Kale SP, Keller NP (2006) Secondary metabolic gene cluster silencing in Aspergillus nidulans. Mol Microbiol 61(6):1636–1645

Bolívar-Anillo HJ, Garrido C, Collado IG (2020) Endophytic microorganisms for biocontrol of the phytopathogenic fungus Botrytis cinerea. Phytochem Rev 19:721–740

Cao R, Liu X, Gao K, Mendgen K, Kang Z, Gao J, Dai Y, Wang X (2009) Mycoparasitism of endophytic fungi isolated from reed on soilborne phytopathogenic fungi and production of cell wall-degrading enzymes in vitro. Curr Microbiol 59:584–592

Cardinale M (2014) Scanning a microhabitat: plant-microbe interactions revealed by confocal laser microscopy. Front Microbiol 5:94

Caruso DJ, Palombo EA, Moulton SE, Zaferanloo B (2022) Exploring the promise of endophytic fungi: a review of novel antimicrobial compounds. Microorganisms 10(10):1990

Chakraborty S, Newton AC (2011) Climate change, plant diseases and food security: an overview. Plant Pathol 60(1):2–14

Checcucci A, Maida I, Bacci G, Ninno C, Bilia AR, Biffi S, Firenzuoli F, Flamini G, Fani R, Mengoni A (2017) Is the plant-associated microbiota of Thymus spp. adapted to plant essential oil? Res Microbiol 168(3):276–282

Chutulo EC, Chalannavar RK (2018) Endophytic mycoflora and their bioactive compounds from Azadirachta indica: a comprehensive review. J Fungi 4(2):42

Collinge DB, Jensen B, Jørgensen HJ (2022) Fungal endophytes in plants and their relationship to plant disease. Curr Opin Microbiol 69:102177

Crandall SG, Gold KM, Jiménez-Gasco MDM, Filgueiras CC, Willett DS (2020) A multi-omics approach to solving problems in plant disease ecology. PLoS One 15(9):e0237975

de Bary A (1866) Morphologie und physiologie der pilze, flechten und myxomyceten, vol 1. Engelmann

Dehne H (1982) Interaction between vesicular-arbuscular mycorrhizal fungi and plant pathogens. Phytopathology 72:1115–1119

Delgado-Baquerizo M, Guerra CA, Cano-Díaz C, Egidi E, Wang J-T, Eisenhauer N, Singh BK, Maestre FT (2020) The proportion of soil-borne pathogens increases with warming at the global scale. Nat Clim Chang 10(6):550–554

Deng Z, Cao L (2017) Fungal endophytes and their interactions with plants in phytoremediation: a review. Chemosphere 168:1100–1106

Deshmukh S, Hückelhoven R, Schäfer P, Imani J, Sharma M, Weiss M, Waller F, Kogel K-H (2006) The root endophytic fungus Piriformospora indica requires host cell death for proliferation during mutualistic symbiosis with barley. Proc Natl Acad Sci 103(49):18450–18457

Deshmukh SK, Verekar SA, Bhave SV (2015) Endophytic fungi: a reservoir of antibacterials. Front Microbiol 5:715

Dirihan S, Helander M, Saloniemi I, Gundel PE, Saikkonen K (2015) Effects of systemic fungal endophytes on the performance of meadow fescue and tall fescue in mixtures with red clover. Grass Forage Sci 70(3):465–473

Freeman E (1904) I.—The seed-fungus of Lolium temulentum, L., the darnel. Philos Trans R Soc London Ser B Contain Papers Biol Charact 196(214–224):1–27

Gao D, Tao Y (2012) Current molecular biologic techniques for characterizing environmental microbial community. Front Environ Sci Eng 6:82–97

Gao F-K, Dai C-C, Liu X-Z (2010) Mechanisms of fungal endophytes in plant protection against pathogens. Afr J Microbiol Res 4(13):1346–1351

Gimenez C, Cabrera R, Reina M, Gonzalez-Coloma A (2007) Fungal endophytes and their role in plant protection. Curr Org Chem 11(8):707–720

Glenn A (2007) Mycotoxigenic Fusarium species in animal feed. Anim Feed Sci Technol 137(3–4):213–240

Glienke-Blanco C, Aguilar-Vildoso CI, Vieira MLC, Barroso PAV, Azevedo JL (2002) Genetic variability in the endophytic fungus Guignardia citricarpa isolated from citrus plants. Genet Mol Biol 25:251–255

Grosch R, Scherwinski K, Lottmann J, Berg G (2006) Fungal antagonists of the plant pathogen Rhizoctonia solani: selection, control efficacy and influence on the indigenous microbial community. Mycol Res 110(12):1464–1474

Hardoim PR, Van Overbeek LS, Berg G, Pirttilä AM, Compant S, Campisano A, Döring M, Sessitsch A (2015) The hidden world within plants: ecological and evolutionary considerations for defining functioning of microbial endophytes. Microbiol Mol Biol Rev 79(3):293–320

Harman GE (2006) Overview of mechanisms and uses of Trichoderma spp. Phytopathology 96(2):190–194

He X, Han G, Lin Y, Tian X, Xiang C, Tian Q, Wang F, He Z (2012) Diversity and decomposition potential of endophytes in leaves of a Cinnamomum camphora plantation in China. Ecol Res 27(2):273–284

Huang L-Q, Niu Y-C, Su L, Deng H, Lyu H (2020) The potential of endophytic fungi isolated from cucurbit plants for biocontrol of soilborne fungal diseases of cucumber. Microbiol Res 231:126369

Ihrmark K, Asmail N, Ubhayasekera W, Melin P, Stenlid J, Karlsson M (2010) Comparative molecular evolution of Trichoderma chitinases in response to mycoparasitic interactions. Evol Bioinforma 6:EBO. S4198

Jacob J, Krishnan GV, Thankappan D, Amma DKBNS (2020) Endophytic bacterial strains induced systemic resistance in agriculturally important crop plants. In: Microbial endophytes. Elsevier, pp 75–105

Kavroulakis N, Ntougias S, Zervakis GI, Ehaliotis C, Haralampidis K, Papadopoulou KK (2007) Role of ethylene in the protection of tomato plants against soil-borne fungal pathogens conferred by an endophytic Fusarium solani strain. J Exp Bot 58(14):3853–3864

Khan RAA, Najeeb S, Mao Z, Ling J, Yang Y, Li Y, Xie B (2020) Bioactive secondary metabolites from Trichoderma spp. against phytopathogenic bacteria and root-knot nematode. Microorganisms 8(3):401

Larran S, Monaco C, Alippi H (2001) Endophytic fungi in leaves of Lycopersicon esculentum Mill. World J Microbiol Biotechnol 17(2):181–184
Le Cocq K, Gurr SJ, Hirsch PR, Mauchline TH (2017) Exploitation of endophytes for sustainable agricultural intensification. Mol Plant Pathol 18(3):469–473
Li J, Zhao J, Xu L, Zhou L, Li X, Wang J (2008) Endophytic fungi from rhizomes of Paris polyphylla var. yunnanensis. World J Microbiol Biotechnol 24:733–737
Liu K, Ding X, Deng B, Chen W (2009) Isolation and characterization of endophytic taxol-producing fungi from Taxus chinensis. J Ind Microbiol Biotechnol 36(9):1171
Maggini V, Mengoni A, Bogani P, Firenzuoli F, Fani R (2020) Promoting model systems of microbiota–medicinal plant interactions. Trends Plant Sci 25(3):223–225
Mahdi T, Mohamed I, Yagi S (2014) Endophytic fungal communities associated with ethno-medicinal plants from Sudan and their antimicrobial and antioxidant prospective. Ecosystems 4:5
Mambaeva A, Świerczyńska I, Pieczul K, Khidirov K, Sadanov A, Shemshura O, Łozowicka B (2019) Biological activity of Trichoderma spp. in greenhouse and in vitro studies Biologiczna aktywność Trichoderma spp. w badaniach szklarniowych oraz in vitro. Progress Plant Protect 59(2):93–98
Mane R, Shinde M, Wagh P, Malkar H (2017) Isolation of Endophytic microorganisms as a source of novel secondary metabolite producers against Tuberculosis. Int J Sci Res Technol 3:1267–1269
Meenambiga S, Rajagopal K (2018) Antibiofilm activity and molecular docking studies of bioactive secondary metabolites from endophytic fungus Aspergillus nidulans on oral Candida albicans. J Appl Pharm Sci 8(3):037–045
Mengoni A, Schat H, Vangronsveld J (2010) Plants as extreme environments? Ni-resistant bacteria and Ni-hyperaccumulators of serpentine flora. Plant Soil 331:5–16
Mishra A, Singh SP, Mahfooz S, Singh SP, Bhattacharya A, Mishra N, Nautiyal C (2018) Endophyte-mediated modulation of defense-related genes and systemic resistance in Withania somnifera (L.) Dunal under Alternaria alternata stress. Appl Environ Microbiol 84(8):e02845–e02817
Moissl-Eichinger C, Pausan M, Taffner J, Berg G, Bang C, Schmitz RA (2018) Archaea are interactive components of complex microbiomes. Trends Microbiol 26(1):70–85
Molinari G (2009) Natural products in drug discovery: present status and perspectives. Pharm Biotechnol 655:13–27
Möller M, Stukenbrock EH (2017) Evolution and genome architecture in fungal plant pathogens. Nat Rev Microbiol 15(12):756–771
Mousa WK, Shearer C, Limay-Rios V, Ettinger CL, Eisen JA, Raizada MN (2016) Root-hair endophyte stacking in finger millet creates a physicochemical barrier to trap the fungal pathogen Fusarium graminearum. Nat Microbiol 1(12):1–12
Moussa M, Ebrahim W, El-Neketi M, Mándi A, Kurtán T, Hartmann R, Lin W, Liu Z, Proksch P (2016) Tetrahydroanthraquinone derivatives from the mangrove-derived endophytic fungus Stemphylium globuliferum. Tetrahedron Lett 57(36):4074–4078
Munir S, Li Y, He P, He P, Ahmed A, Wu Y, He Y (2020) Unraveling the metabolite signature of citrus showing defense response towards Candidatus Liberibacter asiaticus after application of endophyte Bacillus subtilis L1-21. Microbiol Res 234:126425
Nunna SAD, Balachandar D (2022) Authentication of putative competitive bacterial endophytes of rice by re-isolation and DNA fingerprinting assay. J Appl Microbiol 133(3):1808–1820
Ownley BH, Gwinn KD, Vega FE (2010) Endophytic fungal entomopathogens with activity against plant pathogens: ecology and evolution. BioControl 55:113–128
Pandi M, Manikandan R, Muthumary J (2010) Anticancer activity of fungal taxol derived from Botryodiplodia theobromae Pat., an endophytic fungus, against 7, 12 dimethyl benz (a) anthracene (DMBA)-induced mammary gland carcinogenesis in Sprague dawley rats. Biomed Pharmacother 64(1):48–53
Pawlowski ML, Hartman GL (2016) Infection mechanisms and colonization patterns of fungi associated with soybean. Fungal Pathog 12:25–42

Poveda J, Abril-Urias P, Escobar C (2020) Biological control of plant-parasitic nematodes by filamentous fungi inducers of resistance: Trichoderma, mycorrhizal and endophytic fungi. Front Microbiol 11:992

Raimi A, Adeleke R (2021) Bioprospecting of endophytic microorganisms for bioactive compounds of therapeutic importance. Arch Microbiol 203(5):1917–1942

Ram RM, Keswani C, Bisen K, Tripathi R, Singh SP, Singh HB (2018) Biocontrol technology: eco-friendly approaches for sustainable agriculture. In: Omics technologies and bio-engineering. Elsevier, pp 177–190

Ramamoorthy V, Viswanathan R, Raguchander T, Prakasam V, Samiyappan R (2001) Induction of systemic resistance by plant growth promoting rhizobacteria in crop plants against pests and diseases. Crop Prot 20(1):1–11

Ramawat K, Dass S, Mathur M (2009) The chemical diversity of bioactive molecules and therapeutic potential of medicinal plants. In: Herbal drugs: ethnomedicine to modern medicine, pp 7–32

Rao AS, Nayanatara A, Kaup R, Sharma A, Kumar A, Vaghasiya BD, Kishan K, Pai SR (2011) Potential antibacterial and antinfungal activity of aqueous extract of Cynodon dactylon. Int J Pharm Res Dev 2(11):2889–2893

Rodrigues R, White J, Arnold A, Redman R (2008) Fungal endophytes: diversity and function area. Tansley Rev 23:1–17

Rosenblueth M, Martínez-Romero E (2006) Bacterial endophytes and their interactions with hosts. Mol Plant-Microbe Interact 19(8):827–837

Samarakoon MC, Thongbai B, Hyde KD, Brönstrup M, Beutling U, Lambert C, Miller AN, Liu J-K, Promputtha I, Stadler M (2020) Elucidation of the life cycle of the endophytic genus Muscodor and its transfer to Induratia in Induratiaceae fam. nov., based on a polyphasic taxonomic approach. Fungal Divers 101:177–210

Sarsaiya S, Shi J, Chen J (2019) A comprehensive review on fungal endophytes and its dynamics on Orchidaceae plants: current research, challenges, and future possibilities. Bioengineered 10(1):316–334

Shi J, Liu A, Li X, Feng S, Chen W (2011) Inhibitory mechanisms induced by the endophytic bacterium MGY2 in controlling anthracnose of papaya. Biol Control 56(1):2–8

Smith SA, Tank DC, Boulanger L-A, Bascom-Slack CA, Eisenman K, Kingery D, Babbs B, Fenn K, Greene JS, Hann BD (2008) Bioactive endophytes warrant intensified exploration and conservation. PLoS One 3(8):e3052

Strobel G (2018) The emergence of endophytic microbes and their biological promise. J Fungi 4(2):57

Thongsandee W, Matsuda Y, Ito S (2012) Temporal variations in endophytic fungal assemblages of Ginkgo biloba L. J For Res 17(2):213–218

Tiwari K (2015) The future products: Endophytic fungal metabolites. J Biodivers Biopros Dev 2:2–7

Tripathi D, Raikhy G, Kumar D (2019) Chemical elicitors of systemic acquired resistance—salicylic acid and its functional analogs. Curr Plant Biol 17:48–59

Vieira ML, Hughes AF, Gil VB, Vaz AB, Alves TM, Zani CL, Rosa CA, Rosa LH (2012) Diversity and antimicrobial activities of the fungal endophyte community associated with the traditional Brazilian medicinal plant Solanum cernuum Vell.(Solanaceae). Can J Microbiol 58(1):54–66

Wang Y, Zeng QG, Zhang ZB, Yan RM, Wang LY, Zhu D (2011) Isolation and characterization of endophytic huperzine A-producing fungi from Huperzia serrata. J Ind Microbiol Biotechnol 38(9):1267–1278

Wen P, Hu T-G, Linhardt RJ, Liao S-T, Wu H, Zou Y-X (2019) Mulberry: a review of bioactive compounds and advanced processing technology. Trends Food Sci Technol 83:138–158

Wu L, Han T, Li W, Jia M, Xue L, Rahman K, Qin L (2013) Geographic and tissue influences on endophytic fungal communities of Taxus chinensis var. mairei in China. Curr Microbiol 66:40–48

Yadav AN (2020) Plant microbiomes for sustainable agriculture: current research and future challenges. Springer

Zabalgogeazcoa I (2008) Fungal endophytes and their interaction with plant pathogens: a review. Span J Agric Res 6:138–146
Zhang W (2018) Global pesticide use: profile, trend, cost/benefit and more. Proc Int Acad Ecol Environ Sci 8(1):1
Zhang N, O'Donnell K, Sutton DA, Nalim FA, Summerbell RC, Padhye AA, Geiser DM (2006) Members of the Fusarium solani species complex that cause infections in both humans and plants are common in the environment. J Clin Microbiol 44(6):2186–2190
Zimmerman NB, Vitousek PM (2012) Fungal endophyte communities reflect environmental structuring across a Hawaiian landscape. Proc Natl Acad Sci 109(32):13022–13027

Biocontrol Potential of Endophytic Fungi Against Postharvest Grape Pathogens

17

Amr H. Hashem, Amer M. Abdelaziz, Mohamed S. Attia, and Kamel A. Abd-Elsalam

Abstract

The risks of postharvest diseases have not only resulted in huge losses in the global agricultural economy but have also led to the spread of fungal and bacterial toxins in the fruits that humans eat, causing serious diseases that lead to death. Chemical pesticides are common for controlling most diseases, but these pesticides became not effective and affect human health. It is widely recognized as a significant obstacle to ensuring adequate food supply for the increasing global population. Nevertheless, a significant portion of the population continues to rely on chemical fungicides in order to mitigate postharvest losses. However, it is important to note that these fungicides have adverse effects on food quality and can lead to the accumulation of pesticide residues, which in turn constitute a risk to both human health and the surrounding environment. In recent years, there has been a surge in researchers' endeavors to discover biopesticides. Fungal endophytes have become a cost-effective and sustainable method for safely and effectively controlling postharvest infections. Thus, the objective of this chapter is to discuss the use of endophytic fungi to control postharvest diseases of grapes. Additionally, it intends to analyze the patterns of diversity and distribution of endophytic cells in grapes.

Keywords

Fruits · Postharvest pathogens · Endophytes · Biocontrol agents

A. H. Hashem (✉) · A. M. Abdelaziz · M. S. Attia
Botany and Microbiology Department, Faculty of Science, Al-Azhar University, Cairo, Egypt
e-mail: amr.hosny86@azhar.edu.eg

K. A. Abd-Elsalam
Plant Pathology Research Institute, Agricultural Research Center (ARC), Giza, Egypt

© The Author(s), under exclusive license to Springer Nature Singapore Pte Ltd. 2025

K. A. Abd-Elsalam, A. H. Hashem (eds.), *Fungal Endophytes Volume II*,
https://doi.org/10.1007/978-981-97-8804-0_17

17.1 Introduction

The greater costs for horticultural crops as a percentage of land have increased farmers' incomes. These crops provide a significant source of revenue and help to alleviate poverty. Fruits are available year-round in a wide range of varieties, and they are not only delightful to eat but also have good texture, color, flavor, and ease of preparation (Khan et al. 2021). However, in many cases, fruit consumption falls short of the nutritional recommendation target of five to ten portions per day (Singh et al. 2021a). Horticultural produce, particularly fruits, are high in essential elements for a healthy diet (Linling et al. 2013; Singh et al. 2021a, b). These crops give water, carbs, proteins, lipids, fiber, minerals, organic acids, colors, vitamins, and antioxidants to the human body (Yahia et al. 2019). They are an ideal meal replacement and snack because they are little in calories, contain no cholesterol, and are high in carbohydrates (Vincente et al. 2014). Horticultural farming fosters the growth of natural resources, increases land returns, increases land values, increases people's spending power, and thus contributes to overall prosperity (Giller et al. 2021). Grapes are one of the most important fruits in the world (Fahmi et al. 2012). Diseases affect table grapes during postharvest storage, hastening the degeneration of the crop (Lee et al. 2015). Three harmful fungi that might cause postharvest losses in table grapes include *Botrytis cinerea*, *Penicillium* spp., and *Aspergillus* spp. (Solairaj et al. 2020). It is discussed that *Aspergillus* and *Penicillium* species can produce ochratoxin, a carcinogenic mycotoxin that harms human health, particularly the nephridium and hepatic systems (Wang et al. 2019). Controlling fungal diseases in grapes is so critical. Pathogen infections are generally controlled by synthetic chemical fungicides during postharvest techniques (Zhang et al. 2020). Nevertheless, the use of synthetic fungicides has become progressively constrained as a consequence of increased public criticism caused by their toxicological dangers to human health and environmental contamination, as well as the emergence of fungicide-resistant pathogenic strains (Attia et al. 2022a, b). As a result, scientists are looking for alternate methods to control fruit postharvest infections (Wisniewski et al. 2016). Biological control, which employs antagonist microorganisms such as endophytes, has emerged as a potential alternative strategy for controlling fruit and vegetable postharvest diseases (Abdelaziz et al. 2022; Attia et al. 2022b). Endophytes, for example, do not create hazardous secondary metabolites (Naik et al. 2019). As a result, endophytes have significant potential as a biocontrol agent for fruit postharvest deterioration (Morales-Cedeño et al. 2021). Endophytic fungi have shown great potential in controlling the postharvest decay of fruits (Carmona-Hernandez et al. 2019). While many antagonistic endophytic fungi have been found to have a notable impact on the postharvest deterioration of grapes, the efficacy of biocontrol measures also needs further investigation. This chapter aimed to show postharvest pathogens particularly postharvest diseases of grapes. Also, these diseases can be controlled using fungal ensophytes. Additionally, it intends to analyze the patterns of diversity and distribution of endophytic cells in grapes.

17.2 The Concept of Postharvest Damages

Postharvest refers to the deliberate human action of separating the edible commodity from the plant that generated it, with the intention of initiating its journey toward consumption. The postharvest phase concludes when the food is delivered to the ultimate consumer (Sakaguchi et al. 2001). Postharvest management (PHM) can be summarized as "techniques and processes used to extend the shelf life while preserving the quality of fresh or processed horticultural produce" (Sharma et al. 2021). However, Sudheer and Indira (2007) established postharvest degradation of fruits as "the mass of a healthy edible product (exclusive of moisture) that typically gets eaten by humans and which has been purposefully removed by humanity from its current production and development medium and sites in order to be used for human feeding but isn't consumed by humans for any other reason". In addition to both quality and quantity damages, postharvest losses also involve the appearance of horticultural output, which affects its selling price (Kasso and Bekele 2018). Postharvest losses in fruits are commonly believed to be higher compared to cereals and oilseed crops. This is attributed to the fruits' high moisture content, large size and weight, and the fact that their living tissues have a high respiratory rate (Singh et al. 2021a, b). Throughout the marketing process, it is imperative to maintain the vitality and well-being of these tissues. These tissues consist of numerous living cells that require care and maintenance (Ramjan and Ansari 2018). The several types of postharvest losses are addressed here, categorized according to the different stages of postharvesting. (A) Direct loss of produce refers to the physical decay of packaging materials after harvest, resulting in the produce not reaching the consumer. On the other hand, indirect loss occurs at the consumer level and includes factors such as consumer refusal to purchase due to poor quality or infrequent visits to the market (Sommer et al. 1991). (B) Weight loss is defined as the loss caused by a decrease in the moisture content of the produce during storage or transportation (Meng et al. 2008). Excessive moisture absorption can cause an unexpected increase in the weight of a product, leading to potential losses. This can happen when the product is stored in conditions of high relative humidity or when it is exposed to rainfall (Prusky 2011). Grape has gotten a lot of attention because of its economic enhancement as a kind of raisin. Fresh grapes are one of the main export commodities all over the world. The high sugar content of grapes facilitates the proliferation of fungal infections during the transportation and storage of the fruit, serving as a substrate and nourishment for these fungi. Fruit spoilage causes losses ranging from 30% to 40% in rich countries with advanced technology to more than 50% in underdeveloped areas. (Kalia and Gupta 2012) performed a study on postharvest handling and grape losses in the Maharashtra area of Nasik in 2000–2001. They reported a total loss of 1930.9% and fractional losses of 0.001.25%, 5.58.65%, and 12.2516% when grapes were packaged in boxes at the farm, wholesale, and retail levels, respectively. Fruits packed in bamboo baskets arriving in Mumbai market, on the other hand, experienced an average loss of 24.19% owing to rupturing, rotting, and shattering. Grape postharvest loss was estimated to be at 8.30% (Nanda et al.

2012a). Fruit loss was documented at the farm level for 3.21%, 1.93%, and 5.54% of these, respectively, during sorting and grading, shipping, and storage at the farm level. For all, 8.3% of postharvest loss was discovered during several postharvest activity phases. Nevertheless, the produce's sorting and grading process resulted in a maximum loss of 3.2% (Nanda et al. 2012b). Water berries and berry drops were the main causes of fruit losses at the field level. In the Karnataka region of Bijapur, a study was carried out by (Vilas et al. 2011), and they revealed that the most losses were incurred at the farm level as a result of berry mummification, with 13.25% of grapes lost after harvest.

17.3 Endophytic Fungi as Biocontrol Agents

Postharvest diseases pose a significant threat to agricultural production, resulting in losses exceeding 30% and a decline in product quality. These diseases also lead to the production of toxins that can cause poisoning and various severe illnesses in humans (Carlini and Grossi-de-Sá 2002). The Plant Disease Triangle is comprised of these environmental factors. One of these components must be present for biotic illness to develop. Scientists regard to this sequence of incidences as a disease cycle (De Wolf and Isard 2007). Scientists were interested in eliminating different pathogens or inhibiting their effectiveness by using chemical pesticides, whether systemic or nonsystemic, which in turn led to environmental pollution and impact on human health as a direct result of the residual effect of these chemicals used in resistance (Balla et al. 2021). The traditional methods used to confront the biological risks and environmental challenges facing crops depend to a large extent on the use of chemical pesticides. In agricultural production, chemical pesticides are generally used more frequently than herbicides and insecticides, which can have extremely harmful impacts on the environment and consumers, as seen in Fig. 17.1.

Recently, the attention of plant pathologists has turned to thinking of effective modern methods that help in combating disease without harming the environment. The scientists also turned to many modern means of resistance that try to avoid environmental pollution, such as the use of genetically engineered plants that carry the characteristics of resistance, as well as the use of biological control methods (Gerhardson 2002). Finally, attention turned to a new type of resistance, which is known as induced resistance, which is concerned with plant research. Promotion of disease resistance, whether through biotic or abiotic stimuli, for which adding growth regulators to the plant externally is one of its means (Mauch-Mani et al. 2017).

Researchers have shown an interest in postharvest diseases caused by endophytic fungus in recent years (Harman et al. 2004; Hermosa et al. 2012). Nonpathogenic fungi called endophytic fungi are highly effective at stopping the growth of diseases and stimulating plant growth by generating an abundance of antioxidants that shield the plant from conditions where it is exposed to its roots owing to damage. It comes

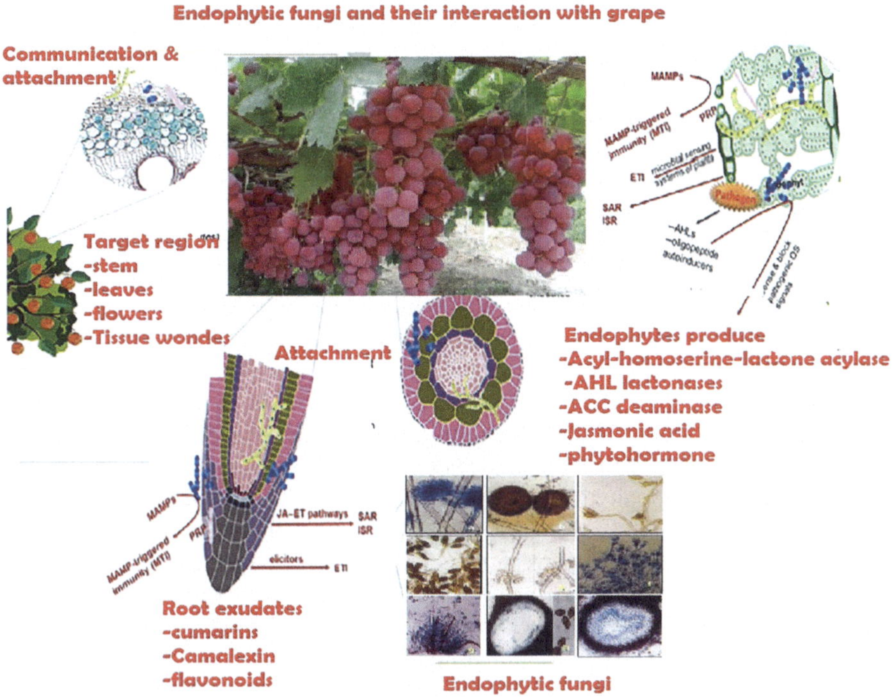

Fig. 17.1 Endophytic fungi and their interaction with grape fruits

additionally to the rivalry with dietary intake and the synthesis of aromatic chemicals to protect plants (Perello et al. 2016). Endophytic fungus is capable of biocontrolling other pathogenic fungi (Attia et al. 2016; Farrag et al. 2017; Veenstra et al. 2019; Zin and Badaluddin 2020). Endophytic fungi have been shown in scientific studies to significantly promote plant growth and improvement by producing secondary metabolites, which play an important role in communication with other species (Poveda et al. 2020). Species such as *Aspergillus*, *Trichoderma*, and *Penicillium* damage plants and natural products with agricultural applications (Alfiky and Weisskopf 2021). The relationships between endophytic fungus and plants have been studied on a number of levels: (1) When the fungal inoculum is near the root, the fungal diffusible chemicals are crucial for plant growth (Kong et al. 2020), (2) once the mycelium gets to the plant root and the two species start interacting physically (Leake et al. 2004), (3) even just connecting with plants by emitting volatile substances (Baldwin et al. 2006), and (4) where the advantages of the fungal inoculation or any of its constituent chemicals have been investigated for biocontrol applications against plant diseases or heterotrophic bacteria in multitrophic interaction systems (Hardoim et al. 2015).

17.4 Establishment Models of Endophytes and Pathogens in the Host Tissue

The host-endophyte connection is complicated and influenced by a variety of intrinsic and external variables (Christian et al. 2020; Patra et al. 2023). The initial phase for any organism to become an endophyte is the institution of microorganisms in plant tissue (Anyasi and Atagana 2019; Taulé et al. 2021). Nevertheless, endophyte colonization is a complicated procedure that is regulated by multiple signals in a series of phases (Dang and Lovell 2016; Fares et al. 2020). Microorganisms were drawn to certain components found in the fluids of plant species. These components, which often consist of sugars, organic acids, amino acids, lipopolysaccharides (LPSs), flavonoids, and proteins, may vary according to the specific microbiological species (Khan et al. 2018). However, bioactive molecules like exopolysaccharides (EPSs), LPSs, cell surface saccharides, and microbial strain cellulase enhance endophyte adhesion on the host surface. Before entering or colonizing the host tissue, the endophytic microbe faces the obstacles of oxidative conditions (Kumari et al. 2022). Several publications have shown the successful adaptation potential of endophytic organisms in this context; for example, *Enterobacter* spp. encodes antioxidant enzymes during poplar plant colonization (Kamran et al. 2022). Pathogenic bacteria release a higher amount of cell wall-degrading enzymes at the sites of infection. Conversely, a limited number of enzymes that break down cell walls were discovered during the process of endophyte colonization. This lack of enzyme activity did not stimulate the plant's defense mechanisms, enabling the endophytes to thrive within the host organ (Khatri et al. 2017; Kikot et al. 2009; Quoc and Bao Chau 2017). Pathogen and endophyte colonization patterns are comparable to some extent. The reaction of plant defense mechanisms, on the other hand, varies and is dependent on the nature of the microorganisms. Likewise, expression sequences in response to oxidative damage differ (Verma et al. 2021). The pathogen employed the salicylate signaling pathway during colonization, but the *Azoarcus* exploited the jasmonate signaling system.

17.5 Host-Endophyte Interaction in Terms of Biocontrol Agents

Throughout plant-microbe connections, microbial strains have been demonstrated to interact with host plants in neutral, commercial, cooperative, or detrimental ways (Manriquez et al. 2021). A variety of factors influence establishment, particularly the genotype of microorganisms or host plants and the surrounding environment (Whipps et al. 2008). Plants have a high level of tolerance to phytopathogen infections. During infection, pathogenic strains release a huge number of biomolecules inside the host (Brown and Hammond-Kosack 2015). After identifying homologous organization via its current pattern recognition receptors (PRRs), the host plant responds correctly and provides immunological activity as the primary line of resistance to control the pathogen (Hou et al. 2019). During coevolution with the host plant, pathogenic strains improved their techniques for reducing MAMP-/PAMP-triggered immunity (Lü et al. 2022). In response, the host plant developed a second line of defense known as effector-triggered resistance (Jacob et al. 2023; Jampílek and Kráľová 2024). The plant mechanism generates receptors that detect or recognize pathogen components. Pathogenic fungus (biotrophic) that relies on the food intake of living cells, for example, may evoke a hypersensitive reaction, culminating in the programmed cell death of plants under attack (Matsumura et al. 2022).

17.6 Endophytes as Biocontrol Agents

Several variables have been regarded relevant in order to investigate endophytes as biological control agents, including survival, stability, storage, application, and marketability (Kashyap et al. 2023). Biopesticides have been developed as an alternative approach for suppressing plant diseases in the field or postharvest (Russi et al. 2024) *Trichoderma*'s chlamydospore structure makes them more appropriate for economic exploitation than other microbial strains due to their stability or survivability under unfavorable conditions (Dutta et al. 2022). The endophytic fungi, on the other hand, can be easily supplied, penetrating and colonizing the host tissue, in contrast to other pathogen (Vandana et al. 2021). However, the efficiency of BCAs against pathogens may be affected by a variety of parameters, including plant development or physiological status, genotype, and ecological situations (Magan 2020). Recent research has revealed the antagonistic activity of a varied spectrum of endophytes originating on the fruit. The presence of diverse bacterial, actinomycete, and fungal species on the fruit surface can impact the occurrence of postharvest diseases (Carmona-Hernandez et al. 2019). *Aureobasidium pullulans* inhibited the growth of *Botrytis cinerea* and *Monilinia laxa* in table grapes, reducing fruit degradation by 10–100% (Schena et al. 2003). *Trichoderma* and *Nodulisporium* are frequently found as fungal biological control agents (BCAs) in the carposphere (Kumari et al. 2022). The use of fungal volatile organic compounds (VOCs) in mycofumigation has lately gained interest as a method to prevent the growth of postharvest pathogens (Kumari et al. 2022).

17.7 Screening of Endophytic Biocontrol Agents

Finding endophyte treatments with the ability to regulate their growth is essential to determining which ones have the best antagonism properties against potential infections (Abdelaziz et al. 2023a; Vurukonda et al. 2018). Detecting these behaviors is dependent on having a higher likelihood of producing microbial endophyte-based biocontrols that are likely to be successful in open field applications rather than only displaying good activities in the laboratory (Kumari et al. 2022). Following that, we go over some tools for detecting and selecting endophytic BCAs. One of the most important procedures is to evaluate antimicrobial agents against diverse phytopathogens. BCAs are often screened based on certain traits such as parasitism, in which BCAs coexist with the host plant, resulting in antagonistic consequences (Ahmed et al. 2020; Hashemi et al. 2022). Moreover, strains possessing the capacity to produce antibiotic or volatile compounds, as well as enzymes like pectinases and cutinases that can disrupt pathogenicity factors or reduce the severity of pathogens, are selected for evaluation as biological control agents (BCAs) (Kumari et al. 2022; Raymaekers et al. 2020). The dual-culture experiment is a commonly used method for directly screening microbial antagonists during in vitro identification based on their phenotypic characteristics (Kumari et al. 2022). BCAs and pathogens were cocultured on semisolid medium in this experiment. Assessing the width of the lesion helps determine the pathogen's antagonistic activity toward BCAs and its level of pathogenicity (Mokiou 2004). Both biological control agents (BCAs) and the infectious agent were generated concurrently on the plates at distinct sites during the assessment, resulting in a notable reduction in the growth of mycelium and fungal spores (Pellan et al. 2021). To assess biocontrol activity, the clear zone surrounding the spotted BCA was measured. The greater the clean zone, the greater the possibility for biocontrol (Abdelaziz et al. 2023b; Dimopoulou et al. 2021). One of the criteria for evaluating biological control is the production of antimicrobial chemicals, which can be either diffusible or volatile, by the microbial endophytic strain. Instances can be evaluated by considering different attributes, such as the size of the lesion, the severity of the disease, or a preestablished disease index (Majumdar 2017; Triantafyllou et al. 2023). The plant's physiological condition is evaluated by measuring water status (including transpiration and stomatal conductance), variations in antioxidant activity (such as levels of enzymatic activity), production of plant defense compounds (such as phytoalexins), and observable indicators of growth (Attia et al. 2022b). The aggressive effects of BCAs vary depending on plant genotype or variety; variances in host genotypes variably affect cellular processes that may modulate the rapidity of disease and host resistance (Yuen et al. 2020). Similarly, the ability of endophytes to colonize plants is influenced by the physiochemical nature of plant exudates, which in turn affects their capacity to control pathogens in a more efficient and effective manner (Kashyap et al. 2023).

17.8 Antifungal Mechanisms

Endophytic fungi contribute in enhancing plant fitness through numerous methods of action. This section covered the general mechanisms that endophytic fungi use. As seen in Fig. 17.2, there are both direct and indirect methods of action. These mechanisms were discussed in detail below. These mechanisms were discussed in detail below. Endophytes have been shown to display several disease defense methods, including direct competition and modifications in the genetic structure of the host plant. Endophytes utilize many techniques to defend against pathogenic attacks on harvested fruits. Endophytic fungi utilize carbon resources to inhibit the development of disease spores. However, the amount of fungi must be controlled based on the type of fruit (Carmona-Hernandez et al. 2019; Wang et al. 2021). *Lactobacillus*

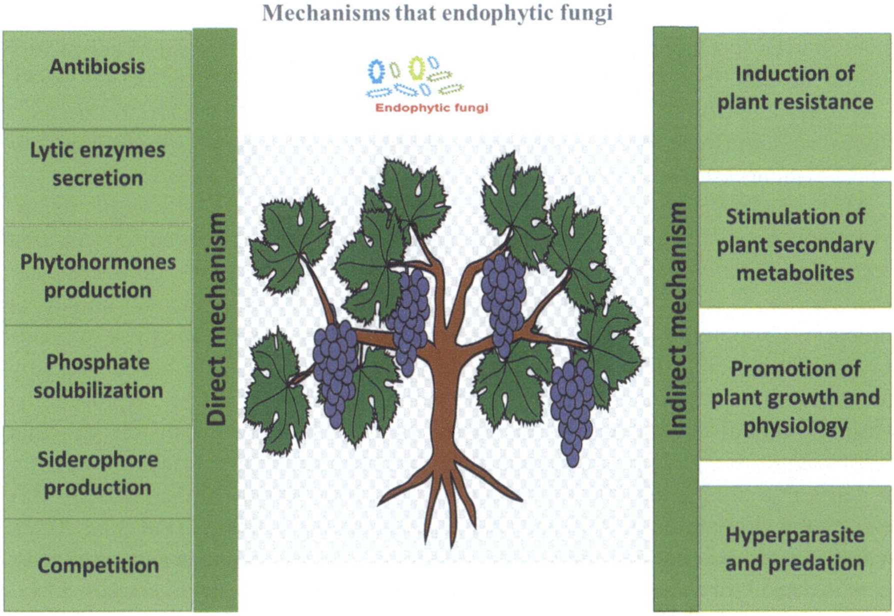

Fig. 17.2 Direct and indirect mechanisms of endophytic fungi in grape growth promotion and disease suppression

plantarum culture effectively inhibited the growth of four fungal infections found in grape berries. Following the cocultivation with *L. plantarum*, the synthesis of mycotoxins by *A. carbonarius* was suppressed by a range of 32–90% (Sadiq et al. 2019).

Iron is an essential mineral for the growth, survival, and harmfulness of pathogens. Endophytes have a higher rate of spreading compared to pathogens in their struggle for survival. They achieve this by producing several types of siderophores, which bind to iron and prevent postharvest pathogens from accessing any iron sources (Kumari et al. 2022). Although the application of endophytic *Trichoderma* spp. for treating fruit postharvest illnesses is still in its early stages, it has the potential to establish a novel and long-lasting method for managing fruit diseases after harvest (Patel et al. 2019). Nevertheless, it is crucial not to disregard the importance of increasing endophyte level and the factors that affect siderophore synthesis in order to optimize the effectiveness of endophytes against postharvest infections (Priyadarshini et al. 2022).

Endophytic fungi employ various strategies, such as the production of cell wall-degrading enzymes and direct parasitism, to combat pathogenic fungi (Rajani et al. 2021). *Choiromyces aboriginum*, an endophytic fungus, effectively inhibited the postharvest pathogen *Pythium* sp. by producing b-1,3-glucanases and disrupting the pathogenic cytoplasm that was entwined around the hyphae (de Carvalho et al. 2020). Just like a microbiologist, *Trichoderma* spp., a fungus that benefits plants, has the ability to hinder the growth of various pathogens through parasitism. As an expert in microbiology, I can provide an example of how a specific strain of

Trichoderma was able to effectively combat the fungal pathogen *F. oxysporum*. This was achieved through the production of a powerful lytic enzyme and by physically coiling around the pathogenic fungal hyphae (Guzmán-Guzmán et al. 2023). Therefore, identifying the ideal media conditions for producing LPSs from endophytes can contribute to the development of effective strategies for long-term control of postharvest fruit infections (Latha et al. 2019).

Most postharvest pathogens have the ability to break through the fruit's defensive mechanism by oxidizing it. As an expert in microbiology, I can tell you that *Penicillium digitatum*, the culprit behind green mold in citrus fruits, produces catalase. This enzyme helps the fungus break down hydrogen peroxide and initiate an infection. Endophytes provide a protective shield against oxidative stress in plants (Lastochkina et al. 2020). Endophytes aid plants in dealing with biotic stress by decreasing lipid peroxidation and proline buildup (Lata et al. 2018). Exploring the world of fungal endophytes, as an expert in microbiology, I can tell you that *Paraburkholderia phytofirmans* strain PsJN has been found to enhance the activity of genes that are responsible for scavenging reactive oxygen species (ROS). This leads to the elimination of ROS and changes in signaling pathways (Pal et al. 2022). While there is existing literature on the relationship between plants and pathogens, there is a need for focused research on how endophytes impact the redox homeostasis of preserved fruits.

Endophytic fungi have the ability to produce various phytohormones, which can help plants become more resilient to both biotic and abiotic stressors after they have been harvested (Kumari et al. 2022). Increased phytohormone production aids in the prevention of stress-induced wilting (Gowtham et al. 2022). Endophytes have the ability to synthesize plant hormones and modify plant-hormone metabolic pathways, resulting in enhanced stress tolerance (Tiwari et al. 2020). The endophytic fungus *Piriformospora indica* plays a crucial role in enhancing stress tolerance in higher plants by facilitating the synthesis of various hormones such as auxins, cytokinin, gibberellins, abscisic acid, ethylene, salicylic acid (SA), jasmonates, and brassinosteroids (Ali et al. 2017).

17.9 Endophytic Fungi for Controlling Mycotoxins

Mycotoxins are a primary cause of qualitative and quantitative fruit loss. Aflatoxins such as patulin, alternariol, and a substance known as are all produced by harmful fungi and may be harmful to both people and their fruit (Rasmussen 2010). Numerous endophytes and secondary substances have demonstrated the ability to regulate mycotoxins (Bacon et al. 2017), but research on their effects on postharvested fruits appears limited. (Bartholomew et al. 2021; Sarrocco and Vannacci 2018) highlighted the importance of using endophytes before harvest to reduce postharvest damage caused by mycotoxins. Endophytic fungus can be included into edible biofilms or used as a packaging element to successfully limit mycotoxins in store fruits (Singh et al. 2023; Zhang et al. 2021). Various endophytic fungi have recently been widely used for successful postharvest pathogen management;

however, techniques using endophytes are quite limited (Díaz-Urbano et al. 2023). By limiting conidia germination in a fungal cell culture, endophytic strains successfully prevent postharvest deterioration (Medina-Romero et al. 2017). During in vivo and in vitro screening, the strain greatly inhibited *Botryosphaeria dothidea* growth and had a robust antagonistic impact against other fungal infections (Zhang et al. 2022). Madbouly et al. (2020) investigated the biocontrol capability of endophytic yeast strains isolated from apple fruits, *Schwanniomyces vanrijiae*, *Galactomyces geotrichum*, and *Pichia kudriavzevii*, against *Monilinia fructigena*. Throughout in vitro testing, all three endophytic yeast strains suppressed conidial germination by 67.6–89.2% and demonstrated inhibitory activity against *M. fructigena*. However, most investigations are still conducted on a laboratory size. Furthermore, we must investigate how the fruit microbiome influences fruit physiology and disease resistance, as well as how the fruit-associated microorganisms changed during the postharvest processes and after BCA application.

17.10 Endophytic Fungi Treatments Versus Pathogenic Fungi of Grape in Postharvest

Grape (*Vitis vinifera* L.) is a highly cultivated fruit plant globally and is prone to illnesses caused by pathogenic fungi (Rienth et al. 2021). Therefore, recently, it has become necessary to reduce the number of plant pathogenic fungi that attack grapes, which is of vital importance in the postharvest phase (Di Canito et al. 2021; Dimkić et al. 2024). Applied microbiology represents an important role for infection prevention and sets new strategic frontiers for maintaining the quality of fruits, especially table and wine grapes, after harvest (Sarrocco and Vannacci 2018; Sellitto et al. 2021). Figure 17.3 shows the mechanism of action of endophytic fungi in improving systemic resistance and suppressing pathogenic microbes. According to several studies, the use of biological control allows for early colonization of fruit surfaces, preventing latent infections that arise after harvest (Morales-Cedeño et al. 2021; Romanazzi et al. 2016; Wenneker and Thomma 2020). Through previous studies, the role and mechanisms of activity of endophytic fungi as biological control against pathogenic fungi have been studied and reported against *Aspergillus* spp. and *Penicillium* spp., *Rhizopus stolonifera* ((Ehrenberg) *Vuillemin*), *Mucor piriformis*, *Colletotrichum acutatum*, *Alternaria* sp., and *Rhizoctonia* sp. (Billar de Almeida et al. 2020; Fontana et al. 2021; Gupta et al. 2022). The results demonstrated the role of endophytic fungi in resistance to postharvest diseases in grape crops (Bolívar-Anillo et al. 2020; Thambugala et al. 2020).

Several endophytic fungi have been studied and exploited as potential antagonists against *Aspergillus carbonarius*, which affects not only grapes but also other fruits (Table 17.1) in recent years (Pantelides et al. 2015; Zhang et al. 2016). *Aspergillus* fungi are often responsible for producing mycotoxins, which are substances that are highly toxic to humans, such as ochratoxin A (Awuchi et al. 2021; Janik et al. 2020). In order to reduce them and avoid their risks, substantial legislation must be enacted regarding them (Regulation (European Commission) No.

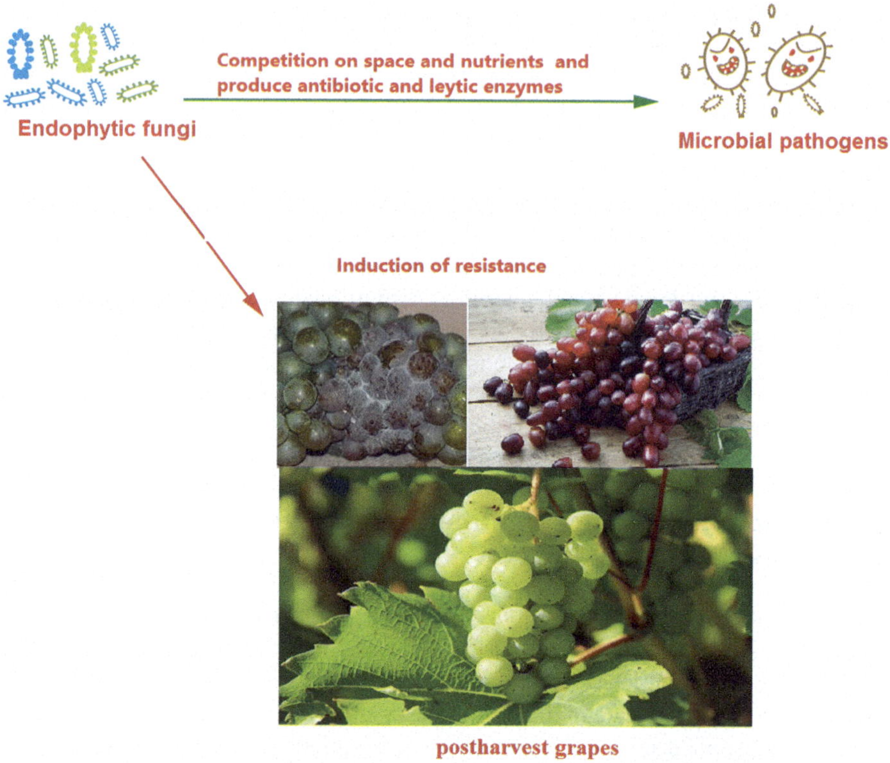

Fig. 17.3 Endophytic fungi and their actions on grape postharvest diseases

401/2006). It is worth noting that many endophytic fungi and yeasts (*Trichosporon mycotoxinivorans* and *Yarrowia lipolytica*) have proven their ability to decompose these toxins and prevent their spread (Doty 2013). In addition, studies have shown that endophytic fungi and yeasts can hinder the development of *A. carbonarius* through competition (Ribeiro et al. 2023; Tsitsigiannis et al. 2012). Furthermore, the effect is through biofilm formation, secretion of hydrolytic enzymes, induction of resistance, and production of lethal toxins and volatile organic compounds (Asma et al. 2022; Mehmood et al. 2023). Endophytic fungi, which produce deadly toxins, can be used to stop the spread of *A. niger* (Gautam et al. 2011).

Penicillium belongs to one of the greatest numbers of fungal species in the environment, and its various species may thrive in a variety of environments. Furthermore, some of them can create mycotoxins (Otero et al. 2020). *P. expansum* produces both patulin (PAT) and citrinin (CIT), both of which are toxic chemicals with potentially serious health consequences (Assaf 2018). Thus, in the grape postharvest phases, it is recommended to select BCAs against specific harmful *Penicillium* species. Numerous non-*Saccharomyces* species have been found to be active against *P. expansum* and *P. digitatum* (De Simone et al. 2020). Because studies on this species' biocontrol activity is considerable, this item goes into greater detail than the

Table 17.1 Endophytic fungi used against postharvest diseases and their hosts (plant species)

Disease	Causative pathogen	Plant species	Endophytic fungi	References
Bunch rot of grape cluster Gray mold of grape	*Botrytis. cinerea*	*Taxus mairei*	*Aspergillus clavatonanicus*	Zhang et al. (2008)
		Melia azeda	*A. fumigatus* LN-4	Li et al. (2012)
		Prunus avium	*Aureobasidium pullulans*	Schena et al. (2003)
		Vitis vinifera L.	*Alternaria* sp. *B. ribis* *Phoma medicaginis* *Bionectria ochroleuca* *Aureobasidium pullulans* *Chaetomium spirochaete*	Cosoveanu et al. (2014)
		Houttuynia cordata Thunb	*Chaetomium globosum*	Pan et al. (2016)
		Picea abies	*Cryptosporiopsis* sp.*Phialocephala spharoides* B.J. Wilson	Terhonen et al. (2016)
		Olea europaea L.	*Daldinia* cf. *concentrica*	Liarzi et al. (2016)
		Aralia elata *Aralia continentalis*	*Drechslera biseptat* *Tricladium splendens*	Paul et al. (2007)
Bunch rot in grapes	*Aspergillus niger*	*Dryopteris uniformis,*	*Candida incommunis* *Candida guilliermondii* *Strain A42*	Zhu et al. (2015), Zahavi et al. (2000)
Fruit decay	*Penicillium* spp.	*Gramineous* plants	*Muscodor suthepensis CMUCib462*	Mao et al. (2019)
Grape downy mildew	*Plasmopara viticola*	*Busseola fusca*	*Beauveria bassiana*	Jaber (2015)

preceding one. *B. cinerea* primarily damages mature or senescent tissues of dicotyledonous hosts, but it typically enters these tissues early in the growth process and stays dormant for a while before rotting tissues quickly in favorable environments since the host biology modifications (Neugebauer et al. 2024; Williamson et al. 2007). As a result, once seemingly healthy crops are harvested and then transported to far-off markets, where the losses are noticeable, significant harm is done. A group of similar *Botrytis* species that are specialized to infect roughly a dozen monocot hosts are also vulnerable to infection by *B. cinerea* on certain monocotyledonous hosts (Valero-Jiménez et al. 2020). Due to its several assault mechanisms, the ability to survive for extended periods of time as sclerotia in crop waste, and the ability to grow on a range of hosts as inoculum sources, *B. cinerea* is a challenging organism to control (Roca-Couso et al. 2021). Due to these factors, it is doubtful that any one control method will be effective, necessitating a deeper comprehension of the host-pathogen relationship, the fungus's operating milieu, and its microbial rivals on the host (Glenn et al. 2013). Because it is currently so expensive to introduce a new fungicide or biological control product to the market, only large crops garner enough attention from the agribusiness sector. *B. cinerea* can produce both phytotoxic proteins and a variety of low molecular weight phytotoxic metabolites (Temme et al. 2012). Botrydial is the phytotoxic metabolite that has been examined the most (Williamson et al. 2007). Furthermore, during infection, *B. cinerea* creates botcinic acid, a polyketide mycotoxin (Dalmais et al. 2011). *B. cinerea* growth is often managed by a fungicide treatment combined with appropriate agronomic techniques. Indeed, while the latter strategy helps to reduce infections, it is insufficient to avoid *B. cinerea* disease in many wine-growing locations (Vezzulli et al. 2022). Mechanisms of yeast action against *B. cinerea* growth have been linked to the development of enzymatic activity, iron or nutrient competition, and the synthesis of VOCs (Roca-Couso et al. 2021). Because of the release of _-1,3-glucanase, these yeasts demonstrated a significant biocontrol activity against *B. cinerea*.

17.11 Conclusions and Future Prospectives

Grapes (*Vitis vinifera* L.) are one of the most important fruits in the world. Due to the adverse climate changes, postharvest diseases have developed, as table grapes are exposed to disease-causing infections during storage after harvest, which leads to rapid deterioration of the fruits. Postharvest diseases pose a significant threat to agricultural production, resulting in substantial losses and a decline in product quality. These diseases can also lead to the production of toxins that cause poisoning and various serious illnesses in humans. This particular fungus is well-known for its ability to cause significant damage to table grapes after they have been harvested. Endophytic fungi have the ability to colonize grape tissue and engage in diverse interactions that help regulate physiological and metabolic pathways. These interactions can then be leveraged to effectively manage illnesses that arise after harvest. In the preharvest and postharvest phases, endophytic fungi strains have been widely used as biocontrol against a variety of grape diseases. Therefore, one of the most important aspects of disease management is the potential colonization efficacy of endophytes. Furthermore, consortia of mixed endophytic fungi demonstrated a useful strategy for managing grape postharvest diseases during application; however, given the greater adaptability and survival of both strains when combined, more research is warranted, especially in conditions with varying environmental factors. Numerous methods have been reported by endophytes to reduce pathogenic development and improve fruit health. While there are many instances of effective bioformulations of microbial endophytic strains that can suppress the pathogenicity of the pest or pathogens under preharvest settings, postharvest pathogen management is still in its infancy as far as these strains are concerned. In postharvest disease management, additional endophytic microbiome application can help lessen or eventually completely eradicate the detrimental reliance on chemical fungicides. In the future, to learn more about the advancements brought about by the usage of endophytic fungi, thorough investigation into genetics and fruit physiology is necessary. This should be considered when looking for various strains of endophytic fungi and studying their metabolism. It is important to investigate the metabolism of endophytic fungi and explore the potential applications of this nutrient as an environmentally friendly biotherapeutic that promotes balanced nutrition and infection prevention. It is necessary to evaluate the biotic and abiotic environmental effects of large-scale applications to create a scientific foundation for the sustainable and safe use of endophytic fungi.

References

Abdelaziz AM, Kalaba MH, Hashem AH, Sharaf MH, Attia MS (2022) Biostimulation of tomato growth and biocontrol of Fusarium wilt disease using certain endophytic fungi. Bot Stud 63(1):34. https://doi.org/10.1186/s40529-022-00364-7

Abdelaziz AM, El-Wakil DA, Hashem AH, Al-Askar AA, AbdElgawad H, Attia MS (2023a) Efficient role of Endophytic aspergillus terreus in biocontrol of Rhizoctonia solani causing damping-off disease of Phaseolus vulgaris and Vicia faba. Microorganisms 11(6):1487

Abdelaziz AM, Sharaf MH, Hashem AH, Al-Askar AA, Marey SA, Mohamed FA et al (2023b) Biocontrol of Fusarium wilt disease in pepper plant by plant growth promoting Penicillium expansum and Trichoderma harzianum. Notulae Botanicae Horti Agrobotanici Cluj-Napoca 51(3):1–23

Ahmed A, Munir S, He P, Li Y, He P, Yixin W, He Y (2020) Biocontrol arsenals of bacterial endophyte: an imminent triumph against clubroot disease. Microbiol Res 241:126565

Alfiky A, Weisskopf L (2021) Deciphering Trichoderma–plant–pathogen interactions for better development of biocontrol applications. J Fungi 7(1):61

Ali S, Charles TC, Glick BR (2017) Endophytic phytohormones and their role in plant growth promotion. In: Functional importance of the plant microbiome: implications for agriculture, forestry and bioenergy, pp 89–105

Anyasi RO, Atagana HI (2019) Endophyte: understanding the microbes and its applications. Pak J Biol Sci 22:154–167

Asma ST, Imre K, Morar A, Herman V, Acaroz U, Mukhtar H et al (2022) An overview of biofilm formation–combating strategies and mechanisms of action of antibiofilm agents. Life 12(8):1110

Assaf CEH (2018) Patulin, main mycotoxin of the apple industry: regulation of its biosynthetic pathway and influence of processing factors in cloudy apple juice production. Institut National Polytechnique de Toulouse-INPT

Attia MS, Younis AM, Ahmed AF, Elaziz A (2016) Comprehensive management for wilt disease Caused by Fusarium oxysporum in tomato plant. IJISET Int J Innov Sci Eng Technol. 4:201

Attia MS, Hashem AH, Badawy AA, Abdelaziz AM (2022a) Biocontrol of early blight disease of eggplant using endophytic aspergillus terreus: improving plant immunological, physiological and antifungal activities. Bot Stud 63(1):26

Attia MS, Salem MS, Abdelaziz AM (2022b) Endophytic fungi Aspergillus spp. reduce fusarial wilt disease severity, enhance growth, metabolism and stimulate the plant defense system in pepper plants. Biomass Convers Biorefinery 14:1–11

Awuchi CG, Ondari EN, Ogbonna CU, Upadhyay AK, Baran K, Okpala COR et al (2021) Mycotoxins affecting animals, foods, humans, and plants: types, occurrence, toxicities, action mechanisms, prevention, and detoxification strategies—a revisit. Food Secur 10(6):1279

Bacon CW, Hinton DM, Mitchell TR (2017) Is quorum signaling by mycotoxins a new risk-mitigating strategy for bacterial biocontrol of Fusarium verticillioides and other endophytic fungal species? J Agric Food Chem 65(33):7071–7080

Baldwin IT, Halitschke R, Paschold A, Von Dahl CC, Preston CA (2006) Volatile signaling in plant-plant interactions: "talking trees" in the genomics era. Science 311(5762):812–815

Balla A, Silini A, Cherif-Silini H, Chenari Bouket A, Moser WK, Nowakowska JA et al (2021) The threat of pests and pathogens and the potential for biological control in Forest ecosystems. Forests 12(11):1579

Bartholomew HP, Bradshaw M, Jurick WM, Fonseca JM (2021) The good, the bad, and the ugly: mycotoxin production during postharvest decay and their influence on tritrophic host–pathogen–microbe interactions. Front Microbiol 12:611881

Billar de Almeida A, Concas J, Campos MD, Materatski P, Varanda C, Patanita M et al (2020) Endophytic fungi as potential biological control agents against grapevine trunk diseases in Alentejo region. Biology 9(12):420

Bolívar-Anillo HJ, Garrido C, Collado IG (2020) Endophytic microorganisms for biocontrol of the phytopathogenic fungus Botrytis cinerea. Phytochem Rev 19:721–740

Brown NA, Hammond-Kosack KE (2015) Secreted biomolecules in fungal plant pathogenesis. In: Fungal biomolecules: sources, applications and recent developments, pp 263–310

Carlini CR, Grossi-de-Sá MF (2002) Plant toxic proteins with insecticidal properties. A review on their potentialities as bioinsecticides. Toxicon 40(11):1515–1539

Carmona-Hernandez S, Reyes-Pérez JJ, Chiquito-Contreras RG, Rincon-Enriquez G, Cerdan-Cabrera CR, Hernandez-Montiel LG (2019) Biocontrol of postharvest fruit fungal diseases by bacterial antagonists: a review. Agronomy 9(3):121

Christian N, Sedio BE, Florez-Buitrago X, Ramírez-Camejo LA, Rojas EI, Mejía LC et al (2020) Host affinity of endophytic fungi and the potential for reciprocal interactions involving host secondary chemistry. Am J Bot 107(2):219–228

Cosoveanu A, Gimenez-Mariño C, Cabrera Y, Hernandez G, Cabrera R (2014) Endophytic fungi from grapevine cultivars in Canary Islands and their activity against phytopatogenic fungi. Int J Agricu Crop Sci 7(15):1497

Dalmais B, Schumacher J, Moraga J, Le Pecheur P, Tudzynski B, Collado IG, Viaud M (2011) The Botrytis cinerea phytotoxin botcinic acid requires two polyketide synthases for production and has a redundant role in virulence with botrydial. Mol Plant Pathol 12(6):564–579

Dang H, Lovell CR (2016) Microbial surface colonization and biofilm development in marine environments. Microbiol Mol Biol Rev 80(1):91–138

de Carvalho JO, Broll V, Martinelli AHS, Lopes FC (2020) Endophytic fungi: positive association with plants. In: Molecular aspects of plant beneficial microbes in agriculture. Elsevier, pp 321–332

De Simone N, Pace B, Grieco F, Chimienti M, Tyibilika V, Santoro V et al (2020) Botrytis cinerea and table grapes: A review of the main physical, chemical, and bio-based control treatments in post-harvest. Food Secur 9(9):1138

De Wolf ED, Isard SA (2007) Disease cycle approach to plant disease prediction. Annu Rev Phytopathol 45:203–220

Di Canito A, Mateo-Vargas MA, Mazzieri M, Cantoral J, Foschino R, Cordero-Bueso G, Vigentini I (2021) The role of yeasts as biocontrol agents for pathogenic fungi on postharvest grapes: a review. Food Secur 10(7):1650

Díaz-Urbano M, Velasco P, Rodríguez VM, Poveda J (2023) Endophytic fungi in postharvest disease management in fresh produce. In: Postharvest management of fresh produce. Elsevier, pp 81–112

Dimkić I, Jelušić A, Kruščić K, Janakiev T (2024) Pathobiome and microbial community shifts associated with vegetable, fruit, and cereal crops. In: Plant pathogen interaction. Springer, pp 237–258

Dimopoulou A, Theologidis I, Benaki D, Koukounia M, Zervakou A, Tzima A et al (2021) Direct antibiotic activity of bacillibactin broadens the biocontrol range of Bacillus amyloliquefaciens MBI600. Msphere 6(4). https://doi.org/10.1128/msphere.00376-00321

Doty SL (2013) Endophytic yeasts: biology and applications. In: *Symbiotic endophytes*. Springer, pp 335–343

Dutta P, Deb L, Pandey AK (2022) Trichoderma-from lab bench to field application: looking back over 50 years. Front Agron 4:932839

Fahmi AI, Nagaty MA, El-Shehawi AM (2012) Fruit quality of Taif grape (Vitis vinifera L.) cultivars. J Am Sci 8(5):590–599

Fares J, Fares MY, Khachfe HH, Salhab HA, Fares Y (2020) Molecular principles of metastasis: a hallmark of cancer revisited. Signal Transduct Target Ther 5(1):28

Farrag A, Attia MS, Younis A, Abd Elaziz A (2017) Potential impacts of elicitors to improve tomato plant disease resistance. Al Azhar Bull Sci 9:311–321

Fontana DC, de Paula S, Torres AG, de Souza VHM, Pascholati SF, Schmidt D, Dourado Neto D (2021) Endophytic fungi: biological control and induced resistance to phytopathogens and abiotic stresses. Pathogens 10(5):570

Gautam AK, Sharma S, Avasthi S, Bhadauria R (2011) Diversity, pathogenicity and toxicology of A. Niger: an important spoilage fungi. Res J Microbiol 6(3):270–280

Gerhardson B (2002) Biological substitutes for pesticides. Trends Biotechnol 20(8):338–343

Giller KE, Delaune T, Silva JV, Descheemaeker K, van de Ven G, Schut AG et al (2021) The future of farming: who will produce our food? Food Secur 13(5):1073–1099

Glenn DM, Kim SH, Ramirez-Villegas J, Läderach P (2013) Response of perennial horticultural crops to climate change. Hortic Rev 41:47–130

Gowtham HG, Singh SB, Shilpa N, Aiyaz M, Nataraj K, Udayashankar AC et al (2022) Insight into recent progress and perspectives in improvement of antioxidant machinery upon PGPR augmentation in plants under drought stress: a review. Antioxidants 11(9):1763

Gupta S, Choudhary M, Singh B, Singh R, Dhar MK, Kaul S (2022) Diversity and biological activity of fungal endophytes of Zingiber officinale Rosc. with emphasis on Aspergillus terreus as a biocontrol agent of its leaf spot. Biocatal Agric Biotechnol 39:102234

Guzmán-Guzmán P, Kumar A, de Los Santos-Villalobos S, Parra-Cota FI, Orozco-Mosqueda MDC, Fadiji AE et al (2023) Trichoderma species: our best fungal allies in the biocontrol of plant diseases—a review. Plan Theory 12(3):432

Hardoim PR, Van Overbeek LS, Berg G, Pirttilä AM, Compant S, Campisano A et al (2015) The hidden world within plants: ecological and evolutionary considerations for defining functioning of microbial endophytes. Microbiol Mol Biol Rev 79(3):293–320

Harman GE, Howell CR, Viterbo A, Chet I, Lorito M (2004) Trichoderma species—opportunistic, avirulent plant symbionts. Nat Rev Microbiol 2(1):43–56

Hashemi M, Tabet D, Sandroni M, Benavent-Celma C, Seematti J, Andersen CB, Grenville-Briggs LJ (2022) The hunt for sustainable biocontrol of oomycete plant pathogens, a case study of Phytophthora infestans. Fungal Biol Rev 40:53–69

Hermosa R, Viterbo A, Chet I, Monte E (2012) Plant-beneficial effects of Trichoderma and of its genes. Microbiology 158(1):17–25

Hou S, Liu Z, Shen H, Wu D (2019) Damage-associated molecular pattern-triggered immunity in plants. Front Plant Sci 10:646

Jaber LR (2015) Grapevine leaf tissue colonization by the fungal entomopathogen Beauveria bassiana sl and its effect against downy mildew. BioControl 60:103–112

Jacob P, Hige J, Dangl JL (2023) Is localized acquired resistance the mechanism for effector-triggered disease resistance in plants? Nat Plants 9(8):1184–1190

Jampílek J, Kráľová K (2024) Fungal volatile organic compounds. In: *Fungal secondary metabolites*. Elsevier, pp 399–426

Janik E, Niemcewicz M, Ceremuga M, Stela M, Saluk-Bijak J, Siadkowski A, Bijak M (2020) Molecular aspects of mycotoxins—A serious problem for human health. Int J Mol Sci 21(21):8187

Kalia A, Gupta RP (2012) Microbiology of fresh and processed fruits. In: Handbook of fruits and fruit processing, pp 51–72

Kamran M, Imran QM, Ahmed MB, Falak N, Khatoon A, Yun B-W (2022) Endophyte-mediated stress tolerance in plants: a sustainable strategy to enhance resilience and assist crop improvement. Cells 11(20):3292

Kashyap N, Singh SK, Yadav N, Singh VK, Kumari M, Kumar D et al (2023) Biocontrol screening of endophytes: applications and limitations. Plan Theory 12(13):2480

Kasso M, Bekele A (2018) Post-harvest loss and quality deterioration of horticultural crops in Dire Dawa Region, Ethiopia. J Saudi Soc Agric Sci 17(1):88–96

Khan MI, Shin JH, Kim JD (2018) The promising future of microalgae: current status, challenges, and optimization of a sustainable and renewable industry for biofuels, feed, and other products. Microb Cell Factories 17(1):1–21

Khan UM, Sameen A, Aadil RM, Shahid M, Sezen S, Zarrabi A et al (2021) Citrus genus and its waste utilization: a review on health-promoting activities and industrial application. Evid Based Complement Alternat Med 2021:1–17

Khatri DK, Tiwari DN, Bariya HS (2017) Chitinolytic efficacy and secretion of cell wall degrading enzymes from Trichoderma spp. in response to phyto-pathological fungi. J Appl Biol Biotechnol 5(6):1–8

Kikot GE, Hours RA, Alconada TM (2009) Contribution of cell wall degrading enzymes to pathogenesis of Fusarium graminearum: a review. J Basic Microbiol 49(3):231–241

Kong W-L, Li P-S, Wu X-Q, Wu T-Y, Sun X-R (2020) Forest tree associated bacterial diffusible and volatile organic compounds against various phytopathogenic fungi. Microorganisms 8(4):590

Kumari M, Qureshi KA, Jaremko M, White J, Singh SK, Sharma VK et al (2022) Deciphering the role of endophytic microbiome in postharvest diseases management of fruits: opportunity areas in commercial up-scale production. Front Plant Sci 13:4390

Lastochkina O, Baymiev A, Shayahmetova A, Garshina D, Koryakov I, Shpirnaya I et al (2020) Effects of endophytic Bacillus subtilis and salicylic acid on postharvest diseases (Phytophthora infestans, Fusarium oxysporum) development in stored potato tubers. Plan Theory 9(1):76

Lata R, Chowdhury S, Gond SK, White JF Jr (2018) Induction of abiotic stress tolerance in plants by endophytic microbes. Lett Appl Microbiol 66(4):268–276

Latha P, Karthikeyan M, Rajeswari E (2019) Endophytic bacteria: prospects and applications for the plant disease management. In: *Plant health under biotic stress: volume 2: microbial Interactions*, pp 1–50

Leake J, Johnson D, Donnelly D, Muckle G, Boddy L, Read D (2004) Networks of power and influence: the role of mycorrhizal mycelium in controlling plant communities and agroecosystem functioning. Can J Bot 82(8):1016–1045

Lee J-S, Kaplunov T, Zutahy Y, Daus A, Alkan N, Lichter A (2015) The significance of postharvest disinfection for prevention of internal decay of table grapes after storage. Sci Hortic 192:346–349

Li X-J, Zhang Q, Zhang A-L, Gao J-M (2012) Metabolites from aspergillus fumigatus, an endophytic fungus associated with Melia azedarach, and their antifungal, antifeedant, and toxic activities. J Agric Food Chem 60(13):3424–3431

Liarzi O, Bar E, Lewinsohn E, Ezra D (2016) Use of the endophytic fungus Daldinia cf. concentrica and its volatiles as bio-control agents. PLoS One 11(12):e0168242

Linling L, Zheng L, Juan H, Hua C, Zhiqin L, Shuiyuan C, Qing T (2013) Isolation and identification of pathogenic fungi causing decay in Luotian chestnut during late storage period and research on pathogenicity. Plant Dis Pests 4(6):10

Lü P, Liu Y, Yu X, Shi C-L, Liu X (2022) The right microbe-associated molecular patterns for effective recognition by plants. Front Microbiol 13:1019069

Madbouly AK, Elyousr KAA, Ismail IM (2020) Biocontrol of Monilinia fructigena, causal agent of brown rot of apple fruit, by using endophytic yeasts. Biol Control 144:104239

Magan N (2020) Importance of ecological windows for efficacy of biocontrol agents. In: How research can stimulate the development of commercial biological control against plant diseases, pp 1–14

Majumdar S (2017) Evaluation of jute rhizospheric bacteria for plant growth promotion and disease suppression

Manriquez B, Muller D, Prigent-Combaret C (2021) Experimental evolution in plant-microbe systems: A tool for deciphering the functioning and evolution of plant-associated microbial communities. Front Microbiol 12:619122

Mao LJ, Chen JJ, Xia CY, Feng XX, Kong DD, Qi ZY et al (2019) Identification and characterization of new Muscodor endophytes from gramineous plants in Xishuangbanna, China. MicrobiologyOpen 8(4):e00666

Matsumura M, Nomoto M, Itaya T, Aratani Y, Iwamoto M, Matsuura T et al (2022) Mechanosensory trichome cells evoke a mechanical stimuli–induced immune response in Arabidopsis thaliana. Nat Commun 13(1):1216

Mauch-Mani B, Baccelli I, Luna E, Flors V (2017) Defense priming: an adaptive part of induced resistance. Annu Rev Plant Biol 68:485–512

Medina-Romero YM, Roque-Flores G, Macías-Rubalcava ML (2017) Volatile organic compounds from endophytic fungi as innovative postharvest control of Fusarium oxysporum in cherry tomato fruits. Appl Microbiol Biotechnol 101:8209–8222

Mehmood N, Saeed M, Zafarullah S, Hyder S, Rizvi ZF, Gondal AS et al (2023) Multifaceted impacts of plant-beneficial pseudomonas spp. in managing various plant diseases and crop yield improvement. ACS Omega 8(25):22296–22315

Meng X, Li B, Liu J, Tian S (2008) Physiological responses and quality attributes of table grape fruit to chitosan preharvest spray and postharvest coating during storage. Food Chem 106(2):501–508

Mokiou S (2004) Ecophysiological approaches to production and formulation of the biocontrol yeast Pichia anomala. Cranfield University

Morales-Cedeño LR, del Carmen Orozco-Mosqueda M, Loeza-Lara PD, Parra-Cota FI, de Los Santos-Villalobos S, Santoyo G (2021) Plant growth-promoting bacterial endophytes as biocontrol agents of pre-and post-harvest diseases: fundamentals, methods of application and future perspectives. Microbiol Res 242:126612

Naik S, Shaanker RU, Ravikanth G, Dayanandan S (2019) How and why do endophytes produce plant secondary metabolites? Symbiosis 78:193–201

Nanda S, Vishwakarma R, Bathla H, Rai A, Chandra P (2012a) Estimation of quantitative harvest and post-harvest losses of major agricultural produce in India. AICRP on Post Harvest Technology, CIPHET, Ludhiana

Nanda S, Vishwakarma R, Bathla H, Rai A, Chandra P (2012b) Harvest and post harvest losses of major crops and livestock produce in India. AICRP, (ICAR)

Neugebauer KA, Perkins JA, Sysak R, Isaacs R, Miles TD (2024) Reducing cluster rots in Michigan wine grapes using combined pathogen and vinegar fly control. Crop Prot 177:106528

Otero C, Arredondo C, Echeverría-Vega A, Gordillo-Fuenzalida F (2020) Penicillium spp. mycotoxins found in food and feed and their health effects. World Mycotoxin J 13(3):323–343

Pal G, Saxena S, Kumar K, Verma A, Sahu PK, Pandey A et al (2022) Endophytic Burkholderia: multifunctional roles in plant growth promotion and stress tolerance. Microbiol Res 265:127201

Pan F, Liu Z-Q, Chen Q, Xu Y-W, Hou K, Wu W (2016) Endophytic fungus strain 28 isolated from Houttuynia cordata possesses wide-spectrum antifungal activity. Braz J Microbiol 47:480–488

Pantelides IS, Christou O, Tsolakidou M-D, Tsaltas D, Ioannou N (2015) Isolation, identification and in vitro screening of grapevine yeasts for the control of black aspergilli on grapes. Biol Control 88:46–53

Patel J, Teli B, Bajpai R, Meher J, Rashid M, Mukherjee A, Yadav SK (2019) Trichoderma-mediated biocontrol and growth promotion in plants: An endophytic approach. In: Role of plant growth promoting microorganisms in sustainable agriculture and nanotechnology. Elsevier, pp 219–239

Patra D, Islam MM, Das P, Sarkar B, Jana SK, Mandal S (2023) Importance of endophytes and mechanisms of their interactions with host-plants. In: Endophytic association: what, Why and how. Elsevier, pp 409–435

Paul NC, Kim W-K, Woo S-K, Park M-S, Yu S-H (2007) Fungal endophytes in roots of Aralia species and their antifungal activity. Plant Pathol J 23(4):287–294

Pellan L, Dieye CAT, Durand N, Fontana A, Strub C, Schorr-Galindo S (2021) Biocontrol agents: toolbox for the screening of weapons against mycotoxigenic Fusarium. J Fungi 7(06):446

Perello C, Llamas E, Burlat V, Ortiz-Alcaide M, Phillips MA, Pulido P, Rodriguez-Concepcion M (2016) Differential subplastidial localization and turnover of enzymes involved in isoprenoid biosynthesis in chloroplasts. PLoS One 11(2):e0150539

Poveda J, Abril-Urias P, Escobar C (2020) Biological control of plant-parasitic nematodes by filamentous fungi inducers of resistance: Trichoderma, mycorrhizal and endophytic fungi. Front Microbiol 11:992

Priyadarshini MS, Panigrahi S, Rath C (2022) *Endophytes: novel microorganisms for plant growth promotion*, vol 1. Darshan Publishers

Prusky D (2011) Reduction of the incidence of postharvest quality losses, and future prospects. Food Secur 3:463–474

Quoc NB, Bao Chau NN (2017) The role of cell wall degrading enzymes in pathogenesis of Magnaporthe oryzae. Curr Prot Pept Sci 18(10):1019–1034

Rajani P, Rajasekaran C, Vasanthakumari M, Olsson SB, Ravikanth G, Shaanker RU (2021) Inhibition of plant pathogenic fungi by endophytic Trichoderma spp. through mycoparasitism and volatile organic compounds. Microbiol Res 242:126595

Ramjan M, Ansari MT (2018) Factors affecting of fruits, vegetables and its quality. J Med Plants 6:16–18

Rasmussen, R. R. (2010). Mycotoxins in maize silage-detection of toxins and toxicological aspects

Raymaekers K, Ponet L, Holtappels D, Berckmans B, Cammue BP (2020) Screening for novel biocontrol agents applicable in plant disease management–a review. Biol Control 144:104240

Ribeiro LS, de Souza ML, Lira JMS, Schwan RF, Batista LR, Silva CF (2023) Volatile compounds for biotechnological applications produced during competitive interactions between yeasts and fungi. J Basic Microbiol. https://doi.org/10.1002/jobm.202200409

Rienth M, Vigneron N, Walker RP, Castellarin SD, Sweetman C, Burbidge CA et al (2021) Modifications of grapevine berry composition induced by main viral and fungal pathogens in a climate change scenario. Front Plant Sci 12:717223

Roca-Couso R, Flores-Félix JD, Rivas R (2021) Mechanisms of action of microbial biocontrol agents against Botrytis cinerea. J Fungi 7(12):1045

Romanazzi G, Smilanick JL, Feliziani E, Droby S (2016) Integrated management of postharvest gray mold on fruit crops. Postharvest Biol Technol 113:69–76

Russi A, Granada CE, Schwambach J (2024) Suppression of Colletotrichum spp. on grape berries, vine leaves, and plants using bacillus velezensis S26 endospores. Sci Hortic 326:112696

Sadiq FA, Yan B, Tian F, Zhao J, Zhang H, Chen W (2019) Lactic acid bacteria as antifungal and anti-mycotoxigenic agents: a comprehensive review. Compr Rev Food Sci Food Saf 18(5):1403–1436

Sakaguchi E, Suzuki M, Favier J, Kawakami S (2001) PH—Postharvest Technology: Numerical simulation of the shaking separation of paddy and brown rice using the discrete element method. J Agric Eng Res 79(3):307–315

Sarrocco S, Vannacci G (2018) Preharvest application of beneficial fungi as a strategy to prevent postharvest mycotoxin contamination: a review. Crop Prot 110:160–170

Schena L, Nigro F, Pentimone I, Ligorio A, Ippolito A (2003) Control of postharvest rots of sweet cherries and table grapes with endophytic isolates of Aureobasidium pullulans. Postharvest Biol Technol 30(3):209–220

Sellitto VM, Zara S, Fracchetti F, Capozzi V, Nardi T (2021) Microbial biocontrol as an alternative to synthetic fungicides: boundaries between pre-and postharvest applications on vegetables and fruits. Fermentation 7(2):60

Sharma D, Malhotra H, Mittal R (2021) Post-harvest management practice (PHMP): A systematic program to attain food security. In: Sustainable Agriculture for Food Security. Apple Academic Press, New York, pp 179–207

Singh D, Sharma RR, Devappa V, Kamil D (2021a) Postharvest handling and diseases of horticultural produce. CRC Press

Singh D, Sharma RR, Kesharwani AK (2021b) Postharvest losses of horticultural produce. In: *Postharvest handling and diseases of horticultural produce*. CRC Press, Boca Raton, pp 1–24

Singh J, Mirza A, Kaur S, Arya SK, Singh G (2023) Application of microbial antagonists for the preservation of fruits: an effective strategy to inhibit the postharvest disease. In: *Relationship between microbes and the environment for sustainable ecosystem services, volume 3*. Elsevier, pp 261–288

Solairaj D, Legrand NNG, Yang Q, Zhang H (2020) Isolation of pathogenic fungi causing postharvest decay in table grapes and in vivo biocontrol activity of selected yeasts against them. Physiol Mol Plant Pathol 110:101478

Sommer A, Katz J, Quigley HA, Miller NR, Robin AL, Richter RC, Witt KA (1991) Clinically detectable nerve fiber atrophy precedes the onset of glaucomatous field loss. Arch Ophthalmol 109(1):77–83

Sudheer K, Indira V (2007) Post harvest technology of horticultural crops, vol Vol. 7. New India Publishing

Taulé C, Vaz-Jauri P, Battistoni F (2021) Insights into the early stages of plant–endophytic bacteria interaction. World J Microbiol Biotechnol 37:1–9

Temme N, Oeser B, Massaroli M, Heller J, Simon A, Gonzalez Collado I et al (2012) BcAtf1, a global regulator, controls various differentiation processes and phytotoxin production in Botrytis cinerea. Mol Plant Pathol 13(7):704–718

Terhonen E, Sipari N, Asiegbu FO (2016) Inhibition of phytopathogens by fungal root endophytes of Norway spruce. Biol Control 99:53–63

Thambugala KM, Daranagama DA, Phillips AJ, Kannangara SD, Promputtha I (2020) Fungi vs. fungi in biocontrol: an overview of fungal antagonists applied against fungal plant pathogens. Front Cell Infect Microbiol 10:604923

Tiwari P, Bajpai M, Singh LK, Mishra S, Yadav AN (2020) Phytohormones producing fungal communities: metabolic engineering for abiotic stress tolerance in crops. In: Agriculturally important fungi for sustainable agriculture: volume 1: perspective for diversity and crop productivity, pp 171–197

Triantafyllou A, Kamou N, Papadopoulou A, Leontidou K, Mellidou I, Karamanoli K (2023) Evaluation of the biocontrol potential of PGPB strains isolated from drought-tolerant tomatoes against fungal pathogens. J Plant Pathol:1–17. https://doi.org/10.1007/s42161-023-01422-0

Tsitsigiannis DI, Dimakopoulou M, Antoniou PP, Tjamos EC (2012) Biological control strategies of mycotoxigenic fungi and associated mycotoxins in Mediterranean basin crops. Phytopathol Mediterr 51:158–174

Valero-Jiménez CA, Steentjes MB, Slot JC, Shi-Kunne X, Scholten OE, van Kan JA (2020) Dynamics in secondary metabolite gene clusters in otherwise highly syntenic and stable genomes in the fungal genus botrytis. Genome Biol Evol 12(12):2491–2507

Vandana UK, Rajkumari J, Singha LP, Satish L, Alavilli H, Sudheer PD et al (2021) The endophytic microbiome as a hotspot of synergistic interactions, with prospects of plant growth promotion. Biology 10(2):101

Veenstra A, Rafudeen MS, Murray SL (2019) Trichoderma asperellum isolated from African maize seed directly inhibits Fusarium verticillioides growth in vitro. Eur J Plant Pathol 153(1):279–283

Verma SK, Sahu PK, Kumar K, Pal G, Gond SK, Kharwar RN, White JF (2021) Endophyte roles in nutrient acquisition, root system architecture development and oxidative stress tolerance. J Appl Microbiol 131(5):2161–2177

Vezzulli S, Gramaje D, Tello J, Gambino G, Bettinelli P, Pirrello C et al (2022) Genomic designing for biotic stress resistant grapevine. In: Genomic designing for biotic stress resistant fruit crops. Springer, pp 87–255

Vilas J, Chinnappa B, Mahadevaiah G (2011) An economic analysis of post-harvest losses of grapes in Karnataka. Mysore J Agric Sci 45(4):905–911

Vincente AR, Manganaris GA, Ortiz CM, Sozzi GO, Crisosto CH (2014) Nutritional quality of fruits and vegetables. In: *Postharvest handling*. Elsevier, pp 69–122

Vurukonda SSKP, Giovanardi D, Stefani E (2018) Plant growth promoting and biocontrol activity of Streptomyces spp. as endophytes. Int J Mol Sci 19(4):952

Wang M, Zhao L, Zhang X, Dhanasekaran S, Abdelhai MH, Yang Q et al (2019) Study on biocontrol of postharvest decay of table grapes caused by Penicillium rubens and the possible resistance mechanisms by Yarrowia lipolytica. Biol Control 130:110–117

Wang H, Zhang R, Duan Y, Jiang W, Chen X, Shen X et al (2021) The endophytic strain Trichoderma asperellum 6S-2: an efficient biocontrol agent against apple replant disease in China and a potential plant-growth-promoting fungus. J Fungi 7(12):1050

Wenneker M, Thomma BP (2020) Latent postharvest pathogens of pome fruit and their management: from single measures to a systems intervention approach. Eur J Plant Pathol 156:663–681

Whipps J, Hand P, Pink D, Bending GD (2008) Phyllosphere microbiology with special reference to diversity and plant genotype. J Appl Microbiol 105(6):1744–1755

Williamson B, Tudzynski B, Tudzynski P, Van Kan JA (2007) Botrytis cinerea: the cause of grey mould disease. Mol Plant Pathol 8(5):561–580

Wisniewski M, Droby S, Norelli J, Liu J, Schena L (2016) Alternative management technologies for postharvest disease control: the journey from simplicity to complexity. Postharvest Biol Technol 122:3–10

Yahia EM, García-Solís P, Celis MEM (2019) Contribution of fruits and vegetables to human nutrition and health. In: *Postharvest physiology and biochemistry of fruits and vegetables*. Elsevier, pp 19–45

Yuen J, Collinge DB, Djurle A, Tronsmo AM (2020) Plant pathology in a changing world 23. In: Plant pathology and plant diseases, p 379

Zahavi T, Cohen L, Weiss B, Schena L, Daus A, Kaplunov T et al (2000) Biological control of botrytis, aspergillus and Rhizopus rots on table and wine grapes in Israel. Postharvest Biol Technol 20(2):115–124

Zhang C-L, Zheng B-Q, Lao J-P, Mao L-J, Chen S-Y, Kubicek CP, Lin F-C (2008) Clavatol and patulin formation as the antagonistic principle of Aspergillus clavatonanicus, an endophytic fungus of Taxus mairei. Appl Microbiol Biotechnol 78:833–840

Zhang H, Apaliya MT, Mahunu GK, Chen L, Li W (2016) Control of ochratoxin A-producing fungi in grape berry by microbial antagonists: A review. Trends Food Sci Technol 51:88–97

Zhang H, Godana EA, Sui Y, Yang Q, Zhang X, Zhao L (2020) Biological control as an alternative to synthetic fungicides for the management of grey and blue mould diseases of table grapes: a review. Crit Rev Microbiol 46(4):450–462

Zhang H, Serwah Boateng NA, Ngolong Ngea GL, Shi Y, Lin H, Yang Q et al (2021) Unravelling the fruit microbiome: the key for developing effective biological control strategies for postharvest diseases. Compr Rev Food Sci Food Saf 20(5):4906–4930

Zhang QQ, Shi J, Shen PY, Xi F, Qian CY, Zhang GH et al (2022) Exploring the efficacy of biocontrol microbes against the fungal pathogen Botryosphaeria dothidea JNHT01 isolated from fresh walnut fruit. Food Secur 11(22):3651

Zhu C, Shi J, Jiang C, Liu Y (2015) Inhibition of the growth and ochratoxin A production by Aspergillus carbonarius and Aspergillus ochraceus in vitro and in vivo through antagonistic yeasts. Food Control 50:125–132

Zin NA, Badaluddin NA (2020) Biological functions of Trichoderma spp. for agriculture applications. Ann Agric Sci 65(2):168–178

18 Biological Control Mechanism of Endophytic Fungi

Feba Sara Koshy, Kamel A. Abd-Elsalam, and Sangilimuthu Alagar Yadav

Abstract

Endophytes are used as biocontrol agents in plant disease management and sustainable agriculture. The numerous endophytic microbes residing in a plant's internal tissues can impact the plant positively or negatively. Beneficial endophytic fungi can directly or indirectly enhance plant health, enabling plants to withstand potential dangers. Plant infections may result from inadequate farming practices and environmental factors. The symbiotic relationship among endophytic fungi and host plants leftovers a vital progressive action in disease control. To address issues with endophyte biocontrol mechanisms, a thorough understanding of the interactions among pathogens and host plants is required, in addition to selecting and using beneficial endophytic fungi. Endophytic fungi, which produce a significant number of biologically active compounds, can be instrumental in managing biotic and abiotic challenges. Consequently, it is increasingly common to search for these microbes that produce beneficial agricultural products.

Keywords

Endophytic fungi · Biocontrol agents · Antibiosis · Competition · Induced stress · Plant diseases

F. S. Koshy · S. A. Yadav (✉)
Department of Biotechnology, Karpagam Academy of Higher Education, Coimbatore, Tamil Nadu, India

K. A. Abd-Elsalam
Plant Pathology Research Institute, Agricultural Research Center (ARC), Giza, Egypt

© The Author(s), under exclusive license to Springer Nature Singapore Pte Ltd. 2025
K. A. Abd-Elsalam, A. H. Hashem (eds.), *Fungal Endophytes Volume II*,
https://doi.org/10.1007/978-981-97-8804-0_18

18.1 Introduction

Apart from food production, plants provide the community with other necessary resources, including wood, fiber, medicines, and green energy. Diseases in plants cause substantial drops in productivity, leading to significant revenue loss, which can result in disastrous social effects (Fontana et al. 2021). Plant diseases currently pose a significant threat to the agricultural business. Diseases and pests cause up to 40% of the yield of commercially significant crops to be lost each year (Adeleke et al. 2021). Disease-related yield losses are influenced by several factors, including the pathogen's virulence, environmental conditions, and the crop's ability to compensate for a reduction in photosynthesis (Alam et al. 2021). The overreliance on certain seeds and insecticides has led to the rise of resistant microbes, creating a need for alternative and sustainable approaches. The shift toward integrated management, combining various strategies, is a smart move. Biological control is a successful case in pathogen control, showcasing the potential of harnessing nature to solve agricultural challenges. It's not just about combating pests and diseases; it's about fostering a healthier, more balanced ecosystem (Suebrasri et al. 2020).

Endophytic fungi establish synergistic relationships with plants by living within their tissues without inflicting visible harm. These fungi can assist their host plants in various ways, including boosting nutrient absorption, increasing tolerance to environmental challenges, and defending against certain pests and diseases. It is a mutual support system where the fungi receive a hospitable environment to live in and the plants gain additional assistance to thrive (Damm et al. 2012). The protection that endophytic fungi provide for plants will be addressed in the article. This chapter highlights some of the most current findings in the field of endophytic fungi used as biological control agents. It also provides reflections to help readers understand the future direction of this field.

18.2 Endophytic Fungi

Microbial endophytes and host plants have a dynamic connection that can be mutualistic or antagonistic, depending on various factors, including genetic compatibility (Harman et al. 2004). Plant growth-promoting endophytes (PGPEs) are a fascinating group of endophytic microbes. These organisms form a mutualistic relationship with plants, offering numerous benefits without causing harm. Some of the key characteristics that make them "plant growth-promoting" include the following:

- Plant growth and development are significantly impacted by phytohormones, which can be produced by PGPEs, including auxins, cytokinins, and gibberellins. These hormones aid in nutrient acquisition by improving the plant's ability to absorb critical components from the soil, often through methods such as mineral solubilization. Secretion of Biocontrol Agents (BCAs): PGPEs may produce chemicals or enzymes that act as biocontrol agents.

- Stress Induction Mechanisms: PGPEs can assist plants in coping with environmental challenges such as drought and salinity by triggering mechanisms that induce stress resistance (Harman et al. 2004).

The complicated interaction between endophytes and their host plants illustrates the complexities of plant-microbe interactions beneath the soil surface. Understanding and leveraging these interactions could significantly impact sustainable agriculture and ecosystem health. This intriguing area of study holds considerable potential for agricultural applications. Examples of endophytes include *Diaporthe phaseolorum*, *Macrophomina phaseolina*, *Daldinia eschscholtzii*, *Trichoderma erinaceum*, and *Trichoderma koningii*. The detailed list of endophytic fungi and their hosts is provided in Table 18.1.

18.2.1 Classification of Endophytic Fungi

Fungal endophytic communities exhibit a high degree of taxonomic diversity, making it challenging to isolate, identify, and characterize every endophytic fungal species studied thus far (Gao et al. 2010). The distinctive symbiotic and ecological relevance of these fungal taxa, as well as their underlying biological issues, can be better understood by classifying and differentiating endophytic fungi according to

Table 18.1 Examples of some endophytic fungi and their host plants

SL. no.	Endophytic fungi	Host	References
1.	*Colletotrichum gloeosporioides*	Citrus spp.	Damm et al. (2012)
2.	*Fusarium oxysporum*	Quercus spp. (oak)	–
3.	*Trichoderma harzianum*	*Gossypium hirsutum* (cotton)	Sun et al. (2008)
4.	*Alternaria alternata*	*Brassica oleracea* (cabbage)	–
5.	*Phomopsis* spp.	*Vitis vinifera* (grapevine)	–
6.	*Acremonium strictum*	*Lolium perenne* (perennial ryegrass)	Schroers et al. (2009)
7.	*Aspergillus niger*	*Zea mays* (maize)	Houbraken et al. (2011)
8.	*Neotyphodium coenophialum*	*Lolium arundinaceum* (tall fescue)	–
9.	*Curvularia lunata*	*Oryza sativa* (rice)	U'Ren et al. (2019)
10.	*Xylaria polymorpha*	*Quercus* spp. (oats)	Hawksworth (2001)
11.	*Phyllosticta capitalensis*	*Musa* spp. (banana)	Aamir et al. (2020)
12.	*Chaetomium globosum*	*Solanum tuberosum* (potato)	Omomowo et al. (2023)
13.	*Penicillium commune*	*Fragaria xananassa* (strawberry)	Segaran and Sathiavelu (2019)
14.	*Epichloë festucae*	*Festuca arundinacea* (tall fescue)	Ting et al. (2010)
15.	*Cladosporium cladosporioides*	*Glycine max* (soybean)	Gao et al. (2010)

their functional characteristics. A variety of factors, including the type and species of the host, the type and density of the inoculum, the developmental stage of the host at the time of inoculation, and the environmental circumstances in the surrounding area, can lead to a wide range of endophytic relationships in plants.

Fungal endophytes are categorized using a number of characteristics. For example, (i) ecology classifies them as either clavicipitaceous or non-clavicipitaceous endophytes; (ii) reproduction classifies them as sexual or asexual endophytes; (iii) mode of transmission classifies them as vertically or horizontally transmitted endophytes; (iv) nutrition source classifies them as biotrophic or necrotrophic endophytes; (v) infection expression classifies them as symptomatic or asymptomatic.

18.3 Biological Control Mechanisms of Endophytic Fungi

Plant pathogen suppression techniques vary, with both closely related and distantly phylogenetic organisms included in the broad category of biological control. Biological control agents encompass a variety of invertebrates, fungi, bacteria, and viruses, with fungi and bacteria being the most commonly used to control plant diseases due to their ability to colonize plants. Regarding mechanisms of action, biological agents can inhibit pathogens by parasitizing them, competing for niches, and secreting harmful or repulsive substances or through predation (Martinuz et al. 2012). Biological control mechanisms via endophytic fungi are a fascinating aspect of plant-microbe interactions. Table 18.2 summarizing the key mechanisms.

18.3.1 Direct Mechanism of Action of Endophytes

Endophytic fungi (EF) have been shown to play an important role in the biological control of plant diseases. EF can inhibit the growth and development of many plant

Table 18.2 Biological control mechanisms via endophytic fungi

Mechanism	Description	Example
Antibiotic production	Endophytic fungi produce chemicals that directly inhibit or kill harmful pathogens	Fungal strains producing penicillin or other antifungal compounds
Competition for space and nutrients	Endophytic fungi colonize the same niche as pathogens, limiting resources and hindering their growth	Endophytes occupying root hairs and preventing fungal pathogens from establishing
Parasitism	Endophytic fungi directly attack and consume pathogenic fungi	Mycoparasites that parasitize and kill other fungal structures
Induced systemic resistance (ISR)	Endophytes trigger plant defense mechanisms, making them more resistant to pathogens	Fungal signals stimulating plant production of protective enzymes and compounds
Enhancement of plant growth	Endophytic fungi improve plant health and vigor, indirectly reducing susceptibility to disease	Increased nutrient uptake, stress tolerance, and root development

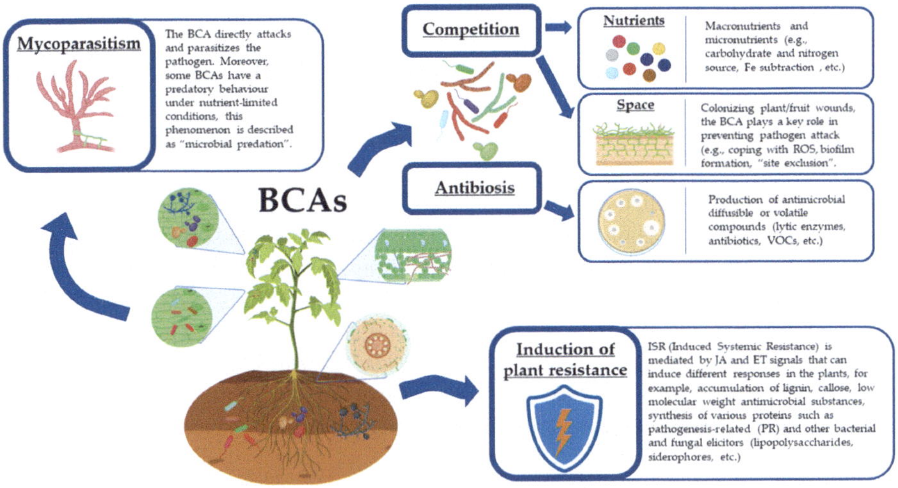

Fig. 18.1 The primary mechanisms of action of biocontrol agents (BCAs) against plant diseases. (Created using Biorender and mentioned in Palmieri et al. 2022)

diseases, such as bacteria, fungus, and nematodes. Endophytic fungi use a variety of ways to exert their biocontrol activity, including both direct and indirect interactions with the host plant and pathogen (Palmieri et al. 2022). Figure 18.1 summarizes the primary ways of action of endophytic fungi. Endophytes prevent infections by competing with pathogens for resources and colonizing sites, enhancing host plant defenses, and producing antimicrobial chemicals that exhibit biocontrol activity similar to other soilborne biocontrol agents (Rodriguez et al. 2009). Organisms are directly involved in mechanisms such as parasitism, competition, and antibiosis during the biocontrol process.

18.3.1.1 Antibiosis

Endophytic fungi use direct antagonism as one of their key strategies for controlling plant diseases. These fungi produce a variety of bioactive secondary metabolites, including antibiotics, enzymes, and volatile organic compounds, that can suppress the development and activity of plant diseases (Khunnamwong et al. 2020). These metabolites have the ability to directly kill or limit pathogen development, preventing disease symptoms in the host plant. Endophytic fungi can also generate cell wall-degrading enzymes, such as chitinases and glucanases, which can degrade pathogen cell walls and disturb structural integrity. This enzyme activity weakens pathogens, making them more vulnerable to the plant's defensive mechanisms and lowering their potential to spread disease.

Antagonism, also known as antibiosis, occurs when bacteria or endophytic fungi develop bioactive secondary metabolites that inhibit the growth or existence of other microbes, particularly pathogens. Endophytic fungi can produce antibiotics or antifungal compounds that are crucial in preventing diseases in the host plant. Numerous antibiotics effective against phytopathogens can be produced by fungal endophytes. *Cassia spectabilis*, a fungal endophyte of *Phomopsis cassiae*, was reported to contain 3,11,12-trihydroxycadalene, a sesquiterpene derivative that exhibited potent antifungal effects against both *Cladosporium sphaerospermum* and *Cladosporium cladosporioides* (Arnold et al. 2003).

18.3.1.1.1 Production of Antimicrobial Compounds

Endophytic fungi produce antimicrobial chemicals as one of their key strategies for biological control (Hussain et al. 2014). These substances can stop the growth and development of plant pathogens such bacteria, fungi, and viruses. Endophytic fungi generate a variety of secondary metabolites, including alkaloids, terpenoids, phenolics, and peptides, all of which have antibacterial characteristics. These substances work by altering pathogens' biological processes, limiting their development, and even causing their death. Endophytic fungi produce antimicrobial chemicals, which assist to restrict plant pathogen growth and minimize disease severity.

18.3.1.2 Parasitism

A fungal parasite is a fungus that feeds on another fungus. Mycoparasitism results from either necrotrophic or biotrophic interactions. Necrotrophic parasites reside on their host's dead cells, while biotrophic parasites obtain nourishment from a living host (Peixoto Neto et al. 2002). The parasitic process involves direct attachment, hyphal puncture, agglomeration around the hyphae, and the production of aberrant hyphae. For instance, the fungus *Geotrichum* sp. parasitizes *T. cucumeris* to gain nutrients. Endophytes isolated from *Phragmites australis* exhibit antagonistic behavior toward eight pathogens, owing to their ability to coil around hyphae and break down the fungal cell wall using chitinases and β-1,3-glucanases (Molitor et al. 2011). Figure 18.1 shows the primary mechanism of action of biocontrol agents against plant diseases.

18.3.1.3 Competitions

Endophytic fungi can potentially exercise biological control by competing with diseases for resources and space in plant tissues. Endophytic fungus, by inhabiting the same ecological niche as infections, can limit the availability of key nutrients such as carbon, nitrogen, and minerals that pathogens require for development and survival (Collinge et al. 2019). This rivalry can lower pathogen proliferation and pathogenicity, resulting in disease suppression. Endophytic fungus can also physically occupy space within plant tissues, preventing diseases from colonizing and establishing themselves. Endophytic fungi's extensive colonization can form a physical barrier that hinders pathogen mobility and dissemination, restricting their capacity to infect and cause disease in host plants. According to Acosta-Martínez et al. (2008), microbial competition is a significant component in determining plant tissue inhabitation and likely how endophytes prevent diseases from colonizing a plant. Within the plant body, endophytes compete with pathogens for resources and space, thus preventing their growth. In contrast to pathogens harbored inside plant tissue, endophytes develop and exploit nutrients more quickly (Rodríguez et al. 2016). Complementary to the pathogen's invasion, endophytic colonization of intracellular plant components is necessary for phytopathological control mechanisms such as competitive exclusion. For instance, *Phytophthora* sp.'s effects on plant leaves were lessened when a sterile seed was treated with an endophytic fungus that was isolated from a cocoa tree (Alencar et al. 2020).

18.3.1.4 Induced Resistance

Endophytic fungi can also induce systemic resistance in the host plant, which is an essential strategy of biological control. When endophytic fungi penetrate plant tissues, they can cause a variety of biochemical and physiological changes, activating the plant's defensive systems. This induced systemic resistance (ISR) improves the plant's capacity to protect against a variety of diseases (Fontana et al. 2021). Endophytic fungi can induce the synthesis of a variety of defense-related chemicals in the host plant, including phytohormones, antimicrobial peptides, and pathogenesis proteins. These chemicals function as signaling molecules, alerting the plant to the presence of pathogens and priming its defenses. As a result, the plant becomes more resistant to disease attacks and can develop a quicker and stronger response.

To boost their defenses against adverse conditions, plants may synthesize substances that affect phytopathogens or alter their structure when endophytic fungi are present. Increased synthesis of phytoalexins and PR proteins, thicker cuticles due to lignin and glucan deposition, and reinforced cell walls are some ways endophytes induce resistance, potentially preventing pathogens from entering and proliferating in the host plant (Tian et al. 2008). The mechanisms by which *S. indica* confers plant resistance to barley powdery mildew were documented (Kavroulakis et al. 2007). The scientists suggested that physiological reactions, such as an increase in local necrosis and barley cilia production, as well as a rise in the expression of the PR gene HvPR17b in the leaves, could be the source of *S. indica*'s resistance to powdery mildew. To begin with, physical contact with the pathogen must be established to validate the efficacy of an endophyte's antimicrobial component against infection. It is difficult to confirm this result, though, because endophytic fungi are embedded in the endosphere of the plant and measuring the rate of metabolite synthesis may be difficult. On the other hand, organic compounds created can easily migrate to the infection site, and plant endophyte metabolites may be translocated through the microbe to the root of these diseases within the plant (Martínez-Medina et al. 2013). Table 18.3 lists some endophytic fungi that induce resistance against plant pathogenic fungi. By understanding how endophytic fungi induce resistance in plants, scientists can develop new strategies for sustainable agriculture that reduce reliance on chemical pesticides. This research is ongoing, but it holds promise for a future where crops are better protected from disease and stress in an eco-friendlier way.

18.3.2 Indirect Effects on Plant Health

In addition to their direct antagonistic and competitive actions, endophytic fungi can indirectly improve plant health and disease resistance. These fungi can help

Table 18.3 Some examples of induced resistance via endophytic fungi

Sl.no.	Endophytic fungi	Hosts	Pathogen targets	References
1.	*Phialomyces macrosporus*	*Coffea arabica*	*Colletotrichum gloeosporioides*	Silva et al. (2011)
2.	*Myrothecium leucotrichum*	*Solanum lycopersicum*	*Alternaria solani*	Ahmed et al. (2000)
3.	*Neotyphodium lolii*	*Lolium perenne*	*Alternaria alternata*	Dastogeer and Wylie (2017)
4.	*Fusarium solani*	*Solanum lycopersicum*	*Fusarium oxysporum*	Zheng et al. (2017)
5.	*Trichoderma harzianum*	*Solanum lycopersicum*	*Alternaria solani*	Fouda et al. (2015)
6.	*Trichoderma harzianum*	*Cucumis sativus*	*Colletotrichum lagenarium*	Spagnoletti et al. (2017)
7.	*Trichoderma virens*	*Solanum lycopersicum*	*Fusarium oxysporum*	Radić and Štrukelj (2012)
8.	*Trichoderma harzianum*	*Capsicum annum*	*Phytophthora capsici*	Silva et al. (2011)

plants absorb and utilize nutrients, resulting in better growth and development. Endophytic fungus can improve the plant's general vigor and resilience by boosting nutrient availability and efficiency, making it less vulnerable to disease. Furthermore, endophytic fungus can influence the physiological processes of plants, such as photosynthesis, water relations, and hormone communication, all of which are critical for plant health and defense (Dastogeer 2018). These fungi can control the plant's hormonal balance by increasing the production of growth hormones and decreasing the production of stress hormones. This hormonal regulation can increase the plant's tolerance to environmental challenges and its capacity to survive pathogen attack (Waqas et al. 2012). Finally, endophytic fungi use a number of techniques to maintain biological control and protect plants against disease. Their direct antagonistic actions, development of systemic resistance, competition for resources, and indirect impacts on plant health all contribute to their biocontrol ability. Understanding these pathways is critical for maximizing the benefits of endophytic fungi in sustainable agriculture and disease management.

18.3.2.1 Stress Tolerance

Endophytic fungi alter certain physiological and biochemical characteristics of drought-tolerant plants, according to research. These alterations include stimulating gaseous exchange, chlorophyll, and phytohormones, thereby accelerating photosynthesis and plant growth, accumulating osmolytes to maintain osmotic balance, reducing ROS production through metabolic activity, supplying water to plants under drought stress conditions, and promoting the production of IAA and GA (Ahmed et al. 2000). Saline stress enhances the biosynthesis of ABA and JA while decreasing the biosynthesis of GAs. The stress mechanism activates the DELLA protein through a XERICO-dependent pathway, which in turn suppresses plant

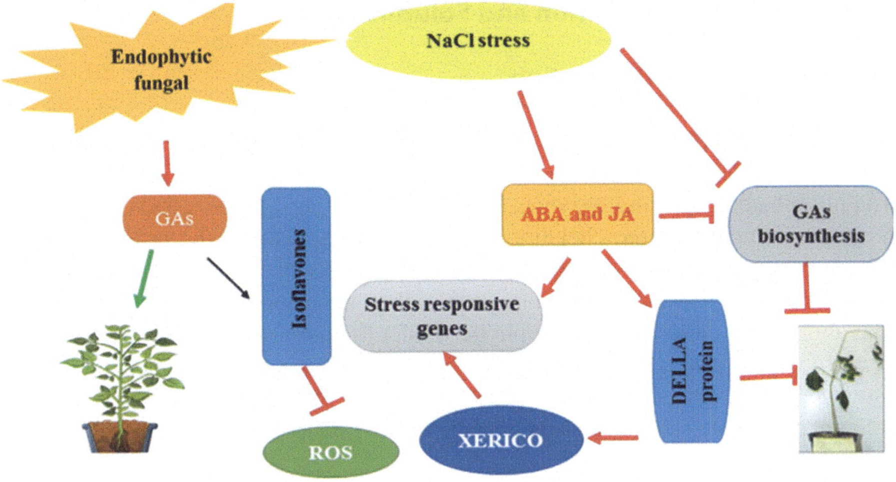

Fig. 18.2 Mechanism of stress tolerance induced by endophytic fungi

growth via a biochemical process. Endophytic fungal inoculation increases GA production, reduces ABA and JA levels, and thus promotes plant growth in stressful environments. The mechanism is explained in Fig. 18.2.

18.3.2.2 Plant Growth Promotion by Endophytic Fungi

Endophytic fungi have a function in both disease control and plant growth and development. These fungi may create plant growth-promoting chemicals such as auxins, cytokinins, and gibberellins, which promote plant growth, root formation, and nutrient uptake. Endophytic fungi indirectly help in disease management by boosting the plant's ability to tolerate pathogen assaults. Many endophytic fungi secrete phytohormones, which promote plant growth, root development, and overall vigor (Silva et al. 2011). Fouda et al. identified three different fungal endophytes from the medicinal plant *Asclepias sinaica*: *Penicillium chrysogenum* Pc_25, *Alternaria alternata* Aa_27, and sterile hyphae Sh_26. They demonstrated that these fungal isolates improved plant development indices by generating ammonia and IAA (Ahmed et al. 2000).

Furthermore, endophytic fungus can help plants withstand abiotic conditions such drought, salt, and heavy metal toxicity. They assist plants in coping with these stress circumstances by creating stress-related proteins and antioxidants that protect plant cells from harm. This improved stress tolerance promotes plant health and minimizes susceptibility to illnesses caused by stress-induced weakening of the plant's defensive mechanisms. Through a variety of ways, endophytic fungi can make a substantial contribution to promoting plant growth. With their host plants, these fungi establish mutualistic relationships that improve the plant's ability to absorb nutrients, withstand environmental stressors, and grow and develop.

18.3.3 Nutrient Acquisition and Solubilization

18.3.3.1 Mineral Solubilization

Endophytic fungi can improve nutrient uptake by solubilizing minerals and increasing their accessibility to the host plant. Phosphate can be dissolved by endophytes through the following methods: utilizing substrates such as calcium, aluminum, and iron phosphate; producing enzymes like phytase and phosphatase; and engaging in processes of acidification, chelation, ion exchange, as well as the production and release of organic acids into the soil. These actions dissolve phosphate complexes, converting them into orthophosphate that is then absorbed and utilized by plants. The ability of a dark-septate endophytic (DSE) fungus, isolated from the roots of *T. aestivum*, *Panicum coloratum*, and *Chloris gayana*, to solubilize phosphates was assessed. The results showed that *Drechslera* sp. had the highest solubility level in calcium phosphate (Dastogeer and Wylie 2017).

18.3.3.2 Siderophore Synthesis

Extracellular siderophore production is a defining property of many microorganisms associated with plants, including endophytic fungi. In the foliar endophytic fungi of Scots pine *Pinus sylvestris* L. and Labrador tea *Rhododendron tomentosum*, *Harmaja ferrirubin*, a PGP siderophore, is produced by the endophyte *Talaromyces pinophilus* in strawberry trees (Kim and Vujanovic 2016). Each of the approximately 300,000 known plant species on Earth hosts a unique community of endophytic microbes. Woody plants may harbor diverse species that are worth studying (Zheng et al. 2017). To create new products, it is necessary to clarify and determine the structures of the most active metabolites. Since the molecules in a crude extract work in concert with other substances produced by the microbe, it is important to remember that individual components of an extract often do not display meaningful microbial activity (Fouda et al. 2015). Endophytic fungi are a potential tool for introduction into plant systems once their mechanisms of action and interactions in plant protection are well understood (Spagnoletti et al. 2017). This can be achieved through direct biological control or by inducing resistance and tolerance to abiotic stressors (Spagnoletti et al. 2021).

18.4 Challenges and Limitations of Endophytic Fungi in Disease Control

While the use of endophytic fungi in the biological control of invasive species shows great promise, there are still challenges that need to be addressed. One of the main challenges is the identification and isolation of effective endophytic fungi. The vast diversity of endophytic fungi makes it challenging to identify those with biocontrol potential. Furthermore, the screening and testing of these fungi for their efficacy and safety require extensive research and resources (Bamisile et al. 2021). Another challenge is the formulation and delivery of endophytic fungi-based biocontrol agents. Developing effective formulations that ensure the survival and establishment of endophytic fungi on the target invasive species is crucial for successful biocontrol. Additionally, the delivery methods need to be optimized to ensure efficient and widespread application (Qaderi and Safaie 2023).

In the future, research efforts should focus on expanding our understanding of the interactions between endophytic fungi, invasive species, and host plants. This knowledge will help in the selection and development of more effective biocontrol strategies. Furthermore, the exploration of novel endophytic fungi from diverse ecosystems can provide a broader range of biocontrol agents for different invasive species (Caruso et al. 2022). While endophytic fungi have shown great potential in disease control, there are several challenges and limitations that need to be addressed to fully harness their benefits. These challenges arise from various factors, including the complexity of the plant-fungus interactions, environmental conditions, and practical considerations in the field. Understanding and overcoming these challenges is crucial for the successful implementation of endophytic fungi as biocontrol agents. In this section, we will explore some of the key challenges and limitations associated with the use of endophytic fungi in disease control.

18.4.1 Complexity of Plant-Fungus Interactions

One of the major challenges in utilizing endophytic fungi for disease control is the complexity of the interactions between the fungi, the host plants, and the pathogens. Endophytic fungi can have both positive and negative effects on plant health, depending on various factors such as the specific fungal species, the plant species, and the environmental conditions. While some endophytic fungi have been found to enhance plant growth and induce resistance against pathogens, others may have no effect or even exacerbate disease symptoms (Mattoo and Nonzom 2021). Understanding the mechanisms underlying these interactions is crucial for effectively utilizing endophytic fungi in disease control. However, the complexity of these interactions makes it challenging to predict the outcomes of introducing specific endophytic fungi into different plant-pathogen systems (Latz et al. 2018). Further research is needed to elucidate the specific mechanisms by which endophytic fungi interact with host plants and pathogens and to identify the factors that influence the outcomes of these interactions.

18.4.2 Environmental Factors

Another challenge in the use of endophytic fungi for disease control is the influence of environmental factors on their efficacy. Environmental conditions such as temperature, humidity, and nutrient availability can significantly impact the growth and activity of endophytic fungi. Some endophytic fungi may thrive under certain environmental conditions but struggle to establish and persist in different environments (Verma et al. 2014). For example, the efficacy of endophytic fungi in disease control may be reduced under conditions of high pathogen pressure or in environments with extreme temperatures or drought. Additionally, the presence of other microorganisms in the plant microbiome can also influence the interactions between endophytic fungi, host plants, and pathogens. Understanding the environmental factors that influence the establishment and activity of endophytic fungi is essential for optimizing their use in disease control strategies.

18.4.3 Practical Considerations

In addition to the biological challenges, there are also practical considerations that need to be addressed when using endophytic fungi for disease control. One of the main challenges is the development of effective and practical delivery methods for introducing endophytic fungi into plants. While various methods such as seed treatment, foliar application, and soil drenching have been explored, each method has its own limitations and may not be suitable for all plant species or disease scenarios (Branine et al. 2019). Furthermore, the production and formulation of endophytic fungi-based products can be challenging and costly (Qaderi and Safaie 2023). The mass production of viable and effective endophytic fungi requires optimized cultivation techniques and quality control measures. The formulation of endophytic fungi into stable and easy-to-use products also presents technical challenges. These practical considerations need to be addressed to ensure the widespread adoption and commercialization of endophytic fungi-based biocontrol products.

18.4.4 Regulation and Safety Concerns

The regulatory framework surrounding the use of endophytic fungi in disease control is another important consideration. Before endophytic fungi-based products can be commercialized and widely used, they need to undergo rigorous testing and evaluation to ensure their safety for humans, animals, and the environment. Regulatory agencies require extensive data on the efficacy, stability, and potential risks associated with the use of these products. Meeting these regulatory requirements can be time-consuming and costly, particularly for small-scale research projects or companies. Additionally, the lack of standardized protocols for testing and evaluating endophytic fungi-based products can further complicate the regulatory process. Addressing these regulatory and safety concerns is essential for the successful commercialization and adoption of endophytic fungi in disease control.

18.4.5 Lack of Knowledge and Awareness

Finally, a significant challenge in the use of endophytic fungi for disease control is the lack of knowledge and awareness among farmers, agronomists, and other stakeholders. Many farmers may not be aware of the potential benefits of endophytic fungi or may have misconceptions about their efficacy and safety. This lack of knowledge and awareness can hinder the adoption of endophytic fungi-based biocontrol strategies in agriculture (Rana et al. 2019).

Educational initiatives and outreach programs are needed to increase awareness and understanding of the benefits and limitations of endophytic fungi in disease control. Providing farmers and other stakeholders with accurate and accessible information about the use of endophytic fungi can help promote their adoption and integration into sustainable disease management practices.

In conclusion, while endophytic fungi hold great promise as biocontrol agents for plant diseases, there are several challenges and limitations that need to be addressed. The complexity of plant-fungus interactions, environmental factors, practical considerations, regulatory concerns, and lack of knowledge and awareness all pose challenges to the effective use of endophytic fungi in disease control. Overcoming these challenges will require further research, technological advancements, and collaborative efforts among researchers, industry, and regulatory agencies. By addressing these challenges, we can unlock the full potential of endophytic fungi in sustainable disease management and contribute to the development of more environmentally friendly and economically viable agricultural practices.

18.5 Recommendations for Future Research

As the field of endophytic fungi research continues to expand, there are several areas that warrant further investigation. These recommendations for future research aim to enhance our understanding of the biological control mechanisms of endophytic fungi and explore their potential applications in various fields.

18.5.1 Exploration of Endophytic Fungal Diversity

One of the key areas for future research is the exploration of endophytic fungal diversity. Although significant progress has been made in identifying and characterizing various endophytic fungi, there is still a vast untapped reservoir of fungal species that remain unexplored (Sun and Guo 2012). By conducting comprehensive surveys and sampling from diverse ecosystems, researchers can uncover novel endophytic fungi with unique biological properties. This exploration will not only contribute to our understanding of fungal diversity but also provide valuable resources for potential biocontrol agents.

18.5.2 Elucidation of Mechanisms of Action

While the mechanisms of action of endophytic fungi in biological control have been partially elucidated, there is still much to learn. Future research should focus on unraveling the intricate interactions between endophytic fungi and their host plants. This includes studying the signaling pathways involved in the induction of plant defense responses, the production of secondary metabolites by endophytic fungi, and the mechanisms by which these metabolites suppress pathogenic organisms (Tiwari and Bae 2022). By gaining a deeper understanding of these mechanisms, researchers can develop more targeted and effective strategies for disease control.

18.5.3 Optimization of Application Methods

To fully harness the potential of endophytic fungi in biological control, it is essential to optimize the application methods. Future research should focus on developing efficient and practical techniques for the inoculation of endophytic fungi into host plants. This includes exploring different delivery systems such as seed treatments, foliar sprays, and root drenches (Branine et al. 2019). Additionally, researchers should investigate the factors that influence the establishment and persistence of endophytic fungi within plant tissues, such as the timing and frequency of application, environmental conditions, and host plant compatibility. By optimizing these application methods, the efficacy of endophytic fungi as biocontrol agents can be maximized.

18.5.4 Integration with Other Control Strategies

Another area of future research is the integration of endophytic fungi with other control strategies. While endophytic fungi have shown promise in disease suppression, their effectiveness can be further enhanced by combining them with other biological control agents, such as beneficial bacteria or mycorrhizal fungi (Parveen and Rashtrapal 2024). Additionally, researchers should explore the potential synergistic effects of combining endophytic fungi with chemical fungicides or cultural practices. By integrating multiple control strategies, a holistic and sustainable approach to disease management can be achieved.

18.5.5 Field Trials and Scale-up Studies

To validate the efficacy of endophytic fungi in real-world conditions, more field trials and scale-up studies are needed (Ganeshan et al. 2021). These studies should evaluate the performance of endophytic fungi in different agricultural systems, including various crops, soil types, and climatic conditions. By conducting long-term field trials, researchers can assess the persistence and stability of endophytic fungi populations over time and determine their impact on disease suppression and crop productivity. Furthermore, scale-up studies can provide valuable insights into the feasibility and practicality of large-scale production and application of endophytic fungi-based products.

18.5.6 Ecological Impacts and Risk Assessment

As with any biological control agent, it is crucial to assess the ecological impacts and potential risks associated with the use of endophytic fungi. Future research should focus on evaluating the nontarget effects of endophytic fungi on beneficial organisms, such as pollinators and natural enemies of pests (van den Brandhof and Wösten 2022). Additionally, risk assessment studies should be conducted to determine the potential for endophytic fungi to become invasive or cause unintended ecological disruptions. By understanding the ecological impacts and risks, appropriate management strategies can be developed to ensure the safe and sustainable use of endophytic fungi in biological control.

18.5.7 Exploration of Nonagricultural Applications

While the majority of research on endophytic fungi has focused on their agricultural applications, there is potential for their use in other fields. Future research should explore the nonagricultural applications of endophytic fungi, such as their role in environmental remediation, biofuel production, and pharmaceutical development. By expanding the scope of research, we can uncover new and innovative uses for endophytic fungi, further highlighting their potential as versatile bioresources (Zhang and White 2021).

In conclusion, the field of endophytic fungi research holds great promise for the development of sustainable and environmentally friendly approaches to disease control. By focusing on the recommendations outlined above, researchers can deepen our understanding of the biological control mechanisms of endophytic fungi and unlock their full potential in various applications. Through continued research and innovation, we can harness the power of endophytic fungi to address the challenges of plant diseases and contribute to a more sustainable future.

18.6 Conclusion

In conclusion, endophytic fungi play a vital role in biological control by employing various mechanisms to suppress plant pathogens and enhance plant health. Through the production of secondary metabolites, competition for resources, enzyme production, and facilitation of nutrient uptake, these fungi contribute to disease suppression and promote plant growth and development. Their ability to induce systemic resistance and enhance plant systemic defense mechanisms further strengthens their role as effective biocontrol agents. Harnessing the potential of endophytic fungi in biological control offers promising prospects for sustainable agriculture and the reduction of chemical pesticide use. Endophytic fungi can trigger a variety of plant defense systems against both biotic and abiotic stressors. These fungi successfully carry out biological control against plant diseases due to their genetic diversity and host plant interactions. They could also be used for bioprospecting new chemicals and plant genetic manipulation. Tolerance to abiotic stresses can be achieved by combining endophytes with target crops. This method shows potential and allows plants to grow in areas where they would not normally thrive. Endophytes produce secondary metabolites with high biological activity, which have the potential to become valuable products. Identifying and characterizing endophytic bacteria from unstudied plants could aid in discovering new species capable of producing chemicals of interest. These chemicals may be used as elicitors in induced resistance, for tolerance induction to abiotic stresses and biological disease prevention.

Acknowledgments The Department of Science and Technology, New Delhi (DST-SHRI-DST/TDT/SHRI/2022/70), Karpagam Academy of Higher Education, and DST-FIST (SR/FST/LS-1/2018/187) are acknowledged by the authors for providing the infrastructure facilities for this research study.

Conflict of Interest There is no conflict of interest, according to the authors.

References

Aamir M, Rai KK, Zehra A, Kumar S, Yadav M, Shukla V, Upadhyay RS (2020) Fungal endophytes: classification, diversity, ecological role, and their relevance in sustainable agriculture. In: Microbial endophytes. Woodhead Publishing, pp 291–323. https://doi.org/10.1016/b978-0-12-818734-0.00012-7

Acosta-Martínez, V., Acosta-Mercado, D. I. M. A. R. I. S., Sotomayor-Ramírez, D., & Cruz-Rodríguez, L. (2008). Microbial communities and enzymatic activities under different management in semiarid soils. Applied Soil Ecology, 38(3), 249–260.

Adeleke BS, Babalola OO, Glick BR (2021) Plant growth-promoting root-colonizing bacterial endophytes. Rhizosphere 20:100433

Ahmed AS, Sánchez CP, Candela ME (2000) Evaluation of induction of systemic resistance in pepper plants (*Capsicum annuum*) to Phytophthora Capsici using *Trichoderma harzianum* and its relation with Capsidiol accumulation. Eur J Plant Pathol 106:817–824. https://doi.org/10.1023/A:1008780022925

Alam B, Lǐ J, Gě Q, Khan MA, Gōng J, Mehmood S, Gŏng W (2021) Endophytic fungi: from symbiosis to secondary metabolite communications or vice versa? Front Plant Sci 12:3060

Alencar MSR, Solino AJDS, Oliveira JSB, Pascholati SF, Schwan-Estrada KRF (2020) Induction of defense mechanisms in tomato plants by saprobic fungi filtrates against early blight diseases. Rev Caatinga 33:671–678

Arnold AE, Mejía LC, Kyllo D, Rojas EI, Maynard Z, Robbins N, Herre EA (2003) Fungal endophytes limit pathogen damage in a tropical tree. Proc Natl Acad Sci 100:15649–15654

Bamisile BS, Siddiqui JA, Akutse KS, Ramos Aguila LC, Xu Y (2021) General limitations to endophytic entomopathogenic fungi use as plant growth promoters, pests and pathogens biocontrol agents. Plan Theory 10(10):2119

Branine M, Bazzicalupo A, Branco S (2019) Biology and applications of endophytic insect-pathogenic fungi. PLoS Pathog 15(7):e1007831

Caruso DJ, Palombo EA, Moulton SE, Zaferanloo B (2022) Exploring the promise of endophytic fungi: a review of novel antimicrobial compounds. Microorganisms 10(10):1990

Collinge DB, Jørgensen HJ, Latz MA, Manzotti A, Ntana F, Rojas EC, Jensen B (2019) Searching for novel fungal biological control agents for plant disease control among endophytes. In: Endophytes for a growing world, vol 31. Cambridge University Press, p 25

Damm U, Cannon PF, Woudenberg JHC, Crous PW (2012) The Colletotrichum acutatum species complex. Stud Mycol 73:37–113

Dastogeer KM (2018) Influence of fungal endophytes on plant physiology is more pronounced under stress than well-watered conditions: a meta-analysis. Planta 248:1403–1416

Dastogeer KMG, Wylie SJ (2017) Plant–Fungi Association: role of fungal endophytes in improving plant tolerance to water stress. In: Singh D, Singh H, Prabha R (eds) Plant-microbe interactions in agro-ecological perspectives. Springer, Singapore

Fontana DC, de Paula S, Torres AG, de Souza VHM, Pascholati SF, Schmidt D, Dourado Neto D (2021) Endophytic fungi: biological control and induced resistance to phytopathogens and abiotic stresses. Pathogens 10(5):570

Fouda AH, Hassan SED, Eid AM, Ewais EED (2015) Biotechnological applications of fungal endophytes associated with medicinal plant Asclepias sinaica (Bioss.). Ann Agric Sci 60(1):95–104

Ganeshan S, Kim SH, Vujanovic V (2021) Scaling-up production of plant endophytes in bioreactors: concepts, challenges and perspectives. Bioresour Bioprocess 8:1–16

Gao FK, Dai CC, Liu XZ (2010) Mechanisms of fungal endophytes in plant protection against pathogens. Afr J Microbiol Res 4(13):1346–1351

Harman GE, Howell CR, Viterbo A, Chet I, Lorito M (2004) Trichoderma species—opportunistic, avirulent plant symbionts. Nat Rev Microbiol 2(1):43–56

Hawksworth DL (2001) The magnitude of fungal diversity: the 1· 5 million species estimate revisited. Mycol Res 105(12):1422–1432. https://doi.org/10.1017/S0953756201004725

Houbraken J, Frisvad JC, Samson RA (2011) Taxonomy of penicillium section citrina. Stud Mycol 70:53–138

Hussain H, Kliche-Spory C, Al-Harrasi A, Al-Rawahi A, Abbas G, Green IR et al (2014) Antimicrobial constituents from three endophytic fungi. Asian Pac J Trop Med 7:S224–S227

Kavroulakis N, Ntougias S, Zervakis GI, Ehaliotis C, Haralampidis K, Papadopoulou KK (2007) Role of ethylene in the protection of tomato plants against soil-borne fungal pathogens conferred by an Endophytic Fusarium Solani strain. J Exp Bot 58:3853–3864. https://doi.org/10.1093/jxb/erm230

Khunnamwong P, Lertwattanasakul N, Jindamorakot S, Suwannarach N, Matsui K, Limtong S (2020) Evaluation of antagonistic activity and mechanisms of endophytic yeasts against pathogenic fungi causing economic crop diseases. Folia Microbiol 65:573–590

Kim SH, Vujanovic V (2016) Relationship between mycoparasites lifestyles and biocontrol behaviors against Fusarium spp. and mycotoxins production. Appl Microbiol Biotechnol 100:5257–5272. https://doi.org/10.1007/s00253-016-7539-z

Latz MA, Jensen B, Collinge DB, Jørgensen HJ (2018) Endophytic fungi as biocontrol agents: elucidating mechanisms in disease suppression. Plant Ecol Diver 11(5–6):555–567

Martínez-Medina A, Fernández I, Sánchez-Guzmán MJ, Jung SC, Pascual JA, Pozo MJ (2013) Deciphering the hormonal Signalling network behind the systemic resistance induced by *Trichoderma harzianum* in tomato. Front Plant Sci 4. https://doi.org/10.3389/fpls.2013.00206

Martinuz A, Schouten A, Sikora R (2012) Systemically induced resistance and microbial competitive exclusion: implications on biological control. Phytopathology 102:260–266. https://doi.org/10.1094/PHYTO-04-11-0120

Mattoo AJ, Nonzom S (2021) Endophytic fungi: understanding complex cross-talks. Symbiosis 83(3):237–264

Molitor A, Zajic D, Voll LM, Pons-Kühnemann J, Samans B, Kogel KH, Waller F (2011) Barley leaf transcriptome and metabolite analysis reveals new aspects of compatibility and Piriformospora indica–mediated systemic induced resistance to powdery mildew. Mol Plant Microbe Interact 24:1427–1439

Omomowo IO, Amao JA, Abubakar A, Ogundola AF, Ezediuno LO, Bamigboye CO (2023) A review on the trends of endophytic fungi bioactivities. Sci Afr 20:e01594

Palmieri D, Ianiri G, Del Grosso C, Barone G, De Curtis F, Castoria R, Lima G (2022) Advances and perspectives in the use of biocontrol agents against fungal plant diseases. Horticulturae 8(7):577

Parveen SS, Rashtrapal PS (2024) Integrated Pest management strategies using Endophytic Entomopathogenic fungi. Plant Sci Today 11(1):568–574

Peixoto Neto PAS, Azevedo JL, Araújo WL (2002) Microrganismos Endofíticos: Interação Com as Plantas e Potencial Biotecnológico. Biotecnol Ciênc Desenvolv 29:62–76

Qaderi S, Safaie N (2023) Endophytic fungi-based wettable powder formulation for Rhizoctonia solani AG-4 biocontrol. Plant Soil 488:1–17

Radić N, Štrukelj B (2012) Endophytic fungi—the treasure chest of antibacterial substances. Phytomedicine 19:1270–1284

Rana KL, Kour D, Sheikh I, Dhiman A, Yadav N, Yadav AN et al (2019) Endophytic fungi: biodiversity, ecological significance, and potential industrial applications. In: *Recent advancement in white biotechnology through fungi: volume 1: diversity and enzymes perspectives*, pp 1–62

Rodriguez RJ, White JF Jr, Arnold AE, Redman RS (2009) Fungal endophytes: diversity and functional roles. New Phytol 182:314–330. https://doi.org/10.1111/j.1469-8137.2009.02773

Rodríguez GAA, de Abreu MS, Pinto FAMF, Monteiro ACA, Núñez ÁMP, de Resende MLV, de Souza JT, de Medeiros FHV (2016) Phialomyces Macrosporus decreases anthracnose severity on coffee seedlings by competition for nutrients and induced resistance. Biol Control 103:119–128

Schroers HJ, O'Donnell K, Lamprecht SC, Kammeyer PL, Johnson S, Sutton DA et al (2009) Taxonomy and phylogeny of the Fusarium dimerum species group. Mycologia 101(1):44–70

Segaran G, Sathiavelu M (2019) Fungal endophytes: a potent biocontrol agent and a bioactive metabolites reservoir. Biocatal Agric Biotechnol 21:101284. https://doi.org/10.1016/j.bcab.2019.101284

Silva VND, Guzzo SD, Lucon CMM, Harakava R (2011) Promoção de crescimento e indução de resistência à antracnose por Trichoderma spp. em pepineiro. Pesq Agrop Brasileira 46:1609–1618

Spagnoletti FN, Tobar NE, Fernández Di Pardo A, Chiocchio VM, Lavado RS (2017) Dark septate endophytes present different potential to solubilize calcium, iron and aluminum phosphates. Appl Soil Ecol 111:25–32. https://doi.org/10.1016/j.apsoil.2016.11.010. ISSN 0929-1393

Spagnoletti FN, Tobar NE, Fernández Di Pardo A, Chiocchio VM, Lavado RS, Fontana DC, de Paula S, Torres AG, de Souza VHM, Pascholati SF, Schmidt D, Dourado Neto D (2021) Endophytic fungi: biological control and induced resistance to phytopathogens and abiotic stresses. Pathogens 10(5):570

Suebrasri T, Harada H, Jogloy S, Ekprasert J, Boonlue S (2020) Auxin-producing fungal endophytes promote growth of sunchoke. Rhizosphere 16:100271

Sun X, Guo LD (2012) Endophytic fungal diversity: review of traditional and molecular techniques. Mycology 3(1):65–76

Sun J, Guo L, Zang W, Ping W, Chi D (2008) Diversity and ecological distribution of endophytic fungi associated with medicinal plants. Sci China Ser C Life Sci 51(8):751–759

Tian P, Nan Z, Li C, Spangenberg G (2008) Effect of the endophyte Neotyphodium Lolii on susceptibility and host physiological response of perennial ryegrass to fungal pathogens. Eur J Plant Pathol 122:593–602

Ting A, Mah S, Tee C (2010) Identification of volatile metabolites from fungal endophytes with biocontrol potential towards Fusarium oxysporum F. sp. cubense race 4. Am J Agric Biol Sci 5(2):177–182

Tiwari P, Bae H (2022) Endophytic fungi: key insights, emerging prospects, and challenges in natural product drug discovery. Microorganisms 10(2):360

U'Ren JM, Lutzoni F, Miadlikowska J, Zimmerman NB, Carbone I, May G, Arnold AE (2019) Host availability drives distributions of fungal endophytes in the imperilled boreal realm. Nat Ecol Evol 3(10):1430–1437

van den Brandhof JG, Wösten HA (2022) Risk assessment of fungal materials. Fungal Biol Biotechnol 9(1):1–20

Verma SK, Gond SK, Mishra A, Sharma VK, Kumar J, Singh DK et al (2014) Impact of environmental variables on the isolation, diversity and antibacterial activity of endophytic fungal communities from Madhuca indica Gmel. at different locations in India. Ann Microbiol 64(2):721–734

Waqas M, Khan AL, Kamran M, Hamayun M, Kang SM, Kim YH, Lee IJ (2012) Endophytic fungi produce gibberellins and indoleacetic acid and promotes host-plant growth during stress. Molecules 17(9):10754–10773

Zhang Q, White JF (2021) Bioprospecting desert plants for endophytic and biostimulant microbes: a strategy for enhancing agricultural production in a hotter, drier future. Biology 10(10):961

Zheng YK, Miao CP, Chen HH, Huang FF, Xia YM, Chen YW, Zhao LX (2017) Endophytic fungi harbored in Panax notoginseng: diversity and potential as biological control agents against host plant pathogens of root-rot disease. J Ginseng Res 41(3):353–360

Hidden Allies: Exploring the Cutting-Edge Insights into the Role of Mycorrhizal Fungi in Plant Health

19

Tikka Dewage Chamarika Priyadarshani, Prathiba Aruni Weerasinghe, Simon Law, and Mohamed Cassim Mohamed Zakeel

Abstract

Sustaining and fortifying plant health is critically important to mitigate poverty, eradicate hunger, safeguard the environment, and foster economic development, as underscored by the United Nations' designation of 2020 as the International Year of Plant Health. Among various approaches to protecting plant health, the use of symbiotic microorganisms, particularly mycorrhizal fungi, has gained substantial appraisal. Mycorrhizal fungi play pivotal roles in protecting and enhancing plant health via sustainable and environmentally friendly mechanisms. Recent advances in methodology, employing cutting-edge technologies, have meticulously dissected the intricate interplay between plants and mycorrhizal fungi, shedding light on the concealed allies within this symbiotic relationship which contributes to overall plant health. The exploration of these hidden allies and their associations with plants not only expands our understanding of the complex mechanisms at play but also paves the way for innovative and effective strategies to harness the benefits of mycorrhizal fungi in agriculture and environmental conservation.

Keywords

AM fungi · Ecosystem resilience · Mycorrhizal symbiosis · Plant-microbe interaction · Symbiotic relationship

T. D. C. Priyadarshani · P. A. Weerasinghe
Department of Plant Sciences, Faculty of Agriculture, Rajarata University of Sri Lanka, Puliyankulama, Anuradhapura, Sri Lanka

S. Law · M. C. M. Zakeel (✉)
CSIRO Agriculture and Food, Black Mountain Science and Innovation Park, Acton, ACT, Australia
e-mail: mohamedzakeel.mohamedcassim@csiro.au

© The Author(s), under exclusive license to Springer Nature Singapore Pte Ltd. 2025

K. A. Abd-Elsalam, A. H. Hashem (eds.), *Fungal Endophytes Volume II*,
https://doi.org/10.1007/978-981-97-8804-0_19

19.1 Introduction

Mycorrhizal fungi have long been recognized as essential contributors to plant health. The term "mycorrhiza" generally refers to the mutually beneficial relationship established between plant roots and fungi colonizing these roots (Bücking et al. 2012). This association creates an interaction that enables the plant to supply the fungus with simple carbohydrates produced through photosynthesis, while the fungi reciprocate by providing scarce mineral nutrients and water to the plant. While generally regarded as a mutualism, the association is in fact dynamic and can span a symbiotic spectrum from obligate mutualism to parasitism, as observed in some orchids (Rasmussen and Rasmussen 2009).

Mycorrhizal fungi play integral roles within ecosystems, allowing plants to thrive in challenging terrestrial environments and contributing to global carbon (C) and nutrient cycling. The vast majority of mycorrhizal fungi engage in obligate symbiotic relationships with plants. Over 90% of all terrestrial plants, including trees, grasses, and numerous agriculturally significant crops, can form mycorrhizal associations (Bianciotto et al. 1996). The capacity to form mycorrhiza has independently evolved multiple times (Bonfante and Genre 2010). Mycorrhizal associations are also observed between the underground organs of gametophytes in many bryophytes and pteridophytes and the sporophytes of most pteridophytes (Smith and Read 2008).

In addition to their crucial role in natural ecosystems, mycorrhizae hold significant potential to improve plant and soil health, contributing to sustainable agricultural production. From a nutritional standpoint, they can expand the effective reach of root systems, enhance mineral nutrient uptake, and increase water absorption and water use efficiency (Huey et al. 2020). Mycorrhiza can indirectly enhance plant productivity by improving the health of the rhizosphere and surrounding bulk soil, e.g., contributing to metal detoxification, stabilizing soil aggregation, minimizing soil erosion, and contributing significantly to the accumulation of stable soil C (Sooriyaarachci et al. 2021; Treseder and Allen 2000). Collectively, these mycorrhizal contributions will enable plants to better tolerate the increasing frequency and magnitude of challenges associated with climate change (Sooriyaarachci et al. 2021).

This book chapter aims to provide an overview of the indispensable role that mycorrhizal fungi play in maintaining and enhancing plant health. We will outline the various types and functions of mycorrhizae, followed by an exploration of nutrient uptake and cycling mechanisms within these interactions. We will also emphasize the importance of mycorrhizae in promoting plant growth and suppressing plant diseases, as well as their influence on abiotic stress tolerance and adaptation to climate change. Additionally, we will discuss recent advances in molecular approaches, highlight the potential of plant-mycorrhizal interactions, and outline the future directions of mycorrhiza research.

19.2 Mycorrhizal Fungi: Types, Structure, and Functioning

A variety of mycorrhizal fungi are found in the environment, each exhibiting distinct structural adaptations. These adaptations enable them to fulfill a range of roles in maintaining ecosystem stability and nutrient acquisition.

19.2.1 Overview of Mycorrhizal Associations

Mycorrhizae play a crucial role in supplying essential nutrients to the plants they associate with and are widely utilized as biofertilizers worldwide (Vosátka et al. 2012). Due to their filamentous structure, these fungi can exploit diverse substrates based on their nutritional strategy (Bonfante and Genre 2010). They are particularly abundant in more decomposed litter and humus, acting as strong C sinks that utilize ample amounts of carbohydrates for growth and metabolism. Furthermore, mycorrhizae play a crucial role in the acquisition of water and minerals.

Mycorrhizal fungi make a significant contribution to plant growth by contributing to hormone and nutrient balance, synthesizing and delivering plant growth regulators (e.g., auxins and gibberellin-like substances), facilitating nutrient solubilization, and triggering resistance against plant pathogens (Huey et al. 2020). Mycorrhizal associations regulate plant growth also by influencing various mechanisms, including the modulation of aquaporin genes, ABA-responsive genes, phytohormone biosynthesis pathways, and transcription factors (Begum et al. 2019).

19.2.2 Types of Mycorrhizal Fungi

Mycorrhizae display a range of diverse forms, influenced by both the host plant's taxonomy and the specific fungal species engaged in the symbiotic relationship (Dighton 2009). The occurrence of these different forms within ecosystems is closely linked to the distribution of host plants, as well as the prevailing climate and soil conditions. Traditionally, mycorrhizae are classified into three main types: ectomycorrhizae (ECM), endomycorrhizae, and ectendomycorrhizae (Zakeel 2017). This classification primarily relies on the positioning of the fungal hyphae in relation to the root tissues of the host plant, with "ecto" denoting hyphae residing primarily outside the root and "endo" indicating hyphae inhabiting not only the interior of root tissues but penetrating the root cells themselves.

ECM provide vital functions in nutrient and C cycling, most notably in forest ecosystems (Anderson and Cairney 2007). These fungi develop a hyphal sheath referred to as the "mantle," which surrounds the root system of the host plant (Fig. 19.1a). These hyphae penetrate the outer layers of the root cortex, establishing a structure known as the Hartig net and extending toward the endodermis of the host plant (Zakeel 2017). ECM fungi are primarily classified within the phylum *Basidiomycota*, with a lesser presence in the *Ascomycota*.

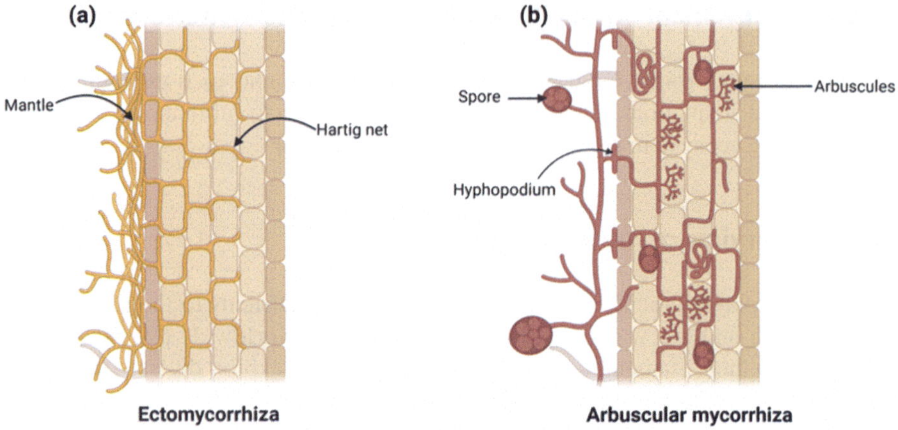

Fig. 19.1 Colonization and structure of (**a**) ectomycorrhizal and (**b**) arbuscular mycorrhizal associations

In contrast, endomycorrhizae instigate the formation of structures entirely contained within the host root without forming external sheaths or mantles (Zakeel 2017). They can be further categorized into three major groups: Arbuscular mycorrhizas (AM), Ericoid endomycorrhizas (EM), and Orchidaceous endomycorrhizas (Moore 2016). AM are primarily associated with a wide range of angiosperms, including tropical forest trees, agricultural crops, and most herbs and grasses (Zakeel 2017). These mycorrhizas form specialized structures called vesicles and arbuscles within root cortical cells, greatly increasing the surface area of the interface separating the symbiotic partners. Ericoid endomycorrhizas, on the other hand, primarily associate with plants in the Ericaceae, Epacridaceae, and Empetraceae families. Initially, the fungi grow on the surface of root hairs and form loose hyphal networks, later penetrating the cortical cell walls of the host to form coils that facilitate nutrient transfer (Wei et al. 2022). Daghino et al. (2022) revealed that Ericoid endomycorrhizas of the *Hyaloscypha* genus can establish symbiotic associations not only with the roots but also with other parts of *Vaccinium myrtillus*, such as stems, leaves, and flowers. This finding highlights the versatility and widespread presence of the symbiosis between Ericoid endomycorrhizas and this plant species.

Orchidaceous endomycorrhizas establish symbiotic relationships with orchids, particularly in their roots and seeds. These mycorrhizas are vital for orchid seed germination as the seeds have minimal energy reserves and rely on C provided by the fungal symbiont. Orchidaceous endomycorrhizas are mainly associated with Basidiomycetes, with the genus *Rhizoctonia* being the most common (Moore 2016). Similar types of mycorrhizal symbiosis, in which mycorrhizae do not derive any advantages from the plants they associate with, are also known as mycoheterotrophic, epiparasitic, or cheating associations (Furman and Trappe 1971; Taylor and Bruns 1999).

Ectendomycorrhizae exhibit characteristics that encompass both endomycorrhizas and ECM. This type of fungal association is predominantly found in the genera

Pinus and, to a lesser extent, *Larix* (Moore 2016). The symbiotic relationship is characterized by the distinct combination of a Hartig net, fungal mantle, and intracellular hyphae, with the latter emerging shortly after the formation of the former (Tu et al. 1996). Through the analysis of the nuclear and mitochondrial genomes of the fungi, it has been confirmed that the majority of ectendomycorrhizae belong to the genus *Wilcoxina*, specifically the species *Wilcoxina mikaloe* and *W. rehmii* (Moore 2016).

19.2.3 Structure and Colonization of Arbuscular Mycorrhizal Fungi

Arbuscular mycorrhizal fungi consist of threadlike structures called hyphae, which form an external network of mycelium that extends into the soil beyond the roots of the host plant. These obligatory symbionts, such as those in the Glomeromycota phylum, colonize host plant roots through spores, extraradical hyphae, or previously colonized roots (van Aarle and Olsson 2003). AM are characterized by structures including arbuscules, vesicles, spores, extraradical hyphae, intraradical hyphae, and auxiliary cells (Fig. 19.1b; Souza 2015). Intraradical hyphae develop from appressoria during the initiation of the symbiotic phase and form infection units (Ramos et al. 2008; Kiriacheck et al. 2009). Once in the cortical zone, intraradical hyphae may differentiate into vesicles, or arbuscular structures or spores once they reach the cortical zone (Cruz et al. 2008).

Unlike intraradical hyphae, extraradical hyphae develop in the pre-symbiotic phase of colonization and have unlimited growth and can extend well beyond the rhizosphere zone (Kiriacheck et al. 2009). While both intra- and extraradical hyphae share similar morphological characters, extraradical hyphae have a wider cell wall composition (Cruz et al. 2008). Extraradical hyphae serve three functions: infection, absorption, and reproduction (Souza 2015).

Arbuscules and vesicles originate from intraradical hyphae, while spores can be produced by both intra- and extraradical hyphae (Ramos et al. 2008). Some species also form auxiliary cells, which originate from extraradical hyphae (Redecker et al. 2013). Two morphological types of hyphae are observed: Arum-type, which colonizes intercellular spaces, and Paris-type, which colonizes cell interiors. Arbuscules are intricately branched fungal structures enclosed by the host's plasma membrane and function as transient sites for efficient nutrient exchange between the symbiotic partners (Smith and Read 1997). As symbiosis progresses, fungi may form storage structures called vesicles in intercellular spaces of the plant root (Nemec 1981). These structures serve as reservoirs of energy-rich lipids and glycogen, which can be called upon to fuel the growth of the fungus.

The development of mycorrhizal colonization is influenced by various abiotic and biotic factors, including mineral nutrient and organic matter availability, pH, pollutants, soil microorganisms, soil moisture content, and temperature (Souza 2015). The symbiotic process begins with the germination of spores in the "asymbiotic phase," independently of the host plant but relying on favorable abiotic and biotic factors (Souza 2015). Transitioning to the "pre-symbiotic" phase, the fungi initiate contact with the host roots and establish the initial mycelium (Kiriacheck et al. 2009). Finally, the "symbiotic" phase commences, during which AM proliferate along the root surface and penetrate the root in response to plant exudates (Souza 2015; Parniske 2008). When fungal hyphae grow through the root tissues, the root plasmalemma invaginates and proliferates around the fungal intrusion (Parniske 2008). Inside the root cells, the hyphae undergo repeated dichotomous branching, forming structures known as arbuscules (Parniske 2008). These branching structures enable the fungus to efficiently exchange nutrients with the plant (Parniske 2008).

19.2.4 Mechanisms and Processes Involved in Mycorrhizal Symbiosis

An important mechanism of mycorrhizal associations is the enhancement of root hydraulic conductivity, leading to increased water uptake by plants (Fernández-Lizarazo and Moreno-Fonseca 2016). Plants infected with mycorrhizal fungi have been observed to exhibit greater water content compared to their non-mycorrhizal counterparts within the same ecological setting (Faber et al. 1991). The increased water absorption in mycorrhizal plants can be ascribed to their extensive growth increasing the root surface area, enabling the redistribution of water from wet soil layers to drier soil layers within the rhizosphere, and even the promotion of lateral and fine root growth, which are essential for water uptake (Lehto and Zwiazek 2011).

Additionally, mycorrhizal associations play a crucial role in nutrient acquisition for plants. The formation of a mantle around the root surface substantially increases the absorptive surface area, providing an efficient interface for nutrient exchange between the fungus and the host plant (Peterson et al. 2004). Particularly, mycorrhizal fungi excel in the absorption of essential nutrients such as nitrogen (N) and phosphorus (P) from the soil. Once assimilated, these nutrients are transported by the fungi to the plant partners, contributing significantly to the plant's nutritional status (Feng 2019). This intricate nutrient exchange mechanism further exemplifies the multifaceted benefits of mycorrhizal symbiosis in facilitating plant growth and health.

Mycorrhizal associations also enhance plant resistance to biotic stresses. For example, AM exhibit anthogenesis capability, allowing them to inhibit the proliferation of soilborne pathogens and enhance the plant's systemic resistance against these harmful microorganisms (Huang et al. 2003). Mycorrhizal fungi residing in the rhizosphere can enhance the physical and chemical characteristics of soil, foster the proliferation of beneficial microbes, and outcompete pathogenic microbes for nutrients and space, thus bolstering the plant's ability to resist diseases (Weng et al. 2022). Mycorrhizal associations can also induce host plants to develop systemic defense systems, increase phytohormone concentrations, regulate gene expression, and enhance protein production, further contributing to disease resistance (Weng et al. 2022; Elliott et al. 2021). The physical barriers formed by mycorrhizal associations, such as mycelial networks, callose, and papillary structures, can further hinder pathogen penetration into host plants (Pozo et al. 2002). Additionally, competition for photosynthetic products between mycorrhizae and pathogens can limit pathogen growth and reproduction (Kuila and Ghosh 2022). Overall, mycorrhizae indirectly reduce pathogen damage and improve disease tolerance by enhancing water and nutrient absorption by plants and disease resistance by activate defense mechanisms in plants (Jung et al. 2012; Ma et al. 2021). However, the mechanisms of disease resistance in mycorrhizal plants may vary and require further investigation.

Mycorrhizal partnerships yield beneficial effects on soil by generating organic acids and glomalin, which serve to shield the soil against erosion, bind heavy metals, enhance C storage, and fortify the stability of soil macroaggregates (Fall et al. 2022). Glomalin, produced by extraradical mycelium, acts as a binding agent, uniting microaggregations within the soil. This enhances water infiltration, reduces runoff and soil erosion, lowers organic matter loss, promotes gas exchange, and augments the retention of both water and minerals (Demenois et al. 2018; Parihar et al. 2020; Lehmann et al. 2020). Mycorrhizal associations also influence microbial activity, diversity, and composition in the soil through their antagonistic or supportive effects (Fall et al. 2022). AM symbiosis elevates the microorganism levels in maize rhizosphere soil, concurrently enhancing the activities of soil urease, protease, and catalase, as indicated by Ye et al. (2015). Furthermore, AM fungi play a pivotal role in interacting with various groups of soil bacteria, exerting influence over the rhizosphere microbial community (Miransari 2011a). The presence of AM fungi, combined with specific plant species, has been demonstrated to influence soil microbial activity. The symbiotic relationship between AM fungi and host plants establishes a networking system akin to a symbiotic framework, interacting with other microorganisms in the soil (Xu et al. 2019).

AM can also be used as a bio-stimulant to enhance stress tolerance in host plants and increase productivity (Rouphael et al. 2015). They contribute to the mitigation of oxidative stress, improve water and nutrient uptake, and modulate the transcription of genes involved in signaling pathways (Abdel Latef and Chaoxing 2011). By increasing radical hyphae, AM confer resistance to conditions like drought and promote the development of efficient root systems to absorb water and increase photosynthetic rates (Singh et al. 2022). Mycorrhizal plants also exhibit tolerance to heavy metals through mechanisms such as immobilization of the heavy metals in cell walls, storage in vacuoles, or chelation in the cytoplasm (Punamiya 2010).

19.3 Nutrient Acquisition and Nutrient Cycling

Mycorrhizal associations demonstrate a remarkable ability to offer nutritional assistance to crop plants, especially in adverse environmental conditions. These mutually beneficial interactions between fungi and plants have significant implications for improving plant nutrient uptake and enhancing crop productivity.

19.3.1 Nutrient Exchange in Mycorrhizal Associations

AM fungi enhance nutrient uptake in plants, improving concentrations of macro- and micronutrients and boosting photosynthesis and biomass production. Particularly beneficial in nutrient-deficient soils, AM aids essential element acquisition. In phosphorus (P)-limited conditions, mycorrhizal associations improve P delivery to host plant roots, as seen in maize (Garcés-Ruiz et al. 2019). Nutrient absorption by AM occurs through two primary pathways: the direct pathway (DP) via the root epidermis and root hairs and the mycorrhizal pathway (MP) through fungal hyphae, transporting nutrients from the soil to the root cells (Smith and Smith 2011).

19.3.2 Enhanced Nutrient Uptake by Mycorrhizal Fungi

The function of mycorrhizal fungi in enhancing the acquisition of mineral nutrients by host plants is investigated across three mycorrhizal groups: ECM, EM, and AM. The interaction between mycorrhizal fungi and host plants can directly impact the mineral nutrition of host plants by boosting the growth of plants via nutrient uptake by fungi (Fig. 19.2a). Additionally, it can indirectly affect the plant by modifying transpiration rates and influencing the composition of rhizosphere microflora.

Mycorrhizal fungi have demonstrated the ability to take up and deliver various nutrients to the host plant. Specifically, they are involved in the acquisition of nutrients such as P, ammonium (NH_4^+), nitrate (NO_3^-), potassium (K), calcium (Ca), sulfate (SO_4^{2-}), copper (Cu), zinc (Zn), and iron (Fe). For instance, under experimental conditions, the external hyphae of AM fungi have been reported to deliver up to 80% of plant P, 25% of plant N, 10% of plant K, 25% of plant Zn, and 60% of plant Cu (Bucking et al. 2012). Conversely, ECM and EM fungi generate secreted enzymes, which grant host plants access to organic forms of N and P that are often

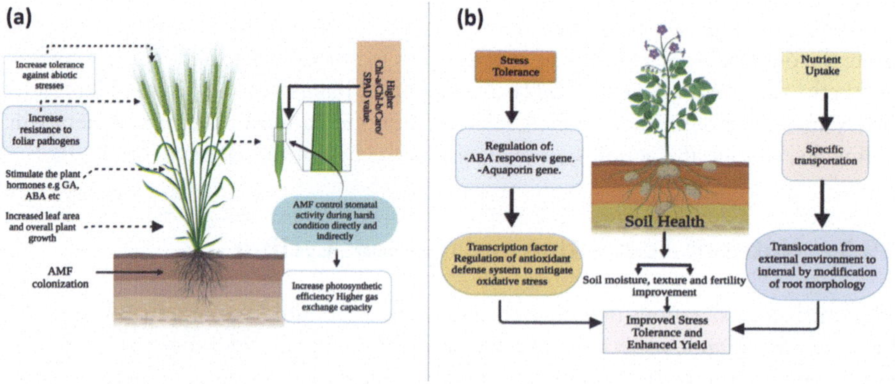

Fig. 19.2 Role of mycorrhizal fungi in (**a**) regulating plant growth and enhancing crop productivity and (**b**) soil health productivity maintenance. (Adapted from Wahab et al. 2023)

inaccessible to AM fungi or non-mycorrhizal roots (Marschner and Dell 1994). However, the specific contributions of these nutrient sources have not been quantified in field studies.

Studies have shown that when *Eucalyptus nutans* is inoculated with *Funneliformis mosseae*, there is a significant increase in plant biomass compared to the non-inoculated treatment (Shi et al. 2021). Observations such as these are likely due to the impact of a common mycorrhizal network, which greatly enhances the efficiency of soil nutrient utilization by increasing the contact area between the root system and soil via a greater number of extraradical hyphae (Begum et al. 2019). Consequently, there is an increase in the concentration of various macronutrients and micronutrients in plants, leading to increased production of photosynthates and ultimately resulting in higher biomass accumulation (Chen et al. 2017).

19.3.3 Role of Mycorrhizal Fungi in Nutrient Cycling and Plant Nutrition

Mycorrhizal fungi play a crucial role in nutrient cycles in terrestrial ecosystems, with approximately 20–50% of photoassimilates allocated to the fungal compartment belowground (Hobbie and Hobbie 2008). Furthermore, mycorrhizal associations play a role in the biogeochemical cycles of soils, particularly in the cycling of P, N, and C (Weng et al. 2022).

19.3.3.1 Key Role of Mycorrhizal Fungi in C Cycle

Mycorrhizal fungi likely play a vital role in the C cycle. Although heavily managed agricultural ecosystems may reduce the abundance of AM, most crops in the field, such as maize, cereals, soybean, and rice grown under non-paddy conditions, are colonized by AM, leading to C allocation to the fungal compartment. Recent observations suggest that mycorrhizal fungi are important regulators of C dynamics, affecting the degradation of fungal residues and increasing C storage in EM-dominated ecosystems compared to AM-dominated ones (Phillips et al. 2013; Averill et al. 2014).

In exchange for various low-diffusive nutrients such as $NH4^+$, P, Zn, and Cu, arbuscular mycorrhizal fungi create a sink demand for C, leading to a 4–20% C drain from plants (Wright et al. 1998; Willis et al. 2013). This additional C, combined with the extensive mycelium of AM, influences the active soil C pool. The dry weight of extraradical hyphae (ERH) constitutes a significant portion of soil microbial biomass (SMB), ranging from ~0.03 to 0.5 mgg^{-1}, contributing to approximately 54–900 kg ha^{-1} of soil organic carbon (SOC) (Miller et al. 1995). This substantial C pool, when incorporated into the soil, could serve as a key factor in mitigating global climate change under appropriate management conditions (Solaiman and Mickan 2014). Despite the considerable amount of C retained by AM, the mean residence time of C in the soil and its physical protection are crucial for its prolonged storage in the soil. Furthermore, under elevated CO_2 conditions, the upregulation of photosynthesis via a greater sink demand and allocation of extra

C to the soil system will impact the soil's inherent microbial community, contributing significantly to terrestrial C flux (Solaiman and Mickan 2014).

19.3.3.2 Key Impact of Mycorrhizal Fungi on N and P Cycles

Mycorrhizal fungi play a pivotal role in providing substantial amounts of N and P to their host plants, particularly in natural ecosystems with limited soil nutrient availability (Fig. 19.2b). While their contribution to plant N nutrition is generally less pronounced than for P, this depends on various factors such as soil type, water content, and pH (Tobar et al. 1994; Hodge and Storer 2014). Additionally, AM can immobilize considerable amounts of N in their mycelia (Hodge and Fitter 2010), while ECM and EM fungi can acquire substantial level of both organic and inorganic N and P, sometimes accounting for up to 80% of plant N and P (Read and Perez-Moreno 2003; Hobbie and Hobbie 2008).

Stable isotope probing has proven invaluable in tracing the movement of N and/or C within mycorrhizal networks (Drigo et al. 2010). While many studies have primarily focused on nutrient uptake, only a limited number have explored the potential of mycorrhizal fungi to mitigate nutrient losses, such as minimizing leaching and denitrification. For example, mycorrhizal fungi have shown a substantial capacity to reduce N (up to 70 kg per ha per year) and P (up to 150 g per ha per year) loss occurring through leaching (Asghari and Cavagnaro 2012; Bender et al. 2015). Notably, even in the absence of a plant growth response, AM have demonstrated an ability to reduce nutrient leaching losses in both organic and inorganic forms (Bender et al. 2015). AM can hydrolyze organic P into inorganic P through the enzyme phosphatase (Shen et al. 2011), and the presence of phosphate-solubilizing bacteria (PSB) can aid in the mineralization of organic P (Etesami et al. 2021). The synergistic association of PSB and AM contributes to the mineralization of insoluble P in the soil, enhancing its availability for plant uptake (Fig. 19.2b; Mahanta et al. 2018).

19.3.4 Mycorrhizal Fungi and Sustainable Agriculture

The surge in agricultural land productivity since the mid-twentieth century, driven by technological advancements and the widespread use of inputs like fertilizers and pesticides, has reached a point where recent years show minimal or no increase in productivity relative to the extent of cultivation (Ortiz-Bobea et al. 2021; Potapov et al. 2022). This stagnation is attributed to unsustainable agricultural practices, emphasizing the pressing need to implement sustainable approaches to address future challenges. Intensive farming, driven by the goal of maximizing productivity, can, however, lead to adverse environmental effects. Crucial to this is the nutritional status of agricultural land, particularly concerning N and P, identified as the most critical and challenging nutrients in agricultural soils.

The challenge with N fertilizers lies in their tight bonding to soil organic matter, fostering competition between plants and soil microorganisms for N absorption postfertilization. Conversely, P fertilizers can precipitate as calcium phosphates in acidic soils when applied at recommended amounts (Delgado and Torrent 2000). Despite the concealed challenges posed by these nutrients in developed countries, the situation contrasts significantly in developing nations, where insufficient fertilizer availability and inadequate agricultural practices contribute to widespread malnutrition and food insecurity. Consequently, a global and urgent call for sustainable farming practices arises, acknowledging the intricate interplay of intensive farming, nutrient challenges, and the need for environmental stewardship.

Various sustainable agricultural practices and techniques can be implemented to address these challenges. A key concept involves harnessing microbes with the capacity to augment nutrient availability and use efficiency. Rhizosphere farming practices—approaches that specifically enhance the health and functionality of the rhizosphere, including reduced tillage, cover cropping, and crop rotation—along with the utilization of beneficial microorganisms in the ecosystem, play pivotal roles in this endeavor. For example, AM fungi boost the utilization of soil-bound N by producing the glycoprotein glomalin, which persists in the soil after hyphal death, contributing to soil physical and biological properties (Treseder and Turner 2007). Glomalin acts as a binding agent for soil particle aggregation and functions as a chelator for heavy metals and toxic pollutants (Singh et al. 2022).

AM have several direct and indirect impacts on organic farming systems. They enhance agricultural land productivity through nutrient acquisition and cycling, promote seedling establishment, and can mitigate extreme environmental conditions. Indirectly, they suppress weed growth, enhance nitrogen fixation ability in legumes, and contribute to soil particle aggregation. Due to these direct and indirect influences, AM are a valuable microbial tool in the sustainable agricultural production systems.

To optimize the benefits of mycorrhizal fungi in sustainable agriculture and enhance overall plant health, farmers can employ a range of practices and interventions. For example, reduced tillage practices minimize soil disturbance, preserving mycorrhizal networks and supporting their establishment (Moitinho et al. 2020; Wilkes et al. 2021). Cover cropping during offseasons ensure a continuous root presence, sustaining mycorrhizal populations (Lehman et al. 2012). Crop rotation diversifies root exudates, promoting a broader range of mycorrhizal associations (Moitinho et al. 2020). Moreover, balancing fertilizer inputs avoids synthetic fertilizer overuse, fostering a healthy symbiotic relationship. Incorporating organic matter into the soil through amendments like compost enhances soil structure and nutrient availability, creating an optimal environment for mycorrhizal development. Efficient water management practices, such as optimal irrigation, maintain ideal soil moisture levels for mycorrhizal activity. Integrated pest management (IPM) strategies safeguard mycorrhizal populations from the negative impacts of chemical pesticides.

19.4 Plant Growth Promotion and Disease Suppression

Plant growth promotion and disease suppression are two significant benefits associated with AM. These fungi can promote plant growth by extending their hyphal networks, which improves nutrient absorption efficiency from the soil. Moreover, they have capacity to suppress plant pathogenic microorganisms.

19.4.1 Promotion of Plant Growth by Mycorrhizal Fungi

Plant growth plays a crucial role in increasing crop productivity. One method to increase plant growth is through the direct addition of nutrients to the soil (fertilization) and the application of plant growth-promoting microorganisms (PGPMs). AM are considered beneficial microbes that positively influence plant growth by improving plant nutrition and water absorption efficiencies and metabolism. They are often referred to as "biological fertilizers" due to their crucial role in nutrient cycling (Mohandas 2013). The extensive hyphal network formed by AM provides them with a greater surface area for accessing nutrients in the soil. They enhance plant nutritional status through two main mechanisms: increasing nutrient availability and facilitating nutrient translocation. These fungi can access larger soil volumes by extending their hyphal networks beyond the nutrient-depleted zones of the rhizosphere (Smith and Smith 2011). Furthermore, they are capable of penetrating smaller pore sizes than plant roots, via the production of thinner fungal hyphae, and can thus uptake nutrients that are otherwise inaccessible (Allen 2007).

AM enhance the soil microenvironment around plant roots and strengthen the plant's resistance to diseases and insect pests. Previous studies have shown the direct positive impact of AM fungi on plant growth in various crop species, including *Solanum lycopersicum* L., *Sorghum bicolor* (L.) Moench, *Withania somnifera* (L.) Dunal, *Cucurbita maxima* Duchesne, *Piper longum* L., *Phaseolus vulgaris* L., *Panicum hemitomon* Schult, and several free fruits, such as *Citrullus lanatus* (Thunb.) Matsum. & Nakai, *Musa acuminata* Colla, and *Prunus cerasifera* L. (Diagne et al. 2020).

AM also play an important role in improving the nutritional quality of food crops (Zhang et al. 2018). For example, in cucumber (*Cucurbita maxima*), AM improve the fat, crude protein, crude fiber, and carbohydrate content in both shoots and roots and increase phytochemical constituents (Diagne et al. 2020). In cereals, such as wheat and barley, they have been shown to modify the concentration and distribution of nutrients within plants, improving protein, Fe, and Zn content (Watts-Williams and Cavagnaro 2018). It is well-known that AM symbiosis specifically induces the expression of transporters, such as AQ genes, P transporters, ammonium transporters, nitrate transporters, sulfur transporters, Zn transporters, C transporters, protein transporters, and more (Zhao et al. 2014; Vinichuk et al. 2013).

19.4.2 Mechanisms of Plant Growth Promotion

Mycorrhizal fungi show great promise as plant growth-promoting rhizosphere organisms for applications in nutrient-deficient and disturbed soils. Among the mechanisms employed by AM to promote plant growth is the biotrophic colonization of the root cortex and development of an extrametrical mycelium, facilitating the acquisition of mineral nutrients and water from the soil (Churchland and Grayston 2014). The external mycelium of AM forms associations with other soil organisms, leading to the formation of water-stable aggregates that are essential for maintaining good soil quality (Azcón-Aguilar and Barea 1997).

Mycorrhizal fungi have the remarkable ability to enhance the nutrient supply of their host plants, especially under nutrient-deficient conditions. They achieve this through efficient soil exploitation and by excreting considerable amounts of phosphatases (Tibbett and Sanders 2002). In unfavorable soil conditions with N and P limitations, ECM fungi have demonstrated mechanisms for adaptation, such as the utilization of seed protein N, as shown by Tibbett et al. (1998). Furthermore, different ECM fungal species exhibit varying abilities to acquire specific nutrients from the soil, which can be seen as a form of niche partitioning (Erland and Taylor 2002).

19.4.3 Suppression of Soilborne Pathogens by Mycorrhizal Fungi

Soil serves as a habitat for numerous soil-inhabiting pathogens, and over the past decades, the intensive application of chemicals has led to the development of pathogen resistance and environmental pollution (Compant et al. 2005). Soilborne root pathogens can cause substantial economic losses during the crop nursery stage, highlighting the urgent need for sustainable control methods. Research has provided evidence that mycorrhizal fungi can contribute to the improvement of soil health and to combating plant pathogens. Numerous reports have demonstrated the alleviation of damage caused by soilborne necrotrophic pathogens. For example, plant colonization by AM has shown positive effects in mitigating the impact of pathogens such as *Fusarium oxysporum* in strawberries (Yanan et al. 2015), cucumbers (Hu et al. 2010), *Armillaria mellea* in grapevines (Nogales García et al. 2008), and *Phytophthora parasitica* in tomatoes (Pozo et al. 2002).

ECM fungi contribute to plant defense through various mechanisms. They provide a mechanical barrier against root pathogens by forming a protective hyphal mantle (Duchesne et al. 1987). Additionally, ECM fungi have the capability to produce antibiotics (Schelkle and Peterson 1997) and can contribute to soil acidification in alkaline environments (Rasanayagam and Jeffries 1992). The synergistic effect of combining bacteria with mycorrhizal fungi has demonstrated significant potential in enhancing plant protection. For example, co-inoculation with *Bacillus subtilis* and ECM fungi has been effective in inhibiting the growth of root pathogens such as *Fusarium* and *Cylindrocarpon* spp. (Schelkle and Peterson 1997). This dual approach takes advantage of the complementary strengths of both microorganisms. Furthermore, both mycorrhizal fungi and bacteria have the ability to produce siderophores (Hrynkiewicz et al. 2010), compounds known to inhibit pathogenic fungi (Neidhardt et al. 1990).

Another key strategy for effectively suppressing soilborne plant pathogens is the induction of plant defense genes during pathogenic infections. Differential and systemic alteration in the expression of defense genes has been observed during the interaction between mycorrhizal beans and *Rhizoctonia solani* (Guillon et al. 2002). The bio-protection ability of AM enhances plant nutritional status, wound repair, root anatomy and longevity alterations, competition with soilborne pathogens for root colonization sites and host photosynthates, activation of plant defenses, and changes in rhizosphere microorganism populations (Harrier and Watson 2004). Mycorrhiza-induced resistance (MIR) in aerial organs can also provide protection against necrotrophic pathogens attacking roots through the quick activation of jasmonate (JA)-dependent signaling pathways in host plants (Pozo and Azcón-Aguilar 2007).

Various research studies have explored the biological control of different plant pathogens. For example, mycorrhizal cotton plants were found to be more resistant to *Thielaviopsis basicola* compared to those with sterile roots (Schönbeck and Dehne 1989). The presence of mycorrhizal fungi has been shown to reduce plant deaths caused by *P. megasper* and the disease index and incidence of *Ralstonia solanacearum* inoculated with *G. rhizogenes* and *G. mossie* (Chou and Schmitthenner 1974). Several factors contribute to the effectiveness of mycorrhizal fungi, including the type of plant disease, the relationship between AM and host plants, the timing and amount of AM fungi inoculation, and environmental factors (Zhou et al. 2019).

19.4.4 Induced Systemic Resistance and Defense Priming

Mycorrhizal fungi play an important role in case of resistance induction of host plants against the pathogens through several mechanisms, and they are capable of rapid and quick activation of the some of the defenses.

19.4.4.1 Induced Systemic Resistance

One of the ways plants defend against plant pathogens is through the activation of induced systemic resistance (ISR). Arbuscular mycorrhiza has shown effectiveness against a diverse range of plant pathogens, including nematodes, oomycetes, bacteria, and fungi and even certain herbivores that attack plants. Plants have the ability to enhance their defensive capabilities when exposed to mycorrhizal infection. This phenomenon is also known as mycorrhiza-induced resistance (MIR) (Fig. 19.3; Cameron et al. 2013). It is suggested that MIR results from both the direct responses of plants to mycorrhizal infection and the indirect immune reactions to rhizobacteria that induce ISR in the mycorrhizosphere (Cameron et al. 2013). In cases of herbivore damage, AM symbionts enhance the release of defense metabolites by interacting with the phytohormone signaling pathway (Meirer and Hunter 2018).

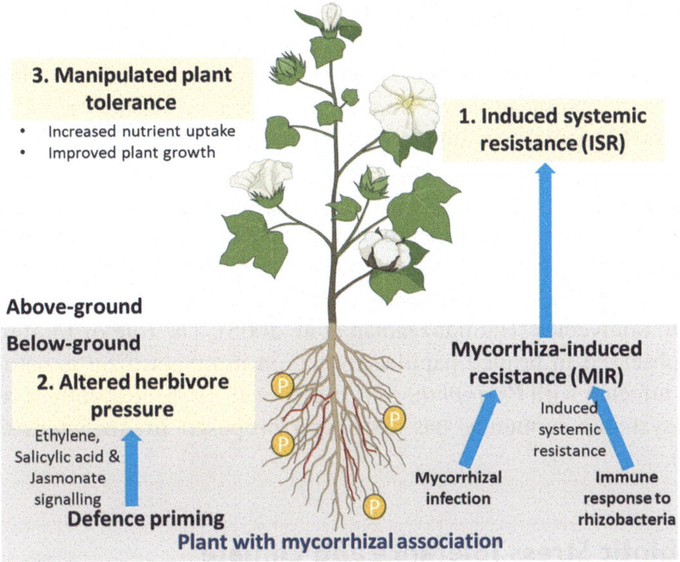

Fig. 19.3 Arbuscular mycorrhizal associations influencing induced systemic resistance, defense priming, and plant tolerance to biotic stress. (Adapted from Hao et al. 2019)

The JA signaling pathway plays a crucial role in both the direct and indirect defense systems of plants against herbivore attacks (Fig. 19.3; Frew et al. 2022).

19.4.4.2 Defense Priming

The rapid and robust activation of defense mechanisms is a factor for successfully controlling herbivores that affect plants (Fig. 19.3). Consequently, preconditioning plant tissues to enable quick and effective defense upon attack is ecologically advantageous and a common feature of the plant immune system (Conrath et al. 2006). This priming boost, known as basal defense, is triggered by AM (Pozo and Azcon-Aguiloar 2007).

Several research studies provide evidence supporting the effective role of AM as a defensive primer. For instance, one study reported that carrot roots transformed by mycorrhizal fungi exhibited stronger defense reactions at challenge sites when faced with *Fusarium* (Benhamou et al. 1994). In the case of infected potato plantlets with *Rhizoctonia*, the presence of AM fungi resulted in increased accumulation levels of the phytoalexins rishitin and solavetivone, while the fungi themselves did not alter the levels of these compounds (Yao et al. 2003). Another study suggested that priming for callose deposition appears to be the mechanism underlying the protection achieved by *G. intraradices* against *Colletotrichum* in cucumber (Lee et al. 2005).

Signaling pathways involving JA, ET, and SA play major roles in plant defense mechanisms (Fig. 19.3). Priming often relies on tissue-sensing signaling molecules (Conrath et al. 2006). SA coordinates defense mechanisms generally effective against biotrophic pathogens, while JA regulates wounding responses and resistance against necrotrophic pathogens (Ton et al. 2002; Glazebrook 2005).

To establish a successful mutualistic interaction upon recognition of the AM fungal partner, a symbiotic program must be activated in the host plant (Reinhardt 2007). Mycorrhizal plants exhibit higher levels of endogenous JA in their roots compared to non-mycorrhizal plants. This increase occurs after the onset of mycorrhization and is likely associated with fully established mycorrhizas (Hause et al. 2002). Elevated levels of basal JA production may be linked to the enhanced resistance of mycorrhizal plants against certain pathogens and insects. Experimental evidence has connected the JA pathway to primed callose deposition and enhanced resistance to oomycetes (Hamiduzzaman et al. 2005). The role of JA signaling has also been observed in primed papillae formation in mycorrhizal tomato root systems when infected with *Phytophthora* (Cordier et al. 1998). Recently, a central role for JA in systemic immunity has also been proposed in *Arabidopsis* (Truman et al. 2007).

19.5 Abiotic Stress Tolerance and Climate Change Adaptation

Mycorrhizal fungi can promote stress tolerance at ecosystem scales, and they can play a pivotal role in case of climate change adaptation.

19.5.1 Mycorrhizal Fungi and Drought Stress

Drought stress is a common abiotic stress that restricts plant growth and development in various ecosystems, particularly in arid and semiarid regions (Boutasknit et al. 2020). However, it has been observed that mycorrhizal fungi, specifically AM, can alleviate drought stress by enhancing water uptake in plants (Bowles et al. 2018). AM hyphae can access water and nutrients in soil pores that are inaccessible to root hairs (Li et al. 2019). Consequently, under drought conditions, these fungi can improve plant productivity and overall plant performance (Augé 2001).

Studies have shown that AM improve water use efficiency through the regulation of stomatal conductance and stomatal aperture (Augé et al. 2014). Additionally, AM can increase the antioxidant defense system in plants to cope with drought-induced oxidative damage (Zou et al. 2021). Boutasknit et al. (2020) confirmed that AM contribute to aquaporin regulation under water scarcity, enabling plants to better tolerate drought stress.

The ability of AM to protect host plants against drought stress can be attributed to several mechanisms, including (i) direct water uptake by extraradical mycelium; (ii) improvement of host plant physiology related to nutrient uptake, osmotic regulations, hormone balance, antioxidant protection, fatty acids, and polyamines; (iii) adaptation of morphology in host plants; (iv) positive impact of mycorrhizal extraradical mycelium on soil structure; and (v) molecular changes (Fig. 19.4; Blee and Anderson 2000; Wu and Zou 2017; Liu et al. 2018; Zhang et al. 2020).

The colonization of mycorrhizal fungi in roots and soil is strongly influenced by drought stress (Wu and Zou 2017). Long-term soil drought can decrease AM colonization, but short-term soil drought does not significantly affect the colonization of AM (Augé 2001). Soil moisture conditions also impact spore germination and development of AM in the soil (Wu and Zou 2017). Spore germination is supported when soil water potential is at or above field capacity, but it decreases when soil water potential drops below field capacity (Daniels and Trappe 1980). Some AM species can adapt quickly to soil drought conditions and make relatively higher root AM colonization (Nasim 2010).

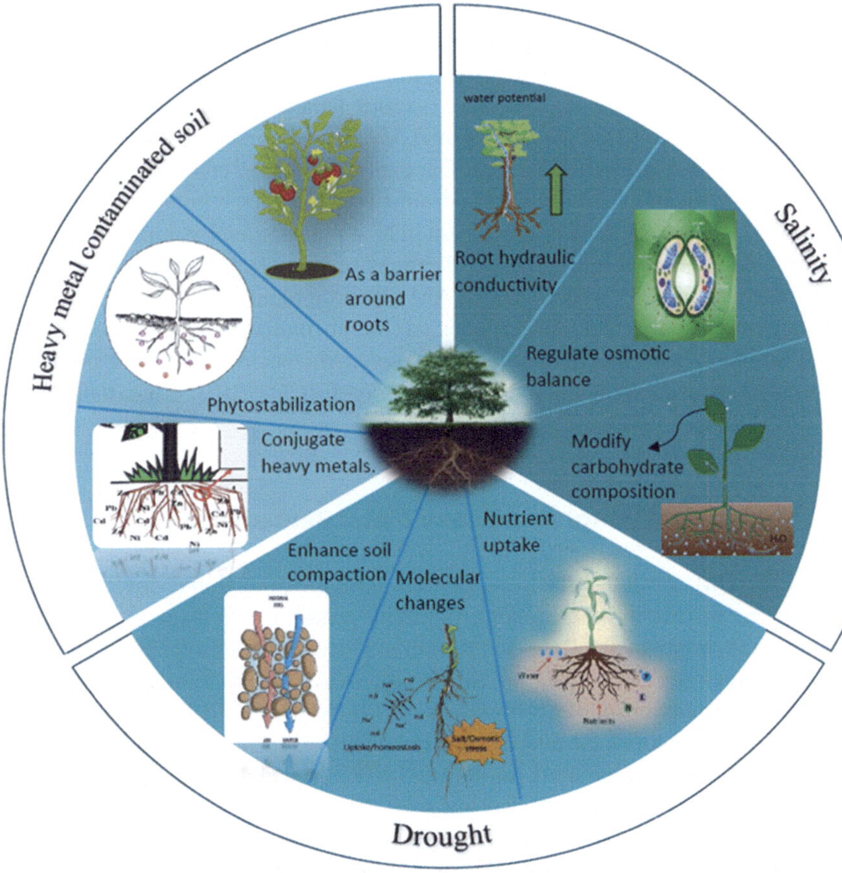

Fig. 19.4 Schematic diagram showing mycorrhizal activities in plants that ameliorate stress due to drought, salinity, and heavy metal contamination in soil

19.5.2 Mycorrhizal Fungi and Salinity Stress

Soil salinization, resulting from factors such as excessive water evaporation, poor drainage, low precipitation, use of wastewater for irrigation, and poor agricultural practices, is a significant environmental problem affecting crops worldwide, especially in arid and semiarid regions (Chourasia et al. 2022). Salinity stress has various impacts on plants, including ion toxicity, osmotic stress, nutrient deficiencies, oxidative stress, reduction in germination, vegetative growth, and reproductive capacity, ultimately leading to decreased crop productivity (Giri et al. 2003; Chinnusamy et al. 2005; Hasanuzzaman et al. 2013). Salinity also affects the photosynthetic process by reducing leaf area index and increasing the production of reactive oxygen species (ROS), which can damage cell organelles, proteins, and lipids in the plasma membrane (Hashem et al. 2014; Ahmad et al. 2018). Furthermore, salinity causes physiological drought and ion imbalances (Feng et al. 2002).

AM symbionts have been found to enhance plant salinity tolerance (Dastogeer et al. 2020). AM can mitigate salinity stress through various mechanisms, including improved nutrient uptake, increased water absorption, modulation of physiological and biochemical properties in host plants, stimulation of antioxidant activities to protect against ROS damage, regulation of osmotic balance, enhancement of photosynthetic rate, and adjustment of hormonal levels to counteract the detrimental effects of salts on crop growth (Fig. 19.4; Dar et al. 2023; Augé et al. 2014; Khalloufi et al. 2017). AM can also increase root hydraulic conductivity, regulate osmotic balance, and modify carbohydrate composition in host plants, thereby improving water absorption capacity and diluting the toxic effects of ions (Ruiz-Lozano and Azcón 2000; Ruiz-Lozano 2003). The degree of AM effect on plant salinity tolerance may vary among studies, and these differences can be attributed to factors such as the extent of salinity, host specificity, types of mycorrhizae involved, and complex interactions with environmental conditions (Porcel et al. 2012; Chandrasekaran et al. 2019; Dastogeer et al. 2020).

19.5.3 Mycorrhizal Fungi and Heavy Metal Remediation

Heavy metal (HM) pollution has emerged as a significant global environmental threat reported in recent years, posing potential risks to natural ecosystems and human health (Kamran et al. 2019). In this context, mycorrhizal fungi have garnered attention for their potential role in mitigating heavy metal contamination and improving the sustainability of agroecosystems (Torres et al. 2018; Balestrini et al. 2018). Beneficial mycorrhizal interactions enhance soil resources and enable efficient responses to environmental limitations. Mycorrhizal fungi not only supply mineral compounds to plants but also improve the soil's physicochemical properties, acting as a physical barrier by forming a sheath around mycorrhizal plants (in the case of ECM) and immobilizing heavy metals in plant roots through processes like phytostabilization (Fig. 19.4; Miransari 2011b; Wu et al. 2017). These fungi are particularly associated with plant roots experiencing severe metal contamination

(Kodre et al. 2017). Apart from being environmentally friendly and cost-effective, mycorrhizal remediation offers secondary benefits such as improved soil nutrition and increased nutrient uptake, as well as toxic metal resistance (Ma et al. 2022).

There are two major pathways of heavy metal phytoremediation involving AM: direct mechanisms through fungal structures and the secretion of glomalin and indirect mechanisms by enhancing the tolerance of host plants. ECM fungi have demonstrated their ability to colonize roots in extreme soil conditions, making them suitable candidates for the restoration of mining sites (Urban et al. 2008; Marx 1975). These fungi can sequester and conjugate heavy metals (Fig. 19.4), reducing their bioavailability in the soil and consequently minimizing their uptake by plants, resulting in improved plant growth. Moreover, ECM fungi thrive in metalliferous soils poor in essential nutrients, enhancing water-plant relationships, nutrient uptake, and overall plant nutrition through extensive extramatrical mycelium (Beniwal et al. 2010; Rajtor and Piotrowska-Seget 2016).

19.5.4 Role of Mycorrhizal Fungi in Climate Change Mitigation and Adaptation

Climate change mitigation involves reducing the release of heat-trapping greenhouse gases into the atmosphere (such as CO_2 and methane) and moderating the impact of rising mean temperatures. To address the harmful impacts of climate change, various actions need to be considered. While extensive research has been conducted on the impacts of climate change and increasing temperatures on fungal populations, there is comparatively less research on the role of mycorrhizal fungi in climate change mitigation and adaption (Deslippe et al. 2011; Morgado et al. 2015).

The integration of mycorrhizal fungi by plant roots offers several beneficial uses in mitigating and tolerating drought conditions. One notable structural adaptation by mycorrhizal fungi is the expansion of the absorbing surface area in the soil through the development and branching of extraradical mycelium (Zhang et al. 2018). They can improve soil aggregate stability by forming a hyphal network that contributes to the formation and maintenance of soil water macroaggregates (Ji et al. 2019), improving water infiltration and storage. Another mechanism contributing to soil aggregate stability is the release of the glycoprotein glomalin by the fungus, which acts as a soil superglue (Cheng et al. 2021).

Furthermore, mycorrhizal colonization can modify root architecture and increase the water- and nutrient-absorbing surface area, particularly under drought conditions (Lehto and Zwiazek 2011). Additionally, they indirectly improve nutrient acquisition in drought stress conditions through the secretion of phosphatase via extraradical hyphae (Sato et al. 2019) or by stimulating root phosphatase exudation, as observed in mycorrhizal citrus trees (Cheng et al. 2021). These mechanisms can enhance the decomposition and acquisition of organic orthophosphate, partially alleviating plant responses to drought stress.

The association with mycorrhizal fungi also triggers various biochemical and physiological modifications in plants, enhancing their drought tolerance. For example, several studies report increased production of enzymatic (He et al. 2020; Huang et al. 2020) and nonenzymatic (Essahibi et al. 2018) antioxidants in mycorrhizal trees such as carob, trifoliate orange, and apple trees under drought conditions. These enzymes and compounds can reduce oxidative damage to carbohydrates, lipids, proteins, and DNA. Mycorrhizal symbiosis has also been found to potentially mitigate osmotic damage through the accumulation of soluble sugars and proline (Huang et al. 2020; Wang et al. 2021).

19.6 Recent Advances in Mycorrhizal Research

19.6.1 Omics Approaches in Studying Mycorrhizal Interactions

Recent developments in omics-based approaches have provided unprecedented insights into mycorrhizal interactions and the agroecosystem services they provide. Integrated multi-omics approaches provide a comprehensive understanding of the genetic background of these microbes, changes that occur in global gene expression in response to perturbation, and the dynamics of molecular messaging that mediate fungal-host coordination. Furthermore, the development of integrative databases for multi-omics data has enabled the identification of regulatory networks that link genome-wide transcriptomics, proteomics, and metabolomics profiles in crops exposed to abiotic stresses (Sandrini et al. 2022). These databases can aid in the identification of key genes, both negative and positive regulators, associated with the biosynthesis of specialized metabolites with functions as far ranging as antimicrobial properties, the initiation of mycorrhization, and mobilization of recalcitrant nutrients (Sandrini et al. 2022). The practical applications of omics approaches are significant, including large-scale genotype screening, exploration of root-associated microbes through metagenomics, screening of microbes for stress tolerance, understanding the mechanisms underlying stress tolerance, identification of novel molecules, and elucidation of mechanistic and functional influences of genes on plant physiology improvement (Sandrini et al. 2022).

Metagenomics and metabarcoding techniques have become common practice in efforts to define the taxonomic structure of fungal communities in terms of richness, evenness, and composition in response to a variety of external pressures, e.g., pollution, climate change, and agrochemical application (Sharma et al. 2020). Improvements in the sensitivity of sequencing-based approaches have also greatly impacted the use of environmental (e) DNA in defining mycorrhizal community composition dynamics at ecosystem scales (Victorino et al. 2020). Recent advances in meta-transcriptomics have built upon this foundation, providing detailed mechanistic insights into the functional dynamics between plant host species and their mycorrhizal partners (Law et al. 2022).

Metabolomics approaches are useful for identifying plant responses to stressful conditions by observing changes in metabolomic profiles. Notable changes have been reported in the metabolomics of mycorrhizal roots (Rivero et al. 2018). In a study utilizing untargeted metabolomics analysis through UPLC-MS on mycorrhizal plant leaves, it was found that mycorrhizal symbiosis had a limited impact on the leaf metabolome in the absence of stress (Rivero et al. 2021). However, it significantly modulated the plant response to herbivory in the damaged area, leading to the local accumulation of defense compounds (Rivero et al. 2021). Transcriptomics also plays a vital role in understanding water stress responses in tomato plants and triggering defense responses against root-knot nematodes (Sandirini et al. 2022).

19.6.2 Molecular Mechanisms of Signal Transduction and Communication

Arbuscular mycorrhizal interactions with plant roots are highly regulated by the plant and fungal partners at the cellular, molecular, and genetic levels during the establishment of symbiosis (Ho-Plagaro et al. 2022). The development of this symbiosis involves several major steps. Prior to physical contact between the fungus and the plant root, molecular cross talk occurs during the pre-symbiotic phase. This communication leads to the formation of hyphopodia on the root surface (Mac Lean et al. 2017). Subsequently, rapid root colonization takes place, facilitated by the peri arbuscular membrane (PAM) developed by the fungal organism, which plays a crucial role in controlling transport and signaling processes between both partners (Bapaume et al. 2012). The life cycle of the arbuscular mycorrhizal fungus within the root is completed by the formation of storage vesicles and spores (Ho-Plágaro and García-Garrido 2022).

At the molecular level, communication between the fungus and the plant occurs before physical contact. Strigolactones (SLs) from the plant side and Myc factors from the fungal side play key roles in this communication (Akiyama et al. 2006). Strigolactones are a group of phytohormones derived from apocarotenoid metabolism that regulate various aspects of plant development (Gomez-Roldan et al. 2008; Lopez-Obando et al. 2015). In terms of pre-symbiotic signaling, when plants secrete SLs into the rhizosphere in response to P deficiency conditions, they serve as signals indicating the presence of a host plant receptive to colonization by AM (Yoneyama et al. 2007). Detection of SLs by AM induces spore germination and activates fungal oxidative metabolism, leading to organelle division, ATP production, and gene expression. These processes provide the energy required for fungal growth and branching, increasing the chances of physical contact with the host root and preparing the fungus for symbiosis establishment (Lopez-Obando et al. 2015).

19.6.3 Role of Mycorrhizal Fungi in Plant-Microbiome Interactions

Microbial communities interact with plants in their natural habitats, and these interactions can have positive or negative effects on ecosystem functions (Fig. 19.5). Mycorrhizal fungi can positively influence the abundance of some of the beneficial

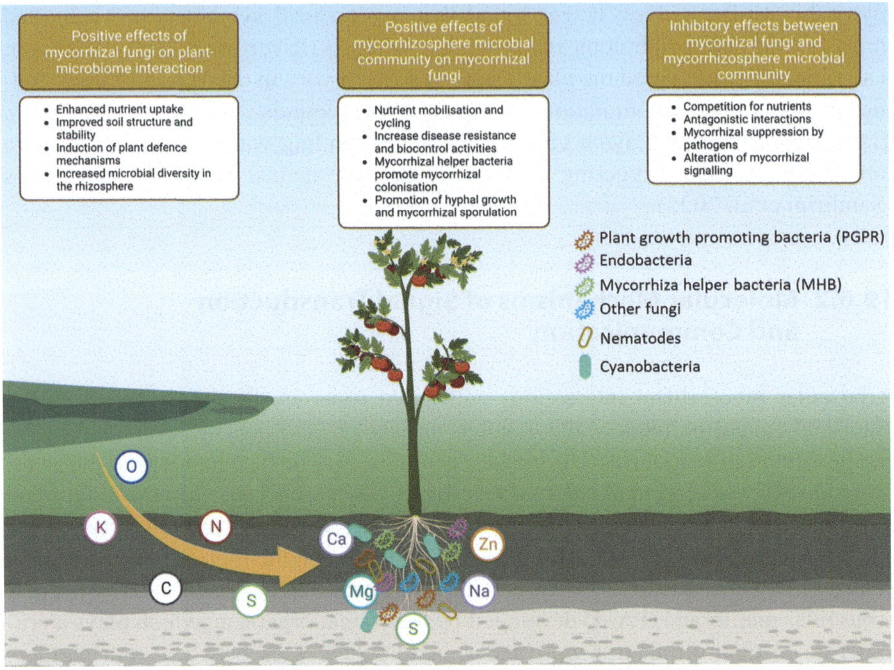

Fig. 19.5 Schematic illustration of interactions by mycorrhizal fungi and their role in plant-microbiome dynamics. (Adapted from Bao et al. 2022)

fungal species while reducing the prevalence of pathogenic bacterial species in the soil (Adomako et al. 2022). Through their synergistic promotion, mycorrhizal fungi can significantly improve plant resistance against pathogens and insect pests (Goicoechea 2020). Their main impacts are observed in the structure and activity of the rhizosphere microbial community under environmental stress conditions (Fig. 19.5; Liu et al. 2023). Arbuscular mycorrhizae increase their biomass, as measured by phospholipid fatty acid analysis (PLFA), but decrease the biomass of microbial bacteria and fungi in the rhizosphere soil (Stamou et al. 2017). Mycorrhizal fungi have direct or indirect effects on plant growth and development, as they produce enzymes, hormones, and toxins that affect mineral nutrition, water absorption, and root development in the soil environment (Fig. 19.5; Shi et al. 2022).

Ectomycorrhizal fungi have been shown to select bacterial communities from the surrounding soil based on symbiotic function or habitat requirements. For example, the short-term contact between soil bacteria and ECM fungal fruiting bodies can directly or indirectly promote the composition of endobacteria (EB) communities within the fruiting bodies (Pent et al. 2017). ECM selectively promote the colonization by specific bacteria that are suitable for their growth, laying the foundation for the establishment and maintenance of the symbiotic relationship (Shi et al. 2022). Bacteria selectively colonize ECM that provide a suitable habitat for their growth and form a new symbiotic relationship, especially under severe environmental stresses (Nazir et al. 2012). The mycelia and fruiting bodies of fungi create a suitable habitat for endobacteria by providing different C sources and protecting them from environmental stresses (Schulz-Bohm et al. 2017).

19.6.4 Emerging Technologies and Future Directions

In the realm of mycorrhizal research, various areas are currently garnering widespread attention and active investigation. These include mycorrhizal associations with host plants, nutrient cycling, ecosystem multifunctionality, mycorrhizal networks, evolution and partner selection, mycorrhizal genomics, and symbiotic molecular cross talk (van der Heijden et al. 2015). Additionally, metagenomics and meta-transcriptomics studies are also proving to be essential areas in advancing mycorrhizal research (Aguiar-Pulidon 2016; Zhao et al. 2023).

There are several key avenues that hold promise for future research in the field of mycorrhizal fungi.

1. Unraveling molecular cross talk: An essential endeavor is gaining a comprehensive understanding of the intricate molecular interactions between plants and mycorrhizal fungi. This exploration can provide novel insights into the genes associated with the establishment and maintenance of mycorrhizal symbiosis. Further investigation of symbiotic gene networks, master transcriptional regulators, and effector proteins, which play central roles in the molecular dialog between plants and fungi, will be crucial (Plett et al. 2014). Additionally, linking molecular data and metabolic pathways with ecophysiological and ecological processes, such as nutrient acquisition and protection against stresses and diseases, will represent a notable advancement in the field (van der Heijden et al. 2015).
2. Comprehensive understanding of mycorrhizal networks and microbiome: With the advancement of sequencing technology and bioinformatics, establishing a comprehensive understanding of the microbiome and mycorrhizal networks is crucial. Delving into the interactions between mycorrhizal fungi and other members of underground food webs, as well as their interactions with other soil biota, will shed light on the intricate fluxes of energy, metabolites, signaling molecules, and nutrients through mycorrhizal networks (Hacquard and Schadt 2014; Tedersoo et al. 2014).
3. Enhanced visualization tools: Developing better visualization tools is essential to study the complex interaction networks of mycorrhizal associations at both the fungal and plant community levels. Network analyses have shown that plant-fungal interactions in major mycorrhizal types can differ in important characteristics such as specificity, nestedness, and modularity (van der Heijden et al. 2015). Understanding coevolutionary processes between plants and mycorrhizal fungi, including the physiological mechanisms responsible for partner choice and the stability of mycorrhizal mutualism, remains an ongoing challenge.

4. Advancements in mycorrhizal technology application: Recent years have witnessed the development of new methods for large-scale production of AM fungi (Ijdo et al. 2011) and seed coating technology with AM fungi (Oliveira et al. 2016). These developments have made their application in horticulture and agriculture more cost-effective and reliable. It is now essential to develop biogeochemical models that can predict the most profitable conditions for the application of mycorrhizal technology.
5. Integration of mycorrhizal associations in global models: Mycorrhizal associations are increasingly being integrated into global models to comprehend C and nutrient cycling (Soudzilovskaia et al. 2019). Improving these models represents a major frontier in mycorrhizal research.

As the scientific community advances in understanding and harnessing the potential of mycorrhizal fungi, the collective efforts of researchers and stakeholders are instrumental in realizing the multitude of benefits that mycorrhizal symbiosis offers in various fields of science, agriculture, and environmental conservation.

19.7 Conclusion

The relationship between mycorrhizal fungi and plants could prove instrumental in protecting plant health via disease suppression, induced systemic resistance, defense priming, and abiotic stress tolerance as well as in promoting plant growth through efficient nutrient cycling and enhanced nutrient absorption, contributing significantly to ecosystem multifunctionality. In the modern scientific landscape, the exploration of this symbiosis has evolved with the aid of state-of the-art tools and technologies, especially focusing on genomics, metagenomics, and the molecular cross talk that facilitates this vital process.

As we advance in research on mycorrhizal associations, it is crucial to understand the intricate networks formed by mycorrhizae and the broader microbiome while dissecting the interplay between the plants and mycorrhizal fungi. To comprehend these complex interactions in greater detail, visualization of these interconnections become important; thus, the development of visualization tools (spatial next-generation sequencing, high-performance liquid chromatography-mass spectrometry, advanced microscopy) stands out as a central aspect of technological advancement. As the significance of mycorrhizal fungi extends beyond individual studies, it serves as a global model for elucidating interactions within nutrient cycles. The recognition of this beneficial microbe as a key player in protecting plant health and shaping ecological dynamics underscores its potential as a cornerstone in sustainable practices for resilient agriculture, environmental conservation, and beyond.

References

Abdel Latef AAH, Chaoxing H (2011) Effect of arbuscular mycorrhizal fungi on growth, mineral nutrition, antioxidant enzymes activity and fruit yield of tomato grown under salinity stress. Sci Hortic 127(3):228–233. https://doi.org/10.1016/j.scienta.2010.09.020

Adomako MO, Roiloa S, Yu FH (2022) Potential roles of soil microorganisms in regulating the effect of soil nutrient heterogeneity on plant performance. Microorganisms 10(12):2399. https://doi.org/10.3390/microorganisms10122399

Aguiar-Pulido V, Huang W, Suarez-Ulloa V, Cickovski T, Mathee K, Narasimhan G (2016) Metagenomics, meta-transcriptomics, and metabolomics approaches for microbiome analysis. Evol Bioinforma 12:EBO-S36436. https://doi.org/10.4137/ebo.s36436

Ahmad H, Hayat S, Ali M, Liu T, Cheng Z (2018) The combination of arbuscular mycorrhizal fungi inoculation (*glomus versiforme*) and 28-homobrassinolide spraying intervals improves growth by enhancing photosynthesis, nutrient absorption, and antioxidant system in cucumber (Cucumis sativus L.) under salinity. Ecol Evol 8(11):5724–5740. https://doi.org/10.1002/ece3.4112

Akiyama K, Hayashi H (2006) Strigolactones: chemical signals for fungal symbionts and parasitic weeds in plant roots. Ann Bot 97(6):925–931. https://doi.org/10.1093/aob/mcl063

Allen MF (2007) Mycorrhizal fungi: highways for water and nutrients in arid soils. Vadose Zone J 6(2):291–297. https://doi.org/10.2136/vzj2006.0068

Anderson IC, Cairney JWG (2007) Ectomycorrhizal fungi: exploring the mycelial frontier. FEMS Microbiol Rev 31(4):388–406. https://doi.org/10.1111/j.1574-6976.2007.00073.x

Asghari HR, Cavagnaro TR (2012) Arbuscular mycorrhizas reduce nitrogen loss via leaching. PLoS One 7(1):e29825. https://doi.org/10.1371/journal.pone.0029825

Augé RM (2001) Water relations, drought and vesicular-arbuscular mycorrhizal symbiosis. Mycorrhiza 11(1):3–42. https://doi.org/10.1007/s005720100097

Augé RM, Toler HD, Saxton AM (2014) Arbuscular mycorrhizal symbiosis alters stomatal conductance of host plants more under drought than under amply watered conditions: a meta-analysis. Mycorrhiza 25(1):13–24. https://doi.org/10.1007/s00572-014-0585-4

Averill C, Turner BL, Finzi AC (2014) Mycorrhiza-mediated competition between plants and decomposers drives soil carbon storage. Nature 505(7484):543–545. https://doi.org/10.1038/nature12901

Azcón-Aguilar C, Barea JM (1997) Arbuscular mycorrhizas and biological control of soil-borne plant pathogens - an overview of the mechanisms involved. Mycorrhiza 6(6):457–464. https://doi.org/10.1007/s005720050147

Balestrini R, Chitarra W, Antoniou C, Ruocco MR, Fotopoulos V (2018) Improvement of plant performance under water deficit with the employment of biological and chemical priming agents. J Agric Sci 156(5):680–688. https://doi.org/10.1017/s0021859618000126

Bao X, Zou J, Zhang B, Wu L, Yang T, Huang Q (2022) Arbuscular mycorrhizal fungi and microbes interaction in rice mycorrhizosphere. Agronomy 12(6):1277

Bapaume L, Reinhardt D (2012) How membranes shape plant symbioses: signalling and transport in nodulation and arbuscular mycorrhiza. Front Plant Sci 3:223. https://doi.org/10.3389/fpls.2012.00223

Begum N, Qin C, Ahanger MA, Raza S, Khan MI, Ashraf M, Ahmed N, Zhang L (2019) Role of arbuscular mycorrhizal fungi in plant growth regulation: implications in abiotic stress tolerance. Front Plant Sci 10:1068. https://doi.org/10.3389/fpls.2019.01068

Bender SF, Conen F, Van der Heijden MGA (2015) Mycorrhizal effects on nutrient cycling, nutrient leaching and N2O production in experimental grassland. Soil Biol Biochem 80:283–292. https://doi.org/10.1016/j.soilbio.2014.10.016

Benhamou N (1994) Resistance responses of mycorrhizal Ri T-DNA-transformed carrot roots to infection by *Fusarium oxysporum* Sp. *chrysanthemi*. Phytopathology 84(9):958. https://doi.org/10.1094/phyto-84-958

Beniwal RS, Langenfeld-Heyser R, Polle A (2010) Ectomycorrhiza and hydrogel protect hybrid poplar from water deficit and unravel plastic responses of xylem anatomy. Environ Exp Bot 69(2):189–197. https://doi.org/10.1016/j.envexpbot.2010.02.005

Bianciotto V, Bandi C, Minerdi D, Sironi M, Tichy HV, Bonfante P (1996) An obligately endosymbiotic mycorrhizal fungus itself harbors obligately intracellular bacteria. Appl Environ Microbiol 62(8):3005–3010. https://doi.org/10.1128/aem.62.8.3005-3010.1996

Blee KA, Anderson AJ (2000) Defense responses in plants to arbuscular mycorrhizal fungi. In: Podila GK, Douds DD (eds) Current advances in mycorrhizae research. The American Phytopathological Society, St Paul, pp 27–44

Bonfante P, Genre A (2010) Mechanisms underlying beneficial plant–fungus interactions in mycorrhizal symbiosis. Nat Commun 1(48):1–11. https://doi.org/10.1038/ncomms1046

Boutasknit A, Baslam M, Ait-El-Mokhtar M, Anli M, Ben-Laouane R, Douira A, El Modafar C, Mitsui T, Wahbi S, Meddich A (2020) Arbuscular mycorrhizal fungi mediate drought tolerance and recovery in two contrasting Carob (Ceratonia siliqua L.) ecotypes by regulating stomatal, water relations, and (in) organic adjustments. Plan Theory 9(1):80. https://doi.org/10.3390/plants9010080

Bowles TM, Jackson LE, Cavagnaro TR (2018) Mycorrhizal fungi enhance plant nutrient acquisition and modulate nitrogen loss with variable water regimes. Glob Chang Biol 24(1):171–182. https://doi.org/10.1111/gcb.13884

Bücking H, Liepold E, Ambilwade P (2012) The role of the mycorrhizal symbiosis in nutrient uptake of plants and the regulatory mechanisms underlying these transport processes. Plant Sci 4:108–132. https://doi.org/10.5772/52570

Cameron DD, Neal AL, van Wees SC, Ton J (2013) Mycorrhiza-induced resistance: more than the sum of its parts? Trends Plant Sci 18(10):539–545

Chandrasekaran M, Chanratana M, Kim K, Seshadri S, Sa T (2019) Impact of arbuscular mycorrhizal fungi on photosynthesis, water status, and gas exchange of plants under salt stress—a meta-analysis. Front Plant Sci 10:457. https://doi.org/10.3389/fpls.2019.00457

Chen W, Li J, Zhu H, Xu P, Chen J, Yao Q (2017) Arbuscular mycorrhizal fungus enhances lateral root formation in Poncirus trifoliata (L.) as revealed by RNA-Seq analysis. Front Plant Sci 8:2039. https://doi.org/10.3389/fpls.2017.02039

Cheng H-Q, Zou Y-N, Wu Q-S, Kuča K (2021) Arbuscular mycorrhizal fungi alleviate drought stress in trifoliate orange by regulating H+-atpase activity and gene expression. Front Plant Sci 12:659694. https://doi.org/10.3389/fpls.2021.659694

Chinnusamy V, Jagendorf A, Zhu J-K (2005) Understanding and improving salt tolerance in plants. Crop Sci 45(2):437–448. https://doi.org/10.2135/cropsci2005.0437

Chou LG, Schmitthenner AF (1974) Effect of rhizobium japonicum and Endogone mosseae on soybean root rot caused by *Phythium ultimum* and *Phytophthora megasperma* var. *Sojae*. Plant Dis Report 58:221–227

Chourasia KN, More SJ, Kumar A, Kumar D, Singh B, Bhardwaj V, Kumar A, Das SK, Singh RK, Zinta G, Tiwari RK, Lal MK (2022) Salinity responses and tolerance mechanisms in underground vegetable crops: an integrative review. Planta 255(3):68. https://doi.org/10.1007/s00425-022-03845-y

Churchland C, Grayston SJ (2014) Specificity of plant-microbe interactions in the tree mycorrhizosphere biome and consequences for soil C cycling. Front Microbiol 5:261. https://doi.org/10.3389/fmicb.2014.00261

Compant S, Duffy B, Nowak J, Clement C, Barka EA (2005) Use of plant growth-promoting bacteria for biocontrol of plant diseases: principles, mechanisms of action, and future prospects. Appl Environ Microbiol 71(9):4951–4959. https://doi.org/10.1128/aem.71.9.4951-4959.2005

Conrath U, Beckers GJM, Flors V, García-Agustín P, Jakab G, Mauch F, Newman M-A, Pieterse CMJ, Poinssot B, Pozo MJ, Pugin A, Schaffrath U, Ton J, Wendehenne D, Zimmerli L, Mauch-Mani B (2006) Priming: getting ready for battle. Mol Plant-Microbe Interact 19(10):1062–1071. https://doi.org/10.1094/mpmi-19-1062

Cordier C, Pozo MJ, Barea JM, Gianinazzi S, Gianinazzi-Pearson V (1998) Cell defense responses associated with localized and systemic resistance to Phytophthora parasitica induced in tomato by an arbuscular mycorrhizal fungus. Mol Plant-Microbe Interact 11(10):1017–1028. https://doi.org/10.1094/mpmi.1998.11.10.1017

Cruz C, Correia PF, Ramos AC, Carvalho L, Bago A, Martins-Loução MA (2008) Arbuscular mycorrhiza in physiological and morphological adaptations of mediterranean plants. Springer eBooks:733–752. https://doi.org/10.1007/978-3-540-78826-3_34

Daghino S, Martino E, Voyron S, Perotto S (2022) Metabarcoding of fungal assemblages in Vaccinium myrtillus endosphere suggests colonization of above-ground organs by some ericoid mycorrhizal and DSE fungi. Sci Rep 12(1):11013. https://doi.org/10.1038/s41598-022-15154-1

Daniels BA, Trappe JM (1980) Factors affecting spore germination of the vesicular-arbuscular mycorrhizal fungus, Glomus epigaeus. Mycologia 72(3):457. https://doi.org/10.2307/3759519

Dar MH, Razvi SM, Singh N, Mushtaq A, Dar S, Hussain S (2023) Arbuscular mycorrhizal fungi for salinity stress: anti-stress role and mechanisms. Pedosphere 33(1):212–224

Dastogeer KMG, Zahan MI, Tahjib-Ul-Arif M, Akter MA, Okazaki S (2020) Plant salinity tolerance conferred by arbuscular mycorrhizal fungi and associated mechanisms: a meta-analysis. Front Plant Sci 11:1927. https://doi.org/10.3389/fpls.2020.588550

Delgado A, Torrent J (2000) Phosphorus forms and desorption patterns in heavily fertilized calcareous and limed acid soils. Soil Sci Soc Am J 64(6):2031–2037

Demenois J, Rey F, Ibanez T, Stokes A, Carriconde F (2018) Linkages between root traits, soil fungi and aggregate stability in tropical plant communities along a successional vegetation gradient. Plant Soil 424(1–2):319–334. https://doi.org/10.1007/s11104-017-3529-x

Deslippe JR, Hartmann M, Mohn WW, Simard SW (2011) Long-term experimental manipulation of climate alters the ectomycorrhizal community of Betula nana in Arctic tundra. Glob Chang Biol 17(4):1625–1636. https://doi.org/10.1111/j.1365-2486.2010.02318.x

Diagne N, Ngom M, Djighaly PI, Fall D, Hocher V, Svistoonoff S (2020) Roles of arbuscular mycorrhizal fungi on plant growth and performance: importance in biotic and abiotic stressed regulation. Diversity 12(10):370. https://doi.org/10.3390/d12100370

Dighton J (2009) Mycorrhizae, Editor(s): Moselio schaechter, encyclopedia of microbiology, 3rd edn. Academic Press, pp 153–162. ISBN 9780123739445

Drigo B, Pijl AS, Duyts H, Kielak AM, Gamper HA, Houtekamer MJ, Boschker HTS, Bodelier PLE, Whiteley AS, Veen JAV, Kowalchuk GA (2010) Shifting carbon flow from roots into associated microbial communities in response to elevated atmospheric CO2. Proc Natl Acad Sci 107(24):10938–10942. https://doi.org/10.1073/pnas.0912421107

Duchesne LC, Peterson RL, Ellis BE (1987) The accumulation of plant-produced antimicrobial compounds in response to ectomycorrhizal fungi: a review. Phytoprotection 68:17–27

Elliott AJ, Daniell TJ, Cameron DD, Field KJ (2021) A commercial arbuscular mycorrhizal inoculum increases root colonization across wheat cultivars but does not increase assimilation of mycorrhiza-acquired nutrients. Plant People Planet 3(5):588–599. https://doi.org/10.1002/ppp3.10094

Erland, S. and Taylor, A. F. S., (2002). Diversity of ectomycorrhizal communities in relation to the abiotic environment. In Mycorrhizal Ecology. Eds. Marcel G A van der Heijden & Ian R Sanders. Ecological studies series. 157. ch 7. Springer-Verlag. Berlin 470.

Essahibi A, Benhiba L, Babram MA, Ghoulam C, Qaddoury A (2018) Influence of arbuscular mycorrhizal fungi on the functional mechanisms associated with drought tolerance in carob (*Ceratonia siliqua* L.). Trees 32(1):87–97. https://doi.org/10.1007/s00468-017-1613-8

Etesami H, Jeong BR, Glick BR (2021) Contribution of arbuscular mycorrhizal fungi, phosphate–solubilizing bacteria, and Silicon to P uptake by plant. Front Plant Sci 12:699618. https://doi.org/10.3389/fpls.2021.699618

Faber BA, Zasoski RJ, Munns DN, Shackel K (1991) A method for measuring hyphal nutrient and water uptake in mycorrhizal plants. Can J Bot 69(1):87–94. https://doi.org/10.1139/b91-012

Fall AF, Nakabonge G, Ssekandi J, Founoune-Mboup H, Apori SO, Ndiaye A, Badji A, Ngom K (2022) Roles of arbuscular mycorrhizal fungi on soil fertility: contribution in the improvement of physical, chemical, and biological properties of the soil. Front Fungal Biol 3. https://doi.org/10.3389/ffunb.2022.723892

Feng G, Zhang FS, Li XL, Tian CY, Tang C, Rengel Z (2002) Improved tolerance of maize plants to salt stress by arbuscular mycorrhiza is related to higher accumulation of soluble sugars in roots. Mycorrhiza 12(4):185–190. https://doi.org/10.1007/s00572-002-0170-0

Feng H, Meng PP, Dou Q, Zhang SX, Wang H-H, Wang CY (2019) Advances in mechanisms of nutrient exchange between mycorrhizal fungi and host plants. Ying Yong Sheng Tai Xue Bao. J Appl Ecol 30(10):3596–3604. https://doi.org/10.13287/j.1001-9332.201910.034

Fernández-Lizarazo JC, Moreno-Fonseca LP (2016) Mechanisms for tolerance to water-deficit stress in plants inoculated with arbuscular mycorrhizal fungi. A review. Agron Colomb 34(2):179–189. https://doi.org/10.15446/agron.colomb.v34n2.55569

Frew A, Antunes PM, Cameron DD, Furman TE, Trappe JM (2022) Plant herbivore protection by arbuscular mycorrhizas: a role for fungal diversity? New Phytol 233(3):1022–1031

Furman TE, Trappe JM (1971) Phylogeny and Ecology of mycotrophic achlorophyllous angiosperms. Q Rev Biol 46(3):219–225. https://doi.org/10.1086/406896

Garcés-Ruiz M, Senés-Guerrero C, Declerck S, Cranenbrouck S (2019) Community composition of arbuscular mycorrhizal fungi associated with native plants growing in a petroleum-polluted soil of the Amazon region of Ecuador. Microbiology 8(4):00703. https://doi.org/10.1002/mbo3.703

Giri B, Kapoor R, Mukerji KG (2003) Influence of arbuscular mycorrhizal fungi and salinity on growth, biomass, and mineral nutrition of Acacia auriculiformis. Biol Fertil Soils 38(3):170–175. https://doi.org/10.1007/s00374-003-0636-z

Glazebrook J (2005) Contrasting mechanisms of defence against biotrophic and necrotrophic pathogens. Annu Rev Phytopathol 43(1):205–227. https://doi.org/10.1146/annurev.phyto.43.040204.135923

Goicoechea N (2020) Mycorrhizal fungi as bio-protectors of crops against Verticillium wilt—a hypothetical scenario under changing environmental conditions. Plan Theory 9(11):1468. https://doi.org/10.3390/plants9111468

Gomez-Roldan V, Fermas S, Brewer PB, Puech-Pagès V, Dun EA, Pillot J-P, Letisse F, Matusova R, Danoun S, Portais J-C, Bouwmeester H, Bécard G, Beveridge CA, Rameau C, Rochange SF (2008) Strigolactone inhibition of shoot branching. Nature 455(7210):189–194. https://doi.org/10.1038/nature07271

Guillon C, St-Arnaud M, Hamel C, Jabaji-Hare SH (2002) Differential and systemic alteration of defence-related gene transcript levels in mycorrhizal bean plants infected with Rhizoctonia solani. Can J Bot 80(3):305–315. https://doi.org/10.1139/b02-015

Hacquard S, Schadt CW (2014) Towards a holistic understanding of the beneficial interactions across the populus microbiome. New Phytol 205(4):1424–1430. https://doi.org/10.1111/nph.13133

Hamiduzzaman MM, Jakab G, Barnavon L, Neuhaus JM, Mauch-Mani B (2005) β-Aminobutyric acid-induced resistance against downy mildew in grapevine acts through the potentiation of callose formation and jasmonic acid signaling. Mol Plant-Microbe Interact 18(8):819–829

Hao Z, Xie W, Chen B (2019) Arbuscular mycorrhizal symbiosis affects plant immunity to viral infection and accumulation. Viruses 11(6):534

Harrier LA, Watson CA (2004) The potential role of arbuscular mycorrhizal (AM) fungi in the bioprotection of plants against soil-borne pathogens in organic and/or other sustainable farming systems. Pest Manag Sci 60(2):149–157. https://doi.org/10.1002/ps.820

Hasanuzzaman M, Nahar K, Fujita M (2013) Plant response to salt stress and role of exogenous protectants to mitigate salt-induced damages. Ecophysiology and Responses of Plants under Salt Stress, pp 25–87. https://doi.org/10.1007/978-1-4614-4747-4_2

Hashem A, Abd-Allah EF, Alqarawi AA, El-Didamony G, Alwhibi Mona S, Egamberdieva D, Ahmad P (2014) Alleviation of adverse impact of salinity on faba bean (*Vicia faba* L.) by arbuscular mycorrhizal fungi. Pak J Bot 46:2003–2013

Hause B, Maier W, Miersch O, Kramell R, Strack D (2002) Induction of jasmonate biosynthesis in arbuscular mycorrhizal barley roots. Plant Physiol 130(3):1213–1220. https://doi.org/10.1104/pp.006007

He J-D, Zou Y-N, Wu Q-S, Kuca K (2020) Mycorrhizas enhance drought tolerance of trifoliate orange by enhancing activities and gene expression of antioxidant enzymes. Sci Hortic 262:108745–108745. https://doi.org/10.1016/j.scienta.2019.108745

Hobbie EA, Hobbie JE (2008) Natural abundance of 15N in nitrogen limited forests and tundra can estimate nitrogen cycling through mycorrhizal fungi: a review. Ecosystems 11(5):815–830. https://doi.org/10.1007/s10021-008-9159-7

Hodge A, Fitter AH (2010) Substantial nitrogen acquisition by arbuscular mycorrhizal fungi from organic material has implications for N cycling. Proc Natl Acad Sci 107(31):13754–13759. https://doi.org/10.1073/pnas.1005874107

Hodge A, Storer K (2014) Arbuscular mycorrhiza and nitrogen: implications for individual plants through to ecosystems. Plant Soil 386(1–2):1–19. https://doi.org/10.1007/s11104-014-2162-1

Ho-Plágaro T, García-Garrido JM (2022) Molecular regulation of arbuscular mycorrhizal symbiosis. Int J Mol Sci 23(11):5960. https://doi.org/10.3390/ijms23115960

Hrynkiewicz K, Baum C, Leinweber P, Weih M, Dimitriou I (2010) The significance of rotation periods for mycorrhiza formation in short rotation coppice. For Ecol Manag 260(11):1943–1949. https://doi.org/10.1016/j.foreco.2010.08.020

Hu X, Wang Y, Wang J, Xue Y, Li Z, Nagao K, Yanagita T, Xue C (2010) Dietary saponins of sea cucumber alleviate orotic acid-induced fatty liver in rats via PPARα and SREBP-1c signaling. Lipids Health Dis 9:1–9. https://doi.org/10.1186/1476-511x-9-25

Huang J, Luo S, Zeng R (2003) Mechanisms of plant disease resistance induced by arbuscular mycorrhizal fungi. J Appl Ecol 14(5):819–822

Huang D, Ma M, Wang Q, Zhang M, Jing G, Li C, Ma F (2020) Arbuscular mycorrhizal fungi enhanced drought resistance in apple by regulating genes in the MAPK pathway. Plant Physiol Biochem 149:245–255. https://doi.org/10.1016/j.plaphy.2020.02.020

Huey CJ, Gopinath SCB, Uda MNA, Zulhaimi HI, Jaafar MN, Kasim FH, Yaakub ARW (2020) Mycorrhiza: a natural resource assists plant growth under varied soil conditions. Biotech 10(5):1–9. https://doi.org/10.1007/s13205-020-02188-3

IJdo M, Cranenbrouck S, Declerck S (2011) Methods for large-scale production of AM fungi: past, present, and future. Mycorrhiza 21(1):1–16. https://doi.org/10.1007/s00572-010-0337-z

Ji L, Tan W, Chen X (2019) Arbuscular mycorrhizal mycelial networks and glomalin-related soil protein increase soil aggregation in Calcaric Regosol under well-watered and drought stress conditions. Soil Tillage Res 185:1–8. https://doi.org/10.1016/j.still.2018.08.010

Jung SC, Martinez-Medina A, Lopez-Raez JA, Pozo MJ (2012) Mycorrhiza-induced resistance and priming of plant defenses. J Chem Ecol 38:651–664

Kamran M, Malik Z, Parveen A, Huang L, Riaz M, Bashir S, Mustafa A, Abbasi GH, Xue B, Ali U (2019) Ameliorative effects of biochar on rapeseed (*Brassica napus* L.) growth and heavy metal immobilization in soil irrigated with untreated wastewater. J Plant Growth Regul 39(1):266–281. https://doi.org/10.1007/s00344-019-09980-3

Khalloufi M, Martínez-Andújar C, Lachaâl M, Karray-Bouraoui N, Pérez-Alfocea F, Albacete A (2017) The interaction between foliar GA 3 application and arbuscular mycorrhizal fungi inoculation improves growth in salinized tomato (*Solanum lycopersicum* L.) plants by modifying the hormonal balance. J Plant Physiol 214:134–144. https://doi.org/10.1016/j.jplph.2017.04.012

Kiriachek SG, de Azevedo LCB, Peres LEP, Lambais MR (2009) Regulação do desenvolvimento de micorrizas arbusculares. Rev Bras Ciênc Solo 33(1):1–16. https://doi.org/10.1590/s0100-06832009000100001

Kodre A, Arčon I, Debeljak M, Potisek M, Likar M, Vogel-Mikuš K (2017) Arbuscular mycorrhizal fungi alter Hg root uptake and ligand environment as studied by X-ray absorption fine structure. Environ Exp Bot 133:12–23

Kuila D, Ghosh S (2022) Aspects, problems and utilization of arbuscular mycorrhizal (AM) application as bio-fertilizer in sustainable agriculture. Curr Res Microb Sci 3:100107. https://doi.org/10.1016/j.crmicr.2022.100107

Law SR, Serrano AR, Daguerre Y, Sundh J, Schneider AN, Stangl ZR, Hurry V (2022) Metatranscriptomics captures dynamic shifts in mycorrhizal coordination in boreal forests. Proc Natl Acad Sci 119(26):e2118852119

Lee CS, Lee YJ, Jeun YC (2005) Observations of infection structures on the leaves of cucumber plants pre-treated with arbuscular mycorrhiza glomus intraradices after challenge inoculation with Colletotrichum orbiculare. Plant Pathol J 21(3):237–243
Lehman RM, Taheri WI, Osborne SL, Buyer JS, Douds DD Jr (2012) Fall cover cropping can increase arbuscular mycorrhizae in soils supporting intensive agricultural production. Appl Soil Ecol 61:300–304
Lehmann J, Hansel CM, Kaiser C, Kleber M, Maher K, Manzoni S, Nunan N, Reichstein M, Schimel JP, Torn MS, Wieder WR, Kögel-Knabner I (2020) Persistence of soil organic carbon caused by functional complexity. Nat Geosci 13(8):529–534. https://doi.org/10.1038/s41561-020-0612-3
Lehto T, Zwiazek JJ (2011) Ectomycorrhizas and water relations of trees: a review. Mycorrhiza 21(2):71–90. https://doi.org/10.1007/s00572-010-0348-9
Li J, Meng B, Chai H, Yang X, Song W, Li S, Lu A, Zhang T, Sun W (2019) Arbuscular mycorrhizal fungi alleviate drought stress in C3 (Leymus chinensis) and C4 (Hemarthria altissima) grasses via altering antioxidant enzyme activities and photosynthesis. Front Plant Sci 10:499. https://doi.org/10.3389/fpls.2019.00499
Liu L, Li J, Yue F, Yan X, Wang F, Bloszies S, Wang Y (2018) Effects of arbuscular mycorrhizal inoculation and biochar amendment on maize growth, cadmium uptake and soil cadmium speciation in Cd-contaminated soil. Chemosphere 194:495–503. https://doi.org/10.1016/j.chemosphere.2017.12.025
Liu L, Ma L, Zhu M, Liu B, Liu X, Shi Y (2023) Rhizosphere microbial community assembly and association networks strongly differ based on vegetation type at a local environment scale. Front Microbiol 14:1129471. https://doi.org/10.3389/fmicb.2023.1129471
Lopez-Obando M, Ligerot Y, Bonhomme S, Boyer F-D, Rameau C (2015) Strigolactone biosynthesis and signaling in plant development. Development 142(21):3615–3619. https://doi.org/10.1242/dev.120006
Ma Y, Zhang H, Wang D, Guo X, Yang T, Xiang X, Walder F, Chu H (2021) Differential responses of arbuscular mycorrhizal fungal communities to long-term fertilization in the wheat rhizosphere and root endosphere. Appl Environ Microbiol 87(17):00349–00321. https://doi.org/10.1128/aem.00349-21
Ma Y, Ankit, Tiwari J, Bauddh K (2022) Plant-mycorrhizal fungi interactions in phytoremediation of geogenic contaminated soils. Front Microbiol 13:843415. https://doi.org/10.3389/fmicb.2022.843415
MacLean AM, Bravo A, Harrison MJ (2017) Plant signaling and metabolic pathways enabling arbuscular mycorrhizal symbiosis. Plant Cell 29(10):2319–2335. https://doi.org/10.1105/tpc.17.00555
Mahanta D, Rai RK, Dhar S, Varghese E, Raja A, Purakayastha TJ (2018) Modification of root properties with phosphate solubilizing bacteria and arbuscular mycorrhiza to reduce rock phosphate application in soybean-wheat cropping system. Ecol Eng 111:31–43. https://doi.org/10.1016/j.ecoleng.2017.11.008
Marschner H, Dell B (1994) Nutrient uptake in mycorrhizal symbiosis. Plant Soil 159(1):89–102. https://doi.org/10.1007/bf00000098
Marx DH (1975) Mycorrhizae and establishment of trees on strip-mined land. Ohio J Sci 75:288–297
Meier AR, Hunter MD (2018) Mycorrhizae alter toxin sequestration and performance of two specialist herbivores. Front Ecol Evol 6:33. https://doi.org/10.3389/fevo.2018.00033
Miller RM, Jastrow JD, Reinhardt DR (1995) External hyphal production of vesicular-arbuscular mycorrhizal fungi in pasture and tallgrass prairie communities. Oecologia 103:17–23. https://doi.org/10.1007/BF00328420
Miransari M (2011a) Interactions between arbuscular mycorrhizal fungi and soil bacteria. Appl Microbiol Biotechnol 89:917–930
Miransari M (2011b) Soil microbes and plant fertilization. Appl Microbiol Biotechnol 92(5):875–885. https://doi.org/10.1007/s00253-011-3521-y

Mohandas S, Poovarasan S, Panneerselvam P, Saritha B, Upreti KK, Kamal R, Sita T (2013) Guava (Psidium guajava L.) rhizosphere glomus mosseae spores harbor actinomycetes with growth promoting and antifungal attributes. Sci Hortic 150:371–376. https://doi.org/10.1016/j.scienta.2012.11.019

Moitinho MR, Fernandes C, Truber PV, Marcelo AV, Corá JE, da Silva Bicalho E (2020) Arbuscular mycorrhizal fungi and soil aggregation in a no-tillage system with crop rotation. J Plant Nutr Soil Sci 183(4):482–491

Moore JA (2016) Plant-microbe interactions influence ecosystem processes: the role of roots and mycorrhizal fungal hyphae on terrestrial carbon dynamics. PhD thesis,. University of Tennessee

Morgado P, Sousa N, Cerqueira JJ (2015) The impact of stress in decision making in the context of uncertainty. J Neurosci Res 93(6):839–847. https://doi.org/10.1002/jnr.23521

Nasim G (2010) The role of arbuscular mycorrhizae in inducing resistance to drought and salinity stress in crops. In: Ashraf M, Ozturk M, Ahmad M (eds) Plant adaptation and phytoremediation. Springer, Dordrecht. https://doi.org/10.1007/978-90-481-9370-7_6

Nazir R, Zhang M, de Boer W, van Elsas JD (2012) The capacity to comigrate with Lyophyllum sp. strain Karsten through different soils is spread among several phylogenetic groups within the genus Burkholderia. Soil Biol Biochem 50:221–233. https://doi.org/10.1016/j.soilbio.2012.03.015

Neidhardt FC, Ingraham JL, Schaechter M (1990) Physiology of the bacterial cell: a molecular approach. Sinauer Associates, Sunderland, p 506

Nemec S (1981) Histochemical characteristics of glomus etunicatus infection of Citrus Limon fibrous roots. Can J Bot 59:609–617

Nogales García A, Camprubí Nieto A, Estaun V, Calvet Pinós M (2008) Mycorrhizal inoculation of grapevines in replant soils: improved field application and plant performance. University of Catalonia, Catalonia

Oliveira RS, Rocha I, Ma Y, Vosátka M, Freitas H (2016) Seed coating with arbuscular mycorrhizal fungi as an ecotechnologicalapproach for sustainable agricultural production of common wheat (*Triticum aestivum* L.). J Toxic Environ Health A 79(7):329–337. https://doi.org/10.1080/15287394.2016.1153448

Ortiz-Bobea A, Ault TR, Carrillo CM, Chambers RG, Lobell DB (2021) Anthropogenic climate change has slowed global agricultural productivity growth. Nat Clim Chang 11(4):306–312

Parihar M, Rakshit A, Meena VS, Gupta VK, Rana K, Choudhary M, Tiwari G, Mishra PK, Pattanayak A, Bisht JK, Jatav SS, Khati P, Jatav HS (2020) The potential of arbuscular mycorrhizal fungi in C cycling: a review. Arch Microbiol 202(7):1581–1596. https://doi.org/10.1007/s00203-020-01915-x

Parniske M (2008) Arbuscular mycorrhiza: the mother of plant root endosymbioses. Nat Rev Microbiol 6(10):763–775

Pent M, Põldmaa K, Bahram M (2017) Bacterial communities in boreal forest mushrooms are shaped both by soil parameters and host identity. Front Microbiol 8:1574. https://doi.org/10.3389/fmicb.2017.00836

Peterson LR, Massicotte HB, Melville LH (2004) Mycorrhizas: anatomy and cell biology. Nrc Research Press; Wallingford, Oxon, Ottawa

Phillips RP, Brzostek E, Midgley MG (2013) The mycorrhizal-associated nutrient economy: a new framework for predicting carbon-nutrient couplings in temperate forests. New Phytol 199(1):41–51. https://doi.org/10.1111/nph.12221

Plett JM, Daguerre Y, Wittulsky S, Vayssières A, Deveau A, Melton SJ et al (2014) Effector MiSSP7 of the mutualistic fungus *Laccaria bicolor* stabilizes the *Populus* JAZ6 protein and represses jasmonic acid (JA) responsive genes. Proc Natl Acad Sci 111(22):8299–8304. https://doi.org/10.1073/pnas.1322671111

Porcel R, Aroca R, Ruiz-Lozano JM (2012) Salinity stress alleviation using arbuscular mycorrhizal fungi. A review. Agron Sustain Dev 32(1):181–200. https://doi.org/10.1007/s13593-011-0029-x

Potapov P, Turubanova S, Hansen MC, Tyukavina A, Zalles V, Khan A, Song X-P, Pickens A, Shen Q, Cortez J (2022) Global maps of cropland extent and change show accelerated cropland expansion in the twenty-first century. Nature Food 3(1):19–28

Pozo MJ, Azcón-Aguilar C (2007) Unraveling mycorrhiza-induced resistance. Curr Opin Plant Biol 10(4):393–398. https://doi.org/10.1016/j.pbi.2007.05.004

Pozo MJ, Cordier C, Dumas-Gaudot E, Gianinazzi S, Barea JM, Azcón-Aguilar C (2002) Localized versus systemic effect of arbuscular mycorrhizal fungi on defence responses to Phytophthora infection in tomato plants. J Exp Bot 53(368):525–534. https://doi.org/10.1093/jexbot/53.368.525

Punamiya P, Datta R, Sarkar D, Barber S, Patel M, Das P (2010) Symbiotic role of glomus mosseae in phytoextraction of lead in vetiver grass [Chrysopogon zizanioides (L.)]. J Hazard Mater 177(1–3):465–474. https://doi.org/10.1016/j.jhazmat.2009.12.056

Rajtor M, Piotrowska-Seget Z (2016) Prospects for arbuscular mycorrhizal fungi (AMF) to assist in phytoremediation of soil hydrocarbon contaminants. Chemosphere 162:105–116. https://doi.org/10.1016/j.chemosphere.2016.07.071

Ramos AC, Façanha AR, Feijó JA (2008) Proton (H+) flux signature for the presymbiotic development of the arbuscular mycorrhizal fungi. New Phytol 178(1):177–188

Rasanayagam S, Jeffries P (1992) Production of acid is responsible for antibiosis by some ectomycorrhizal fungi. Mycol Res 96(11):971–976. https://doi.org/10.1016/s0953-7562(09)80600-x

Rasmussen HN, Rasmussen FN (2009) Orchid mycorrhiza: implications of a mycophagous life style. Oikos 118(3):334–345

Read DJ, Perez-Moreno J (2003) Mycorrhizas and nutrient cycling in ecosystems—a journey towards relevance? New Phytol 157(3):475–492. https://doi.org/10.1046/j.1469-8137.2003.00704.x

Redecker D, Schüßler A, Stockinger H, Stürmer SL, Morton JB, Walker C (2013) An evidence-based consensus for the classification of arbuscular mycorrhizal fungi (Glomeromycota). Mycorrhiza 23(7):515–531. https://doi.org/10.1007/s00572-013-0486-y

Reinhardt D (2007) Programming good relations—development of the arbuscular mycorrhizal symbiosis. Curr Opin Plant Biol 10(1):98–105

Rivero J, Álvarez D, Flors V, Azcón-Aguilar C, Pozo MJ (2018) Root metabolic plasticity underlies functional diversity in mycorrhiza-enhanced stress tolerance in tomato. New Phytol 220(4):1322–1336

Rivero J, Lidoy J, Llopis-Giménez Á, Herrero S, Flors V, Pozo MJ (2021) Mycorrhizal symbiosis primes the accumulation of antiherbivore compounds and enhances herbivore mortality in tomato. J Exp Bot 72(13):5038–5050

Rouphael Y, Franken P, Schneider C, Schwarz D, Giovannetti M, Agnolucci M, Pascale SD, Bonini P, Colla G (2015) Arbuscular mycorrhizal fungi act as biostimulants in horticultural crops. Sci Hortic 196:91–108. https://doi.org/10.1016/j.scienta.2015.09.002

Ruiz-Lozano JM (2003) Arbuscular mycorrhizal symbiosis and alleviation of osmotic stress. New perspectives for molecular studies. Mycorrhiza 13(6):309–317. https://doi.org/10.1007/s00572-003-0237-6

Ruiz-Lozano JM, Azcón R (2000) Symbiotic efficiency and infectivity of an autochthonous arbuscular mycorrhizal glomus sp. from saline soils and Glomus deserticola under salinity. Mycorrhiza 10(3):137–143. https://doi.org/10.1007/s005720000075

Sandrini M, Nerva L, Sillo F, Balestrini R, Chitarra W, Zampieri E (2022) Abiotic stress and belowground microbiome: the potential of omics approaches. Int J Mol Sci 23(3):1091

Sato T, Hachiya S, Inamura N, Ezawa T, Cheng W, Tawaraya K (2019) Secretion of acid phosphatase from extraradical hyphae of the arbuscular mycorrhizal fungus Rhizophagus clarus is regulated in response to phosphate availability. Mycorrhiza 29(6):599–605. https://doi.org/10.1007/s00572-019-00923-0

Schelkle M, Peterson RL (1997) Suppression of common root pathogens by helper bacteria and ectomycorrhizal fungi in vitro. Mycorrhiza 6(6):481–485. https://doi.org/10.1007/s005720050151

Schönbeck F, Dehne HW (1989) VA-mycorrhiza and plant health. In: Developments in soil science, vol 18. Elsevier, pp 83–91. https://doi.org/10.1016/s0166-2481(08)70200-9
Schulz-Bohm K, Martín-Sánchez L, Garbeva P (2017) Microbial volatiles: small molecules with an important role in intra- and inter-kingdom interactions. Front Microbiol 8:2484. https://doi.org/10.3389/fmicb.2017.02484
Sharma MP, Grover M, Chourasiya D, Bharti A, Agnihotri R, Maheshwari HS, Pareek A, Buyer JS, Sharma SK, Schütz L, Mathimaran N, Singla-Pareek SL, Grossman JM, Bagyaraj DJ (2020) Deciphering the role of trehalose in tripartite symbiosis among rhizobia, arbuscular mycorrhizal fungi, and legumes for enhancing abiotic stress tolerance in crop plants. Front Microbiol 11:509919. https://doi.org/10.3389/fmicb.2020.509919
Shen J, Yuan L, Zhang J, Li H, Bai Z, Chen X, Zhang W, Zhang F (2011) Phosphorus dynamics: from soil to plant. Plant Physiol 156(3):997–1005. https://doi.org/10.1104/pp.111.175232
Shi J, Zhao B, Zheng S, Zhang X, Wang X, Dong W, Xie Q, Wang G, Xiao Y, Chen F, Yu N, Wang E (2021) A phosphate starvation response-centered network regulates mycorrhizal symbiosis. Cell 184(22):5527–5540.e18. https://doi.org/10.1016/j.cell.2021.09.030
Shi J, Zhao B, Jin R, Hou L, Zhang X, Dai H, Yu N, Wang E (2022) A phosphate starvation response-regulated receptor-like kinase, OsADK1, is required for mycorrhizal symbiosis and phosphate starvation responses. New Phytol 236(6):2282–2293
Singh AK, Zhu X, Chen C, Wu J, Yang B, Zakari S, Jiang XJ, Singh N, Liu W (2022) The role of glomalin in mitigation of multiple soil degradation problems. Crit Rev Environ Sci Technol 52(9):1604–1638. https://doi.org/10.1080/10643389.2020.1862561
Smith SE, Read DJ (1997) Mycorrhizal symbiosis, 2nd edn. Academic Press, London. [Online] Available at: https://www.scirp.org/(S(i43dyn45teexjx455qlt3d2q))/reference/ReferencesPapers.aspx?ReferenceID=1300286. Accessed 27 Oct 2023
Smith SE, Read DJ (2008) Mycorrhizal Symbiosis, 3rd edn. Academic Press, London
Smith FA, Smith SE (2011) What is the significance of the arbuscular mycorrhizal colonisation of many economically important crop plants? Plant Soil 348(1–2):63–79. https://doi.org/10.1007/s11104-011-0865-0
Solaiman ZM, Mickan B (2014) Use of mycorrhiza in sustainable agriculture and land restoration. In: Solaiman Z, Abbott L, Varma A (eds) Mycorrhizal fungi: use in sustainable agriculture and land restoration. Soil biology, vol 41. Springer, Berlin, Heidelberg. https://doi.org/10.1007/978-3-662-45370-4_1
Sooriyaarachchi ND, Zakeel MCM, Safeena MIS, Abhayapala KMRD (2021) Role of rhizosphere and endophytic microbes in alleviation of biotic and abiotic stress in plants. Springer eBooks, pp 195–235. https://doi.org/10.1007/978-981-16-2922-8_9
Soudzilovskaia NA, van Bodegom PM, Terrer C, van't Zelfde M, McCallum I, McCormack ML et al (2019) Global mycorrhizal plant distribution linked to terrestrial carbon stocks. Nat Commun 10:5077. https://doi.org/10.1038/s41467-019-13019-2
Souza T (2015) Handbook of arbuscular mycorrhizal fungi. Springer
Stamou GP, Konstadinou S, Monokrousos N, Mastrogianni A, Orfanoudakis M, Hassiotis C, Menkissoglu-Spiroudi U, Vokou D, Papatheodorou EM (2017) The effects of arbuscular mycorrhizal fungi and essential oil on soil microbial community and N-related enzymes during the fungal early colonization phase. Aims Microbiol 3(4):938–959. https://doi.org/10.3934/microbiol.2017.4.938
Taylor DL, Bruns TD (1999) Community structure of ectomycorrhizal fungi in a Pinus muricata forest: minimal overlap between the mature forest and resistant propagule communities. Mol Ecol 8(11):1837–1850
Tedersoo L, Bahram M, Põlme S, Kõljalg U, Yorou NS, Wijesundera R, Villarreal Ruiz L, Vasco-Palacios AM, Thu PQ, Suija A, et al. (2014). Fungal biogeography. Global diversity and geography of soil fungi. Science 346(6213). https://doi.org/10.1126/science.1256688
Tibbett M, Sanders F (2002) Ectomycorrhizal symbiosis can enhance plant nutrition through improved access to discrete organic nutrient patches of high resource quality. Ann Bot 89(6):783–789

Tibbett M, Grantham K, Sanders FE, Cairney J (1998) Induction of cold active acid phosphomonoesterase activity at low temperature in psychrotrophic ectomycorrhizal Hebeloma spp. Mycol Res 102(12):1533–1539. https://doi.org/10.1017/s0953756298006753

Tobar R, Azcón R, Barea JM (1994) Improved nitrogen uptake and transport from15N-labelled nitrate by external hyphae of arbuscular mycorrhiza under water-stressed conditions. New Phytol 126(1):119–122. https://doi.org/10.1111/j.1469-8137.1994.tb07536.x

Ton J, Van Pelt JA, Van Loon LC, Pieterse CMJ (2002) Differential effectiveness of salicylate-dependent and jasmonate/ethylene-dependent induced resistance in Arabidopsis. Mol Plant-Microbe Interact 15(1):27–34. https://doi.org/10.1094/mpmi.2002.15.1.27

Torres N, Antolín MC, Goicoechea N (2018) Arbuscular mycorrhizal symbiosis as a promising resource for improving berry quality in grapevines under changing environments. Front Plant Sci 9:00897. https://doi.org/10.3389/fpls.2018.00897

Treseder KK, Allen MF (2000) Mycorrhizal fungi have a potential role in soil carbon storage under elevated CO2 and nitrogen deposition. New Phytol 147(1):189–200

Treseder KK, Turner KM (2007) Glomalin in ecosystems. Soil Sci Soc Am J 71(4):1257–1266

Truman W, Bennett MK, Kubigsteltig I, Turnbull C, Grant M (2007) Arabidopsis systemic immunity uses conserved defense signaling pathways and is mediated by jasmonates. Proc Natl Acad Sci USA 104(3):1075–1080. https://doi.org/10.1073/pnas.0605423104

Tu CC, Hsieh TF, Chang Y-C (1996) Vegetable diseases incited by Rhizoctonia spp. In: Rhizoctonia species: taxonomy, molecular biology, ecology, pathology and disease control. Springer, Dordrecht, pp 369–377. https://doi.org/10.1007/978-94-017-2901-7_34

Urban A, Puschenreiter M, Strauss J, Gorfer M (2008) Diversity and structure of ectomycorrhizal and co-associated fungal communities in a serpentine soil. Mycorrhiza 18:339–354

van Aarle IM, Olsson PA (2003) Fungal lipid accumulation and development of mycelial structures by two arbuscular mycorrhizal fungi. Appl Environ Microbiol 69(11):6762–6767. https://doi.org/10.1128/aem.69.11.6762-6767.2003

van der Heijden MGA, Martin FM, Selosse M-A, Sanders IR (2015) Mycorrhizal ecology and evolution: the past, the present, and the future. New Phytol 205(4):1406–1423. https://doi.org/10.1111/nph.13288

Victorino ÍMM, Berruti A, Orgiazzi A, Voyron S, Bianciotto V, Lumini E (2020) High-throughput DNA sequence-based analysis of AMF communities. In: Ferrol N, Lanfranco L (eds) Arbuscular mycorrhizal fungi. Methods in molecular biology, vol 2146, Humana, New York. https://doi.org/10.1007/978-1-0716-0603-2_9

Vinichuk M, Mårtensson A, Rosén K (2013) Inoculation with arbuscular mycorrhizae does not improve 137Cs uptake in crops grown in the Chernobyl region. J Environ Radioact 126:14–19

Vosátka M, Látr A, Gianinazzi S, Albrechtová J (2012) Development of arbuscular mycorrhizal biotechnology and industry: current achievements and bottlenecks. Symbiosis 58(1–3):29–37. https://doi.org/10.1007/s13199-012-0208-9

Wahab A, Muhammad M, Munir A, Abdi G, Zaman W, Ayaz A, Khizar C, Reddy SPP (2023) Role of arbuscular mycorrhizal fungi in regulating growth, enhancing productivity, and potentially influencing ecosystems under abiotic and biotic stresses. Plan Theory 12(17):3102

Wang J, Zhang H, Gao J, Zhang Y, Liu Y, Tang M (2021) Effects of ectomycorrhizal fungi (Suillus variegatus) on the growth, hydraulic function, and non-structural carbohydrates of Pinus tabulaeformis under drought stress. BMC Plant Biol 21(1):1–13. https://doi.org/10.1186/s12870-021-02945-3

Watts-Williams SJ, Cavagnaro TR (2018) Arbuscular mycorrhizal fungi increase grain zinc concentration and modify the expression of root ZIP transporter genes in a modern barley (Hordeum vulgare) cultivar. Plant Sci 274:163–170

Wei X, Zhang W, Zulfiqar F, Zhang C, Chen J (2022) Ericoid mycorrhizal fungi as biostimulants for improving propagation and production of ericaceous plants. Front Plant Sci 13:102739. https://doi.org/10.3389/fpls.2022.1027390

Weng W, Yan J, Zhou M, Yao X, Gao A, Ma C, Cheng J, Ruan J (2022) Roles of arbuscular mycorrhizal fungi as a biocontrol agent in the control of plant diseases. Microorganisms 10(7):1266

Wilkes TI, Warner DJ, Edmonds-Brown V, Davies KG, Denholm I (2021) Zero tillage systems conserve arbuscular mycorrhizal fungi, enhancing soil glomalin and water stable aggregates with implications for soil stability. Soil Systems 5(1):4

Willis A, Rodrigues BF, Harris PJ (2013) The ecology of arbuscular mycorrhizal fungi. Crit Rev Plant Sci 32(1):1–20

Wright SF, Upadhyaya A, Buyer JS (1998) Comparison of N-linked oligosaccharides of glomalin from arbuscular mycorrhizal fungi and soils by capillary electrophoresis. Soil Biol Biochem 30(13):1853–1857

Wu Q-S, Zou Y-N (2017) Arbuscular mycorrhizal fungi and tolerance of drought stress in plants. Arbuscular Mycorrhizas and Stress Tolerance of Plants, pp 25–41. https://doi.org/10.1007/978-981-10-4115-0_2

Wu HH, Zou YN, Rahman MM, Ni QD, Wu QS (2017) Mycorrhizas alter sucrose and proline metabolism in trifoliate orange exposed to drought stress. Sci Rep 7(1):42389. https://doi.org/10.1038/srep42389

Xu H, Shao H, Lu Y (2019) Arbuscular mycorrhiza fungi and related soil microbial activity drive carbon mineralization in the maize rhizosphere. Ecotoxicol Environ Saf 182:109476

Yanan W, Xusheng Z, Baozhong Y, Wenchao Z, Jintang G (2015) Biochemical defences induced by mycorrhizae fungi glomus mosseae in controlling strawberry Fusarium wilt. Open Biomed Eng J 9(1):301–304. https://doi.org/10.2174/1874120701509010301

Yao Y-F, Zhang YM, Zhou P, Zhang B, Qiu WY, Tseng SCG (2003) Therapeutic penetrating keratoplasty in severe fungal keratitis using cryopreserved donor corneas. Br J Ophthalmol 87(5):543–547. https://doi.org/10.1136/bjo.87.5.543

Ye S, Yang Y, Xin G, Wang Y, Ruan L, Ye G (2015) Studies of the Italian ryegrass–rice rotation system in southern China: arbuscular mycorrhizal symbiosis affects soil microorganisms and enzyme activities in the Lolium mutiflorum L. rhizosphere. Appl Soil Ecol 90:26–34

Yoneyama K, Xie X, Kusumoto D, Sekimoto H, Sugimoto Y, Takeuchi Y, Yoneyama K (2007) Nitrogen deficiency as well as phosphorus deficiency in sorghum promotes the production and exudation of 5-deoxystrigol, the host recognition signal for arbuscular mycorrhizal fungi and root parasites. Planta 227(1):125–132. https://doi.org/10.1007/s00425-007-0600-5

Zakeel MCM (2017) Agricultural microbiology. S. Godage and Brothers (pvT) ltd, p 192

Zhang F, Zou Y-N, Wu Q-S (2018) Quantitative estimation of water uptake by mycorrhizal extraradical hyphae in citrus under drought stress. Sci Hortic 229:132–136. https://doi.org/10.1016/j.scienta.2017.10.038

Zhang F, Zou Y-N, Wu Q-S, Kuča K (2020) Arbuscular mycorrhizas modulate root polyamine metabolism to enhance drought tolerance of trifoliate orange. Environ Exp Bot 171:103926. https://doi.org/10.1016/j.envexpbot.2019.103926

Zhao X, Zhang J, Chen C, Yang J, Zhu H, Liu M, Lv F (2014) Deep sequencing–based comparative transcriptional profiles of Cymbidium hybridum roots in response to mycorrhizal and non-mycorrhizal beneficial fungi. BMC Genomics 15:1–22

Zhao W, Chen Z, Yang X, Sheng L, Mao H, Zhu S (2023) Metagenomics reveal arbuscular mycorrhizal fungi altering functional gene expression of rhizosphere microbial community to enhance Iris tectorum's resistance to Cr stress. Sci Total Environ 895:164970. https://doi.org/10.1016/j.scitotenv.2023.164970

Zhou Y, Zhou B, Pache L, Chang M, Khodabakhshi AH, Tanaseichuk O, Benner C, Chanda SK (2019) Metascape provides a biologist-oriented resource for the analysis of systems-level datasets. Nat Commun 10(1):1523. https://doi.org/10.1038/s41467-019-09234-6

Zou YN, Wu QS, Kuča K (2021) Unravelling the role of arbuscular mycorrhizal fungi in mitigating the oxidative burst of plants under drought stress. Plant Biol 23:50–57. https://doi.org/10.1111/plb.13161

Endophytic Fungi Secondary Metabolites and Their Antimicrobial Activity

20

Prajakta Y. Pachorkar and Vaishnavi R. Pawar

Abstract

Endophytic fungi are considered as the most beneficial and important components of plant ecosystems; they live in plant tissues without causing any harmful symptoms or harmful changes on the ecosystem—they inhabit the internal tissues without negative impacts. They live in symbiotic association with plants, aiding their growth, improvement, and development and enhancing tolerance to certain biotic and abiotic factors such as stress and damage. They mitigate these negative impacts by synthesizing a wide range of bioactive compounds which exhibit various antigenic activities such as antimicrobial, antimalarial, anti-inflammatory, anti-hypercholesterolemic, anticancer, antioxidant, and biocontrol properties. They actively participated in bioremediation, nutrient cycling, and biodegradation, which makes them an important source for the pharmaceutical industry in terms of antibiotics and drugs, as well as for the agricultural sector. Endophytic fungi play an important role in antibiotic production that enables them to survive and withstand in competitive habitats with other microorganisms and protect their host plants against fungal and bacterial pathogens. Nowadays, pharmaceutical industries rely on plant-based antibiotics, which leads to the diminishment of natural flora due to deforestation. Thus, there is a need to develop complementary and subsidiary sources for the pharmaceutical industry. Endophytic fungi serve as an alternative antimicrobial source, possessing compounds similar to those found in plants, and can replace traditional methods. Large-scale cultivation of endophytic fungi produces many natural products while keeping the overall cost reasonable. The biochemical diversity of endophytic fungi suggests that there are still many unknown active compounds which are not known for its beneficial role in pharmaceutical aspects. This chapter

P. Y. Pachorkar (✉) · V. R. Pawar
P.G. Department of Microbiology, K. R. T. Arts, B. H. Commerce and A. M. Science College, Nashik, Maharashtra, India

© The Author(s), under exclusive license to Springer Nature Singapore Pte Ltd. 2025

K. A. Abd-Elsalam, A. H. Hashem (eds.), *Fungal Endophytes Volume II*,
https://doi.org/10.1007/978-981-97-8804-0_20

focuses mainly on the secondary metabolites produced by endophytic fungi and their pharmaceutical aspects for the prevention of various diseases, in light of environmental and biotechnological advances.

Keywords

Endophytic fungi · Host-plant interactions · Antimicrobial activity · Bioactive compounds · Pharmacological activity · Nanotechnology · Antiparasitic activity · Bioactive molecules · Biocontrol · Endophyte

20.1 Introduction

Endophytic fungi are beneficial and have a mutual relationship with their host plants, providing several positive effects. They produce substances that promote plant growth and possess antibacterial, antioxidant, and antitumor properties. Research on endophytic fungi is crucial for plant protection, biopesticide development, biopharmacology, and the discovery of new medicinal resources. These areas are critical for the advancement of agriculture, forestry, and medicine. Recent advances in multi-omics approaches have enabled the identification and characterization of metabolites through various advanced techniques and the comparative analysis of extensive omics datasets. This also elucidates the role of diverse endophytic fungi associated with key crops, which is essential for the sustainability of agriculture (Jha et al. 2023). In this chapter, we will explore the analysis of new bioactive components derived from mangrove endophytic fungi. These components are widely used in the production and packaging of natural compounds important to the pharmaceutical, industrial, and agricultural sectors. It has been discovered that all plant species can synthesize endophytes. Endophytic biology studies plants in conditions likely to encounter interesting endophytic fungi across various settings and ecosystems, including mangrove areas. We will use research from studies conducted by Suryanarayanan et al. (2012), Ananda et al. (2002), and Bibi et al. (2020) to aid our analysis of the antimicrobial activity of these components in pharmacology, antiparasitic activity, and other related fields. Plants flourish with a diverse range of microorganisms such as bacteria and fungi, collectively known as phytomicrobiomes. Thus, the relationship among the phytomicrobiomes is crucial for maintaining ecological balance. Through photosynthesis, plants convert atmospheric carbon into food, making them autotrophic. Fungal species then utilize this fixed carbon to provide nutrients from the soil rhizosphere which harbors the plant growth promoting microorganisms to the host plant (Wu et al. 2012).

Endophytic fungi colonize various plant parts such as buds, cells, hyaline cells, weed inflorescences, leaf segments, petioles, roots, stems, seeds, and fruits to fulfill their roles. This symbiotic association exists in two forms: mycorrhizal and endophytic fungi. The presence of special features like cellular contact with arbuscles, specific coordination between the fungal and plant partners, and mutual benefit are the three characteristics that distinguish mycorrhizal fungi from other groups. The structure of endophyte communities is influenced by various biotic and abiotic factors such as plant-microbe interactions, biogeography, plant species, and soil texture and other conditions, which impact species diversity, richness, and relative abundance (Deng et al. 2017).

The interactions between fungi and plants have both direct and indirect mechanisms. Few examples of these mechanisms include the production of secondary metabolites that shield plants from diseases and herbivory and facilitate plant growth through various strategies. Endophytic fungi also help to protect plants from a wide range of pathogens by colonizing sites, competing with pathogens for nutrients, producing antibiotics, and inducing systemic resistance against pathogens (Ownley et al. 2010). To protect plants from herbivores, plants use various methods such as synthesizing poisons that prevent plant-pathogen contact, thus minimizing the detrimental effects on plants, and tackle them from their attackers.

The endophytic fungus can improve the growth of the host plant in different ways, which includes production of various phytohormones or increasing the plant"s resistance to certain stresses (Khan et al. 2015). Promoting plant growth is a crucial part of agriculture, where traditional methods are being complemented by modern techniques. Among these techniques, the use of microorganisms in plant growth and development stands out. While they help to accelerate plant growth and sustainability, they also pose challenges that require attention for proper plant growth. According to (Conti et al. 2012), these chemicals play a crucial role in maintaining social interaction and the ecology of productive species. Secondary metabolites are sources of novel bioactive natural products that can be used in various industries, including agriculture, medicine, and pharmaceuticals. The use of microorganisms in these sectors can help prevent deforestation and preserve ecological flora (Rönsberg et al. 2013).

There are many different varieties of fungi on Earth, with around 1.5 million species identified so far. Most of these species are important for the environment and biotechnology. However, only about 300,000 plant species host various fungal species. These microorganisms that live inside the plant"s body without harming them are called "endophytic." According to researchers, the relationship between endophytic fungi and plants can be mutually beneficial, leading to positive effects such as increased disease resistance, tolerance to stressors like herbivorous attacks, and greater nutrient availability and biodiversity. This occurs because the endophytic fungus affects the plant"s metabolic pathways, causing it to produce different phytochemical substances (as documented by Strobel and Daisy in 2003, Bibi et al. 2020). Endophytic fungi are recognized as to produce a diverse range of therapeutic substances that include vitamins, immunosuppressants, immunological modulators, anticancer agents, and insecticides. According to various researchers such as (Sebastianes et al. 2017) and (Thatoi et al. 2013), these fungi are a great source of antimicrobial agents. Endophytes produce secondary metabolites that have a wide range of pharmacological potential, which facilitates drug development. They can also be used to synthesize organic acids, enzymes, surfactants, and emulsifiers (Dezam et al. 2017; Maroldi et al. 2018; Martinho et al. 2019). Thus, endophytic fungi are a promising resource for the development of new drugs and other useful substances.

Endophytic fungi perform a wide range of functions, including protecting plants from disease, enhancing plant resistance, inhibiting the growth of plant pathogens, producing biofertilizers, developing pesticides, acting as antiviral and antibacterial agents, and synthesizing volatile antibiotics. The discovery of new isolates of these fungi is crucial for promoting environmental and ecosystem health, as well as expanding seed variety. Fungal endophytes are known for synthesizing and producing bioactive compounds against microbial diseases instead of chemicals. Additionally, expand the application of fungal endophytes and develop methods of formulation to ensure the application is safer, easier, cheaper, and more effective (Hashem et al. 2023).

This chapter is focused on endophytic fungi, covering their biodiversity, antibacterial activity, and economic impact. It also discusses the different methods used to synthesize secondary metabolites, and the importance of taxonomy in pharmacological interventions and secondary metabolite shows interactions with host plants. Close examination of endophytic fungi within the host plant is necessary to mitigate and suppress their harmful effects. The impact of external factors such as biotic and abiotic variables, as well as climate change, on the isolation of bioactive components like terpenoids, polyphenols, and alkaloids and their pharmacological benefits are also considered.

20.2 Classification

Endophytic fungi can be mainly categorized into two primary classes based on a variety of factors, including ecological functions, colonization and transmission patterns, taxonomy, and tissue specialization (Rodriguez et al. 2009). According to Márquez et al. 2012, non-clavicipitaceous endophytes are referred to as NC endophytes, while clavicipitaceous endophytes are referred to as C endophytes and primarily inhabit grasses. Additional classification is based on the mechanisms of activity to protect plants from various influences, which can be either direct or indirect.

20.2.1 Class I Clavicipitaceous Endophytes

Ascomycota is a group of fungi that includes this class. They are primarily associated with grasses, with some being independent and others living symbiotically with insects, rushes, grasses, and sedges. Members of this class of fungi belong to the *Ascomycota* group and are mainly associated with grasses. Some live in symbiotic association with insects, rushes, grasses, and sedges. Certain members derived from the seeds of *Lolium linicolum*, *L. temulentum*, *L. arvense*, and *L. remotum* are responsible for producing alkaloids that have toxic effects on both humans and animals. The intercellular gaps of rhizomes plants of various locations such as surfaces of leaf blades, leaf sheaths, and culms are the primary locations for Class I endophyte mycelium. Reports suggest that these endophytes have antifungal and antinematodal properties. According to Hasanuzzaman et al. (2018), they also confer tolerance to the host plant against abiotic stresses such as drought.

20.2.2 Class II Non-clavicipitaceous Endophytes

This fungal species exhibits remarkable diversity among endophytic fungi. It is associated with tropical trees (leaves), seedless vascular plants (club mosses), conifers (gymnosperms), nonvascular plants (liverworts and mosses), and herbaceous angiosperms, rather than grasses. The majority of these fungi belong to the *Ascomycota* family, while very few are basidiomycetes. Some endophytes are categorized as dark septate endophytes (DSE) due to their deeply melanized septa. It boosts biomass, offers resistance against abiotic stress, and guards against fungal infections.

20.3 Mechanism of Plant Protection by Endophytes

Lack of nutrients is a significant obstacle to plant growth in agriculture. However, various fungi participate in the geochemical nutrition cycle and provide access to plants and the microbial community through different methods. The use of plant growth promoting rhizobacteria such as microorganisms as bio-inoculants can enhance soil nutrient availability, reduce the usage of chemical fertilizers, decrease environmental pollution, and promote sustainable agriculture by supplying a range of macronutrients such as nitrogen, phosphorus, potassium, magnesium, calcium, and sulfur, as well as micronutrients such as manganese and zinc. Plant-associated microbes control plant development and morphogenesis or enhance their defenses by releasing phytohormones and small molecules.

Microorganisms in agriculture offer numerous benefits and have the potential to revolutionize the industry. Figure 20.1 and Table 20.1 depict the mechanism of plant protection by use of endophytes.

20.4 Relationship Between Endophytes Within Host Plants

Microorganisms in agriculture offer numerous benefits and have the potential to revolutionize the industry. Endophytic fungi are specific to their host plants or groups of plants, a specificity determined by the fungus being confined to a particular host or group of hosts. This is crucial for identifying and establishing specific interactions between plants and fungi, which aid in the colonization of multiple hosts by the fungus.

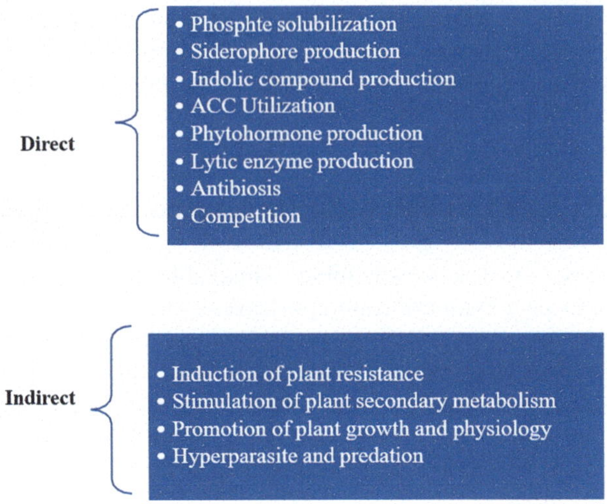

Fig. 20.1 Mechanism of plant protection

- In nature, plants and endophytes exist in a symbiotic relationship. A breakdown in their interaction can lead to the stimulation of host breakdown by the endophytic fungi.
- The host plant and endophyte fungi maintain a mutual relationship with each other.
- The host plant can break down into products from the endophyte microbes.

Endophyte microbes can break down secondary compounds from host plants, offering the potential for biochemical transformations.

20.4.1 Factors Influencing the Endophytic Fungi

Numerous species and strains of endophytic genera have the ability to produce medicinal compounds found in their host plants. Examples of these genera include *Phomopsis*, *Fusarium*, *Pestalotiopsis*, and *Aspergillus*. Several medicinal agents, such as Taxol, Brefeldin A, Camptothecin, and Podophyllotoxin, are produced on a large scale after optimization. However, the potential of fungal endophytes to produce host-like medicinal compounds remains largely unexplored due to factors affecting the screening, isolation, and development processes. Rainfall, humidity, and seasonal cycles are some of the factors that affect the ability of endophytic fungi to function. (Mishra et al. 2012) noted that there are fewer isolates during the winter compared to other seasons. This could be because certain plants accumulate non-structural biomolecules, primarily carbohydrates, when under water stress.

Table 20.1 Endophytic fungi as a source of secondary metabolites

Sr. no.	Type of mechanism	Particular	Importance	Genera	References
1	Direct	Phosphate solubilization	Phosphorus is one of the essential nutrients for optimum plant growth and development after nitrogen Inorganic phosphate solubilization, through microorganisms, is one of the major mechanisms involved in plant growth The potential of endophytic fungi for their ability to solubilize insoluble phosphates in the presence of tricalcium (TCP)	*Penicillium*, *Aspergillus*, *Piriformospora*, *Curvularia* are nonsymbiotic and symbionts such as Arbuscular mycorrhiza (AM) fungi	Mehta et al. (2019), Sujatha et al. (2020)
2		Siderophore production	Plants and microorganisms in soil require micronutrient such as nickel, copper, zinc, and iron but the bioavailability of these elements but its inadequate under environmental conditions. Plants and microbial species have developed a chelation approach to stimulate metal availability during stress situations. The fungal species in soil and plant endophytes synthesize hydroxamate and carboxylate type siderophores of several classes—coprogens, fusarinines	*Penicillium chrysogenum*, *Aspergillus sydowii*, *Aspergillus terreus*	Satpute et al. (2019), Babu et al. (2015)
3		Indole acetic acid	Root-associated endophytes are known to promote plant growth through nutrient supplementation Phytohormones Endophytes can synthesize IAA de novo to influence the IAA homeostasis in plants It is involved in nearly each plant process such as cell growth, root initiation, tropism, fruit ripening, or senescence and influences the plant metabolism	*Penicillium* sp., *Fusarium* sp., *Co niochaeta* sp., *Al ternaria* sp., *Cladosporium* sp.	Jahn et al. (2021)

(continued)

Table 20.1 (continued)

Sr. no.	Type of mechanism	Particular	Importance	Genera	References
4		ACC utilization	ACC mainly help plants to tolerate various stresses like salinity, acidity, alkalinity, heavy metals, extreme temperature	*Penicillium aurantiogriseum*, *Aspergillus flavus*, *Aspergillus ochraceus*, *Fusarium oxysporum*, *Trichoderma asperellum*, *Ceriporia lacerate*, *Alternaria brassicae*,	Wang et al. (2022a, b), Zheng et al. (2023)
5		Phytohormone production	Endophytic fungi produce auxins, gibberellins (GAs), and cytokinins that help by increasing plant height, fresh and dry matter of shoots and roots, and chlorophyll content in leaf and apical meristem	*Homa glomerata* and *P enicillium* sp., *Penicillium citrinum*	Wang et al. (2022a, b)
6		Lytic enzyme production	The hydrolytic enzymes viz. chitinase, cellulase produced by phytopathogens help plants to overcome harmful effects caused by other phytopathogens by antibiosis, parasitism and competition	*Trichoderma asperellum*, *Curvularia chiangmaiensis*, *Fusarium solani*	Putri et al. (2022), Akram et al. (2023)
7	Indirect	Induction of plant resistance	The resistance and biocontrol potential of endophytic microbes can be achieved by the ability to colonize plant tissues, produce hydrogen cyanide and exopolysaccharide, and stimulate novel genes involved in secretion systems and secondary metabolite Secretions	*Trichoderma asperellum*, *Curvulariachian gmaiensis*, *Fusariumsolani*	Singh et al. (2022), Adeleke et al. (2022)

8		Stimulation of plant secondary metabolism	Secondary metabolites (SMs) are the products formed by interactions with the environment during plant growth and development. SMs mainly include alkaloids, terpenoids, peptides, phenols, sterols, and SMs regulating Plant growth and biological defense, and they are also involved in the plant response abiotic stresses, such as drought, low temperature, salinity, and metals	*Alternaria alternata*, *Aspergillus fumigatus*, *Penicillium urticae*	Tian et al. (2017), Jin et al. (2017), Xie et al. (2018), Li et al. (2023)
9		Promotion of plant growth and physiology	Symbiotic relationship between plants and soil microbiota and the synergistic mode of action show positive interaction fertilizers to develop more sustainable agriculture	Class I *Metarhizium*, *Claviceps* Class II *Ascomycota* *Penicillium*, *Aspergillus*, *Fusarium*, *Colletotrichum*, *Trichoderma*	
10		Hyperparasite and predation	A hyperparasite is an organism capable of parasitizing another parasite	*Acrodontium crateriforme*, *Cladosporium oxysporum*	Bermúdez-Cova et al. (2023)

20.5 Endophytic Fungi as a Source of Secondary Metabolites

For the isolation of secondary metabolites initially medicinal plants were used to isolate secondary metabolites and develop bioactive molecules with various industrial and commercial applications. The extensive use of these plants has threatened many species, and extinction of rare species of plants and extracting bioactive molecules from them is laborious, costly, and time-consuming. Therefore, there is an urgent need for alternative sources and strategies to produce bioactive molecules similar to those of plant origin. Endophytic fungi are a promising source of secondary metabolites that synthesize bioactive products with complex molecular structures, playing a crucial role in biotechnological applications. These fungi exhibit various activities, including antifungal, antialgal, antiprotozoal, anticancer, antimalarial, antibiotic, and immunosuppressive activities, which are beneficial for plant growth and development (Pelo et al. 2020; Anand et al. 2023).

Endophytic fungi are natural sources that can be extracted from plants and used commercially to produce bioactive metabolites at a low cost without causing pollution. In recent years, endophytic fungi have become the main source for extracting secondary metabolites known for their antimicrobial potential. These metabolites include alkaloids, terpenoids, phenols, phenylpropanoids, quinones, lactones, and peptides. Moreover, biomolecules such as polysaccharides, polypeptides, unsaturated fatty acids, and glycoproteins act as elicitors, accelerating the production of signaling molecules. This process is stimulated by the accumulation of bioactive phytochemicals in plants, providing defenses against disease-causing organisms. The bioactive compounds produced by endophytic fungi are depicted in Table 20.2.

Table 20.2 Bioactive compounds produced by endophytic fungi

Endophytes	Bioactive compounds	References
Chloridium sp.	Javanicin	Mulyani et al. (2023)
Cryptosporiopsis quercina	Saadamycin	Balkrishna et al. (2022)
Cytonaema sp.	Cytonic acids A and B	
Ganoderma boninense	Cyclododecane, petalostemumol, rapamycin	
Fusarium sp.	Xularosides	Chouhan et al. (2022)
Fusarium proliferatum	Beauvericin, kakadumycin, methylcoumarin	
Nigrospora sp.	Saadamycin	Choudhary et al. (2022)
Streptomyces sp.	Hypericin, kakadumycin A	Harshitha et al. (2023)
S. coelicolor	Munumbicins	
S. hygroscopicus	Clethramycin	

20.5.1 Alkaloids

Endophytic fungi produce nitrogen-containing bioactive compounds that exhibit various biochemical activities, such as antifungal, anticancer, and antiviral properties. This makes them vital in the discovery of novel drugs and developmental processes. Amines and amides are commonly isolated secondary metabolites from endophytic fungi. For instance, phomoenamide was acquired from the *Phomopsis* sp. and shows antimicrobial activity against *Mycobacterium tuberculosis*. Another example is 7-amino-4-methylcoumarin, which was extracted from *Xylaria*, an endophytic fungal species. This compound exhibits a broad spectrum of antimicrobial properties, demonstrating strong antibacterial and antifungal activities against a range of microorganisms. These include *Aeromonas hydrophila*, *Escherichia coli*, *Salmonella typhi*, *S. enteritidis*, *S. typhimurium*, *Shigella* spp., *Staphylococcus aureus*, *Vibrio anguillarum*, *V. parahaemolyticus*, and *Yersinia* spp., as well as *Aspergillus niger* and *Candida albicans*.

20.5.2 Steroids

Many steroids derived from endophytes exhibit moderate antimicrobial activity. For instance, *Colletotrichum* sp. produces several steroids, including 6β-trihydroxyergosta-7,22-diene, 5α-dihydroxy-6β-phenyl-acetoxyergosta-7,22-diene, 3-oxoergosta-4,6,8,22-tetraene, and 3-oxoergosta-4-ene. Meanwhile, *Nodulisporium* sp. has shown strong inhibitory activity against *B. megaterium* and *M. violaceum*. Additionally, an endophytic fungus exhibited antifungal, antialgal, and antibacterial activities against *Microbotryum violaceum*, *Chlorella fusca*, *B. megaterium*, and *E. coli* (Zhang et al. 2012).

20.5.3 Quinones

Many steroids derived from endophytes exhibit moderate antimicrobial activity. For instance, *Colletotrichum* sp. produces several steroids. *Nodulisporium* sp. has shown strong inhibitory activity against *B. megaterium* and *M. violaceum*. Additionally, an endophytic fungus has demonstrated antifungal, antialgal, and antibacterial activities against *Microbotryum violaceum*, *Chlorella fusca*, *B. megaterium*, and *E. coli* (Zhang et al. 2012).

20.5.4 Peptides

Antimicrobial peptides are garnering significant attention as they are small molecules composed of 6–100 amino acids. These peptides differ in amino acid composition, charge, mass, and three-dimensional structure and exhibit antimicrobial properties against a variety of pathogens, including fungi and bacteria. They are also present in fungal endophytes such as *Acremonium* sp., which produce a group of antifungal-anticancer peptides known as leucinostatins. Among them, leucinostatin A has demonstrated antifungal activity against the oomycetous plant-pathogenic fungus *Pythium ultimum*. Antibacterial compounds, specifically cyclo-(Pro-Thr) and cyclo-(Pro-Tyr), extracted from endophytic *Penicillium* sp. associated with *Acrostichum aureum* (a mangrove plant), exhibit potential against *C. albicans* and *S. aureus*.

20.5.5 Phenols

Phenols are acidic phenolic compounds derived from various endophytic fungal species, including *Pestalotiopsis mangiferae*. These compounds exhibit strong antimicrobial properties, inhibiting the growth of bacteria such as *E. coli*, *Klebsiella pneumoniae*, *Pseudomonas aeruginosa*, *Bacillus subtilis*, and *Micrococcus luteus*, as well as the fungus *Candida albicans*. Furthermore, two phenolic compounds, pestalachloride A and B, have demonstrated excellent antifungal activity against a range of plant pathogens.

20.5.6 Terpenoids

Terpenoids are chemical compounds extracted from endophytes, including diterpenoids, triterpenoids, and sesquiterpenes. These compounds possess antimicrobial properties and offer several biological benefits, such as anticancer, antimicrobial, and antiprotozoal activities. The *Acremonium zeae* endophyte found in maize produces pyrrocidines that exhibit significant antifungal activity against *Aspergillus flavus* and *F. verticillioides*. Additionally, they have antibacterial potential against both Gram-positive and drug-resistant bacteria (Yu et al. 2010; Wicklow et al. 2005).

20.5.7 Polyketides

Polyketides are a diverse group of fungal metabolites with significant biological activity, making them important in drug discovery. They exhibit antibiotic, anticancer, antifungal, antiparasitic, and immunosuppressant properties. Two polyketides, 8-dihydroxy-6-methoxy-3-methyl-3,4-dihydromellein and 4,4-dihydro-1H-isochromen-1-one, have been isolated from fungal endophytes and shown inhibitory activity against *A. niger* and *B. subtilis*.

20.5.8 Acids

Endophytes are known for producing various pharmacologically active compounds, including torreyanic acid, fusaric acid, and their derivatives. These compounds exhibit antimicrobial potential, as reported by Deshmukh et al. in 2015. Additionally, a study conducted by Li et al. in 2016 found that salvianolic acid extracted from *Phoma glomerata* also possesses antimicrobial properties.

20.6 Broad Spectrum Uses of Endophytes as Biologically Active Substances

20.6.1 Biomedical Active Substances

Mankind is impacted by the development of antibiotic-resistant bacteria and viruses, but these species are difficult to manage; thus, there is a need for the generation of novel sources which are safe and ecologically friendly active chemicals that have a powerful and effective effect against a wide variety of pathogenic bacteria (Hashem et al. 2023). Endophytic fungi are mainly known for their bioactive and biomedical applications. Thus, endophytes are considered a new category of microbial source that can produce a variety of biological components, have major values for study, and have broad prospects for development (Zhang et al. 2021). The endophytes helps to control a wide range of human health issues, such as numerous microbial pathogens, such as fungi, bacteria, and viruses, through various mechanisms, such as the production of antibiotics, antifungal, antiviral, anticancer, lytic enzymes, and degradation of toxins. Endophytic fungi have metabolic variety and metabolite pathways, which will play important role in metabolite isolation and applications in medicine (Peng et al. 2021). Thus, some biological activities of endophytic fungi constitute an important source of biologically active substances of medical importance—extracts of these isolates were characterized based on varieties of their antioxidant activity, which encourages research on biologically active secondary receptors that would solve many health problems in humanity. Utilizing fungal endophytes as a sustainable alternative to chemical-based methods for producing bioactive compounds, can effectively combat microbial diseases. To expand the applications of fungal endophytes, it is crucial to develop novel methods for formulating biologically active products that harness the microbial wealth within these organisms. This will accelerate the discovery of new biomolecules, enabling the development of effective treatments for life-threatening diseases and ultimately preserving human health through the identification of active compounds. Endophytes offer a promising source of novel biologics for medical and pharmaceutical applications, presenting a safer, more cost-effective, and efficient alternative.

There is an interrelationship between plants and endophytic fungi inhabiting the tissues, and they both live with symbiotic association with each other which shows potential implications for health and growth. Fungal bioactive compounds offer great potential for human health, necessitating research on their colonization of various plant parts to understand their roles in boosting plant defense and growth. This involves exploring their widespread presence and diversity across plant species and addressing the challenges and importance of isolating and extracting these valuable compounds (Ibrahim et al. 2021). It focuses on the health benefits of these bioactive compounds, including their capacity to exhibit antioxidant, anti-inflammatory, antimicrobial, and anticancer effects. The health-enhancing attributes of bioactive compounds derived from plant endophytic fungi shine as a beacon of promise for the future of healthcare and medicine within the intricate interplay of nature and science (Kumari et al. 2021).

20.6.2 Ecological and Physiologically Active Substances

Endophytic fungi are naturally found within plant tissues and cause no detectable disease symptoms in the plant. However, it is believed to possess supporting ecological and physiological benefits for the plant. The fungi confer to their host plant and show resistance to various pathogenic, parasitic, and pest attacks by augmenting the plant's growth rate and improving adaptability and tolerance to pests and pathogens. Furthermore, the relationships between fungal endophytes and their host plant help synthesize varieties of biologically important secondary metabolites. Many of these metabolites have shown antibiotic and pest-repellant properties. Others show plant's adaptive properties to withstand environmental stress such as drought, significant variation in temperature, salinity, and heavy metal accumulation. In recent years, the interest in fungal endophytes has shifted to their propensity to produce new unique compounds with bioactive properties. These properties include anticancer, antimicrobials, immunosuppressant, and antioxidant functions. These fungi could become the future of medicine, agriculture, and industrialization revolutions, based on all of the possible biochemicals produced by endophytes.

The protective nature of endophytes and their host plant against pathogens and other unfavorable environmental factors improved plant survival rate. Endophytic fungi can be harnessed to provide agricultural benefits, replacing agrochemicals, and pharmaceutical applications, offering an alternative to chemical-based medicines. Their bioactive metabolites can help solve pressing issues in agriculture, environmental health, and medicine. More so, with the latest technology like genetic engineering, various endophytic fungi can be manipulated in a better way to explore some of the unexplored areas and gain more information regarding their hidden benefits (Omomowo et al. 2023).

20.6.3 Endophytic Fungi as a Source of Nanoparticles

Nanotechnology is a branch of science with important applications in various fields such as medicine and electronics, positively impacting these areas (Sandhu et al. 2017). This field has great potential in biomedical applications, as living systems are complex networks of nanomachines. The physical and chemical characteristics of nanoparticles at the nanoscale make them highly useful in different areas, including chemical signaling, information storage, catalysis, electronics, drug discovery and labeling, ecological treatments, and high commercial value as mentioned in Fig. 20.2. Currently, nanoparticles produced by various microorganisms such as fungi, bacteria, and yeasts are being exploited (Ibrahim 2015; Ingale and Chaudhari 2013).

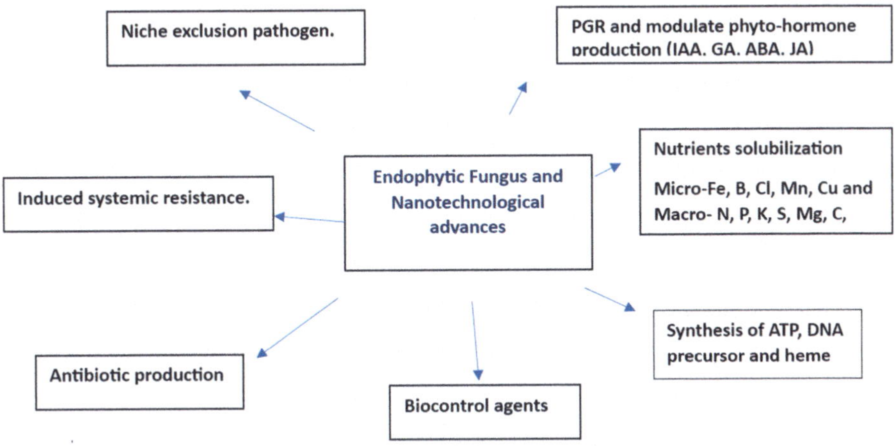

Fig. 20.2 Plant growth promotion by nanoparticles synthesized from endophytic fungus

Fungi are considered the most ideal candidates for metal nanoparticle synthesis due to their antimicrobial properties, which have a wide range of applications in nutrition, health, agriculture, and medicine (Das et al. 2014; Khalil et al. 2019). Silver nanoparticles (AgNPs) are among the most widely used nanoparticles, synthesized from various fungi such as *Fusarium oxysporum*, *F. incarnatum*, *F. solani*, *Alternaria alternata*, *Aspergillus niger*, *A. rouxii*, *A. terreus*, *A. versicolor*, *Coriolus versicolor*, *Penicillium funiculosum*, *Humicola* sp., *Schizophyllum commune*, *Phanerochaete chrysosporium*, *Trichoderma viride*, and *Puccinia graminis*, as well as actinomycetes (Ludwaba et al. 2024; Prabhu et al. 2024).

20.7 Conclusion

Endophytic fungi have emerged as intriguing microorganisms capable of creating a wide range of secondary metabolites with powerful antibacterial properties in plants. Through their symbiotic association with host plants, these fungi have coevolved to produce complicated disease control and plant protection systems. Endophytic fungi create secondary compounds that impede the growth and development of plant pathogens, providing a natural defense system against microbial diseases. These compounds" antimicrobial action can be related to their capacity to disrupt infections" crucial cellular processes, such as cell wall breakdown, enzyme inhibition, and interference with critical metabolic pathways.

Furthermore, these metabolites frequently display pathogen specificity, making them useful tools in disease management efforts. However, further study is needed to completely understand the biosynthesis processes, regulation, and mechanism of action of these secondary metabolites. Exploring the potential of endophytic fungi in biocontrol formulations, as well as their interactions with other beneficial microbes, holds promise for the development of long-term and ecologically acceptable plant health management solutions. The research of endophytic fungus secondary metabolites and their antimicrobial action in plants continues to reveal these organisms" enormous potential for altering agricultural methods and furthering our understanding of microbial ecology.

References

Adeleke BS, Ayilara MS, Akinola SA, Babalola OO (2022) Biocontrol mechanisms of endophytic fungi. Egypt J Biol Pest Control 32(1):1–17

Akram S, Ahmed A, He P, He P, Liu Y, Wu Y et al (2023) Uniting the role of Endophytic fungi against plant pathogens and their interaction. J Fungi 9(1):72

Anand K, Kumar V, Prasher IB, Sethi M, Raj H, Ranjan H et al (2023) Bioactive molecules from fungal endophytes and their applications in pharmaceutical industries: challenges and future scope. J Basic Microbiol 63:690

Ananda K, Sridhar KR (2002) Diversity of endophytic fungi in the roots of mangrove species on the west coast of India. Can J Microbiol 48(10):871–878

Babu AN, Jogaiah S, Ito SI, Nagaraj AK, Tran LSP (2015) Improvement of growth, fruit weight and early blight disease protection of tomato plants by rhizosphere bacteria is correlated with their beneficial traits and induced biosynthesis of antioxidant peroxidase and polyphenol oxidase. Plant Sci 231:62–73

Balkrishna A, Joshi R, Arya ADV (2022) Endophytic fungi: a comprehensive review on their secondary metabolites, pharmacological interventions and host plant interactions. Endophytic Fungi, p 1

Bermúdez-Cova MA, Haelewaters D, de Bekker C, Piepenbring M, Schoutteten N, Quandt CA (2023) Hyperparasitic fungi—definitions, diversity, ecology, and research. Authorea Preprints

Bibi SN, Gokhan Z, Rajesh J, Mahomoodally MF (2020) Fungal endophytes associated with mangroves–chemistry and biopharmaceutical potential. S Afr J Bot 134:187–212

Choudhary J, Mishra SK, Rao Z, Shukla S (2022) Endophytic microbes: potential source of antibiotic production. Microbial Biotechnology: Role in Ecological Sustainability and Research, pp 89–106

Chouhan S, Agrawal L, Prakash A (2022) Amelioration in traditional farming system by exploring the different plant growth-promoting attributes of endophytes for sustainable agriculture. Arch Microbiol 204(2):151

Conti R, Guimarães DO, Pupo MT (2012) Aprendendo com as interações da natureza: microrganismos simbiontes como fontes de produtos naturais bioativos. Ciência e Cultura 64(3):43–47

Das VL, Thomas R, Varghese RT, Soniya EV, Mathew J, Radhakrishnan EK (2014) Extracellular synthesis of silver nanoparticles by the Bacillus strain CS 11 isolated from industrialized area. 3 Biotech 4:121–126

Deng Z, Cao L (2017) Fungal endophytes and their interactions with plants in phytoremediation: a review. Chemosphere 168:1100–1106

Deshmukh SK, Verekar SA, Bhave SV (2015) Endophytic fungi: a reservoir of antibacterials. Front Microbiol 5:715

Dezam APG, Vasconcellos VM, Lacava PT, Farinas CS (2017) Microbial production of organic acids by endophytic fungi. Biocatal Agric Biotechnol 11:282–287

Harshitha K, Nair AR, Pillai PP (2023) Overview of bioactive metabolite (s) produced by endophytes and future perspectives on epigenetic modification/regulation of cryptic biosynthetic pathways. Phytochem Lett 53:116–131

Hasanuzzaman M, Mahmud JA, Anee TI, Nahar K, Islam MT (2018) Drought stress tolerance in wheat: omics approaches in understanding and enhancing antioxidant defense. In: Abiotic stress-mediated sensing and signaling in plants: an omics perspective, pp 267–307

Hashem AH, Attia MS, Kandil EK, Fawzi MM, Abdelrahman AS, Khader MS et al (2023) Bioactive compounds and biomedical applications of endophytic fungi: a recent review. Microb Cell Fact 22(1):1–23

Ibrahim HM (2015) Green synthesis and characterization of silver nanoparticles using banana peel extract and their antimicrobial activity against representative microorganisms. J Radiat Res Appl Sci 8(3):265–275
Ibrahim M, Oyebanji E, Fowora M, Aiyeolemi A, Orabuchi C, Akinnawo B, Adekunle AA (2021) Extracts of endophytic fungi from leaves of selected Nigerian ethnomedicinal plants exhibited antioxidant activity. BMC Complement Med Ther 21(1):1–13
Ingale AG, Chaudhari AN (2013) Biogenic synthesis of nanoparticles and potential applications: an eco-friendly approach. J Nanomed Nanotechol 4(165):1–7
Jahn L, Hofmann U, Ludwig-Müller J (2021) Indole-3-acetic acid is synthesized by the endophyte Cyanodermella asteris via a tryptophan-dependent and-independent way and mediates the interaction with a non-host plant. Int J Mol Sci 22(5):2651
Jha P, Kaur T, Chhabra I, Panja A, Paul S, Kumar V, Malik T (2023) Endophytic fungi: hidden treasure chest of antimicrobial metabolites interrelationship of endophytes and metabolites. Front Microbiol 14
Jin Z, Gao L, Zhang L, Liu T, Yu F, Zhang Z et al (2017) Antimicrobial activity of saponins produced by two novel endophytic fungi from Panax notoginseng. Nat Prod Res 31(22):2700–2703
Khalil NM, Abd El-Ghany MN, Rodríguez-Couto S (2019) Antifungal and anti-mycotoxin efficacy of biogenic silver nanoparticles produced by Fusarium chlamydosporum and Penicillium chrysogenum at non-cytotoxic doses. Chemosphere 218:477–486
Khan M, Damalas CA (2015) Farmers' knowledge about common pests and pesticide safety in conventional cotton production in Pakistan. Crop Prot 77:45–51
Kumari P, Singh A, Singh DK, Sharma VK, Kumar J, Gupta VK et al (2021) Isolation and purification of bioactive metabolites from an endophytic fungus Penicillium citrinum of Azadirachta indica. S Afr J Bot 139:449–457
Li Z, Xiong K, Wen W, Li L, Xu D (2023) Functional endophytes regulating plant secondary metabolism: current status, prospects and applications. Int J Mol Sci 24(2):1153
Ludwaba B, Jimoh MO, Laubscher CP, Nchu F (2024) Insecticidal and antioxidant potential of volatile compounds and nanoparticles from Tulbaghia violacea Harv. Inoculated with endophytic fungus Beauveria bassiana (Bals.-Criv.) Vuill. S Afr J Bot 165:246–256
Maroldi MMC, Vasconcellos VM, Lacava PT, Farinas CS (2018) Potential of mangrove-associated endophytic fungi for production of carbohydrolases with high saccharification efficiency. Appl Biochem Biotechnol 184:806–820
Márquez SS, Bills GF, Herrero N, Zabalgogeazcoa I (2012) Non-systemic fungal endophytes of grasses. Fungal Ecol 5(3):289–297
Martinho V, dos Santos Lima LM, Barros CA, Ferrari VB, Passarini MRZ, Santos LA et al (2019) Enzymatic potential and biosurfactant production by endophytic fungi from mangrove forest in southeastern Brazil. AMB Express 9(1):1–8
Mehta P, Sharma R, Putatunda C, Walia A (2019) Endophytic fungi: role in phosphate solubilization. Advances in Endophytic Fungal Research: Present Status and Future Challenges, pp 183–209
Mishra A, Gond SK, Kumar A, Sharma VK, Verma SK, Kharwar RN, Sieber TN (2012) Season and tissue type affect fungal endophyte communities of the Indian medicinal plant Tinospora cordifolia more strongly than geographic location. Microb Ecol 64:388–398
Mulyani Y, Sinaga SE, Supratman U (2023) Phytochemistry and biological activities of endophytic fungi from the meliaceae family. Molecules 28(2):778
Omomowo IO, Amao JA, Abubakar A, Ogundola AF, Ezediuno LO, Bamigboye CO (2023) A review on the trends of endophytic fungi bioactivities. Sci Afr 4:e01594
Ownley BH, Gwinn KD, Vega FE (2010) Endophytic fungal entomopathogens with activity against plant pathogens: ecology and evolution. BioControl 55:113–128
Pelo S, Mavumengwana V, Green E (2020) Diversity and antimicrobial activity of culturable fungal endophytes in Solanum mauritianum. Int J Environ Res Public Health 17(2):439
Peng XY, Wu JT, Shao CL, Li ZY, Chen M, Wang CY (2021) Co-culture: stimulate the metabolic potential and explore the molecular diversity of natural products from microorganisms. Marine Life Sc Technol 3:1–12

Prabhu N, Deepika K, Nandhini SM, Begam SA (2024) Mycofabrication of silver nanoparticles: synthesis, characterization and its biological applications. In: Biogenic nanomaterials for environmental sustainability: principles, practices, and opportunities. Springer International Publishing, Cham, pp 189–206

Putri ND, Sulistyowati L, Aini LQ, Muhibuddin A, Trianti I (2022) Screening of endophytic fungi as potential antagonistic agents of Pyricularia oryzae and evaluation of their ability in producing hydrolytic enzymes. Biodiversit J Biol Divers 23(2)

Rodriguez RJ, White JF Jr, Arnold AE, Redman ARA (2009) Fungal endophytes: diversity and functional roles. New Phytol 182(2):314–330

Rönsberg D, Debbab A, Mándi A, Wray V, Dai H, Kurtán T et al (2013) Secondary metabolites from the endophytic fungus Pestalotiopsis virgatula isolated from the mangrove plant Sonneratia caseolaris. Tetrahedron Lett 54(25):3256–3259

Sandhu SS, Kumar S, Aharwal RP, Nozawa M (2017) Endophytic fungi: eco-friendly future resource for novel bioactive compounds. Endophytes: Biol Biotechnol 1:303–331

Satapute P, Kamble MV, Adhikari SS, Jogaiah S (2019) Influence of triazole pesticides on tillage soil microbial populations and metabolic changes. Sci Total Environ 651:2334–2344

Sebastianes FLS, de Azevedo JL, Lacava PT (2017) Diversity and biotechnological potential of endophytic microorganisms associated with tropical mangrove forests. Diversity and Benefits of Microorganisms from the Tropics, pp 37–56

Singh P, Chauhan PK, Upadhyay SK, Singh RK, Dwivedi P, Wang J et al (2022) Mechanistic insights and potential use of siderophores producing microbes in rhizosphere for mitigation of stress in plants grown in degraded land. Front Microbiol 13:898979

Sujatha E, Gunaswetha K, Bramhachari PV (2020) Current perspectives on phosphate-solubilizing endophytic fungi: ecological significances and biotechnological applications. Plant Microbiomes for Sustainable Agriculture:79–96

Suryanarayanan TS, Thirunavukkarasu N, Govindarajulu MB, Gopalan V (2012) Fungal endophytes: an untapped source of biocatalysts. Fungal Divers 54(1):19–30

Thatoi H, Behera BC, Mishra RR (2013) Ecological role and biotechnological potential of mangrove fungi: a review. Mycology 4(1):54–71

Tian N, Liu F, Wang P, Zhang X, Li X, Wu G (2017) The molecular basis of glandular trichome development and secondary metabolism in plants. Plant Gene 12:1–12

Wang L, Huang X, Li J, Huang J, Bao S, He C et al (2022a) Metabolites of zearalenone and phytohormones secreted by endophytic fungus strain TH15 regulating the root development in Tetrastigma hemsleyanum. PCTOC 150(3):683–694

Wang X, Tian Z, Xi Y, Guo Y (2022b) Identification of endophytic fungi with ACC deaminase-producing isolated from halophyte Kosteletzkya Virginica. Plant Signal Behav 17(1):2152224

Wicklow DT, Roth S, Deyrup ST, Gloer JB (2005) A protective endophyte of maize: Acremonium zeae antibiotics inhibitory to Aspergillus flavus and Fusarium verticillioides. Mycol Res 109(5):610–618

Wu B, Maruyama H, Teramoto M, Hogetsu T (2012) Structural and functional interactions between extraradical mycelia of ectomycorrhizal Pisolithus isolates. New Phytol 194(4):1070–1078

Xie J, Wu YY, Zhang TY, Zhang MY, Peng F, Lin B, Zhang YX (2018) New antimicrobial compounds produced by endophytic Penicillium janthinellum isolated from Panax notoginseng as potential inhibitors of FtsZ. Fitoterapia 131:35–43

Yu H, Zhang L, Li L, Zheng C, Guo L, Li W, Sun P, Qin L (2010) Recent developments and future prospects of antimicrobial metabolites produced by endophytes. Microbiol Res 165(6):437–449

Zhang H, Liu R, Yang Y, Li H, Zhou F (2021) One strain-many compounds method for production of diverse metabolic profiles using Chaetomium sp. from Astragalus membranaceus. Chem Nat Compd 57:1177–1180

Zhang Y, Han T, Ming Q, Wu L, Rahman K, Qin L (2012) Alkaloids produced by endophytic fungi: a review. Nat Prod Commun 7(7):1934578X1200700742

Zheng J, Xie X, Li C, Wang H, Yu Y, Huang B (2023) Regulation mechanism of plant response to heavy metal stress mediated by endophytic fungi. Int J Phytoremediation 25:1–18

The manufacturer's authorised representative in the EU is Springer Nature Customer Service Centre GmbH, Europaplatz 3, 69115 Heidelberg, Germany. If you have any concerns regarding our products, please contact ProductSafety@springernature.com

Printed and bound by CPI Group (UK) Ltd, Croydon, CR0 4YY
01/10/2026
02225961-0019